The Authors

After graduating with first class honours in mathematics and philosophy from the University of Glasgow in 1971, EPHRAIM BOROWSKI pursued his study of the philosophy of mathematics at The Queen's College, Oxford, where he achieved Distinction in the BPhil degree in 1973. He was then awarded a Senior Scholarship by Hertford College, Oxford, and taught both there and for The Queen's College. In 1976 he returned to Glasgow where he was successively Lecturer in Logic, Senior Lecturer, and Head of the Department of Philosophy. For many years Mr Borowski also taught a number of arts courses for the Open University, and was editor of *Open Mind*. He has published in professional journals, chiefly on the subject of identity, but also on a diverse range of topics relating to logic and language. He was a major contributor on mathematics, logic, and philosophy to the *Collins English Dictionary* and to a number of other reference books. Having retired from the University, he is now Principal Examiner in Philosophy for the Scottish Qualifications Authority and sits on a number of government and other committees, principally concerned with promoting community relations and social justice in Scotland.

A native of St Andrews in Fife, JONATHAN BORWEIN was educated there and in London, Ontario, before graduating in mathematics from the University of Western Ontario in 1971. The recipient of an Ontario Rhodes Scholarship, he completed a DPhil in optimization theory in 1974 at Jesus College, Oxford. Since then he has taught at Carnegie-Mellon, Dalhousie, and Waterloo Universities. From 1993 to 2003 he was Shrum Professor of Science at Simon Fraser University, and was the founding Director of the Centre for Experimental and Constructive Mathematics. In 2004 he rejoined the Faculty of Computer Science at Dalhousie as a Canada Research Chair in Distributed and Collaborative Research. A respected researcher and expositor, his research interests encompass optimization theory, functional analysis and computational mathematics. He has published extensively in these and other areas of pure and applied mathematics and has co-authored eight books including a graduate-level text *Pi and the AGM* (1986), *A Dictionary of Real Numbers* (1991), *Pi – a Sourcebook* (1997) and *Convex Analysis and Nonlinear Optimization* (2000). Dr Borwein is a Fellow of the Canadian Royal Society (1994) and of the American Association for the Advancement of Science (2002). He received an honorary Doctorate from Limoges in 1999. Other distinctions include the 1987 Coxeter-James Lecture of the Canadian Mathematical Society, the 1993 Chauvenet Prize of the Mathematical Association of America, and the presidency of the Canadian Mathematical Society (2000-2002). He is also a founder of a software house that produces *The MathResource*, an interactive version of this dictionary (www.mathresources.com) among other products.

Collins
Web-linked
Dictionary *of*
Mathematics

Collins
Web-linked
Dictionary *of*
Mathematics

E. J. Borowski & J. M. Borwein

Collins

An imprint of HarperCollins*Publishers*

HarperCollins*Publishers*
Westerhill Road, Bishopbriggs,
Glasgow G64 2QT

www.collins.co.uk

First published 1989
Second edition 2002
Reissued 2005

ISBN-10: 0-06-085179-1 (in the United States)
ISBN-13: 978-0-06-085179-8

Originally published in the U.S. copyright © 1991 as
HARPERCOLLINS DICTIONARY OF MATHEMATICS.

FIRST COLLINS U.S. EDITION 2006
HarperCollins books may be purchased for educational, business,
or sales promotional use. For information in the United States, please
write to: Special Markets Department, HarperCollins*Publishers*,
10 East 53rd Street, New York, NY 10022.

Typeset by Davidson Pre-Press Graphics Ltd, Glasgow
Printed and bound in Great Britain by Clays Ltd, St Ives plc.

05 06 07 08 09 10 9 8 7 6 5 4 3 2 1

In memory of our grandparents.

ת" נ" צ" ב" ה"

Contents

Preface

It is the immodest hope of the authors that this Dictionary will not only prove valuable as a reference book for students of mathematics at all levels from secondary school to master's degree, but also offer much to interest a more general readership. We have no illusions that a reference book can supplant authoritative textbooks, but we nonetheless believe that the Dictionary will serve a distinct, complementary, and useful purpose. We have thus attempted to include anything we consider of interest or mnemonic value to the reader: there are formal accounts of certain terms of elementary arithmetic (in terms unavoidably reminiscent of Tom Lehrer's 'New Math'), informal accounts of a number of common logical paradoxes, and less than hagiographic biographies of the luminaries in the history of mathematics.

Our more serious aspiration was the inclusion of any term that an undergraduate might encounter not only within, but also in reading around, any course at any college or university, and we have also deliberately set out to tailor the explanation of each term to the mathematical knowledge of the reader who is likely to consult it. But this ideal overreaches the practicalities, since syllabuses vary greatly in scope, level, and order of presentation, and there are also various demarcation disputes, particularly at the intersection of applied mathematics and physics. In resolving these conflicts, we gave some weight to the existence of the other volumes of the present series; for that reason there is little computing or economics, but a considerable amount of logic. However, our prime consideration remained the likelihood of a mathematics student encountering a given term, so that our comparative liberality towards mechanics and statistics is a consequence of the prominence of these subjects in many undergraduate syllabuses. Thus in general we believe we have erred on the side of inclusiveness, albeit that some lacunae undoubtedly remain, however inadvertently, and that for some advanced terms we have taken refuge in mere indications of the context in which the term occurs, and limitations of space and the demands of readability have dictated that some definitions are relatively informal.

In accordance with our principle of relative inclusiveness, we have attempted to define every term used in our definitions. However, just as we have tried to set the level of each entry for the reader who is likely to have recourse to it, so we have only explicitly signalled a CROSS-REFERENCE where we judge it likely to assist that reader, and for synonymous terms we have attempted to cross-refer the less common to the more common, although a degree of arbitrariness (to say nothing of subjectivity) is inevitable in such judgements. We have likewise attempted, in addition to the substantive biographical entries, to give biographical information under every headword that includes a personal name, but we have not always succeeded, and the absence of such information should not be taken to reflect on the standing of the individual; whenever our research was successful, the biographical matter appears as a *subsidiary entry* following the first substantive eponym. More generally, *italics* denote a term which has no entry of its own, other than (unless the two entries would in any event be adjacent in the lexical ordering) a cross-reference. Cross-references are always to the identical form, except that singular and plural are not distinguished; in particular, a sequence of words in this style always corresponds to a single entry except in a very few cases in which it was impossible to avoid ambiguity without violence to syntax or sense.

A major difficulty faced by a reference book of this kind is the existence of distinct – and inconsistent – usages. We have sought to record the contradictions and ambiguities between recognizable mathematical 'dialects', as in the terminology of orderings and metric spaces, or with respect to the commutativity of certain algebraic objects, but the reader must remain on guard against the unbounded possibility of idiosyncrasy. Use of this handbook should thus be subject to the caveat that authors are authoritative about their own usage, and we should not be taken to be promulgating a unique definitively correct usage where none exists.

*****

The Dictionary originated in 1984–85, when I was asked successively to review the logic, philosophy, and mathematics entries in the *Collins English Dictionary*, and it therefore affords me an opportunity to express my appreciation for the encouragement of its editors, Patrick Hanks and Bill McLeod. It soon became apparent that the only English language dictionaries of post-school mathematics in print were out of date, poorly written, prohibitively expensive, or, in the classic case, all three. Jon Borwein readily agreed to fulfil my immediate need for a consultant's consultant, and this work has grown out of that collaboration, following us both around the world.

Many of our friends, students, and colleagues were sucked into the project, and we are happy to acknowledge their unstinting assistance; Margaret Jones in particular was always ready to lend a hand. The entire text was read and commented on at various stages by Professor Robin Knops of Heriot-Watt University and Dr John Bowers of Leeds, on behalf of the publishers, and by Glasgow students Michael McQuillan,

Andrew Robertson, Martin Hendry, and John Lamb; Karen Chandler and Todd Cardno in Halifax also assisted us with research. By virtue of the addictive nature of lexicography, John Bowers, Andrew Robertson (now at Oxford), and Michael McQuillan (now at Harvard) became increasingly involved in the project, and we are grateful for their many suggestions which we have included in the present text. In addition, Peter Breeze of Glasgow University was particularly helpful in steering us away from statistical solecism.

We must also acknowledge material help – substantial in both senses – from our respective families, from whom I must single out my mother for her help with the early clerical chores. Our respective universities and departments enabled us to consult not only by granting us leave of absence but also by more technological means; I must also express my particular gratitude to the Open University for freeing my time for this work by relieving me from all my teaching and administrative commitments. I wish also to thank the Borwein family for making me welcome in both Limoges and Halifax, and Il Conte Paolo Sassoli de Bianchi in Venice and the Schnellers and Newmans in Jerusalem, for the hospitality which enabled me to combine work and pleasure in these most magnetically atmospheric cities.

One other contribution which cannot pass unacknowledged is that of the Apple Macintosh computer, on which all the words, figures, and formulae were created, assembled, and printed, and I am therefore glad of the opportunity to thank Pat Docherty and his staff at Scotsys Computer Systems for their generous help with hardware and advice. Dr N Smythe of the Australian National University generously provided a pre-release version of his ANUgraph program, Gerry McCauley wrote raw PostScript programmes to draw some of the graphs, and John Milne of Bureau–Graphics Ltd converted much of this material into a more usable form. At Collins Publishers, as well as my editors, Jim Carney and Eddie Moore, who devoted incredible time and precision to this book, I must thank Alan Macfarlane, Ian Crofton, Charlie Ranstead, and Peter Harrison for allowing me to persuade them down an untrodden path towards desk-top publishing, for their patience when unforeseen complexities threatened to overwhelm us all, and for their support in the aftermath. I must also express my gratitude to John Laing and Ted Carden in Collins' design department for making available the space and the hardware that enabled me to undertake the final reassembly of all the diverse elements of text, formulae and graphics; it was here that the Macintosh was indispensible, for I know of no other machine which could have so enhanced the author's control of his own material. My greatest debt, however, is undoubtedly to Jim Carney for his commitment and support in overcoming all the obstacles that fate and technology set before us.

When, inevitably, disagreements arose amongst our far–flung community of authors and collaborators, the blurring by the computer of the roles of author, editor, and typesetter conspired with geography, particularly as deadlines approached – and passed – to dictate that I take responsibility for making final editorial decisions; I must therefore exculpate Jon Borwein from any barbarisms I may have perpetrated in his name. In consolation, we recall that no smaller a lexicographer than Dr Johnson, asked to justify his erroneous definition of 'pastern', replied "ignorance, madam, pure ignorance". We humbly follow his example, and invite readers to draw our own pasterns to our attention.

Ephraim Borowski
Glasgow, 1988

Preface to second edition

Although we took the opportunity of successive reprintings during the 1990s to correct the few 'pasterns' which have been brought to our attention, this is the first major revision, with the addition of some 200 entries. Our thanks are due to Becci DeCamillis of Simon Fraser University for her conscientious review of the entire text, and to Eddie Moore of HarperCollins for his long-suffering patience as we developed the new asymptotic theory of deadlines!

The authors welcome comments and suggestions for the next edition. Please e-mail mathdict@beeb.net.

EJB JMB
February 2002

a, *abbrev. for* ATTO–, used in symbols for fractions of the physical units of the SYSTEME INTERNATIONAL.

A, *n.* the number 10 in HEXADECIMAL notation.

a-, *prefix.* **1.** not; for example, an aperiodic function is one that is not periodic. However, some terms have more than one correlative negative term with different meanings; for example, the classes of ASYMMETRIC and NON-SYMMETRIC relations are disjoint.

2. *abbrev. for* ARC, ANTI–, or ARGUMENT, indicating the inverse of the named function, in expressions such as alog for the antilogarithm, atan for the arctangent, etc. These functions are often symbolized by writing ' –I ' superscript to the name of the function, as in sin^{-1} for arcsin, the inverse sine, and coth^{-1} for arc-coth, the inverse hyperbolic cotangent.

abacus, *n.* a counting device containing rods on which a number of beads are free to move in such a way that each configuration of beads represents a unique whole number by a form of PLACE–VALUE NOTATION. For example, if each rod holds nine beads, the successive rods could represent units, tens, hundreds, etc., in the decimal counting system, and the number represented by a particular configuration would depend upon how many beads on each rod are moved towards the horizontal bar. However, the Chinese abacus, as shown in Fig. 1, uses a more complex system involving a mixture of base 5 and base 10.

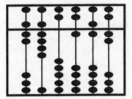

Fig 1. An **abacus** showing the number 790 162.

Abel, Niels Henrik (1802–29), Norwegian mathematician who made significant contributions to both algebra and analysis, and particularly to the study of GROUPS and of INFINITE SERIES. He proved the insolubility of the QUINTIC equation at the age of 19.

Abelian group, *n.* a GROUP on which the defined binary operation is COMMUTATIVE; that is, if *a* and *b* are members of an Abelian group, then *ab* = *ba*. All CYCLIC GROUPS, such as the integers under addition modulo *n*, are Abelian, while the SYMMETRIC GROUP on more than two letters is not.

Abelian theorem, see TAUBERIAN condition.

Abel's limit theorem, *n.* the result that the ABEL SUMMATION method is REGULAR in that for a CONVERGENT series the limit it assigns exists and agrees with its sum. Compare CESARO SUMMATION.

1

Abel's partial summation formula, *n.* the formula of SUMMABILITY THEORY that for two arbitrary sequences $\{a_n\}$ and $\{b_n\}$, and with

$$\sum_{k=1}^{n} a_k = A_n \,,$$

one has

$$\sum_{k=m}^{n} a_k b_k = \sum_{k=m}^{n} A_k(b_k - b_{k+1}) + A_n b_{n+1} - A_{m-1} b_m.$$

Abel summation, *n.* the method of SUMMABILITY THEORY that computes the sum of a possibly DIVERGENT series of complex numbers as the limit, as z approaches 1 from below, of the POWER SERIES with the given sequence as coefficients. This assumes that the series has RADIUS OF CONVERGENCE equal to 1. Compare ABEL'S LIMIT THEOREM.

Abel's test, *n.* **1.** a test for the CONVERGENCE of INFINITE SERIES that states that if $\{a_n\}$ is a bounded MONOTONIC sequence, and if $\sum b_n$ is a convergent series, then $\sum a_n b_n$ converges. In many applications, $\{a_n\}$ actually converges to zero.
2. Abel's test for uniform convergence. a test for the UNIFORM CONVERGENCE of INFINITE SERIES that states that if $\{a_n(z)\}$ and $\{b_n(z)\}$ are sequences of complex functions on a compact set K, such that $a_n(z)$ is bounded in K, then

$$\sum \left| a_n(z) - a_{n+1}(z) \right|$$

is convergent and has a sum that is bounded in K, and $\sum b_n(z)$ is uniformly convergent in K, then $\sum a_n(z) b_n(z)$ is uniformly convergent on K.

above, *prep.* greater than. The limit of a function *from above* is the ONE–SIDED LIMIT where x is restricted to values greater than a, that is, the RIGHT–HAND LIMIT, written variously

$$\lim_{x \downarrow a} f(x) = \lim_{x \to a+} f(x) = f(a+).$$

abscissa, *n.* the horizontal or x-coordinate of a point in a two-dimensional system of CARTESIAN COORDINATES, equal to the distance of the point from the y-axis measured parallel to the x-axis; for example, in Fig. 2 the abscissa of the point P is 4. Compare ORDINATE.

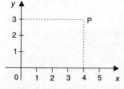

Fig. 2. The **abscissa** of P is 4.

absolute or **numerical,** *adj.* **1.** having a magnitude but no sign; see ABSOLUTE VALUE.
2. not RELATIVE, or not CONDITIONAL. For example, π is an absolute constant.
absolute error, *n.* the magnitude (ignoring sign) of the deviation of an observation from its true or predicted value. See also ERROR, RELATIVE ERROR.
absolute frequency, see FREQUENCY.
absolute geometry, *n.* EUCLIDEAN GEOMETRY without the PARALLEL POSTULATE.

absolutely continuous, *adj.* **1.** (of a real function) continuous by virtue of being defined as the DEFINITE INTEGRAL from *a* to *x* of another function that is merely LEBESGUE INTEGRABLE, where *x* is the independent variable of the given function. This property is stronger than BOUNDED VARIATION.
2. (more generally, of a MEASURE with respect to a second measure) assigning measure zero to any set that is null for the second measure, written $\eta \ll \mu$; that is, $\eta \ll \mu$ if $\eta(E) = 0$ whenever $\mu(E) = 0$. For example, the zero measure is absolutely continuous with respect to any Lebesgue measure.

absolutely convergent, *adj.* **1.** (of a SERIES) such that the series of ABSOLUTE VALUES of its terms CONVERGES; then $\sum a_i$ is said to converge absolutely to *s*, where *s* is the sum of the terms, a_i, of the given series, rather than of their absolute values, $|a_i|$. For example,

$$\sum_{n=1}^{\infty} \frac{(-1)^{n-1}}{n^2} = 1 - \frac{1}{4} + \frac{1}{9} - \frac{1}{16} + \cdots$$

is absolutely convergent, since

$$\sum_{n=1}^{\infty} \left| \frac{(-1)^{n-1}}{n^2} \right| = \sum_{n=1}^{\infty} \frac{1}{n^2} = \frac{\pi^2}{6}$$

but

$$\sum_{n=1}^{\infty} \frac{(-1)^{n-1}}{n} = 1 - \frac{1}{2} + \frac{1}{3} - \frac{1}{4} + \cdots$$

is not absolutely convergent, since

$$\sum_{n=1}^{\infty} \left| \frac{(-1)^{n-1}}{n} \right| = 1 + \frac{1}{2} + \frac{1}{3} + \frac{1}{4} + \cdots$$

diverges. See also COMPARISON TEST.
2. (of an infinite product) such that the logarithms of the successive terms form an absolutely convergent series.
Compare CONDITIONALLY CONVERGENT.

absolutely normal number, *n.* a real number that is a NORMAL NUMBER with respect to every BASE.

absolutely summable, *adj.* (of an infinite series) having a CONVERGENT series of ABSOLUTE VALUES.

absolutely symmetric, see SYMMETRIC FUNCTION.

absolute retract, see RETRACT.

absolute temperature, *n.* (*Statistical physics*) a measure of the heat energy possessed by a system. It may be defined by the relation

$$T = \frac{1}{k} \left(\frac{\partial U}{\partial \log g} \right),$$

where *k* is a constant that relates the mean KINETIC ENERGY and absolute temperature of a system, *U* is the total energy of the system and *g* is the number of possible STATES accessible to the system.

absolute value, *n.* **1.** the positive real number equal to a given real but disregarding its sign; written $|x|$. Where *r* is positive,

$$|r| = r = |-r|.$$

2. another term for MODULUS (sense 1).

absorbing set, *n.* a subset of a VECTOR SPACE over a field of numbers with the property that, for any point *x* in the space, *tx* belongs to the set whenever *t* is sufficiently small and positive. For example, the unit disk is an absorbing set in the Cartesian plane.

absorbing state, *n.* a STATE of a MARKOV CHAIN from which the probability of exit is zero; a singleton ERGODIC SET.

absorption laws, *n.* the laws stating that for all sets A and B (subsets of some UNIVERSAL SET),

$$A \cap (A \cup B) = A \cup (A \cap B) = A$$

abstract, *n.* (*Logic*) an expression, usually denoting a class or property, formed by ABSTRACTION. For example, $\hat{x}(Fx)$ denotes the class of things with the property *F*.

abstract algebra, *n.* that part of ALGEBRA concerned with the study of GROUPS, SEMI–GROUPS, RINGS, MODULES, FIELDS, and similar structures.

abstraction, *n.* (*Logic*) **1.** the process of formulating a generalized concept of a common property by disregarding the differences between a number of particular instances. On such an account, we acquired the concept *red* by recognizing it as common to, and so abstracting it from the other properties of, those individual objects we were originally taught to call red.
2. an operator that forms a class name or predicate from any given expression. See LAMBDA–CALCULUS.

abstract machine, *n.* any hypothetical computing machine defined in terms of the operations it performs rather than its internal physical structure. See AUTOMATA THEORY, TURING MACHINE.

abundant number, *n.* a natural number the distinct PROPER FACTORS of which have a sum exceeding the number. For example, 12 is abundant since its distinct proper integral factors are 1, 2, 3, 4, and 6, whose sum is 16. Compare DEFICIENT NUMBER, PERFECT NUMBER.

acceleration, *n.* **1.** the rate of change of VELOCITY with respect to time; this is a VECTOR quantity that may be either instantaneous or average depending upon the context. The standard units are metres per second per second (abbrev. ms^{-2}).
2. (*Continuum mechanics*) a generalization of the foregoing, the MATERIAL DERIVATIVE of the VELOCITY of a point of a BODY, evaluated at that point.

accumulation point, *n.* another name for CLUSTER POINT.

accuracy, *n.* a measure of the precision of a numerical value of some quantity, such as the number of SIGNIFICANT DIGITS or DECIMAL PLACES, or the range of possible error stated either in absolute or proportional terms. Thus one speaks of an accuracy of $\pm 5\%$ to mean that the true value lies between 95% and 105% of the stated value; or one may specify a time as 9.30 am $\pm$ 5 minutes. See also PRECISION.

accurate or **correct,** *adj.* (of a TRUNCATED decimal number) **1. accurate to *n* significant digits.** correctly representing the first *n* digits of the given number following the first non-zero digit, but approximating to the nearest digit in the final position. For example, since $\pi = 3.14159...$, the approximation 3.1416 is accurate to 5 significant digits.

2. accurate to *n* decimal places. giving the first *n* digits after the decimal point without further approximation. For example, $\pi = 3.1415$ is in this sense accurate to 4 decimal places. This usage is less common than the preceding, and may arise by confusion with such expressions as 'π to 4 decimal places equals 3.1415', where there is no suggestion of maximizing accuracy.

Achilles' paradox, *n.* **1.** also called **racecourse paradox.** the classical paradox of Achilles and the tortoise that argues that motion can never be completed. Since the tortoise is given a head start, before Achilles can overtake it he must first reach its initial position, but by then the tortoise has farther advanced. This argument repeats indefinitely, so that before he can overtake the tortoise, Achilles has first to cover an infinite number of distinct distances. See ZENO'S PARADOXES.

2. the paradox of deduction, published in 1895 by the Oxford mathematician Charles Dodgson under his literary pseudonym Lewis Carroll, that shows the need to distinguish between AXIOMS and RULES OF INFERENCE, and the indispensability of the latter to logic. In his parable, the tortoise attempts to persuade Achilles of a conclusion, *Q*, from *if P then Q* and *P*, by MODUS PONENS; however, Achilles refuses to accept that the rule licenses the detachment of the consequent, so the tortoise advances another premise,

$$\textit{if P and if P then } Q \textit{, then } Q \textit{,}$$

in an attempt to bridge the gap, but this still only permits the inference of the conclusion by detachment. The conversation thus iterates *ad infinitum* (or in Carroll's version of the fable, *ad nauseam*).

acnode, *n.* another word for ISOLATED POINT, a point not lying on a curve but satisfying its equation.

acos, *abbrev. and symbol for* the inverse COSINE function. See ARC–COSINE.

acosh, *symbol for* the inverse HYPERBOLIC COSINE function. See ARC–COSH.

acosec, *abbrev. and symbol for* the inverse COSECANT function. See ARC–COSECANT.

acosech, *symbol for* the inverse HYPERBOLIC COSECANT function. See ARC–COSECH.

acot, *abbrev. and symbol for* the inverse COTANGENT function. See ARC–COTANGENT.

acoth, *symbol for* the inverse HYPERBOLIC COTANGENT function. See ARC–COTH.

act, *vb.* (of a GROUP) to have an ACTION defined (on a non-empty set).

action, *n.* **1.** (of a GROUP G on a non-empty set S) a HOMOMORPHISM Φ from the group to the group of PERMUTATIONS on S. For *g* in G and *s* in S the image $\Phi(g)(s)$ may be written gs, sg, or $s(g\Phi)$. Thus there is an ASSOCIATIVE MAPPING with an IDENTITY from $G \times S$ into S, under which the product of any member of the group with any member of the set is a member of the set, such that for each *g, h* in G and *s* in S,

$$g(hs) = (gh)s \quad \text{and} \quad 1s = s,$$

where 1 is the identity. For example, the action of the permutation group S_n on the polynomial ring $\mathbb{Z}[t_1,..., t_n]$ is given by

$$\alpha f(t_1,..., t_n) = f(t_{\alpha(1)},..., t_{\alpha(n)}).$$

2. (*Mechanics*) the definite integral of the LAGRANGIAN of a given BODY or of a discrete set of PARTICLES, from some reference time to the current time. See also HAMILTON'S PRINCIPLE OF LEAST ACTION.

active, *adj.* (of a constraint) see BINDING.

acsc, *abbrev. and symbol for* the inverse COSECANT function. See ARC–COSECANT.

acsch, *symbol for* the inverse HYPERBOLIC COSECANT function. See ARC–COSECH.

actn, *abbrev. and symbol for* the inverse COTANGENT function. See ARC–COTANGENT.

actnh, *symbol for* the inverse HYPERBOLIC COTANGENT function. See ARC–COTANH.

acute, *adj.* **1.** (of an angle) smaller than a RIGHT ANGLE.

2. (of a triangle) having all its angles acute, such as the example in Fig. 3. Compare OBTUSE.

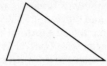

Fig. 3. An **acute** triangle; all its angles are acute.

acyclic, *adj.* not cyclic, having no CYCLES.

add, *vb.* **1.** to combine numbers or quantities by calculating the total number of units contained in them. One may speak of adding a row of figures, adding a and b, or adding a to b. Formally, the addition of two natural numbers is defined recursively by

$$a + 0 = a; \quad a + (n + 1) = (a + n) + 1;$$

it can also be defined in terms of the CARDINALITY of the DISJOINT UNION of sets whose cardinalities are the given numbers. Addition of numerical quantities such as integers, rationals, and reals is defined as an extension of this operation.

2. to apply the OPERATION defined on a GROUP, or any analogous operation written with the ADDITION sign. See also SUM.

addend, *n.* any of a sequence of numbers that are to be added. Compare SUM.

adding machine, *n.* a device that performs simple arithmetical operations, usually neither hand-held nor providing a printout. Compare CALCULATOR.

addition, *n.* **1.** the operation, process, or action of calculating the SUM of two or more numbers or quantities.

2. any operation represented by the *addition sign*, + (usually read *plus*). See also ADD.

addition formula, *n.* **1.** any of a number of IDENTITIES used to express TRIGONOMETRIC FUNCTIONS of a sum or difference of angles as a sum or difference of products of the functions of the individual angles. The formulae for the sine and cosine of a sum of angles are

$$\sin(A + B) = \sin A \cos B + \cos A \sin B$$
$$\cos(A + B) = \cos A \cos B - \sin A \sin B$$

from which the formulae for tangent and other functions, and for functions of differences of angles can easily be derived.

2. any analogous formula for a function f, that gives the value of $f(x + y)$ in terms of $f(x)$, $f(y)$, and related functions. See also ALGEBRAIC ADDITION THEOREM.

additive, *adj.* **1.** (of a function between semigroups) DISTRIBUTING over addition, so that

$$f(x + y) = f(x) + f(y).$$

The only continuous or measurable additive functions on the real line are of the form $f(x) = cx$. Compare MULTIPLICATIVE.

2. (of a SET FUNCTION on a class of sets) DISTRIBUTING over addition, so that, for disjoint sets whose UNION is in the class,

$$f(A \cup B) = f(A) + f(B).$$

See also MEASURE.

additive identity, *n.* an IDENTITY ELEMENT under an ADDITIVE operation; a ZERO.

additive inverse, *n.* (in a RING or GROUP) the element that is INVERSE to a given element with respect to the addition operation.

adherent point, *n.* (*Topology*) a point of the CLOSURE of a set.

ad infinitum, *adv.* (*Latin*) endlessly, repeating indefinitely, generating an infinite series of terms, often with the implication of an INFINITE REGRESS or of fruitless CYCLING; however, the term is sometimes used when an infinite sequence or series is described by listing an initial segment rather than by giving a recursion formula.

adj., *abbrev. for* ADJOINT (especially sense 2).

adjacency matrix, *n.* (*Graph theory*) a matrix of which the rows and columns correspond to vertices of a graph, and of which the i,j th entry is 1 if vertex i is adjacent to vertex j and 0 otherwise. More generally, one counts the number of arcs passing between two vertices in a DIGRAPH or other structure. This is usually a more efficient representation of a graph than is the INCIDENCE MATRIX.

adjacent, *adj.* **1.** (*Graph theory*) **a.** (of a pair of vertices in a graph) joined by a common edge. Thus in Fig. 4, the pair of vertices A and D are adjacent, but vertices A and B are not.

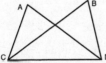

Fig. 4. AC and CD are **adjacent** edges;
A and D are adjacent vertices.

b. (of a pair of edges in a graph) meeting at a common vertex; in Fig. 4, AC and BC are adjacent, but AC and BD are not.

2. (*as substantive*) (*Geometry*) the side of a right-angled triangle other than the hypotenuse that forms an arm of a given angle. For example, in Fig. 5, the adjacent side to θ is BC. Compare OPPOSITE.

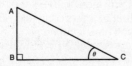

Fig. 5. The **adjacent** of θ is BC.

adjacent angles, *n.* any pair of angles formed between two intersecting lines, and lying on the same side of one of them, such as θ and ϕ in Fig. 6; such angles are SUPPLEMENTARY. Compare OPPOSITE ANGLES.

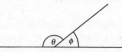

Fig. 6. **Adjacent angles**.

adjoint, *n.* **1.** also called **Hermitian conjugate. a.** the matrix that is the TRANSPOSE of the matrix of the COMPLEX CONJUGATES of the entries of a given matrix, often denoted A^*, A' or $A^\perp$.
b. also called **dual.** the operator A^* that is CONJUGATE to a given linear operator A between NORMED SPACES X and Y; it is defined by

$$\langle Ax, y \rangle = \langle x, A^*y \rangle.$$

where $\langle \ , \ \rangle$ represents a pairing between a space and its DUAL, and A^* maps Y* into X*. (See also sense 4.)
2. also called **adjugate.** (*Matrix theory*) the matrix the entries of whose TRANSPOSE are the COFACTORS of the given square matrix. If the given matrix is invertible, its adjoint is obtained by multiplying its INVERSE MATRIX by its DETERMINANT.
3. the ADJOINT EQUATION of a linear differential or control equation.
4. Hilbert space adjoint. the OPERATOR A^* that is CONJUGATE to a given linear operator A; it is defined on a HILBERT SPACE by

$$\langle Ax, y \rangle = \langle x, A^*y \rangle.$$

In this case $(cA)^* = \bar{c}A^*$, while if A^* is viewed as a mapping between dual spaces, as in sense 1b, then $(cA)^* = cA^*$.

adjoint equation, *n.* (*Differential equations*) the vector equation constructed from a given LINEAR EQUATION SYSTEM via the negative of the ADJOINT of the matrix involved, that is by replacing the original equation $y' = Ay$ by the equation $z' = -A^*z$. There is a corresponding adjoint for a scalar n^{th} order equation described with LAGRANGE'S IDENTITY.

adjugate, *n.* a less common term for ADJOINT (sense 2).

admissible variation, (*Calculus of variations*) see VARIATION.

a.e., (*Measure theory*) *abbrev. for* ALMOST EVERYWHERE.

affine, *adj.* (*Geometry*) characterizing, or involving AFFINE TRANSFORMATIONS.

affine geometry, *n.* the study of AFFINE space; a geometry of generality between EUCLIDEAN GEOMETRY and PROJECTIVE GEOMETRY, obtained by choosing a line at infinity in a projective geometry, in which distances are only compared on parallel lines, so that there is no notion of perpendicularity.

affine hull, *n.* the set of all elements derived from a given set by AFFINE TRANSFORMATIONS.

affinely independent set, *n.* a minimal subset with a given AFFINE SPAN.

affine manifold or **affine subspace,** *n.* a subset of a VECTOR SPACE, usually over the field of real numbers that contains all lines between points of the subset; equivalently, a TRANSLATE of a vector subspace. A non-trivial affine manifold in three-space must be a point, line, or plane.

affine plane, *n.* **1.** an AFFINE GEOMETRY of two dimensions.

2. (*Combinatorics*) a SUBDESIGN of a finite projective plane that is itself a finite projective plane.

affine span, *n.* the smallest AFFINE MANIFOLD containing a given subset of a vector space.

affine transformation or **affinity,** *n.* a transformation that preserves COLLINEARITY, and hence parallelism and straightness, especially in classical geometry; in particular TRANSLATION, ROTATION, and REFLECTION in an axis are affinities. Formally, an affine transformation is a change of variables in which the new variables are affine combinations of the original variables. See AFFINE MANIFOLD.

affinity, *n.* another term for AFFINE TRANSFORMATION.

affirmative, *adj.* (*Logic*) **1.** (of a categorial proposition) affirming the satisfaction by the subject of the predicate, for example *all birds have feathers*, and *some men are married*.

2. not containing NEGATION.

3. (*as substantive*) an affirmative proposition.

Compare NEGATIVE.

a fortiori, *adv.* (*Latin:* from the stronger) by reason of a preceding stronger statement or result; for example, since the number 7 is prime, it is *a fortiori* not divisible by 3.

agm, *abbrev. for* ARITHMETIC–GEOMETRIC MEAN.

agonic, *adj.* another word for SKEW.

agree, *vb.* (of two functions) to have the same values for the same arguments; that is, *f* and *g* agree on the set S if $f(x) = g(x)$ for all *x* in S. The sets of measurable functions that agree almost everywhere form equivalence classes under this relation.

Airy function, *n.* the solution of the DIFFERENTIAL EQUATION $\phi'' - t\phi = 0$, namely

$$\phi(t) = \frac{1}{\sqrt{\pi}} \int_0^\infty \cos(tx + \frac{x^3}{3})\, dx.$$

Akerman function, *n.* one of a class of RECURSIVE FUNCTIONS which grow so rapidly that they challenge our usual notion of COMPUTABLE. For example, if

$$A(1, j) = 2^j \qquad \text{for } j > 1$$
$$A(i, 1) = A(i-1, 2) \qquad \text{for } i > 1$$
$$A(i, j) = A(i-1, A(i, j-1)) \qquad \text{for } i, j > 1,$$

then $A(1,3) = 8$, $A(2,3) = 65\,536$, and $A(2,4)$ has 19 729 digits!

AKS primality test, *n.* the first deterministic POLYNOMIAL TIME ALGORITHM that tests whether a given number is PRIME or not, named after its discoverers in 2002, Agarwal, Saxena and Kayal.

Alaoglu's theorem, *n.* another name for the BANACH–ALAOGLU THEOREM.

aleph, *n.* any infinite CARDINAL NUMBER, usually denoted by the Hebrew letter ℵ (aleph). See CONTINUUM HYPOTHESIS.

aleph-null, aleph-nought, or **aleph-zero,** *n.* the smallest ALEPH, defined as the CARDINALITY of the positive integers, and also that of the rationals and of the algebraic numbers, but not of the reals. The usual symbol is $\aleph_0$.

alethic, *adj.* (*Logic*) **1.** (of a MODALITY) pertaining to truth and falsehood, such as *possibly true* and *necessarily true*.

2. (of a MODAL LOGIC) formalizing, or having as its intended interpretation, these concepts.

Compare EPISTEMIC, DEONTIC.

Alexander's sub-base theorem, *n.* the theorem that asserts that if every open COVER of a topological space by elements of a SUB-BASE has a finite sub-cover, then the space is COMPACT. (Named after the US topologist and algebraist, *James Waddell Alexander* (1888–1971), who worked on functions of complex variables and on the theory of KNOTS.)

Alexandroff compactification, *n.* another name for ONE–POINT COMPACTI-FICATION. (Named after *Pavel Alexandroff* (1896–1982), Russian topologist and set theorist.)

alg., *abbrev. for* ALGEBRA or ALGEBRAIC.

algebra, *n.* **1a.** the branch of elementary mathematics that generalizes arithmetic by using VARIABLES to range over numbers, for example in arithmetical IDENTITIES such as $x + y = y + x$.

b. in particular, the use of symbols standing for unknown quantities in order to determine their value by the ELEMENTARY OPERATIONS of arithmetic.

2. also called **abstract algebra.** the study of systems, such as RINGS, GROUPS, and FIELDS, endowed with FINITARY OPERATIONS with specific properties.

3. any FORMAL CALCULUS used to model and study the properties of the entities that are the intended interpretation of their symbols, such as the algebra of logic, and the algebra of classes; thus one might construct an algebra of colour properties.

4. (more specifically) a BOOLEAN ALGEBRA, or SIGMA–ALGEBRA (σ-algebra), and in particular, the ALGEBRA OF SUBSETS or ALGEBRA OF PROPOSITIONS.

5. any formal system with only functions and constants, but no relations except possibly identity.

6. a RING that is a MODULE over a FIELD. See also ALGEBRA OVER A FIELD.

See also LINEAR ALGEBRA.

algebraic, *adj.* **1.** of or relating to ALGEBRA.

2. relating only to finite numbers, operations and expressions; constructible using only FINITARY methods. See ALGEBRAIC FUNCTION. Compare TRANSCENDENTAL.

3a. consisting of or relating to roots of a polynomial equation with rational coefficients. See ALGEBRAIC NUMBER. Compare TRANSCENDENTAL NUMBER.

b. algebraic over a field. consisting of or relating to the roots of a polynomial equation with coefficients that are members of the given field. See also ALGEBRAIC EQUATION.

4. (of an expression) not containing any individual terms or quantifiers, expressed only in terms of variables, and intended to be interpreted as being true for all members of some set. For example,

$$x(y + z) = xy + xz$$

is an algebraic formulation of the distribution law for arithmetic multiplication over addition.

algebraic addition theorem, *n.* (*Analysis*) any theorem or IDENTITY that gives an ADDITION FORMULA for a function f in terms of a polynomial P in three complex variables such that

$$P(f(x), f(y), f(x + y)) = 0$$

for all complex arguments. A MEROMORPHIC FUNCTION has such an addition theorem if and only if it is rational, trigonometric or elliptic. For example,

$$\exp(x + y) = \exp(x) \exp(y)$$

is an addition theorem for the exponential with $P(x, y, z) = xy - z$.

algebraically soluble, *adj.* (of a first-order ORDINARY DIFFERENTIAL EQUATION) such that, where u and t are respectively the dependent and independent variables, either it can be solved for $p = du/dt$ and the resulting first-degree equation in p can be solved; or it can be solved for u, giving an equation that can be differentiated to give a linear first-order equation in the depen dent variable t and independent variable p; or it can be solved for t, giving an equation that can be differentiated with respect to u to give a linear first-order equation in the dependent variable u and the independent variable p, after writing $dt/dp = 1/p$.

algebraic closure, *n.* the extension of a given set, field, etc. to one containing all the roots of all polynomials whose coefficients are members of the given set. A set is *algebraically closed* if it coincides with its algebraic closure. Thus neither the reals nor the rationals are algebraically closed, since they do not contain the roots of the polynomial $x^2 + 1$, but the complex field is algebraically closed and is the closure of both subfields.

algebraic cycle, see HODGE CONJECTURE

algebraic equation, *n.* an equation of the form $p(x) = 0$ where p is a polynomial of degree n with coefficients in a given base field, usually the rationals; n is then also the degree of the algebraic equation.

algebraic extension or **algebraic extension field,** *n.* an EXTENSION FIELD of a base field with the property that every element of the extension is ALGEBRAIC (in sense 3b) over the base. An algebraic extension of the rationals is an ALGEBRAIC NUMBER FIELD. A non-algebraic extension is a *transcendental extension*.

algebraic function, *n.* any function generated by ALGEBRAIC operations alone; any function that can be constructed in a finite number of steps from the ELEMENTARY OPERATIONS and the inverses of any function already constructed. Precisely, f is algebraic over a given base field if there is a two-variable polynomial P over the field such that $P(x, f(x)) = 0$. Compare TRANSCEN-DENTAL FUNCTION.

algebraic geometry, *n.* **1.** the study of geometry by algebraic methods, especially the study of algebraic AFFINE or PROJECTIVE GEOMETRY. This has arisen by generalization from the original study of points on curves and families of curves on a surface, and from the project of classifying all ALGEBRAIC VARIETIES. These methods have been applied, for example, to NUMBER THEORY.

2. in particular, the study of COMMUTATIVE RINGS with IDENTITY, viewed as a ring of 'regular functions'.

3. an abstract geometry that consists of a set with a LINEAR DEPENDENCE relation that is preserved by a BIJECTION onto the set of one-dimensional subspaces of a VECTOR SPACE; the properties of the geometry are those properties of the set that are invariant under the action of a given subgroup of linear transformations.

algebraic independence, *n.* **1.** (*Number theory*) the failure of a set of numbers to satisfy any non-trivial polynomial with rational or ALGEBRAIC coefficients. **2.** the LINEAR INDEPENDENCE of a set of complex numbers viewed as a vector space over the field of algebraic numbers.

algebraic integer, *n.* **1.** an ALGEBRAIC NUMBER that is the root of an IRREDUCIBLE polynomial with integer coefficients where that of the highest power is 1. **2.** an ALGEBRAIC NUMBER that is a polynomial with integer coefficients in a finite number of SURDS, and with leading coefficient 1; an element of the INTEGRAL DOMAIN in a FINITE EXTENSION of the field of rationals that is generated by the integers together with the elements in the extension that are not rational.

algebraic number, *n.* any number that is the root of a polynomial equation with coefficients drawn from a given field, in particular the rationals; in that case, $\sqrt{2}$ is an algebraic number, but π is not. The algebraic numbers form a field. Compare ALGEBRAIC EQUATION, TRANSCENDENTAL NUMBER.

algebraic number field, *n.* a subfield of the complex numbers that arises as a finite degree ALGEBRAIC EXTENSION FIELD of the rational field.

algebraic number theory, *n.* that part of NUMBER THEORY that employs ALGEBRAIC methods.

algebraic structure, *n.* a STRUCTURE (sense 2) such as a GROUP or FIELD in which the predicates or relations are of an ALGEBRAIC nature.

algebraic system, *n.* a set together with a family of OPERATIONS and a family of RELATIONS.

algebraic topology, *n.* **1.** another name for TOPOLOGY (sense 2). **2.** those parts of topology employing group-theoretic and other algebraic methods. The main problem is the classification of TOPOLOGICAL SPACES as classes of HOMEOMORPHIC spaces, or, as this problem is very difficult, the simpler problem of classifying spaces by HOMOTOPY. See BROUWER'S THEOREM, HOMOLOGY, JORDAN CURVE THEOREM, HAUSDORFF GROUP.

algebraic variety, *n.* a subset of an *n*-dimensional ALGEBRAIC GEOMETRY (in sense 2), consisting of all points $(x_1, ..., x_{n+1})$ that satisfy a system of polynomial equations

$$P_\alpha(x_1, ..., x_{n+1}) = 0, \quad \alpha \in A.$$

algebra of propositions, *n.* the BOOLEAN ALGEBRA whose intended interpretation is PROPOSITIONAL CALCULUS.

algebra of sets, **algebra of subsets** or **field of sets,** *n.* the BOOLEAN ALGEBRA whose intended interpretation is SET THEORY.

algebra over a field, *n.* a RING that is also a VECTOR SPACE in which the SCALARS are members of a FIELD that satisfies the condition that if **x** and **y** are any members of the ring, and *a* and *b* are any scalars, then

$$(a\mathbf{x})(b\mathbf{y}) = (ab)(\mathbf{xy});$$

more generally, a ring that is also a MODULE over a commutative unitary ring. The continuous or differentiable functions on an interval form an algebra, with multiplication defined pointwise.

algorithm or **algorism,** *n.* **1.** a step-by-step procedure by which an operation can be carried out without any exercise of intelligence, and so, for example, by a machine; formally, a RECURSIVE specification of a procedure by which a given type of problem can be solved in a finite number of mechanical steps. Simple algorithms familiar in elementary arithmetic are those used in the extraction of square roots and in long division. The problem of how much of mathematics can be described in such terms is the subject of COMPUTABIL-ITY THEORY, and HILBERT'S PROGRAMME was in essence an attempt to demonstrate that all of mathematics was recoverable from algorithms that operate upon strings of mathematical symbols. See AUTOMATA THEORY. See also GÖDEL'S THEOREM, TURING MACHINE.

2. a recursive definition enabling any member of an infinite sequence of terms to be generated by its repeated application.

aliorelative, *adj.* (*Logic*) another word for IRREFLEXIVE.

aliquant part, *n.* a number or quantity that is not an exact divisor of a given number or quantity. For example, 5 is an aliquant part of 12. Compare ALIQUOT PART.

aliquot part, *n.* a number or quantity that is an exact divisor of a given number or quantity. For example, 4 is an aliquot part of 12. Often it is required that the divisor be proper. Compare ALIQUANT PART.

almost all or **almost everywhere,** *adv.* often written **a.e.** or **p.p.** (of a property) holding for all values except on a set of zero MEASURE, especially in LEBESGUE MEASURE. For example, if $f(x) = 1$ for all real x, and $g(x) = 1$ for all irrational x and 0 for all rational x, then f and g agree almost everywhere, since the rationals have measure zero.

almost disjoint, *adj.* (of a collection of subsets) such that the intersections of all pairs of distinct members of the collection are finite.

almost surely, *adv.* another term for ALMOST EVERYWHERE, especially in probability theory.

aln, *n. abbrev. and symbol for* the ANTILOGARITHM of a NATURAL LOGARITHM.

alog, *n. abbrev. and symbol for* ANTILOGARITHM. If no BASE is specified, it may be taken to be 10.

alphabet, *n.* the set of symbols from which an n-tuple is drawn to form a WORD of a CODE.

alpha-beta theorem, see SCHNIRELMANN DENSITY.

alternant, *n.* **1.** (*Logic*) another word for DISJUNCT.

2. a DETERMINANT constructed from n functions and n points (not necessarily distinct) by taking the i,j^{th} entry to be the value of the ith function at the jth point, or vice versa. The VANDERMONDE DETERMINANT is an alternant, as is a WRONSKIAN.

alternate angles, *n.* either pair of angles contained between two given lines and a TRANSVERSAL, and lying on opposite sides of the transversal, such as θ and ϕ in Fig. 7 overleaf. These angles are equal if and only if the given lines are parallel.

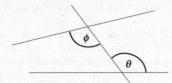

Fig. 7. **Alternate angles**.

alternating form, see MULTILINEAR FUNCTION.

alternating group, *n.* the subgroup of the SYMMETRIC GROUP consisting of all EVEN PERMUTATIONS of *n* objects; it is denoted A_n, has order $n!/2$, and has index 2 in S_n, the symmetric group of degree *n*. For $n \geq 5$, A_n is the only proper non-trivial NORMAL subgroup of S_n, and is itself SIMPLE. See also GENERATE.

alternating multilinear function, see MULTILINEAR FUNCTION.

alternating series, *n.* a series the terms of which are alternately positive and negative, such as

$$1 - \frac{1}{2} + \frac{1}{3} - \frac{1}{4} + \frac{1}{5} \ldots$$

(which converges to log 2).

alternating series test or **Leibniz's alternating series test,** *n.* the result that to establish CONDITIONAL CONVERGENCE of an ALTERNATING SERIES one need only verify that the terms decrease monotonically to zero in absolute value. In this case the error after adding *n* terms is always less than the magnitude of the next term. This is a special case of DIRICHLET'S TEST. For example,

$$1 - \frac{1}{3} + \frac{1}{5} - \frac{1}{7} + \frac{1}{9} \ldots$$

is convergent, since

$$|a_n| = \left| \frac{(-1)^n}{2n+1} \right| = \frac{1}{2n+1}$$

decreases monotonically to zero as *n* tends to infinity; its sum is $\pi/4$.

alternation, *n.* (*Logic*) another word for DISJUNCTION.

alternation theorem, *n.* the result, for continuous functions $g_1, \ldots, g_n$ on $[a, b]$ that satisfy the HAAR CONDITION (as holds for $1, x, \ldots, x^n$), that in order for a GENERALIZED POLYNOMIAL P to be the BEST APPROXIMATION in CHEBYSHEV NORM to a continuous function *f* it is necessary and sufficient that the error function $r = f - P$ possesses at least $n+1$ *alternations*, points at which the

$$r(x_i) = -r(x_{i-1}) = \pm \| r \|_\infty.$$

See UNICITY, VANDERMONDE DETERMINANT.

alternative hypothesis, *n.* (*Statistics*) any hypothesis that given data do not conform with a given NULL HYPOTHESIS; the alternative is only accepted if the value of a TEST STATISTIC is sufficient, at a chosen SIGNIFICANCE LEVEL, to reject the null. See HYPOTHESIS TESTING.

alternative theorem, *n.* any theorem stating that of two systems of equations or inequalities one or the other always has a solution. See FARKAS' LEMMA, FREDHOLM ALTERNATIVE.

altitude, *n.* **1.** any line segment between a VERTEX and a side of a POLYGON, that is PERPENDICULAR to the side and is the longest such perpendicular through that vertex; especially, any perpendicular from a vertex to the opposite side of a triangle. For example, in Fig. 8, AD is an altitude of triangle ABC, and FH is an altitude of heptagon ABCDEFG. (Note that where there are re-entrant angles, as in the latter case, the perpendicular from the vertex meets the opposite side outside the figure.)

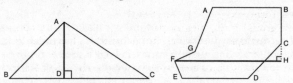

Fig. 8. AD is an **altitude** of the triangle, and FH of the polygon.

2. the length of such a line.

ambiguous case, *n.* the situation in which, when the lengths of two sides and one of the angles of a triangle other than that between the given sides are known, if the known angle is less than 90°, it may not be possible to solve for the third side or for the other two angles.

amicable numbers, *n.* a pair of integers each of which is the sum of the distinct PROPER FACTORS of the other. For example, 220 and 284 are amicable since 284 has factors 1, 2, 4, 71, and 142, which add up to 220, while 220 has factors 1, 2, 4, 5, 10, 11, 20, 22, 44, 55, and 110, which sum to 284. Compare PERFECT NUMBERS. See also SIGMA FUNCTION.

amp, *n. abbrev. and symbol for* AMPLITUDE.

amplitude, *n.* **1.** the maximum difference between the value of a PERIODIC FUNCTION and its MEAN. Thus, for example, the amplitude of $y = \sin x + 0.5$ is 1; in Fig. 9 this is represented by the vertical lines from a maximum or minimum to the mean line $y = 0.5$.

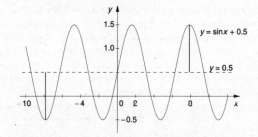

Fig. 9. The vertical lines represent the **amplitude** of the function.

2. also called **argument**, **anomaly**, or **azimuth.** the angle between the positive real axis and the vector representing a given complex number in the ARGAND DIAGRAM; if the point (x, y), representing the number $x + iy$, has POLAR COORDINATES (r, θ), its amplitude is θ, i.e.

$$\text{amp}(x + iy) = \arctan (y/x),$$

and $r(\cos\theta + i\sin\theta)$ equals the given number. For example, θ in Fig. 10 is the amplitude of $-4 + 3i$, which is approximately 2.5 radians (143°). Compare MODULUS. See also PHASE.

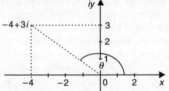

Fig. 10. θ is the **amplitude** of the complex number $3i-4$.

analog device or **analogue device,** *n.* a mechanical or electrical device in which a continuously varying quantity such as voltage is used to represent another quantity to which it is related by a continuous BIJECTION; for example, the hands of a conventional clock give an analogue representation of the passage of time. Such devices are used in monitoring systems, simulation processes, etc., but seldom, since the development of the DIGITAL COMPUTER, for arithmetical operations; for example, a slide rule is an analog device for arithmetic computation, since the numbers to be operated upon are represented by physical distances on the rule that are proportional to the logarithms of the given numbers.

analysis, *n.* the branch of mathematics principally concerned with LIMITS of functions, sequences and series, and with other infinite processes applied to them. It largely grows out of CALCULUS, and is now often divided into classical REAL ANALYSIS and COMPLEX ANALYSIS, and more abstract FUNCTIONAL ANALYSIS and the theory of LINEAR OPERATORS. See also NON–STANDARD ANALYSIS.

analysis of variance, *n.* (*Statistics*) any of a number of techniques for resolving the observed VARIANCE between sets of data into COMPONENTS, especially to determine whether the difference between two or more samples is explicable as random sampling variation within the same underlying population.

analysis situs, *n.* a former name for TOPOLOGY.

analyst, *n.* a student of ANALYSIS.

analytic, *adj.* **1.** also called **regular** or **holomorphic.** (of a complex function) having a complex derivative at every point of its domain, and in consequence possessing derivatives of all orders and agreeing with its TAYLOR SERIES locally. For example, exp$z$ is analytic on the complex plane, and log$z$ is analytic on $\mathbb{C} \setminus (-\infty, 0]$.

2. also called **real analytic.** (of a real function) possessing derivatives of all orders and agreeing with its TAYLOR SERIES locally.

3. (*Logic*) (of a proposition) **a.** true by virtue of the meanings of the words alone without reference to the facts; for example, *all spinsters are unmarried* is analytic.

b. true or false by virtue of meaning alone; for example, *all spinsters are married* is analytically false.

Compare SYNTHETIC.

analytical engine or **difference engine,** *n.* a mechanical precursor of the modern digital computer, including a punched-card reader and a memory storage device, of which the principle was described by Charles Babbage in 1834, but which was never completed.

analytic continuation, *n.* **1.** the construction of an analytic function the RESTRICTION of which to a given domain is a given analytic function.

2. the unique analytic function that extends a given analytic function to a larger domain. For example, sinz on the complex plane is the analytic continuation of sinz on the real line, since there is no other analytic function on ℂ of which the restriction to the reals is the sine function.

analytic geometry, *n.* that part of geometry based on COORDINATE GEOMETRY.

analytic proof or **analytic method,** *n.* proof by algebraic construction, as contrasted with SYNTHETIC PROOF, that is, deduction from axioms.

analytic set, *n.* another term for SOUSLIN SET.

analytic structure, *n.* a covering of a TOPOLOGICAL SPACE by sets that are homeomorphic to open sets in a fixed EUCLIDEAN SPACE, such that, whenever two members of the cover overlap, the coordinate transformations in both directions are ANALYTIC in their intersection; equivalently, a C^ω DIFFERENTIAL STRUCTURE. See MANIFOLD.

ancestral, *n.* (*Logic*) **1.** a relation, derived from a given relation, that holds between two elements of its field whenever there is a chain of instances of the given relation leading from the first element to the second. Thus the ancestral of *parent of* is *ancestor of*, since *x* is an ancestor of *y* if and only if there is a sequence of individuals each of whom is the parent of the next, and of whom the first is *x* and the last *y*.

2. the set of elements that have the ancestral of a given relation to a given element of its domain. Thus the ancestral of 5 under the successor relation on the positive integers is the class {1, 2, 3, 4}. In a tree, a node A is a member of the ancestral of a node B if and only if there is a path from the root to B that includes A.

anchor ring, *n.* another term for a TORUS.

and, *conj.* usual English reading of the CONJUNCTION operator.

angle, *n.* **1.** the figure formed by two line segments that extend from a common point, or by regions of two planes that extend from a common line.

2. a measure of such a shape as the divergence of one line or plane from the other, the difference between their directions being measured, usually in DEGREES or RADIANS, by the amount of rotation required to superimpose one upon the other. The angle between two planes is taken to be the angle between two lines, one lying in each plane, drawn perpendicular to the intersection of the planes at the same point; for example, in Fig. 11 below, the angle between the planes ABCD and CDEF is the angle YXZ, where X is any point on the line CD, and YX and ZX lie in the respective planes and are perpendicular to CD. If directions are distinguished, the anticlockwise direction is conventionally taken to be positive.

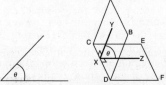

Fig. 11. **Angle**.

3. the space between two such lines or planes.

angle brackets, *n.* the paired symbols, ⟨ ⟩, often used to indicate that the terms lying between them form a SEQUENCE or ORDERED *n*-TUPLE, or to denote the INNER PRODUCT of two vectors. They may also denote the SUBGROUP or IDEAL generated by a subset of a given group or ring. In some spoken usages the left and right bracket are read as *bra* and *ket* respectively.

angle of declination or **angle of depression,** see DECLINATION.

angle of elevation or **angle of inclination,** see INCLINATION.

angular, *adj.* of, relating to, or measured in terms of, angles. See ANGULAR ACCELERATION, ANGULAR MOMENTUM, ANGULAR VELOCITY.

angular acceleration, *n.* the RATE OF CHANGE of ANGULAR VELOCITY.

angular momentum or **moment of momentum,** *n.* **1.** a measure of the momentum of a body due to its motion around an axis of rotation, equal, in the case of circular motion, to the product of its mass and its ANGULAR VELOCITY; more generally, if a particle has momentum **m** and position vector **x**, then its angular momentum is the vector product **x** × **m**. For a RIGID BODY, the angular momentum is $A\omega_1 + B\omega_2 + C\omega_3$, where A, B, C are the PRINCIPAL MOMENTS OF INERTIA, and ω_1, ω_2, ω_3 are the angular velocities about the PRINCIPAL AXES.

2. (*Continuum mechanics*) (around a point P) the generalization of the foregoing, the integral with respect to velocity:

$$\mathbf{H}(\mathbf{R}_t; P) = \int \rho \, (\mathbf{x} - \mathbf{p}) \times \mathbf{v} \, dv,$$

over the volume of the CONFIGURATION of a given SUB–BODY **R** at time *t*, where **x** is the position and **v** the velocity of the points of **R**, ρ is the DENSITY, and **p** is the position vector of P.

angular velocity, *n.* **1.** a measure of the rate of rotation of a rigid body or point around a fixed axis, given by a vector parallel to the axis of rotation with magnitude equal to the RATE OF CHANGE of the angle formed at that fixed point between successive POSITION RADII of the rotating object and some fixed direction; it is measured in the anticlockwise direction. The usual symbol is ω.

2. (*Continuum mechanics*) the AXIAL VECTOR of the BODY SPIN of a given BODY, or equivalently, one half of the VORTICITY.

anharmonic ratio, *n.* another word for CROSS RATIO.

Ann, *abbrev. for* ANNIHILATOR (sense 3).

annihilator, *n.* **1.** the collection of all functions of some prescribed type that take value zero on every member of a given set.

2a. in particular, the linear subspace of all continuous linear functionals whose value is zero on every member of a given set in a NORMED SPACE.

b. the ORTHOGONAL COMPLEMENT of a set in HILBERT SPACE.

See POLAR SET.

3. the set of elements of a ring whose product with every member of a given subset of a MODULE over the ring is the zero element of the ring; this set is an IDEAL of the ring, and is denoted Ann X, where X is the given subset.

annulus, *n.* **1.** also called **ring.** the region enclosed between two concentric circles; its area is $\pi(R^2 - r^2)$, where R and r are the radii of the larger and smaller circle respectively, as shown in Fig. 12, opposite.

Fig. 12. The shaded region is an **annulus**.

2. early twentieth-century word for RING (sense 2).

anomaly, *n.* another word for AMPLITUDE in polar coordinates.

antecedent, *n.* (*Logic*) the hypothetical clause, the clause that implies the other, in a CONDITIONAL STATEMENT. For example, *Fafner is a dragon* is the antecedent of *Fafner breathes fire if he is a dragon*. Compare CONSEQUENT.

anti-, *prefix.* indicating the INVERSE of a function. For example, the *antitrigonometric* and the *antihyperbolic* functions are the respective inverses of the TRIGONOMETRIC and the HYPERBOLIC functions. These functions are sometimes symbolized by prefixing 'a-' to the name of the function, or more usually by writing ' $^{-1}$ ' superscript, as in $\sin^{-1}$ for the inverse sine, or $\coth^{-1}$ for the inverse hyperbolic cotangent. See also ANTILOGARITHM.

anticlastic, *adj.* (of a surface) having CURVATURES of opposite signs in two perpendicular directions at a given point; saddle-shaped. For example, in the surface shown in Fig. 13, X is a minimum between A and B, but a maximum between C and D. Compare SYNCLASTIC. See also SADDLE POINT.

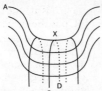

Fig. 13. An **anticlastic** surface.

anticlockwise or **counterclockwise,** *adj.* or *adv.* (of rotation, an angle, etc.) in a direction opposite to that conventionally taken by the hands of a clock. If a horizontal line segment has its left-hand end, A, fixed, its right-hand end, B, moves anticlockwise when it attempts to move upwards, as depicted in Fig. 14. For the purpose of measuring angles, fixing points by POLAR COORDINATES, etc., this is conventionally taken to be the positive direction. Compare CLOCKWISE.

Fig. 14. B turns **anticlockwise** about A.

antiderivative or **primitive,** *n.* a function whose DERIVATIVE is a given function; for example, $\log x$ is an antiderivative for $1/x$. See ANTIDIFFERENTIATE.

antidesignated, (*Logic*) see DESIGNATED.

antidifferentiate, *vb.* to find an ANTIDERIVATIVE for a given function; especially when deriving an INDEFINITE INTEGRAL, or when evaluating a DEFINITE INTEGRAL using the FUNDAMENTAL THEOREM OF CALCULUS rather than as an infinite sum of infinitesimal elements. Compare INTEGRATE.

antilog, *n. abbrev. for* ANTILOGARITHM. If no BASE is specified, it is usually taken to be 10.

antilogarithm, *n.* a number the LOGARITHM of which to a given BASE is a given number; that is, the result of raising the given base to the given power, usually written *alog* or *antilog*, or, in the case of NATURAL LOGARITHMS, *aln*. The antilogarithm of 2 to the base 8, written antilog$_8$2, alog$_8$2 or log$_8^{-1}$2, is $8^2 = 64$; the *common antilogarithm* of x is usually written without a subscript as alog$x$, log^{-1}x or antilog$x$, and equals alog$_{10}x$; and the *natural antilogarithm* of x, written aln$x$, is exp$x$. See EXPONENTIAL, LOGARITHM, NATURAL LOGARITHM.

antiparallel, *adj.* (of a pair of lines) **1.** cutting two given PARALLEL lines in such a way that the interior opposite angles of the QUADRILATERAL so formed total two right angles. For example, in Fig. 15, given a pair of parallel lines AB and CD, and a TRANSVERSAL, XY, making an angle of θ with them, MN is antiparallel to XY since it makes an angle of θ with the opposite directions of the given parallels.

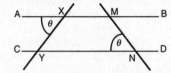

Fig. 15. The bold lines are **antiparallel** with respect to the others.

2. (of a pair of DIRECTED LINES) having the same DIRECTION but opposite SENSE, so that given any line segment AB, the directed lines $\overrightarrow{AB}$ and $\overrightarrow{BA}$ are antiparallel.

antipodal points or **antipodes,** *n.* two points at opposite ends of a diameter, such as A and B in the sphere in Fig. 16.

Fig. 16. **Antipodal points**.

antiprism, *n.* the solid, first recognised by Kepler, obtained from a PRISM by rotating the two identical faces relative to one another so that the vertices of one correspond to the sides of the other, and then joining the vertices of the two faces alternately. The lateral faces are then all triangular, and if these triangles are all EQUILATERAL, the antiprism is said to be SEMI-REGULAR.

antisymmetric, *adj.***1.** (*Logic*) (of a relation) never holding between a pair of elements in one order when it holds between them in the other, except when $x = y$; that is, ordered pairs of elements $\langle x, y \rangle$ and $\langle y, x \rangle$ can not both have the relation unless the relata are identical. For example, *less than or equal to* ($\leq$), and *no younger than* are antisymmetric relations. Any ASYMMETRIC relation is *a fortiori* antisymmetric, as is any empty relation. Compare NON-SYMMETRIC, SYMMETRIC.

2. (of a matrix) another term for SKEW-SYMMETRIC.

antitone, *n.* another term for MONOTONE DECREASING.

Apery's theorem, *n.*, the result recently proven by the French mathematician *Apery* (1916–) that the value of the ZETA FUNCTION at 3 is IRRATIONAL.

apex, *n.* the highest vertex in a given orientation of a polygon, especially a triangle; the vertex opposite the BASE. For example, A is the apex in the orientation of the pentagon shown in Fig. 17.

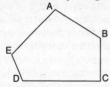

Fig. 17. A is the **apex** in this orientation.

Apollonian packing, *n.* a packing of the interior of an equilateral curvilinear triangle, V, with closed disks B_i. It can be shown by FRACTAL theory that the dimension of

$$V \setminus \bigcup_{i=1}^{\infty} B_i$$

is greater than one.

Apollonius of Perga, (c. 255–170 BC) Greek geometer who wrote widely on pure and applied mathematics, and improved on Aristotle's approximation for π; his only surviving work is his treatise on CONICS.

Apollonius' circle, *n.*, the circle consisting of the LOCUS of points the ratio of the distances of which from two given points is a fixed number; the locus of the apexes of all triangles on a given base the other two sides of which are in a fixed proportion. It is a circle the diametral points of which on the extension of the base are HARMONIC POINTS.

a posteriori, *adj.* **1.** (*Logic*) empirical; not able to be known entirely independently of experience. For example, the fact that all cats are agile is known *a posteriori*, while the fact that all cats are mammals is known A PRIORI; this is because a cat's agility is merely a matter of observable fact, whereas its being a mammal is a matter of definition. This is an epistemological property and so is distinct from the logical property of being SYNTHETIC.
2. (*Statistics*) another term for POSTERIOR. See EMPIRICAL PROBABILITY.

apothem, *n.* **1.** a line from the centre of a regular polygon perpendicular to one of its sides, such as the bold line in the regular hexagon of Fig. 18.
2. the length of such a line.

Fig. 18. The bold line is an **apothem**.

application, *n.* (*Logic*) **1.** the process of determining the value of a function for a given argument.
2. (*Combinatory logic*) the primitive binary function $(x, y) = x(y)$, equivalent to LAMBDA–CONVERSION.

applied, *adj.* related to or put to practical use, as in APPLIED MATHEMATICS. Compare PURE.

applied mathematics, *n.* the branch of mathematics that is concerned to describe (or MODEL), natural, social, or technological processes in mathematical terms, and that therefore has practical application to influence or predict the effects of any phenomena that are amenable to mathematical description and analysis. Although economics, linguistics, music, etc., are all thus within the subject matter of applied mathematics, the term is often used more restrictively to refer only to topics arising within physics and its technological applications, or even more narrowly to mechanics. The boundary between pure and applied mathematics is not sharp, since any practical problem can be considered in the abstract, and conversely any topic of pure enquiry may have unforeseen applications. It is thus not the content but the intent of the practitioner that determines what counts as applied mathematics.

approximate, *vb.* to find an expression for (some quantity) ACCURATE to a specified degree.

approximate line search, see LINE SEARCH METHOD.

approximation, *n.* **1.** an estimate of the value of some quantity, ACCURATE to a desired degree.

2. any expression in simpler terms than, and approximately equivalent to, a given expression; for example, a function or sequence asymptotic to a given function or sequence.

a priori, *adj.* **1a.** relating to or involving deductive reasoning.

b. (*Logic*) strictly, able to be known without appeal to experience. For example, any matter of definition, such as that all cats are mammals, is knowable *a priori*. However, this is an epistemological property and so distinct from the logical property of being ANALYTIC; in fact Kant claimed that mathematics is *a priori* but not analytic. Compare *A POSTERIORI*.

2. (*Statistics*) another term for PRIOR. See MATHEMATICAL PROBABILITY.

apse or **apsidal point,** *n.* any point at which the direction of motion of a point moving round a closed curve is perpendicular to its radius vector. Every point of a circle is thus an apse, and the apsidal points of an ellipse are the endpoints of its axes.

Arabic numerals, *n.* the sequence of symbols *0, 1, 2, 3, 4, 5, 6, 7, 8, 9* that represent the successive units of the PLACE–VALUE counting system with base 10 (the DECIMAL SYSTEM). They reached the West in the Middle Ages through translations of Arabic mathematical texts (although they are believed to have originated as sanskrit numerals in India), and the ease of calculation that resulted from the adoption of a place-value system revolutionized Western mathematics. Compare ROMAN NUMERALS.

arbitrary constant, *n.* a non-numerical symbol that represents an unspecified CONSTANT, typically in generalized expressions. For example, in the general linear equation $y = ax + b$, a and b are arbitrary constants, while x and y are VARIABLES, and y is thus not conceived of as a function of a or b. See also CONSTANT OF INTEGRATION, PARAMETER.

arc, *n.* **1a.** any continuous section of a curve, graph or geometric figure; formally, the image of the unit interval under a continuous function. See also PATH, CONNECTED.

b. in particular, a section of the circumference of a circle lying between two points on the circle. Any line that intersects a circle thus divides it into two arcs; the longer is the *major arc* (ADC in Fig. 19), and the shorter is the *minor arc* (ABC).

Fig. 19. ABC is the minor **arc**, and ADC the major **arc**.

c. (for some authors) more restrictively, a HOMEOMORPHIC image of the unit interval. In complex analysis it is fairly usual to require that an arc is also SMOOTH.

2. an EDGE of a NETWORK or DIGRAPH.

3. see MINUTE OF ARC.

arc-, *prefix.* denoting the INVERSE function of a given TRIGONOMETRIC or HYPERBOLIC FUNCTION, usually written arcsin, asin, or sin⁻¹, arctanh, atanh, or tanh⁻¹, etc. Thus, for example, $x = \text{arcsech} y$ if and only if $y = \text{sech} x$.

arc-connected or **arcwise connected**, *adj.* (of a TOPOLOGICAL SPACE)
1. another word for PATH–CONNECTED.

2. (more rigorously, in those usages that distinguish) having the property that any two points can be connected by ARCS rather than by PATHS, that is by HOMEOMORPHIC images of the unit interval.

arc-cosecant, *n.* written **cosec⁻¹**, **csc⁻¹**, or **acsc.** the inverse of the COSECANT function, so that its value for any argument is an angle in radians whose cosecant equals the given argument; that is, $y = \text{cosec}^{-1} x$ if and only if $x = \text{cosec} y$. It is defined for arguments less than −1 or greater than 1, and its PRINCIPAL VALUES (often written *Cosec⁻¹y*) are by convention taken to lie between $-\pi/2$ and $\pi/2$. The graph of these principal values is shown in Fig. 20, and it can be seen that the function can never take the value 0.

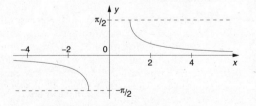

Fig. 20. **Arc-cosecant**. Graph of the principal values of the inverse cosecant.

(Some authors instead give the range of the principal values as the two intervals $-\pi < y \le -\pi/2$ and $0 < y \le \pi/2$.) The derivative of $\text{cosec}^{-1} x$ is

$$\frac{-1}{x\sqrt{x^2 - 1}},$$

and an antiderivative or indefinite integral is

$$x \, \text{cosec}^{-1} x + \ln[x + \sqrt{(x^2 - 1)}].$$

arc-cosech, *n.* written **cosech**$^{-1}$, **csch**$^{-1}$, or **acsch.** the inverse of the HYPERBOLIC COSECANT function, so that for any argument the hyperbolic cosecant of its value equals the given argument; that is, $y = \text{cosech}^{-1}x$ if and only if $x = \text{cosech}y$. It is defined for all non-zero arguments, and its graph is shown in Fig. 21. The derivative of $\text{cosech}^{-1}x$ is

$$\frac{-1}{|x|\sqrt{1+x^2}},$$

and an antiderivative or indefinite integral is

$$x\,\text{cosech}^{-1}x + \frac{x}{|x|}\sinh^{-1}x.$$

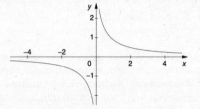

Fig. 21. **Arc-cosech.** Graph of the inverse hyperbolic cosecant function.

arc-cosh, *n.* written **cosh**$^{-1}$, **ch**$^{-1}$, or **acosh.** the inverse of the HYPERBOLIC COSINE function, so that $y = \cosh^{-1}x$ if and only if $x = \cosh y$. It is defined for arguments greater than or equal to 1, when

$$\cosh^{-1}x = \ln[x + \sqrt{(x^2 - 1)}];$$

and its PRINCIPAL VALUES (often written $Cosh^{-1}y$) are by convention taken to be the non-negative ones shown in the graph in Fig. 22. The derivative of $\cosh^{-1}x$ is

$$\frac{-1}{\sqrt{x^2 - 1}},$$

and an antiderivative or indefinite integral is

$$x\,\cosh^{-1}x - \sqrt{x^2 - 1}.$$

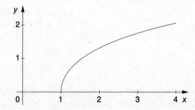

Fig. 22. **Arc-cosh.** Graph of the principal values of the inverse hyperbolic cosine.

arc-cosine, *n.* written **cos**$^{-1}$, **arccos**, or **acos.** the inverse of the COSINE function, so that for any argument the value of the arc-cosine function is an angle in radians whose cosine equals the given argument; that is, $y = \cos^{-1}x$ if and only if $x = \cos y$. It is defined for arguments between –1 and 1, and its PRINCIPAL VALUES (often written $Cos^{-1}y$) are by convention taken to be those between

0 and π. The graph of these principal values is shown in Fig. 23.

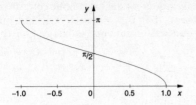

Fig. 23. **Arc-cosine**. Graph of the principal values of the inverse cosine.

The derivative of $\cos^{-1}x$ is

$$\frac{-1}{\sqrt{1-x^2}},$$

and an antiderivative or indefinite integral is

$$x\cos^{-1}x - \sqrt{1-x^2}.$$

arc-cotangent, *n.* written **cotan⁻¹**, **cot⁻¹**, **ctn⁻¹**, or **actn.** the inverse of the COTANGENT function, so that its value for any argument is an angle in radians whose cotangent equals the given argument; that is, $y = \cotan^{-1}x$ if and only if $x = \cotan y$. It is defined for all real arguments and its PRINCIPAL VALUES (often written $Cotan^{-1}y$, $Cot^{-1}y$, etc.) are by convention taken to be those strictly between 0 and π. These principal values is shown in Fig. 24.

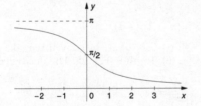

Fig. 24. **Arc-cotangent**. Graph of the principal values of the inverse cotangent.

The derivative of $\cotan^{-1}x$ is

$$\frac{-1}{1+x^2},$$

and an antiderivative or indefinite integral is

$$x\cotan^{-1}x + \tfrac{1}{2}\ln(1+x^2).$$

arc-cotanh or **arc-coth,** *n.* written **cotanh⁻¹**, **coth⁻¹**, or **acoth.** the inverse of the HYPERBOLIC COTANGENT function, so that for any argument the hyperbolic cotangent of its value equals the given argument; that is, $y = \cotanh^{-1}x$ if and only if $x = \cotanh y$. It is defined for arguments less than –1 or greater than 1, and its graph is shown in Fig. 25 overleaf. The derivative of $\cotanh^{-1}x$ is

$$\frac{1}{1-x^2},$$

and an antiderivative or indefinite integral is

$$x\cotanh^{-1}x + \tfrac{1}{2}\ln(1-x^2).$$

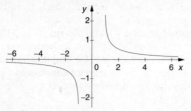

Fig. 25. **Arc-cotanh**. Graph of the inverse hyperbolic tangent.

Archimedean property, *n.* **1.** the ORDER AXIOM for the real line that states that if *a* and *b* are real numbers such that $a < b/n$ for all natural numbers *n* then $a \leq 0$, or equivalently, that for any positive *a* and *b* there is a positive integer *n* such that $a < nb$, and thus that every real number is less than some natural number. This is equivalent to the assertion that the real numbers are CON-DITIONALLY COMPLETE. An INFINITESIMAL is non-Archimedean as it is less than any positive non-zero number. See also COFINAL, DENSE, NON–STANDARD ANALYSIS.

2. the corresponding property of a PARTIAL ORDER on an ORDERED VECTOR SPACE. This fails in the LEXICAL ORDER on Euclidean 2-space.

Archimedean solid, *n.* one of the thirteen SEMI-REGULAR polyhedra, including the ICOSIDODECAHEDRON. Compare PLATONIC SOLID.

Archimedean (or **Archimedes'**) **spiral,** *n.* a SPIRAL with polar equation $r = a\theta$; its graph is shown in Fig. 26.

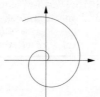

Fig. 26. **Archimedean spiral**.

Archimedes, (c. 287–212 BC) Greek mathematician, physicist, and inventor, generally regarded as the greatest mathematician of antiquity. His rigorous geometrical technique of measuring curved lines, areas, and surfaces anticipated modern calculus, and he laid the foundations of mechanics, statics and hydrostatics.

Archimedes' method, *n.* the method of estimating π (PI) by computing the area or perimeter of INSCRIBED and CIRCUMSCRIBED polygons for a unit circle using more and more sides. Archimedes estimated that

$$3 \tfrac{10}{71} < \pi < 3 \tfrac{1}{7}$$

from computations with polygons with 96 sides. This gives 3.14...; that is, it is accurate to two decimal places. Compare EUDOXUS' AXIOM.

Archimedes' spiral, *n.* another name for ARCHIMEDEAN SPIRAL.

arc length, *n.* the LENGTH (in sense 2) of an ARC of a curve.

arcograph, *n.* another name for CYCLOGRAPH.

arc-secant, *n.* written **sec⁻¹**, **arcsec**, or **asec.** the inverse of the SECANT function, so that for any argument the value of the arc-secant is an angle in radians whose secant equals the given argument; that is, $y = \sec^{-1}x$ if and only if $x = \sec y$. It is defined for arguments between –1 and 1, and its PRINCIPAL VALUES (often written $Sec^{-1}\theta$) are by convention taken to lie between 0 and π. The graph of these principal values is shown in Fig. 27, and it can be seen that the function never takes the value $\pi/2$. (Some authors instead give the range of the principal values as the two intervals $-\pi \leq \theta < -\pi/2$ and $0 \leq \theta < \pi/2$.) The derivative of $\sec^{-1}x$ is

$$\frac{1}{x\sqrt{x^2 - 1}},$$

and an antiderivative or indefinite integral is

$$x\sec^{-1}x - \ln[x + \sqrt{x^2 - 1}].$$

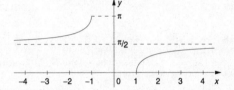

Fig. 27. **Arc-secant.** Graph of the principal values of the inverse secant.

arc-sech, *n.* written **sech⁻¹** or **asech.** the inverse of the HYPERBOLIC SECANT function, so that for any argument the hyperbolic secant of its value equals the given argument; that is, $y = \operatorname{sech}^{-1}x$ if and only if $x = \operatorname{sech} y$. It is defined for arguments between 0 and 1, and its PRINCIPAL VALUES (often written $Sech^{-1}y$) are by convention taken to be the positive ones shown in the graph in Fig. 28. The derivative of $\operatorname{sech}^{-1}x$ is

$$\frac{-1}{x\sqrt{1 - x^2}},$$

and an antiderivative or indefinite integral is

$$x\operatorname{sech}^{-1}x + \sin^{-1}x.$$

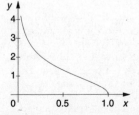

Fig. 28. **Arc-sech.** Graph of the principal values of the inverse hyperbolic secant.

arc-sine, *n.* written **sin⁻¹**, **arcsin**, or **asin.** the inverse of the SINE function, so that its value for any argument is an angle in radians whose sine equals the given argument; that is, $y = \sin^{-1}x$ if and only if $x = \sin y$. It is defined for arguments between –1 and 1, and its PRINCIPAL VALUES (often written $Sin^{-1}y$) are by convention taken to be those between $-\pi/2$ and $\pi/2$. The graph of these principal values is shown in Fig. 29.

The derivative of $\sin^{-1}x$ is

$$\frac{1}{\sqrt{1 - x^2}},$$

and an antiderivative or indefinite integral is

$$x \sin^{-1}x + \sqrt{1 - x^2}.$$

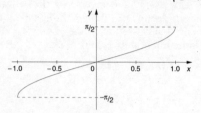

Fig. 29. **Arc-sine**. Graph of the principal values of the inverse sine.

arc–sinh, *n.* written **sinh⁻¹**, **sh⁻¹**, or **asinh.** the inverse of the HYPERBOLIC SINE function, so that $y = \sinh^{-1}x$ if and only if $x = \sinh y$, defined for all real x by

$$\sinh^{-1}x = \ln[x + \sqrt{(x^2 + 1)}];$$

its graph is as shown in Fig. 30.

The derivative of $\sinh^{-1}x$ is

$$\frac{1}{\sqrt{x^2 + 1}},$$

and an antiderivative or indefinite integral is

$$x \sinh^{-1}x - \sqrt{1 + x^2}.$$

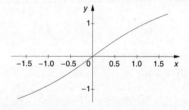

Fig. 30. **Arc-sinh**. Graph of the inverse hyperbolic sine.

arc-tangent, *n.* written **tan⁻¹**, **arctan** or **atn.** the inverse of the TANGENT function, so that its value for any argument is an angle in radians whose tangent equals the given argument; that is, $y = \tan^{-1}x$ if and only if $x = \tan y$. It is defined for all real arguments and its PRINCIPAL VALUES (often written $Tan^{-1}y$) are by convention taken to be those strictly between $-\pi/2$ and $\pi/2$. The graph of these principal values is shown in Fig. 31 opposite.

The derivative of $\tan^{-1}x$ is

$$\frac{1}{1 + x^2},$$

and an antiderivative or indefinite integral is

$$x \tan^{-1}x - \tfrac{1}{2} \ln(1 + x^2).$$

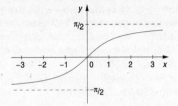

Fig. 31. **Arc-tangent**.
Graph of the principal
values of the inverse
tangent.

arc-tanh, *n.* written **tanh⁻¹, th⁻¹**, or **atanh.** the inverse of the HYPERBOLIC TANGENT function, so that for any argument the hyperbolic tangent of its value equals the given argument; that is, $y = \tanh^{-1}x$ if and only if $x = \tanh y$. It is defined for arguments between -1 and 1, and its graph is shown in Fig. 32. The derivative of $\tanh^{-1}x$ is

$$\frac{1}{1 - x^2},$$

and an antiderivative or indefinite integral is

$$x\tanh^{-1}x + \tfrac{1}{2}\ln(1 - x^2).$$

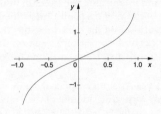

Fig. 32. **Arc-tanh**.
Graph of the inverse
hyperbolic tangent.

area, *n.* **1a.** part of a two-dimensional surface enclosed within a specified boundary or geometric figure.
b. the measure or extent of such a or part of a surface.
2. the two-dimensional extent of the surface of a solid or of some part thereof, especially one bounded by a closed curve. Thus one speaks of the area of a sphere, meaning the area of its surface. See SURFACE AREA.

arg., *n. abbrev. and symbol for* the ARGUMENT of a complex number. The PRINCIPAL VALUE is usually taken to be in the range $-\pi < \theta \leq \pi$ and denoted *Arg*.

Argand diagram or **Gaussian plane,** *n.* a diagram in which COMPLEX NUMBERS are represented by the points in the CARTESIAN PLANE the coordinates of which are respectively the REAL and IMAGINARY PARTS of the given number, so that the complex number $a + ib$ is represented by the point (a, b), or by

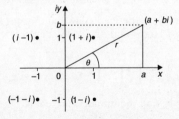

Fig. 33. **Argand diagram.**
Point (a,b) represents the
complex number $a + bi$.

the corresponding POSITION VECTOR $\langle a, b \rangle$. If, as shown in Fig 33, the POLAR COORDINATES of (a, b) are (r, θ), r is the MODULUS and θ the AMPLITUDE of $a + ib$; also shown are the points $(\pm 1, \pm 1)$, representing the four complex numbers $1+i$, $1-i$, $i-1$, and $-1-i$. (Named after the Swiss accountant and amateur mathematician *Jean Argand* (1768–1822).)

argument, *n.* **1.** an element to which an operation, function, predicate, etc. applies; especially the INDEPENDENT VARIABLE of a function.

2. another word for the AMPLITUDE or PHASE of a complex number. Compare MODULUS.

3. (*Logic*) **a.** a process or instance of INDUCTIVE or DEDUCTIVE reasoning that purports to show its conclusion to be true.

b. a sequence of statements one of which is the CONCLUSION and the remainder the PREMISES.

c. formally, an ordered pair the first element of which is a set of statements (the premises), and the second element of which is a single statement (the conclusion).

Aristotle (384–322 BC), Greek philosopher and scientist who studied under Plato, tutored Alexander the Great, and established the Lyceum in opposition to the Platonic Academy. Partly because of his profound influence on medieval Christianity and the incorporation of his doctrines into those of the Church, a large number of his works have come down to us, on subjects ranging from syllogistic logic, theories of meaning, and metaphysics to zoology, cosmology, and aesthetics.

Aristotelian logic, *n.* the logical theories of Aristotle, especially as developed in the Middle Ages, concerned mainly with the principle of the SYLLOGISM; traditional as contrasted with modern, SYMBOLIC or MATHEMATICAL LOGIC.

arith., *abbrev. for* ARITHMETIC or ARITHMETICAL.

arithmetic, *n.* **1.** the branch of mathematics concerned with numerical calculations, such as addition, subtraction, multiplication, division and the extraction of roots.

2. the higher arithmetic. another term for NUMBER THEORY.

arithmetic or **arithmetical,** *adj.* involving or pertaining to ARITHMETIC.

arithmetic function, *n.* (*Number theory*) any function defined on the NATURAL NUMBERS or integers; equivalently, a SEQUENCE viewed functionally. For example, the Euler phi function is an arithmetic function.

arithmetic-geometric mean (abbrev. **agm**), *n.* the common limit of the series of ARITHMETIC MEANS and GEOMETRIC MEANS obtained by ARITHMETIC-GEOMETRIC MEAN ITERATION.

arithmetic-geometric mean inequality, *n.* the inequality that the ARITHMETIC MEAN of a set of numbers is always larger than the GEOMETRIC MEAN of the numbers, that is,

$$\frac{1}{n} \left[\sum_{i=1}^{n} a_i \right] \geq \left[\prod_{i=1}^{n} a_i \right]^{\frac{1}{n}}$$

with equality occurring if and only if all the numbers are equal.

arithmetic-geometric mean iteration, *n.* the two-term iteration in which one repeatedly computes both the ARITHMETIC MEAN and the GEOMETRIC MEAN of two positive numbers: $a_0 = a$, $b_0 = b$ and

$$a_{n+1} = \tfrac{1}{2}(a_n + b_n); \quad b_{n+1} = \sqrt{(a_n b_n)}$$

This converges quadratically to a common limit $M(a, b)$, which was discovered by Gauss to satisfy

$$M(1, b) = \frac{\pi}{2K\sqrt{1 - b^2}}$$

when normalized so that $a = 1 > b$. Here K is the COMPLETE ELLIPTIC INTEGRAL of the first kind. This leads to a very rapid method of computing elliptic integrals.

arithmetic mean or **mean,** *n.* the average of a set of numbers or quantities calculated by dividing their sum by the number of terms. For example, the arithmetic mean of 3, 4 and 8 is 5. Compare GEOMETRIC MEAN.

arithmetic progression, *n.* a SEQUENCE of numbers or quantities, each differing by a constant amount (the *common difference*) from its predecessor; for example, the sequence 3, 6, 9, 12, 15,.... The nth term of a progression, whose first term is a and whose common difference is d, is $a + (n-1)d$. See also ARITHMETIC SERIES. Compare GEOMETRIC PROGRESSION.

arithmetic sequence, another term for ARITHMETIC SEQUENCE.

arithmetic series, *n.* the sum of the terms of an ARITHMETIC PROGRESSION, for example $3 + 6 + 9 + 12 +....$. The sum of the first n terms of such a series, whose first term is a and whose common difference is d, is

$$na + \tfrac{1}{2}n(n - 1)d.$$

arity, *n.* the number of ARGUMENTS of a function or relation. See also N-ARY.

arm, *n.* **1.** either of the lines forming an ANGLE.

2. either of the sides of a RIGHT-ANGLED TRIANGLE other than the HYPOTENUSE.

Armijo's method, *n.* a popular method of LINE SEARCH METHOD, in which for fixed positive quantities s, α and β, one moves from x to

$$x - s\,\beta^m\,\nabla f(x)$$

where m is the smallest non-negative integer for which

$$f(x) - f(x - s\,\beta^m\,\nabla f(x)) \geq \alpha s\,\beta^m \|\nabla f(x)\|^2.$$

arrangement, *n.* **1.** (*Combinatorics*) a PERMUTATION (*ordered arrangement*) or COMBINATION (*unordered arrangement*) of a set of objects.

2. (*Algebra*) any ordered sequence of elements. In this sense, 3,1,2 is an arrangement, whereas the permutation 1,2,3 → 3,1,2 is an operation upon arrangements.

array, *n.* **1.** a two-dimensional arrangement of numbers or symbols in rows and columns in such a manner that two arrays are identical if and only if they have the same number of rows and the same number of columns, and the entries which occupy the same position in the respective arrays are equal; for example, a MATRIX, COLUMN VECTOR, or DETERMINANT.

2. a higher-dimensional analogue of the above.

arrow or **morphism,** *n.* a generalization in CATEGORY THEORY of the concept of a MAPPING. See also DIAGRAM OF ARROWS.

arrow paradox, *n.* the classical paradox that the motion of an arrow is illusory, since an object in flight always occupies a space equal to itself, but what occupies a space equal to itself is not moving, and so the arrow is always at rest. See ZENO'S PARADOXES.

Arrow's impossibility theorem, *n.* the celebrated result that there is no way of consistently aggregating the distinct PREFERENCES of more than two indivi- duals in such a way as to satisfy four conditions each of which by itself seems intuitively reasonable, when the aggregation is required to yield a COMPLETE, TRANSITIVE, and REFLEXIVE ordering of their collective preferences. The four conditions are: that the collective ordering must be applicable in all cases; that a preference shared by all the individuals should be reflected in the joint preference; that the collective ordering should be independent of irrele- vant alternatives (that is, alternatives not available in the given choice); and that no individual has dictatorial power (that is, no one ordering invariably determines the joint ordering). (Named after the American economist *Kenneth Arrow,* who was awarded the Nobel prize in 1972.)

artificial variable, see SLACK VARIABLE.

Artinian module, *n.* a MODULE that satisfies the DESCENDING CHAIN CONDITION, so that every strictly descending (decreasing) chain of submodules is finite; this is equivalent to the satisfaction of the MINIMUM CONDITION. Every Artinian module is also a NOETHERIAN MODULE, but not necessarily vice versa; for example, the integers are a Noetherian but not an Artinian $\mathbb{Z}$-module. (Named after *Emil Artin* (1898–1962), German-born American algebraist and group theorist.)

Artinian ring, *n.* a RING that, regarded as a (left or right) R-MODULE, is an a ARTINIAN MODULE.

Artin's conjecture on primitive roots, *n.* a quantitative form of the conjecture that every non-square integer is a PRIMITIVE ROOT of infinitely many primes. The quantitative conjecture is known to follow from an extended form of the RIEMANN HYPOTHESIS.

Arzela–Ascoli theorem, *n.* the complex case of the ASCOLI THEOREM.

ascending chain condition, *n.* the condition on SUBMODULES that no *ascending chain*

$$I_1 \subseteq I_2 \subseteq I_3 \subseteq \dots ,$$

of which each member is contained in the next, has more than a finite number of distinct members, equivalent to the MAXIMUM CONDITION that every non-empty set of submodules has a maximal element. See NOETHERIAN MODULE. Compare DESCENDING CHAIN CONDITION.

ASCII, *acronym for* American Standard Code for Information Interchange; a BINARY CODE used in computing to represent letters, numbers and other standard symbols.

Ascoli's theorem, *n.* the result that a family of functions that is POINT–WISE BOUNDED and EQUI–CONTINUOUS on a compact space is TOTALLY BOUNDED in the SUPREMUM NORM. Every sequence in such a family will contain a norm convergent subsequence. In the complex case, this is known as the ARZELA–

ASCOLI THEOREM. (Named after the Italian analyst *Giulio Ascoli* (1843–96).) See NORMAL FAMILY.

asec, *abbrev. and symbol for* ARC–SECANT, the inverse SECANT function.

asech, *symbol for* the inverse HYPERBOLIC SECANT function. See ARC–SECH.

asin, *abbrev. and symbol for* ARC–SINE, the inverse SINE function.

asinh, *symbol for* the inverse HYPERBOLIC SINE function. See ARC–SINH.

assignment, *n.* **1.** (*Logic*) a function that associates specific elements of a domain with each free variable in a FORMAL CALCULUS. Compare INTERPRETATION. See also MODEL, VALUATION.

2. (*Computing*) a statement in a program that allocates a value to a variable, usually written as in

$$x := y + z.$$

If the new value is a function of the previous value of that same variable this notation is still used; thus

$$x := x + 1$$

is an instruction to increase the value of x by one.

assignment problem, *n.* one of a class of COMBINATORIAL ANALYSIS, QUADRATIC PROGRAMMING, or LINEAR PROGRAMMING problems derived from the problem of matching individuals and tasks, often in order to maximize job satisfaction or some other measure of appropriateness.

associate numbers, *n.* any pair of elements of an INTEGRAL DOMAIN that are UNIT multiples of each other, such as $3 + i$ and $3i - 1$ in the GAUSSIAN INTEGERS.

associative, *adj.* **1.** (of a binary operation) having the property that the bracketing of its arguments may be disregarded since

$$(a \bullet b) \bullet c = a \bullet (b \bullet c)$$

where $\bullet$ is the operator. For example, conjunction and multiplication are associative, but vector product is not.

2. (of an algebraic structure) possessing an associative operator.

associative law, *n.* the theorem or axiom of any particular calculus or mathematical system that a given operation is ASSOCIATIVE.

assumption, *n.* (*Logic*) a statement that is taken to be true for the purposes of a particular argument and is used as a premise in order to infer its consequences, but that may not be otherwise accepted. Compare AXIOM.

astroid or **star curve,** *n.* a HYPOCYCLOID with four CUSPS; a curve such as that in Fig. 34, with parametric equations $x = \cos^3 t$, $y = k \sin^3 t$.

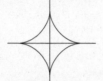

Fig. 34. **Astroid**.

asymmetric, *adj.* **1.** not SYMMETRIC.

2. (*Logic*) (of a relation) never holding between a pair of elements in one order when it holds between them in the other; that is, ordered pairs of elements $\langle x, y \rangle$ and $\langle y, x \rangle$ can never both have the relation. For example the relation *father of* is asymmetric, but *brother of* is not, since one may be one's brother's brother, but never one's father's father. Compare SYMMETRIC, ANTI–SYMMETRIC, NON–SYMMETRIC.

asymptote, *n.* **1.** (*Euclidean geometry*) a straight line whose perpendicular distance from a curve decreases to zero as the distance from the origin increases without limit. Often it is required that the line actually be tangent to the curve at infinity. For example, the curve $y = 1/x$, shown in Fig. 35, has a vertical asymptote at zero and a horizontal asymptote at infinity. The rotated hyperbola $x^2 - y^2 = 1$ has 45° asymptotes at plus and minus infinity.

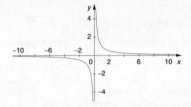

Fig. 35. $y = 1/x$ has the axes as **asymptotes**.

2. (*Augmented Euclidean geometry, affine geometry*) a tangent to a curve at its intersection with the line at infinity.

asymptotic, *adj.* **1.** (of a function, series, etc.) approaching a given value (its ASYMPTOTE) arbitrarily closely, as the independent variable, or some expression containing a variable, approaches a limit or tends to infinity. Graphically, the perpendicular distance between a curve and its asymptote tends to zero as the distance from the origin tends to a limit or infinity; in Fig. 36 this limit is zero.

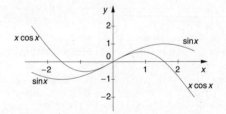

Fig. 36. Curves can be **asymptotic** at infinity or at a point.

2. (of a pair of functions) approaching one another infinitely closely as their arguments tend to infinity or some other value; this is often written as $f(x) \sim g(x)$, when

$$\lim_{x \to \infty} \frac{f(x)}{g(x)} = 1$$

For example, as shown in Fig. 36 above,

$$\sin x \sim x \cos x$$

as x tends to zero, and

$$2 \cosh x \sim \exp x$$

as x tends to infinity. See ORDER NOTATION. See also PRIME NUMBER THEOREM.

asymptotically stable, see STABLE.

asymptotic density, *n.* (of a sequence of positive integers) see SCHNIRELMANN DENSITY.

asymptotic direction, *n.* (at a point on a surface) any direction in which NORMAL CURVATURE vanishes, the contact of the tangent plane being of at least third order.

asymptotic expansion, *n.* (for a function f) any divergent series of the form

$$S_n(z) = a_0 + \frac{a_1}{z} + \dots + \frac{a_n}{z^n} + \dots$$

and such that for all n

$$\lim_{z \to \infty} z^n \left[S_n(z) - f(z) \right] = 0$$

which may then be written

$$f(z) \sim a_0 + \frac{a_1}{z} + \dots .$$

Estimates of varying degrees of accuracy thus produced play an important role in analysis, especially with $n = 1$. Two functions may have the same asymptotic expansion; for example, $e^{1/z}$ and $e^{1/z} + e^{-z}$ both have asymptotic expansion, namely

$$1 + \frac{1}{z.1!} + \frac{1}{z.2!} + \dots$$

for $|\arg z| < \pi/2$. More generally, if f/g has an asymptotic expansion, then f is asymptotic to its product with g. See STIRLING'S FORMULA.

asymptotic stability, see STABLE.

atan or **atn,** *abbrev. and symbol for* ARC–TANGENT, the inverse TANGENT function.

atanh, *symbol for* the inverse HYPERBOLIC TANGENT function. See ARC–TANH.

atlas, *n.* a collection of CHARTS that cover a MANIFOLD. It is said to be a $C^{(r)}$-*atlas* if each pair of charts has a $C^{(r)}$-OVERLAP.

atom, *n.* **1.** (*Measure theory*) a set, often a point, in a MEASURE SPACE with strictly positive measure, and such that any subset of the set either has equal measure or has zero measure.

2. (*Lattice theory*) a minimal non-zero element in a BOOLEAN ALGEBRA.

atomic, *adj.* **1.** (*Logic*) (of a sentence, formula, etc.) having no internal structure at the appropriate level of analysis; for example, in predicate calculus, *Fa* is an *atomic sentence*, and *Fx* an *atomic predicate*. Compare COMPOUND STATEMENT.

2. also called **purely atomic.** (of a MEASURE) having a SUPPORT a countable subset of which has complement of zero measure. Compare NON-ATOMIC.

attainable set, *n.* another term for REACHABLE SET.

atto- (symbol **a**), *prefix* denoting a fraction of 10^{-18} of the physical units of the SYSTEME INTERNATIONAL.

attractor, *n.* (*Fractal theory*) an equilibrium state or set of states to which a DYNAMICAL SYSTEM converges. Formally, a closed set E such that for a given mapping f, $f(E)$ is contained in E, and for all x in some given set containing E, the distance from $f^m(x)$ to E tends to zero as m tends to infinity. It is usually required that the ORBIT of f is dense in E for some value of x. If the attractor, E, is a FRACTAL set, then it is said to be *strange*; for example, in the *Julia sets*, $z \to z^2 - \mu$ makes E strange for some values of μ.

augend, *n.* a number or quantity to which another, the ADDEND, is added.

augmentation, *n.* the enlargement of a set of equations or a matrix in linear programming, matrix theory, or control theory. See AUGMENTED MATRIX. Compare BORDERING.

augmented Euclidean geometry, *n.* a EUCLIDEAN GEOMETRY to which have been added a notional LINE AT INFINITY on which parallel lines meet and (unless the geometry is real) *complex points* that have complex coordinates in any system of CARTESIAN COORDINATES; it can be represented by all non-zero complex (or real) triples, where two triples represent the same point if they are complex multiples of one another. A point is IMPROPER if $z = 0$ or the ratio of x/z to y/z is complex; this yields POINTS AT INFINITY and complex points. Other points are PROPER POINTS with Cartesian coordinates $(x/z, y/z)$.

augmented Lagrangian, *n.* any of a number of combinations of a PENALTY FUNCTION term (usually quadratic) with a LAGRANGIAN, the intent being to produce algorithms that exploit the best features of both. For example, given the simple differentiable problem of minimizing $f(x)$ subject to the conditions $h_1(x) = 0, ..., h_n(x) = 0$, a standard augmented Lagrangian is

$$L(x, \lambda, \alpha) = f(x) + \sum_{i=1}^{n} \lambda_i h_i(x) + \alpha \sum_{i=1}^{n} h_i(x)^2,$$

for a positive parameter α, and real numbers λ_i. If the function $L(x, \lambda^*, \alpha)$ has a local minimum at x^*, while x^* and λ_i^* satisfy the KUHN–TUCKER CONDITIONS, then x^* must be a local minimum of the original problem. Conversely, under reasonable conditions, given a local minimum x^* of the original problem, and corresponding Lagrange multipliers λ_i^*, there will be a value α^* such that for $\alpha > \alpha^*$ the function $L(x, \lambda^*, \alpha)$ has a local minimum at x^*.

augmented matrix, *n.* a matrix derived from a given matrix by adjoining a constant vector to its columns, for example when using GAUSSIAN ELIMINATION to invert the matrix. More generally, any matrix of which the given matrix is a submatrix.

aut, *conj.* (*Logic*) another word for the EXCLUSIVE DISJUNCTION operator; it is the Latin word for 'or' in this sense, as distinct from VEL.

Aut, *abbrev. for* AUTOMORPHISM; AutS is the set of automorphisms of an algebraic structure, S.

autocorrelation or **serial correlation,** *n.* (*Statistics*) the condition occurring when successive items in a sequence are correlated so that their CORRELATION is not zero and they are not INDEPENDENT. Compare AUTOCOVARIANCE.

autocovariance, *n.* (*Statistics*) the condition occurring when successive items in a sequence are correlated so that their COVARIANCE is not zero and they are not INDEPENDENT. Compare AUTOCORRELATION.

automata theory, *n.* the mathematical study and modelling of certain ABSTRACT MACHINES and their capacity to solve various types of problem by means of the algorithms available to them. See TURING MACHINE.

automorphic function, *n.* an ANALYTIC function, *f*, on a domain D, such that for a GROUP of MÖBIUS TRANSFORMATIONS, T, T(*z*) is in D and *f*(T(*z*)) = *f*(*z*) for all *z* in D.

automorphism, *n.* an ISOMORPHISM the domain and range of which are identical, such as a PERMUTATION of a set.

autonomous, *adj.* **1.** (of a system of ORDINARY DIFFERENTIAL EQUATIONS) not explicitly depending on the variable of differentiation (often time); that is, such that there is no explicit occurrence of the independent variable in the equation $dy/dt = f(y)$. For example, the equation $dx/dt = x$ is autonomous, but the equation $dx/dt = t$ is not.
2. (of an ORDINARY DIFFERENTIAL EQUATION of the SECOND ORDER) able to be reduced to a FIRST–ORDER equation by writing

$$\frac{du}{dt} = p, \quad \frac{d^2u}{dt^2} = p\,\frac{dp}{du},$$

and regarding *u* as the independent variable, where *u* is the dependent variable, and *t* the independent variable of the given equation.

auxiliary equation, *n.* any simpler equation that assists the solution of a given more difficult one, usually obtained by taking transforms; especially an equation of the same form as a given DIFFERENTIAL EQUATION but with scalar variables replacing the derivatives. For example, the auxiliary equation of

$$\frac{d^2x}{dy^2} + b\,\frac{dx}{dy} + cy = 0$$

is $D^2 + bD + c = 0$.

average, *n.* **1.** the usual term for the ARITHMETIC MEAN of discrete quantities.
2. *adj.* (of a continuously variable ratio such as speed) derived as the ratio of the differences between the initial and final values of the two quantities constituting the ratio. Thus an average speed of 60 miles per hour is attained by travelling a total number of miles in as many minutes, whatever speeds may have been attained during the journey.

average deviation, *n.* (*Statistics*) another term for MEAN DEVIATION.

axial vector, *n.* (for a given skew-symmetric second order CARTESIAN TENSOR, **W**) the unique three-dimensional Euclidean vector ω that satisfies the equation **Wx** = ω × **x**, for all other such vectors, **x**.

axiom, *n.* a statement that is stipulated to be true for the purpose of constructing a theory in which THEOREMS may be derived by its rules of inference; a PRIMITIVE statement of a DEDUCTIVE FORMAL SYSTEM. Compare ASSUMPTION.

axiomatic probability, *n.* the study of PROBABILITY in terms of PROBABILITY MEASURE.

axiomatic set theory, *n.* the presentation of set theory as a formal set of UNINTERPRETED axioms and rules of inference, rather than as the formalization of an antecedent body of knowledge. Compare NAIVE SET THEORY.

axiomatic system, *n.* (*Logic*) any logical system with a set of explicitly stated AXIOMS from which THEOREMS are derived by TRANSFORMATION RULES. Compare NATURAL DEDUCTION.

axiom of choice, *n.* the axiom of set theory stating that from every family of DISJOINT sets, a set can be constructed containing exactly one element of each of the given family of sets. It is independent of the other axioms, and is rejected by INTUITIONISM because of the NON-CONSTRUCTIVE definition of the choice set. See also ZORN'S LEMMA, TRANSFINITE INDUCTION, WELL-ORDERING PRINCIPLE, HAUSDORFF'S MAXIMALITY THEOREM.

axiom of inaccessibility, see INACCESSIBLE CARDINAL.

axiom of infinity, *n.* the axiom of set theory that specifies an algorithm for constructing infinitely many distinct sets.

axis, *n.* **1.** one of the lines used to locate a point in a COORDINATE SYSTEM in terms of its perpendicular or angular distance from them. In a CARTESIAN COORDINATE system, the number of axes equals the DIMENSION of the space, and all axes are mutually perpendicular. The axes are usually labelled x, y, z, ... For example, in Fig. 37 the point P has Cartesian coordinates a and b, which are the directed lengths of the projections of its position vector on the x- and y-axes respectively (so that a is here negative); P is identified with respect to POLAR COORDINATES by the length of its position vector (r in Fig. 37) and the angle (θ) between it and the positive direction of the x-axis. See COORDINATE GEOMETRY.

2. an AXIS OF SYMMETRY or AXIS OF ROTATION.

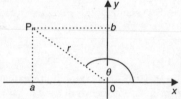

Fig. 37. 0x and 0y are respectively the x- and y- **axes**.

axis of perspectivity, see PERSPECTIVE.

axis of rotation, *n.* a line around which some body or curve rotates. For example, the cylinder in Fig. 38 opposite is formed by rotating the line segment AB around the axis of rotation XY; this line is thus an AXIS OF SYMMETRY of any cross-section of the resulting surface that includes it. See SURFACE OF REVOLUTION.

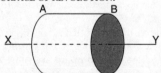

Fig. 38. XY is the **axis of rotation** of the cylinder.

axis of symmetry, *n.* a line around which a geometric figure is symmetrical, in that for every point P of the figure there is another, P', such that the perpendiculars from the two points to the line are coincident and of equal length. For example, a REGULAR hexagon has six axes of symmetry, namely all the BISECTORS of its angles and sides; the graph of cos$x$, as shown in Fig. 39 opposite, has the y-axis as an axis of symmetry. Compare CENTRE OF SYMMETRY.

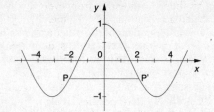

Fig. 39. The *y*-axis is the **axis of symmetry**.

azimuth, *n*. another word for AMPLITUDE in polar coordinates.

b

B, *n.* the number 11 in HEXADECIMAL notation.

Babbage's engine, *n.* the ANALYTICAL ENGINE invented by the English analyst and statistician *Charles Babbage* (1792–1871), who was a founder of both the Royal Statistical and the Royal Astronomical Societies, and also founded a society with the objective of supplanting Newton's notation with Leibniz's for the differential calculus.

B*-algebra, *n.* a BANACH ALGEBRA with a conjugate-linear anti-automorphic INVOLUTION, *, such that

$$x^{**} = x, \qquad x^* + y^* = (x+y)^*,$$
$$x^*y^* = (yx)^*, \qquad (cx)^* = \bar{c}(x)^*,$$

and that satisfies

$$\| xx^* \| = \| x \|^2.$$

The prototype is the ADJOINT of a matrix or an operator on Hilbert space. A B*-algebra of operators with this involution is called a *C*-algebra*.

back-substitution, *n.* the process of solving a sequence of LINEAR EQUATIONS is in ECHELON FORM by first solving the last equation for the first unknown appearing in it, any other unknowns being set equal to PARAMETERS taking arbitrary values, then substituting this solution into the previous equation, which can then likewise be solved for the first unknown appearing in it, and so on.

backward difference, see DIFFERENCE QUOTIENT, DIFFERENCE SEQUENCE.

backward error analysis, *n.* (*Numerical analysis*) (for an ALGORITHM) an analysis of the error in computing an approximation of an exact quantity that views the computed quantity as an exact solution of a perturbed problem. This distinguishes rounding error from truncation error, and leads to estimates that allow one to determine that the algorithm is numerically stable. Compare FORWARD ERROR ANALYSIS.

backward induction, *n.* the form of INDUCTION in which the inductive step is an argument to the effect that what fails at step $n+1$ must fail at or before step n. This is closer than classical induction to BAR INDUCTION as required by INTUITONISTS. See INFINITE DESCENT. Compare *REDUCTIO AD ABSURDUM*.

Baire category, *n.* a measure of the size of sets in a topological space. The countable union of NOWHERE DENSE sets is said to be *first category* (sometimes called *meagre*) and any other set is *second category*. The complement of a first category set is called *residual*. The rationals form a first category subset of the reals, as does the CANTOR TERNARY SET. (Named after the French analyst *René Baire* (1874–1932).)

Baire category theorem, *n.* the theorem that every COMPLETE METRIC SPACE is a BAIRE SPACE.

Baire set, see BOREL MEASURE.

Baire space, *n.* a TOPOLOGICAL SPACE with the property that the intersection of a countable family of open DENSE subsets of the space is itself dense in the space. In particular, such a space is second category in itself. For example, any regular locally compact space is a Baire space. See also BAIRE CATEGORY THEOREM.

baker's transformation, *n.* the transformation of the unit square with LEBESGUE MEASURE that is given analytically by

$$T(x, y) = (2x, y/2) \qquad \text{for } 0 \le x < \tfrac{1}{2}$$
$$T(x, y) = (2x - 1, \tfrac{1}{2}[y+1]) \qquad \text{for } \tfrac{1}{2} \le x < 1.$$

This corresponds to first deforming the unit square into a rectangle twice as wide and half as tall,

$$[0, 2] \times [0, \tfrac{1}{2}],$$

and then cutting this rectangle along the line $x = 1$, and placing the right-hand part above the left, to reform a square. This is called the baker's transformation because of its similarity to a method of kneading dough.

balanced, *adj.* (of a set) being a subset B of a VECTOR SPACE with the property that tx belongs to the set B whenever x belongs to B and $|t| \le 1$ (in modulus). For example, the unit disk in the Cartesian plane is a balanced set.

balanced block design, see BLOCK DESIGN.

ball, *n.* a set in a METRIC SPACE consisting of all the points the distance of which from a given point is, in an *open ball,* less than, or, in a *closed ball,* less than or equal to, a given constant. An open ball is an OPEN SET in the metric space, and the open ball centred on a with radius ε is often denoted $N(a, \varepsilon)$ or $N_\varepsilon(a)$; a closed ball is a CLOSED SET and is denoted $B(a, \varepsilon)$ or $B_\varepsilon(a)$, or otherwise. A ball is sometimes called a *disk,* especially in the complex plane, or a *sphere,* although this latter term is sometimes used only for the frontier of the ball. Compare NEIGHBOURHOOD.

Banach, Stefan (1892–1945), Polish mathematician who founded FUNCTIONAL ANALYSIS. He defined NORMED LINEAR SPACES, explored BANACH SPACES, proved the HAHN–BANACH THEOREM and the BANACH–STEINHAUS THEOREM, and originated other fundamental concepts and theorems in functional analysis and studied their applications. He became a full professor at the University of Lvov in 1927 and was dean of the faculty there from 1939 to 1941. During the German occupation of Lvov between 1941 and 1944, his health deteriorated rapidly, and he died shortly after its liberation.

Banach–Alaoglu theorem, *n.* the theorem that in a DUAL BANACH SPACE the unit ball is WEAK-STAR COMPACT, and, more generally, that the POLAR of a neighbourhood of the origin in a TOPOLOGICAL VECTOR SPACE is also weak-star compact.

Banach algebra, *n.* an ALGEBRA over the real or complex field that is also a COMPLETE NORMED SPACE, and that satisfies the inequality

$$\| xy \| \le \| x \| \cdot \| y \|$$

for all elements of the space. For example, the continuous functions on a compact set are a Banach algebra in the supremum norm as are all bounded operators on Hilbert space in operator norm.

Banach contraction mapping theorem, see CONTRACTION MAPPING THEOREM.

Banach limit, *n.* any translation-invariant positive LINEAR FUNCTIONAL on the vector space of all bounded sequences that sends a constant sequence to that constant value. Such limits must assign each convergent sequence its correct limit, and can be shown to exist in various non-constructive fashions.

Banach space, *n.* a COMPLETE NORMED SPACE. The vector space of continuous functions on a compact set in CHEBYSHEV NORM is a Banach space. See also HILBERT SPACE, L_p-SPACE.

Banach–Steinhaus theorem, see UNIFORM BOUNDED PRINCIPLE.

Banach–Tarski theorem, *n.* the seemingly paradoxical result that if A and B are bounded subsets of Euclidean space of three or more dimensions and both sets have INTERIOR POINTS, then A can be finitely decomposed and re-assembled using RIGID MOTIONS to form a set CONGRUENT to B. In particular, a solid sphere may be so transformed into two spheres each the size of the original.

bang-bang principle, *n.* the principle applicable to linear time problems in CONTROL THEORY that asserts that the optimal solution is 'bang-bang' in that the control mechanism is either fully on or fully off and has only finitely many switchings. These bang-bang CONTROLS arise as EXTREME POINTS of the feasible controls.

bar, *n.* **1.** the small superscript symbol, ⁻, as in $\bar{x}$, used to distinguish entities otherwise indicated by the same letter, such as VECTORS and SCALARS, or to indicate the COMPLEX CONJUGATE of a complex number, the CLOSURE of a topological set, or a statistical MEAN.

2. (in INTUITIONIST logic) a subset, S, of a SPREAD in a finite-width tree, such that every continuation of the infinite sequence assigned to a given node has an element in S; a bar for a tree is a bar for the root of the tree. Intuitively, S constitutes a bar to progress up the tree from the given node in the sense that there is no branch that avoids S. See BAR INDUCTION.

Barcan formula, *n.* (*Logic*) the expression of MODAL LOGIC

$$(\forall x) \, \Box \, Fx \rightarrow \Box \, (\forall x) \, Fx$$

that asserts that if everything has some necessary property, then it is necessary that everything has that property, whence, since

$$\Box \, (\forall x) \, Fx \equiv - \Diamond \, (\exists x) \, -Fx,$$

it is not even possible that something (else) might exist that would lack that property. (Here $\Box$ is the 'necessity' and $\Diamond$ the 'possibility' operator.) It, or equivalently,

$$\Diamond \, (\exists x) \, Fx \rightarrow (\exists x) \, \Diamond \, Fx,$$

is an axiom in some modal systems, but is not provable in others, and its intuitive acceptability is disputed on the grounds that it permits modalities *DE RE* to be derived from modalities *DE DICTO*. (Named after the American logician *Ruth Barcan* (Mrs Ruth Marcus).)

bar chart or **bar graph,** *n.* (*Statistics*) a diagram consisting of a sequence of vertical or horizontal bars or rectangles, each of which represents an equal interval of the values of a variable, and has height proportional to the quantities of the phenomenon under consideration in that interval. For example, in Fig. 40, each bar represents a ten-year age span, and its height is proportional to the proportion of that age-group in full-time employment. A bar chart may also be used to illustrate discrete data, in which case each bar represents a distinct circumstance. Compare HISTOGRAM.

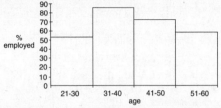

Fig. 40. A **bar chart** showing the proportion in full-time work by age.

bar induction, *n.* an intuitionistically valid form of INDUCTION based on the principle for finitary SPREADS that if Q is a subset of the spread containing a BAR P, and if, whenever all immediate descendants of a sequence *a* belong to Q then so does *a*, then the empty sequence also belongs to Q. See also BACKWARD INDUCTION, INTUITIONISM.

barrel, *n.* a closed, CONVEX, ABSORBING, BALANCED subset of a NORMED SPACE or a TOPOLOGICAL VECTOR SPACE.

barrelled or **barreled space,** *n.* a TOPOLOGICAL VECTOR SPACE in which every BARREL contains a neighbourhood of the origin. Every BANACH SPACE and every FRECHET SPACE is barrelled.

barrier function, *n.* another term for (interior) PENALTY FUNCTION.

barycentre, *n.* the CENTROID of a set. When the set is a *k*-dimensional SIMPLEX, the barycentre has all BARYCENTRIC COORDINATES equal to $1/(k+1)$.

barycentric coordinates, *n.* the unique set of non-negative coefficients, λ_i with $\sum_i \lambda_i = 1$, that identifies a given point, *x*, in a SIMPLEX of $n+1$ points, p_i, that do not all lie in the same hyperplane, as a CONVEX COMBINATION $x = \sum_i \lambda_i p_i$.

base, *n.* **1a.** a side of a polygon, especially a triangle, usually the one at the bottom in a given orientation. For example, DC is the base of the polygon in Fig. 41 in the orientation shown, but any other side may also be referred to as a base.

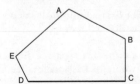 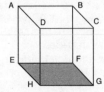

Fig. 41. DC is the **base** of the polygon, and EFGH the **base** of the cube.

b. a face of a solid, especially a cone, cylinder, pyramid or prism, usually that on which it stands in a given orientation. For example, EFGH is the base of the cube in Fig. 41 above in the orientation shown, but any other face may also be referred to as a base.

c. any cross-section of an infinite cone such that each point in the cone is a unique multiple of a point in the cross-section.

2. also called **radix. a.** the number of distinct single-digit numbers (including zero) in a counting system. For example, the binary system has base 2, as it has only two digits, written 1 and 0, so that the binary numeral 101 represents the decimal number

$$(1 \times 2^2) + (0 \times 2^1) + (1 \times 2^0) = 5.$$

The base of a system of notation is thus the number represented by the numeral 10 in that notation. Where it is necessary to be explicit, the base is written subscript after the numeral; so, for example, one writes $101_2 = 5_{10}$. See also PLACE VALUE.

b. the number in terms of which a given number is expressed as a LOGARITHM or EXPONENTIAL. For example, since $1000 = 10^3$, the logarithm of 1000 to base 10 is 3. Where it is necessary to be explicit, the base is written subscript to the symbol; so, for example, one writes $\log_{10} 1000 = 3$.

3. the number of distinct RESIDUES in a system of MODULAR ARITHMETIC.

4a. a substructure of a given mathematical structure from which the whole structure can be generated.

b. base for a topology. in particular, a collection of open sets such that every member of the topology is a union of members of the collection.

c. base at a point or **local base.** more particularly, a subcollection of neighbourhoods of the given point with the property that every neighbourhood of the point contains a member of the subcollection. This is also called a *base for the neighbourhood system.*

See also BASIS.

base clause, or **base,** *n.* the initial instance from which a generalization is proved by MATHEMATICAL INDUCTION; the statement that defines the first element of the infinite sequence generated by the induction. See RECURSIVE DEFINITION.

base field, *n.* the underlying FIELD of a vector space or other structure. For example, one speaks of polynomials over a given base field.

base period, *n.* (*Statistics*) the period used as a standard for comparison for some variable, such as, for example, consumer prices; 100 is usually taken as the INDEX NUMBER for the variable in the base period, so that an index of 150 for a given period indicates that prices then were one and a half times those of the base period.

basic feasible solution, *n.* a FEASIBLE SOLUTION in LINEAR PROGRAMMING that corresponds to an EXTREME POINT of the FEASIBLE SET. The term is used because the solution corresponds to a BASIS in the SIMPLEX TABLEAUX.

basic variables, see SIMPLEX METHOD.

basis, *n.* **1.** any set of vectors that determines a space as the set of sums of their multiples. It is also called a *Hamel basis,* especially when the basis vectors are orthogonal.

2a. in a EUCLIDEAN SPACE, a maximal set of mutually ORTHOGONAL VECTORS, in terms of which all the elements of the space are uniquely expressible, and the number of which is the DIMENSION of the space; thus the vectors $\mathbf{x}$, $\mathbf{y}$, and $\mathbf{z}$, in the positive directions of the coordinate axes, form a basis of the three-dimensional space all members of which can be written as LINEAR COMBINATIONS $a\mathbf{x} + b\mathbf{y} + c\mathbf{z}$.

b. any LINEARLY INDEPENDENT subset of a VECTOR SPACE that GENERATES the space. The CARDINALITY of such a subset is the DIMENSION of the space. For example, the dimension of the vector space of all polynomials over a field is $\aleph_0$, and the set with elements $1, x, x^2, \ldots, x^n, \ldots$, forms a basis; the vectors $(1,0,0), (0,1,0), (0,0,1)$ are a basis for three-dimensional Euclidean space.
c. (in a FREE MODULE) any linearly independent set that SPANS the module.
3. also called **Schauder basis.** in a SEPARABLE NORMED SPACE, a sequence of vectors $\{x_i\}$ in terms of which each element can be expressed uniquely as an infinite combination:

$$x = \sum_{i=1}^{\infty} v_i \, x_i \, .$$

See SCHAUDER BASIS PROBLEM.

basis theorem, *n.* the theorem that any LINEARLY INDEPENDENT set of d vectors is a BASIS of a vector space of finite DIMENSION d.

Bayesian, *adj.* (*Statistics*) (of a theory) presupposing known PRIOR PROBABILITIES (that may be subjectively assessed); these can be revised in the light of experience in accordance with BAYES' THEOREM. An hypothesis is thus confirmed by an experimental observation that is likely given the truth of the hypothesis, and unlikely given its falsehood. (Named after the English probability theorist and theologian *Thomas Bayes* (1702–61), who also published a defence of Newton's calculus against the criticisms of the philosopher Berkeley.) Compare MAXIMUM LIKELIHOOD.

Bayes' theorem, *n.* (*Statistics*) the fundamental result that expresses the CONDITIONAL PROBABILITY $P(E|A)$ of an event E given an event A as

$$P(E|A) = P(A|E) \, \frac{P(E)}{P(A)} \, ;$$

more generally, where E_n is one of a set of E_i that constitute a PARTITION of the sample space,

$$P(E_n|A) = \frac{P(A|E_n) \, P(E_n)}{\sum_i [P(A|E_i) \, P(E_i)]} \, .$$

This enables PRIOR estimates of probability to be continually revised in the light of observations.

bcd, *abbrev. for* BINARY CODED DECIMAL.

Begriffsschrift, *n.* (*German*: concept script) the original and idiosyncratic notation for the PREDICATE CALCULUS, of which examples are shown in Fig. 42, devised by Frege.

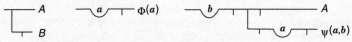

Fig. 42. **Begriffsschrift**.
Frege's notation for $-B \rightarrow A$, for $(\forall x)-Fx$, and for $(\forall y)[(\exists x)Rxy \,\&\, A]$.

behavioural variable, *n.* another term for STATE VARIABLE.
Bellman's principle of optimality, *n.* the fundamental principle of DYNAMIC PROGRAMMING stating that the optimal solution to an n-step dynamic process

must come from an optimal solution of the $(n-1)$-step process that commences with the optimal outcome of the first step. This principle can be extended to permit the RECURSIVE solution of many dynamic programming problems. Bellman's principle of optimality should not be confused with Lewis Carroll's 'Bellman's principle' that what is said thrice is true.

bell-shaped curve, see NORMAL CURVE.

below, *prep.* less than. The limit of a function *from below* is the LEFT–HAND LIMIT written variously as

$$\lim_{x \uparrow a} f(x) = \lim_{x \to a-} f(x) = f(a-),$$

the ONE–SIDED LIMIT where x is restricted to values less than a.

bending moment, *n.* (*Mechanics*) the TORQUE of the COUPLE that, together with the TENSION and the SHEARING FORCE, is equivalent to the total force at a point in a thin elastic beam.

Bernoulli, Jakob or **Jacques** (also known as **James**) (1654–1705), Swiss analyst, probability theorist, and physicist, after whom a large variety of results in analysis and statistics are named. He was the most outstanding of a large family of mathematicians, amongst whom his brother *Johann* or *Jean* (also known as *John*) (1667–1748) and his nephew *Nikolaus* (1687–1759) were also eminent. The dynasty was founded by his father *Nikolaus* (1623–1708), whose family had fled from Antwerp to Basel to escape religious persecution, and John's descendants continued to make significant contributions to mathematics for the succeeding three generations.

Bernoulli equation, *n.* a DIFFERENTIAL EQUATION of the form

$$dy/dx + \phi y = \psi y^n,$$

where ϕ and ψ are functions of x alone. It may be written in linear form by the change of variable $z = y^{1-n}$, and is closely related to the JACOBI EQUATION.

Bernoulli number, *n.* the sequence $\{B_n\}$ of coefficients of the power series defined by

$$\frac{z}{e^z - 1} + \frac{z}{2} = \sum_{m=0}^{\infty} B_{2m} \frac{z^{2m}}{(2m)!}$$

for even indices, and with $B_1 = -\frac{1}{2}$ and all other odd terms zero. These numbers allow evaluation of even values of the ZETA FUNCTION:

$$\zeta(2m) = (-1)^{m+1} B_{2m} \frac{(2\pi)^{2m}}{2(2m)!}$$

This gives $\zeta(6) = \frac{\pi^6}{945}$.

Bernoulli's theorem, *n.* (*Probability*) a form of the WEAK LAW OF LARGE NUMBERS for a sequence of random variables.

Bernoulli trial, *n.* (*Statistics*) one of a sequence of INDEPENDENT repetitions of an experiment with two possible outcomes (often termed success and failure) whose probabilities do not vary between repetitions; for example, a sequence of throws of a die where success is defined as throwing either a one or a six. A sequence of any fixed number of such trials is a *binomial experiment* and the collective outcome of a sequence of Bernoulli trials is described by a BINOMIAL DISTRIBUTION.

Bernstein polynomials, *n.* the sequence of polynomials defined for a given continuous function *f* on the interval [0,1] by the formula

$$B_n(f)(x) = B_n(x) = \sum_{k=0}^{n} f\left(\frac{k}{n}\right) \binom{n}{k} x^k (1-x)^{n-k}.$$

The polynomials $B_n(f)$ converge in UNIFORM NORM to *f*. This yields a proof of the WEIERSTRASS APPROXIMATION THEOREM. (Named after the Russian analyst *Sergei Natanovich Bernstein* (1880–1968).)

Bernstein's theorem, see SCHRÖDER–BERNSTEIN THEOREM.

Berry's paradox, *n.* the semantic paradox discovered by the English librarian *G.G.Berry,* that results from classifying the positive integers in terms of the smallest number of syllables of ordinary English needed to describe them (for example, 3 628 800 is describable in only five syllables as factorial ten). There will then be a least integer not describable using less than 19 syllables, that is, a least integer that is not a member of the first 18 classes. However, the description "the least integer not describable using less than 19 syllables" itself identifies that number using only 18 syllables, thereby contradicting itself. This is a simplified version of RICHARD'S PARADOX, and Russell's solution was to distinguish levels of language by his THEORY OF TYPES, so the paradoxical description could only count ordinary numerical expressions for integers and not descriptions which quantify over other descriptions. (On this basis he asserted that the least such integer is 111 777.) See also LIAR PARADOX, GRELLING'S PARADOX, RUSSELL'S PARADOX.

Bertrand's postulate, *n.* the conjecture that, for any integer *n* greater than 3, there is always a prime between *n* and $2n-2$. This is in fact so, as Chebyshev established. Indeed, for any positive number ε and for sufficiently large integers there is always a prime between *n* and $(1+\varepsilon)n$. (Named after the French geometer and analyst *Joseph Louis Bertrand* (1822–1903).)

Bessel function, *n.* one of a class of SPECIAL FUNCTIONS related to the HYPERGEOMETRIC functions, that arise as solutions of BESSEL'S EQUATION, and for which tables of values are available. The simplest of these is the *Bessel function of the first kind* of order zero:

$$J_0 = \sum_{k=0}^{\infty} \frac{(-1)^k (x/2)^{2k}}{(k!)^2}.$$

(Named after the German astronomer *Friedrich Wilhelm Bessel* (1784–1846), who calculated the orbit of Halley's comet while a 20-year-old warehouseman. He later made the first accurate measurements of stellar positions, and became Professor of Astronomy at Königsberg.)

Bessel's equation, *n.* the second-order differential equation

$$x^2 y'' + xy' + (\lambda^2 x^2 - \nu^2)y = 0.$$

The BESSEL FUNCTION and the NEUMANN FUNCTION are independent solutions for this equation.

Bessel's inequality, *n.* the FOURIER SERIES inequality stating that the sum of the squares of the moduli of the FOURIER COEFFICIENTS of a function *f* on the interval $[0, 2\pi]$ is dominated by the integral of the squared function, and so satisfies

$$\sum_{n=0}^{\infty} |c_n|^2 \le \int_{0}^{2\pi} f(x)^2 \, dx.$$

More generally, if $\{f_y\}$ is an ORTHONORMAL set in a Hilbert space H, and $f \in$ H, then

$$\sum_y |(f, f_y)|^2 \le \|f\|^2.$$

See also PARSEVAL'S THEOREM.

best approximation, *n.* (in a METRIC SPACE) a point of a specified set that is closest to a given point that usually lies outside the set. For example, in the simplest Chebyshev approximation one seeks the polynomial closest in Chebyshev norm to a prescribed continuous function.

beta function, *n.* the function

$$B(p, q) = \int_{0}^{1} x^{p-1}(1-x)^{q-1} \, dx$$

that has the relation to the GAMMA FUNCTION that

$$B(p, q) = \frac{\Gamma(p)\,\Gamma(q)}{\Gamma(p+q)}.$$

In particular, for integral arguments m and n, the beta function is related to the BINOMIAL COEFFICIENT by

$$B(m+1, n+1) = \frac{n!\, m!}{(n+m+1)!} = \frac{1}{\binom{n+m+1}{m}}.$$

The beta function is one of the most important SPECIAL FUNCTIONS.

between, *prep.* (of one element of an ORDERING with respect to two others) being a member of a CHAIN the first and last elements of which are respectively the two given elements. For example, the integer a lies between b and c if and only if either

$$b < a < c \quad \text{or} \quad c < a < b;$$

a point A lies between two others, B and C, if and only if they are ordered B A C by some suitable relation, such as *to the right of.* More generally, there may be other elements of the chain from B to A or from A to C. If it is necessary to specify that the intermediate element is not permitted to be identical with either of the endpoints of the chain, a is said to be *strictly between b* and *c.*

between-subjects design, *n.* (*Statistics*) an experimental design that is concerned to measure the value of a dependent variable for distinct and unrelated groups subjected to each of the experimental conditions. Compare WITHIN-SUBJECTS DESIGN, MATCHED–PAIRS DESIGN.

Bezout's lemma or **Bezout's identity,** *n.* the generalization, to polynomials over a field, of a result known to Euclid for integers, that if d is the GREATEST COMMON DIVISOR of f and g, it can be written as $d = af + bg$ for two other polynomials a and b. (Named after the French geometer and analyst, *Étienne Bezout* (1730–83).)

Bezout's theorem, *n.* the result that two algebraic plane curves with degree *m* and *n* respectively and with no common component have exactly *mn* points of intersection, counting multiplicity and points at infinity.

bi-, *prefix denoting* two; for example, the *bidual* is the normed DUAL of the dual of a normed space. See also BINORMAL, BILINEAR, BINARY.

bias, *n.* (*Statistics*) **1.** an extraneous latent influence on, unrecognized conflated variable in, or selectivity in the choice of, a sample, that influences its distribution and so renders it unable correctly to reflect the desired population parameters.

2. the EXPECTED VALUE of $(T - \theta)$, for T an ESTIMATOR of the parameter θ.

biased, *adj.* (of a SAMPLE) having a distribution that is not determined only by the population from which it is drawn, but also by some property that influences the distribution of the sample. For example, an opinion poll might be biased by geographical location.

bicompact, *adj.* a former term for COMPACT as contrasted with SEQUENTIALLY COMPACT.

biconditional, *n.* (*Logic*) another word for EQUIVALENCE, a statement of the form *A if and only if B*, or for the symbol representing this binary relation. See also CONDITIONAL.

bicontinuous, *adj.* (of a function) CONTINUOUS and possessing a continuous INVERSE. A continuous bijection whose domain is compact is necessarily bicontinuous, when the range is a HAUSDORFF SPACE.

Bieberbach's conjecture, *n.* the conjecture, proved by Louis de Branges in 1985, that if S is the class of normalized injective HOLOMORPHIC functions, so that S consists of one-to-one holomorphic functions from the unit disk with POWER SERIES of the form

$$z + a_2 z^2 + \ldots + a_n z^n + \ldots$$

for $|z| < 1$, then for every function in S, the coefficients satisfy $|a_n| \le n$ for all *n*.

bifurcation, *n.* (in a DYNAMICAL SYSTEM) a value of a parameter at which the ATTRACTOR changes.

bijection, *n.* a ONE–TO–ONE correspondence; a function or mapping that associates two sets in such a way that one and only one member of its range is paired with each member of its domain as shown in Fig. 43. For example, the pairing of married men with the women to whom they are married is a bijection between the sets of married men and married women if and only if society is monogamous. A bijection is both INJECTIVE and SURJECTIVE, and has an inverse.

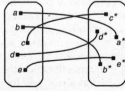

Fig. 43. **Bijection**.

bijective, *adj.* (of a function, relation, etc.) constituting a BIJECTION, both INJECTIVE and SURJECTIVE.

bilateral shift, *n.* the linear operator defined on a space of (square summable) doubly infinite sequences

$$\{x_n\}_{n=-\infty}^{\infty}$$

by

$$(Sx)_n = x_{n-1}.$$

Compare UNILATERAL SHIFT.

bilateral symmetry, *n.* SYMMETRY of order 2, such as exhibited by the human body around some central plane or axis.

bilinear, *adj.* of or relating to a function of two variables that is LINEAR with respect to each variable independently, as $f(x, y) = xy$.

bilinear functional, *n.* a complex-valued function, f, on the CARTESIAN PRODUCT of two VECTOR SPACES over the complex numbers that satisfies

$$f(\alpha\mathbf{u} + \beta\mathbf{v}, \mathbf{w}) = \alpha f(\mathbf{u}, \mathbf{w}) + \beta f(\mathbf{v}, \mathbf{w}),$$

and

$$f(\mathbf{v}, \gamma\mathbf{w} + \delta\mathbf{x}) = \overline{\gamma}f(\mathbf{v}, \mathbf{w}) + \overline{\delta}f(\mathbf{v}, \mathbf{x}),$$

where $\alpha, \beta, \gamma, \delta$ are scalars, and $\mathbf{u}, \mathbf{v}, \mathbf{w}, \mathbf{x}$ are vectors.

billion, *n.* **1.** (*in UK and Germany*) one million million, 10^{12}.

2. (*in USA and France*) one thousand million, 10^9.

bimodal, *adj.* (*Statistics*) (of a distribution) having two distinct peaks of frequency, as shown in Fig. 44. For example, the incidence of certain accidents relative to age is bimodal, since they occur more frequently among children and the elderly than they do in the rest of the population.

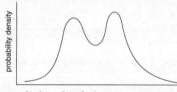

Fig. 44. A **bimodal** distribution.

binary, *adj.* **1.** of, relating to, or expressed in, BINARY NOTATION or BINARY CODE.

2. (*Logic*) also called **dyadic**. (of a relation, expression, operation, etc.) having two arguments; applying to two elements of its domain.

binary code, *n.* (*Computing*) the representation of each letter, number or element of some symbol set, and so of any sequence of such symbols, as a unique sequence of BINARY DIGITS, as in ASCII.

binary coded decimal, *n.* (abbrev. **bcd**) a number in BINARY CODE, but representing DECIMAL PLACE–VALUE NOTATION. Such a number is written in groups of four BITS, each group being the binary number equal to the corresponding digit of the given decimal number. Thus 0110 1001 0011 represents 693, since $0110_2 = 6_{10}$, $1001_2 = 9_{10}$, and $0011_2 = 3_{10}$.

binary digit, *n.* either of the numbers 0 and 1, of the BINARY SYSTEM, usually abbreviated to BIT.

binary line search, *n.* another term for DICHOTOMOUS LINE SEARCH.

binary notation, *n.* the PLACE–VALUE NOTATION to base 2, in which numbers are expressed by sequences of the digits 0 and 1. This system is the basis of all digital computing since these two digits can be represented as the *on* and *off* states of an electrical switch.

binary number, *n.* a number expressed in PLACE–VALUE NOTATION to base 2. For example, 101.01 to base 2, written 101.01_2, represents

$$(1 \times 2^2) + (0 \times 2^1) + (1 \times 2^0) + (0 \times 2^{-1}) + (1 \times 2^{-2}),$$

that is, $4 + 0 + 1 + 0 + \frac{1}{4} = 5\frac{1}{4}$.

binary operation, *n.* an operation that applies to two numbers, quantities or expressions.

binary relation, *n.* a RELATION explicitly involving ordered pairs.

binary system, *n.* arithmetic using BINARY NOTATION.

binary tree, *n.* a TREE in which every node has at most two SUCCESSORS, as illustrated in Fig. 45.

Fig. 45. **Binary tree**.

binary word, *n.* a WORD using only two letters, usually 0 and 1.

bind, *vb.* (*Logic*) to bring (a variable) within the scope of an appropriate QUANTIFIER. See BOUND (sense 4).

binding or **active,** *adj.* (of a CONSTRAINT given by a WEAK INEQUALITY) satisfied as equality by a given point. For example, the constraint $x^2 + y^2 \leq 2$ is binding at (1, 1), since $1^2 + 1^2 = 2$, but is not binding at (0, 1).

Binet-Cauchy formula, another term for CAUCHY-BINET FORMULA.

binomial, *n.* a mathematical expression consisting of two terms, such as $2x + 3y$. See also BINOMIAL EXPANSION.

binomial coefficient, *n.* **1.** any of the numerical factors that multiply the successive terms in the expansion of an expression of the form $(x + a)^n$, for integral n, in accordance with the BINOMIAL THEOREM. These are any terms of the form

$$\frac{n!}{(n-k)!\ k!},$$

which is the $(k+1)^{\text{th}}$ coefficient in the expansion of $(x + a)^n$, written

$$\binom{n}{k}, \ ^nC_k, \text{ or } C_k^n.$$

This is equal to the number of distinct COMBINATIONS of k items selected from a pool of n without replacement. For each n, the sum of all the binomial coefficients, that is the sum of the entries in the nth row of PASCAL'S TRIANGLE, is 2^n. More generally, for any α, real or complex, and non-negative integral k, the binomial coefficient is defined analogously by

$$\binom{\alpha}{0} = 1; \quad \binom{\alpha}{k} = \frac{\alpha\,(\alpha-1)\,(\alpha-2)\,...\,(\alpha-k+1)}{k!}.$$

2. q-binomial coefficient. see Q-BINOMIAL THEOREM.

binomial distribution, *n.* a statistical DISTRIBUTION giving the probability of obtaining a specified number of successes in a BINOMIAL EXPERIMENT; written $\text{Bi}(n, p)$, where n is the number of trials, and p the probability of success in each. It has probability distribution function

$$\binom{n}{x} p^x (1-p)^{n-x},$$

and so has mean np and variance $np(1-p)$. See also NEGATIVE BINOMIAL DISTRIBUTION.

binomial expansion, *n.* the expansion in accordance with the BINOMIAL THEOREM of some power of a BINOMIAL expression. For example, the binomial expansion of $(x+a)^3$ is

$$x^3 + 3x^2a + 3xa^2 + a^3,$$

where the BINOMIAL COEFFICIENTS are given by PASCAL'S TRIANGLE.

binomial experiment, *n.* (*Statistics*) an experiment consisting of a fixed number of BERNOULLI TRIALS.

binomial series, *n.* the series

$$1 + \frac{\alpha}{1!} x + \frac{\alpha(\alpha-1)}{2!} x^2 + \ldots + \frac{\alpha(\alpha-1) \ldots (\alpha-n+1)}{x!} x^n + \ldots.$$

which is the MACLAURIN SERIES for the function $(1 + x)^\alpha$, in which the coefficients of the powers of x are the BINOMIAL COEFFICIENTS. In general it is valid for $-1 < x < 1$ and if α is a non-negative integer, the expansion is a polynomial equal to $(1 + x)^\alpha$ for all x.

binomial theorem, *n.* **1.** the theorem that gives the form of the expansion of any positive integral power of a BINOMIAL, $(x+a)^n$, as a POLYNOMIAL with $n+1$ terms, namely

$$x^n + nx^{n-1}a + \frac{n(n-1)}{2} x^{n-2}a^2 + \binom{n}{k} x^{n-k}a^k + \ldots + a^n,$$

or more generally, for any real α and complex z with modulus strictly less than 1,

$$(1 + z)^\alpha = \sum_{k=0}^{\infty} \binom{\alpha}{k} z^k,$$

where $\binom{n}{k}$ and $\binom{\alpha}{k}$ are the appropriate BINOMIAL COEFFICIENTS.

2. see Q-BINOMIAL THEOREM.

binormal, *n.* the VECTOR perpendicular to both the tangent and the normal to a curve at a point in three-space, given by the VECTOR PRODUCT $\mathbf{B} = \mathbf{T} \times \mathbf{N}$, where $\mathbf{T}$ is the TANGENT VECTOR and $\mathbf{N}$ is the PRINCIPAL NORMAL vector. See also FRENET'S FORMULAE.

bi-orthogonal, *adj.* (of two sequences (a_n) and (b_n) in HILBERT SPACE) such that $\langle a_n, b_m \rangle$ is unity when $m = n$, and zero otherwise. Compare ORTHOGONAL.

bipartite, *adj.* **1.** divided in two distinct parts.

2. (of a graph) having the property that the vertices can be PARTITIONED into two sets such that every edge has one vertex in each set. Compare MATCHING.

bipolar set, *n*. the set of vectors, denoted S^{00} or S^{oo}, that are polar to the POLAR SET of a given set, S, of vectors in a HILBERT SPACE; this coincides in the real setting with the CONVEX HULL of S and zero.

biquadrate, *adj*. quartic, raised to the fourth power.

biquadratic, *adj*. **1.** quartic, of or relating to the fourth power.

2. (*as substantive*) an equation in which the highest order term is of the fourth power and in which only even powers occur with non-zero coefficients, such as $x^4 + 3x^2 - 5 = 0$. Such equations may thus be solved by the QUADRATIC FORMULA.

Birch and Swinnerton–Dyer Conjecture, *n*. the conjecture (one of the MILLENNIUM PRIZE PROBLEMS) that, when the solutions in whole numbers to ALGEBRAIC EQUATIONS are the points of an abelian variety, the size of the group of rational points is related to the behavior of an associated zeta function $z(s)$ near the point $s = 1$. In particular this amazing conjecture asserts that if $z(1)$ is equal to 0 then there are an infinite number of rational points (solutions), and conversely, if $z(1)$ is not equal to 0, then there is only a finite number of such points. This is a special case of the tenth HILBERT PROBLEM, which was shown in 1970 by Matiyasevich to have no general solution; i.e. there is no general method for determining when algebraic equations have a solution in the whole numbers.

Birkhoff, George David (1884 – 1944), American analyst and topologist who was President of both the American Mathematical Society and the Association for the Advancement of Science, and influenced an entire generation of American mathematicians. Although his own major work was in the application of analysis to dynamics, he also contributed to the study of difference equations, constructed a relativistic theory of gravity independently of Einstein, and devised a mathematical theory of 'aesthetic measure'.

Birkhoff (or **strong** or **pointwise**) **ergodic theorem,** *n*. the theorem that, for any MEASURE–PRESERVING TRANSFORMATION T on a measure space and any integrable f, the CESARO MEANS of $f(T^n x)$ converge almost everywhere to an invariant function f^* for which $f^*(Tx) = f^*(x)$; when the underlying space has finite measure, f and f^* have the same integrals. This is sometimes called the *pointwise* (or *strong*) *ergodic theorem*, to distinguish it from the *mean* (or *weak*) *ergodic theorem*, due to von Neumann, in which one only obtains CONVERGENCE IN MEAN SQUARE.

Birkhoff's theorem, *n*. the theorem that every DOUBLY–STOCHASTIC matrix can be expressed as a CONVEX COMBINATION of PERMUTATION MATRICES.

bisect, *vb*. to divide (a geometrical figure) into two equal parts.

bisection method, see DICHOTOMOUS LINE SEARCH.

bisector or **bisectrix,** *n*. a straight line or plane that BISECTS a given angle or line. For example, the perpendicular bisectors of the sides of any triangle are concurrent. See Fig. 46.

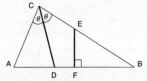

Fig. 46. EF is a **bisector** of AB, and CD a **bisector** of the angle ACB.

Bishop–Phelps theorem, see SUPPORT POINT.

bit, *n. abbrev. for* BINARY DIGIT. **1.** a single digit of BINARY NOTATION, represented by either a 0 or a 1.

2. the smallest unit of information, indicating either the presence or the absence of a single feature.

3. a unit of capacity of a computer, capable of storing a single unit of information, consisting of an element of its physical structure capable of being in either of two states, such as a switch with *on* and *off* positions, or a microscopic magnet capable of alignment in two directions. Compare BYTE.

bitangent, *n.* a line TANGENT to a curve or surface at two different points.

bivariate, *adj.* (*Statistics*) (of a distribution) involving two RANDOM VARIABLES, not necessarily independent of one another.

block design, *n.* **1.** (*Statistics*) a DESIGN in which groups of subjects are regarded as sufficiently homogeneous to behave in the same way, so that comparison of the application of different EXPERIMENTAL CONDITIONS to subjects in the same group is meaningful. It is a *balanced block design* if all the blocks are of equal size and all treatments occur equally often; it is a *completely balanced block design* if in addition each treatment occurs equally often in each block, that is, if the block size is a multiple of the number of treatments. **2.** (*Combinatorics*) a family of subsets (*blocks*) of a given finite set (of VARIETIES or points) such that each block has the same number of members and such that each pair of points belongs to the same number of blocks. For example,

$$\{ \{1,2,4\}, \{2,3,5\}, \{3,4,6\}, \{4,5,7\}, \{5,6,1\}, \{6,7,2\}, \{7,1,3\} \}$$

is a block design on P = {1, ..., 7} with each pair in exactly one block. The simplest examples occur as FINITE GEOMETRIES such as the seven-point FINITE PROJECTIVE PLANE. A block design on a set of v points with k blocks in which each point belongs to exactly λ blocks is called a (v, k, λ)-design; thus the above example is a (7, 3, 1) design. In general the construction of a design for given values of v, k, and λ is a non-trivial problem. (This example is, in statistical terms, a balanced, incomplete block design.)

See also ERROR CORRECTING CODES.

block-diagonal matrix, *n.* a matrix the only non-zero elements of which lie in a sequence of square matrices arranged along the main diagonal; the block-diagonal matrix, C, with two blocks, one an $m \times m$ matrix A, and the other an $n \times n$ matrix B, is denoted diag[A, B], and has entries $c_{ij} = a_{ij}$ for i and $j \le n$, and $c_{ij} = b_{i-n, j-n}$ for i and $j > n$, all other entries being zero. For example, where A and C are 2×2 matrices and B is a 3×3 matrix, diag[A,B,C] is as shown in Fig. 47.

$$\begin{bmatrix} a_{11} & a_{12} & 0 & 0 & 0 & 0 & 0 \\ a_{21} & a_{22} & 0 & 0 & 0 & 0 & 0 \\ 0 & 0 & b_{11} & b_{12} & b_{13} & 0 & 0 \\ 0 & 0 & b_{21} & b_{22} & b_{23} & 0 & 0 \\ 0 & 0 & b_{31} & b_{32} & b_{33} & 0 & 0 \\ 0 & 0 & 0 & 0 & 0 & c_{11} & c_{12} \\ 0 & 0 & 0 & 0 & 0 & c_{21} & c_{22} \end{bmatrix} = \begin{bmatrix} A & 0 & 0 \\ 0 & B & 0 \\ 0 & 0 & C \end{bmatrix}$$

Fig. 47. **Block diagonal matrix**.

block multiplication, *n.* multiplication of matrices the entries of which are themselves matrices rather than field entries. This allows one to exploit the structure of the matrix. See SCHUR COMPLEMENT.

body, *n.* **1.** a subset of a VECTOR SPACE that has a non-empty INTERIOR.

2. (*Continuum mechanics*) a volume of continuously distributed physical matter, such as an expanse of fluid or an elastic band; formally, a SMOOTH three-dimensional MANIFOLD that is HOMEOMORPHIC to the closure of a CONNECTED open subset of EUCLIDEAN POINT-SPACE.

body force, *n.* (*Continuum mechanics*) a FORCE experienced by the points of a BODY other than those due to their contact with other points of the body or contact with external boundaries, such as the effects of self-gravitation or gravitation due to an external source. Formally, the effect of the body forces on a SUB-BODY is given by the integral

$$\int \rho \, \mathbf{b} \, dv$$

over the volume of its current CONFIGURATION, where **b** is the BODY FORCE DENSITY, and ρ is the DENSITY. Compare CONTACT FORCE, BODY TORQUE.

body force density, *n.* (*Continuum mechanics*) a VECTOR FIELD that represents the BODY FORCES per unit MASS (or per unit volume) acting on a BODY, for example, the downward acceleration equal to the LOCAL GRAVITATIONAL CONSTANT.

body spin, **spin tensor** or **vorticity tensor,** *n.* (*Continuum mechanics*) the skew symmetric part of the VELOCITY GRADIENT; if Ω is the body spin and **L** is the VELOCITY GRADIENT then

$$\Omega = \tfrac{1}{2} (\mathbf{L} - \mathbf{L}^{T}).$$

This is the local ANGULAR VELOCITY expressed in tensor form. Compare EULERIAN STRAIN RATE.

body torque, *n.* (*Continuum mechanics*) the TORQUE experienced by the points of a BODY other than those due to their contact with other points of the body or contact with external boundaries, such as the effects of self-gravitation or gravitation due to an external source. Formally, the effect of the body torque on a SUB-BODY is given by the integral

$$\int \rho \, \mathbf{x} \times \mathbf{b} \, dv$$

over the volume of its current CONFIGURATION, where **b** is the BODY FORCE DENSITY, and ρ is the DENSITY at the point with POSITION VECTOR **x**. Compare CONTACT TORQUE, BODY FORCE.

Bolzano's theorem or **intermediate value theorem,** *n.* the theorem that if a real function f is continuous on a closed bounded interval $[a, b]$, then it takes every value between $f(a)$ and $f(b)$ for at least one argument between a and b. This intermediate value property, which derivatives also possess by virtue of the MEAN-VALUE THEOREM, is also called the *Darboux property*. (Named after the Czech analyst *Bernhard Bolzano* (1781–1848).)

Bolzano–Weierstrass theorem, *n.* the theorem that every bounded infinite subset of EUCLIDEAN SPACE possesses a CLUSTER POINT, and so that every bounded infinite sequence possesses a convergent subsequence. See also COMPACT, HEINE–BOREL COVERING THEOREM.

Bolza's problem, *n.* (*Calculus of variations*) the general problem of determining the arc, from a given class, that minimizes a function of the form

$$g(a, y(a), b, y(b)) + \int_a^b f(t, y(t), y'(t)) \, dt,$$

subject to constraints. Note that the objective function involves explicitly both an integral and an evaluation at the endpoints. (Named after the German-born American analyst *Oskar Bolza* (1857–1942).)

Bonnet's mean-value theorem, see MEAN–VALUE THEOREM.

Boole, George (1815–64), British mathematician who is best known for his innovatory work in formal logic. Although he had little formal education, he also made contributions to analysis, differential equations, algebra and probability theory, was elected a Fellow of the Royal Society and held the Chair of Mathematics at Cork.

Boolean, *adj.* (of a variable, function, operator, etc.) taking either of the values *true* and *false*, typically used in computing to record the result of a test.

Boolean algebra, *n.* **1.** a DISTRIBUTIVE LATTICE with a ZERO and a UNITY in which every member has a COMPLEMENT that is itself a member.

2. the ALGEBRA OF CLASSES devised by George Boole, in which the operations of COMPLEMENTATION, UNION, and INTERSECTION are defined. It is isomorphic with the SENTENTIAL CALCULUS, and the term is thus sometimes applied to the internal logic of a digital computer. See also SET THEORY.

Boolean ring, *n.* **1.** a RING in which every member is IDEMPOTENT.

2. less abstractly, a class of sets that is closed under finite UNION and RELATIVE COMPLEMENT. A Boolean ring with a largest element is identifiable with a BOOLEAN ALGEBRA.

bordering, *n.* the enlargement of a matrix or determinant annexing a column and a row, especially when all the entries of the annexed row and column are 0 except for a 1 in the common entry, so that the value of the determinant does not change. Compare AUGMENT.

Borel, Félix Édouard Justin Émile (1871–1956), French measure theorist and probability theorist, who, together with Lebesgue and Baire, founded the theory of real-valued functions, and also contributed to the development of game theory. He was also a member of the Chamber of Deputies and for fifteen years was Navy Minister until his imprisonment by the Vichy regime; he then joined the French Resistance. He was appointed to a Chair specially created for him at the Sorbonne in 1909, received the *Croix de Guerre* after the First World War, and the Resistance Medal and Grand Cross of the *Légion d'Honneur* after the Second, and was the first gold medallist of the French National Centre for Scientific Research in 1955.

Borel–Cantelli lemma, *n.* the result that, given an infinite sequence of events in a PROBABILITY SPACE for which the sum of the individual probabilities is finite, then the probability that infinitely many events occur is zero. If the events are independent and the sum of the individual probabilities is infinite, then the probability that infinitely many events occur is unity. More generally, if $\{A_n\}$ is a sequence of MEASURABLE sets in a measure space whose measures $\mu(A_n)$ have a finite sum, then $\mu(\limsup A_n) = 0$; that is, the set of points that are in an infinite number of the given sets has measure zero.

Borel measurable function, see MEASURABLE.

Borel measure, *n.* any MEASURE defined on the SIGMA–ALGEBRA generated by all the OPEN (or equivalently compact) subsets of a COMPACT topological space (the *Borel field*), especially on the unit interval. When the space is only locally compact, one distinguishes the sigma-algebra generated by the compact sets (the *Borel sets*) from that generated by the compact G_δ subsets (the *Baire sets*) that are expressible as finite intersections of open sets.

Borel set or **Borel measurable set,** *n.* any set derived from the intervals on the real line by repeated application of countable union and intersection; the Borel sets constitute a sigma-algebra. Any such set is MEASURABLE; see also BOREL MEASURE.

Borromean rings, *n.* an arrangement of three circles that are interwoven in such a way that, if any one of them is removed, then the other two are no longer linked. Named after the wealthy 15th century Milanese banking family of Borromeo, whose family crest incorporated the design.

borrow, *vb.* (in the algorithm for subtraction in elementary arithmetic) to redistribute a number between PLACE–VALUES to enable subtraction in each place-value position to be effected within the natural numbers. For example, to subtract 25 from 73, we first try to take 5 from 3 in the units column; as this is impossible within the natural numbers, we 'borrow' 10 from the tens column as shown in Fig. 48, treating 73 not as 7 tens and 3 units but as 6 tens and 13 units. Then 5 from 13 is 8 in the units and 2 from 6 is 4; so $73 - 25 = 48$. Compare CARRY.

$$
\begin{array}{llll}
& \underline{\text{first attempt}} & \underline{\text{second attempt}} & \\
73 = & (7 \times 10) + 3 & = \ (6 \times 10) + 13 & \\
-25 = & -(2 \times 10) \ \underline{- \ 5} & = -(2 \times 10) \ \underline{- \ 5} & \\
& \text{fails} & (4 \times 10) + \ 8 \ = \ 48 &
\end{array}
$$

Fig. 48. **Borrow**. See main entry.

Borsuk–Ulam theorem, *n.* the result that there does not exist any continuous odd mapping of the unit *n*-sphere into the unit $(n{-}1)$-sphere.

bottleneck problems, *n.* a class of NETWORK optimization problems involving restrictions (*bottlenecks*) on the NETWORK FLOWS.

bound, *n.* **1.** a number that is greater than all the numbers in a given set of numbers (an *upper bound*), or less than all the numbers in the given set (a *lower bound*). If the bound holds uniformly, usually for every member of a sequence, it is a *uniform bound*. See SUPREMUM, INFIMUM.

2. more generally, an element of an ORDERING that has the same ordering relation to all the members of a given subset; for example, since it is a subset of every set the empty set is a bound on any family of sets ordered by weak inclusion.

3. whence, an estimate of the extent of a given set.

4. (*Logic*) (of a variable) occurring within the SCOPE of a QUANTIFIER that indicates the degree of generality of the OPEN SENTENCE in which the variable occurs; for example, in

$$(x)(Fx \rightarrow Gxy),$$

x is bound, but y is not. Compare FREE.

boundary, *n.* another word for FRONTIER.

boundary condition, *n.* a condition imposed on the solution of a DIFFERENTIAL EQUATION so as to obtain the desired PARTICULAR SOLUTION. This is often an INITIAL CONDITION.

boundary hyperplane, see SUPPORT POINT.

boundary point, *n.* a member of the FRONTIER of a set.

bounded, *adj.* **1.** (of a set) having a BOUND, especially where there is a MEASURE in terms of which all the elements of the set, or the differences between all pairs of members, are less than some value, or else all its members lie within some other well-defined set. Thus the open unit interval is bounded while the real line is not. A set is bounded in a metric space exactly when its DIAMETER is finite.

2. (of an operator, function, etc.) having a bounded set of values. A bounded real function must be bounded both above and below.

3. (of a LINEAR OPERATOR) sending bounded sets to bounded sets; between NORMED SPACES this is equivalent to continuity of the linear operator.

bounded above, *adj.* having an UPPER BOUND.

bounded away from zero, *adj.* having a LOWER BOUND strictly greater than zero, or correspondingly an UPPER BOUND strictly less than zero.

bounded below, *adj.* having a LOWER BOUND.

bounded variation, *n.* the property of a real-valued function that its VARIATION is BOUNDED; it is then expressible as the difference of two MONOTONE nondecreasing functions. See TOTAL VARIATION.

Bourbaki, Nicolas (fl. 1939–), the author of, to date, 36 volumes of comprehensive texts covering most areas of mathematics in a rigorous axiomatic manner. In fact the name is that of a junior Napoleonic officer, and is the collective pseudonym of a changing and secret group of mathematicians, most of them French, who have collaborated since the 1930s with the intention of achieving a complete and definitive compilation of mathematical knowledge. Their work is distinguished not only by its rigour but also by the idiosyncrasy of their terminology and their classification of the various areas of mathematical endeavour in terms of structure rather than subject.

bow compass or **bow spring compass,** *n.* (*Geometry*) a COMPASS in which the legs are joined by a flexible metal bow–shaped spring rather than by a hinge, the angle being adjusted by a screw.

Bowditch curves, *n.* another term for LISSAJOUS FIGURES.

box, *n.* a set in $\mathbb{R}^n$ consisting of the n–fold CARTESIAN PRODUCT of intervals of the form $[a, b)$, $[a, \infty)$, $(-\infty, b)$, or $\mathbb{R}$

bra, see ANGLE BRACKET.

brace, *n.* either of the pair of BRACKETS, { }, used to indicate that the expression between them is to be evaluated and treated as a single unit in the evaluation of the whole; they are usually only used in an expression that already contains both PARENTHESES and SQUARE BRACKETS, and they have lower priority than either (that is, the contents of the brackets or parentheses are evaluated before those of the braces). Such brackets are also often employed for set definitions; thus one writes $\{a, b, c\}$ for the set whose members are a, b, and c, and $\{x : Fx\}$ for the class of elements that have the property F.

brachistochrone or **brachystochrone problem,** *n.* the classical and motivating problem of the CALCULUS OF VARIATIONS that sought the path taken by a constrained weighted particle falling in the shortest time between two points not in the same vertical under gravity. That the solution is a CYCLOID was first discovered by Johann Bernoulli.

bracket, *n.* either of any pair of symbols used to enclose a number of items that are to be regarded as constituting a single expression, or to indicate that the expression between them is to be evaluated before the remainder of the formula and treated as a single unit in the evaluation of the whole. For example, to evaluate $2 + (4 \times 3)$ we first calculate $4 \times 3 = 12$ and then add this result to 2. It is sometimes conventional to use PARENTHESES before SQUARE BRACKETS, and the latter before BRACES, and, in an expression containing all of these, this is the order in which they are to be evaluated. If these provide insufficient levels of bracketing, a VINCULUM is used, and it takes precedence. The ANGLE BRACKETS in the expression $\langle a_1, a_2, a_3 \rangle$ and the braces in the expression $\{a_1, a_2, a_3\}$ indicate that the terms between them are to be treated together as a sequence and a set respectively.

braid, *n.* a number of strings plaited together. The study of the different ways in which a number of strings can be braided is related to GROUP theory.

branch, *n.* **1.** a continuous section of a curve with an endpoint at which it meets another branch so that it is continuous but not differentiable at that point. The graph in Fig. 49 has two branches and a CUSP at B. See also OSCULATION.

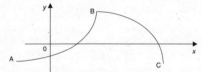

Fig. 49. AB and BC are distinct **branches** of this curve.

2. a continuous SELECTION from an ANALYTIC SET–VALUED FUNCTION such as the LOGARITHM.

3. a PATH in a TREE that either is infinite or has an end point as its final member.

branch-and-bound method, *n.* a TREE-based HEURISTIC search method that avoids exhaustive search by using one branch of the tree to set a BOUND to the desired quantity and eliminating the other branches as soon as they contravene that bound.

branch point, *n.* a point at which one can switch from one BRANCH of an ANALYTIC FUNCTION to another. See also RIEMANN SURFACE.

Brianchon's theorem, *n.* (*Projective geometry*) the theorem that if a hexagon is circumscribed about a CONIC then its diagonals are concurrent; this is the DUAL of PASCAL'S MYSTIC HEXAGRAM THEOREM.

Briggsian logarithm, *n.* a less common name for the COMMON LOGARITHM. (Named after the English mathematician and Oxford professor *Henry Briggs* (1561–1630), who proposed that logarithms to base 10 would be more useful than NAPERIAN LOGARITHMS; he published his first tables after consultation with Napier, and later published tables of logarithms to 14 decimal places, of sines to 15 places, and of tangents to 10 places.)

broken line, *n.* a line consisting of a number of concatenated straight line segments. A smooth curve can be approximated by a broken line, which approaches the given curve as the number of segments tends to infinity.

Brouwer's form of the degree, see DEGREE (sense 4).

Brouwer's theorem, *n.* a FIXED–POINT THEOREM stating that a continuous mapping of a COMPACT CONVEX SET into itself has a FIXED POINT; so, for example, in the complex numbers any mapping of the unit disk to itself has a fixed point. This theorem was shown by Schauder and Tychonoff to remain valid for a normed space or a locally convex space. (Named after the Dutch logician *Luitzen Egbertus Jan Brouwer,* who was the principal theoretician of mathematical INTUITIONISM, and the founder of modern topology.) See also CONTRACTION MAPPING THEOREM.

Brouwer's theorem (**on domain invariance**), see INVARIANCE OF DOMAIN.

Brownian motion, see WIENER PROCESS.

Bruck–Ryser–Chowla theorem, *n.* the theorem, proved in 1950, that if D is a SYMMETRIC BLOCK DESIGN on v points, with k points belonging to each block and with each pair of points contained in λ blocks, then if v is even, $v - \lambda$ is square, and if v is odd then the equation

$$x^2 = (k - \lambda)y^2 + (-1)^{\left[\frac{v-1}{2}\right]}\lambda z^2$$

has no non-trivial solution. The converse has not been proved.

Buffon's needle, see NEEDLE PROBLEM.

bundle, *n.* (*Geometry*) a family of lines or planes all passing through a single point. See also PENCIL, SHEAF.

Buniakovski's inequality, *n.* a less common name for the CAUCHY–SCHWARZ INEQUALITY. (Named after the Russian probability theorist *Viktor Jakovlevich Buniakovski* (1804–99).)

Burali-Forti paradox, *n.* (*Logic*) the PARADOX that the ORDINAL of the set of all ordinals must necessarily be larger than any of the members of the set, and so would be an ordinal which was not contained in the set of all ordinals, whence the set of all ordinals cannot itself be assigned an ordinal, and in fact, the set of all ordinals is not an allowable set. (Named after the Italian mathematician, *Cesare Burali-Forti* (1861–1931).) Compare CANTOR'S PARADOX, RUSSELL'S PARADOX.

Buridan's ass, *n.* (*Logic*) the medieval paradox of the ass who is equidistant from two equal heaps of grain of identical quality but starves to death because there is no ground for preferring one heap to the other. In a more modern setting, Buridan's fireman is unable to decide which of two properties to save first and so loses both. This is not to be confused with Nero's fiddling; one did not care, while the other cares too much. The paradox shows that choice cannot be determined by reasoned preference. (Named after the French philosopher and physicist, *Jean Buridan* (c.1295–1356), who was a pupil of Ockham, became Rector of the University of Paris, and is credited with founding the University of Vienna.) The paradox is first found in Aristotle, and since it does not appear in Buridan's writings, the name seems to arise from its role as a counterexample to his determinism.)

byte, *n.* (*Computing*) **1.** a sequence of BITS, usually eight or sixteen, encoding a single piece of data and processed together, as the successive four-bit bytes of a binary coded decimal. Half a byte is informally referred to as a *nibble*, and half a nibble as a *crumb*.
2. a unit of information, equivalent to a single character.
3. a unit of capacity of a computer, capable of storing a single character.

C

c, *abbrev. for* CENTI–, used in symbols for fractions of the physical units of the SYSTEME INTERNATIONAL.

C, *n.* **1.** the number 12 in HEXADECIMAL notation.

2. the ROMAN NUMERAL for 100.

ℂ or 𝒞, *n.* the set of COMPLEX NUMBERS. Compare ℝ, ℚ, ℤ.

C*-algebra, see B*-ALGEBRA.

C$^{(r)}$, *modifier.* (of a mapping from $\mathbb{R}^m$ to $\mathbb{R}^n$) r-times CONTINUOUSLY DIFFE-RENTIABLE, where r is a positive integer or ω. A mapping is C^ω if it is real ANALYTIC. See also DIFFERENTIAL STRUCTURE, ATLAS, OVERLAP.

calculate, *vb.* to compute; to determine (a number or value) from given information by a mathematical procedure or ALGORITHM.

calculation, *n.* the act or process of calculating, or a record of the steps of such a process.

calculator or **calculating machine,** *n.* a device, usually hand-held and electronic, for performing arithmetical operations or other evaluations. See also COMPUTER.

calculus or **infinitesimal calculus,** *n.* **1.** the branch of mathematics, developed largely by Newton and, independently, by Leibniz, that was originally con-ceived in terms of the effects on a function of an INFINITESIMAL change in the value of the independent variable, and is now understood in terms of LIMITS of real functions. The *differential calculus* concerns the RATE OF CHANGE of the dependent variable, and so the SLOPE of a curve; the *integral calculus* extends the notion of the sum of a finite number of discrete values of a function to a continuous function, and allows one to obtain the area under a curve. See also DEFINITE INTEGRAL, INDEFINITE INTEGRAL, DIFFERENTIAL.

2. (*Logic*) an UNINTERPRETED FORMAL SYSTEM, consisting of a vocabulary of PRIMITIVE TERMS, and sets of FORMATION RULES and TRANSFORMATION RULES. Compare FORMAL LANGUAGE.

3. any formal theory or set of rules for calculation. In this sense one speaks of, for example, FUNCTIONAL CALCULUS, PREDICATE CALCULUS, and LAMBDA-CALCULUS.

calculus of variations or **variational calculus,** *n.* the extension of CALCULUS concerned with the MAXIMA or MINIMA of DEFINITE INTEGRALS, and hence with finding functions that maximize or minimize a given function of those functions; this is analogous to the differential calculus, in which values of a function are found that maximize or minimize a given function of those values. In the simplest form, one attempts to minimize

$$I(y) \ = \int\limits_a^b f(y, y', t) \ \mathrm{d}t$$

over a class of piece-wise smooth arcs, the values at the endpoints of which are fixed or that satisfy other relevant constraints. For example, a typical problem seeks the shortest distance between two points on some surface. The calculus of variations was first developed by Euler in 1744, although both Newton and Jakob Bernoulli had solved problems involving variational methods; it has since become one of the major branches of ANALYSIS. See also CONTROL THEORY, EULER–LAGRANGE EQUATIONS, OPTIMIZATION THEORY, BRACHISTOCHRONE PROBLEM.

cancel, *vb.* to eliminate terms from an expression thereby leaving it in a simpler form. For example, ratios such as $^2/_8$ and $2a^2/ab$ can be simplified by dividing their numerator and denominator by a common factor to give $^1/_4$ and $2a/b$ respectively. Similarly, identical terms can be removed from both sides of an equation by subtraction or division, so that $3x + y = y + 6$ can be simplified to $3x = 6$, and further to $x = 2$ by a further cancellation of a common factor of both sides. See also SIMPLIFY, ELIMINATE.

cancellation law, *n.* **1.** (in an additive algebraic structure) a law asserting that whenever $a + b = a + c$ then it follows that $b = c$. In a GROUP this is an immediate consequence of the existence of INVERSE elements.
2. (in an multiplicative algebraic structure) a law asserting that whenever $a \times b = a \times c$ then $b = c$ follows. A commutative RING is an INTEGRAL DOMAIN exactly when the cancellation law holds for the ring multiplication.

canonical, *adj.* (of an expression, etc.) expressed in a standard form, such as the simplest form of equation of a type of curve derived from a given equation by a suitable change of variables; often a canonical expression is unique. For example, the canonical decomposition of an integer is as a product of powers of its prime factors; the canonical equation of a circle is $x^2 + y^2 = r^2$, where the centre is taken to be the origin. See also NATURAL TRANSFORMATION, JORDAN NORMAL FORM.

canonical basis or **standard basis,** *n.* the set of ORTHONORMAL vectors, $(1,0,0,...,0)$, $(0,1,0,...,0)$, ..., $(0,...,0,1)$, whose kth element has a 1 in the kth place and zeroes elsewhere, that is a BASIS for n-dimensional EUCLIDEAN SPACE.

Cantor, Georg Ferdinand Ludwig Philip (1845–1918), German mathematician, noted as the founder of set theory and for his basic contributions to classical analysis and topology. He originated the definition of real numbers as equivalence classes of CAUCHY SEQUENCES of rational numbers, the definition of open and closed sets, and the theory of TRANSFINITE NUMBERS. He began teaching at the University of Halle in 1869, became full professor in 1879, and remained there until his death in 1918 after a long mental illness.

Cantor–Bendixson theorem, see DERIVED SET.

Cantor–Bernstein theorem, see SCHRÖDER–BERNSTEIN THEOREM.

Cantorian set theory, *n.* (*Logic*) another name for NAIVE SET THEORY.

Cantor's diagonal theorem, *n.* the theorem of set theory that the POWER SET of any set, finite or infinite, cannot be put in one-to-one correspondence without remainder with the members of the given set; that is, any set has strictly more subsets than members. This is proved by a DIAGONAL PROCESS. See also CANTOR'S PARADOX.

Cantor set, see CANTOR'S TERNARY SET.

Cantor's intersection theorem, *n.* the theorem that in a COMPLETE METRIC SPACE, any NESTED sequence of sets the DIAMETERS of which decrease to zero contains a unique point of intersection.

Cantor's paradox, *n.* (*Logic*) the paradox derived in CANTORIAN SET THEORY from the supposition of an all-inclusive infinite set: every subset of such a set would be a member of it, but by CANTOR'S DIAGONAL THEOREM, every set has strictly more subsets than it has members. There is thus no largest cardinal number. Compare BURALI–FORTI'S PARADOX, RUSSELL'S PARADOX.

Cantor's ternary set or **Cantor set,** *n.* the subset of the interval [0, 1] formed by iteratively removing the open middle third, then the open middle third of each of the remaining intervals, and so on; this is the set of points in the interval whose TERNARY representations contain no 1s. This produces an UNCOUNTABLE PERFECT set of zero LEBESGUE MEASURE that has many applications in MEASURE THEORY and TOPOLOGY.

cap, n. the symbol for set INTERSECTION, written $S \cap T$ or $\cap_i S_i$.

capacity, see NETWORK.

Caratheodory, Constantin (1873–1950), German analyst who worked as an engineer in Egypt before studying mathematics, and later taught in Germany, Poland, and Greece. He saved the library of the new Greek University of Smyrna from the Turks and transferred it to Athens. His most important work was in the CALCULUS OF VARIATIONS, but he also made significant contributions to the theory of functions of several variables, measure theory, thermodynamics, and relativity.

Caratheodory measurable, *adj.* (of a set A with respect to an OUTER MEASURE μ^*) having the property that for all B,

$$\mu^*(B) = \mu^*(B \cap A) + \mu^*(B \setminus A).$$

Caratheodory outer measure, see OUTER MEASURE.

Caratheodory's extension theorem, *n.* the theorem that if μ is a MEASURE on an algebra A and μ^* is an OUTER MEASURE of μ, then the collection A* of μ^*-CARATHEODORY MEASURABLE sets is a SIGMA–ALGEBRA, and the restriction of μ^* to A* is a measure that extends μ. This theorem makes it possible to obtain Lebesgue measure from length measure on the half-open intervals of the real line.

Caratheodory's theorem, *n.* the theorem that in EUCLIDEAN SPACE every point in the CONVEX HULL of a given set S lies in some SIMPLEX with vertices in S. Compare EXTREME POINT.

Cardano, Girolamo (1501–76), Italian mathematician and physician who, having declined several posts of court physician, made the most significant contribution of his generation to both subjects. He wrote an early work on probability, solved both the CUBIC (building on the work of Tartaglia) and (with his servant Ferrari) the QUARTIC, was the first to identify typhus, and wrote popular works on science, philosophy, and astrology. His son was executed in 1560, and he himself was imprisoned for heresy in 1570 and deprived of his post. Although forbidden to publish, he completed his autobiography shortly before his death.

Cardano's formula, *n.* the name commonly given to the formula, due to Ferro and Tartaglia, for the SOLUTION BY RADICALS of the general (normalized) CUBIC equation,

$$x^3 + rx^2 + sx + t = 0.$$

On substituting x = $y - {}^1\!/_3\, r$, one obtains the *reduced form of the cubic*,

$$y^3 + py + q = 0,$$

in which the quadratic term is missing. Here

$$p = s - \frac{r^2}{3} \quad \text{and} \quad q = \frac{2r^3}{27} - \frac{sr}{3} + t.$$

The DISCRIMINANT of the cubic is

$$\Delta^2 = \left[\frac{q^2}{4} + \frac{p^3}{27} \right]$$

so that there are repeated roots if and only if $\Delta^2 = 0$; when Δ^2 is positive the unique real solution to the reduced cubic is

$$\left[-\frac{q}{2} + \Delta \right]^{1/3} + \left[-\frac{q}{2} - \Delta \right]^{1/3},$$

and when Δ^2 is negative there are three real solutions best expressed trigonometrically. There is a corresponding solution to the general QUARTIC equation, associated with the names of Ferrari and Cardano, that proceeds by producing a cubic resolvent equation to which the previous formula applies; the discriminant of the quartic is also the discriminant of the AUXILIARY EQUATION of the cubic.

cardinal, *n.* a CARDINAL NUMBER.

cardinality, *n.* the CARDINAL NUMBER associated with a given class; since two sets are defined to have the same cardinality if their members can be put in ONE-TO-ONE CORRESPONDENCE, this is an EQUIVALENCE RELATION, and the cardinality of any given finite class is taken to be the largest member of the initial sequence of natural numbers (beginning with 1) that so corresponds to it; for example, the cardinality of {knife, fork, spoon}, is

$$|\,\{\text{knife, fork, spoon}\}\,| \;=\; |\,\{1, 2, 3\}\,| \;=\; 3.$$

This property may then be used to define arithmetical operations in terms of set operations (see ADDITION). See also ALEPH.

cardinal number, *n.* 1. a measure of the size of a set that does not take into account the order of its members. This can be defined in terms of the CARDINALITY of a RECURSIVELY GENERATED sequence of classes and is a wider concept than NATURAL NUMBER.

2. any particular number having this function. For example, one, zero, and ALEPH–NULL are cardinals.

3. precisely, the smallest ORDINAL NUMBER that is EQUIPOLLENT to a given set. See also INFINITY. Compare ORDINAL NUMBER.

cardioid, *n.* a heart-shaped curve generated by a fixed point on a circle as it rolls round another circle of equal radius; in Fig. 50 overleaf, P_1 is the initial position of this point, and P_2 another position. It has equation

$$r = a(1 - \cos \phi),$$

where *a* is the common radius of the two circles, and ϕ is the polar angle. The term is sometimes also used of similar heart-shaped curves.

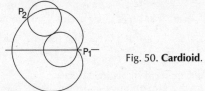

Fig. 50. **Cardioid**.

Carleson's theorem, *n.* the 1966 result that the partial sums of the FOURIER SERIES of a square-integrable function converge almost everywhere to the given function. This remains true in L_p-SPACE for $p > 1$. For $p = 1$, an example due to Kolmogorov shows that the series may diverge everywhere. (Named after the Swedish mathematician *Lennart Axel Edvard Carleson* (1928–).)

Carmichael number, *n.* another term for PSEUDO-PRIME (sense 1).

carrier, *n.* another name for CORRESPONDENCE or SET–VALUED FUNCTION.

carry, *vb.* in addition and multiplication, to transfer a digit or digits from a lower to a higher PLACE–VALUE column. For example, in adding 19, 27 and 48, first the 9, 7 and 8 are added, giving 24, which is 4 units and 2 tens; we thus write 4 in the unit position of the sum and *carry* the 2 tens to add to the 1, 2, and 3 tens of the addends, giving a sum of 94, as shown in Fig. 51.

$$
\text{add} \begin{cases} 19 = (1 \times 10) + 9 \\ 27 = (2 \times 10) + 7 \\ 48 = (4 \times 10) + \underline{8} \\ \overset{\frown}{(24)} \end{cases}
$$

$$
\underline{(2 \times 10) + 4} \\ = (9 \times 10) + 4 = 94
$$

Fig. 51. **Carry**. See main entry.

Cartesian, *adj.* derived from or relating to the works of DESCARTES, especially the algebraic representation of planar geometry. See also CARTESIAN CO-ORDINATES, CARTESIAN PLANE, CARTESIAN PRODUCT.

Cartesian coordinates or **rectangular coordinates,** *n.* a system for the representation of a point in space in terms of its distance, measured along a set of mutually perpendicular AXES, from a given ORIGIN: on the CARTESIAN PLANE the point $\langle a, b \rangle$ is located by measuring *a* units along the *x*-axis and *b* units along the *y*-axis, and then finding the point of intersection of the perpendiculars to the axes at those points, as shown in Fig. 52 opposite; *a* is then the *abscissa* and *b* the *ordinate*. By convention, the positive directions of the axes point to the right and upwards, so that the four points $(\pm 1, \pm 1)$ are placed as shown; by convention, the first QUADRANT is that in which both quantities are positive, and the other quadrants are numbered anticlockwise from the first. This system extends naturally to three or more dimensions; in three dimensions the convention is to use a RIGHT–HANDED set of axes, and many standard formulae presuppose this. See COORDINATE GEOMETRY.

Cartesian distance, *n.* another name for EUCLIDEAN DISTANCE.

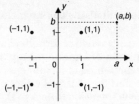

Fig. 52. **Cartesian coordinates**.
See main entry.

Cartesian distance, n. another name for EUCLIDEAN DISTANCE.

Cartesian plane, n. the two-dimensional space the points of which are identified by their CARTESIAN COORDINATES; the CARTESIAN SQUARE of the REAL LINE. See COORDINATE GEOMETRY.

Cartesian product, n. **1.** also called **cross product.** the set of ORDERED n-TUPLES the elements of which are respectively members of any sequence of given sets. The product $A \times B \times C$ is the set of all triples $\langle a, b, c \rangle$ where a is a member of A, b is a member of B, and c is a member of C.

2. another term for EXTERNAL DIRECT PRODUCT. See also DIRECT SUM.

Cartesian space, n. another term for EUCLIDEAN SPACE.

Cartesian square, n. the CARTESIAN PRODUCT of any set with itself. For example, Cartesian coordinates are the Cartesian square of the real numbers.

Cartesian tensor, n. a tensor defined on a VECTOR SPACE with an ORTHO-NORMAL BASIS.

Casorati-Weierstrass theorem, n. a weak form of PICARD'S THEOREM, stating that an analytic function comes arbitrarily close to all values in any neighbourhood of an ESSENTIAL SINGULARITY; that is, that the image of any ball centred on the singularity is dense in the complex numbers.

Cassini ovals, see OVALS OF CASSINI.

casting out nines, n. a method of checking arithmetical calculations that relies on the fact that, MODULO 9, the sum of the digits of a sum or product equals the sum or product of the sums of the digits of the numbers added or multiplied, where the process of adding digits is repeated where necessary until the result is less than nine. For example, we can determine that 365×248 cannot be 91520, since adding the digits of each of the multiplicands yields 14, whose digits sum to 5, so that their product should have the same *nines-complement* as $5 \times 5 = 25$, that is 7, whereas similarly casting out nines from the supposed product gives 8. However, since this is merely a necessary and not a sufficient condition, the process can only be used to detect errors, but can neither determine nor verify a result.

Catalan numbers, n. the integers of the form

$$\frac{(2n)!}{n!(n+1)!} = \frac{1}{n+1}\binom{2n}{n},$$

the first four being 2, 3, 10, 420.

Catalan's constant, n. the sum of the alternating series

$$\sum_{n=0}^{\infty} (-1)^n (2n+1)^{-2} = 1 - \frac{1}{9} + \frac{1}{25} - \frac{1}{49} + \dots,$$

which is approximately equal to 0.915965. It is not known whether this constant is rational or irrational.

catastrophe theory, *n.* **1.** the mathematical theory that classifies SURFACES up to DIFFEOMORPHISM according to their form.

2. the popular application of the theory to the explanation of abruptly changing phenomena, by, for example, the discontinuity of a line on the topmost fold of a folded surface. This can be simply represented in two dimensions, as in Fig. 53: y is a continuous function of x, but the observed values of x, considered in terms of progression along the y-dimension from A to D, are those indicated by the bold line, and there is a discontinuity between B and C indicated by the dotted line.

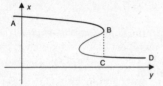

Fig. 53. **Catastrophe theory**.
See main entry.

categorial, *adj.* **1.** pertaining to CATEGORY theory.

2. also called **categorical.** (of a statement) consisting of a subject, a predicate, and a quantifier that asserts a relationship between two classes, such as *all A are B*, or *some A are not B*. See SYLLOGISM.

category, *n.* **1.** a structure consisting of a class of objects, denoted ObC, and a class of ARROWS (or MORPHISMS), denoted ArC, such that disjoint classes of morphisms are associated with each pair of objects and the morphisms are ASSOCIATIVE with an IDENTITY under COMPOSITION. Categories are useful as an abstract model for the study of structures and mappings that preserve these structures. For example, in the DIAGRAM OF ARROWS of Fig. 54, the objects are groups G and A, and the morphisms represented by the arrows are homomorphisms θ and ϕ, and the identity function ψ. When the categories are small and the morphisms are functions, it is called a classical category or *kittygory*. See also FUNCTOR.

Fig. 54. **Category**. See main entry.

2. short for BAIRE CATEGORY.

category theory, *n.* an abstraction from the study of structures and structure-preserving MAPPINGS, such as groups and their homomorphisms or topologies and their homeomorphisms. See CATEGORY.

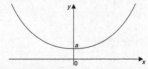

Fig. 55. **Catenary**.

catenary, *n.* the curve described by a uniform heavy flexible cord hanging freely between two points; when symmetrical about the y-axis, as in Fig. 55, its equation is $y = a\cosh(x/a)$, with a the point of intersection with the y-axis.

catenoid, *n.* the geometrical surface generated by rotating a CATENARY around its axis of symmetry.

Cauchy, Augustin Louis, Baron (1789–1857), French mathematician and physicist whose work had great impact on, and introduced rigour to, virtually all branches of mathematics. In particular, he laid the foundations of modern analysis in terms of limits and continuity, and developed the theory of functions of complex variables. After serving as an engineer in the force preparing Napoleon's abortive invasion of Britain, he was encouraged to pursue a mathematical career by Laplace (whom he had met when his family fled from the Reign of Terror) and by Lagrange. He became Professor at the *École Polytechnique*, the Sorbonne, and the *Collège de France*. Because of his political and religious views, he refused to take the oath of allegiance to Louis-Philippe in 1830 and followed Charles X into exile; the University of Turin appointed him to a specially created Chair, but he left to tutor the grandson of Charles X. He published a total of 789 works, including monographs on definite integrals and on wave propagation, and papers on geometry, number theory, elasticity, the theory of error, astronomy, and optics.

Cauchy-Binet formula or **Binet-Cauchy formula,** *n.* the formula that expresses the DETERMINANT of a product AB of two MATRICES A and B, where A is $m \times n$ and B is $n \times m$ $(m \leq n)$, as the products of all possible MINORS of the maximal order m of A with the corresponding minors of the same order of B.

Cauchy condensation test, *n.* the test for the CONVERGENCE of a series that relies on the fact that if $\{p_n\}$ are decreasing positive terms, then $\sum p^n$ and $\sum 2^n p_2 n$ converge or diverge together.

Cauchy condition or **Cauchy criterion,** *n.* **1.** the necessary and sufficient condition for an infinite sequence to CONVERGE that the absolute difference between successive terms with sufficiently large indices tends to zero.

2. more generally, any of a number of conditions establishing the convergence of a SEQUENCE, SERIES, PRODUCT, etc., by verifying that one obtains a CAUCHY SEQUENCE and then appealing to the COMPLETENESS of the underlying metric space. For example, if $\{a_i\}$ is a Cauchy sequence of elements of a normed space, then the associated series converges whenever for every $\varepsilon > 0$ there is an N such that

$$\left\| \sum_{k=m}^{j} a_k \right\| < \varepsilon \quad \text{for all } j > m > N.$$

3. an INITIAL CONDITION for a second-order PARTIAL DIFFERENTIAL EQUATION in which u and $\partial u / \partial t$ are given at $t = 0$, where u is a function of the variable t.

Cauchy form of the remainder, see LAGRANGE FORM OF THE REMAINDER.

Cauchy–Green deformation tensors, *n.* (*Continuum mechanics*) the TENSORS $\mathbf{F}^T\mathbf{F}$ (the *right Cauchy–Green deformation tensor*) and $\mathbf{F}\mathbf{F}^T$ (the *left Cauchy–Green deformation tensor*) for a given DEFORMATION GRADIENT $\mathbf{F}$.

Cauchy–Hadamard theorem, *n.* the theorem that the RADIUS OF CONVERGENCE of a TAYLOR SERIES with coefficients $\{a_i\}$ is

$$\frac{1}{\limsup\limits_{z \to \infty} \sqrt[n]{|a_n|}}.$$

Cauchy–Kowalewska theorem, n. the theorem that if $\mathbf{S}_r$ is a system of r PARTIAL DIFFERENTIAL EQUATIONS in r unknown functions, $v_1,...,v_r$, of $p+1$ real variables, $x_1,..., x_p+1$, of the form

$$\frac{\partial v_j}{\partial x_{p+1}} = H_j\left(x_1, \ldots, x_{p+1}, v_1, \ldots, v_r, \frac{\partial v_1}{\partial x_1}, \frac{\partial v_2}{\partial x_2}, \ldots, \frac{\partial v_r}{\partial x_p}\right),$$

where there are no derivatives with respect to x_{p+1} on the right-hand side and the H_j are real ANALYTIC functions, then there exists a neighbourhood of the origin in $(p+1)$-dimensional Euclidean space in which $\mathbf{S}_r$ may be uniquely solved for analytic functions $v_1,...,v_r$. This is the only general theorem for PARTIAL DIFFERENTIAL EQUATIONS, but most systems are not of this kind.

Cauchy principal value, n. the evaluation of an IMPROPER INTEGRAL on the interval $[-\infty, \infty]$ as the symmetric (two-sided) limit of the integral on intervals of the form $[-n, n]$. This may well converge even if the sum of the two ordinary improper integrals over $[-\infty, a]$ and $[a, \infty]$ does not. This is the case for ODD functions such as x or $\sin x$.

Cauchy product, n. the CONVOLUTION of two sequences $\{a_n\}$ and $\{b_n\}$ given by

$$c_n = \sum_{k=0}^{n} a_{n-k}b_k.$$

If the series $\sum a_n$ sums to A and $\sum b_n$ to B, then the Cauchy product series sums to AB if one of the series is absolutely convergent. See MERTEN'S THEOREM.

Cauchy–Riemann equations, n. the equations linking the PARTIAL DERIVATIVES of the real and imaginary parts of an ANALYTIC function of z with real part x and imaginary part y. When $f = u + iv$, with u and v real-valued, these are

$$\frac{\partial u}{\partial x} = \frac{\partial v}{\partial y} \quad \text{and} \quad \frac{\partial u}{\partial y} = -\frac{\partial v}{\partial x}.$$

When the partial derivatives are continuous, these equations are also sufficient for analyticity, when the domain of analyticity is a REGION.

Cauchy–Schwarz inequality or **Buniakovski's inequality,** n. the inequality, valid for any INNER PRODUCT, that

$$\langle x, y \rangle \leq \| x \| \cdot \| y \|.$$

In a EUCLIDEAN SPACE this can be written as CAUCHY'S INEQUALITY.

Cauchy sequence or **fundamental sequence,** n. an infinite sequence of points or values the distances between which tend to zero as their indices tend to infinity; $\{a_i\}$ is a Cauchy sequence in a metric space if, for every $\varepsilon > 0$, there is an N such that

$$d(a_i, a_j) < \varepsilon \quad \text{for all } i,j > N.$$

For example, $\{1/n\}$ is a Cauchy sequence. See COMPLETE.

Cauchy's inequality, n. the special case of the CAUCHY–SCHWARZ INEQUALITY in a EUCLIDEAN SPACE:

$$\sum_{i=1}^{n} a_i b_i \leq \sqrt{\left(\sum_{i=1}^{n} a_i^2\right)\left(\sum_{i=1}^{n} b_i^2\right)}.$$

This follows from the COSINE LAW.

Cauchy's integral formula, *n.* the identity, for an ANALYTIC function *f* on a STAR–LIKE region G, that

$$f(c)\, n(\Gamma, c) \;=\; \frac{1}{2\pi i} \int_\Gamma \frac{f(z)}{z - c} \; dz,$$

where Γ is a curve in G, *c* is a point of G not on Γ, and $n(\Gamma, c)$ is the WINDING NUMBER of Γ. This fundamental result is most useful for a positively-oriented simple closed curve surrounding *c*, when $n(\Gamma, c) = 1$.

Cauchy's integral theorem, *n.* the theorem of complex analysis stating that the CONTOUR INTEGRAL of an ANALYTIC function around a SIMPLE CLOSED CURVE is zero. See also GREEN'S THEOREM, RESIDUE THEOREM OF CAUCHY.

Cauchy's lemma, *n.* the result that if G is a finite GROUP and *p* is a prime number that divides the ORDER of G, then G contains an element of order *p*. See also SYLOW SUBGROUP.

Cauchy's mean-value theorem or **generalized mean-value theorem,** *n.* the theorem that if *f* and *g* are differentiable in an interval (a, b) and continuous on $[a, b]$, then

$$f'(c)\,[g(b) - g(a)] \;=\; g'(c)\,[f(b) - f(a)]$$

for some point *c* in the open interval.

Cauchy's ratio test, see RATIO TEST.

Cauchy's residue theorem, *n.* see RESIDUE THEOREM OF CAUCHY.

Cauchy's root test, see ROOT TEST.

Cauchy's stress principle, *n.* the fundamental axiom of CONTINUUM MECHANICS that postulates that the STRESS VECTOR at a point on a surface of a BODY depends continuously on the unit outward NORMAL to the surface at that point.

Cauchy's stress theorem, *n.* (*Continuum mechanics*) the theorem that states that the STRESS VECTOR at a point on a given surface of a BODY is given by σn, where σ is a symmetric second order CARTESIAN TENSOR and **n** is the unit outward NORMAL to the surface at that point.

Cauchy's vorticity formula, *n.* (*Continuum mechanics*) the relation, for a given BODY executing a CIRCULATION PRESERVING MOTION, of the VORTICITIES ω_r and ω_t respectively in the reference and current CONFIGURATIONS; this states that

$$\omega_t \;=\; \frac{\mathbf{F}\omega_r}{\det \mathbf{F}} \,,$$

where F is the DEFORMATION GRADIENT.

Cavalieri's principle, *n.* the principle that solids with the same height and with cross-sections of the same area have the same volume; in particular, this applies to PRISMS and CYLINDERS with equal bases and heights. This is easily established using the INTEGRAL CALCULUS. (Named after the Italian mathematician and physicist Francesco Bonaventura Cavalieri (1598–1647), whose work in some respects anticipated the integral calculus. He postponed its publication in deference to Galileo, whom he regarded as his master.)

Cayley, Arthur (1821–95), English algebraist and analyst, who, despite precocious mathematical ability, was obliged to earn a living as a lawyer for 14 years before being appointed to the Sadlerian Chair of Mathematics at Cambridge. He published more than 900 papers touching on most branches of mathematics, but his particular contributions were to the theory of matrices, algebraic invariance, and multidimensional geometry, work soon to be of significance in the development of relativity theory and quantum mechanics.

Cayley algebra, *n.* a non-associative non-commutative DIVISION ALGEBRA of dimension 8 over the real numbers.

Cayley–Hamilton theorem, *n.* the result that a SQUARE MATRIX satisfies its own CHARACTERISTIC equation.

Cayley representation theorem, *n.* the result that every GROUP is ISOMORPHIC to a group of PERMUTATIONS

cdf, (*Statistics*) *abbrev. for* CUMULATIVE DISTRIBUTION FUNCTION.

ceiling or **least integer function,** *n.* (*Computing*) the smallest integer not less than a given real number. Compare FLOOR.

centesimal, *adj.* hundredth, or pertaining to hundredth parts.

centi- (*symbol* c), prefix denoting a fraction of 100th of the physical units of the SYSTEME INTERNATIONAL.

centile, *n.* (*Statistics*) another word for PERCENTILE.

central angle, *n.* any angle the vertex of which is the centre of a given circle; for example in Fig. 56, angles AOB, AOC, and BOC are all central angles.

Fig. 56. The angles at O are **central angles**.

central conic, *n.* a CONIC with a CENTER OF SYMMETRY, and thus either an ELLIPSE or a HYPERBOLA. The conic with equation $ax^2 + 2hxy + by^2 + 2gx + 2fy + c = 0$ is central if and only if $ab \neq h^2$.

central difference, see DIFFERENCE QUOTIENT, DIFFERENCE SEQUENCE.

central dilatation, see DILATATION.

centralizer, *n.* the subgroup, denoted $C_G(x)$, consisting of elements that COMMUTE with a given element or subset of a GROUP, G. See also CENTRE. Compare NORMALIZER.

central limit theorem, *n.* the fundamental statistical result that if a sequence of INDEPENDENT IDENTICALLY DISTRIBUTED RANDOM VARIABLES each has finite VARIANCE, then as their number increases, their sum (or, equivalently, their arithmetic mean) approaches a NORMALLY distributed random variable. Hence, in particular, if sufficiently many samples are successively drawn from any population, the sum or mean of the sample values can be thought of, approximately, as an outcome from a normally distributed random variable.

central moment, *n.* another term for MOMENT ABOUT THE MEAN.

centre, *n.* **1a.** the point from which all the points on the circumference of a circle are equidistant.

b. the point of intersection of the AXES of an ellipse or a hyperbola.

2. see CENTRE OF SYMMETRY, CENTROID.

3. the set of elements of a GROUP, G, that COMMUTE with every member of the group. It is denoted Z(G) and is equal to the intersection of the CENTRALIZERS of the elements of the group. See also INNER AUTOMORPHISM.

4. the POLE of the LINE AT INFINITY with respect to a CONIC in affine or Euclidean geometry.

5. (*Geometry*) the invariant point of various PERSPECTIVITIES.

6. (*Complex analysis*) the point about which a POWER SERIES is evaluated; the centre of the CIRCLE OF CONVERGENCE of the given series.

centre of curvature, *n.* the centre of the CIRCLE OF CURVATURE of a curve at a given point.

centre of mass, *n.* (*Continuum mechanics*) **1.** the weighted mean position of a finite discrete set of PARTICLES with POSITION VECTORS $\mathbf{x}_1,...,\mathbf{x}_n$ and masses $m_1,...,m_n$; its position vector is

$$\overline{\mathbf{x}} = \frac{\sum_{i=1}^{n} m_i\,\mathbf{x}_i}{\sum_{i=1}^{n} m_i}.$$

2. (*Mechanics*) (for a continuous BODY) the position vector given by the corresponding ratio of integrals over the volume of the CONFIGURATION of the body at time *t*, that is,

$$\overline{\mathbf{x}}(t) = \left(\int \rho\,\mathbf{x}\,dv\right) / \left(\int \rho\,dv\right)$$

where the DENSITY is ρ at the point with position vector $\mathbf{x}$.

centre of perspectivity, see PERSPECTIVE.

centre of similitude, see SIMILITUDE.

centre of symmetry or **centre,** *n.* a point around which a curve is SYMMETRICAL, so that for every point P on the curve there is another P' such that the directed lines joining the centre to P and to P' are of equal length and opposite sense. For example, the sine curve in Fig. 57 has a centre of symmetry about the origin, O, with respect to which P and P' are images of one another, but it has no axis of symmetry. Compare AXIS OF SYMMETRY.

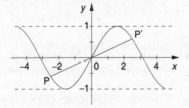

Fig. 57. O is the **centre of symmetry** of this curve.

centrifugal force, *n.* an apparent force acting outwards along a radius that opposes a real force such as the tension in a cord connecting a body to a fixed point around which it travels in a circular path; this is given by $m\,\omega \times (\omega \times \mathbf{x})$ for a PARTICLE of mass *m* and position vector $\mathbf{x}$, where ω is the ANGULAR VELOCITY of a given ROTATING FRAME OF REFERENCE. Like the CORIOLIS FORCE this is not in fact a force, but notional compensation for the rotating axes.

centroid, *n.* **1.** (of a triangle) the point of coincidence of the MEDIANS, as shown in Fig. 58.

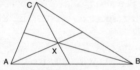

Fig. 58. X is the **centroid** of the triangle ABC.

2. also called **centre.** **a.** a point whose coordinates are the mean values of the coordinates of the points in a given set.

b. the CENTRE OF MASS of an object of uniform density or of a geometric figure.

Cesaro summation, *n.* the computation, in SUMMABILITY THEORY, of the limit of a possibly DIVERGENT sequence of numbers as the limit, as *n* approaches infinity, of the MEANS of the first *n* numbers. The method is REGULAR in that for a CONVERGENT sequence this limit exists and agrees with the original limit. This method assigns the limit of $\frac{1}{2}$ to the sequence 1,0,1,0,1,.... (Named after the Italian analyst and geometer *Ernesto Cesaro* (1859–1906).) Compare ABEL SUMMATION.

ceteris paribus, *adv.* (*Latin*) other things being equal; that is, in the absence of any other change (in the conditions of a theorem, for example).

Ceva's theorem, *n.* the result that three CEVIANS, one through each vertex of a triangle, are concurrent if and only if the product of the ratios in which they divide the opposite sides is unity. For example, the concurrence of the medians shown in Fig. 58 is a special case. MENELAUS' THEOREM is the DUAL of this result.

cevian, *n.* any line segment joining a vertex of a triangle to a point on the opposite side (or on its extension).

cgs, *abbrev. for* the METRIC SYSTEM, now superseded, that uses the centimetre, the gram, and the second as units of length, mass, and time respectively.

ch, *symbol for* the hyperbolic cosine function, COSH.

ch⁻¹, symbol for the inverse hyperbolic cosine function, ARC–COSH.

chain, *n.* **1.** any set that has a LINEAR ORDERING; it may be a subset of a PARTIALLY ORDERED set on which the ordering is CONNECTED.

2. a collection of NESTED sets.

3. Markov chain. another term for MARKOV PROCESS.

4. (*Graph theory*) any PATH joining two vertices in a GRAPH. In particular, an Eulerian chain uses each EDGE exactly once; a Hamiltonian chain uses each VERTEX exactly once.

chain condition, *n.* either an ASCENDING CHAIN CONDITION or an DESCENDING CHAIN CONDITION.

chain rule, *n.* a theorem that may be used in the DIFFERENTIATION of a function of a function. It states that

$$\frac{dy}{dx} = \frac{dy}{dt} \times \frac{dt}{dx} ,$$

where *y* is a differentiable function of *t*, and *t* is a differentiable function of *x*. This enables a function $y = f(x)$ to be differentiated by finding a suitable function u, such that f is a COMPOSITION of y and u, y is a differentiable function of u, and u is a differentiable function of x. Similarly, for PARTIAL DIFFERENTIATION,

$$\frac{\partial f}{\partial x} = \left(\frac{\partial f}{\partial u} \times \frac{\partial u}{\partial x} \right) + \left(\frac{\partial f}{\partial v} \times \frac{\partial v}{\partial x} \right),$$

where f is a function of u and v, each of which is a function of x.

Champernowne's number, n. the number 0.123 456 789 101 112… whose digits are those of all the natural numbers in succession. This is known to be a NORMAL NUMBER in base 10.

chance variable, n. another term for RANDOM VARIABLE.

change of observer or **change of reference,** n. (*Mechanics*) a mapping that corresponds to the different perception of physical events by different observers; formally, a mapping gf^{-1}, where f and g are OBSERVERS. In classical mechanics it is required that the distance and time between events is invariant under such transformations; that is, is independent of the choice of observer.

change of variables, n. a TRANSFORMATION in which new expressions are substituted for the variables of a given expression, especially where such substitutions are LINEAR and the effect is to change the position of a curve relative to the coordinate axes.

channel, n. (*Information theory*) a route by which discrete pieces of information, MESSAGES, are transmitted from sender to receiver, defined by the INPUT SET, the OUTPUT SET, and the PROBABILITY LAW for the channel.

chaos, n. apparently random but recurrent behaviour in a deterministic system; formally, a DYNAMICAL SYSTEM whose ATTRACTOR is a FRACTAL set.

chaotic, *adj*. of or relating to CHAOS.

character, n. **1.** a multiplicative functional from a GROUP to the complex numbers; more formally and generally, the mapping from a given group, G, to a field, that corresponds to the REPRESENTATION, R, of G, under which the image of an element, x, is the TRACE of $R(x)$.

2. see FINITE CHARACTER.

characteristic, n. **1.** the integral part of a COMMON LOGARITHM, representing the order of magnitude but not the digits of the given number, and equal to the exponent of the greatest power of 10 less than it. For example, the characteristic of log 450 is 3, and of log 4.5 is 1. Compare MANTISSA.

2. *adj*. of or relating to the CHARACTERISTIC FUNCTION of a matrix.

characteristic curve, n. (*Partial differential equations*) the curve determined by the CHARACTERISTIC EQUATION.

characteristic equation, n. **1.** the equation

$$\det[A - tI] = 0,$$

derived from a given square matrix A, where I is the conformable unit matrix. See CHARACTERISTIC POLYNOMIAL.

2. the second order PARTIAL DIFFERENTIAL EQUATION for which

$$a\left(\frac{dy}{dx}\right)^2 - b\left(\frac{dy}{dx}\right) + c = 0,$$

(where

$$au_{xx} + bu_{xy} + cu_{yy} + du_x + eu_y + fu = h$$

is the general form of a second-order partial differential equation), the solutions to which give the *characteristic curves* of the partial differential equation, allowing it to be written in a simpler canonical form.

characteristic function, *n.* **1.** (of a set) the function taking the value 1 for arguments that are members of the given set, and 0 otherwise. Compare INDICATOR FUNCTION.

2. another term for the CHARACTERISTIC POLYNOMIAL of a matrix.

3. (*Statistics*) a function derived from the PROBABILITY DISTRIBUTION FUNC-TION that in particular enables the distribution of the sums of random variables to be analysed since two distributions have the same characteristic function only if they are identical almost everywhere.

characteristic of a field, *n.* the smallest positive natural number n such that 0 is the sum of the unit element with itself n times. If no such n exists, the field is said to be of characteristic zero. See MODULAR FIELD.

characteristic polynomial or **characteristic function,** *n.* the polynomial $\det[A - tI]$ derived from a given square matrix A, where I is the unit matrix and t is a scalar variable; the roots of this polynomial are the LATENT ROOTS (or EIGENVALUES), λ, of A, for which there is a column matrix, the EIGENVECTOR, **X** such that A**X** = λ**X**. For example, the characteristic polynomial of

$$\begin{bmatrix} 2 & 1 \\ -1 & 1 \end{bmatrix}$$

is $t^2 - 3t + 1$. See also QUADRATIC FORM.

characteristic root, value, or **number,** *n.* other terms for LATENT ROOT or EIGENVALUE. See CHARACTERISTIC POLYNOMIAL.

characteristic subset or **subgroup,** *n.* a subset or subgroup of a group that is mapped by all AUTOMORPHISMS of the group onto itself. The DERIVED SUB-GROUP and the CENTRE of a group are characteristic, and any characteristic subgroup is a NORMAL SUBGROUP.

characteristic vector, *n.* another term for EIGENVECTOR. See CHARACTERISTIC POLYNOMIAL.

Charpit's method, *n.* a method for solving a first-order PARTIAL DIFFERENTIAL EQUATION of the form

$$F\left(x, y, z, \frac{\partial z}{\partial x}, \frac{\partial z}{\partial y}\right) = 0,$$

in which a solution of a specimen of LAGRANGE'S LINEAR EQUATION is used to provide a second first-order partial differential equation

$$f\left(x, y, z, \frac{\partial z}{\partial x}, \frac{\partial z}{\partial y}\right) = 0,$$

that has the property that solving these equations for $\partial z/\partial x$ and $\partial z/\partial y$ produces an integrable TOTAL DIFFERENTIAL EQUATION

$$dz = \frac{\partial z}{\partial x}\,dx + \frac{\partial z}{\partial y}\,dy.$$

The GENERAL SOLUTION of this ordinary differential equation is a COMPLETE SOLUTION of the equation F = 0.

chart, *n.* **1.** another word for GRAPH (sense 1).

2. (*Differential geometry*) also called (**local**) **coordinate system.** a neighbourhood of a point in a MANIFOLD together with its mapping into Euclidean *n*-space; formally, a pair $(U_\lambda, \phi_\lambda)$ where U_λ is an element of a COVER of the manifold and ϕ_λ is a HOMEOMORPHISM that maps it to an open subset of $\mathbb{R}^n$. A collection of charts that cover the manifold is called an *atlas.*

Chebyshev (or **Chebychev, Chebysev, Chebycheff, Tchebychev,** etc.)**, Pafnuti Lvovich** (1821–94), Russian algebraist, analyst, probability theorist, and number theorist.

Chebyshev approximation, *n.* the problem of finding the polynomial closest in CHEBYSHEV NORM to a given continuous function.

Chebyshev norm, supremum norm, or **uniform norm,** *n.* the NORM placed on continuous or bounded functions on a set S that assigns to each function the SUPREMUM of the moduli of the values of the function on the set:

$$\|f\|_\infty = \sup\{\,|f(x)\,| : x \in S\}.$$

So endowed, the real or complex continuous functions on a compact set S form a BANACH SPACE that is denoted C(S).

Chebyshev polynomials (**of the first kind**), *n.* the ORTHOGONAL polynomials defined by $T_n(x) = \cos(n \arccos x)$. These arise as the polynomials of degree *n* of smallest CHEBYSHEV NORM on $[-1, 1]$, with leading coefficient unity. See BEST APPROXIMATION.

Chebyshev's inequality, *n.* **1.** (*Statistics*) **a.** the fundamental theorem that the probability of a random variable differing from its mean by more than *k* standard deviations is less than or equal to $1/k^2$.

b. more generally, the result that, for all $\varepsilon > 0$,

$$P(|X - c| > \varepsilon) \leq \frac{1}{\varepsilon^2} E\left[(X - c)^2\right]$$

where X is a random variable and *c* is a constant.

2. the inequality for two non-increasing sequences of real numbers, $a_1, .., a_n$ and $b_1, .., b_n$, that

$$\frac{1}{n^2} \sum_{j=1}^{n} a_j \sum_{i=1}^{n} b_i \leq \frac{1}{n} \sum_{k=1}^{n} a_k b_k ,$$

with identity if and only if all the a_i are equal and all the b_i are equal.

Chebyshev's theorem, *n.* (*Statistics*) a form of the WEAK LAW OF LARGE NUMBERS.

check digit, *n.* a bit added to each word in a BINARY CODE in order to detect errors. Consider a code in which all the words of length *n* are possible code words. Such a code cannot be error-detecting, but if an additional bit is added to each code word so that the number of 1s in each derived code word is even, the new code is error-detecting. For example, from the code with codewords 00, 01, 10, and 11, the new code with codewords 000, 011, 101, and 110 would be obtained. This simple example is a PARITY check, and more complicated examples are often used for the purposes of detecting errors.

chief series, *n.* another term for PRINCIPAL SERIES.

Chinese postman problem, *n.* the problem of finding the closed WALK in a weighted graph that includes each edge exactly once and has least WEIGHT. This is analogical to the city postman who wishes to visit all the streets in his area to deliver his letters and return to his starting point, having covered the least possible distance. See also TRAVELLING SALESMAN PROBLEM.

Chinese remainder theorem, *n.* the basic result in number theory that if a set of integers, m_i, are pairwise RELATIVELY PRIME, then the CONGRUENCES $x \equiv a_i$ (mod m_i) have a unique solution for x, MODULO the product of all the m_i.

chi-square distribution or **χ^2-distribution,** *n.* (*Statistics*) a continuous single-parameter distribution derived as a special case of the GAMMA DISTRIBUTION; it is used especially to measure GOODNESS OF FIT, and to test hypotheses and obtain CONFIDENCE INTERVALS for the VARIANCE of a NORMALLY distributed random variable. Its probability distribution function is

$$\chi^2(v) = \frac{x^{(\frac{v}{2}-1)} e^{(-\frac{x}{2})}}{2^{(\frac{v}{2})} \Gamma(\frac{v}{2})},$$

where the single parameter is known as the number of DEGREES OF FREEDOM.

chi-square test or **χ^2-test,** *n.* (*Statistics*) a test derived from the CHI-SQUARE DISTRIBUTION and used to compare the goodness of fit of theoretical and observed frequency distributions, or to compare NOMINAL DATA derived from unmatched groups of subjects.

Cholesky decomposition or **factorization,** *n.* the factorization of a POSITIVE DEFINITE matrix A as $LL^* = R^*R$, where L is LOWER-TRIANGULAR, R is UPPER-TRIANGULAR, and L^* and R^* are their respective TRANSPOSES. The matrix R is sometimes called a *Cholesky factor* or 'square root' of A and can be computed directly from element-by-element comparison, starting in the first row. Compare L-U DECOMPOSITION.

chord, *n.* a straight line segment connecting two points on a curve or surface and lying between them.

chromatic number, *n.* the maximum number of colours, denoted $\chi(G)$, that must be used to colour the edges (or, dually, the vertices) of a GRAPH (or map) so that edges meeting at a common vertex are distinct colours. A graph with $\chi(G)$ equal to k is said to be k-colourable. All BIPARTITE GRAPHS are two-colourable; and all PLANAR GRAPHS are four-colourable as a consequence of the FOUR COLOUR THEOREM.

Church's theorem, *n.* (*Logic*) the result that no decision procedure exists for arithmetic. (Named after the US logician Alonzo Church (1903–), who held chairs in both mathematics and philosophy at UCLA.) See also GÖDEL'S THEOREM.

Church's thesis, *n.* (*Logic, computing*) the hypothesis that a function is RECURSIVE if and only if it is EFFECTIVELY COMPUTABLE. See also TURING MACHINE.

cipher or **cypher,** *n.* **1.** an obsolete name for ZERO.

2. any of the ARABIC NUMERALS, 1, ..., 9, or the Arabic counting system as a whole.

cir. or **circ.,** *abbrev. for* CIRCLE, CIRCULAR, or CIRCUMFERENCE.

circle, *n.* **1a.** a closed plane curve every point of which is equidistant from a given fixed point, the centre. Its equation is

$$(x-h)^2 + (y-k)^2 = r^2,$$

where r is the RADIUS, and (h, k) is the CENTRE; its parametric equations are

$$x = r\cos\theta; \quad y = r\sin\theta$$

(compare ELLIPSE). It has CIRCUMFERENCE of length $2\pi r$.

b. the figure enclosed by such a curve, with area πr^2. Fig. 59 shows such a figure together with some of its more important elements.

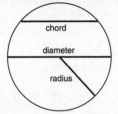
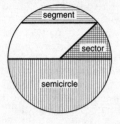

Fig. 59. Some main elements of a **circle**.

2. see GREAT CIRCLE.

3. (*Logic*) see VICIOUS CIRCLE.

circle of convergence, *n.* a circle on the ARGAND DIAGRAM such that a given POWER SERIES converges at all points within the circle and diverges at all points outside it. For real series the term interval of convergence is preferred. Formally, for any power series $\sum_i c_i (z-a)^i$, there is an R such that the series converges, if $R \neq 0$, for all z such that $|z-a| < R$, and diverges for all z such that $|z-a| > R$. The radius R may also be infinite, in which case the circle of convergence is the entire plane, or zero, in which case it is a single point, and is the absolute value of the limit of the ratio of each term to the next. The series may either converge or diverge at points for which $|z-a|$ equals the radius; that is, that lie on the circumference of the circle. For example, the series $\sum z^n/n$ has circle of convergence $|z| = 1$, shown in Fig. 60; it converges absolutely inside the circle (the shaded area), and diverges outside, but it converges conditionally everywhere on the circle of convergence except at $z = 1$ (the point P), where it diverges. See also RADIUS OF CONVERGENCE.

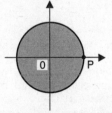

Fig. 60. **Circle of convergence**. See main entry.

circle of curvature or **osculating circle,** *n.* the circle with the same tangent and CURVATURE as a given curve at a given point; its radius, called the RADIUS OF CURVATURE, is NORMAL to the concave side of the curve at the point, and is equal to the reciprocal of its CURVATURE. For example, Fig. 61 (overleaf) shows the graph of $y = x^2$, and its circle of curvature at $x = 0$; its centre, C, is the *centre of curvature* at this point. The function has curvature 2 at this point, and its radius of curvature, shown by the line CO, therefore has length $1/2$ and is in the direction of the y-axis.

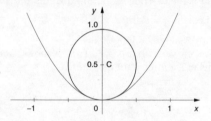

Fig. 61. **Circle of curvature**. C is the centre of curvature at 0.

circuit, *n.* a SIMPLE CLOSED CHAIN in a GRAPH. An *Eulerian circuit* uses each EDGE exactly once; a *Hamiltonian circuit* uses each VERTEX exactly once.

circulant, *n.* a DETERMINANT or matrix in which the elements of a given row are the elements of the previous row moved cyclically one to the right. All entries on the MAIN DIAGONAL are therefore identical.

circular, *adj.* **1.** (of an argument) proving a conclusion that was already among the premises.

2. (of an explanation or construction) given in terms of something that is itself explained in terms of, or constructed from, the very thing that is supposedly being explained or constructed.

See VICIOUS CIRCLE. Compare INFINITE REGRESS.

circular measure, *n.* the measurement of the size of an angle in RADIANS.

circular point, see UMBILIC.

circular function, *n.* another name for TRIGONOMETRIC FUNCTION.

circular triangle, *n.* a triangle constructed from three intersecting arcs of circles. The sum of the angles between the tangents to the arcs at the points of intersection can take any value between 0 and 1080° (6π radians). If only minor arcs are considered, the upper bound is 4π radians; a limiting case in which these arcs are semicircles is shown in Fig. 62. Compare SPHERICAL TRIANGLE.

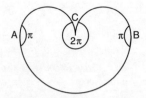

Fig. 62. **Circular triangle**. See main entry.

circulating decimal, *n.* another term for RECURRING DECIMAL.

circulation, *n.* (*Continuum mechanics*) the curvilinear integral $C(\Gamma, t)$ defined as the integral $\int_{\Gamma} \mathbf{v} \cdot d\mathbf{x}$, around a curve Γ in the CONFIGURATION of a BODY at time t, where $\mathbf{v}$ is the VELOCITY at the point with POSITION VECTOR $\mathbf{x}$.

circulation preserving motion, *n.* (*Continuum mechanics*) a MOTION in which CIRCULATION is independent of time. See also CAUCHY'S VORTICITY FORMULA.

circumcentre, *n.* the centre of the CIRCUMCIRCLE of a given figure. The circumcentre of a triangle is the point of intersection of the perpendicular bisectors of its sides, as shown in Fig. 63 opposite. Compare INCENTRE.

Fig. 63. X is the **circumcentre** of triangle ABC.

circumcircle, *n*. a circle that CIRCUMSCRIBES a given polygon (where this is possible), passing through all its vertices. For example, the circumcircle of the triangle ABC in Fig. 63 is as shown; X is its CIRCUMCENTRE. Compare INCIRCLE.

circumference, *n*. **1.** the boundary of a specific region or geometric figure, especially of a circle.

2. the length of a closed curve or of the boundary of such a geometric figure.

circumscribe, *vb*. to draw a specified geometric figure around another in such a way that they are in contact but do not intersect: a *circumscribed polygon* has sides which are tangential to the given figure, or which pass through the endpoints of its sides. For example, in Fig. 64 the regular hexagon UVWXYZ circumscribes the quadrilateral ABCD. See also CIRCUMCIRCLE. Compare INSCRIBE.

Fig. 64. The hexagon **circumscribes** the quadrilateral.

cis, *abbrev. for* cos + *i*sin. Thus, by EULER'S FORMULA (sense 2), cis$\theta = e^{i\theta}$.

cissoid, *n*. a geometric curve the two branches of which meet at a CUSP at the origin and are ASYMPTOTIC to a line parallel to the *y*-axis. Its equation is

$$y^2(2a - x) = x^3,$$

where $2a$ is the distance from the *y*-axis of the asymptote. If O is a fixed point, OD is the diameter of a circle with radius a, and C is a point that describes that circle, then the cissoid is the locus of another point, P, that moves in

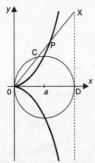

Fig. 65. **Cissoid**.

such a way that its distance, OP, from the fixed point equals the distance between C and the tangent to the circle at D; this locus is shown in Fig. 65 opposite by the bold curve. Compare SISTROID.

Clairaut's equation, *n*. the DIFFERENTIAL EQUATION

$$xy' - y + f(y') = 0.$$

Clairaut's form, *n*. a FIRST–ORDER ordinary or partial DIFFERENTIAL EQUATION

$$z = \sum_{i=1}^{n} x_i \frac{\partial z}{\partial x_i} + f\left(\frac{\partial z}{\partial x_1}, \frac{\partial z}{\partial x_2}, \dots, \frac{\partial z}{\partial x_n}\right)$$

where f is a differentiable function. If $n > 1$, the equation has the COMPLETE SOLUTION

$$z = \sum_i a_i x_i + f(a_1, a_2, \dots, a_n),$$

where $a_1, a_2, \dots, a_n$ are arbitrary constants; the GENERAL SOLUTION for $n = 1$ has the same form. The equation may also have a SINGULAR SOLUTION or a solution that is not obtainable from the complete solution.

clamped boundary condition, see SPLINE–FITTING.

Clarke generalized directional derivative, *n*. the function associated with a real-valued locally LIPSCHITZ FUNCTION f on a normed space X, by the formula

$$f^0(z; h) = \limsup_{x \to t,\, t \to 0+} \frac{f(y + th) - f(y)}{t}$$

The function $f^0(x; \ )$ is always SUBLINEAR and coincides with the ordinary directional derivative for a convex or a continuously differentiable function. The *Clarke generalized gradient* denoted $\partial f(x)$ is defined to be the set of linear functionals with

$$\phi(h) \le f^0(x; h) \quad \text{for all } h \text{ in X.}$$

This set is non-empty, WEAK–STAR compact, and convex. It coincides with the SUBGRADIENT for a convex function, and with the GRADIENT for a continuously differentiable function. If the space is finite dimensional it can be realized as the closed convex hull of all limit points of sequences of gradients of the function at arguments approaching the point.

class, *n*. **1.** another name for a SET, especially a finite set.

2. also called **proper class.** in certain formulations of set theory, a set that cannot itself be a member of other sets. Adopting this restriction makes it impossible to speak of the class of all classes, and so avoids RUSSELL'S PARADOX.

class equation, *n*. the equation

$$|G| = |Z(G)| + \sum_i |cl(x_i)|$$

where G is a finite GROUP, Z(G) is its CENTRE, and $cl(x_1) \dots cl(x_n)$ are all the distinct non-singleton CONJUGACY CLASSES of G.

classical, *adj.* **1.** (of a theory) distinguished from some, usually later, variant with a more complex structure; often the term is used to distinguish a variant that the author regards as no longer interesting.

2. (of a logical or mathematical system) having the law of EXCLUDED MIDDLE as an axiom or theorem, so that every statement is known to be either true or false even although it may not be known which. Compare INTUITIONIST.

3. (of some entity) well-behaved, in terms of some classical theory.

classical category, see CATEGORY.

classical eigenvalue problem, see GENERALIZED EIGENVALUE PROBLEM.

classical probability, *n.* another name for MATHEMATICAL PROBABILITY.

class interval, *n.* (*Statistics*) one of the intervals into which the range of a variable of a distribution is divided, especially one of the divisions of the base line of a BAR CHART or HISTOGRAM. For example, in Fig. 66, the base line is divided into unequal class intervals, and the frequency of each class is proportional to the area of the bar.

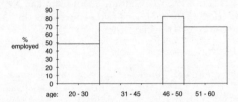

Fig. 66. The age range consists of four unequal **class intervals**.

class mark, *n.* (*Statistics*) a value within a CLASS INTERVAL, often its midpoint or the nearest integral value, used to represent the interval for computational convenience. For example, in Fig. 66 we might use 25, 38, 48, and 55 as the class marks corresponding to each of the intervals shown.

class number, *n.* the finite number h_F of equivalence classes of EQUIVALENT IDEALS in the ring of INTEGERS, D, of an ALGEBRAIC NUMBER FIELD, F. The class number is 1 if and only if D is a PRINCIPAL IDEAL DOMAIN.

clique, *n.* a completely CONNECTED SUBGRAPH of a GRAPH.

clock arithmetic, *n.* arithmetic to a given finite MODULUS, analogous to the numbers on a clock face, for which $12 + 1 \equiv 1 \pmod{12}$. See CONGRUENCE.

clockwise, *adj.* or *adv.* (of rotation, an angle, etc.) in the same direction as that conventionally taken by the hands of a clock. If a horizontal line segment has its left-hand end, A, fixed, its right-hand end, B, moves clockwise when it moves downwards, as depicted in Fig. 67. By convention, for the purpose of measuring angles, fixing points by POLAR COORDINATES, etc., this is taken to be the negative direction. Compare ANTICLOCKWISE.

Fig. 67. Clockwise.

clopen, *adj.* (of a set in a TOPOLOGY) both OPEN and CLOSED. The space itself is always clopen.

closed, *adj.* 1. (of a set under an operation) containing all members of the set produced by that operation acting on members of the set. For example, the positive integers are closed under addition but not under subtraction, since $n + m$ is also a positive integer for any positive integers n and m, but $n - m$ may be negative or zero and so not in the set.

2. (of a curve or surface) completely enclosing an area or volume. See CLOSED CURVE.

3. (of a set in a TOPOLOGY) containing all of its LIMIT POINTS, being the complement of an OPEN SET. See also CLOSED INTERVAL.

4. (of a set) being the ALGEBRAIC CLOSURE of some set.

5. (of a function or MULTIVALUED FUNCTION) possessing a GRAPH that is topologically closed.

6. (of a function between two TOPOLOGICAL SPACES) sending closed sets to closed sets.

7. (of a path in a GRAPH) having the same VERTEX at each end.

8. (of a DIFFERENTIAL FORM) having EXTERIOR DIFFERENTIAL equal to zero. Compare EXACT (sense 3).

9. (of a branch in a SEMANTIC TABLEAU) containing inconsistent propositions. If every branch is closed, then the tableau is said to be closed; this fact shows the given set of propositions to be inconsistent.

10. see ORBIT.

closed ball, see BALL.

closed curve, *n.* a curve that completely encloses an area, having no endpoints. Formally, a closed curve is the continuous image of a CLOSED SET; it has COORDINATE FUNCTIONS such that each coordinate is a continuous function $f_i(t)$ on the interval $[0,1]$ of the real line, and $f_i(0) = f_i(1)$. A closed curve is simple if it does not intersect with itself; thus the curve in Fig. 68 is closed but not simple.

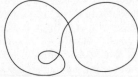

Fig. 68. **Closed curve**.

closed form, *n.* an expression for a given function or quantity, especially an integral, in terms of known and well understood quantities, such as the evaluation of

$$\int_{-\infty}^{\infty} \exp(-x^2) \, dx$$

as $\sqrt{\pi}$.

closed graph theorem, *n.* a theorem asserting that a given function (or MULTIVALUED FUNCTION) with a closed graph is continuous (or SEMICONTINUOUS). The classical closed graph theorem proves that a linear operator between BANACH or FRECHET SPACES is continuous if and only if it has a closed graph.

closed interval, *n.* an interval the complement of which is OPEN, a set of real numbers lying between and including its endpoints; written $[x, y]$ where x is both the MINIMUM and the INFIMUM of the interval and y both its MAXIMUM and its SUPREMUM. $[x, \infty[$ and $]-\infty, x]$ are regarded as closed. Compare OPEN INTERVAL.

closed mapping, *n.* a function or MULTIVALUED FUNCTION between TOPOLOGICAL SPACES that sends closed sets to closed sets. Compare OPEN MAPPING.

closed neighbourhood, see NEIGHBOURHOOD.

closed path, *n.* a PATH whose terminal point is identical with its initial point.

closed sentence, *n*. (*Logic*) an expression that contains no FREE occurrence of any variable, so that all its variables are BOUND by QUANTIFIERS. Compare OPEN SENTENCE.

closed set, *n*. **1.** the complement of an OPEN SET.

2. (*Topology*) a set that contains all its CLUSTER POINTS.

closure, *n*. **1.** the smallest CLOSED SET containing a given set, equal to the intersection of all closed sets containing that set. For example, the closure of the positive integers under subtraction is the set of all integers. See CLOSED (sense 1). See also HULL, ALGEBRAIC CLOSURE.

2. the set of points in a space every NEIGHBOURHOOD of which has a non-empty intersection with a given set. The closure of A is written Ā or Cl A. For example, the closure of the open interval (0, 1) is the closed interval [0, 1]; the closure of the rationals is the reals. Compare INTERIOR. See also CLUSTER POINT.

3. (*Logic*) the CLOSED SENTENCE formed by prefixing QUANTIFIERS to a given OPEN SENTENCE to BIND all its FREE VARIABLES; especially the universal closure of the given sentence, formed by binding its free variables with universal quantifiers. Mathematical IDENTITIES written without quantifiers are abbreviations for their universal closures; thus one writes $a + b = b + a$ as the commutative law for addition, indicating that the result of adding any two elements is independent of order.

4. the operation forming such a set or sentence.

cluster, *n*. (*Statistics*) a naturally occurring subgroup of a population used in STRATIFIED SAMPLING.

cluster point, limit point, or **accumulation point,** *n*. a point every PUNCTURED NEIGHBOURHOOD of which has a non-empty intersection with a given set; a point where neighbourhoods intersect with the set other than at the point itself. Compare CLOSURE.

cn, see JACOBIAN ELLIPTIC FUNCTIONS.

cnf, abbrev. for CONJUNCTIVE NORMAL FORM.

coarser, *adj*. (of a TOPOLOGY) strictly contained in another topology. For example, the topology of open intervals on the reals is coarser than the DISCRETE TOPOLOGY; the coarsest topology is the INDISCRETE TOPOLOGY.

coaxial, *adj*. having the same AXIS. For example coaxial cable consists of a number of concentric cylinders of wire mesh round an axial central wire.

Cobb–Douglas function, *n*. a commonly used PRODUCTION FUNCTION of the form $AL^a K^b M^c$ where L, K, and M respectively measure the quantities of labour, capital, and raw material consumed for a given output rate; constants A, a, b, c, are parameters. Increasing return occurs when $a + b + c > 1$.

code, *n*. **1.** (*Computing*) a PROGRAM or a fragment of a program written in a programming language, or the language itself. See also BINARY CODE, ASCII.

2. (*Information theory*) a set of *n*-tuples of elements drawn from an *alphabet*, S; where each such string is a *word*, and *n* is the *word-length* of the code, the code is thus a subset of S^n.

coding theory, *n*. the mathematical theory of the encryption of messages, which is used to ensure the security of transmitted data, and can be applied to recover corrupted data. Some of the most secure codes depend on the decomposition of a very large number into prime factors; this is simple if one factor is known, but extremely difficult otherwise (an example of the P VERSUS NP PROBLEM). See also ERROR-CORRECTING CODES.

codimension or **deficiency,** n. (for a subspace of a vector space) the dimension of the algebraic FACTOR SPACE, or of a vector space COMPLEMENT. A HYPER-PLANE through the origin has codimension 1.

codomain, n. a set within which the VALUES of a function lie, as opposed to the set of values that the function actually takes (the RANGE). For example, the function $y = 1/x$ for integral values of x can be said to have the interval $[-1, 1]$ as codomain, although 0 is not a value of the function for any argument; its range is in fact only the set of rationals of form $1/x$. Compare DOMAIN.

coefficient, n. **1.** a numerical or constant multiplier of the variables in an algebraic term. For example, the coefficient of $3xyz$ is 3; the coefficient of $\cos x$ in $5\cos x$ is 5. Pascal's triangle is an array of coefficients in this sense. **2.** the multiplier of a given power of a variable in an expression. For example, in this sense the coefficient of x in $3xyz + zx^2$ is $3yz$; this is equivalent to the preceding sense when y and z are regarded as temporarily fixed, as when calculating partial derivatives.

coefficient functionals, n. the maps $f_\gamma : x \mapsto a_\gamma$, such that

$$\sum_{\gamma \in \Gamma} a_\gamma x_\gamma$$

is the representation of x with respect to a BASIS $\{x_\gamma\}_{\gamma \in \Gamma}$ of a VECTOR SPACE.

coefficient matrix, n. a MATRIX whose entries are the COEFFICIENTS of a system of LINEAR EQUATIONS.

coefficient of kinetic friction, see FRICTION.

coefficient of probability, n. (*Statistical physics*) the single-valued real function, P, determining the probability that a member of a system lies within an elemental volume, dv, of PHASE SPACE. This probability may be represented by an expression of the form $dp = P\,dv$. P will also fulfil the condition $\int P\,dv = 1$, where the integral is over all phase space. Otherwise P is arbitrary.

coefficient of restitution, n. (*Mechanics*) for a set of objects involved in a collision, the empirically determined ratio, e, of the total MOMENTUM after the collision to the total momentum before it. For standard billiard balls, e is almost unity, but for balls made of putty, e would be very close to zero.

coefficient of static friction, see FRICTION.

coercive, *adj*. (of a function) tenting to infinity as the values of the independent variable tend to infinity.

cofactor or **signed minor,** n. a DETERMINANT that is derived from a given matrix or determinant by the deletion of the rows and columns containing a specified entry or submatrix; the (i, j)-*cofactor* of a matrix A, is the number

$$\hat{A}_{i,j} = (-1)^{i+j} \det(A_{i,j}),$$

where $A_{i,j}$ is the original matrix with the ith row and j^{th} column deleted; $\hat{A}_{i,j}$ is thus positive or negative according as $i + j$ is even or odd. For example, deleting the row and column containing 2 in

$$M = \begin{bmatrix} 1 & 2 & 3 \\ 4 & 5 & 6 \\ 7 & 8 & 9 \end{bmatrix}$$

and then taking the determinant of the result yields

$$\hat{A}_{i,j} = \begin{vmatrix} 4 & 6 \\ 7 & 9 \end{vmatrix}$$

so that the cofactor of 2 in M is

$$(-1)^{1+2}[(4 \times 9) - (6 \times 7)] = 6.$$

Any determinant can be expanded as the sum of the products of the entries in any one row or column and their respective cofactors, and the series of the products of the elements in one row or column with the cofactors of the elements in another is zero. With sign abstracted $A_{i,j}$ is sometimes called a *minor*.

cofinal, *adj.* (of a subset D of a set E with a PARTIAL ORDERING ≥) having the property that for any x in E there exists y in D with $y \geq x$.

cofinite subset, *n.* any set whose complement is finite.

cofunction or **complementary function,** *n.* a TRIGONOMETRIC FUNCTION whose values for any argument are equal to the given function of the COMPLEMENTARY ANGLE; thus sine and cosine are cofunctions, since $\sin\theta = \cos(\pi/2 - \theta)$.

coincident, *adj.* (of configurations) having all points in common.

colatitude, *n.* given a point P in SPHERICAL COORDINATES, the angle from the POLAR AXIS to the line segment joining the origin to P.

collinear, *adj.* (of a set of points) lying in the same straight LINE.

collineation, *n.* a BIJECTION of a PROJECTIVE GEOMETRY onto another or itself, that maps lines onto lines. See AFFINE TRANSFORMATION, CORRELATION.

cologarithm (abbrev. **colog**.), *n.* the LOGARITHM of the reciprocal of a number, equal to the additive inverse of its logarithm. For example,

$$\text{colog } 100 = \log 0.01 = \log 10^{-2} = -2 = -\log 100.$$

colourable, *adj.* (of a graph or map) able to be coloured with finitely many colours; having finite CHROMATIC NUMBER. See FOUR COLOUR PROBLEM.

column, *n.* **1.** a vertical linear array of numbers or terms, for example in a MATRIX or the array representation of a DETERMINANT; an $n \times 1$ array such as

$$\begin{bmatrix} a \\ b \\ c \end{bmatrix},$$

whether regarded itself as matrix or as a part of a larger matrix such as

$$\begin{bmatrix} a & b & c \\ d & e & f \\ g & h & i \end{bmatrix}.$$

2. (*as modifier*) operating upon, or relating to, the columns of a matrix, and contrasted with a corresponding ROW operation. For example, ELEMENTARY OPERATIONS on the columns of a matrix are *elementary column operations*.

column equivalence, *n.* the relation that holds between a pair of matrices when one is obtainable from the other by a finite sequence of ELEMENTARY MATRIX OPERATIONS on its COLUMNS. Compare ROW EQUIVALENCE.

column rank, *n.* the RANK of the COLUMN SPACE of a matrix.

column-reduced echelon form, see REDUCED ECHELON FORM.

column space, *n.* the VECTOR SPACE generated by the COLUMNS of a matrix. The dimension of this space is called the *column rank* and coincides with both the ROW–RANK and the RANK of the matrix.

column-stochastic, see STOCHASTIC.

column vector, *n.* an *n*-TUPLE of quantities written as an $n \times 1$ matrix, that is, as a COLUMN.

combination, *n.* **1.** also called **unordered arrangement.** a selection of a subset of objects from a set without regard to order. If repetitions are not permitted, the number of distinct combinations selecting *r* objects out of *n* is

$$\binom{n}{r} = \frac{n!}{(n-r)!\, r!}$$

(also written $_nC_r$ or ^nC_r). For example, the distinct combinations of two objects selected from $\{a, b, c, d\}$ are *ab, ac, ad, bc, bd, cd,* since *db* counts as the same selection as *bd*; there are $15!/(12! \times 3!) = 455$ ways of selecting 12 jurors from a panel of 15. Compare ARRANGEMENT, PERMUTATION. See also BINOMIAL THEOREM, PASCAL'S TRIANGLE.

2. in a structure such as a vector space, any finite sum of appropriate multiples of given elements, such as a LINEAR COMBINATION, AFFINE combination, or CONVEX combination.

combinatorial analysis or **combinatorics,** *n.* the branch of mathematics concerned with the theory of enumeration, COMBINATIONS, and PERMUTATIONS in order to solve problems about the possibility of constructing arrangements of objects to satisfy specified conditions.

combinatorial logic or **combinatorics,** *n.* the formal study of functions regarded in terms of the binary operation of APPLICATION; this is equivalent to the LAMBDA–CALCULUS.

commensurable, *adj.* (of two quantities) in rational proportion, both being integral multiples of the same quantity; expressible in common units. For example, minutes and seconds are commensurable, but days and light years are not. To say that log 3 and log 2 are not commensurable is to assert the irrationality of $\log_3(2)$.

common, *adj.* (of a CYCLOID, EPICYCLOID or HYPOCYCLOID) described by a point lying on, rather than outside or within, the circumference of a circle as it rolls without slipping round some other figure; Fig. 69 shows the generation of a common cycloid. Compare CONTRACTED, EXTENDED.

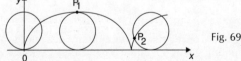

Fig. 69. A **common** cycloid.

common denominator, *n.* an integer that is exactly divisible by all the denominators of a set of fractions; a common multiple of the denominators. For example, all multiples of 12 are common denominators of $\frac{1}{2}$, $\frac{1}{4}$ and $\frac{1}{8}$, and these fractions may be uniformly expressed as $\frac{6}{12}$, $\frac{3}{12}$, and $\frac{2}{12}$.

common difference, *n.* the difference between the successive terms of an ARITHMETIC PROGRESSION.

common factor or **common divisor,** *n.* a number, polynomial, or quantity that is a FACTOR of each of a given set. For example, 5 is a common factor of 15 and 20. See also HIGHEST COMMON FACTOR.

common fraction, *n.* another name for SIMPLE FRACTION.

common logarithm, *n.* a LOGARITHM to base 10; that is, the power to which 10 has to be raised to equal a given number. The common logarithm of x is usually written $\log x$, or, if the base is significant, $\log_{10} x$. See also ANTI-LOGARITHM. Compare NATURAL LOGARITHM.

common multiple, *n.* an integer, polynomial, or quantity that is an integral MULTIPLE of each in a given set. For example, 20 is a common multiple of 2, 4, 5, and 10, while $x^3 - x^2 - x + 1$ is a common multiple of $x^2 - 1$ and $(x-1)^2$.

common tangent, *n.* a line that is a TANGENT to two or more curves; for example, in Fig. 70, the line AB is tangent to both the sine curve and the circle.

Fig. 70. AB is a **common tangent**.

commutative or **permutable,** *adj.* **1.** (of an operator) giving the same result irrespective of the order of arguments. For example, in the reals, addition is commutative since $a + b = b + a$, but subtraction is not since $a - b \neq b - a$. **2.** (of a structure) possessing a commutative operator. A commutative group is called ABELIAN.

commutative diagram, *n.* (*Algebra*) a DIAGRAM OF ARROWS in which one asserts that all directed paths between any two given vertices, yield the same COMPOSITION arrow. This is basic to CATEGORY THEORY. Thus, for the diagram of functions between sets in Fig. 71, to say that the diagram *commutes* is to assert that $g \circ f = j \circ h = k$, where k is the resultant from X to Z.

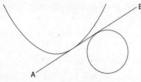

Fig. 71. **Commutative diagram**.

commutative law, *n.* an axiom or theorem of a particular mathematical or formal system stating that a given operator is COMMUTATIVE. For example, the commutative law for set union is the set theoretic axiom $A \cup B = B \cup A$.

commutative ring, *n.* a RING in which the multiplication is COMMUTATIVE. For example, the integers are a commutative ring, but 2×2 matrices are not.

commutator, *n.* **1.** (of two elements of a GROUP) the quantity

$$[x, y] = x^{-1}y^{-1}xy,$$

for x and y in the group. The subgroup of a group generated by all commutators is the DERIVED SUBGROUP.

2. (*Operator theory*) the operator PQ–QP, where P and Q are given operators.

commute, *vb.* to obey a COMMUTATIVE LAW, especially of a group, semi-group, or ring operation.

commuting indeterminate, *n.* an element that COMMUTES with every element of a given RING and is not the root of any polynomial equation over the ring. See POLYNOMIAL RING.

compact, *adj.* **1a.** (of a TOPOLOGICAL SPACE) having the property that every collection of OPEN SETS the union of which is the whole space has a finite subcollection with the same property. This is equivalent to the FINITE INTERSECTION PROPERTY. In particular, in n-dimensional EUCLIDEAN SPACE a set is compact if and only if it is closed and bounded. For example, the closed interval $[0, 1]$ is compact, but the open interval $(0, 1)$ is not, as

$$\left\{ \left(\frac{1}{n}, n \right) \right\}_{n \in N}$$

is a cover of $(0, 1)$ that has no finite subcover. See also LINDELÖF SPACE. Compare SEQUENTIALLY COMPACT.

b. (of a subspace) such that every cover in the INDUCED TOPOLOGY has a finite subcover.

2. (of a relation) having the property that for any pair of elements such that a is related to b, there is some element c such that a is related to c and c is related to b. For example, *less than* is compact on the rational numbers, since for any pair of rationals, a and b, $\frac{1}{2}(a + b)$ is a rational between a and b.

3. (of a mapping between topological vector spaces, especially BANACH SPACES) having the property that the image of any bounded set has a compact closure. See also COMPLETELY CONTINUOUS.

compactification, *n.* a COMPACT TOPOLOGICAL SPACE containing a given topological space. A *one-point compactification* adds a single point, designated ∞, to a HAUSDORFF SPACE. The extended interval $[0, \infty]$ is a one-point compactification of $[0, \infty[$ in which all complements of bounded intervals are NEIGHBOURHOODS of ∞.

compactum, *n.* a TOPOLOGICAL SPACE that is COMPACT and METRIZABLE.

compactness theorem, *n.* (*Logic*) the theorem that a FORMULA is VALID in a THEORY T if and only if it is valid in some finitely axiomatized part of T. Consequently, by the COMPLETENESS THEOREM, a theory has a MODEL if every finitely axiomatized part has. Compare LÖWENHEIM–SKOLEM THEOREM.

companion matrix, *n.* a matrix constructed from a given polynomial

$$p(x) = x^n - a_{n-1}x^{n-1} - \ldots - a_1 x - a_0$$

by placing the negatives of the coefficients of the polynomial in increasing order of degree in the last row of a unit SUPER–DIAGONAL MATRIX. For example, a four-dimensional companion matrix has the form

$$\begin{bmatrix} 0 & 1 & 0 & 0 \\ 0 & 0 & 1 & 0 \\ 0 & 0 & 0 & 1 \\ a_0 & a_1 & a_2 & a_3 \end{bmatrix}.$$

This procedure is used in the construction of CANONICAL forms such as the JORDAN FORM.

comparable, *adj.* (of two elements *a* and *b* of a PARTIALLY ORDERED SET) related by the ordering so that either $a \leq b$ or $b \leq a$.

comparison test, *n.* the test that establishes the ABSOLUTE CONVERGENCE of a series by verifying that the terms are DOMINATED by the terms of a convergent series with positive terms. The first series is said to be subordinate to or MAJORIZED by the second series. For example, since

$$\sum_{n=1}^{\infty} \frac{1}{n(n+1)}$$

is convergent, and

$$\frac{2}{n(n+1)} \geq \frac{1}{n^2} ,$$

it follows by the comparison test that

$$\sum_{n=1}^{\infty} \frac{1}{n^2}$$

converges. See also RATIO TEST, ROOT TEST.

comparison tolerance, *n.* (*Numerical analysis*) a number used as one of several measures of equality in numerical COMPUTATION. One common method is to consider two numbers *a* and *b* to be equal if the RELATIVE ERROR

$$|a-b| / |b|$$

is less than the comparison tolerance ε, so that the comparison tolerance is an upper bound to the allowable relative errors.

compass and straight-edge construction, see CONSTRUCT.

compasses, *n. pl.* a geometrical instrument with two legs hinged to one another, of which usually one is pointed and the other holds a pen, used to trace out a circle the radius of which is the distance between the tips of the legs and the centre of which is the fixed end. Compare DIVIDERS.

competitive equilibrium, *n.* (*Mathematical economics, game theory*) an EQUILIBRIUM state arrived at without cooperation between agents or players.

complement, *n.* **1a.** intuitively, the class of all things that are not members of a given SET. Because it is not relative to a universe, however, this attempted definition is too all-inclusive, and so gives rise to the contradictions of RUSSELL'S PARADOX and CANTOR'S PARADOX.

b. properly, the class of all members of a given UNIVERSE OF DISCOURSE that are not members of a given class, often written C(A) or A′, where A is the given set. For example, if some given UNIVERSAL SET is represented by the rectangle U in Fig. 72, then the shaded region S′ is the complement of the unshaded region S (and vice versa).

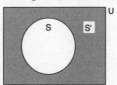

Fig. 72. S′ is the **complement** of S in U.

c. more generally, the RELATIVE COMPLEMENT of one set in another; the complement of a set in the preceding sense is its complement in the understood universal set.

2. the difference between some given value and a fixed total value, especially the COMPLEMENTARY ANGLE of a given angle.

3. (in a VECTOR SPACE) a subspace disjoint from a given subspace that when added to it gives the whole space.

4. in general, any element of a structure that is COMPLEMENTARY to a given element, such as orthogonal vectors or elements in a lattice of which the meet is zero.

complementarity problem, *n.* an optimization model including LINEAR and INTEGER PROGRAMMING, and having applications in FIXED POINT THEORY. Given a function *f* on EUCLIDEAN SPACE, a vector *x* is sought such that *x* and *f(x)* are non-negative and ORTHOGONAL; equivalently, *x* and *f(x)* are COMPLEMENTARY in the lattice of non-negative vectors. Compare VARIATIONAL INEQUALITY.

complementary, *adj.* **1a.** constituting a COMPLEMENT to one another.

b. MUTUALLY EXCLUSIVE and EXHAUSTIVE so that each is the complement of the other.

2. more generally, related in a way definable in such terms. For example, sine and cosine are called complementary functions (COFUNCTIONS) since, for any value of *x*, arcsin*x* and arcos*x* are complementary angles.

3. (of two vectors in Euclidean space) each having coordinates of one sign and being ORTHOGONAL to one another. Thus, (1, 2, 0) and (0, 0, –3) are complementary. More generally, one allows one vector to lie in a given convex cone, the other to lie in the POLAR CONE, and the inner product of the two to be zero.

complementary angle, *n.* the angle that is the difference between a given angle and a right angle. Thus, as shown in Fig. 73, the acute angles of a right-angled triangle are complementary.

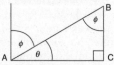

Fig. 73. θ and ϕ are **complementary angles**.

complementary function, *n.* **1.** (for a LINEAR DIFFERENTIAL EQUATION) an arbitrary LINEAR COMBINATION of *n* LINEARLY INDEPENDENT solutions of the corresponding HOMOGENEOUS equation, where *n* is the ORDER of the given linear equation.

2. see COFUNCTION.

complementary minor, *n.* the MINOR of a given square matrix or determinant obtained by deleting from it the elements of a given minor.

complementary modulus, *n.* the quantity $\sqrt{(1-k^2)}$, where *k* is the MODULUS of a complete ELLIPTIC INTEGRAL or ELLIPTIC FUNCTION.

complementary slackness, *n.* (*Optimization*) the condition that arises in the KUHN–TUCKER THEOREM and other similar results, in which the MULTIPLIER and the constraint vector are COMPLEMENTARY VECTORS to one another.

Equivalently, only BINDING CONSTRAINTS have non-zero multipliers, and all multipliers for inequality constraints are restricted in sign.

complementation, *n.* the process of taking COMPLEMENTS, especially in set theory.

complete, *adj.* **1.** (of a METRIC SPACE) having the property that every CAUCHY SEQUENCE converges; for example, the real numbers are complete but the rationals are not, where the metric is the absolute difference between the numbers.

2. also called **order complete** or **Dedekind complete**. (of a PARTIALLY ORDERED set) such that every subset has a SUPREMUM and INFIMUM: for example, the reals are not complete but the interval [0,1] is.

3. (of a GRAPH) containing all possible EDGES between its VERTICES; thus the hexagon in Fig. 74 is a complete graph, since every pair of vertices is joined by an edge. See also COMPLETE QUADRILATERAL.

Fig. 74. The **complete** graph of a regular hexagon.

4. (of a logical THEORY) having the property that every (semantically) VALID formula can be proved (syntactically) from the AXIOMS. See also STRONG COMPLETENESS. Compare CONSISTENT.

5. (of a SUFFICIENT STATISTIC for a parameter θ) having the property that if the EXPECTED VALUE of a function of the statistic is zero for all values of the parameter, then the function is identically zero.

6. (of a GROUP) having a trivial CENTRE, and isomorphic to the group of its own AUTOMORPHISMS.

7. (of an ORTHONORMAL set) MAXIMAL.

8. (of a PARTIALLY ORDERED set) another, more ambiguous, word for CONNECTED in the sense of a TOTAL ORDER.

complete elliptic integral, *n.* any ELLIPTIC INTEGRAL expressible in terms of the functions K, the *complete elliptic integral of the first kind*, and E, the *complete elliptic integral of the second kind*. The first and second complete elliptic integral are related by LEGENDRE'S IDENTITY: for any $0 < k < 1$

$$K(k)\,E[\sqrt{(1-k^2)}] + E(k)\,K[\sqrt{(1-k^2)}] - K(k)\,K[\sqrt{(1-k^2)}] = \pi/2.$$

See K, E.

complete induction, **second-kind induction**, or **general induction,** *n.* INDUCTION in which the inductive step is from all integers less than n to all less than $n + 1$, rather than from the single integer n to $n + 1$. Compare FIRST–KIND INDUCTION.

completely balanced block design, see BLOCK DESIGN.

completely continuous, *adj.* (of a mapping between BANACH SPACES) mapping weakly convergent to norm-convergent sequences. When the domain is reflexive this coincides with the mapping being COMPACT.

completely monotone function, *n.* an infinitely differentiable function

$$f: \,]0, \infty[\, \to \, \mathbb{R}$$

for which the sequence of derivatives satisfies

$$(-1)^n f^{(n)}(x) \geq 0.$$

completely normal topological space, see NORMAL TOPOLOGICAL.

completely regular, see REGULAR.

complete matching, see MATCHING.

complete measure, *n.* a MEASURE such that all subsets of a NULL set are measurable. For example, Lebesgue measure is complete, but Borel measure is not.

completeness postulate, *n.* the axiom that the real line is ORDER–COMPLETE. This is equivalent to the ARCHIMEDEAN PROPERTY.

completeness theorem or **Godel's completeness theorem,** *n.* the result that a theory is CONSISTENT if and only if it has a MODEL.

complete quadrangle, *n.* the CONFIGURATION of four points, with no three COLLINEAR, and the six joining lines; the lines joining the vertices of a complete quadrangle thus have segments that are the sides and diagonals of the quadrilateral defined by the vertices, as shown in Fig. 75. The DUAL CONFIGURATION replacing points by lines is a COMPLETE QUADRILATERAL.

Fig. 75. **Complete quadrangle**.
See main entry.

complete quadrilateral, *n.* the CONFIGURATION of four lines, of which no three are COINCIDENT, and their six points of pair-wise intersection. The DUAL CONFIGURATION replacing lines by points is a COMPLETE QUADRANGLE. Thus, in Fig. 76, the quadrilateral ABCD consists of only four of the points of intersection of the lines of which its sides are segments; the complete quadrilateral includes also the points E and F, and there are three diagonals, AD, BC, and EF.

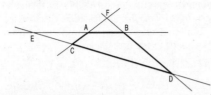

Fig. 76. **Complete quadrilateral.**
See main entry.

complete regularity, see REGULAR.

complete residue system, *n.* any set of representatives chosen one from each RESIDUE CLASS. This is sometimes called a *complete residue class*, but as it is clearly not a residue class this usage is at best confusing.

complete solution, *n.* a solution of a PARTIAL DIFFERENTIAL EQUATION of order *n* that contains *n* arbitrary constants.

complete symmetric group, see SYMMETRIC GROUP.

complete the square, *vb.* to solve QUADRATIC equations by replacing the quadratic expression $x^2 + bx + c$ by

$$(x + b/2)^2 + c - (b/2)^2,$$

and so obtaining a purely quadratic equation with no linear term.

completion, *n.* **1.** (of a METRIC SPACE) the minimal COMPLETE metric space containing the given space.

2. a MEASURE that extends a given measure to a COMPLETE MEASURE.

complex, *adj.* **1.** of, involving, or of the form of a COMPLEX NUMBER, consisting of a REAL and an IMAGINARY PART, either of which may be zero.

– *n.* **2.** another term for a SIMPLICIAL COMPLEX. Sometimes, however, less is required of a complex than of a simplicial complex.

3. (*Group theory*) an archaic term for a subset of a GROUP.

complex analysis, *n.* the study of COMPLEX functions, and especially the study of ANALYTIC FUNCTIONS since it is the properties of complex differentiation that distinguish complex from REAL ANALYSIS.

complex conjugate, *n.* the COMPLEX NUMBER whose IMAGINARY PART is the negative of that of a given complex number, their REAL PARTS being equal. Thus $a - ib$ is the complex conjugate of $a + ib$.

complex fraction, *n.* another term for COMPOUND FRACTION.

complex function, *n.* a function whose DOMAIN and CODOMAIN are both sets of COMPLEX NUMBERS. Compare REAL FUNCTION.

complexification, *n.* the treatment of the DIRECT PRODUCT of a given real VECTOR SPACE with itself as a complex vector space. This is entirely analogous to the identification of the complex numbers with the real plane. All complex vector spaces can be so reconstructed from the underlying real vector space, with multiplication restricted to real scalars.

complexity, *n.* any of various measures of the difficulty of a given DECISION PROBLEM, computational method, or ALGORITHM; for example, the total number of bits, flops, or operations used may be regarded as approximately a function of the size of the problem, or the amount of work involved in its solution. See also FAST FOURIER TRANSFORM, POLYNOMIAL TIME ALGORITHM, NP COMPLETE.

complexity theory, *n.* the branch of mathematics concerned with classifying computational methods and determining their degree of COMPLEXITY.

complex number, *n.* any number of the form $a + ib$, where a and b are real numbers and i is the square root of -1; a and ib are respectively the REAL and the IMAGINARY PART of the number. Setting a or b equal to zero gives an imaginary and a real number respectively.

complex plane, *n.* the complex numbers considered as identified with the infinite two-dimensional space defined by the REAL and IMAGINARY AXES of the ARGAND DIAGRAM; for example, as shown in Fig. 77, the point (a, b) of the complex plane represents the complex number $a + ib$. Compare EXTENDED PLANE.

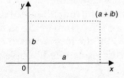

Fig. 77. **Complex plane.** $a + ib$ is represented by the point (a, b).

complex point, see AUGMENTED EUCLIDEAN GEOMETRY.

complex velocity potential, *n.* (*Continuum mechanics*) a HOLOMORPHIC function that describes an INCOMPRESSIBLE and IRROTATIONAL two-dimensional STEADY MOTION, the COMPLEX CONJUGATE of the derivative of which is the VELOCITY expressed in terms of complex numbers; the real part is the VELOCITY POTENTIAL and the imaginary part is the STREAM FUNCTION.

component, *n.* **1a.** one of the elements of the ORDERED SET that represents a VECTOR, such as 2 in $\langle 1, 2, 3 \rangle$.

b. more generally, one of a set of two or more VECTORS of which the RESULTANT is a given vector, especially one of such a set of ORTHOGONAL vectors lying in a specified direction such as parallel to a coordinate axis. Thus in this sense, the second component of $\langle 1, 2, 3 \rangle$ is $\langle 0, 2, 0 \rangle$.

2. the effect in a specified direction of a physical quantity such as a force. If a vector represents the magnitude and direction of the physical quantity, its component in any direction is represented by the PROJECTION of the vector onto a line in the specified direction. For example, in Fig. 78, the component of the vector $\overrightarrow{AB}$ in the direction of $\overrightarrow{CD}$ is $\overrightarrow{XY}$.

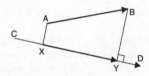

Fig. 78. $\overrightarrow{XY}$ is a **component** of $\overrightarrow{AB}$.

3. (of a TOPOLOGICAL SPACE) a maximal CONNECTED subset of the given space. For example, the set of positive x and y with $xy = 1$ is a component of the corresponding hyperbola.

4. (*Statistics*) any one of the variables of a MULTIVARIATE DISTRIBUTION.

5. (*Logic*) any of the ATOMIC statements making up a COMPOUND STATEMENT.

component transformation law, *n.* the identity that relates TENSORS defined over VECTOR SPACES with different bases of the same dimension:

$$T^{a'_1 \ldots a'_r}_{b'_1 \ldots b'_s} = X^{a'_1}_{d} \ldots X^{a'_r}_{\sigma} X^{d_1}_{b'_1} \ldots X^{d_s}_{b'_s} T^{c_1 \ldots c_r}_{d_1 \ldots d_s},$$

where T^r_s has basis

$$\left\{ \mathbf{e}^{b_1 \ldots b_s}_{a_1 \ldots a_r} \right\},$$

and $X^{a'}_c$ and $X^d_{b'}$ are the matrices of the change of basis of T.

composite, *adj.* (of a number, polynomial, function, etc.) having PROPER FACTORS. For example, $y = x(x + 2)$ is a composite function.

composite number, *n.* an integer, *m*, that is the product of two or more integers other than ± 1 and $\pm m$; any integer (other than 0 or ± 1) that is not a PRIME NUMBER. For example, $365 = 73 \times 5$ is composite.

composition, *n.* **1.** the operation that forms a single function from two given functions by applying the second function to the value of the first for any argument; it is only defined when the RANGE of the first is contained in the DOMAIN of the second. Repeated composition is denoted by a superscript numeral as $f^{(n)}$; so, for example, $f \circ f \circ f \circ f = f^{(4)}$.

2. the function so formed. In analysis, the composition of f with g is often written $f \circ g$ or fg where g is the function applied first; that is,

$$f \circ g(x) = f(g(x)).$$

For example, the composition of $x + 3$ with x^2 is $x^2 + 3$; that of x^2 with $x + 3$ is $(x + 3)^2$. In some subjects, such as group theory or topology, mappings are written after their argument, and the notation for composition preserves this order, so that the result of applying τ after σ is written $\sigma \circ \tau$.

composition factors, n. the NORMAL FACTORS of a COMPOSITION SERIES.

composition series, n. (for a group) a NORMAL SERIES in which every NORMAL FACTOR is a non-trivial SIMPLE group. Every finite group has a composition series; for example, $(S_n, A_n, 1)$ is a composition series for the symmetric group S_n for $n \geq 5$, where A_n is the alternating group. See also JORDAN-HOLDER THEOREM, SOLUBLE GROUP.

compound fraction or **complex fraction,** n. a FRACTION of which the numerator or denominator contains fractions, such as

$$\frac{\frac{5}{7}}{1 + \frac{3}{5}}.$$

See also CONTINUED FRACTION.

compound interest, n. the interest accumulated over a given period at a given rate when each successive interest payment is aggregated with the principal sum for the purpose of calculating the next interest payment. For example, capital of \$$C$ invested at i % *per annum* compound yields

$$\$C(1 + i/100)^n$$

after n years. More generally, the interest on P units at i % compounded n times per period yields

$$P(1 + i/(100n))^n$$

units at the end of the period, which is equivalent to receiving (i/n) % interest for n periods and reinvesting the principal. As the interest becomes INSTANTANEOUS, with n tending to infinity, the sum approaches

$$P \times \exp(i/100).$$

Compare SIMPLE INTEREST. See EXPONENTIAL FUNCTION.

compound number, n. a quantity expressed in different but related units, such as 3 minutes and 10 seconds.

compound statement, n. (*Logic*) a statement formed from simple statements by the use of words such as 'and', 'or', 'not', 'if ... then ...' or their corresponding symbols. The simple statements involved are the components of the compound statement. For example, $(p \& \neg q) \lor r$ is a compound statement built up from the components p, q and r.

comprehension axiom, n. the axiom of set theory that states that for any property there exists a set consisting of all elements with the given property.

compressible, *adj.* (*Continuum mechanics*) (of a BODY) such that its DENSITY is not independent of time for all possible motions.

compressive normal stress, n. a NORMAL STRESS that is opposite in direction to the outward normal at a point on a surface. Compare TENSILE NORMAL STRESS.

computability theory, *n.* the study of ALGORITHMS, and in particular their power and limitations, often conceived of in terms of TURING MACHINES. This grew out of HILBERT'S PROGRAMME, which was ultimately proved impossible by GÖDEL'S THEOREM. See also CHURCH'S THESIS, AUTOMATA THEORY.

computable, *adj.* able to be carried out by the operation of an ALGORITHM.

computation, *n.* **1.** a calculation, especially of a number or value from given information by use of an ALGORITHM.

2. any step-wise calculation, especially one that could be followed by a suitably programmed computer.

compute, *vb.* to calculate (an answer, result, etc.), especially by a stepwise procedure; whence, in particular, to use a COMPUTER.

computer, *n.* an electronic device that carries out arithmetical and logical operations in accordance with a precise sequence of instructions (a PROGRAM) and so can process data or carry out any task that can be expressed in such a way. Although there are ANALOGUE DEVICES that qualify as computers, it is typically a DIGITAL COMPUTER with a number of separate parts: an input device such as a keyboard, a central processing unit (cpu) consisting of a large number of LOGIC GATES, memory units such as disks or magnetic tapes, and output devices such as a visual display unit and a printer. AUTOMATA THEORY and COMPLEXITY THEORY are concerned not with the physical properties of actual computers, but with the theoretical capabilities of ABSTRACT MACHINES, defined in terms of their programs.

concatenate, *vb.* **1.** to adjoin one symbol or string of symbols to the end of another, thereby forming the symbol for a new mathematical object from the symbols for the given objects. For example, concatenation of ordered pairs and ordered triples forms ordered quintuples; more properly, concatenating pairs $\langle a, b \rangle$ and triples $\langle c, d, e \rangle$ yields ordered pairs of ordered sets, $\langle \langle a, b \rangle, \langle c, d, e \rangle \rangle$, but these are isomorphic to the quintuples $\langle a, b, c, d, e \rangle$.

2. (*Logic*) to adjoin one quoted expression to another in such a way as to form a single quoted expression. For example, a quoted conjunction, 'P & Q', is the concatenation of the first conjunct, the symbol '&', and the second conjunct, which is written

$$P \wedge \text{`\&'} \wedge Q,$$

often abbreviated to the QUASI–QUOTATION $\ulcorner P \& Q \urcorner$.

concave, *adj.* **1.** (of a polygon) having an interior angle greater than 180°, as in Fig. 79.

2. (of a real–valued function, or surface) **a.** having the property that the chord joining any two points on its graph lies below the graph. Thus if Fig. 79 has the usual orientation with respect to the coordinate axes, both paths from A to B are concave.

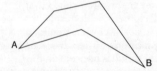

Fig. 79. **Concave**. A concave polygon.

b. formally and more generally, such that for arguments x and y in the appropriate abstract space, and t in the interval [0, 1],

$$tf(x) + (1-t)f(y) \leq f(tx + (1-t)y).$$

See QUASI–CONCAVE.

concave down, *adj.* (of a curve on an interval) having a decreasing derivative as the independent variable increases as between A and B in Fig. 80; having negative second derivative and CURVATURE; CONCAVE.

concave up, *adj.* (of a curve on an interval) having an increasing derivative as the independent variable increases as between B and C in Fig. 80; having positive second derivative and CURVATURE; CONVEX.

Fig. 80. **Concave down**;
concave up. See main entries.

concavity, *n.* the fact of whether the graph of a curve is CONCAVE or CONVEX. See CONCAVE UP, CONCAVE DOWN.

concentrated, *adj.* (of a MEASURE, μ, on a set, B) such that the measure of any measurable set, E, is the measure of its intersection with B; that is,

$$\mu(E) = \mu(B « E).$$

concentric, *adj.* (of a family of geometrical figures, especially circles) having a common centre.

Fig. 81. **Concentric** circles.

conchoid, *n.* a plane curve consisting of two branches situated about a line to which they are ASYMPTOTIC, so that a line from a fixed point (the *pole*) intersecting both branches is of constant length between asymptote and either branch. Its equation is

$$(x - a)^2 (x^2 + y^2) = b^2 x^2,$$

or in polar coordinates

$$r\cos\theta = a \pm b\cos\theta,$$

where a is the distance between the pole and a vertical asymptote, and b is the length of the constant segment; the curve for $b > a$ is shown in Fig. 82.

Fig. 82. **Conchoid**.

conclusion, *n.* (*Logic*) **1.** a statement that purports to follow from another or others (the PREMISES) by means of an ARGUMENT or PROOF.
2. a statement that does in fact validly follow from given premises. See VALID.

concomitant matrix, *n.* the matrix of coefficients in LAGRANGE'S IDENTITY that involves an nth-order linear differential equation and its adjoint.

concrete number, *n.* a number that counts a particular set of objects; for example, three dogs or ten men. See also NUMERICAL QUANTIFIER.

concurrence, *n.* **1.** the fact that three or more lines coincide.

2. the point at which they do so.

concurrent, *adj.* having a point in common. For example, a number of concurrent lines all pass through the same point, as in Fig. 83.

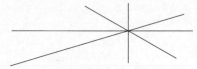

Fig. 83. **Concurrent** lines.

concyclic, *adj.* (of a number of points) such that there is a circle that passes through all of the points.

condensation point, *n.* (*Topology*) a point of a given set such that every neighbourhood of the point is UNCOUNTABLE. In EUCLIDEAN SPACE the set of condensation points of any set is PERFECT and omits only countably many points of the set.

condition, *n.* **1.** a presupposition, especially a restriction on the domain, that is indispensable to the proof of a theorem and stated as part of it; a restriction on the truth of the stated result, so that the latter only holds if the stated precondition is satisfied.

2. (*Logic*) a statement related to another by IMPLICATION ; it is either a NECESSARY CONDITION if its truth is required for the truth of the other, or a SUFFICIENT CONDITION if its truth guarantees that of the other.

3. (*Statistics*) short for EXPERIMENTAL CONDITION (sense 1).

conditional, *adj.* **1.** (of an EQUATION or INEQUALITY) true only for certain values of the variable for which it can be solved. For example, $x^2 - 1 = x + 1$ is a conditional equation, as it is true only for $x = 2$ and $x = -1$. Compare IDENTITY (sense 3).

2. (*Statistics*) with respect to some RANDOM VARIABLE of which the value is taken to be fixed. See CONDITIONAL PROBABILITY, CONDITIONAL EXPECTATION, CONDITIONAL DISTRIBUTION.

3a. (*Logic*) also called **hypothetical.** (of a statement, proposition, etc.) having IMPLICATION as its main connective; consisting of two component propositions purporting to be so related that the second (the CONSEQUENT) cannot be true if the first (the ANTECEDENT) is false, so that the compound statement is false only when its components do have these values. The usual English for this relation is *if... then...*, where *if* is followed by the antecedent, and *then* by the consequent; however, *if P then Q, only if Q then P, Q if P,* and *P only if Q* are all equivalent, and all of these forms are usually symbolized $P \rightarrow Q$, or $p \supset q$.

b. (*as substantive*) a conditional statement.

4. (of a property) holding only under certain CONDITIONS or restrictions. See CONDITIONALLY COMPLETE, CONDITIONALLY CONVERGENT.

conditional completeness, *n.* see CONDITIONALLY COMPLETE.

conditional convergence, *n.* see CONDITIONALLY CONVERGENT.

conditional distribution, *n.* the PROBABILITY DISTRIBUTION of a subset of COMPONENTS of a random vector, conditional upon the values taken by another subset of components.

conditional expectation, *n.* (*Statistics*) **1.** the EXPECTED VALUE of a RANDOM VARIABLE, X, or any function of it, given that an event, B, is known to have occurred; written $E(X|B)$. This is the sum or integral of the products of all the possible values of the random variable or function and the respective CONDITIONAL PROBABILITIES of each. Where y_i is a value of the discrete random variable Y, $E(X)$, the expected value of X, is the sum of the products of the conditional expectations $E(X|Y=y_i)$ and the probabilities of each of the y_i respectively. See also BAYES' THEOREM, MARGINAL EXPECTATION, RADON-NIKODYM THEOREM.

2. (of a random variable *f*, given *x*) formally, the function

$$e_{(x)}(f) = d\sigma/d\alpha$$

on a product probability space $(X \times Y, \mu)$, defined as the RADON-NIKODYM DERIVATIVE of σ with respect to α, where

$$\sigma(A) = \int_{A \times Y} f \, d\mu$$

and

$$\alpha(A) = \mu(A \times Y).$$

3. (of X, given $X_1,...,X_n$) rigorously, any random variable *g* that is measurable with respect to the sigma-field **D** generated by all inverse images of Borel sets B_k, $\{X_k \in B_k\}$, and that satisfies

$$\int_D g \, dP = \int_D X \, dP$$

for all D in **D**. This is written $E(X|X_1,...,X_n)$ or $E(X|\mathbf{D})$, and the definition holds for any sigma-subfield **D**. One can then define

$$P(A|X) = E(\chi_A|X).$$

If X and Y have a joint density $f(x, y)$, then X has density $f(x)$ and

$$E(Y|X=x) = \int y \frac{f(x,y)}{f(x)} \, dy.$$

conditionalization, *n.* (*Logic*) **1.** the process of deriving a CONDITIONAL statement from an argument by taking the conjunction of the premises of the given argument as its ANTECEDENT and its conclusion as its CONSEQUENT. **2.** the statement so derived; this is true if and only if the argument is VALID. For example, the conditionalization of

> *All men are mortal.*
> *Socrates is a man.*
> So *Socrates is mortal.*

is

> *If all men are mortal and Socrates is a man, then Socrates is mortal.*

See DEDUCTION THEOREM.

conditionally complete or **relatively complete,** *adj.* (of a PARTIALLY ORDERED SET) such that every set with an UPPER or LOWER BOUND has a SUPREMUM or INFIMUM, respectively. Compare COMPLETE (sense 3).

conditionally convergent, *adj.* CONVERGENT but not ABSOLUTELY CONVERGENT, so that although the given series converges, the series of its absolute values does not. For example, the convergent series

$$\sum_{n=1}^{\infty} \frac{(-1)^{n-1}}{n} = 1 - \frac{1}{2} + \frac{1}{3} - \frac{1}{4} + \frac{1}{5} \cdots$$

is only conditionally convergent, since

$$\sum_{n=1}^{\infty} \left| \frac{(-1)^{n-1}}{n} \right| = 1 + \frac{1}{2} + \frac{1}{3} + \frac{1}{4} + \frac{1}{5} \cdots$$

diverges. See also COMPARISON TEST.

conditional probability, *n.* (*Statistics*) **1.** the PROBABILITY, $P(A|B)$, of one event, A, occurring, given that another, B, is already known to have occurred; this is defined by

$$P(A|B) = P(A \& B)/P(B).$$

Where x_i and y_j are values of the discrete random variables X and Y, and p_{ij} is the JOINT PROBABILITY of $X = x_i$ and $Y = y_j$, the conditional probability is given by

$$P(x_i | y_j) = p_{ij}/(\Sigma_i p_{ij}).$$

See also BAYES'S THEOREM, RADON–NIKODYM THEOREM.

2. (of a set E in $X \times Y$, given x) formally, the function

$$\mu_{(x)}(E) = e_{(x)}(\chi_E)$$

where $e_{(x)}(f)$ is the CONDITIONAL EXPECTATION, given x, of the RANDOM VARIABLE; this is defined in terms of the RADON–NIKODYM DERIVATIVE with respect to α, where

$$\alpha(A) = \mu(A \times Y)$$

and χ_E is the characteristic function of E. Here $\mu_{(x)}$ behaves much like a measure, in that, given a countable disjoint family of measurable sets $\{E_n\}$,

$$\mu_{(x)}(\cup_n E_n) = \Sigma_n \mu_{(x)}(E_n)$$

for almost all x with respect to α.

condition number, *n.* (*Numerical analysis*) any of a number of measures of the stability of computational problems. In particular, the *relative condition number* of evaluating a real function $f(x)$ when a perturbation $x + \delta x$ is introduced is

$$\frac{|f(x+dx) - f(x)|}{|f(x)|} \times \frac{|x|}{|\delta x|},$$

which for small displacements behaves like

$$\frac{x f'(x)}{f(x)}.$$

More generally, the *condition number of a linear system*, $Ax = b$, is usually taken to be

$$\| A \| \ \| A^{-1} \|,$$

and is denoted $\text{cond}(A)$, where the norm on matrices is at the choice of the user, although often it is taken as the maximum of the sums of absolute values of rows, that is,

$$\| A \| = \max_j \left| \sum_{j=i}^{n} a_{ij} \right|.$$

See also WELL–CONDITIONED, ILL–CONDITIONED.

cone, *n.* **1.** also called **nappe. a.** a solid with a plane base bounded by a closed curve every point of which is joined to a fixed point (the *vertex*) lying outside the plane of the base. A conical surface is swept out by a line segment, such as VA in Fig. 84, with one end, V, fixed at the vertex and the other, A, tracing the curve. If no qualification is stated, the base is usually understood to be circular or elliptical; in the figure it is the plane ellipse ABCD. The volume of a circular cone is $\frac{1}{3} \pi r^2 h$, where r is the radius of the base and h is the perpendicular height of the cone. A *right circular cone* has its vertex perpendicularly above or below the centre of a circular base. See also FRUSTUM.

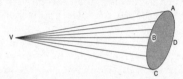

Fig. 84. **Cone**.

b. either of the infinite solids swept out by the infinite lines of which these segments are part. Two cones are thus generated, of which one has ABCD as a cross-section, and the other is a reflection of the first in the vertex V.
2. the infinite solid bounded by the locus of a line passing through a fixed point (the vertex) as it sweeps out a plane closed curve; this constructs two cones in the preceding sense, joined at the vertex.
3. in a VECTOR SPACE, a TRANSLATE of any set that is closed under positive multiplication. Often a cone is required to contain the origin, and to be CONVEX.

confidence interval or **interval estimate,** *n.* (*Statistics*) an interval of values bounded by CONFIDENCE LIMITS derived by sampling, within which the true value of a population parameter is stated to lie with specified probability in the sense that $[F, G]$ is a 95% confidence interval of some parameter if F and G are both functions of a random sample from the given distribution, and 95% of such samples yield intervals that include the true value of the parameter. Compare POINT ESTIMATE.

confidence level, *n.* (*Statistics*) a measure of the reliability of a result. For example, a confidence level of 95% or 0.95 means that there is a probability of 5% that the result is unreliable; less strictly, it is often taken to mean that the probability of error is less than 5%. Compare SIGNIFICANCE LEVEL.

confidence limits, *n.* the endpoints of a CONFIDENCE INTERVAL, between which the true value of a population parameter lies in at least the specified percentage of estimates of the limits.

configuration, *n.* **1.** (*Geometry*) a finite set of points and lines such that each point lies on the same number of lines and each line passes through the same number of points. Any configuration has a DUAL in which points and lines are interchanged. For example, a COMPLETE QUADRILATERAL and a COMPLETE QUADRANGLE are dual configurations; DESARGUE'S THEOREM gives rise to a self-dual configuration. Compare BLOCK DESIGN, FINITE GEOMETRY.

2. another term for BLOCK DESIGN, especially one for which one specifies the sizes of the components.

3. (*Mechanics*) an abstraction from the notion of a body having a particular position and orientation at a time; a representation of a BODY by a three-dimensional geometric figure. For example, the natural configuration for a solid cube would be the unit cube. More formally, a configuration is a BIJECTION from a given BODY to three-dimensional EUCLIDEAN POINT SPACE; an arbitrary configuration is chosen as the *reference configuration* and the *current configuration* of a SUB–BODY is its image at a given time under a MOTION. See also MATERIAL DESCRIPTION, SPATIAL DESCRIPTION.

4. (*Statistical physics*) a distribution of energy among the particles of a system with more than two particles in which the particles are not considered distinguishable; for example, if the system consists of three particles, A, B, C, and the total energy of the system is one unit, then the three STATES $A = 1, B = 0, C = 0$; $A = 0, B = 1, C = 0$; and $A = 0, B = 0, C = 1$ correspond to the same configuration $(1, 0, 0)$.

confirm, *vb.* (of an experiment) to make (an hypothesis) more likely to be true. By BAYES' THEOREM, if an experimental result is more likely under one hypothesis than another, the occurrence of that result increases the probability of the former hypothesis. For example, every single sighting of a white swan confirms the hypothesis that all swans are white, since it tends to increase the probability of its truth; however, no sequence of such sightings, however many, are ever sufficient to prove its truth. See INDUCTION.

confirmation paradox, see HEMPEL'S PARADOX.

confocal, *adj.* having a common FOCUS or common foci; for example, Fig. 85 shows three confocal ellipses.

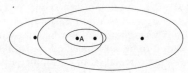

Fig. 85. **Confocal**. A is a focus of all three ellipses.

conformable, *adj.* (of a sequence of matrices) related so that the number of elements in each row of the first matrix is equal to the number of elements in each column of the second (that is, the number of columns of the first equals the number of rows of the second), and so on, so that they may be multiplied in that order. For example, the matrices

$$A = \begin{bmatrix} a_{11} & a_{12} & a_{13} \\ a_{21} & a_{22} & a_{23} \end{bmatrix}$$

and

$$\mathbf{B} = \begin{bmatrix} b_{11} & b_{12} & b_{13} & b_{14} \\ b_{21} & b_{22} & b_{23} & b_{24} \\ b_{31} & b_{32} & b_{33} & b_{34} \end{bmatrix}$$

may be multiplied in that order, since the INNER PRODUCT of the rows of A with the columns of B is well-defined. However, B is not conformable with A, since the rows of B and the columns of A have different numbers of elements, and so their inner products are not well-defined.

conformal, *adj.* **1.** (of a TRANSFORMATION) preserving both the size and the sign of angles, in the sense that, if two arcs meet at a given angle, their images meet at the same angle. An ANALYTIC function is conformal at any point where its derivative is non-vanishing. In addition, any conformal mapping of a complex variable that possesses continuous partial derivatives is analytic.

2. (of a parameter) relating to such a transformation.

congruence, *n.* **1.** the relationship between two integers relative to another (the MODULUS of the congruence) when the difference between the given numbers is an exact multiple of the modulus, usually written

$$x \equiv y \pmod{m}.$$

For example, $8 \equiv 2 \pmod 3$, since $8 - 2 = 6 = 2 \times 3$. It is an equivalence relation fundamental to number theory. In general a *congruence equation* seeks to solve

$$f(x) \equiv 0 \pmod m$$

where f is an INTEGRAL POLYNOMIAL. See also MODULAR ARITHMETIC, FERMAT'S LITTLE THEOREM, LINEAR CONGRUENCE, QUADRATIC CONGRUENCE.

2. (*Group theory*) either of the relations of *left* or *right congruence* between two elements, x and y, of a GROUP, G, relative to a subgroup, H, of G. The elements are *left congruent* modulo H, written $x \equiv_l y \pmod{\text{H}}$, if $x^{-1}y$ is in the subgroup; they are *right congruent* modulo H, written $x \equiv_r y \pmod{\text{H}}$, if yx^{-1} is in the subgroup. Congruence mod m in the integers is a special case of this, with $\text{G} = \mathbb{Z}$, and $\text{H} = \{0, 1,..., m-1\}$.

3. (*Geometry*) the fact or relation of being CONGRUENT, an ISOMETRY.

4. (*Logic*) any EQUIVALENCE RELATION preserved by every operation in the given structure, so that if $x \equiv y$, then $f(x) \equiv f(y)$ for every operation f.

congruence class, *n.* a set of elements each of which is congruent to every other in the class; an EQUIVALENCE CLASS under a CONGRUENCE relation.

congruent, *adj.* **1.** (*Geometry*) (of a set of figures) having identical size and shape so that they may be exactly superimposed. Thus in Fig. 86, the corresponding sides of the two polygons are equal in length and the angles enclosed between corresponding sides are also equal; the figures thus differ only in orientation. Compare SIMILAR. See also EQUIVALENT.

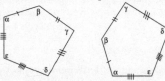

Fig. 86. **Congruent** polygons.

2a. congruent modulo m. (of a pair of integers) differing by an exact multiple of m, related by CONGRUENCE to the given MODULUS.

b. congruent modulo H. (of a pair of elements of a GROUP) related by a (left or right) CONGRUENCE relation (in sense 2).

3. (of two matrices A and B) having the property that $B = PAP^t$, where P^t is the TRANSPOSE of a non-singular matrix P. Every real symmetric matrix of RANK r is congruent (over the reals) to a diagonal matrix with r entries equal to ± 1 and all others 0. Compare CONJUNCTIVE. See SIMILAR, EQUIVALENT.

congruential method, n. one of a number of methods used by a RANDOM NUMBER GENERATOR to produce PSEUDO–RANDOM numbers by taking congruences modulo a large number.

conic, adj. **1.** of or relating to a CONE.

2. (*as substantive*) any expression that represents a CONIC SECTION; any second-degree equation. Thus

$$ax^2 + bxy + cy^2 + dx + ey + f = 0$$

is the general conic in Cartesian coordinates; it can also be expressed as

$$(p^2 + q^2)\,[\,(x - \alpha)^2 + (y - \beta)^2\,] = e^2(px + qy + r)^2,$$

where e is the ECCENTRICITY, (α, β) the FOCUS, and $px + qy + r$ is the equation of the DIRECTRIX of the conic. In VERTEX FORM the equation is

$$y^2 = 2px - (1 - \varepsilon^2)\,x^2,$$

where $2p$ is the *parameter* of the conic, that is, the length of its LATUS RECTUM, which in the ELLIPSE and HYPERBOLA equals b^2/a (where a and b are the lengths of the semi-axes of the conic), and ε is the NUMERICAL ECCENTRICITY, e/a; there are many other equivalent descriptions. Fig. 87 shows the graphs of these curves for specific values of ε, which is constant for any family of similar curves: $0 < \varepsilon < 1$ for an ELLIPSE; $\varepsilon = 1$ for a PARABOLA ; $\varepsilon > 1$ for an HYPERBOLA; and $\varepsilon = 0$ for a circle (when p is the radius).

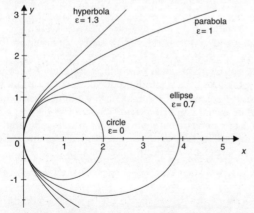

Fig. 87. **Conics** with various eccentricities.

Lines and points are *degenerate conics*; in augmented Euclidean geometry all degenerate conics are pairs of lines or repeated lines, but in Euclidean geometry, points, for example, are determined by equations such as $x^2 + y^2 = 0$, and there are a large number of distinct cases.

3. (*as substantive*) another word for CONIC SECTION.

conics, *n.* (*functioning as singular*) the branch of geometry and coordinate geometry concerned with the properties of the CONIC SECTIONS.

conic section or **conic,** *n.* one of a group of curves formed by the intersection of a plane and a right circular cone, as shown in Fig. 88. This curve is either a CIRCLE, if the plane is parallel to the base of the cone; an ELLIPSE, if it is at any other angle at which the intersection is a closed curve; a PARABOLA, if it is parallel to any line joining the vertex of the cone to a point on its base; or an HYPERBOLA, if it is at any other angle. Lines and points are *degenerate conics* obtained when the intersecting plane includes the vertex of the cone. Conic sections can be conceived geometrically as the loci of points satisfying certain distance relations from a given point, the FOCUS, and a given line, the DIRECTRIX ; the ECCENTRICITY is then defined as the ratio of these distances, which is constant for a particular family of similar curves; these properties are described algebraically by the CONIC equations.

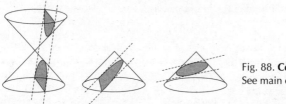

Fig. 88. **Conic sections**.
See main entry.

conjecture, *n.* a statement which may be true, but for which a proof (or disproof) has not been found.

conjugacy class, *n.* the set, cl(a), of all elements in a GROUP that are CONJUGATE to a given element, a, in the group. The conjugacy classes PARTITION the group, and all elements conjugate to one another have the same ORDER. The order of the conjugacy class in a finite group, is the INDEX of the CENTRALIZER of one such element in the group.

conjugacy problem, *n.* the problem of finding an ALGORITHM that decides whether the elements represented by two WORDS are CONJUGATE in a GROUP generated by a given ALPHABET. There exists a finitely presented semigroup for which this problem is insoluble.

conjugate, *adj.* **1.** (of two angles) having a sum of $360°$.

2. (of two complex numbers) differing only in the sign of their IMAGINARY PARTS. For example, $4 + 3i$ and $4 - 3i$ are conjugate.

3. (of two ALGEBRAIC NUMBERS) **a.** being roots of the same irreducible algebraic equation with rational coefficients. For example, $3 + 2\sqrt{2}$ and $3 - 2\sqrt{2}$ are conjugate roots of the equation $x^2 - 6x + 1 = 0$.

b. related by an automorphism that leaves the base field fixed, so that one algebraic number is the image of the other under the automorphism.

4. (of two elements of a matrix) interchanged when the rows and columns are interchanged; if a_{ij} is the element at the intersection of the ith row and the jth column, then a_{ji} is its conjugate entry. See TRANSPOSE.

5a. (of two lines) such that one passes through the POLE of the other.

b. (of two points) such that one lies on the POLAR of the other.

6. (of two POTENTIAL functions) being the real and imaginary parts of a given ANALYTIC function. The pair of functions are then called *conjugate harmonics*.

7. (of vectors with respect to a SYMMETRIC matrix) being LINEARLY INDEPENDENT pairwise ORTHOGONAL with respect to the given matrix: $\mathbf{d}_1$ and $\mathbf{d}_2$ are said to be *H-conjugate* directions if $\langle \mathbf{d}_1, \mathbf{Hd}_2 \rangle = 0$.

8. (*Group theory*) **a.** (of a pair of elements, x and y) related as $y = a^{-1}xa$ for some a in the group.

b. (of a pair of subsets, S and T) such that there is an element a of the group for which $S = a^{-1}Ta$.

– *n.* **9.** also called **Fenchel conjugate.** (of a CONVEX function) the convex function defined in HILBERT SPACE X by the formula:

$$f^*(y) = \sup\{\langle y, x \rangle - f(x) : x \in X\}.$$

This definition can be extended to any locally convex space. The definition yields the fundamental *Young's inequality*

$$f^*(y) + f(x) \geq \langle y, x \rangle.$$

10. another term for the ADJOINT of a LINEAR OPERATOR.

11. another term for a DUAL NORMED SPACE or a more general dual space.

conjugate axis, *n.* the axis of a HYPERBOLA that does not pass through the FOCI. Compare TRANSVERSE AXIS.

conjugate exponents, *n.* (*Measure theory*) any two positive numbers whose reciprocals sum to unity; 1 and ∞ are also regarded as a conjugate pair.

conjugate gradient methods, *n.* any of a class of QUASI–NEWTON METHODS for minimizing a differentiable function of n variables by sequentially generating CONJUGATE directions. See also DESCENT METHOD.

conjugate-linear functional, *n.* a complex functional on a complex vector space that is ADDITIVE and conjugate HOMOGENEOUS; equivalently, f is the CONJUGATE of a complex linear functional

$$g: f(x) = \bar{g}(x).$$

conjugate pairs of points, see HARMONIC POINTS.

conjugate ruled surface, *n.* a RULED SURFACE whose generators are tangent to those of a given ruled surface.

conjugate surd, see SURD.

conjugate variable, *n.* an ADJOINT variable of a DIFFERENTIAL EQUATION.

conjunct, *n.* (*Logic*) either of the component statements or propositions of a CONJUNCTION.

conjunction, *n.* **1.** (*Logic*) also called **logical product. a.** the binary TRUTH-FUNCTIONAL sentential connective that forms a compound sentence from two given sentences, and corresponds to the English *and*. Its TRUTH–TABLE is shown in Fig. 89.

P	Q	P & Q
T	T	T
T	F	F
F	T	F
F	F	F

Fig. 89. **Conjunction**.
Truth table for conjunction.

b. a sentence so formed, usually written *P* & *Q*, *P* ∧ *Q*, or *P.Q*, where *P* and *Q* are the component sentences (the CONJUNCTS); a conjunctive statement is true if and only if its conjuncts are both true.

c. any generalization of this operation or relation, or any sentence formed thereby, such as an operator the argument of which is a set of statements and the value of which is a single statement that is true when and only when all the elements of the given set are true.

2. (*Grammar*) any expression that combines sentences to form more complex sentences; the ordinary language equivalents of sentential CONNECTIVES. Thus conjunction in the preceding sense is not the only conjunction in this sense; confusion may result.

conjunction elimination, *n.* (*Logic*) the ELIMINATION RULE of SENTENTIAL CALCULUS, or any of its extensions, that permits the inference of either conjunct from a given CONJUNCTION.

conjunction introduction, *n.* (*Logic*) the INTRODUCTION RULE of SENTENTIAL CALCULUS, or any of its extensions, that permits the inference, given two statements as premises, of their CONJUNCTION as conclusion.

conjunctive, *adj.* (of two matrices A and B) such that B = PAP* with P* the HERMITIAN TRANSPOSE or ADJOINT of a non–singular matrix P. Every matrix of RANK *r* is conjunctive (over the complex field) to a diagonal matrix with *r* entries equal to ±1 and all others zero. Compare CONGRUENT. See SIMILAR, EQUIVALENT.

conjunctive normal form (abbrev. **cnf**), *n.* the form to which every statement of SENTENTIAL CALCULUS can be reduced, consisting of a conjunction of disjunctions each of the disjuncts of which is either an atomic formula or the negation of one. Because of the associativity of both disjunction and conjunction, no brackets are necessary, as it is understood that conjunction has wider scope. For example, ((P → Q) & P) → Q reduces to the cnf

$$(Q \vee -P \vee P) \ \& \ (Q \vee -Q \vee -P).$$

The *conjunctive normal form theorem* states the direct consequence of this definition, that a well-formed formula is a TAUTOLOGY if and only if every one of the disjunctions in its cnf contains both an atomic variable and its negation; the preceding example is thus a tautology. Compare DISJUNCTIVE NORMAL FORM.

connected, *adj.* **1.** (of a relation) such that either it or its CONVERSE holds between any two members of the domain.

2. (of an ordering) constituting a TOTAL ORDERING.

3. (of a set of real numbers) having the property that if *a* and *b* are in the set and *c* lies between them, then *c* is also in the set, constituting a CONTINUUM.

4. (of a TOPOLOGICAL set) not able to be PARTITIONED into two non-empty OPEN subsets each of which has no points in common with the CLOSURE of the other. For example, the rationals are not connected, although the reals are. The set is called *path-*, *arc-*, *pathwise*, or *arcwise connected* if each pair of points is joined by a PATH in the set; a connected set is not necessarily pathwise connected.

5. (of a GRAPH) having the property that there is a PATH between every pair of vertices along a sequence of edges in the graph.

connectedness or **connectivity,** *n.* the property of being CONNECTED.

connective, *n.* (*Logic*) a function, or the symbol for it in a FORMAL LANGUAGE, forming compound sentences from simple ones, corresponding to English conjunctions such as *or* and *not*.

conoid, *n.* a geometrical surface or solid formed by rotating a CONIC SECTION around one AXIS. See PARABOLOID, HYPERBOLOID, ELLIPSOID.

consequence, *n.* (*Logic*) **1.** a conclusion reached by reasoning, the conclusion of an argument from given premises.

2. logical consequence. VALID deducibility, the relation between the conclusion and the premises of a valid argument.

consequent or **succedent,** *n.* (*Logic*) the resultant clause in a CONDITIONAL sentence; the clause that is implied by the other. For example, *Fafner breathes fire* is the consequent of *Fafner breathes fire if he is a dragon.* Compare ANTECEDENT.

conservative, *adj.* (of a force) describable by a CONSERVATIVE VECTOR FIELD, such that the WORK DONE when the point of action moves from A to B does not depend on the path taken.

conservative extension, see EXTENSION (sense 2).

conservative vector field, *n.* a VECTOR FIELD whose CURL is zero. Its coordinates arise as the GRADIENT of a POTENTIAL ; for example, gravity is a conservative vector field. Compare EXACT (sense 3).

conservative summability method, *n.* a method of summation that assigns a finite limit to every convergent sequence, but that may assign a number different from the limit. Compare REGULAR. See ABEL SUMMATION.

consistency theorem, *n.* (*Logic*) the result that an OPEN THEORY is CONSISTENT exactly when no disjunction of negations of instances of its non-logical axioms is a QUASI–TAUTOLOGY. Compare COMPLETENESS THEOREM.

consistent, *adj.* **1.** (*Logic*) also called **sound. a.** (of a set of statements) capable of all being true at the same time in the same circumstances or under the same INTERPRETATION.

b. (of a FORMAL SYSTEM) not enabling the deduction of a contradiction from the axioms, or, more generally, not having an ATOMIC sentence as a theorem. Compare COMPLETE (sense 4).

2a. (of a congruence or equation) possessing a solution.

b. more generally, (of a system of relations or equations, particularly a linear system) able to be simultaneously satisfied; possessing a solution.

3. (*Statistics*) (of a sequence of tests) such that, as the sample size increases, the probability of accepting a fixed ALTERNATIVE HYPOTHESIS when it is true tends to unity.

constant, *n.* **1a.** a numerical expression that is a part of an algebraic expression. For example, in $x + 2$, 2 is a constant.

b. an unspecified numerical value. For example, if a is proportional to b, then a/b is a constant.

2. a quantity that is regarded as fixed and unchanging for the purposes of a particular calculation. For example, in $y = mx + c$, the general equation of a line, m and c are constants, while x and y are VARIABLES.

3. a specific invariant quantity whose value is determined *a priori*, such as π or e (the base of the NATURAL LOGARITHMS).

4. the value of a specific physical quantity that is determined by the laws of nature and the choice of units, such as c (the speed of light) or γ (the universal gravitational constant; see GRAVITY).

5. see LOGICAL CONSTANT.

constant matrix, $n.$ a matrix all the entries of which are constants; sometimes, in particular, one in which all the entries are the same constant.

constant of integration, $n.$ the arbitrary constant term in the expression of the INDEFINITE INTEGRAL of a function (a consequence of the MEAN–VALUE THEOREM from which it follows that the only functions with zero derivative are constants). For example,

$$\int \sin x = -\cos x + c$$

for any constant c; here c is the constant of integration.

constitutive equation, $n.$ (*Continuum mechanics*) an equation describing the form of the STRESS TENSOR and other quantities for a given body. For example, the constitutive equation for the stress in an incompressible Newtonian viscous fluid is

$$\sigma = -p(\mathbf{x}, t)\,\mathbf{I} + 2\eta\,[\Sigma - \tfrac{1}{3}\,(\text{tr}\Sigma)\,\mathbf{I}\,],$$

where p is a scalar function of the density (pressure), η is a constant (the viscosity), and Σ is the EULERIAN STRAIN RATE. See also ELASTIC.

constraint or **side-condition,** $n.$ a condition that restricts the range of applicability or interest of some statement or result, typically written as a functional equation or inequality.

constrained optimization, $n.$ OPTIMIZATION subject to CONSTRAINTS, as, for example, in LINEAR PROGRAMMING.

constraint qualification, $n.$ a REGULARITY CONDITION upon CONSTRAINTS imposed in order to guarantee that some NECESSARY CONDITION obtains, as with LAGRANGE MULTIPLIERS. Two of the most common of such conditions are to suppose LINEAR INDEPENDENCE of the derivatives of equality constraints, and to suppose SLATER'S CONDITION for convex inequality constraints.

construct, *vb.* **1.** to draw (a line, angle or figure) that satisfies certain specified requirements; especially, in classical geometry to do so without measuring devices, using only a straight-edge and compasses. See CONSTRUCTION.

2. to define any mathematical entity in terms of simpler entities and operations, such as those of set theory.

constructible, *adj.* **1.** able to be CONSTRUCTED in finitely many steps with ruler and compass. This corresponds to determining the *constructible numbers* or quantities whose SOLUTION BY RADICALS involves only quadratic surds. A celebrated result of Gauss's is that the only constructible REGULAR POLYGONS are those with $2^m p_1 p_2 ... p_n$ sides with each p_k a distinct FERMAT PRIME. Compare DOUBLING THE CUBE, SQUARING THE CIRCLE, TRISECTING THE ANGLE.

2. more generally, having a CONSTRUCTIVE proof of its existence.

construction, *n.* **1.** the drawing of a line, angle or figure that accords with certain specified conditions, used in solving a geometrical problem or proving a theorem. *Ruler and compass* (or *compass and straight-edge*) *constructions* in classical geometry do not permit the use of measuring devices.

2. more generally, the specification of a structure, such as a topology or an algebra, satisfying prescribed conditions.

constructive, *adj.* (of a proof or definition) not merely asserting the existence of some entity, but specifying how it might be constructed, as required by INTUITIONIST logic. Usually a constructive proof is taken to be FINITARY. For example, the AXIOM OF CHOICE is not constructive since it does not indicate how a choice set is to be constructed, while the AXIOM OF INFINITY is constructive as it specifies an algorithm that generates infinitely many objects.

constructive dilemma, *n.* see DILEMMA.

constructivism, *n.* the philosophical doctrine that mathematical entities do not exist independently of our construction of them. Compare FINITISM, INTUITIONISM.

contact force, *n.* (*Continuum mechanics*) a FORCE experienced by the points of a BODY due to their contact with other points of the body or with external boundaries, such as by the application of an external pressure to its boundary or (if ELASTIC) by its need to return to its original position. Formally, the effect of the contact forces on a SUB–BODY is given by the integral

$$\int \mathbf{t}(\mathbf{x}, \partial \mathbf{R}_t) \, da$$

over the surface, $\partial \mathbf{R}_t$, of its current CONFIGURATION, where $\mathbf{t}(\mathbf{x}, \partial \mathbf{R}_t)$ is the *contact force density* (or *load, traction,* or *stress vector*) on the surface. Compare BODY FORCE, CONTACT TORQUE.

contact torque, *n.* (*Continuum mechanics*) the TORQUE experienced by the points of a BODY due to their contact with other points of the body or contact with external boundaries, for example, due to the internal magnetic interreactions of its material points. Formally, the effect of the contact torque on a SUB–BODY, $\mathbf{R}$, is given by the sum of the integrals

$$\int \mathbf{x} \times \mathbf{t}(\mathbf{x}, \partial \mathbf{R}_t) \, da + \int \mathbf{c}(\mathbf{x}, \partial \mathbf{R}_t) \, da$$

over the surface, $\partial \mathbf{R}_t$, of its current CONFIGURATION, where $\mathbf{t}(\mathbf{x}, \partial \mathbf{R}_t)$ and $\mathbf{c}(\mathbf{x}, \partial \mathbf{R}_t)$ are respectively the CONTACT FORCE density and the *contact torque density* on the surface, and $\mathbf{x}$ is the POSITION VECTOR of a point of $\mathbf{R}$. Although there are theories in which the existence of a non-zero contact torque density is postulated, no such body has yet been found in nature. Compare BODY TORQUE.

contain, *vb.* to have as a SUBSET; a set *strictly contains* its PROPER SUBSETS.

content, Jordan content, or **Jordan measure,** *n.* a form of MEASURE especially useful in SURFACE INTEGRATION; it is to the RIEMANN INTEGRAL what measure is to the LEBESGUE INTEGRAL, and provides the most practical definition of area. See INNER MEASURE and OUTER MEASURE.

contextual definition, see DEFINITION.

contingency, *n.* (*Logic*) **1.** the state of being CONTINGENT.

2. a contingent statement or proposition; one whose TRUTH–TABLE contains both *true* and *false*. For example, *the moon is made of green cheese* is a contingency. Compare TAUTOLOGY, INCONSISTENCY.

contingency table, *n.* (*Statistics*) an array showing the frequency of occurrence of certain events in each of a number of samples.

contingent, *adj.* (*Logic*) (of a statement or proposition) true under certain conditions and false under others; neither necessarily true nor necessarily false. Compare TAUTOLOGOUS, INCONSISTENT.

continued fraction, *n.* a number that is an integer plus a fraction whose denominator is itself an integer plus a fraction, and so on. For example, the GOLDEN MEAN is

$$\frac{1+\sqrt{5}}{2} = 1 + \cfrac{1}{1 + \cfrac{1}{1 + \cfrac{1}{1 + \cfrac{1}{1 + \cdots}}}}$$

It can be shown that every irrational number has a representation as a continued fraction; these are much used in the solution of DIOPHANTINE EQUATIONS. See CONVERGENTS. Compare COMPOUND FRACTION.

continued product, see PRODUCT.

continuity equation, *n.* another term for SPATIAL EQUATION OF CONTINUITY.

continuous, *adj.* **1a.** (of a function) informally, having a value that changes gradually as the independent variable or variables change, so that at every value, *a*, of the independent variable, the difference between $f(x)$ and $f(a)$ approaches zero as *x* approaches *a*. More formally, a real function $y = f(x)$ is continuous at a point *a* if and only if it is defined at $x = a$ and

$$\lim_{x \to a} f(x) = f(a),$$

that is, precisely, if

for every $\varepsilon > 0$ there exists a $\delta > 0$ such that
$|f(x) - f(a)| < \varepsilon$ for all *x* such that $|x - a| < \delta$.

A function is *continuous on the left* at the point if the above condition holds only for values of *x* less than *a*, and *continuous on the right* if it holds for values greater than *a*; it is continuous at a point if and only if it is both continuous on the left and continuous on the right at that point. The function itself is said to be continuous if it is continuous at all points. The function is said to be *uniformly continuous* on a set if the value of δ depends only on ε and not on the point *a* in the set.

b. (of a curve) representing a continuous function.

2. (of a function defined between metric spaces) having the analogous property that $y = f(x)$ is continuous at a point *p* if and only if it is defined at $x = p$ and

for every $\varepsilon > 0$ there exists a $\delta > 0$ such that
$d(f(x), f(p)) < \varepsilon$ for all *x* such that $d(x, p) < \delta$,

or in NEIGHBOURHOOD notation, if

for every $\varepsilon > 0$ there exists a $\delta > 0$ such that

$f(x) \in N(\varepsilon, f(p))$ for all x such that $x \in N(\delta, p)$.

If for all p in some set, the value of δ depends only on ε and not on the specific point p, the function is said to be *uniformly continuous* on the set. Every continuous function defined on a COMPACT set is uniformly continuous thereon.

3. (of a function f between topological spaces, at a point p) more generally, having the property that, given any NEIGHBOURHOOD V of $f(p)$, there exists a neighbourhood U of p with $f(U)$ inside V. A function is then continuous at every point exactly if the INVERSE IMAGE of any OPEN set is open (and of every CLOSED set is closed). This reduces to the previous definition when the topological spaces are metric. See also LIMIT, DIFFERENTIABLE.

4. (*Statistics*) (of a RANDOM VARIABLE or RANDOM VECTOR) not DISCRETE; having a continuum of possible values so that its distribution requires integration rather than summation to determine its CUMULATIVE PROBABILITY.

5. (of a MEASURE or MEASURE RING) another word for NON-ATOMIC.

continuous deformation, see DEFORMATION.

continuously differentiable, *adj.* (of a function) possessing a CONTINUOUS DERIVATIVE; a mapping from $\mathbb{R}^m$ to $\mathbb{R}^n$ is said to be $C^{(r)}$ if it is r-times continuously differentiable.

continuous multifunction, see SEMICONTINUOUS.

continuous spectrum, see SPECTRUM.

continuum, *n.* **1.** a COMPACT CONNECTED set.

2. the continuum. the set of all REAL NUMBERS.

3. a continuous distribution of matter. See CONTINUUM MECHANICS.

continuum hypothesis, *n.* the hypothesis that the CARDINALITY of the CONTINUUM is the smallest NON-DENUMERABLE cardinal. This has been shown to be UNDECIDABLE in that both it and its negation are consistent with the standard axioms of set theory. The *generalized continuum hypothesis* states that for any infinite cardinal the next greater cardinal is that of its POWER SET.

continuum mechanics, *n.* the study of the properties of idealized fluids and other materials, regarded as a *continuum*, that is, a continuous distribution of matter with no empty space, so that molecular structure is ignored and average pressure, velocity, etc. are considered; hence for these purposes, a PARTICLE is not a physical molecule but an INFINITESIMAL element of a BODY.

contour, *n.* a PIECEWISE SMOOTH CURVE in COMPLEX ANALYSIS.

contour integral, *n.* a CURVILINEAR INTEGRAL in COMPLEX ANALYSIS, typically around a SIMPLE CLOSED CURVE or SIMPLE CLOSED CONTOUR.

contour line, *n.* (on a surface) a line consisting of points of equal value for some given function. It is possible to represent a function of two variables, or a three-place relation, in this way. For example, to represent the function $z = x^2 + y^2$ we can draw a series of plane curves for different values of z; as shown in Fig. 90 opposite, each of these is the shape of a cross section of the surface $z = x^2 + y^2$ at that value of z, and is in fact a circle with radius $\sqrt{z}$, so that the surface is a paraboloid with its vertex at the origin.

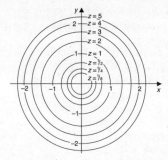

Fig. 90. A **contour** representation of $z = x^2 + y^2$.

contract curve, *n.* the locus of points in an EDGEWORTH BOX at which the respective INDIFFERENCE CURVES of the two consumers are tangential to one another. This curve is optimal in the sense that if the consumers are at a point off the contract curve, neither loses and at least one benefits from moving to a point on the curve.

contracted, *adj.* (of a CYCLOID, EPICYCLOID or HYPOCYCLOID) described by a point attached to but lying within, rather than on or outside, the circumference of a circle as it rolls round some other figure without slipping; for example, Fig. 91 shows a contracted cycloid. Compare COMMON, EXTENDED.

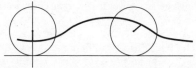

Fig. 91. A **contracted** cycloid.

contraction, *n.* **1.** also called **contraction mapping.** a mapping on a METRIC SPACE that uniformly shrinks distances. That is, T is a contraction if there exists $k < 1$ such that $d(Tx, Ty) \leq kd(x, y)$ for all x and y in the space. See CONTRACTION MAPPING THEOREM.

2. an operation on TENSORS that yields a tensor of type $(r-1, s-1)$ from one of type (r, s), by setting a superscript equal to a subscript.

contraction mapping theorem, *n.* the theorem that a CONTRACTION MAPPING defined on a COMPLETE METRIC SPACE must have a unique FIXED POINT. Compare BROUWER'S THEOREM, TARSKI FIXED POINT THEOREM.

contradiction, *n.* **1.** a statement or proposition that is CONTRADICTORY to a given statement.

2. a necessary falsehood; a statement that is false under all circumstances. Compare CONTINGENCY, TAUTOLOGY.

contradictory, *adj.* (*Logic*) **1.** (of a pair of statements) unable both to be true or both to be false at the same time, under the same circumstances or in the same INTERPRETATION.

2a. (of a single statement) not able to be true when a given statement is true, or false when it is false.

b. (*as substantive*) a statement that is contradictory to a given statement.

3. another word for SELF–CONTRADICTORY.

Compare CONTRARY, SUBCONTRARY.

contragredient matrix, *n.* the INVERSE of the TRANSPOSE of a given matrix.

contraposition, *n.* (*Logic*) **1.** the logical principle that if *p* implies *not q* then *q* implies *not p*.

2. the derivation of its CONTRAPOSITIVE from a given statement.

contrapositive, *n.* (*Logic*) a statement derived from a given statement by interchanging its subject and predicate terms and altering each from positive to negative or vice versa. For example, the contrapositive of *all cats are mammals* is *all non-mammals are non-cats*; the contrapositive of *if it's Thursday, this can't be Belgium* is *if this is Belgium, it can't be Thursday.*

contrary, *adj.* (*Logic*) **1.** (of a pair of statements) such that they cannot both be true at the same time, under the same circumstances or in the same INTERPRETATION (although they may both be false).

2a. (of a single statement) not able to be true when a given statement is true.

b. (*as substantive*) a statement that is contrary to a given statement.

Compare CONTRADICTORY, SUBCONTRARY.

contravariant, see FUNCTOR.

contravariant metric tensor, see METRIC TENSOR.

contravariant tensor, *n.* an element of the TENSOR PRODUCT,

$$T^r = T \otimes \ldots \otimes T,$$

of a VECTOR SPACE with itself *r* times; such a tensor is said to be contravariant of order *r*. Compare COVARIANT TENSOR.

control, 1. *vb.* (*Statistics*) to rule out the effect of irrelevant variables by designing an experiment in which their effect is nullified; this may be done, for example, by random allocation of subjects to experimental conditions or by selection of matched pairs of subjects. Thus one might control for a learning effect in repeated testing by applying the tests to each subject in randomly determined order, or one might control for an irrelevant variable by ensuring that it has the same value for all one's subjects.

– *n.* **2.** a CONTROL CONDITION, or the group of subjects to which it is applied.

3. another term for CONTROL VARIABLE.

control chart, *n.* (*Statistics*) a chart on which observed values of a variable are plotted, usually relative to its EXPECTED VALUE under some condition and its allowable deviation, so that excessive variations can be detected.

control condition, *n.* (*Statistics*) the condition in which the subjects of an experiment are not subjected to the treatment whose effectiveness is under investigation; the condition before experimental intervention, such as the course of a disease without therapy, with which the effect of a treatment is to be compared. Compare EXPERIMENTAL CONDITION .

control theory or **optimal control,** *n.* the branch of mathematics, growing out of the CALCULUS OF VARIATIONS, that studies methods of solution of OPTIMIZATION problems subject to constraints expressed by DIFFERENTIAL EQUATIONS; it is especially applicable to control mechanisms and cost effectiveness. See PONTRYAGIN'S MAXIMUM PRINCIPLE.

control variable or **control,** *n.* (*Control theory*) one of the governing variables in a CONTROL THEORY problem, as contrasted with the STATE VARIABLES.

converge, *vb.* **1.** (of an infinite sequence) to be CONVERGENT to a finite LIMIT as the number of terms tends to infinity. An infinite series converges if the sequence of its partials sums converges as the number of terms tends to infinity. Compare OSCILLATE.

2. (of an IMPROPER INTEGRAL) to possess a finite value.

3. (of a NET) see NET CONVERGENCE.

See CONVERGENT. Compare DIVERGE.

convergence, *n.* the property that, or manner in which, a sequence, series, or integral, is CONVERGENT to a finite limit.

convergent, *adj.* **1.** (of an infinite SEQUENCE of numbers or vectors) having a finite limit, so that if a_n is the nth element of the sequence, there exists an L (the limit) such that for every $\varepsilon > 0$ there is an N such that

$$|a_n - L| < \varepsilon \text{ for all } n > N.$$

A corresponding definition applies in a metric space setting.

2. (of an infinite SERIES) having a finite sum, generating a sequence of PARTIAL SUMS that has a finite limit. If the series $a_0 + a_1 + a_2 + \ldots$, that is, the sequence of partial sums

$$\langle a_0, a_0 + a_1, a_0 + a_1 + a_2, \ldots \rangle,$$

converges, then the sequence $\langle a_0, a_1, a_2, \ldots \rangle$ must converge to zero, but not necessarily conversely. For example, the sequence

$$\langle 1, \tfrac{1}{2}, \tfrac{1}{3}, \tfrac{1}{4}, \tfrac{1}{5}, \ldots \rangle$$

is convergent, but the series

$$1 + \tfrac{1}{2} + \tfrac{1}{3} + \tfrac{1}{4} + \tfrac{1}{5} + \ldots$$

is not.

3. (of an IMPROPER INTEGRAL) having a finite value that is defined as the limit of the proper integrals as the limits of integration tend to a limit.

4. more generally, of any function, approaching a limit.

5. (of a sequence or NET in a topological space) EVENTUALLY in every neighbourhood of the point. See NET CONVERGENCE.

See ABSOLUTE CONVERGENCE, CONDITIONAL CONVERGENCE, CONVERGENT IN MEAN, CONVERGENT IN MEASURE. See also RATE OF CONVERGENCE, LINEAR CONVERGENCE, UNIFORM CONVERGENCE, POINTWISE CONVERGENT. Compare DIVERGENT, OSCILLATING.

convergent in mean, *adj.* (of a sequence of INTEGRABLE functions on a set) having the property that the integral of the absolute value of the differences of the functions from the limit function tends to zero. Thus, the sequence $\{f_n\}$ converges to f in mean on an interval $[a, b]$ if

$$\int_a^b |f_n(x) - f(x)| \ \mathrm{d}x \ \to \ 0$$

as n tends to ∞. *Convergence in mean square* instead requires that

$$\int_a^b |f_n(x) - f(x)|^2 \ \mathrm{d}x \ \to \ 0.$$

Compare CONVERGENT IN MEASURE, POINTWISE CONVERGENT.

convergent in measure, *adj.* (of a sequence, $\{f_n\}$, of MEASURABLE functions) convergent with respect to some MEASURE, P, in the sense that, for every $\varepsilon > 0$

$$P(\{\, x : \mid f_n(x) - f(x) \mid \, > \varepsilon \,\})$$

tends to zero as n tends to infinity; f is then the *limit* of the sequence. Compare CONVERGENT IN MEAN, POINTWISE CONVERGENT.

convergents, *n. pl.* (of a continued fraction) the rational numbers obtained by truncation of a CONTINUED FRACTION. If the fraction is *simple*, in that all the numerators are unity and all the denominators are positive integers, then the nth *convergent* is

$$\frac{p_n}{q_n} = \left[a_0, a_1, \ldots, a_n\right] = a_0 + \cfrac{1}{a_1 + \cfrac{1}{a_2 + \cfrac{1}{a_3 + \cdots \cfrac{}{\cdots \frac{1}{a_n}}}}}$$

the limit of which exists and defines a simple CONTINUED FRACTION. Here p_n and q_n satisfy

$$p_0 = a_0; \quad p_1 = 1 + a_1 a_0; \quad p_n = a_n p_{n-1} + p_{n-2};$$
$$q_0 = 1; \quad q_1 = a_1; \quad q_n = a_n q_{n-1} + q_{n-2}.$$

converse, *n.* (*Logic*) **1.** a relation that holds of an ordered pair of elements, $\langle x, y \rangle$ if and only if a given relation holds of the ordered pair $\langle y, x \rangle$; that is, x has the converse relation to y if and only if y has the given relation to x. For example, on the domain of males, *father of* is the converse of *son of*. The converse of a given relation Rxy is often written Ryx.

2a. in ARISTOTELIAN LOGIC, a proposition derived from another by interchanging its subject and predicate terms. Thus, for example, *all men are liars* can be derived from *all liars are men*; this, however, is clearly not a valid form of argument.

b. analogously, a conditional statement derived from another by interchanging antecedent and consequent, such as

if John missed the meeting, then his train was late

from

if John's train was late, then he missed the meeting.

This is not a valid form of argument unless *if* is taken to represent the biconditional.

convert, *vb.* **1.** to change the units of a quantity. For example, to convert from miles to kilometres, one multiplies by 1.61.

2. to obtain the CONVERSE of a given proposition or relation.

convex, *adj.* **1.** (of a polygon) having no interior angle greater than 180°, so that all lines joining any pair of points on the boundary of the figure lie wholly inside it. Thus, in Fig. 92 opposite, the pentagon ADBFE is convex, but ACBFE is not.

2. (of a function) **a.** having the property that the chord joining any two points on its graph lies above the graph. Thus, with the usual orientation of the coordinate axes in Fig. 92, both ACB and AEFB are convex, although the polygon is not itself convex; ADB is not convex.

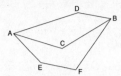

Fig. 92. **Convex.** See main entry.

b. formally and more generally, such that for arguments x and y in the appropriate abstract space, and t in the interval $[0, 1]$,

$$tf(x) + (1-t)f(y) \geq f(tx + (1-t)y).$$

3. (of a set of points in a real VECTOR SPACE) having the property that if two points are in the set then so are all the points on the line segment joining them; that is, if x and y are any two points in the set, then so is $tx + (1-t)y$, for all t between 0 and 1.

Compare CONCAVE.

convex combination, *n.* a LINEAR COMBINATION, $\sum t_i a_i$, of a finite set of elements, a_i, such that all the coefficients, t_i, of the elements are non-negative and $\sum t_i = 1$.

convex hull, *n.* the intersection of all CONVEX sets containing a subset A of a real VECTOR SPACE; equivalently, the set of all CONVEX COMBINATIONS of elements of A.

convex polyhedron, see POLYHEDRON.

convex polytope, see POLYHEDRON.

convex quadrangle, see QUADRANGLE.

convolution, *n.* **1a.** a function or series derived from two given functions or series by integration: the *integral convolution* of $f(x)$ and $g(x)$ is

$$\int_0^x f(t)\, g(x-t)\, \mathrm{dt},$$

while the convolution of two series produces the CAUCHY PRODUCT.

b. analogously, the *infimal convolution* of $f(x)$ and $g(x)$ is

$$(f \square g)(x) = \inf_t f(t) g(x-t),$$

and is a convex function if the two given functions are convex.

c. the process of constructing such a function.

2. (*Statistics*) a method of determination of the sum of two random variables by integration or summation.

convolve, *vb.* to form the CONVOLUTION of a pair of functions.

coordinate, *n.* **1.** (*in plural*) a set of numbers that uniquely identifies the position of a point relative to a set of fixed reference points, lines, directions, etc., that constitute a COORDINATE SYSTEM. See CARTESIAN COORDINATES, POLAR COORDINATES, SPHERICAL COORDINATES, CURVILINEAR COORDINATES, HOMOGENEOUS COORDINATES.

2. (*with qualifier*) that one of such a set of numbers that relates to the specified direction, angle, etc.; for example, the x-coordinate is the distance from the ORIGIN to the foot of the perpendicular from the point to x-axis, measured along that axis. See also ORDINATE, ABSCISSA.

coordinate change or **coordinate transformation**, *n.* (*Differential geometry*) a mapping

$$\phi\psi^{-1} : \psi(U \cap V) \to \phi(U \cap V),$$

where (U, ϕ) and (V, ψ) are CHARTS.

coordinate function, *n.* a function that defines one COORDINATE of a curve in terms of a PARAMETER; if $y = f(x)$ is satisfied by the set of points $(u(t), v(t))$, then $x = u(t)$ and $y = v(t)$ are the coordinate functions. See also PARAMETRIC EQUATIONS.

coordinate geometry or **analytic geometry**, *n.* the branch of mathematics in which geometrical points and figures are described using algebraic notation in terms of their position in a coordinate system. See also CARTESIAN COORDINATES.

coordinate plane, *n.* a two-dimensional system of coordinates.

coordinate space, *n.* (*Statistical physics*) a space of s dimensions representing a system with s DEGREES OF FREEDOM in which the rectangular coordinates specify the positions of the points within the system.

coordinate system, *n.* **1.** any system for locating points by their COORDINATES with respect to some set of reference points, lines, directions, etc.

2. (*Differential geometry*) another word for CHART.

coplanar, *adj.* lying in the same plane; for example, coplanar lines.

coprime, *adj.* another word for RELATIVELY PRIME.

core, *n.* **1.** (of a set in a VECTOR SPACE) the points of the set such that an OPEN LINE SEGMENT may be constructed in the set containing the point. Compare ABSORBING SET.

2. (*Game theory*) a solution concept that seeks PARETO OPTIMAL outcomes in games in which coalitions are allowed between players.

3. (*Group theory*) the intersection of all the CONJUGATES of a subgroup, H, in a given group, G; it is the largest NORMAL subgroups of G contained in H, and is denoted core H.

Coriolis force, *n.* an apparent force experienced by a body that moves along a radius of a ROTATING FRAME OF REFERENCE, and opposing the rotation of the body relative to the stationary frame of reference; like the CENTRIFUGAL FORCE this is not in fact a force, but notional compensation for the rotating axes. It is given by $2m\,\omega\,\mathbf{v}$ for a PARTICLE of mass m moving with velocity $\mathbf{v}$ with respect to a rotating frame of reference with ANGULAR VELOCITY ω.

corollary, *n.* a proposition that follows directly from the statement or the proof of another proposition; a subsidiary theorem.

correct to n decimal places, see ACCURATE (sense 1).

correction, *n.* a number or quantity added to or subtracted from the result of a calculation or observation in order to increase its accuracy. For example, when weighing goods, one has to *correct* for the weight of any container.

correlation, *n.* **1.** (*Statistics*) **a.** the extent of correspondence between the ordering of two RANDOM VARIABLES. It is a *positive correlation* when each variable tends to increase or decrease as the other does, and a *negative* or *inverse* *correlation* if one tends to increase as the other decreases; thus, for example, in Fig. 93 there is a high negative correlation between values of *x* and *y* in the first example, and a low positive one in the second.

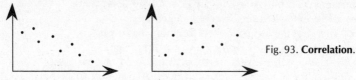

Fig. 93. **Correlation**.

(a) high negative correlation.   (b) low positive correlation.

b. also called **correlation coefficient.** any of a number of STATISTICS that measure the degree of correlation between two random variables, for example, by dividing their COVARIANCE by the square root of the product of their VARIANCES. The closer the absolute value of the correlation coefficient, which is usually denoted $\rho(X, Y)$, is to 1, the greater the correlation; a coefficient of 0 is necessary but not sufficient for the random variables to be independent. See also PEARSON'S CORRELATION COEFFICIENT, SPEARMAN'S RANK–ORDER COEFFICIENT.

2. (*Projective geometry*) a BIJECTION of the set of points of a projective geometry onto the set of highest-dimension SUBGEOMETRIES of another projective geometry or itself.

correlation matrix, *n.* (*Statistics*) the square $n \times n$ matrix of which the entries are the pairwise CORRELATIONS of the variables of a RANDOM VECTOR of length n; the $(i,j)^{th}$ entry is the correlation between the ith and the jth variables. Compare VARIANCE–COVARIANCE MATRIX.

correspond, *vb.* (of a pair of numbers, objects or quantities) to be related by a ONE–TO–ONE CORRESPONDENCE, so that one is a member of the domain and the other its image under the mapping.

correspondence, *n.* **1.** a ONE–TO–ONE CORRESPONDENCE.

2. less frequently, any mapping or relation between the members of discrete sets, whether one-to-one, ONE–MANY, MANY–ONE, or even many–many, as in Fig. 94.

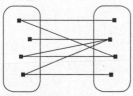

Fig. 94. A many–many **correspondence**.

3. another word for SET–VALUED MAPPING especially when the image set is always non-empty.

correspondence theorem, *n.* (*Algebra*) the result that there is a BIJECTION from the set of subgroups of the image of a GROUP under a HOMOMORPHISM onto the set of subgroups of the group that contain the KERNEL of the homomorphism; similar results hold for RINGS, MODULES, etc.

corresponding angles, *n.* angles that are on the same side of a TRANSVERSAL that cuts two lines, and on the same side of each line.

corresponding sides, *n.* in CONGRUENT polygons, the pairs of sides which can be superimposed on one another. In SIMILAR polygons, the ratio of the length of a side on the larger polygon to the length of its corresponding side on the smaller polygon is the same for all the sides.

cos, *abbrev. and symbol for* the COSINE function.

cos⁻¹, *symbol for* the inverse COSINE function, ARC–COSINE.

cosec, *abbrev. and symbol for* the COSECANT function.

cosec⁻¹, *symbol for* the inverse COSECANT function, ARC–COSECANT.

cosecant (abbrev. **cosec** or **csc**), *n.* a TRIGONOMETRIC FUNCTION that in a right-angled triangle is equal to the ratio of the length of the hypotenuse to that of the side opposite the given angle; the reciprocal of SINE. Its graph is shown in Fig. 95. If θ is the angle measured in radians from the x-axis of a coordinate system swept out in an anticlockwise direction by a radius of length r centred on the origin, then $\operatorname{cosec}\theta = r/y$, where y is the ORDINATE of the end of the radius. The derivative of the cosecant function is

$$- \operatorname{cosec} x \cot x,$$

and an antiderivative or indefinite integral is

$$\ln | \operatorname{cosec} x - \cot x |.$$

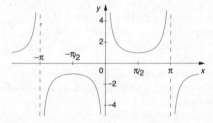

Fig 95. The graph of the **cosecant** function.

cosech, *symbol* for the HYPERBOLIC FUNCTION, hyperbolic cosecant; the reciprocal of the hyperbolic sine function, SINH. Its graph is shown in Fig. 96. The derivative of cosech$x$ is

$$- \operatorname{cosech} x \cotanh x,$$

and an antiderivative or indefinite integral is

$$\log \tanh (x/2).$$

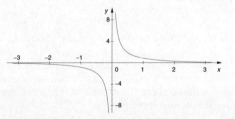

Fig. 96. **Cosech.** Graph of the hyperbolic cosecant function.

cosech⁻¹, *symbol for* the inverse HYPERBOLIC COSECANT function, ARC–COSECH .

coset, *n.* a subset of a given GROUP, written *aH* or *Ha*, whose members are the products of a given element, *a*, of the group with a given subgroup, *H*. The cosets of *H* in *G* are disjoint and form a PARTITION of *G*. The left and right cosets, *aH* and *Ha*, may be distinct in a non-commutative group, and if *aH* = *Ha* for all *a* in *G*, then *H* is said to be *normal* in *G*. See also TRANSVERSAL.

cosh or **ch,** *symbol for* the HYPERBOLIC FUNCTION hyperbolic cosine, related to the COSINE function by the identity $\cosh z = \cos iz$, for z a complex number, and $i = \sqrt{-1}$. It can be defined in terms of the EXPONENTIAL FUNCTION as

$$\cosh z = \tfrac{1}{2}(e^z + e^{-z}).$$

It is an EVEN function of which both the derivative and an antiderivative (or indefinite integral) is SINH, the hyperbolic sine function; its graph is shown in Fig. 97. The functions $\cosh z$ and $\sinh z$ together satisfy

$$\cosh^2 z - \sinh^2 z = 1$$
$$\cosh(2z) = \cosh^2 z + \sinh^2 z.$$

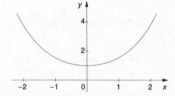

Fig. 97. **Cosh.** The graph of the hyperbolic cosine function.

cosh⁻¹, *symbol for* the inverse HYPERBOLIC COSINE function, ARC–COSH.

cosine (abbrev. **cos**), *n.* a TRIGONOMETRIC FUNCTION that in a right-angled triangle is equal to the ratio of the side adjacent the given angle to the hypotenuse. If θ is the angle measured in radians from the x-axis of a co-ordinate system swept out in an anticlockwise direction by a radius of length r centred on the origin, then $\cos\theta = x/r$ where x is the ABCISSA of the end of the radius. It is an EVEN function of which an antiderivative (or indefinite integral) is the SINE function, sin, and the derivative is $-\sin$; its graph is shown in Fig. 98.

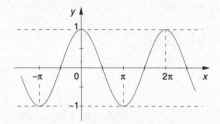

Fig. 98. Graph of the **cosine** function.

The functions $\cos z$ and $\sin z$ together satisfy

$$\cos^2 z + \sin^2 z = 1$$
$$\cos(2z) = \cos^2 z - \sin^2 z.$$

123

It is best defined as a complex function by the power series:

$$\cos z = \sum_{n=0}^{\infty} \frac{(-1)^n z^{2n}}{(2n)!}.$$

See also DE MOIVRE'S FORMULAE.

cosine law, *n.* the relationship that holds between the lengths and angles of a triangle with sides of length *a*, *b*, and *c*, namely

$$c^2 = a^2 + b^2 - 2ab\cos C,$$

where, as in Fig. 99, *C* is the angle opposite the side of length *c*. In a EUCLIDEAN SPACE this becomes the vector identity

$$\| x - y \|^2 = \| x \|^2 + \| y \|^2 - 2\langle x, y \rangle,$$

valid for any INNER PRODUCT SPACE, by virtue of the definition of the INNER PRODUCT. Compare CAUCHY–SCHWARZ INEQUALITY.

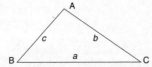

Fig. 99. **Cosine law**. See main entry.

cosine series, *n.* **1.** the POWER SERIES expansion for the COSINE FUNCTION:

$$\cos x = 1 - \frac{x^2}{2!} + \frac{x^4}{4!} - \frac{x^6}{6!} + \dots$$

This is valid for all *x*.

2. A series in which all the terms are cosine functions. See FOURIER SERIES.

cot or **cotan**, *abbrev. and symbol for* the COTANGENT function.

cot⁻¹ or **cotan⁻¹**, *symbol for* the inverse COTANGENT function, ARC–COTANGENT.

cotangent (abbrev. **cot** or **cotan**), *n.* a TRIGONOMETRIC FUNCTION that in a right-angled triangle is equal to the ratio of the length of the side adjacent to the given angle to that of the side opposite it; it is the reciprocal of the TANGENT function, equal to the ratio of COSINE to SINE. Its graph is shown in Fig. 100. If θ is the angle from the *x*-axis of a coordinate system swept out in an anti-clockwise direction by a radius centred on the origin, then $\cot\theta = x/y$, where *x* is the ABSCISSA and *y* the ORDINATE of the end of the radius. Its derivative is $-\csc^2 x$, and an antiderivative (or indefinite integral) is $\log \sin x$.

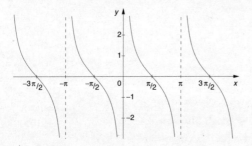

Fig. 100. Graph of the **cotangent** function

cotanh or **coth,** *symbol for* the HYPERBOLIC FUNCTION hyperbolic cotangent, the reciprocal of hyperbolic tangent function TANH, equal to the ratio of COSH to SINH ; its graph is shown in Fig. 101. Its derivative is $-\mathrm{cosech}^2 x$, and an antiderivative (or indefinite integral) is $\log \sinh x$.

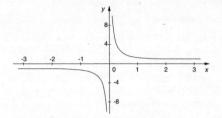

Fig. 101. **Cotanh**. Graph of the hyperbolic cotangent function.

cotanh⁻¹ or **coth⁻¹,** *symbol for* the inverse HYPERBOLIC COTANGENT function, ARC–COTANH.

coterminal angles, *n.* angles of rotation with the same initial and terminal arms; for example, 60° is coterminal with 420°. Such angles differ by whole multiples of 360°.

count, *vb.* **1.** to enumerate; to check (some collection of objects) in order to ascertain their CARDINAL number. In this sense, to count a set of objects is to put them in one-to-one correspondence with an initial segment of the natural numbers. Counting is contrasted with calculating; so one might say that someone's arithmetic was so weak he had to count on his fingers.

2a. to recite numbers in ascending order (up to a stated maximum); for example, to count up to a thousand.

b. to recite in ascending order the multiples of a given number. For example, to count in threes is to count '3, 6, 9, 12, …'.

countable, *adj.* (of a set of objects) able to be put into ONE–TO–ONE CORRES-PONDENCE with a subset of the natural numbers; a countable set may be either finite or DENUMERABLE.

countably additive, *adj.* (of a SET FUNCTION, S, defined on a class of sets) having the property that, for any COUNTABLE pairwise disjoint family of sets $\{A_n\}_{n \in N}$ in the class,

$$S\left(\bigcup_{n \in N} A_n\right) = \sum_{n \in N} S(A_n) .$$

whenever the union lies in the given class. For example, by definition, a MEASURE is countably additive on its sigma-algebra. Compare ADDITIVE.

countably infinite, *adj.* another term for DENUMERABLE.

counterclockwise, another term for ANTICLOCKWISE.

counter-domain, *n.* (of a relation) the set of objects for which there is something that has the given relation. If the relation is regarded as a function, the counter-domain is the RANGE rather than the CODOMAIN of the function. Compare DOMAIN.

counterexample, *n.* an example that disproves a general statement, or that shows an argument to be invalid by satisfying its antecedent or premises while rendering its consequent or conclusion obviously false. For instance,

125

a counterexample to the truth of *all mortals are men* might be any (mortal) cat. A counterexample to the inference of *P* from *if P then Q* together with *Q*, would be any substitution for *P* and *Q* that would make the premises true and the conclusion false: to show this, *P* may be taken to be '2 + 2 = 5', and *Q* to be '2 + 2 > 3'; then *Q* is clearly true, and so is *if P then Q*, that is

$$\text{if } 2 + 2 = 5, \text{ then } 2 + 2 > 3,$$

since whatever is equal to 5 must be greater than 3; but the conclusion *P* is clearly false, so that this cannot be a valid argument as it is capable of leading from true premises to a false conclusion. See VALID.

counterfactual or **counterfactual conditional,** *n.* a CONDITIONAL STATEMENT, such as *had the US not entered the war, Germany would have won*, whose ANTECEDENT is in fact false, and that is usually interpreted modally in terms of the truth of the consequent in the nearest POSSIBLE WORLD in which the antecedent is true. See also COUNTERPART.

counter-harmonic mean, see NEO–PYTHAGOREAN MEANS.

counterimage, **inverse image** or **pre-image,** *n.* the set of elements whose IMAGE under a mapping lies in a given set; the counter-image is denoted $f^{-1}(S)$ or $f^{-l}(S)$, and is well-defined even if the inverse mapping is not; for example, the counterimage of the square root function on the non-negative reals is all the reals, even though this is a set-valued function. Compare IMAGE.

counterpart, *n.* (*Logic*) the object in one POSSIBLE WORLD that most closely resembles a given object in another, and that therefore figures in certain accounts of the semantics of COUNTERFACTUAL statements about the latter. On this account, counterparts of the same object need not be identical, and indeed if a counterfactual statement is true they will differ in exactly that respect. For example, *Nelson might have been celibate* is true if and only if there is some possible world (which may be the actual world) in which the predicate *celibate* is satisfied by that world's counterpart of Nelson.

counting measure, *n.* the MEASURE function whose value for each finite subset of a given set is its CARDINALITY. Note that the measure can be viewed as defined on the SIGMA–ALGEBRA of all countable subsets, or on the entire POWER SET.

counting number, see NATURAL NUMBER.

couple, *n.* (*Mechanics*) a pair of parallel FORCES of equal magnitude but opposite direction acting along different lines. The TORQUE of the couple about any point in space is a vector with direction perpendicular to the plane of the lines along which they act and with magnitude equal to the product of the magnitude of either force and the distance between the lines. Two couples are equivalent if they have the same torque.

coupled, *adj.* (of a pair of equations) interdependent or interrelated in some fashion.

Courant penalty function, see PENALTY FUNCTION METHODS. (Named after the US analyst, complex variable theorist, and applied mathematician *Richard Courant* (1888–1972).)

cov, (*Statistics*) *abbrev. and symbol for* COVARIANCE.

covariance (abbrev. **cov**), *n.* (*Statistics*) a measure, denoted Cov(X, Y), of the association between two RANDOM VARIABLES, X and Y, equal to the EXPECTED

VALUE of the product of their DEVIATIONS from the mean. It may be estimated by the sum of products of deviations from the sample mean for the associated values of the two variables, divided by the number of sample points.

covariance matrix, *n.* another term for the VARIANCE–COVARIANCE MATRIX. Compare CORRELATION MATRIX.

covariant, *adj.* see FUNCTOR.

covariant tensor, *n.* **1.** an element of the TENSOR PRODUCT,

$$T_s = T^* \otimes \ldots \otimes T^*,$$

of the DUAL of a VECTOR SPACE with itself *s* times; such a tensor is said to be covariant of order *s*.

2. a MULTILINEAR FUNCTION ; the covariant tensor is said to be of rank *r* if the function is of degree *r* and has its domain in the *r*-fold product of Euclidean *n*-space. Compare CONTRAVARIANT TENSOR .

covector, *n.* an alternating COVARIANT TENSOR of rank *r*.

cover or **covering,** *n.* **1.** a collection of sets whose union contains a given set.

2. (of a graph) see KONIG'S THEOREM.

covers, *abbrev. for* COVERSED SINE.

coversed sine, *n.* a trigonometric function equal to $1 - \sin x$. See SINE.

Cramer's rule, *n.* a method for solving SIMULTANEOUS EQUATIONS by the use of matrices: given *n* equations in *n* unknowns, of the form

$$a_{i,1}x_1 + a_{i,2}x_2 + \ldots + a_{i,n}x_n = b_i;$$

these can be written as $AX = B$, where A is the matrix of coefficients $a_{i,j}$, **X** the column of unknowns, and **B** the column of constants; then if A is non-singular, the system of equations has a unique solution,

$$x_i = \frac{\Delta_i}{\det A},$$

where Δ_i is the DETERMINANT of the matrix derived from A by replacing its ith column by **B**. (Named after the Swiss mathematician and physicist, *Gabriel Cramer* (1704–52).)

critical point, *n.* **1.** another name (*mainly USA*) for a STATIONARY POINT.

2. a point at which a function has a first DERIVATIVE that is infinite, so that the curve has a vertical tangent. For example, the function $y = \sqrt{(x + 2)}$, whose graph is shown in Fig. 102, has a critical point at $x = -2$; the bold line is the tangent at X. Compare STATIONARY POINT.

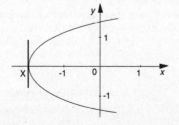

Fig. 102. X is a **critical point** of the function.

127

critical region, *n.* (*Statistics*) the set of values of a TEST STATISTIC for which the NULL HYPOTHESIS would be rejected at a given SIGNIFICANCE LEVEL. See HYPOTHESIS TESTING.

cross-cap, *n.* a NON–ORIENTABLE surface, made by deforming a MÖBIUS STRIP, or by pulling part of a sphere through a slit on its surface. Compare HANDLE.

cross-correlation, *n.* (*Statistics*) the CORRELATION between two sequences of random variables in a TIME–SERIES.

cross-cut, *n.* a simple ARC between two distinct points on a SURFACE.

crossed quadrangle, see QUADRANGLE.

cross-multiply, *vb.* to simplify an equation involving fractions by multiplying the numerator of each side by the denominator of the other, since $a/b = c/d$ if and only if $ad = bc$.

cross product, *n.* **1.** another name for VECTOR PRODUCT.

2. another name for CARTESIAN PRODUCT.

cross ratio, *n.* (*Geometry*) (of four collinear points) the product of certain ratios obtained as follows: let distinct points A, B, C, D lie in some order on a line and compute the DIRECTED RATIO $(AC.BD)/(AD.BC)$, denoted by $(A,B; C,D)$ or $\{AB, CD\}$. Equivalently, if A, B, C, D have parameters a, b, c, d, then

$$(A, B; C, D) = \frac{(a-c)(b-d)}{(a-d)(b-c)}.$$

If no ordering of the points will produce an HARMONIC RATIO, there will be six possible cross ratios from four points. A PROJECTIVITY is one that maintains the cross ratio. See DIVISION OF A SEGMENT. See also HARMONIC POINTS.

cruciform, *n.* a geometric curve shaped like a cross that has four similar branches that are ASYMPTOTIC to two mutually perpendicular pairs of lines, as in Fig. 103. Its equation is

$$x^2y^2 - a^2x^2 - a^2y^2 = 0,$$

where $x = \pm a$ and $y = \pm a$ are the four lines.

Fig. 103. The four branches of a **cruciform** with their asymptotes.

crude, *adj.* (of statistical data) not analysed, consisting simply of the collected values of the variables.

crumb, *n.* (*informal*) half a NIBBLE; a BYTE of two BITS.

crunode, *n.* a point at which two branches of a curve intersect and each has a distinct tangent; for example, X in Fig. 104. Compare SPINODE, OSCULATION.

Fig. 104. X is a **crunode** of the curve.

csc, *abbrev. and symbol for* the COSECANT function.

csc⁻¹, *symbol for* the inverse COSECANT function. See ARC–COSECANT.

csch, *symbol for* the HYPERBOLIC COSECANT function. See COSECH.

csch⁻¹, *symbol for* the inverse HYPERBOLIC COSECANT function. See ARC–COSECH.

ctn, *abbrev. and symbol for* the COTANGENT function.

ctn⁻¹, *symbol for* the inverse COTANGENT function. See ARC–COTANGENT.

ctnh, *symbol for* the HYPERBOLIC COTANGENT function. See COTANH.

ctnh⁻¹, *symbol for* the inverse HYPERBOLIC COTANGENT function. See ARC–COTANH.

cubage, *n.* the volume of a solid body or figure.

cubature, *n.* the calculation of the volume of a solid, or the volume itself. See QUADRATURE.

cube, *n.* **1.** a solid with six identical square sides that are mutually perpendicular, as shown in Fig. 105.

Fig. 105. **Cube**.

2a. the result of multiplying a number, quantity, or expression by itself three times; the third power of a number, quantity, etc. For example, the cube of 2 is $2 \times 2 \times 2 = 8$; the cube of x is x^3.

b. *vb.* to raise a number, quantity, or expression to its third power.

cube root, *n.* the number, quantity, or expression whose CUBE is a given number, quantity, or expression. For example, the cube root of 8 is 2. The real cube root of x is usually written $^3\sqrt{x}$; every non-zero real has one real and two complex cube roots.

cubic, *adj.* **1.** having the shape of a CUBE.

2. of or relating to the third power.

3. (of an algebraic expression, equation, etc.) containing a variable raised to the third power, and none raised to a higher power; of the third degree.

4. (of an algebraic expression, equation, etc.) containing a term in which the variables have EXPONENTS the sum of which equals 3, and no term in which the exponents have a higher sum; of the third degree.

5. denoting a measure of three-dimensional space that is derived from a linear measure by raising it to the third power; thus a cubic metre is the volume enclosed by a cube each of the sides of which is one metre long.

6. (*as substantive*) a cubic equation, function, expression or term.

cubical, *adj.* of, involving, or related to volume or to the third power.

cubic resolvent equation, *n.* an intermediate equation arising in the solution of the general QUARTIC. See CARDANO'S FORMULA.

cubiform, *adj.* having the shape of a CUBE.

cuboctahedron, *n.* one of the ARCHIMEDIAN SOLIDS, with 14 faces and 12 vertices. It can be obtained from either a CUBE or an OCTOHEDRON by cutting off the corners in such a way that the new vertices lie at the midpoints of the original edges.

cuboid, *adj*. **1**. shaped in some respect like a CUBE, especially having rectangular but not necessarily square faces.

2. (*as substantive*) a geometric solid of which the six faces are mutually perpendicular rectangles; a rectangular PARALLELEPIPED.

cumulant, *n*. (*Statistics*) any of the coefficients, κ_n, of $t^n/n!$ in the TAYLOR SERIES of the natural logarithm of the MOMENT GENERATING FUNCTION, $M_X(t)$, of a random variable. κ_1 is the MEAN and κ_2 is the VARIANCE.

cumulative distribution function (abbrev. **cdf**), *n*. (*Statistics*) a function defined on the SAMPLE SPACE of a distribution and taking as its value at each point the probability of the random variable being less than or equal to that argument; the function $F(x) = P(X \leq x)$, where X is the random variable, that is the sum or integral of the PROBABILITY DENSITY FUNCTION of the distribution.

cumulative frequency, *n*. (*Statistics*) the frequency of occurrence of all values less than each given value of a random variable, equal to the sum of the frequencies of each value of the variable less than that given value.

cup, *n*. the symbol for set UNION, written $S \cup T$ or $\bigcup_i S_i$. Compare CAP.

curl or **rotation,** *n*. a vector quantity, written $\nabla \times \mathbf{A}$, curl$\mathbf{A}$, or rot$\mathbf{A}$, associated with a VECTOR FIELD that is the VECTOR PRODUCT of the operator

$$\nabla = \mathbf{i}\frac{\partial}{\partial x} + \mathbf{j}\frac{\partial}{\partial y} + \mathbf{k}\frac{\partial}{\partial z}$$

with a three-dimensional vector function $\mathbf{A}$, where $\mathbf{i}, \mathbf{j}, \mathbf{k}$ are mutually perpendicular unit vectors, and $\partial/\partial x$, etc. are the PARTIAL DERIVATIVES of $\mathbf{A}$.

current, *n*. (*Electromagnetism*) the derivative, with respect to time, of the quantity of charge crossing a surface. When the rate at which charge is flowing varies over the surface then one may define the *current density* by the relationship

$$i = \int \mathbf{j} \cdot d\mathbf{S},$$

where i is the current and $\mathbf{j}$ the current density through a surface $\mathbf{S}$.

current configuration, *n*. (*Continuum mechanics*) the CONFIGURATION of a BODY under a MOTION at a given time. See also SPATIAL DESCRIPTION.

current density, see CURRENT.

curtate trochoid, see TROCHOID.

curvature, *n*. **1**. the rate of change of inclination of the tangent to a curve relative to the length of arc; the change per unit length measured as the limit as that length tends to zero. If $y = f(x)$, the curvature of $f(x)$ is

$$\frac{y''}{\left(1 + (y')^2\right)^{3/2}}$$

where y' and y'' are respectively the first and second derivatives of the function. The curvature is positive if the curve is *concave up*, and negative if it is *concave down*.

2. also called **first curvature**. (at a point of a space curve) the magnitude of the CURVATURE VECTOR, denoted ρ or κ.

curvature vector, *n*. (of a space curve) the derivative of a UNIT TANGENT VECTOR with respect to ARC LENGTH; the second derivative of the position vector of the curve when parametrized by arc length; the product of the CURVATURE and the unit vector in the NORMAL direction. See FRENET FORMULAE.

curve, *n.* another word for ARC (sense 1a), especially when arcs are required to be CONTINUOUSLY DIFFERENTIABLE.

curvilinear, *adj.* **1.** consisting of, or characterized by, a curved line.

2. (of a set of coordinates) determined by or determining a system of three usually ORTHOGONAL surfaces; in a system of Cartesian coordinates, these are the coordinate planes.

curvilinear integral or **line integral,** *n.* **1a. curvilinear integral of the first kind.** the INTEGRAL denoted

$$\int_C \phi(\mathbf{x})\, dx_i,$$

of a SCALAR FIELD, $\phi(\mathbf{x})$, along a CURVE C in n–dimensional EUCLIDEAN SPACE, with respect to the ith position variable; if C can be represented as $\mathbf{x}(t)$, in terms of a PARAMETER t in the interval $[a, b]$, then it is equal to

$$\int_a^b \phi(\mathbf{x}(t))\, x_i(t)\, dt.$$

For example, in two dimensions, the curvilinear integral of a function $F(x, y)$ along $y = f(x)$ from C to D is written

$$\int_{CD} F(x, y)\, dx,$$

and is equal to

$$\int_a^b F[x, f(x)]\, dx.$$

where C is the point $(a, f(a))$ and D is $(b, f(b))$; similarly for y or for more variables.

b. curvilinear integral of the second kind. a LINEAR COMBINATION of curvilinear integrals of the first kind; for example, if C is a curve and ϕ and $\mathbf{F}$ are a SCALAR FIELD and a VECTOR FIELD respectively, then the standard notations for curvilinear integrals of the second kind are defined as follows:

$$\int_C \phi\, d\mathbf{x} = \sum_{i=1}^n \mathbf{e}_i \int_C \phi\, dx_i;$$

$$\int_C \mathbf{F} \cdot d\mathbf{x} = \sum_{i=1}^n \int_C F_i\, dx_i;$$

$$\int_C \mathbf{F} \times d\mathbf{x} = \sum_{i=1}^3 \sum_{j=1}^3 \sum_{k=1}^3 \mathbf{e}_k\, \varepsilon_{ijk} \int_C F_i\, dx_j;$$

$$\int_C \mathbf{F} \otimes d\mathbf{x} = \sum_{i=1}^n \sum_{j=1}^n \mathbf{e}_i \otimes \mathbf{e}_j \int_C F_i\, dx_j.$$

2a. the INTEGRAL of a SCALAR FIELD ϕ along a curve C with respect to ARC-LENGTH; that is if C can be represented as $\mathbf{x}(s)$, for $0 \le s \le l$, where s is the arc–length parameter, then the CURVILINEAR INTEGRAL

$$\int_C \phi \, ds$$

of ϕ with respect to s is defined as

$$\int_0^1 \phi \, (x(s)) \, ds.$$

For example, in two dimensions,

$$\int_{CD} F(x, y) \, ds \;=\; F[x, f(x)] \sqrt{1 + \left(\tfrac{dy}{dx}\right)^2} \; dx.$$

Compare SURFACE INTEGRAL.

cusp, *n.* a point at which two branches of a curve meet, and at which the limits of the tangent on each branch coincide. It is a *cusp of the first kind* if the two branches are at opposite sides of the common tangent as in (a) in Fig. 106, and a *cusp of the second kind* if they are on the same side, as in (b). It is a *double cusp* or point of OSCULATION if the curves extend on both sides of the cusp.

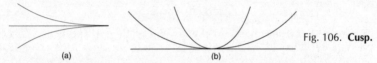

Fig. 106. **Cusp.**

(a)   (b)

(a) A simple cusp of the first kind.   (b) A double cusp of the second kind.

cut, *vb.* **1.** to remove a portion of the complex plane, leaving a *cut plane*, so that some function has an ANALYTIC BRANCH in the remaining region; thus the PRINCIPAL VALUE of the LOGARITHM lies in the cut plane excluding $]-\infty, 0]$.

2. (*Topology*) to remove a point, a *cut point*, from a set so that the remainder of the set is DISCONNECTED. For example, the interval $(-1, 1)$ is disconnected by cutting zero. See also CONNECTED.

3. (*Optimization*) to remove a portion of the FEASIBLE SET by adding a linear inequality, called a *cutting plane*, as in INTEGER PROGRAMMING so as to remove the current estimate of the solution while maintaining all possible solutions.

$-n.$ **4.** see DEDEKIND CUT.

5. (*Logic*) a rule of ELIMINATION in SEQUENT CALCULUS, according to which A, C $\vdash$ D can be derived from A $\vdash$ B and B, C $\vdash$ D.

6. (in a network) see NETWORK CUT.

cycle, *n.* **1.** a PERMUTATION in which one subset of elements are moved cyclically and the remainder are not moved, such as

$$\langle\, 1, 2, 3, 4, 5 \,\rangle \;\to\; \langle\, 1, 4, 3, 2, 5 \,\rangle.$$

If γ is a cycle that permutes exactly l elements, then l is the *length* of the cycle, and γ has order l, that is $\gamma^l = e$, where e is the identity permutation. Every permutation has a unique factorization as a product of disjoint cycles. See also ALTERNATING GROUP, PERMUTATION.

2. a SIMPLE CLOSED PATH in a graph.

cycle pattern, *n.* a representation of the unique decomposition of a PERMUTATION as a product of disjoint CYCLES. If γ is a permutation of n symbols, and the decomposition consists of λ_j cycles of length j for each j between 1 and n, then the cycle pattern is written

$$1^{\lambda_1} 2^{\lambda_2} \ldots n^{\lambda_n}.$$

Two elements of the permutation group S_n are conjugate if and only if they have the same cycle pattern, and the number of permutations with a given cycle pattern is

$$\frac{n!}{(1^{\lambda_1})(2^{\lambda_2})\ldots(n^{\lambda_n})(\lambda_1)!(\lambda_2)!\ldots(\lambda_n)!}$$

cyclic, *adj.* **1.** having CYCLES. Compare ACYCLIC.

2. (of a polygon) such that all the vertices lie on a circle. See for example CYCLIC QUADRILATERAL.

cyclic group, *n.* a GROUP all of the members of which are powers of a given element, the *generator*, such as the integers modulo n under addition, for which $n-1$ is a generator. Any subgroup of a cyclic group is also cyclic.

cyclic permutation, *n.* a PERMUTATION that advances all the elements of a finite sequence by the same number of places MODULO the length of the sequence.

cyclic quadrilateral, *n.* a quadrilateral inscribed in a circle, so that all its vertices lie on the circumference. A cyclic quadrilateral has opposite angles that are supplementary, so that in Fig. 107, for example, the angles at A and C total 180°.

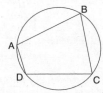

Fig. 107. **Cyclic quadrilateral**.

cycling, *n.* (*Numerical analysis*) the behaviour of an algorithm when the same approximate solution recurs repeatedly. It can happen in NEWTON'S METHOD for finding a zero of a real function that every second value produced is the same. See the SIMPLEX METHOD.

cycloid, *n.* the curve described by a point on (or rigidly connected to) the circumference of a circle as it rolls without slipping on a straight line. For example, a *common cycloid* is drawn by a point, such as O in Fig. 108 overleaf, that lies on the circumference of the circle; P_1 and P_2 are other positions of the point. The cycloid is *extended* if the generating point is outside the circle, and *contracted* if it is inside. If the term is used without qualification, usually the common cycloid is intended; the terms TROCHOID and *roulette* are sometimes used for the more general concept, and *prolate* and *curtate trochoids* are extended and contracted cycloids (or vice versa: there is no consistency about which is which). A cycloid has parametric equations

$$x = r(\theta - \sin\theta); \quad y = r(1 - \cos\theta)$$

Compare EPICYCLOID, HYPOCYCLOID.

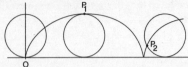

Fig. 108. A common **cycloid**.

cyclometric function, n. another word for TRIGONOMETRIC FUNCTION.

cyclosymmetric, see SYMMETRIC FUNCTION.

cyclotomic, *adj.* of or concerning n^{th} ROOTS OF UNITY. The *cyclotomic equation* is $z^n - 1 = 0$. The nth *cyclotomic polynomial* is the polynomial the roots of which are all the primitive n^{th} roots of unity, and thus, for prime n, is the cyclotomic equation of degree n. A *cyclotomic field* is an EXTENSION FIELD of the rationals obtained by adjoining a root of unity.

cylinder, n. **1.** commonly, a solid bounded by two parallel planes and by the locus of a line that moves round a fixed closed curve at a fixed angle to the planes, as in Fig. 109; usually, if nothing else is specified, it is a *right circular cylinder*; that is, the curves are circles and the line is perpendicular to the parallel planes. The volume of any cylinder is the area of its base multiplied by the perpendicular distance between the planes.

2. geometrically, a surface formed by a line segment moving round a closed plane curve at a fixed angle to its plane.

3. (in a CARTESIAN PRODUCT) a set that is the direct product of a set and a non-trivial vector space.

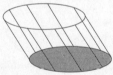

Fig. 109. **Cylinder**.

cylindrical coordinates, n. a set of coordinates that locate a point in space by the SPHERICAL COORDINATES of the foot of a perpendicular from it to a coordinate plane and its height above that plane; for example, the point (x, y, z) has cylindrical coordinates (r, θ, z), where (r, θ) are the polar coordinates of (x, y). In Fig. 110, Q is the foot of the perpendicular from P to the x–y plane; the coordinates of P are then the length of the radius vector of Q, the anticlockwise angle between the x-axis and this vector, and the directed length of QP (that is, the height of P above the x–y plane). Compare SPHERICAL COORDINATES.

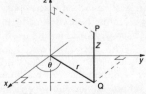

Fig. 110. The **cylindrical coordinates** of P are (r, θ, z).

cylindroid, n. a CYLINDER with ELLIPTICAL cross-section.

cypher, n. alternative UK spelling of (US) CIPHER.

d

d, 1. *symbol* concatenated with a variable or functional symbol, and originally denoting an INFINITESIMAL INCREMENT in that variable or function. The notation now survives in compound expressions: in particular, if $y = f(x)$ denotes a real function, then dy/dx denotes the DERIVATIVE of y with respect to x, which is defined as the limit of the ratio of a small increment δx of x to the corresponding increment $\delta y = f(x + \delta x) - f(x)$ of y, as δx tends to zero (see Fig. 111). The notation dx for a DIFFERENTIAL, and $\int f(x)$ dx for INTEGRATION, where dx denotes an ELEMENT are also derived in this way. The corresponding notation for PARTIAL DERIVATIVES is ∂x. See also DIFFERENTIATION. Compare DELTA.

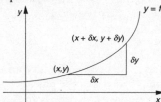

Fig. 111. dx/dy is the limit of $\delta x/\delta y$.

2. *abbrev. for* DECI–, used in symbols for fractions of the physical units of the SYSTEME INTERNATIONAL.

3. (*Number theory*) the DIVISOR FUNCTION, $d(n)$.

D, *n.* **1.** the number 13 in HEXADECIMAL notation.

2. the ROMAN NUMERAL for 500.

da, *abbrev. for* DEKA–, used in symbols for multiples of the physical units of the SYSTEME INTERNATIONAL.

D'Alembert's ratio test, see RATIO TEST.

damped oscillation, *n.* an OSCILLATION in which the AMPLITUDE decreases over time, as in

$$f(x) = e^{-x/2} (\sin 2\pi x - \cos 2\pi x)$$

the graph of which is shown in Fig. 112 below; note that the period of the oscillation in this example does not change.

Fig. 112. **Damped oscillation.**

135

Daniell integration, *n.* an approach to LEBESGUE INTEGRATION using extensions of positive LINEAR FUNCTIONALS. (Named after *P. J. Daniell* (1886–1946).)

Darboux integral, *n.* the limit of a DARBOUX SUM as the lengths of the subintervals tends to zero. (Named after *Jean Gaston Darboux* (1842–1917), French analyst and analytical geometer.) See UPPER INTEGRAL, LOWER INTEGRAL.

Darboux property, see BOLZANO'S THEOREM.

Darboux sum, *n.* an UPPER SUM or LOWER SUM for an integral.

data, *n. plural.* the information, usually numerical values, gathered from an experiment, observation, survey or other study, for example of a RANDOM SAMPLE drawn from an underlying POPULATION. Data are DISCRETE if the underlying population is finite or countably infinite and are CONTINUOUS if the underlying population is a CONTINUUM such as an interval on the real line. NOMINAL DATA are date that are descriptive rather than numerical or quantitative, and have no natural order.

datum, *n.* **1.** a single item of DATA; the information obtained from a single observation.

2. a given value, such as the origin of a coordinate system, from which other data are measured. For example sea level is a geographical datum. (From Latin, *datum* = given.)

decade, *n.* any sequence of ten successive terms.

decagon, *n.* a polygon with ten sides.

decahedron, *n.* a solid with ten plane faces. No decahedron is REGULAR.

deci- (symbol **d**), *prefix* denoting a tenth part of any of the physical units of the SYSTEME INTERNATIONAL.

decidable, *adj.* (*Logic*) **1a.** (of a statement) able to be shown either to be true or to be false.

b. (of a WELL–FORMED formula of a given theory) either provable, or having a provable negation, in the given theory; equivalently, either the formula or its negation is a theorem. For example, the CONTINUUM HYPOTHESIS has been shown not to be decidable.

2. (of a formal theory) having the property that each theorem is RECURSIVE; equivalently, by CHURCH'S THESIS, that it is possible to determine by a mechanistic procedure whether or not any given WELL–FORMED FORMULA is a theorem. For example, sentential calculus is decidable, but predicate calculus is not. See DECISION PROCEDURE. See also COMPLETE.

decile, *n.* (*Statistics*) any one of the nine values of a RANDOM VARIABLE that divide its DISTRIBUTION into ten equal parts, so that the probability of a variable having a value between one decile and the next is $1/10$; the cumulative relative frequency of the nth decile is $10n\%$. The ninth decile is the value below which 90% of the population lie. See PERCENTILE.

decimal, *adj.* **1.** relating to or using powers of 10 or the base 10.

2. relating to or expressed in the PLACE–VALUE NOTATION with base 10. See DECIMAL SYSTEM.

3. (*as substantive*) a DECIMAL FRACTION.

4. (*informal*) relating to or expressed in PLACE–VALUE NOTATION similar to that of decimal fractions but to a different base. For example, $2\frac{5}{8}$ can be expressed as a binary 'decimal' as 10.101. See RADIX.

decimal fraction or **decimal,** *n*. a FRACTION written in the PLACE–VALUE NOTA-
TION with base 10. It is preceded by a dot, after which each successive digit
indicates a multiple of the successive negative powers of 10; thus 0.435 can
be expanded as

$$(4 \times 10^{-1}) + (3 \times 10^{-2}) + (5 \times 10^{-3}),$$

or, in VULGAR FRACTIONS,

$$\frac{4}{10} + \frac{3}{100} + \frac{5}{1000},$$

which equals 435 thousandths. In general, a decimal fraction is equal to the
simple fraction of which the numerator is the integer consisting of the digits following the decimal point and the denominator is 10 raised to the
power of the number of digits after the point. See also RECURRING DECIMAL.

decimalize, *vb*. to change (a number, quantity, etc.) to the DECIMAL SYSTEM
(sense 2). See also METRICATION.

decimal notation, *n*. **1**. another name for the DECIMAL SYSTEM (sense 1).

2. a loose term for PLACE–VALUE NOTATION. See DECIMAL (sense 4).

decimal place, *n*. **1**. the position of a digit after the DECIMAL POINT in a DECI-
MAL FRACTION, so that the nth digit has a PLACE VALUE of 10^{-n}. For example, in
0.025, 5 is in the third decimal place.

2. the number of digits after the DECIMAL POINT in a DECIMAL FRACTION. For
example, 0.025 is expressed to three decimal places. See also ACCURATE
(sense 2).

decimal point, *n*. the dot placed between the integral and fractional parts of
a number expressed in the notation of the DECIMAL SYSTEM.

Note: In accordance with international convention, the decimal point should
be represented by either a dot or a comma on the line (as, for example, in
4.5 or 4,5), and digits should be grouped in threes without further punctua-
tion on either side; thus one should write, for example, 12 345 678.901 23.
However, some people, especially in the UK, have the practice of writing
commas between the groups (as, for example, in 12,345.67). In some European countries the precisely contrary convention prevails, so that '12,345.67'
would there represent $12 + (34567 \times 10^{-5})$.

decimal system, *n*. **1**. the PLACE–VALUE system with BASE 10 that constitutes
the numerical notation in general use, in which numbers are expressed as
sequences of ARABIC NUMERALS 0 to 9 in which each successive digit to the
left and right of the unit position indicates a multiple of the successive
(respectively positive and negative) powers of 10; thus 123.45 can be expanded as

$$(1 \times 10^2) + (2 \times 10^1) + (3 \times 10^0) + (4 \times 10^{-1}) + (5 \times 10^{-2}).$$

2. a system of measurement such as the METRIC SYSTEM, in which the units
are related by multiples of ten.

decision problem, *n*. (*Logic*) the problem, for any given theory, of whether
there exists a DECISION PROCEDURE for questions such as whether an expression is a theorem. See also SATISFIABILITY PROBLEM, TURING MACHINE.

decision procedure, *n*. (*Logic*) an algorithm able to determine mechanistically whether or not any given WELL–FORMED FORMULA of a formal theory is

a theorem. Not all CONSISTENT theories possess such a procedure; for example, truth-tables provide a decision procedure for sentential calculus, but predicate calculus can be proved to have no such procedure. Those for which one does exist are called DECIDABLE.

decision theory, *n.* (*Statistics*) the study of strategies for decision making under conditions of uncertainty in such a way as to maximize the EXPECTED UTITLITY. See also GAME THEORY.

declination or **depression,** *n.* the angle measured in a clockwise direction from the positive direction of the *x*-axis to a given line; for example the declination of OP in Fig. 113 is θ. The declination of a line thus has opposite sign to the conventional anticlockwise measurement of angles. Compare INCLINATION.

Fig. 113. θ is the **declination** of OP.

decompose, *vb.* to give a DECOMPOSITION for a given object or quantity.

decomposition, *n.* the expression of a given object or quantity in terms of a number of simpler components. For example, a number has a decomposition as a product of primes, a set as a canonical union of appropriate disjoint subsets, a vector as the resultant of orthogonal components, and a signed measure as a difference of positive measures.

decreasing, *adj.* (of a function of a single variable) having the property (locally or globally) that when $x > y$, then the value of the function at x is less than its value at y, $f(x) \leq f(y)$. If $f(x) < f(y)$ the function is *strictly decreasing*. The function shown in Fig. 114 is decreasing on the closed interval $[a, b]$, and strictly decreasing on the open interval $]a, b[$. See also MONOTONE.

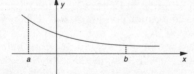

Fig. 114. A **decreasing** function.

decrement, *n.* a negative INCREMENT.

decrypt, *vb.* to recover data from a code. See CODING THEORY.

decryption, *n.* recovering data from a code. See CODING THEORY.

Dedekind, Julius Wilhelm Richard (1831–1916), German mathematician who studied under Gauss and Dirichlet. He defined the real numbers by means of the DEDEKIND CUT, and originated the concepts of RING (in the sense of DEDEKIND RING) and UNIT, and the definition of IDEAL in algebra.

Dedekind-complete, *adj.* (of a PARTIALLY ORDERED set) another term for COMPLETE (sense 2).

Dedekind cut, *n.* a PARTITION of a sequence into two disjoint subsequences, all the members of one of which are less than all those of the other. This device is used to define the IRRATIONAL NUMBERS in terms of pairs of sequences of rationals, as opposed to using a metric COMPLETION. For example √2 is defined as the pair

$$\langle \{ x : x^2 > 2 \}, \{ x : x^2 < 2 \} \rangle.$$

The corresponding construction in a lattice gives rise to its *Dedekind completion*. See COMPLETION.

Dedekind ring, *n.* an INTEGRAL DOMAIN in which every non-zero ideal is INVERTIBLE; it follows that every non-zero ideal can be written uniquely as a product of PRIME IDEALS.

de dicto, *adj.* (*Logic*) relating to the expression of a belief, possibility, etc., rather than to the entities referred to. For example, *the number of planets is the number of satellites of the Sun*, is necessary *de dicto*, as its truth is not dependent upon which number in fact that is. Compare *DE RE*.

deducibility, *n.* the property of being able to be deduced as the conclusion of a VALID argument within the appropriate system.

deduction, *n.* **1.** the process of reasoning typical of mathematics and logic, in which a CONCLUSION follows necessarily from given PREMISES so that it cannot be false when the premises are true. See also SYNTHETIC PROOF.
2. a systematic method of constructing ARGUMENTS of this type, especially one amenable to FORMALIZATION and study by the science of LOGIC.
3. an argument that is or purports to be of this type. In this sense one says, for example, 'his deduction of John's guilt was faulty'.
4. the conclusion of such an argument. It is in this sense that one says 'his deduction was that John was guilty'.
Compare INDUCTION. See also VALIDITY.

deduction theorem, *n.* (*Logic*) the property of many formal systems that if an argument is VALID, then its premises imply its conclusion; that is, it is true that if the premises of the given argument are true then so is its conclusion. Formally, the conditional statement is derived from an argument by taking the conjunction of the premises as antecedent and the conclusion as consequent. This means that to prove the truth of a conditional statement it is sufficient to prove its consequent as a conclusion from its antecedent as premises. For example, to prove the truth of

If all men are mortal and Socrates is a man, then Socrates is mortal.

it suffices to prove the validity of the argument:

All men are mortal.
Socrates is a man.
So Socrates is mortal.

See also CONDITIONALIZATION.

deductive, *adj.* of or relating to DEDUCTION.

def, *abbrev. for* DEFINITION.

defect, *n.* (of a SPHERICAL TRIANGLE) the difference between the sum of the internal angles of the given triangle and 3π.

defective number, another term for DEFICIENT NUMBER.

deferred approach to the limit, another term for RICHARDSON EXTRAPOLATION.

deficiency, *n.* (of a vector sub-space) another term for CODIMENSION.

deficient number or **defective number,** *n.* any natural number that exceeds the sum of its proper divisors. Clearly, any prime number is deficient. Compare ABUNDANT NUMBER, PERFECT NUMBER.

definiendum, *n.* (*Latin*) the expression of which a DEFINITION is required or given. Compare DEFINIENS.

definiens, *n.* (*Latin*) the expression in terms of which a DEFINITION, especially an explicit definition, is given. Compare DEFINIENDUM.

definite, see POSITIVE SEMI–DEFINITE.

definite description, *n.* (*Logic*) **1a.** an expression capable of having a unique reference; for example, 'the woman in white' or 'Rosemary's baby'.

b. an analogous plural expression; for example, 'the dogs of war'.

2. theory of definite descriptions. the analysis, proposed by Bertrand Russell, of singular definite descriptions in which a sentence of the form *the F is G* is said to be equivalent to

there is one and only one F and it is G,

which is abbreviated $G[(\imath x)Fx]$. This can be defined in terms of the UNIQUE QUANTIFIER as

$$G[(\imath x)Fx] \equiv (\exists! x)Fx \ \& \ (\forall x)(Fx \to Gx)$$

or in terms of the EXISTENTIAL QUANTIFIER as

$$G[(\imath x)Fx] \equiv (\exists x)((Fx \ \& \ (\forall y)(Fy \to x = y)) \ \& \ Gx).$$

definite integral, *n.* **1.** (informally) the expression for the evaluation of the INDEFINITE INTEGRAL of a positive function between two LIMITS OF INTEGRATION, representing the area between the graph of the given function and the *x*-axis between these values of *x*. If the given limits of integration are *a* and *b*, and the interval $[a, b]$ is divided into *n* equal subintervals of width δx, then, as shown in Fig. 115, the lightly shaded region is the limit as *n* tends to infinity of the sum of the areas of the rectangles constructed on each subinterval with height $f(x)$ for some *x* in that subinterval; these are known as *elements* of area, of which the deeper shaded rectangle is an example. This limit is written

$$\int_a^b f(x) \ \mathrm{d}x \ \text{ or } \ \int_a^b f(x) \ \mathrm{d}x$$

and is evaluated as $F(b) - F(a)$, where $f(x)$ is the given function, $x = a$ and $x = b$ are the limits of integration, and $F(x)$ is the indefinite integral $\int f(x) \ \mathrm{d}x$. See also FUNDAMENTAL THEOREM OF CALCULUS.

2. the actual value of such an expression.

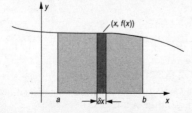

Fig. 115. **Definite integral**. See main entry.

3. (more properly) **a.** the definite integral is said to exist in the sense of the RIEMANN INTEGRAL, or otherwise, if the appropriate limit of DARBOUX SUMS exists. The definite integral of a continuous positive function between a and b then gives the area enclosed by the curve and axis between those bounds. For general continuous functions, the integral is the algebraic sum of the enclosed areas above and below the x-axis, where the latter has negative sign.

b. more generally, the definite integral of a function is said to exist if the LEBESGUE INTEGRAL of the product of the function with the CHARACTERISTIC FUNCTION of that interval exists; that is, provided that the function can be approximated by measurable simple functions. If the function is continuous, then this gives the Riemann integral as the limit of DARBOUX SUMS.

definition (abbrev. **def**, **defn** or **df**), *n.* a precise statement of the meaning of an expression (the *definiendum*) in terms equivalent to it. An *explicit definition* is an identity between the definiendum and another expression (the *definiens*) all the terms of which are already well understood; for example, the empty set can be defined in terms of negation, identity, and set membership by the definition

$$\varnothing =_{df} \{x : x \neq x\}.$$

This permits the definiendum to be replaced by the definiens wherever it occurs. An *implicit* or *contextual definition* is an identity between complex expressions in only one of which the definiendum occurs, such as

$$G[(\iota x)Fx] =_{df} (\exists x)((Fx \mathbin{\&} (\forall y)(Fy \rightarrow x = y)) \mathbin{\&} Gx)$$

in Russell's theory of definite descriptions, and so does not specify a definiens that may be directly substituted for the definiendum. Definitions are often written using the EQUALS SIGN either with 'df' subscript to the right, with a colon to the left, with a small triangle superscript, or simply with the word 'definition' or its abbreviation 'defn' in the same line; alternatively, the equivalence sign, $\equiv$, is used. The definiendum is usually written on the left and the definiens on the right. See also ASSIGNMENT.

defn, *abbrev. for* DEFINITION.

deformation or **continuous deformation,** *n.* a transformation the effect of which is to change the shape of a figure by stretching but not tearing: $T(p)$ from A to B is a deformation if there is a CONTINUOUS function $F(p, t)$, for t between 0 and 1, such that $F(p, 0) = p$ and $F(p, 1) = T(p)$. T is then said to *deform* A into B.

deformation gradient, *n.* (*Continuum mechanics*) a measure of the extent of the DEFORMATION caused by the MOTION of a BODY; formally, the GRADIENT, with respect to the reference CONFIGURATION, of the positions of the points of the current configuration of a body. Compare DISPLACEMENT GRADIENT.

degeneracy, *n.* **1.** (*Statistical physics*) the number of DEGENERATE STATES of a system with a given total energy.

2. see SIMPLEX METHOD.

degenerate, *adj.* a limiting case of some type of entity that is equivalent to some simpler type, often obtained by setting some coefficient or parameter to zero. For example, a point is a degenerate circle.

degenerate conic, *n.* a CONIC SECTION obtained when the intersecting plane includes the vertex of the cone; these are a point, a (repeated) line, and a pair of intersecting lines. Equivalently, the degenerate conics have equations that, when regarded as QUADRATIC FORMS in x and y equated to zero, have SINGULAR matrices.

degenerate state, *n.* (*Statistical physics*) one of a set of distinct STATES of a system that are independent solutions of the equation of the system for a given value of the total energy of the system. The number of such solutions is known as the *degeneracy* of that total energy.

degree, *n.* **1. degree of arc.** a measure of angle equal to one 360th part of the angle traced out by one full revolution of a line segment around one of its endpoints, written °. One degree is divided into 60 MINUTES, or 3600 SECONDS. Compare RADIAN.

2. the highest power or sum of powers in any term of a given polynomial or algebric equation, or the sum of powers in any one term. For example, $x^4 + 3x^2 - x$ and xy^2z are both of the fourth degree. See also QUADRATIC, CUBIC, QUARTIC, etc.

3. the greatest power of the derivative of highest order in a DIFFERENTIAL EQUATION. For example, $D_3^2 + D_2^3 + D_1^4 = 0$, where D_i is the ith derivative, is a second-degree differential equation. Compare ORDER.

4. (of a REPRESENTATION of a GROUP) the degree of the GENERAL LINEAR GROUP over a FIELD into which the representation is a HOMOMORPHISM from the given group.

5. (of a VERTEX in GRAPH) the number of coincident EDGES at the given vertex. In a NETWORK or DIGRAPH, entering arcs (the *in-degree*) and exiting arcs (the *out-degree*) are counted separately.

6. (*Topology*) another word for GENUS.

7. topological degree. (for a continuously differentiable function, *f*, on Euclidean space) the excess of the number of points of a given region, *G*, in $f^{-1}(a)$ at which the JACOBIAN is positive over those at which it is negative; this is referred to as the *degree of f at a in region G*, and written $D[a, G, f]$. More generally, this extends to a number (known as *Brouwer's form of the degree* for a continuous function *f* defined on *G*) that is a TOPOLOGICAL INVARIANT that, when non-zero, ensures that $f(x) = a$ has solution in *G*.

8. (of an EXTENSION FIELD with respect to a BASE FIELD) the dimension of the extension viewed as a VECTOR SPACE over the base field.

9. (of membership of a set) see FUZZY SET THEORY.

degrees of freedom, *n.* **1a.** the minimum number of parameters necessary to describe completely a state or property of a system.

b. (*Mechanics*) the minimum number of position variables required to describe a given continuous or discrete set of PARTICLES. A particle has three degrees of freedom, and a rigid body has six.

2. (*Statistics*) the number of independent unrestricted RANDOM VARIABLES constituting a STATISTIC; the number of degrees of freedom is usually one less than the number of variables.

3a. the only parameter of the family of CHI–SQUARE DISTRIBUTIONS, T–DISTRIBUTIONS, F–DISTRIBUTIONS, etc.

b. the number of observations in a GOODNESS OF FIT test STATISTIC less the number of PARAMETERS estimated in the model.

c. formally, the RANK of the matrix associated with a SUM OF SQUARES. See SUM OF SQUARES THEOREM.

deka- (symbol **da**), *prefix* denoting a multiple of ten times any of the physical units of the SYSTEME INTERNATIONAL.

del, see DIFFERENTIAL OPERATOR.

deleted neighbourhood, *n.* another name for PUNCTURED NEIGHBOURHOOD.

Delian altar problem, *n.* see DOUBLING THE CUBE.

delta, *n.* **1.** see EPSILON–DELTA NOTATION.

2. a finite increment, δx or Δx, in the value of a variable, as contrasted with the infinitesimal increment dx. See also d, DIFFERENTIATION.

3. the symbol $\partial/\partial v$ for a PARTIAL DERIVATIVE.

4. see KRONECKER'S DELTA.

delta function, see DIRAC DELTA FUNCTION.

deltoid, *n.* a non-convex quadrilateral with two pairs of adjacent equal sides. Compare KITE.

de Moivre, Abraham (1667–1754), French-born analyst and probability theorist who was educated in Belgium and settled in England after fleeing from French persecution of the Huguenots. He worked with Halley and Newton, and was elected to the Royal Society of London, and the Paris and Berlin Academies, but never obtained a permanent post; indeed his interest in probability was not unrelated to his work as a consultant on both insurance and gambling.

de Moivre's formulae, *n.* the identities,

$$(\cos x + i\sin x)^n = \cos nx + i\sin nx,$$

valid for all complex x and n, that are an immediate consequence of EULER'S FORMULA, $\exp(ix) = \cos x + i\sin x$. Equating REAL PARTS of these expressions provides a simple way of expressing $\cos nx$ as a polynomial in $\sin x$ and $\cos x$.

demonstration, *n.* a PROOF, especially one making explicit all the assumptions, rules, and steps of the derivation of a mathematical theorem.

De Morgan, Augustus (1806–71), Indian-born British analyst, probability theorist and logician, who was the first Professor of Mathematics at University College, London, and the first President of the London Mathematical Society. He also addressed philosophical questions, clarified the nature of mathematical INDUCTION, and generalized the notion of ALGEBRA, as well as initiating the revision of traditional ARISTOTELIAN LOGIC.

De Morgan's laws, *n.* the theorems in a BOOLEAN ALGEBRA, such as set theory or sentential calculus, that state the DUALITY of the binary operations; in sentential logic the laws are:

$$-(P \,\&\, Q) \;\dashv\vdash\; -P \vee -Q$$
$$-(P \vee Q) \;\dashv\vdash\; -P \,\&\, -Q,$$

and in set theory:

$$C(S \cap T) = C(S) \cup C(T)$$
$$C(S \cup T) = C(S) \cap C(T),$$

where $C(S)$ denotes the COMPLEMENT of S, or their equivalents.

denary, *adj.* calculated or based on ten; DECIMAL.

denominator, *n.* the DIVISOR in a simple fraction indicating the size of each of the parts whose number is given by the NUMERATOR. In a/b, the denominator is b.

dense, *adj.* **1.** (of a set in an ORDERED space) having the property that between any two COMPARABLE elements a third can be interposed. Thus the rational numbers are dense, since for any rationals a and b, $\frac{1}{2}(a + b)$ lies between them and is also a rational.

2. (of a set in a TOPOLOGY) having a CLOSURE that contains a given set. More simply, one set is *dense in another* if the second is contained in the closure of the first. For example, the rationals are dense in the reals, since the latter are contained in the closure of the former.

3. (of a matrix) see SPARSE.

dense in itself, *adj.* (of a set in a TOPOLOGICAL SPACE) such that every PUNCTURED NEIGHBOURHOOD of every element of the set intersects the set.

density, *n.* **1.** the property of being DENSE.

2. also called **density function.** (*Statistics*) another name for PROBABILITY DENSITY FUNCTION.

3. (of a sequence of positive integers) see SCHNIRELMANN DENSITY, UNIFORM DISTRIBUTION.

4. (*Continuum mechanics*) **a.** the MASS per unit volume; more formally, the SCALAR FIELD $\rho(\mathbf{x}, t)$, unique almost everywhere, such that the MASS of a SUB-BODY is given by the integral

$$\int \rho(\mathbf{x}, t) \, dv,$$

over the volume of the current CONFIGURATION of the sub-body.

b. see INTERNAL ENERGY DENSITY, BODY FORCE DENSITY. See also CONTACT FORCE, CONTACT TORQUE.

density of a point, see METRIC DENSITY.

denumerable, **enumerable**, or **numerable,** *adj.* capable of being put into ONE-TO-ONE CORRESPONDENCE with the positive integers; COUNTABLE but infinite. For example, the rational numbers are denumerable, but the reals are *non-denumerable.*

deontic logic, *n.* the branch of MODAL LOGIC that seeks to represent the relationships of the concepts of obligatoriness and permissibility. Compare ALETHIC, EPISTEMIC.

dependent, *adj.* **1.** determined by another value. See DEPENDENT VARIABLE.

2. (of a system of linear equations, vectors, etc.) having the property that any one may be expressed as a LINEAR COMBINATION of the remainder. See LINEAR DEPENDENCE.

3. see STATISTICAL DEPENDENCE.

dependent variable, *n.* **1.** a variable whose value is determined by that taken by the INDEPENDENT VARIABLES; for example, in $y = f(x)$, y is the dependent variable.

2. (*Statistics*) also called **response variable, predicted variable.** the variable whose values are observed for different values of the INDEPENDENT VARIABLE.

depression, another term for DECLINATION.

derangement, *n.* a PERMUTATION such that no element occurs in its original position. The number of derangements of *n* objects is precisely

$$n! \left[1 - 1 + \frac{1}{2} - \frac{1}{6} + \dots + \frac{(-1)^n}{n!} \right].$$

For example, there are nine derangements of four objects. Asymptotically, the proportion of derangements approaches $1/e$.

de re, *adj.* (*Logic*) relating to an actual individual mentioned, rather than to the expression of a belief, possibility, etc., concerning it. For example, *the number of planets is a perfect square*, is necessary *de re*, since its truth is dependent upon which number that in fact is. Compare DE DICTO.

derivation, *n.* **1.** the act or process of deducing some expression from given others, or a record of the steps of this process.

2. the operation of finding the DERIVATIVE of a function.

3. an ADDITIVE mapping on a COMMUTATIVE RING that satisfies the analogue of the PRODUCT RULE.

derivation rules, *n.* the TRANSFORMATION RULES or RULES OF INFERENCE of a system of FORMAL LOGIC by which THEOREMS may be recursively derived from AXIOMS. Compare FORMATION RULES.

derivative or **differential coefficient,** *n.* **1.** (for a function $f(x)$ at the argument x) the LIMIT of the DIFFERENCE QUOTIENT

$$\frac{f(x + \Delta x) - f(x)}{\Delta x}$$

as the INCREMENT Δx tends to 0. For functions of a single variable, if the left- and right-hand limits exist and are equal, it is the GRADIENT of the curve at x, and is the limit of the gradient of the chord joining the points $(x, f(x))$ and $(x + \Delta x, f(x + \Delta x))$, as shown in Fig. 116.

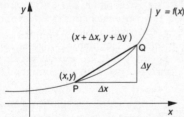

Fig. 116. The **derivative** dx/dy is the limit of $\Delta x/\Delta y$ as Q approaches P.

The function of x defined as this limit for each argument x is the *first deriva tive* of $y = f(x)$; it is the RATE OF CHANGE of the value of the function with respect to the independent variable, and is written

$$\frac{dy}{dx}, \ f'(x), \ \text{or} \ D_x f(x),$$

while the ratio of differences of which this is the limit is written $\delta y/\delta x$. The process of extracting the derivative is called DIFFERENTIATION. For example, the first derivative of ax^n is anx^{n-1}. The *second derivative* is the first derivative of the first derivative, and is written

$$\frac{d^2 y}{dx^2}, \ f''(x), \ D^2 f(x), \ \text{or} \ D_{xx} f(x).$$

These definitions are readily extended to functions of several variables; see PARTIAL DERIVATIVE, COMPLETE DERIVATIVE.

2. the GATEAUX DERIVATIVE or FRECHET DERIVATIVE of a vector space mapping.

derivative test, see FIRST DERIVATIVE TEST and SECOND DERIVATIVE TEST.

derive, *vb.* to obtain a function by DIFFERENTIATION.

derived series, *n.* (*Group theory*) the series defined inductively starting with a given GROUP, G, by taking each $G^{(n+1)}$ to be the DERIVED SUBGROUP of the preceding member $G^{(n)}$. See also SOLUBLE GROUP.

derived set, *n.* the set of all the CLUSTER POINTS of a given set. The *second derived set* is the derived set of the derived set. This process can be continued transfinitely. The *Cantor–Bendixson theorem* asserts that on the real line this process terminates COUNTABLY, in that every closed set is the DISJOINT UNION of a PERFECT set and a FINITE or DENUMERABLE set.

derived subgroup, *n.* the subgroup generated by the set of COMMUTATORS of a given GROUP. The derived subgroup is a CHARACTERISTIC SUBGROUP, and is denoted G′. See also DERIVED SERIES.

Desargues' theorem, *n.* the theorem that if the lines joining corresponding vertices of two triangles in three-dimensional space pass through a common point (that is, if the triangles are perspective PROJECTIONS of one another) then the points of intersection of corresponding sides are COLLINEAR. In Fig. 117 (below), the two given triangles are ABC and KLM, and O is the common point of intersection of the lines AK, BL and CM; if one now extends the sides opposite the paired angles, their respective points of intersection are Y, Z and X, which are collinear. This does not hold if any pair of sides is parallel; in that case the two remaining points of intersection determine a line parallel to the parallel sides. It was the recognition that the generality of the theorem could be maintained by adding POINTS AT INFINITY to Euclidean space that led to the development of PROJECTIVE GEOMETRY. The DUAL form of this is exactly the converse: in this case X, Y and Z are the collinear points of intersection of the pairs of sides AB and KL, BC and LM, and AC and KM respectively; the lines joining the vertices opposite the paired sides are now again respectively CM, AK, and BL, which are coincident. This leads to a self-dual CONFIGURATION with ten points and ten lines, as shown in Fig. 117.

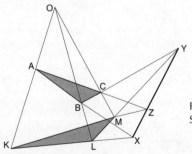

Fig. 117. **Desargues' theorem**.
See main entry.

(Named after the French pioneer of projective geometry, *Gerard Desargues* (1591–1661). He was an engineer who advised Cardinal Richelieu, met

Descartes during the siege of La Rochelle, was a member of Mersenne's circle in Paris, and taught Pascal. He wrote mainly on projective geometry (as well as a manual of musical composition), but in an obscure notation, so that the importance of his work was not appreciated, and indeed it was lost until a manuscript was rediscovered in the middle of the 19th century.)

Descartes, René (1596–1650), French scientist, philosopher, and mathematician who founded ANALYTIC GEOMETRY, and introduced EXPONENTIAL NOTATION, CARTESIAN COORDINATES, and methods of solving POLYNOMIAL EQUATIONS to mathematics. His work as a whole was ruled by the desire to systematize all knowledge as resting only on what is clearly self-evident, on the axiomatic model of Euclid's geometry, and thereby to achieve certainty. His method involved the suspension of belief in anything that could conceivably be doubted, and he then founded his edifice on the argument that, while doubting, one cannot doubt that one doubts, and therefore thinks; he expressed this single self-justifying proposition in his famous Latin phrase '*Cogito, ergo sum*' ('I think, therefore I am'). The adjective *Cartesian* is derived from the archaic spelling *Des Cartes*.

Descartes' rule of signs, *n.* a rule discovered by Descartes that the number of positive roots of a polynomial, counting multiplicity, is equal in parity to, and no greater than, the number of changes of sign of the coefficients of the polynomial; the number of negative roots is similarly related to the polynomial in $-x$. Zeros in the coefficient string are ignored. For example, $x^3 + x^2 - x - 1$ has one positive root; replacing x by $-x$ yields $-x^3 + x^2 + x - 1$ and shows that the original polynomial has two negative roots. Compare STURM SEQUENCE.

descendant, *n.* an element related to a given one by a chain of instances of a given RELATION; equivalently, in a TREE, a node further from the root than a given node, and such that there is a branch including both. For example, 5 is a descendant of 3 under the successor relation on the whole numbers.

descending chain condition, *n.* the condition on SUBMODULES that no *descending chain*

$$M_1 \supseteq M_2 \supseteq M_3 \supseteq \dots$$

has more than a finite number of distinct members; that is, for every such chain there is an n such that $M_n = M_m$ for all $m \geq n$. Equivalently, every nonempty set of submodules has a minimal element. Similar conditions are defined for RINGS, GROUPS, etc. See also ARTINIAN MODULE. Compare ASCENDING CHAIN CONDITION, MINIMUM CONDITION.

descent methods, *n.* the class of numerical optimization methods that proceed iteratively in order to reduce continually the value of some given func-tion, called the *descent function*, often in a predetermined direction, called the *descent direction*. See STEEPEST DESCENT, CONJUGATE GRADIENTS, QUASI–NEWTON METHOD.

describe, *vb.* to draw or to follow the shape of a line, curve or figure. For example, in a non-resistant medium, a projectile describes a parabola.

description, *n.* **1.** the act of drawing a line or figure, the tracing of such a figure by the path of some object.

2. an expression containing a predicate and capable of replacing a name as the subject of a sentence. See also DEFINITE DESCRIPTION.

descriptive geometry, *n.* the study of the PROJECTION of three-dimensional solids onto a plane surface in order to solve spatial problems by graphical methods.

descriptive statistics, *n.* the use of STATISTICS to summarize a set of known data in a clear and concise manner, such as in terms of its mean and variance, or diagrammatically by a histogram. Compare STATISTICAL INFERENCE.

design or **experimental design,** *n.* (*Statistics*) (for an experiment in which the experimenter has control over the EXPERIMENTAL CONDITIONS) a formal description of the constraints on the allocation of subjects to conditions. See also BLOCK DESIGN.

designated, *adj.* (*Logic*) (of a TRUTH VALUE in a VALUATION SYSTEM) functioning as the analogue of truth in a two-valued system. It is often convenient to regard all the designated values as species of truth and all the *anti-designated* values as species of falsehood, at which level the law of EXCLUDED MIDDLE may then hold, or else there may still remain a TRUTH–VALUE GAP between designated and anti-designated values. See MANY–VALUED LOGIC.

destructive dilemma, see DILEMMA.

detach, *vb.* (*Logic*) to derive (an unconditional statement) by MODUS PONENS; given the truth of a conditional together with that of its antecedent, to infer the unconditional truth of its consequent. For example, given

> *If Joan is on that train, she'll be late for her meeting*

and

> *Joan is on that train*

one can infer by *detachment* that

> *Joan will be late for her meeting.*

detachment, see MODUS PONENS.

determinant (abbrev. **det**), *n.* a SCALAR quantity representing a certain defined alternating sum of products of elements of a SQUARE MATRIX, one from each row and column. Determinants have the property that

$$\det(AB) = \det(A) \times \det(B).$$

If A is the 2 x 2 matrix

$$\begin{bmatrix} a & b \\ c & d \end{bmatrix}$$

then the determinant of A, denoted det(A) and often written

$$\begin{vmatrix} a & b \\ c & d \end{vmatrix}$$

is equal to $ad - bc$. A determinant of rank n can be evaluated in terms of determinants of rank $n-1$ (its MINORS or COFACTORS); for example, the 3×3 determinant

$$\begin{vmatrix} a & b & c \\ d & e & f \\ g & h & i \end{vmatrix} = a \begin{vmatrix} e & f \\ h & i \end{vmatrix} - b \begin{vmatrix} d & f \\ g & i \end{vmatrix} + c \begin{vmatrix} d & e \\ g & h \end{vmatrix}.$$

Formally, if $A = [a_{ij}]$ is an $n \times n$ matrix with entries in a commutative UNITARY RING, then

$$\det(A) = \sum_{\sigma \in S_n} \varepsilon(\sigma)\, a_{1\sigma_1}\, a_{2\sigma_2} \ldots a_{n\sigma_n},$$

where S_n is the set of PERMUTATIONS of the integers 1 to n, $\varepsilon(\sigma)$ is the SIGNATURE of the permutation σ, and σ_i is the ith member of the permutation σ. A matrix is invertible if and only if its determinant is non-zero, and determinants may be used in the solution of simultaneous equations, etc. by matrix methods, although GAUSSIAN ELIMINATION or related techniques are almost always preferred in actual computation for matrices larger than 3×3. Compare PERMANENT.

determine, *vb.* to be sufficient for the unique specification of (some entity). For example, any two points determine a straight line; a definite integral is determined up to a constant.

develop, *vb.* **1.** to expand (a function or expression) in the form of a SERIES.
2. to PROJECT (a surface) onto a plane without stretching or shrinking any element; this is equivalent to rolling it out on the plane. For example, a half-cone is *developable*, but a sphere is not.

deviation, *n.* (*Statistics*) the difference between any one of a sequence of observed values of a variable and some value, such as the MEAN. Compare DISPERSION.

deviatoric, *adj.* (of second-order CARTESIAN TENSOR) having zero TRACE. The *deviatoric part* of a second-order Cartesian tensor, **T**, is $\mathbf{T} - \frac{1}{3}(\operatorname{tr} \mathbf{T})\mathbf{I}$, where **I** is the identity.

df, 1. *abbrev. for* DEFINITION, especially written subscript to the EQUALS SIGN as in '$=_{\text{df}}$', or in the margin, to indicate that an identity is true by definition.
2. *abbrev. for* DEGREES OF FREEDOM.

diagonal, *n.* **1.** a line joining any two VERTICES of a POLYGON that are not joined by any of its edges; that is, that are not adjacent. For example, AC and BD are diagonals of the Euclidean quadrangle ABCD in Fig. 118; in the complete quadrilateral ABCDXY, XY is the third diagonal.

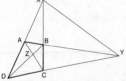

Fig. 118. AC, BD and XY are the **diagonals** of quadrilateral ABCDXY.

2. a line joining any two vertices of a polyhedron that are not in the same face, for example, AG and BH in the cuboid of Fig. 119; however, AC and AH are not diagonals of the solid.

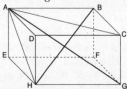

Fig. 119. AG and BH are **diagonals** of a cuboid.

3a. also called **main diagonal.** the sequence of elements of a square matrix that lie between the top left and bottom right of the array; that is, the elements a_{ii} where a_{jk} is the element in the jth row and k^{th} column. See DIAGONAL MATRIX.

b. off diagonal. the sequence of elements of a square matrix lying between the top right and bottom left elements.

c. any other related sequence of elements of a matrix, such as the elements above the main diagonal. See SUPERDIAGONAL.

diagonalizable matrix, *n.* a matrix which is SIMILAR to a DIAGONAL MATRIX.

diagonal matrix, *n.* a MATRIX of which all the elements other than those on its MAIN DIAGONAL are zero.

diagonally dominant matrix, *n.* a SYMMETRIC MATRIX in which every element on the main DIAGONAL is larger than the sum of the absolute values of the remaining elements in its row.

diagonal point, *n.* any of the three points at which non-adjacent sides of a COMPLETE QUADRANGLE intersect.

diagonal process, *n.* the technique of constructing a new member of a set from a list of given members by making its nth term differ from the nth term of the nth member, so that the new member is distinct from every member on the given list, and the set that includes the new member must have strictly larger cardinality than the original list. This method is used in the proof of CANTOR'S DIAGONAL THEOREM, and to show the uncountability of any proper interval of the real line.

diagram, *n.* **1.** a pictorial or graphical representation of some entities and their relations, such as, for example, an Argand diagram or a Venn diagram.
2. diagram of arrows (*Algebra*) the graphical representation of functions between structured sets, using arrows to specify the desired relationships, used especially in CATEGORY THEORY. See COMMUTATIVE DIAGRAM.

diameter, *n.* **1.** a straight line passing through the centre of a CLOSED geometric figure, especially a circle or sphere, for example AB in Fig. 120.

Fig. 120. AB is a **diameter** of the sphere.

2. the length of the segment of such a line of which the end-points are on the perimeter or surface of the figure.

3. (of a set in a METRIC SPACE) the SUPREMUM of distances between pairs of points in the set.

diametral, *adj.* located on or forming a DIAMETER; for example, a diametral plane is one that includes a diameter of a given circle.

diametrical, *adj.* related to or along a DIAMETER.

diamond, *n.* another term for RHOMBUS.

dichotomous line search, binary line search, or **bisection method,** *n.* the iterative method of finding the maximum of a UNIMODAL FUNCTION that proceeds at each iteration by excluding one half of the remaining interval by testing two new function values, both near the midpoint of the present interval, one above and one below.

dichotomy, *n.* **1.** division into two equal parts, as in DICHOTOMOUS LINE SEARCH.

2. the classical PARADOX that motion can never be initiated because before a body can traverse a distance it must first complete the first half of the distance, and before that the first quarter, etc., and so that a runner cannot start until he has made the last of this infinite sequence of steps. See ZENO'S PARADOXES.

dictionary, *n.* a book that acts as a guide and *aide-memoire*, but is no substitute for a text-book. This entry could only list the contents of an adequate dictionary of mathematics if it were of infinite length, so that the word is not included in this dictionary in order to avoid certain PARADOXES.

Dido's problem, *n.* the classical name for the ISOPERIMETRIC PROBLEM, named after Dido, first Queen of Carthage. According to Virgil, Dido was offered whatever area she could enclose with an ox hide on which to found her city. Her solution was to work the ox-hide into cord and use this to enclose a large circle, since a circle maximizes the area enclosed by a closed curve of fixed length.

diffeomorphism, *n.* (*Topology*) a DIFFERENTIABLE mapping that has a differentiable inverse. Two sets are *diffeomorphically equivalent* if there is a diffeomorphism of one onto the other. For example, the reals and the interval $]0, \infty[$ are diffeomorphically equivalent, since the diffeomorphism

$$f : \mathbb{R} \mapsto]0, \infty[\ : \ f(x) = e^x$$

has an inverse

$$g : \ ]0, \infty[\ \mapsto \ \mathbb{R} \ : \ g(x) = \log x.$$

This is a stronger relation than HOMEOMORPHISM as there are pairs of sets that are homeomorphic but not diffeomorphic.

difference, *n.* **1.** the result of the subtraction of one number or quantity from another; the number or quantity requiring to be added to one to yield the other.

2. (of two sets) **a.** the set of members of the first that are not members of the second, written A \ B. See RELATIVE COMPLEMENT.

b. see SYMMETRIC DIFFERENCE.

difference engine, another term for BABBAGE'S ENGINE.

difference equation, *n.* a RECURRENCE RELATION, especially one written in the notation of DIFFERENCE SEQUENCES. Sometimes the term is used for an initial calculation of terms of a difference sequence in the hope of finding a recognizable recurrence relation.

difference of squares, *n.* any expression of the form $a^2 - b^2$, which can be factorized as $(a + b)(a - b)$.

difference polynomial, *n.* the POLYNOMIAL, usually denoted d, over the RING, $\mathbb{Z}[t_1, \ldots, t_n]$ defined as the product

$$\prod_{i < j} (t_i - t_j).$$

Under the natural ACTION of the PERMUTATION GROUP S_n on $\mathbb{Z}[t_1, \ldots, t_n]$,

α in S_n maps $f(t_1,\dots,t_n))$ to $f(\alpha(t_1),\dots,\alpha(t_n))$; $\alpha d = \pm d$, and α is even or odd according as αd equals d or $-d$.

difference quotient, *n.* a ratio of the form

$$\frac{f(x+\Delta x) - f(x)}{\Delta x}.$$

This is known as a *forward* or *backward difference quotient* according as Δx is positive or negative. The average of these produces the *central difference quotient*

$$\frac{f(x+\Delta x) - f(x-\Delta x)}{2\Delta x}.$$

Higher-order difference quotients are similarly defined. These differences are important in numerical estimation of DERIVATIVES.

difference sequence, *n.* a sequence of numbers constructed as the differences between the successive terms of a given sequence. Higher-order differences of a sequence $\{x_k\}$ are defined recursively. Thus the mth *forward difference sequence* at x_k is given by

$$\Delta^m x_k = \Delta^{m-1} x_{k+1} - \Delta^{m-1} x_k,$$

where $\Delta^0 x_k = x_k$. Then, for example,

$$\Delta^3 x_k = x_{k+3} - 3x_{k+2} + 3x_{k+1} - x_k,$$

in accord with PASCAL'S TRIANGLE. *Backward* and *central differences* are similarly defined. See also DIVIDED DIFFERENCE.

differentiable, *adj.* **1.** (of a function or operator) possessing a well-defined DERIVATIVE. See also MANIFOLD.

2. (of a function of the real variables $x_1,\dots, x_n$) having PARTIAL DERIVATIVES such that near the point $(a_1,\dots, a_n)$,

$$f(x_1, \dots, x_n) - f(a_1, \dots, a_n) = \sum_{i=1}^{n} f_{x_i}(a_1, \dots, a_n)(x_i - a_i) + \varepsilon\left(\sum_{i=1}^{n} |x_i - a_i|\right),$$

where ε tends to zero as $(x_1,\dots, x_n)$ tends to $(a_1,\dots, a_n)$, and f_{x_i} is the partial derivative of f with respect to x_i.

differential, *adj.* **1.** of, pertaining to, or containing DERIVATIVES.

2. (*as substantive*) an INCREMENT in a given function, expressed as the product of the DERIVATIVE of that function and the corresponding increment of the independent variable; if $F(x)$ is the given function, then

$$dF = \frac{dF}{dx} \times dx.$$

(However, when dx is an increment in x, dF is not in general the increment in F.)

3. (*as substantive*) an increment in a given function of two or more variables, expressed as the sum of the products of each PARTIAL DERIVATIVE and the increment in the corresponding variable; if $F(x_1,\dots, x_n)$ is the given function, then

$$dF = \sum_{i=1}^{n} \frac{\partial F}{\partial x_i} \times \partial x_i.$$

(However, when ∂x_i are the increments in x_i, dF is not in general the increment in F.)

4. (*as substantive*) a mapping, *df*, derived from a given mapping, *f*, between two NORMED VECTOR SPACES, such that

$$\lim_{\|h\| \to 0} \frac{\| f(x+h) - f(x) - df(x)h \|}{\| h \|} = 0.$$

See also FRECHET DIFFERENTIAL, GATEAUX DIFFERENTIAL.

differential calculus, *n.* the branch of CALCULUS that studies DERIVATIVES and DIFFERENTIALS and their evaluation and use. Compare INTEGRAL CALCULUS.

differential coefficient, *n.* another name for the DERIVATIVE or its value for a given argument. See also PARTIAL DERIVATIVE.

differential equation, *n.* an equation containing DERIVATIVES or DIFFEREN-TIALS of a function. A *partial differential equation* (PDE) contains the PARTIAL DERIVATIVES of a function of more than one variable; otherwise the equation is an *ordinary differential equation* (ODE). FIRST–ORDER partial differential equations are reducible to systems of ODEs. Readily soluble first-order differential equations of the first DEGREE include EXACT, SEPARABLE, HOMO-GENEOUS, and LINEAR DIFFERENTIAL EQUATIONS. Readily soluble first-order differential equations of higher degrees are those which are ALGEBRAICALLY SOLUBLE in the first derivative or either variable, and CLAIRAUT'S FORM. Second-order partial differential equations are fundamental to physics and include the WAVE EQUATION, the HEAT EQUATION, and LAPLACE'S EQUATION. The general *quasi-linear* equation of the second order is

$$A(x,y)\,u_{xx} + 2B(x,y)\,u_{xy} + C(x,y)\,u_{yy} + F(x,y,u,u_x,u_y) = 0,$$

where *A*, *B*, and *C* can also be functions of *u*, u_x, or u_y; it is *hyperbolic, parabolic*, or *elliptic* depending upon whether $B^2 - AC$ is negative, zero, or positive. The definitions can be extended to functions of more variables, but are of increasing complexity. There are methods for finding COMPLETE SOLUTIONS of non-linear first-order PARTIAL DIFFERENTIAL EQUATIONS if the equation explicitly contains either no dependent variable or no independent variable, or if it either has the form

$$f\left(\frac{\partial z}{\partial x}, x \right) = g\left(\frac{\partial z}{\partial x}, y \right)$$

for some functions *f* and *g*, or has CLAIRAUT'S FORM. See CHARPIT'S METHOD, JACOBI'S METHOD. See also LAGRANGE'S LINEAR EQUATION.

differential form, *n.* a part of the formalization of the notion of SURFACE INTEGRATION that is central to a modern treatment of STOKES'S THEOREM in which one talks of integrating *k-forms* over *k-surfaces*. A typical 1-form is $xdy - ydx$. More precisely, a differential form of degree *r* in *n* variables is a mapping from a domain in *n*-space into the set of *r*-COVECTORS.

differential geometry, *n.* **1.** the study of GEOMETRY by the methods of CALCULUS; for example, in order to determine the area of a surface.

2. the study of DIFFERENTIAL MANIFOLDS with an induced structure, the structure typically originating from problems of geometry or mechanics. See LORENZ GROUP.

differential manifold, see MANIFOLD (sense 2).

differential operator, *n.* **1.** the operator DEL used in VECTOR ANALYSIS and

defined as

$$\nabla = \mathbf{i}\,\frac{\partial}{\partial x} + \mathbf{j}\,\frac{\partial}{\partial y} + \mathbf{k}\,\frac{\partial}{\partial z},$$

where $\mathbf{i}$, $\mathbf{j}$, $\mathbf{k}$ are unit vectors in the direction of the x-, y-, and z-axes, respectively, and $\partial/\partial x$, $\partial/\partial y$, and $\partial/\partial z$ are the respective PARTIAL DERIVATIVES of the given function of x, y and z. See also GRADIENT, DIVERGENCE, CURL, LAPLACE OPERATOR.

2. any operator involving DERIVATIVES.

differential structure, *n.* a maximal CONTINUOUSLY DIFFERENTIABLE ATLAS; if it is r times continuously differentiable, the atlas is called a $C^{(r)}$ differential structure. See MANIFOLD (sense 2).

differentiate, *vb.* to find the FIRST DERIVATIVE of (a function).

differentiation, *n.* the operation or process of determining the FIRST DERIVATIVE of a function.

diffusion equation, *n.* another term for HEAT EQUATION.

digamma function or **psi function,** *n.* the LOGARITHMIC DERIVATIVE of the GAMMA FUNCTION, defined by

$$\Psi(z) = \frac{\Gamma'(z)}{\Gamma(z)}.$$

digit, *n.* **1.** also called **figure.** any of the ten ARABIC NUMERALS, *0, 1, 2, 3, 4, 5, 6, 7, 8, 9,* of the decimal system.

2. by analogy, any of the single symbols used to represent the numbers from 0 to $b-1$ in counting to the base b. For example, in base 12, the extra digits T and E are used; in base 16, the extra digits are A, B, C, D, E, F.

digital, *adj.* in numerical form; for example, digital data are tables of numerical values, as opposed to, for example, a graph, and a digital watch has a numerical display, as contrasted with the ANALOG display of a conventional clock-face.

digital computer, *n.* a COMPUTER, usually electronic, that can be programmed to carry out a variety of functions on an input string of data to produce an output string. Both data and program are represented internally in BINARY notation, and the program applies elementary binary operations to the input data to generate the output data. Compare ANALOG DEVICE.

digraph, *n.* (*Graph theory*) a generalization of the notion of a GRAPH in which EDGES are DIRECTED and so go from one VERTEX to another. These play an important role in NETWORK optimization problems.

dihedral, *adj.* **1.** having or formed by two intersecting planes.

2. (*as substantive*) also called **dihedron** or **dihedral angle.** the figure formed by two intersecting planes and the line in which they meet, such as that shown in Fig. 121. If a measure of this angle is required, it is taken to be that

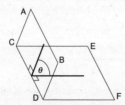

Fig. 121. θ is the angle between the planes of the **dihedral.**

between any pair of lines, one in each plane, perpendicular to the common line at the same point. Fig. 121 shows a dihedral angle ABCDEF; θ is the size of this angle.

dihedral group, *n.* the GROUP of SYMMETRIES of a regular polygon, denoted D_n or dih(n), where n is the number of sides of the polygon. For example, D_4 is the dihedral group of a square, which has order 8; its members are shown in Fig. 122, where the corners are marked merely to distinguish the positions of the regular figure.

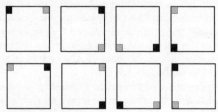

Fig. 122. The **dihedral group** of a square.

dihedron, see DIHEDRAL.

dilatation or **dilation,** *n.* (*Geometry*) **1.** a transformation that takes parallel lines into parallel lines, a direction-preserving SIMILARITY. A dilatation that is not merely a TRANSLATION is called a *central dilatation* because all lines joining corresponding points of a figure and its image are concurrent. Compare HOMOTHETY.

2. dilatation at P with ratio *k*. a mapping of a space into itself that is defined by

$$f(x) = kx + (1 - k)p,$$

where k is a non-zero real number, and p is the set of coordinates of the point P, the *centre* of the dilatation. The dilatation is said to be positive or negative according as k is positive or negative; for example, in Fig. 123, XYZ is a positive dilatation of ABC with respect to the origin; ABC is a negative dilatation of XYZ.

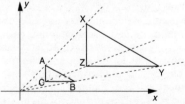

Fig. 123. ABC and XYZ are **dilatations** of one another.

dilemma, *n.* (*Logic*) a form of argument of which one of the premises is the conjunction of two conditional statements and the other affirms the disjunction of their antecedents, and the conclusion is the disjunction of their consequents; its form is

> *if P then Q, and if R then S;*
> *either P or R;*
> so *either Q or S.*

This is sometimes called *constructive dilemma*, as distinct from *destructive dilemma* in which the second premise is the disjunction of the negations of the consequents and the conclusion is the disjunction of the negations of the antecedents of the first premise:

> *if P then Q, and if R then S;*
> *either not-Q or not-S;*
> so *either not-P or not-R.*

dilogarithm, see POLYLOGARITHM.

dimension, *n*. **1.** each of a set of independent and mutually perpendicular, or ORTHOGONAL, directions in which a EUCLIDEAN SPACE may be measured. **2.** also called **Hamel dimension.** the minimal number of mutually INDEPENDENT vectors that generate the given space, that is, in terms of LINEAR COMBINATIONS of which every element of the space can be canonically expressed; the cardinality of a BASIS of the space. **3.** in particular, the number of coordinates required to locate a point in a space; for example, the space of ordinary experience is three-dimensional, and a flat surface is two-dimensional. **4.** various topological measures of size determined by COVERING properties of the space in question. See also HAUSDORFF DIMENSION, TOPOLOGICAL DIMENSION.

Dini derivatives, *n*. (of a real-valued function on real *n*-space) the four directional quantities defined, for a point, **x**, and a direction, **h**, as the LIMITS SUPERIOR and the LIMITS INFERIOR, as *t* tends to 0 both from above and from below, of the quotient

$$\frac{f(x + th) - f(x)}{t}.$$

All four limits coincide with the DIRECTIONAL DERIVATIVE if the latter exists. (Named after *Ulise Dini* (1845–1918), Italian analyst.)

Dini's theorem, *n*. the result that a monotone decreasing sequence of continuous functions defined on a compact set that converges pointwise to a continuous limit actually exhibits UNIFORM CONVERGENCE.

Diophantine equation, *n*. a polynomial equation in several unknowns, with integral coefficients, to be solved or proved insoluble in integers, such as PYTHAGORAS' THEOREM or FERMAT'S LAST THEOREM. In 1970, Matiyasevich proved that no general algorithm exists for determining whether a given Diophantine equation is soluble, thereby answering Hilbert's tenth problem. (Named after the 3rd century BC Greek mathematician, *Diophantus of Alexandria,* of whose life all that is known are the ages of his marriage and death, inferred from an arithmetic riddle. Only six of the supposed 13 books of his *Arithmetica* are extant, but these introduced the earliest known algebraic notation, and deal with the algebraic solution, in the rationals, of a wide range of number theoretic and geometric problems.)

Dirac delta function, *n*. the function, $\delta(x)$, defined to be zero for all non-zero real *x*, and infinite for $x = 0$. Naively,

$$\int_{-\infty}^{\infty} \delta(x) \ dx \ = \ 1.$$

This can be made rigorous by use of DISTRIBUTION theory, and is of use in quantum mechanics and the study of partial differential equations.

direct, *adj.* (of a relationship) relating two variables in such a way that an increase in the value of one is associated with an increase in the value of the other. Compare INVERSE.

directed, *adj.* (of a number, line, angle, etc.) having an orientation or direction distinguished from an opposite orientation or direction, usually by the use of PLUS and MINUS signs. Thus two points A and B determine not one but two *directed lines*, $\overrightarrow{AB}$ and $\overrightarrow{BA}$, lying in the same position (its *direction*), but with opposite orientation (its *sense*); but any two points can be used to distinguish the two directed lines by specifying which precedes the other. In this case, $|\overrightarrow{AB}| = |\overrightarrow{BA}|$ and $\overrightarrow{AB} = -\overrightarrow{BA}$; thus a line, AB, is not a directed line, as it has only direction and not sense. For example, in Fig. 124, the directed lines $\overrightarrow{BA}$ and $\overrightarrow{CD}$ have the same length and direction, but opposite sense; regarded as vectors they have resultant zero. Inclination and declination are directed angles with opposite sense, as, for example, are angles XOY and XOZ in the figure.

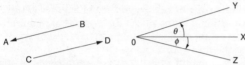

Fig. 124. **Directed** lines and angles.

directed number, *n.* **1.** another term for a SIGNED NUMBER or INTEGER that may be positive, negative or zero.

2. more generally, any numerical quantity that may be positive or negative.

directed ratio, *n.* a RATIO of DIRECTED quantities, as opposed to their absolute magnitudes.

directed set, *n.* a set endowed with a TRANSITIVE and REFLEXIVE relation, $\geq$, such that, given any two points a and b in the set, there exists another point c in the set with $a \geq c, c \geq b$. The relation then *directs* the set. For example, the finite subsets of an infinite set are directed by inclusion. Any TOTAL ORDERING is directed.

direction, *n.* the ORIENTATION of a line in space, distinguished from its SENSE in the case of a DIRECTED LINE.

directional derivative, *n.* the LIMIT of the DIVIDED DIFFERENCES of a given function in a given direction, **h**:

$$f'(x,h) = \lim_{t \to 0} \frac{f(x + th) - f(x)}{t}$$

for **x** in real n-space; one-sided directional derivatives are also defined for t tending to zero from above or below. When the GRADIENT of the function exists continuously this limit coincides with $\langle \nabla f(\mathbf{x}), \mathbf{h} \rangle$, and in Euclidean space **h** may always be taken as a set of DIRECTION COSINES. See also CLARKE GENERALIZED DIRECTIONAL DERIVATIVE, DINI DERIVATIVES.

direction angles, *n.* the triple of angles, usually designated α, β, γ, that a line in space or a vector makes with the positive directions of the x-, y- and z-coordinate axes respectively, and that are sufficient to determine the orientation of the line or vector. Thus, for example, in Fig. 125, the direction angles of P are the angles xOP, yOP, and zOP respectively.

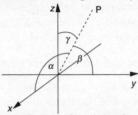

Fig. 125. α, β, γ are the **direction angles** of P.

direction cosines, *n.* the triple of cosines of the DIRECTION ANGLES of a line or vector, that uniquely determine its orientation.

direction field, *n.* (of a first-order differential equation) the set of triples or *lineal elements* consisting of independent variable, dependent variable and derivative at points in the domain of the governing function. If the equation is of the form $p = dy/dx = f(x, y)$ the lineal elements are (x, y, p). The CONTOUR lines with p constant are called *isoclines.*

direction numbers or **direction ratios,** *n.* a sequence of numbers determining the orientation of a line or vector relative to a set of axes, by virtue of the fact that they are proportional to the DIRECTION COSINES of that line.

direct product, *n.* **1.** a construction analogous to the DIRECT SUM of groups, but with multiplication as the operation.

2. a construction analogous to that of a DIRECT SUM of infinitely many subspaces when the sums are permitted, but not required, to be infinite; thus, for example,

$$\bigotimes_{i=1}^{\infty} \mathbb{R}_i$$

represents the set of all finite or infinite sequences of real numbers.

direct proof, *n.* a PROOF proceeding from premises to conclusion by successive steps of deriving intermediate conclusions from preceding steps, rather than by eliminating the possibility of the falsehood of the hypothesis, as in an INDIRECT PROOF.

direct proportion or **direct variation,** *n.* the relation between two variable quantities the corresponding values of which are a constant multiple of one another. Compare INVERSE PROPORTION.

directrix, *n.* a fixed line, on the convex side of a CONIC SECTION, in terms of which, together with the FOCUS and the ECCENTRICITY, the locus of points that constitute the conic is defined. For example, in Fig. 126, the ellipse is defined as the locus of points whose distance from the focus F is in some fixed proportion to their perpendicular distance from the line XYZ; P, Q, and R are three such points. See CONIC.

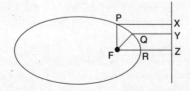

Fig. 126. XYZ is the **directrix** of the ellipse.

direct sum, n. (of VECTOR SPACES, ABELIAN GROUPS, etc.) the decomposition of the vector space or group as a sum of subspaces or subgroups such that each element has a unique representation as a finite sum of elements, one from each subspace or subgroup. One writes

$$X = \bigoplus_{i=1}^{n} X_i,$$

so that

$$\mathbb{R}^n = \bigoplus_{i=1}^{n} \mathbb{R}_i.$$

A finite direct sum may also be written as

$$\bigotimes_{i=1}^{n} X_i$$

and may be identified with the CARTESIAN PRODUCT of sets that are disjoint apart from identity (although it is not in general identifiable with the Cartesian product). When infinitely many spaces are involved the use of the term DIRECT PRODUCT indicates that one does not require the sums to be finite; in this case

$$\bigoplus_{i=1}^{\infty} \mathbb{R}_i$$

represents all finite sequences of real numbers, and

$$\bigotimes_{i=1}^{\infty} \mathbb{R}_i$$

all real sequences, finite or infinite. Some authors distinguish between 'direct' for the finite, and 'Cartesian' for the unrestricted case, rather than between 'sum' and 'product'.

2. see INTERNAL DIRECT SUM, EXTERNAL DIRECT SUM.

direct variation, n. another term for DIRECT PROPORTION.

Dirichlet, Peter Gustav Lejeune (1805–59), French-born German mathematician who became Professor at Berlin. His *Vorlesungen über Zahlentheorie* contained important results about IDEALS and provided a lucid exposition of Gauss' results in number theory. He made important advances in the fields of number theory, complex analysis, mechanics, and the study of FOURIER SERIES, and introduced the modern concept of a FUNCTION as a many–one relation.

Dirichlet's condition, n. the condition, as extended by Jordan, that a periodic function be of BOUNDED VARIATION in a neighbourhood of a point. This suffices to guarantee that the FOURIER SERIES of the function converges pointwise to the average of the limits of the function from the right and left, and so, if the function is continuous, to the function value at the point. A related result (*Fejer's condition*) shows that if the function is merely integrable then pointwise CESARO convergence is achieved, and that when the function is continuous the CESARO AVERAGES of the PARTIAL SUMS of the Fourier series converge uniformly to the function.

Dirichlet series, n. any one of a class of series of the form

$$\sum_{n=1}^{\infty} a_n n^{-s}$$

that are of great importance in number theory. Typically the coefficients satisfy divisibility or other number theoretic conditions. The most important of these series are the *Dirichlet L-series* of which the simplest are the ZETA FUNCTION and the primitive L-series modulo three:

$$1^{-s} - 2^{-s} + 4^{-s} - 5^{-s} + \ldots.$$

Dirichlet's kernel, *n.* the KERNEL, important in FOURIER ANALYSIS, that is defined as the sum

$$\frac{1}{2} + \sum_{k=1}^{n} \cos kt = \frac{\sin \frac{(2n+1)t}{2}}{2 \sin \frac{t}{2}},$$

for t not a multiple of 2π.

Dirichlet's principle, *n.* another name for the PIGEON–HOLE PRINCIPLE.

Dirichlet's problem, *n.* the partial differential equation problem that looks for solutions to LAPLACE'S EQUATION in a region subject to BOUNDARY CONDITIONS. These are typically either requirements of *Dirichlet type* that the solutions agree with a given continuous function on the boundary of the region, or of *Neumann type* that require the NORMAL DERIVATIVE to satisfy a boundary condition. PONTRYAGIN'S MAXIMUM PRINCIPLE implies the uniqueness of the solution in many cases.

Dirichlet's test, *n.* **1.** a test for the CONVERGENCE of an INFINITE SERIES: if $\{a_n\}$ and $\{b_n\}$ are sequences such that $\sum a_n$ has bounded partial sums, and $\{b_n\}$ is strictly decreasing and converges to zero, then $\sum a_n b_n$ is convergent. This test is often useful in testing the convergence of a POWER SERIES on the boundary of its CIRCLE OF CONVERGENCE. The ALTERNATING SERIES TEST is a special case of this result.

2. a test for the UNIFORM CONVERGENCE of an infinite series: if $\{a_n(z)\}$ and $\{b_n(z)\}$ are sequences of complex functions on a compact set K, such that $\sum a_n(z)$ has partial sums that are uniformly bounded in K, $\sum \{b_n(z) - b_{n+1}(z)\}$ is uniformly absolutely convergent in K, and $\{b_n(z)\}$ converges uniformly to zero in K, then $\sum a_n(z) b_n(z)$ is uniformly convergent in K.

Dirichlet's theorem, *n.* the result that if $f(x)$ is a BOUNDED PERIODIC function containing at most a finite number of maxima and minima and a finite number of discontinuities in each period, then where f is continuous the FOURIER SERIES for f converges to f, and where f is discontinuous it converges to the average of the right and left limits of f at the discontinuity.

Dirichlet-type boundary conditions, see DIRICHLET'S PROBLEM.

disc, variant spelling of DISK.

disconnected, *adj.* not topologically or graphically CONNECTED. For example, any punctured neighbourhood in the reals is disconnected, as is the cut interval $(-1,1) \setminus \{0\}$.

discontinuity, *n.* **1.** a point or value of the independent variable at which the value of a function is not equal to its limit as the value of the independent variable approaches that point, or where it is not defined. For example, $y = (x^2 - 4)^{-1}$ has discontinuities at $x = 2$ and $x = -2$, as shown in Fig. 127 opposite; these are in fact SINGULARITIES.

2. the property of being DISCONTINUOUS.

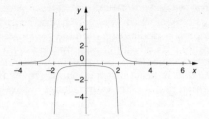

Fig. 127. $y = 1/(x^2-4)$ has **discontinuities** at ± 2.

discontinuous, *adj.* (of a function or curve) not everywhere CONTINUOUS, having a DISCONTINUITY.

discrete, *adj.* **1.** (of a function or random variable or vector) taking a succession of distinct values. Formally, a discrete random variable is one with a COUNTABLE number of possible outcomes. For example, Fig. 128 is a graph showing the proportion of a class who attend lectures on each day of the week; it would make no sense to join the dots, as this would give the impression that the independent variable is continuous, but there are clearly no intermediate values between Monday and Tuesday. Compare CONTINUOUS. **2.** (of a TOPOLOGICAL SET) having no CLUSTER POINTS, so that every point is ISOLATED. For example, the integers are discrete, but the rationals are not, since they are DENSE in the reals.

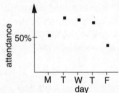

Fig. 128. A **discrete** function.

discrete Fourier transform, *n.* another term for the FINITE FOURIER TRANSFORM.

discrete topology, *n.* the TOPOLOGY on a given space that consists of its entire POWER SET.

discretization, *n.* (*Numerical analysis*) a DISCRETE approximation to a continuous or other non-discrete object, often for the purposes of computation, such as replacing DIFFERENTIALS by DIFFERENCE QUOTIENTS or QUADRATURE.

discriminant, *n.* an algebraic expression, related to the coefficients of a polynomial equation (or to a number field), that gives information about the roots of the polynomial; principally, the discriminant is non-zero if and only if the roots are distinct. For example,

$$D = b^2 - 4ac$$

is the discriminant of the quadratic equation $ax^2 + bx + c = 0$; D is positive exactly when the equation has distinct real roots, and is zero exactly when it has equal real roots. More precisely, the discriminant of a polynomial p of degree n over a given field is the quantity

$$D(p) = (-1)^{n(n-1)/2} R(p, p')$$

where R is the RESOLVENT of p and p'. See also CARDANO'S FORMULA.

discriminatory, *adj.* (of a statistical test) see UNBIASED (sense 3).

disjoint, *adj.* (of two sets) having no members in common, having an INTERSECTION that is empty; for example, the odd integers and the even integers are disjoint sets. Two sets are disjoint if and only if the properties of membership are MUTUALLY EXCLUSIVE, and in a VENN DIAGRAM, such as Fig. 129, the overlapping region is shaded to indicate that it is empty. (In EULER'S CIRCLES, they would be represented by non-intersecting circles.)

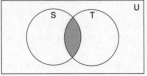

Fig. 129. Venn diagram of **disjoint sets** S and T.

disjoint union, *n.* a binary operator that constructs the set of all elements of a pair of given sets, when all those elements are regarded as distinct; this can be done by first indexing the elements of the sets to ensure they are disjoint and then taking the union of the indexed sets. The disjoint union of S and T is written

$$S \cup^* T = (S \times \{0\}) \cup (T \times \{1\}),$$

and this construction is shown in Fig. 130; although there is no necessity to form these particular products, and one might, for example, instead take $A^* = \{\langle a, 1 \rangle : a \in A\}$ and $B^* = \{\langle 1, b \rangle : b \in B\}$, the former construction is preferable, as it permits direct extension to disjoint unions of more than two sets. The cardinality of a disjoint union is always the sum of the cardinalities of the given sets, and this may therefore be used to define addition in set-theoretic terms.

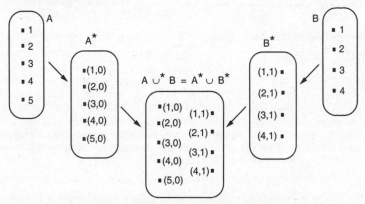

Fig. 130. One possible construction of the **disjoint union**.

disjunct or **alternant,** *n.* (*Logic*) either of a pair of propositions or formulae related or operated upon by DISJUNCTION.

disjunction, **alternation**, or **logical sum,** *n.* (*Logic*) **1.** the binary TRUTH–FUNCTIONAL sentential connective that forms a compound sentence from two given sentences and corresponds to the English *or*. *Inclusive disjunction* assigns the value true whenever either or both of its arguments is true, and false only when both are false; *exclusive disjunction* assigns the value true if and only if one but not both of its disjuncts is true; it is thus sometimes called *non-equivalence*. The TRUTH–TABLES for both functions are shown in Fig. 131.

2. a sentence so formed. The inclusive disjunction of P and Q is usually written P ∨ Q and sometimes read 'P *vel* Q'; the less common exclusive disjunction has no standard symbol, but is sometimes written P ⊻ Q and read 'P *aut* Q'. Since exclusive disjunction is equivalent to non-equivalence it is also sometimes written P ≢ Q.

P	Q	P ∨ Q	P ⊻ Q
T	T	T	F
T	F	T	T
F	T	T	T
F	F	F	F

Fig. 131. Truth-tables for inclusive and exclusive **disjunction**.

disjunctive normal form (abbrev. **dnf**), *n.* (*Logic*) the form to which every statement of SENTENTIAL CALCULUS can be reduced, consisting of a disjunction of conjunctions each of the conjuncts of which is either an atomic formula or its negation. For example,

$$(P \rightarrow Q) \ \& \ (P \ \& - Q)$$

has disjunctive normal form

$$(P \ \& - P \ \& - Q) \lor (Q \ \& \ P \ \& - Q).$$

(Associativity permits the omission of brackets round the separate conjunctions, and where the context permits it, juxtaposition may be used for the conjunctions.) The *disjunctive normal form theorem* states the direct consequence of this definition that a well-formed formula is a CONTRADICTION if and only if every one of the disjuncts contains the conjunction of some atomic variable and its negation; for example, the preceding example is a contradiction. Compare CONJUNCTIVE NORMAL FORM.

disjunctive syllogism, *n.* (*Logic*) a form of ARGUMENT one of the PREMISES of which is the DISJUNCTION of two statements and the other of which is the NEGATION of one of these statements; its form is

either P or Q		*either P or Q*
not P	or	*not Q*
so Q.		*so P.*

disk or **disc,** *n.* an OPEN or CLOSED BALL in a METRIC SPACE, usually the complex plane.

dispersion, *n.* (*Statistics*) the degree to which the values of a FREQUENCY DISTRIBUTION are scattered around some central point, usually the ARITHMETIC MEAN or MEDIAN.

displacement, *n.* **1.** also called **displacement vector.** a VECTOR representing the difference between the POSITION VECTORS of two positions; especially the distance and direction by which a point or figure is moved in a TRANSLA-

163

TION to another position. For example, in Fig 132 below, the vectors $\overrightarrow{AX}$ and $\overrightarrow{CZ}$ both represent the translation of triangle ABC to XYZ; if the coordinates of A and X are respectively $(1, 2)$ and $(5, -2)$, then the displacement vector is $\langle 4, -4 \rangle$. The position vector of a point is the vector of its displacement from the origin.

2. (*Continuum mechanics*) in particular, the difference between the POSITION VECTORS of the images of a given PARTICLE under its reference and its current CONFIGURATIONS.

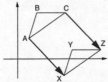

Fig. 132. $\overrightarrow{AX}$ is the **displacement** vector of A.

displacement gradient, *n.* (*Continuum mechanics*) the GRADIENT, with respect to the REFERENCE CONFIGURATION, of the DISPLACEMENT of a given body; thus the displacement gradient is equal to the DEFORMATION GRADIENT minus the identity.

dissect, *vb.* to divide (an interval) into a number of subintervals the union of which is the given interval and the only common points of which are the endpoints of the subintervals. For example, $[0, \frac{1}{3}], [\frac{1}{3}, 1]$ is a dissection of $[0,1]$. See also PARTITION (sense 4).

dissymmetry, *n.* **1.** a lack of SYMMETRY.

2. the relation between two objects when one is the mirror image of the other, that is, when one is a reflection of the other in some AXIS OF SYMMETRY.

distance, *n.* **1.** (between two points in EUCLIDEAN SPACE) the length of the shortest line segment joining the given points, measured as the square root of the sums of the squares of the differences between the coordinates of the two points; if A and B are respectively the points (a_1, a_2) and (b_1, b_2) in the Cartesian plane, then the length

$$| AB | = \sqrt{(a_1 - b_1)^2 + (a_2 - b_2)^2},$$

since, as shown in Fig. 133, AB is the hypotenuse of a right-angled triangle whose other sides are the differences of the coordinates of A and B. In *n*-dimensional Euclidean space,

$$| AB | = \sqrt{\sum_{i=1}^{n} (a_i - b_i)^2}.$$

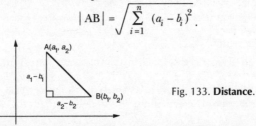

Fig. 133. **Distance.**

2. any length measured along a curve, especially straight LINE or GREAT CIRCLE. See also GEODESIC.

3. (from a point to a line) see PERPENDICULAR DISTANCE.

4a. (in a TOPOLOGICAL SPACE) a METRIC. See also NORM.

b. (between sets in a METRIC SPACE) the INFIMUM of the distance between points in one set and points in the other.

distinct, *adj.* (of a pair of entities) not NUMERICALLY IDENTICAL.

distinctiveness ratio, *n.* (*Statistics*) the ratio of the RELATIVE FREQUENCY of some event in a given sample to that in the general population or another relevant sample.

distribute, *vb.* to apply, or (of an operator) to obey, a DISTRIBUTIVE LAW; for example, multiplication distributes over addition.

distribution, *n.* **1.** (*Statistics*) the set of possible values of a random variable, or points in a SAMPLE SPACE, considered in terms of their theoretical or observed FREQUENCY.

2. also called **generalized function.** a generalization of the concept of a function, defined as continuous linear FUNCTIONALS over spaces of infinitely differentiable functions, introduced so that all continuous functions possess partial *distributional derivatives* (also called *Schwartzian derivatives*) that are again distributions. This leads to so-called *weak solutions* of differential equations and is of importance in the theory of PARTIAL DIFFERENTIAL EQUATIONS.

distribution function, see CUMULATIVE DISTRIBUTION FUNCTION.

distributive law, *n.* an axiom or theorem of a particular formal system that states for a given pair of operators that one *distributes* over the other; that is, that an expression in which one has as an argument a term containing the other is equivalent to an expression in which the latter has wider scope and the former operates directly on each of what were the terms of the latter. For example,

$$a(b + c) = ab + ac$$

is the distributive law for for arithmetic multiplication over addition. The converse does not hold for these operators, but set union and intersection, for example, distribute over one another in either order.

div, *abbrev. for* DIVERGENCE.

diverge, *vb.* **1a.** (of an infinite sequence or series) not having a finite limit.

b. in some contexts, neither having a finite limit nor an absolute bound as the number of terms tends to infinity; this usage therefore excludes functions that OSCILLATE boundedly.

2. (of an IMPROPER INTEGRAL) not having a finite value.

3. (of an INFINITE PRODUCT of non-zero complex numbers) see DIVERGE TO ZERO.

Compare CONVERGE.

divergence, *n.* **1.** (for a VECTOR-valued function, **A**) the scalar quantity $\nabla \cdot \mathbf{A}$ or div**A**, defined in terms of the DIFFERENTIAL OPERATOR as

$$\nabla \cdot \mathbf{A} = \left[\mathbf{i} \frac{\partial}{\partial x} + \mathbf{j} \frac{\partial}{\partial y} + \mathbf{k} \frac{\partial}{\partial z} \right] \cdot \mathbf{A},$$

where **i**, **j**, **k** are unit vectors in the directions of the *x*-, *y*-, and *z*-axes respectively; if **A** = (u, v, w), then

$$\nabla \cdot \mathbf{A} = \frac{\partial u}{\partial x} + \frac{\partial v}{\partial y} + \frac{\partial w}{\partial z}.$$

Compare GRADIENT, CURL.

2. more generally, for a CARTESIAN TENSOR

$$T_{ijk...}\,e_i \otimes e_j \otimes e_k \otimes \ldots \,,$$

the quantity

$$\frac{\partial}{\partial x_i}\,(T_{ijk...})\,e_i \otimes e_j \otimes e_k \otimes \ldots .$$

divergence theorem or **Gauss' theorem**, *n.* (in VECTOR ANALYSIS) the theorem that the triple integral of the DIVERGENCE of a function over a region, G, is equal to the SURFACE INTEGRAL of the normal component, **A.n**, of the function over the boundary of the region:

$$\iiint\limits_{G} \text{div}\,\mathbf{A}\;dV \;=\; \iint\limits_{\partial G} \mathbf{A}\,.\,\mathbf{n}\;dS,$$

where **n** is the external UNIT NORMAL to the surface. Compare STOKE'S THEOREM, GREEN'S THEOREM.

divergent, *adj.* **1a.** (of an infinite sequence) not having a finite limit.

b. in some contexts, neither having a finite limit nor oscillating boundedly.

2. (of an infinite series) not having a finite sum, so that it generates a sequence of PARTIAL SUMS that does not CONVERGE. If $\langle a_0, a_1, a_2, \ldots \rangle$ is a divergent sequence then the series

$$a_0 + a_1 + a_2 + \ldots,$$

that is, the sequence

$$\langle\, a_0,\, a_0 + a_1,\, a_0 + a_1 + a_2,\, \ldots \rangle$$

must also diverge; but not necessarily conversely: the series

$$1 + \tfrac{1}{2} + \tfrac{1}{3} + \tfrac{1}{4} + \tfrac{1}{5} + \ldots$$

is divergent, but the sequence of terms is not. Compare CONVERGENT.

diverge to zero, *vb.* (of an INFINITE PRODUCT of non-zero complex numbers) having partial products tending to a zero limit as *n* tends to infinity. If a sequence is finitely zero, one determines its convergence by considering convergence of the non-zero tail, although the value of the product will be zero in both the convergent and the divergent case. These conventions allow one to convert products securely into series by taking logarithms.

diversity, *n.* (*Logic*) the relation that holds between two entities if and only if they are not IDENTICAL; the property of being NUMERICALLY DISTINCT.

divide, *vb.* **1.** to calculate the multiplier of a given number required to yield a product equal to another number; loosely, the number of times the former is contained in the latter. One may speak of dividing 50 by 10, or dividing 10 into 50. The DIVISOR divides the DIVIDEND to yield a QUOTIENT.

2. (of a number) to have as an exact multiple another number. In this sense one says, for example, that 10 divides 50.

divided difference or **first divided difference sequence,** *n.* another name for a sequence of DIFFERENCE QUOTIENTS of the form:

$$\frac{f(x_{k+1}) \;-\; f(x_k)}{x_{k+1} \;-\; x_k}.$$

The *second* and *higher divided differences* are defined recursively: the $(n+1)^{th}$ divided difference sequence is the first divided difference sequence of the nth divided difference sequence with respect to the same points. See also DIFFERENCE SEQUENCE.

dividend, *n.* a number or quantity that is to be divided by another number or quantity. Compare DIVISOR.

dividers, *n.* an instrument with two arms hinged together, each ending with a point, used to transfer measurements from one place to another. For example, by placing the points of the dividers on two points of a map and then transferring them to a scale, one can read off the distance between them. Compare COMPASSES.

divisibility, *n.* the capacity of one number or quantity to be exactly divided by another.

divisible, *adj.* (of a number) capable of being exactly divided (by another number).

division, *n.* **1.** the inverse operation of multiplication, calculating the multiplier of a given number that yields a product equal to another number; thus if *a* (the DIVIDEND) is divided by *b* (the DIVISOR), the result *q* (the QUOTIENT), written $a \div b$, a/b, $^a/_b$, or $\frac{a}{b}$, has the property that $bq = a$.
2. see LONG DIVISION.

division algebra, *n.* an ALGEBRA OVER A FIELD in which all non-zero elements have multiplicative inverses. The only commutative and associative division algebras over the real field are the reals (of dimension 1) and the complex numbers (of dimension 2); the QUATERNIONS are a non-commutative associative four-dimensional division algebra, and the CAYLEY ALGEBRA is a non-commutative and non-associative algebra of dimension 8. Compare DIVISION RING. See also FROBENIUS' THEOREM.

division algorithm, *n.* the fundamental result in number theory that for any two natural numbers *a* and *b*, there are two unique others, *q* and *r*, such that $a = qb + r$ and $r < b$. This also holds in a EUCLIDEAN DOMAIN by virtue of the existence of a GAUGE.

division of a segment, *n.* (*Geometry*) the construction of a point that divides a line segment in some desired proportion. See INTERNAL DIVISION, EXTERNAL DIVISION, INTERNAL AND EXTERNAL DIVISION.

division ring, *n.* a RING in which every non-zero element, *a*, has an inverse a^{-1}, such that

$$aa^{-1} = e = a^{-1}a,$$

where *e* is the (multiplicative) IDENTITY element. A commutative division ring is a field; the division ring $\mathbb{H}$ of QUATERNIONS is a non-commutative division ring. Compare QUOTIENT RING.

division sign, *n.* the symbol '$\div$' placed between two numbers or quantities to indicate that the former is to be DIVIDED by the latter; for example, $124 \div 31 = 4$.

divisor, *n.* **1.** the number or quantity that is to be divided into another number or quantity (the DIVIDEND).
2. another word for FACTOR.

divisor function, *n.* (*Number theory*) the function, $d(n)$, that counts the number of divisors of n, including 1 and n. When p is prime, $d(p^a)$ is $a + 1$, and since d is MULTIPLICATIVE, the value for any other argument can be easily computed from its prime factorization.

dn, see JACOBIAN ELLIPTIC FUNCTIONS.

dnf, *abbrev. for* DISJUNCTIVE NORMAL FORM.

dodecagon, *n.* a polygon with twelve sides.

dodecahedron, *n.* a polyhedron with twelve plane faces. All the faces of a *regular dodecahedron* are regular pentagons, and it is one of the five PLATONIC SOLIDS.

domain, *n.* **1a.** also called **essential domain.** the set of values of the independent variables of a given function, partial function, or multi-valued function; the set of all the first members of the ordered pairs that constitute the function. In this sense the domain of the real-valued square root function cannot exceed the non-negative reals, and the mapping f of Fig. 134 has the set S′ as its domain, as only these members have an image under the mapping. Compare RANGE.

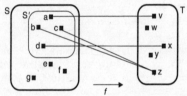

Fig. 134. S is the maximal **domain** and S′ the essential domain of f.

b. also called **maximal domain.** the set on which a given function is defined; the set from which the first members of the ordered pairs that formally constitute the function are drawn. In this sense the domain of the real-valued square root function may be taken to be the real numbers, or only the positive reals, or any other set appropriate to the context; for example, the mapping f defined by the diagram of Fig. 134 has S as its domain. Compare CODOMAIN.

2. an INTEGRAL DOMAIN. See also EUCLIDEAN DOMAIN.

3. a CONNECTED OPEN SET. Compare REGION.

domain of definition, *n.* another term for RANGE OF SIGNIFICANCE.

domain of discourse, *n.* (*Logic*) another term for UNIVERSE OF INTERPRETATION.

dominate, *vb.* see DOMINATED.

dominated, *adj.* **1.** (of a subset in a PARTIAL ORDER) possessing an upper bound that is then said to *dominate* the subset.

2a. (of a sequence of positive terms) such that each element is less than the corresponding member of a given second sequence; that is, $\{a_i\}$ is dominated by $\{c_i\}$ if for every i, $a_i \leq c_i$.

b. more generally, such that the sequence of ABSOLUTE VALUES of the terms of a given sequence of real numbers, or MODULI of the terms of a given sequence of complex numbers, is dominated in the preceding sense by a given second sequence.

dominated convergence theorem, *n.* the theorem of LEBESGUE INTEGRATION that allows one to evaluate the limit of the integrals of a sequence of functions as the integral of the pointwise limit of the functions as soon as the sequence of functions is DOMINATED in absolute value by an integrable function. For example, if $\{f_n\}$ is a sequence of integrable functions, convergent almost everywhere to f, and if there is an integrable g such that $|f_n| < g$ for all n, then f is integrable, and

$$\int f \, d\mu = \lim_{n \to \infty} \int f_n \, d\mu.$$

This result can be generalized to an arbitrary complete measure space. Compare MONOTONE CONVERGENCE.

dot, *n.* the symbol '.' representing the DECIMAL POINT, CONJUNCTION, or MULTIPLICATION. In some of these usages it is often written above the line, as in '$x \cdot y$', although there is an international convention is that decimal points (or commas with the same meaning) are always written on the line. A dot is also the symbol for FLUXIONS, and in this sense is always written superscript centrally above the letter representing the function to be differentiated, as in $\dot{x}, \ddot{x}$.

dot product, *n.* another name for SCALAR PRODUCT.

double-angle formula, *n.* any formula for a TRIGONOMETRIC or HYPERBOLIC function expressing the value of the function at twice one value in terms of that value. For example,

$$\cos 2z = 2\cos^2 z - 1$$

$$\sinh 2x = 2\sinh x \cosh x$$

Compare HALF–ANGLE FORMULA, ADDITION FORMULA.

double false position, see FALSE POSITION.

double integral, *n.* an INTEGRAL of a function with respect to two variables, written

$$\iint f(x, y) \, dx \, dy \quad \text{or} \quad \iint f(x, y) \, dA.$$

Double integrals can be used in the evaluation of single integrals, as, for example, in

$$\int_{-\infty}^{\infty} \exp(-x^2) \, dx = \left[\int_{-\infty}^{\infty} \int_{-\infty}^{\infty} \exp(-(x^2 + y^2)) \, dx \, dy \right]^{1/2} = \sqrt{\pi}.$$

See MULTIPLE INTEGRATION.

double negation, *n.* (*Logic*) the principle, axiom, or rule of inference, according to which a statement is equivalent to, or can be derived from, the negation of its negation. For example, *It is not the case that John is not here* and *John is here* are related in this way. INTUITIONIST logic denies that this relationship holds in both directions: it permits the derivation of its double negation from a given sentence, but not vice versa.

double ordinate, *n.* a line segment between two points on a curve, and parallel to a coordinate axis.

double point, *n.* a point at which a curve intersects itself, such as a CRUNODE. Compare SINGULAR POINT.

double precision, see PRECISION.

double root, *n.* one of a pair of equal roots of the same polynomial or equation; this occurs when the polynomial has $(x-a)^2$ as a factor, where a is the double root.

double ruled surface, see RULED SURFACE.

double sequence, *n.* a doubly-indexed SEQUENCE, such as

$$a_{n,m} = (-1)^{n+m}(n+m).$$

double series, *n.* a doubly-indexed SERIES, such as

$$\sum_{n,m=0}^{\infty} \frac{1}{n^2 + m^2}.$$

If the absolute values sum in any way, then $\sum_n \sum_m$ and $\sum_m \sum_n$ converge and diverge together, and when they converge their sums are equal.

double tangent, *n.* **1.** a line that is TANGENT to a curve at two distinct points. **2.** a pair of distinct but COINCIDENT tangents at the same point of a curve, for example at a cusp.

doubling the cube or **Delian altar problem,** *n.* the traditional geometric problem of constructing, using a straight-edge and compasses only, a cube whose volume is double that of a given cube, named after the oracle at Delos which prescribed doubling the altar as a means of halting the plague of 428 BC. The problem was solved using conics by Apollonius in 3rd century BC, but it was not shown to be impossible using Euclidean constructions until the 18th century, since $\sqrt[3]{2}$ is not a CONSTRUCTIBLE number.

doubly periodic, see PERIODIC FUNCTION.

doubly stochastic, *adj.* (of a matrix) STOCHASTIC in both rows and columns; such that both columns and rows are non-negative and sum to unity. See BIRKHOFF'S THEOREM.

doxastic logic, *n.* the branch of MODAL LOGIC that studies the concept of belief. Compare DEONTIC LOGIC.

dragon curve, see FRACTAL.

drawer principle, *n.* another term for PIGEON–HOLE PRINCIPLE.

dual, *adj.* **1.** (of a pair of algebraic structures) ISOMORPHIC with one another under an exchange of certain operators and perhaps constants, usually involving the distribution of negation over another operator.

2. (of a pair of operators) interchangeable in this way.

3. (of a pair of theorems) derived from one another by such an exchange.

4. (of an operator) another term for ADJOINT (sense 1b).

– *as substantive.* **5.** an entity related to another in one of these ways. For example, the dual of conjunction is disjunction; the dual of the proposition that $P \cup P' = U$ is that $P \cap P' = \varnothing$; and the dual of a given Boolean algebra is another in which union and intersection and the null and universal sets are interchanged.

6. (of a VECTOR SPACE) the vector space of LINEAR FUNCTIONALS on the given vector space; the dual of the dual is ISOMORPHIC to the original space. The dual of a vector space T is often written T*.

7. the vector space of all continuous linear functionals on a given TOPO-LOGICAL VECTOR SPACE.

8. see PRIMAL–DUAL METHODS.

dual isomorphism, *n.* an ISOMORPHISM between a space and its DUAL, especially important in GALOIS THEORY.

duality gap, see STRONG DUALITY.

duality, *n.* the interchangability of two types of entity in a given theory. For example, points and lines in projective geometry, intersection and union in set theory, or existential and universal quantifiers in predicate calculus.

duality theory of linear programming, *n.* the assertion that a *dual pair* of LINEAR PROGRAMS is in STRONG DUALITY, if both are feasible. When the *primal linear program* is written

$$p = \max \{ \langle c, x \rangle : Ax \leq b, x \leq 0 \}$$

with inequalities taken coordinate-wise, then the *dual linear program* is

$$d = \min \{ \langle b, y \rangle : A^*y \geq c, y \geq 0 \},$$

where A^* is the transpose of the original matrix. The strong duality assertion is that the two optimal values (p and d) agree and are attained. These terms are used somewhat analogously to relate non-linear programs.

dual linear program, see DUALITY THEORY OF LINEAR PROGRAMMING.

dual norm, *n.* the NORM placed on the DUAL of a given normed space by

$$\|f\| = \sup \{ |f(x)| : \|x\| = 1 \},$$

for each continuous linear functional. This is the OPERATOR NORM on the dual space. This space is necessarily complete, and so is a BANACH SPACE.

dummy suffix convention, *n.* another term for SUMMATION CONVENTION.

dummy variable, *n.* a variable occurring in a function, but on which the value of the function does not depend. For example, the variable with respect to which a function is integrated, and the variable that indexes an infinite series are dummy variables.

dump, see TRANSSHIPMENT PROBLEM.

duodecimal, *adj.* **1.** of or relating to the number system with base 12.

2. (*as substantive*) a number expressed in PLACE–VALUE NOTATION with base 12, using as digits the numerals 0 to 9 and the letters T or A (ten) and E or B (eleven); for example,

$$2E4_{12} = (2 \times 12^2) + (11 \times 12) + (4 \times 12^0)$$

(= 424 in decimal notation).

duplication of the cube, *n.* another term for DOUBLING THE CUBE.

dyad, *n.* a pair of vectors written without an operator being indicated, and which may form part of either a SCALAR PRODUCT or a VECTOR PRODUCT; if **uv** is the dyad, **uv.w** is defined to be **u**(**v.w**), and **uv**×**w** is **u**(**v**×**w**). The use of dyads has been largely superseded by that of TENSORS.

dyadic, *adj.* **1.** with two as BASE; two-fold. For example, the dyadic rationals are those of which the reduced forms have a power of 2 as denominator.

2. another word for BINARY; for example, a dyadic tree. Compare MONADIC, POLYADIC.

3a. relating to DYADS.

b. (*as substantive*) the sum of two DYADS.

dyadic product, *n.* another term for TENSOR PRODUCT.

dynamical (or **dynamic) system,** *n.* any natural process, such as a feedback loop, or any mathematical model in which each successive state is a function of the preceding state; a STOCHASTIC PROCESS that can be described by a system of differential equations. The *orbit* of the system is the sequence of the repeated compositions of this function, starting from an initial state or *seed*. If the system reaches an equilibrium or cycles between a number of states, the equilibrium points (or sets of equilibrium points) are called the *attractors* of the system. If the initial seed determines the equilibrium state, a system with two or more attractors divides the complex plane into equivalence classes, and the boundary between such regions is typically a FRACTAL curve. In many natural processes, such as turbulence or rapid eye movements, apparently random ('*chaotic*') behaviour is nonetheless deterministic and may be modelled by a dynamical system in which the nature of the attractor depends upon minute variation of some seemingly irrelevant parameter of the system; a point at which the attractor changes is called a *bifurcation*. In a chaotic system, bifurcation may be observed to recur at all levels of analysis or magnification, showing the self-symmetry typical of fractals; in fact such a system has an attractor that bifurcates infinitely often and is a fractal set (a *strange attractor*).

dynamic programming, *n.* the study of discrete or continuous multi-step RECURSIVE optimization problems to which BELLMAN'S PRINCIPLE is applicable.

dynamics, *n.* (*functioning as singular*) **1.** also called **kinetics.** the branch of MECHANICS concerned with the forces that change or produce the motions of bodies. Compare STATICS, KINEMATICS.

2. less commonly, the branch of mechanics that includes STATICS and KINETICS.

3. loosely, any branch of science concerned with forces.

4. the governing differential equations in a CONTROL problem.

e

e, *n.* **1.** also called **Euler number.** the TRANSCENDENTAL NUMBER, the approximate value of which is 2.718281828..., that is defined either as the value of the EXPONENTIAL FUNCTION for *x* = 1, or directly as

$$e = \lim_{n \to \infty} \left(1 + \frac{1}{n}\right)^n$$

or

$$e = \sum_{n=0}^{\infty} \frac{1}{n!} = 1 + 1 + \frac{1}{2} + \frac{1}{6} + \frac{1}{24} + \dots .$$

See also COMPOUND INTEREST.

2. a common notation for the IDENTITY element of a GROUP.

E, *n.* **1.** the number 14 in HEXADECIMAL notation.

2. the number 11 in DUODECIMAL notation.

3. the **complete elliptic integral of the second kind**, a SPECIAL FUNCTION, the formula for which is

$$\int_0^{\pi/2} (1 - k^2 \sin^2\theta)^{1/2} \, d\theta,$$

from which one may compute the arc length of an ellipse. See ELLIPTIC INTEGRAL.

4. (*Statistics*) the operator that yields the EXPECTED VALUE of a RANDOM VARIABLE. See also MEAN, VARIANCE.

5. *abbrev. for* EXA-, used in symbols for multiples of the physical units of the SYSTEME INTERNATIONAL.

Eberlein–Smulian theorem, *n.* the theorem that asserts the equivalence of COMPACTNESS and SEQUENTIAL COMPACTNESS in the WEAK TOPOLOGY for a BANACH SPACE.

eccentric, *adj.* **1.** (of geometric figures) not having a common centre; not CONCENTRIC. For example, the circles in Fig. 135 are eccentric.

2. (of an ellipse or ellipsoid) having widely separated FOCI, and so having ECCENTRICITY close to unity.

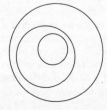

Fig. 135. **Eccentric** circles.

eccentricity, *n.* **1.** also called **linear eccentricity.** a parameter, often denoted *e,* that identifies the shape of a CONIC SECTION as a LOCUS of points such that it is the ratio between the distance of any such point from a given fixed point (the FOCUS) and its distance from a given fixed line (the DIRECTRIX). Clearly this constant is independent of the position, orientation and size of the curve, and so identifies a family of similarly shaped curves.

2. also called **numerical eccentricity.** the ratio of the linear eccentricity to half the length of the MAJOR AXIS of a conic, often denoted ε, that is constant for a family of similar curves. If the equation of the curve is given in VERTEX FORM as

$$y^2 = 2px - (1 - \varepsilon^2)x^2$$

with $2p$ the length of the LATUS RECTUM, then, as shown in Fig. 136, if ε = 0, the curve is a CIRCLE; if ε < 1, the curve is an ELLIPSE; if ε = 1, it is a PARABOLA; and if ε > 1, it is a HYPERBOLA.

Compare ELLIPTICITY.

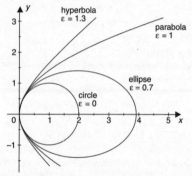

Fig. 136. **Eccentricity**. Graphs of $y^2 = 2x - (1 - \varepsilon^2)x^2$ for ε shown.

ecentre, *n.* another word (mainly US) for EXCENTRE.

echelon form, *n.* a matrix derived from another by a finite sequence of ELEMENTARY OPERATIONS, and, in a *row echelon form,* having the properties that all the non-zero rows precede all the zero rows, and that the first non-zero element in a non-zero row appears to the right of the first non-zero element in the preceding row; in a *column echelon form,* the conditions attaching to rows and columns are interchanged. The relations between such derived echelon forms and the given matrix are respectively ROW EQUIVALENCE and COLUMN EQUIVALENCE. See also REDUCED ECHELON FORM.

ecircle, *n.* another word (mainly US) for ESCRIBED CIRCLE.

economy, *n.* a mathematical model of an economic system in which, commonly, there are m producers, each with a production set P_j, and n consumers, each with a consumption set C_i, and an associated preference ordering $\leq_i$. These sets lie in a EUCLIDEAN SPACE, the dimension of which corresponds to the number of goods in the economy. There is a total resource level w. The producers each produce a vector y_j in P_j and the consumers have a demand x_i in C_i. These $m + n$ elements define a state of the economy. The *excess demand* is

$$\sum x_i - \sum y_j - w,$$

and a *market equilibrium* occurs if excess demand is zero. Individual consumers will try to maximize their preference satisfaction; this leads to the study of equilibria that occur as a market equilibrium together with a pricing of resources such that every producer maximizes profit and every consumer optimizes satisfaction.

edge, *n.* **1.** a line along which two FACES of a solid or two surfaces meet, as shown in Fig. 137.

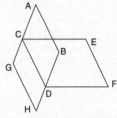

Fig. 137. CD is an **edge** (sense 1).

2. (*Graph theory*) a line segment, which may or may not be DIRECTED, joining two nodes or vertices of a graph.

Edgeworth box, *n.* a diagrammatic representation of a simple two-good exchange ECONOMY in which two consumers' INDIFFERENCE CURVES are drawn in a box with the satisfaction of the first increasing from bottom left to top right corner and that of the second correspondingly decreasing. (Invented by the Irish economist *Francis Ysidro Edgeworth* (1845–1926).) See also CONTRACT CURVE.

Edgeworth–Pareto point, *n.* another term for EFFICIENT POINT.

effective, *adj.* (of a procedure) able to be carried out mechanistically by following a finite number of steps of an algorithm; in particular, a value may be *effectively computable*. Context may determine whether this condition is to be met in principle or in practice. See CHURCH'S THESIS.

efficient code, *n.* (*Information theory*) an ERROR–CORRECTING CODE containing a high ratio of INFORMATION to WORD-length.

efficient point, Pareto optimal point, or **Edgeworth–Pareto point,** *n.* a MINIMAL (or MAXIMAL) point; a non-DOMINATED point.

Egoroff's theorem, *n.* (*Measure theory*) the result that if a sequence of almost everywhere finite and measurable functions converges almost everywhere to a finite limit, on a set of finite measure, E, then there is, for each positive ε, a measurable subset F with measure at most ε such that UNIFORM CONVERGENCE occurs on E \ F. (Named after the Russian analyst, *D. F. Egoroff* (1869–1931).)

Egyptian fraction, *n.* a fraction of the form $1/n$, where n is an integer, so called because the Egyptians largely considered fractions in this form.

eigenfunction, *n.* **1.** another term for CHARACTERISTIC FUNCTION of a matrix.
2. an EIGENVECTOR that is itself a function.

eigenvalue or **eigenroot,** *n.* other terms for LATENT ROOT. See CHARACTERISTIC POLYNOMIAL.

eigenvector, **latent vector**, or **characteristic vector**, *n*. (for a matrix or linear operator, A) a non-zero vector **X**, such that A**X** = λ**X**, so that λ is a LATENT ROOT of A.

Einstein, Albert (1879–1955), German-born American physicist who developed RELATIVITY theory and is popularly considered to be one of the greatest physicists of all time. He published four entirely original papers in 1905, the year he obtained his doctorate; these were on special relativity, on the equivalence of mass and energy, on the particular nature of light, and on Brownian motion. In 1921 he won the Nobel Prize in Physics for his work on the photoelectric effect. He also worked on gravitation, the general theory of relativity, and unified field theory. He fled Nazi persecution and obtained a post at Princeton in 1933, where his work laid the theoretic foundations for the invention of the atomic bomb, and so to the Allied victory over Japan; he was a public advocate of the Zionist movement, and in 1953 he was invited to become President of the recently established State of Israel.

Eisenstein's criterion, *n*. a sufficient condition for a polynomial with integer coefficients to be IRREDUCIBLE (over the rationals or integers) that states that it suffices to find a prime *p* that does not divide the leading coefficient of the polynomial and that does divide all other coefficients, but is such that p^2 does not divide the constant coefficient. For example, the polynomial $x^n - 3$ is irreducible for all *n*, but $x^n - 9$ need not be. The corresponding result holds in a PRINCIPAL IDEAL DOMAIN. (Named after the German number theorist, algebraist and analyst *Ferdinand Gotthold Max Eisenstein* (1823–52).)

elastic, *adj*. **1.** (of a function) having ELASTICITY greater than unity. In economics, demand for a good is said to be elastic if an increase in the price results in an increase in revenue. Compare INELASTIC.

2. (*Continuum mechanics*) (of a BODY) resuming its original shape and size without supply of energy, after an arbitrary deformation; formally, such that the STRESS TENSOR is given by a symmetric tensor-valued function of the DEFORMATION GRADIENT. Compare GREEN–ELASTIC BODY.

elasticity, *n*. **1.** the quantity *e* defined at a point *x* for a function *f* as the absolute value of the derivative of log *f*(*x*) with respect to log *x*. The function is ELASTIC when *e* > 1, and INELASTIC when *e* < 1. In economic terms, if a good is subject to demand *d*(*p*) that is dependent upon the price *p* and so yields revenue *pd*(*p*), then if *d*(*p*) is elastic, the revenue increases as the price *p* increases; if it is inelastic, the revenue decreases as *p* increases.

2. the study of ELASTIC bodies.

electric field, *n*. (*Electromagnetism*) a VECTOR FIELD created by a stationary charge that describes the interactions of that charge with others in its neighbourhood.

electromagnetic field tensor, see MAXWELL'S LAWS.

electromagnetic potentials, *n*. (*Electromagnetism*) a SCALAR FIELD, φ, and a VECTOR FIELD, **A**, in terms of which the ELECTRIC FIELD and MAGNETIC FIELD, **E** and **B** respectively, may be expressed completely. This results in a degree of simplification in the use of MAXWELL'S EQUATIONS, particularly in their relativistic formulation. The basic forms of **A** and φ are given by

$$\mathbf{B} = \nabla \times \mathbf{A}, \ \ \mathbf{E} = -\nabla\phi - \frac{1}{c}\frac{\mathbf{A}}{t},$$

where c is the velocity of light. **A** and φ are not uniquely determined by these conditions and are often chosen to satisfy the *Lorentz condition* that

$$\nabla \cdot \mathbf{A} + \frac{1}{c}\frac{\phi}{t} = 0.$$

electromagnetic radiation, *n.* (*Electromagnetism*) a flow of ENERGY produced when a body which carries charge undergoes ACCELERATION. In a vacuum, all electromagnetic radiation is propagated with constant velocity; this is the velocity of light, which is invariant with respect to the relative velocity of an observer of such radiation. See also ELECTROMAGNETIC WAVE, MAXWELL'S EQUATIONS.

electromagnetic wave, *n.* (*Electromagnetism*) a model of the method of propagation of ELECTROMAGNETIC RADIATION, consisting of an ELECTRIC FIELD and MAGNETIC FIELD oscillating in planes normal to each other and to the direction of propagation of the wave.

element, *n.* **1.** a point, line, plane or part of a geometric figure.

2. also called **member.** one of the objects or numbers that together constitute a set or class; if there is a structure imposed on the set, it may be a group, ring, field, etc. The *p-elements* of a group are those of order p^{α}, where p is a prime, and α is any positive integer.

3. any of the terms in the array constituting a determinant or matrix.

4a. any one of the rectangles summed in an INTEGRAL and represented by the expression following the integral sign; in

$$\int_{a}^{b} f(x) \ \mathrm{d}x,$$

$f(x) \, \mathrm{d}x$ is an element of area; originally, an integral was conceived of as the limit of an infinite sum of INFINITESIMAL elements. See DEFINITE INTEGRAL.

b. (*Continuum mechanics*) in particular, the small quantity of a body surrounding the given point; formally the PARTICLE of a BODY at a given point.

elementarily equivalent, *adj.* (of MODELS) such that every sentence true in one model is true in the other. This is a weaker relation than isomorphism; that is, ISOMORPHIC models are elementarily equivalent, but not necessarily vice versa. For example, the non-standard real numbers and the ordinary real numbers are elementarily equivalent.

elementary, *adj.* **1.** (*Logic*) (of a theory) able to be formalized in FIRST–ORDER PREDICATE CALCULUS. For example, *elementary Peano arithmetic* is the extension of first-order predicate calculus that is obtained by adding to its axioms expressions of the PEANO AXIOMS as well-formed expressions of that theory.

2. pertaining to or consisting of ELEMENTARY FUNCTIONS or OPERATIONS.

3. (*Number theory*) not using techniques involving complex numbers, such as complex analysis. Elementary proofs are not necessarily simple.

elementary column operation, *n.* an ELEMENTARY MATRIX OPERATION on the COLUMNS of a matrix.

elementary divisor, *n.* any of the distinct linear factors of the CHARACTERISTIC POLYNOMIAL of a matrix.

elementary function, *n.* a member of the class of functions built up from the exponential and trigonometric functions and their inverses by RECURSION upon the ELEMENTARY OPERATIONS and COMPOSITION. For example,

$$\log[\tan^{-1}\sqrt{(\exp(x^2) + 1)}]$$

is constructed by successively applying each of the following operations to the result of the preceding: squaring, exponentiating, adding 1, extracting the square root, finding the inverse of the sine and taking the logarithm. Some definitions admit more functions by permitting recursion also upon inversion.

elementary matrix (abbrev. **E-matrix**), *n.* a matrix obtained by applying an ELEMENTARY MATRIX OPERATION to the appropriate identity matrix. Any non-singular matrix is a product of such matrices, and each *elementary row* (or *column*) *matrix* represents respectively an elementary row or column matrix operation.

elementary matrix operation (abbrev. **E-operation**), *n.* **1.** one of the operations of multiplying a row or column of a matrix by a scalar, adding a scalar multiple of another row or column to a given one, or transposing two rows or columns; these are known as *elementary row operations* and *elementary column operations* respectively. If pre- or post-multiplication by a matrix is equivalent to such a transformation, it is called an ELEMENTARY MATRIX; for example, if θ is the elementary row operation of interchanging the first and second rows, and if A is any given matrix, then $\theta(A) = EA$, where, for an identity matrix I conformable with A,

$$E = \theta(I) = \begin{bmatrix} 0 & 1 & 0 & 0 & \dots \\ 1 & 0 & 0 & 0 & \dots \\ 0 & 0 & 1 & 0 & \dots \\ \dots & \dots & \dots & \dots \end{bmatrix}.$$

2. the derivation of one matrix from another by any such operation.

elementary operation , *n.* **1.** one of the mathematical operations of addition, subtraction, multiplication, division and extraction of integral roots.

2. see ELEMENTARY MATRIX OPERATION.

elementary reduction, *n.* (*Group theory*) the operation that forms a WORD from another by replacing any term of the form xx^{-1} or $x^{-1}x$ by the EMPTY WORD; or the word so formed.

elementary row operation, *n.* an ELEMENTARY MATRIX OPERATION on the ROWS of a matrix.

elementary symmetric function, see SYMMETRIC FUNCTION.

elevation, another word for INCLINATION.

eliminant, *n.* another term for RESULTANT.

eliminate, *vb.* **1.** to remove one or more variables from consideration in a system of SIMULTANEOUS EQUATIONS by using ELEMENTARY OPERATIONS to derive another system of the same number or fewer equations in which those variables do not occur. For example, the variable y can be eliminated from the pair of equations

$$x + y = 3$$
$$x - 2y = 5$$

by multiplying the former by 2 and adding it to the latter to give the single equation $3x = 11$. See also GAUSSIAN ELIMINATION.

2. to remove a common element from the numerator and denominator of a fraction, or from both sides of an equation, thereby simplifying it and yielding an equivalent expression. For example, we can eliminate y from

$$3x + 2y = 4z + 2y$$

to give $3x = 4z$. See also CANCEL.

elimination rule, *n.* (*Logic*) any syntactic rule specifying the conditions under which a formula or statement containing a specified operator may permit the valid derivation of others that do not contain it. For example, *conjunction elimination* is the rule permitting the derivation of either P alone or Q alone from P & Q; *universal elimination* permits the derivation of *Fa* from $(\forall x)Fx$. Compare INTRODUCTION RULE.

ellipse, *n.* a closed geometric figure shaped like an elongated circle and symmetric about two axes of different lengths (the MAJOR and MINOR AXES); the CONIC SECTION with ECCENTRICITY less than 1. In Fig. 138 the axes of symmetry are both shown, and the eccentricity is the ratio PF/PX, where F is a FOCUS and X the foot of the perpendicular from the variable point P to the DIRECTRIX DE.

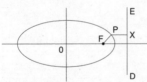

Fig. 138. An **ellipse** with focus F and directrix DE.

An ellipse is formed by the intersection of a bounded nappe of a right circular CONE with a plane that does not cut its base; that is, it is the projection of a circle onto another non-parallel plane, and it is the locus of points for which the sum of the distances from the two foci is constant. The canonical equation of an ellipse is

$$\frac{x^2}{a^2} + \frac{y^2}{b^2} = 1,$$

when the ellipse is symmetrical about the origin as shown, and intersects the axes at the points $(\pm a, 0)$ and $(0, \pm b)$; its parametric equations are

$$x = a\cos\theta, \quad y = b\sin\theta.$$

It has its foci at $(\pm ae, 0)$, where e is the eccentricity, and

$$ae = \sqrt{a^2 - b^2}.$$

The area is then πab, but there is no general closed form for the perimeter without using ELLIPTIC INTEGRALS.

ellipsoid or **spheroid,** *n.* **1a.** a geometrical surface or solid that is symmetrical about its three axes, and the plane sections of which are circles or ellipses. When its axes coincide with the coordinate axes its equation is

$$\frac{x^2}{a^2} + \frac{y^2}{b^2} + \frac{z^2}{c^2} = 1,$$

where the points $(\pm a, 0, 0)$, $(0, \pm b, 0)$ and $(0, 0, \pm c)$ are the intercepts with the *x*-, *y*-, and *z*-axes respectively.
b. the corresponding multidimensional surface or solid. In standard form an ellipsoid arises as a LEVEL SET

$$\{ x : \langle x, Ax \rangle \leq 1 \}$$

where A is a POSITIVE DEFINITE matrix.

ellipsoid method, *n.* (*Optimization*) one of a class of POLYNOMIAL–TIME ALGORITHMS for LINEAR PROGRAMMING that find a feasible point for a system of linear inequalities by generating a uniformly shrinking sequence of multidimensional ELLIPSOIDS, the centres of which will ultimately be feasible if any feasible point exists, and which diagnose infeasibility otherwise. See KHACHIYAN'S METHOD.

ellipsoid of revolution, see SURFACE OF REVOLUTION.

elliptical, *adj.* relating to or having the shape of an ELLIPSE.

elliptic curve, *n.* a CURVE defined by a POLYNOMIAL equation with rational coefficients, with GENUS 1 and containing at least one point with rational coordinates.

elliptic equation, *n.* (of a second-order PARTIAL DIFFERENTIAL EQUATION) having DISCRIMINANT $b^2 - 4ac$ negative, where

$$au_{xx} + bu_{xy} + cu_{yy} + du_x + eu_y + fu = h$$

is the general form of a second-order partial differential equation.

elliptic function, *n.* a non-ELEMENTARY TRANSCENDENTAL FUNCTION that can be defined as the inverse of certain ELLIPTIC INTEGRALS; a DOUBLY PERIODIC MEROMORPHIC function. Gauss was led to the theory of elliptic functions by his determination of the arc length of the LEMNISCATE. See also JACOBIAN ELLIPTIC FUNCTIONS, WEIERSTRASS ELLIPTIC FUNCTION.

elliptic geometry or **Riemannian geometry,** *n.* a NON–EUCLIDEAN GEOMETRY in which a line has no parallels through any given point; in the simplest setting, it has a model on the surface of a sphere, on which lines are represented by GREAT CIRCLES. Compare LOBACHEVSKIAN GEOMETRY.

elliptic integral, *n.* a DEFINITE INTEGRAL not usually evaluable in CLOSED FORM by ANTIDIFFERENTIATION. More precisely, an elliptic integral is an integral of the form

$$\int_c^u R(x, y) \, dx,$$

where R is a rational function of x and y, where y^2 is a QUARTIC polynomial in x, and where c is a fixed constant. The integral is called a *complete elliptic integral* if the range of integration is maximal; otherwise, it is an *incomplete elliptic integral*. The two most fundamental elliptic integrals are

$$K(k) = \int_0^{\pi/2} (1 - k^2 \sin^2\theta)^{-1/2} \, d\theta$$

(the *complete elliptic integral of the first kind*), and

$$E(k) = \int_0^{\pi/2} (1 - k^2 \sin^2\theta)^{1/2} \, d\theta$$

(the *complete elliptic integral of the second kind*).
For $0 < u < \pi/2$

$$F(k, u) = \int_0^u (1 - k^2 \sin^2\theta)^{-1/2} \, d\theta$$

is an incomplete elliptic integral of the first kind. When $u = \pi/2$, this coincides with $K(k)$. Elliptic functions can be used to evaluate the period of a pendulum with amplitude α and length L as

$$4 \sqrt{\frac{L}{g}} \; K(\sin\frac{\alpha}{2}),$$

which yields the SIMPLE HARMONIC APPROXIMATION $2\pi\sqrt{(L/g)}$ for small amplitudes. Similarly, the length of the circumference of an ellipse in standard form is $4aE(e)$, where a is the length of the major axis, and e is the eccentricity. See ARITHMETIC–GEOMETRIC MEAN ITERATION.

ellipticity, *n.* the degree of deviation from a circle or sphere of an ellipsoidal shape or path, measured as the ratio of the major to minor axes. See also ECCENTRICITY.

elliptic paraboloid, see PARABOLOID.

E-matrix, *abbrev. for* ELEMENTARY MATRIX.

embedding, *n.* another word for INJECTION.

empirical, *adj.* **1.** derived from experience rather than from logical principles alone. Compare *A PRIORI*.

2. (*Logic*) strictly, not able to be known independently of experience. This is thus an epistemological property, and hence distinct from the logical property of being SYNTHETIC.

empirical probability, *n.* (*Statistics*) the POSTERIOR PROBABILITY of an event derived on the basis of its observed frequency in a sample. For example, suppose we have obtained 52 heads in 100 tosses of coin A and 43 heads in 100 tosses of coin B; if we now toss one of these coins without knowing whether it is A or B, and the outcome is tails, then, by BAYES' THEOREM, the empirical probability that it is coin B is approximately 54.3%. Compare MATHEMATICAL PROBABILITY.

empty, *adj.* **1.** (of a set or class) having no members. There is in fact only one empty set, since identity for sets is defined solely in terms of their having the same members. Consequently, since there are neither chimeras nor unicorns, both sets are empty and so are identical; whence all chimeras are unicorns. The symbol '$\emptyset$' denotes the empty set. See also NULL.

2. (*Logic*) (of a name or description) vacuous, having no reference.

empty word, *n.* (*Group theory*) any entity that is not a NON–EMPTY WORD; multiplication is extended to the empty word, denoted **1**, by providing that **1**.$u = u$, for any word u.

enantiomorphic, *adj.* (of a pair of asymmetric figures in Euclidean geometry) such that each is the mirror image of the other; such that the given figures are not SUPERPOSABLE, but a reflection of either is CONGRUENT to the other. Such a pair are said to be of opposite *handedness*; for example, a pair of gloves, and a left- and right-handed trihedral are pairs of *enantiomorphs*. A pair of enantiomorphic figures in n dimensions are congruent in $n+1$ dimensions.

encrypt, *vb.* to transform data into code. See CODING THEORY.

encryption, *n.* transforming data into code. See CODING THEORY.

End, *abbrev. for* the set of ENDOMORPHISMS of an algebraic structure.

endecadic, *adj.* of or relating to eleven.

endomorphism (abbrev. **End**), *n.* a HOMOMORPHISM of a structure into itself.

endow, *vb.* to define a relation or function (on a given structure) in order to consider them together as a single mathematical entity. For example, the metric space (S, δ) consists of the set S endowed with the metric δ.

endpoint, *n.* **1.** a MAXIMAL or MINIMAL point of a SEGMENT on a line or INTERVAL; for example, the endpoints of the interval $[0, 1]$ are the points 0 and 1. **2.** more generally, either of the ends of a curvilinear ARC; that is, the image of either of the endpoints (in the preceding sense) of the unit interval under a given mapping.

energy balance equation, *n.* (*Continuum mechanics*) the result that the rate of change of the sum of the KINETIC ENERGY and INTERNAL ENERGY of a SUB-BODY is equal to the sum of the POWER and the HEATING of that sub-body.

energy principle, *n.* (*Mechanics*) the consequence of the LAWS OF MOTION that in a system of discrete PARTICLES with all forces CONSERVATIVE or doing no WORK, the sum of the KINETIC ENERGY and POTENTIAL ENERGY is constant.

ennea- or **nona-,** *prefix denoting* nine. For example, an *ennead* is a sequence of nine elements; an *enneagon* is a polygon with nine sides; and an *ennea-hedron* is a polyhedron with nine faces.

entail, *vb.* (*Logic*) to have as a necessary consequence.

entailment, *n.* (*Logic*) **1.** a statement to the effect that one statement is a necessary consequence of another. For example, although *Vivian is a non-male human* implies that she is a woman, it does not entail that conclusion, since it is merely an accidental fact rather than a necessary truth that all humans are either male or female. Such a proposition is usually symbolized $P \rightarrow Q$, where P and Q are the component statements; the symbol '$\rightarrow$' is often informally called *fish-hook*. **2.** the relationship that holds between the two statements of which this is true.
Compare IMPLICATION.

entire, *adj.* (of a complex function) ANALYTIC at all points of the finite complex plane; for example, $f(z) = e^z$ is an entire function.

entire surd, see SURD.

entropy, *n.* **1.** the quantity, for a measurable partition ξ of a probability space with probability P,

$$H(\xi) = - \sum_{C \in \xi} P(C) \log_2 P(C),$$

where H(ξ) is taken to be infinite if the partition is uncountable. For a discrete random variable X whose i^{th} outcome has probability p_i, this becomes

$$H(X) = -\sum_{i=0}^{k} p_i \log_2 p_i,$$

where the choice of the base of the logarithm is only one of convenience. This corresponds to the expected value of the INFORMATION FUNCTION of the partition.

2. (*Statistical physics*) the macroscopic variable representing the degree of disorder within a system and corresponding to a statistical description of the system by the relation that the entropy is equal to the EXPECTED VALUE of $k\log P$, where k is a constant that relates the mean KINETIC ENERGY and ABSOLUTE TEMPERATURE of a system, and P is the COEFFICIENT OF PROBABILITY of the system.

entry, *n.* one of the elements constituting a matrix, determinant, vector, or array, usually regarded in terms of its position, so that a_{ij} is the entry at the intersection of the ith row and the jth column.

enumerable, *adj.* another word for DENUMERABLE.

envelope, *n.* a curve or surface that is tangential to each of a family of curves or surfaces. Fig. 139 shows an annulus as the envelope of the family of circles with radius a and centre at a distance b from a given fixed point P.

Fig. 139. The annulus is the **envelope** of the family of circles.

E-operation, *abbrev. for* ELEMENTARY MATRIX OPERATION.

epi, *adj.* (*Category theory*) (of an ARROW, $h: a \rightarrow b$) having the property that its compositions with distinct arrows are necessarily distinct; that is, for any arrows g_1 and g_2: $b \rightarrow c$, if $g_1 \circ h = g_2 \circ h$ then $g_1 = g_2$.

epicycle, *n.* **1.** a circle that rolls around the inside or outside of another circle, thereby generating an EPICYCLOID (as shown in Fig. 140 overleaf) or HYPOCYCLOID.

2. (*Ptolemaic astronomy*) a small circle the centre of which moves around a larger circle. When the centre of the larger circle was taken to be the Earth, these were thought to represent the paths of planets through the heavens. Epicycles were used to construct extremely accurate descriptive mechanical models of the solar system and thereby to predict the position of the planets relative to the Earth. However, when the heliocentric model of the solar system succeeded in describing the same data with inherently greater simplicity, in terms of elliptical paths one of whose foci was the Sun, both the theory and the associated cosmology were abandoned.

epicycloid, *n.* the curve described by a point on, or rigidly attached to, the circumference of a circle (the *epicycle*) as this circle rolls round the outside of another fixed coplanar circle. A COMMON epicycloid, such as the CARDI-

OID, is one in which the point that describes the curve is on, rather than within or outside, the circumference of the circle. For example, the curve of which a section is shown in Fig. 140, is a common epicycloid; P_1 is the initial position of the generating point, and P_2 is another position. Compare CYCLOID, HYPOCYCLOID.

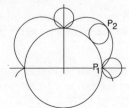

Fig. 140. Part of a common **epicycloid**.

epigraph, *n.* the set of points that lie on or above the graph of a real-valued function; the set of points (x, y) such that $y \geq f(x)$. For example, Fig. 141 shows the epigraph of $y = x^2$. A real-valued function is CONVEX precisely if the corresponding epigraph is convex. Compare LEVEL SET.

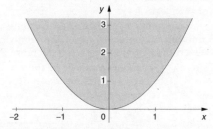

Fig. 141. The **epigraph** of $y = x^2$.

epimorphism, *n.* a SURJECTIVE HOMOMORPHISM. See MORPHISM. See also NATURAL EPIMORPHISM. Compare ISOMORPHISM, MONOMORPHISM .

epistemic logic, *n.* the branch of MODAL LOGIC that seeks to represent the relationships of the concepts of knowledge, belief and ignorance. Compare ALETHIC.

epsilon, *n.* **1.** the symbol, 'ε', conventionally used for a small but strictly positive quantity. See EPSILON–DELTA NOTATION.
2. a small but strictly positive quantity, especially the maximum difference between any pair of members of a set. For example, one writes ε-*sequence*, ε-*neighbourhood*. See CONTINUOUS. See also EPSILON NET.
3. the symbol

$$\varepsilon_{i_1, i_2, \ldots, i_k},$$

for the SIGNATURE of a PERMUTATION; this is a generalization of KRÖNECKER'S DELTA.

epsilon-delta notation, *n.* the standard notation used in the definitions of LIMITS, CONTINUITY, and related concepts; the principal notion is that a function tends to a limit at a given point if its value lies within some small ε (*epsilon*) of the limit whenever the independent variable is within a small δ (*delta*) of the given argument. Formally, a real function $f(x)$ has a limit L at a point p, at which it is defined, if it satisfies the condition:

for every $\varepsilon > 0$ there exists a $\delta > 0$ such that

$|f(x) - L| < \varepsilon$ for all x such that $|x - p| < \delta$.

More generally, a function has a limit at a point p in a METRIC SPACE at which it is defined if it satisfies the same condition where the absolute differences $|x - p|$ and $|f(x) - L|$ are replaced by the corresponding metric distances, or in NEIGHBOURHOOD notation, if

for every $\varepsilon > 0$ there exists a $\delta > 0$ such that

$f(x) \in N(\varepsilon, L)$ for all x such that $x \in N(\delta, p)$.

In each case, the function is then said to be CONTINUOUS at p if its limit at p exists and equals $f(p)$.

epsilon neighbourhood, see NEIGHBOURHOOD.

epsilon net, *n.* a finite or infinite set of points in a METRIC SPACE such that each point of the space is within DISTANCE ε of some point in the set.

equal, *vb.* **1.** (of an algebraic or arithmetic expression) to have as its value, the latter being specified as some other arithmetic or algebraic expression. Thus, the result of a computation, or the solution of an EQUATION are written as, for example, $7 + 5 = 12$ and $x = 5$ respectively.

2. (of two mathematical expressions) to refer to the same quantity or entity; for example $x^2 + 2y = 2x + y^2$ when x equals y. See EQUATION, IDENTITY.

equality, *n.* **1.** the condition or state of being EQUAL.

2. a statement indicating that the quantities on either side of the equals sign are equal in value, or that the expressions on either side of the sign have the same reference; usually a CONDITIONAL EQUATION rather than an IDENTITY. Compare INEQUALITY.

equals sign or **equal sign,** *n.* the symbol '=' used between two expressions to indicate the identity of their references or values. DEFINITIONS are often written using this symbol either with 'df' subscript to the right, with a small triangle superscript, or simply with the word 'definition' or its abbreviation 'defn' in the same line; the equals sign written with a colon to the left is used both for definitions, and in computing for the ASSIGNMENT of values to variables, where

$$x := x + y$$

means that the new value of the variable labelled x is to be the sum of the previous values of x and y.

equate, *vb.* to form an EQUATION by joining two expressions with the equals sign; to assert that two expressions are equal in value.

equate coefficients, *vb.* to conclude from the fact that two POLYNOMIALS are identical that the COEFFICIENTS of the terms must be equal, whence the ROOTS of the polynomial equation can be derived. For example, if the roots of $x^2 + ax + b = 0$ are α and β, then

$$x^2 + ax + b = (x - \alpha)(x - \beta) = x^2 - (\alpha + \beta)x + \alpha\beta.$$

Since the coefficients of x and the constant terms (the coefficients of x^0) must be the same in both expressions, we can conclude that $\alpha + \beta = -a$ and $\alpha\beta = b$.

equation, *n.* a formula that asserts that two expressions have the same value; it is either an *identical equation* (usually called an IDENTITY), which is true for any values of the variables, or a *conditional equation*, which is only true for certain values of the variables (the ROOTS of the equation). For example, $x^2 - 1 = (x+1)(x-1)$ is an identity, and $x^2 - 1 = 3$ is a conditional equation with roots $x = \pm 2$.

equation of continuity, see SPATIAL EQUATION OF CONTINUITY.

equations of motion, *n.* **1.** (for a BODY) see EULER'S LAWS OF MOTION.
2. (for a PARTICLE) see NEWTON'S LAWS OF MOTION.
3. (for a set of PARTICLES) see HAMILTON'S PRINCIPLE OF LEAST ACTION.

equator, *n.* a circle dividing a sphere or other surface into two equal symmetrical parts, such as the horizontal circle in Fig. 142.

Fig. 142. **Equator**.

equi-, *prefix* denoting equality.

equiangular, *adj.* **1.** (of a geometrical figure) having all angles equal. The figure of Fig. 145, opposite, is equiangular, but those of Fig. 143 are not.
2. (of a pair of figures) each having angles that are equal to the corresponding angles of the other; thus the trapezia in Fig. 143 are equiangular in this sense but not in the preceding.

Fig. 143. These figures are **equiangular** in sense 2 but not sense 1.

equiangular spiral, *n.* another term for LOGARITHMIC SPIRAL.

equicontinuous, *adj.* **1.** (of a family, F, of real functions between METRIC SPACES, at a point, *c*) such that for every $\varepsilon > 0$ there is a single $\delta > 0$ such that for all functions f in F

$$|f(x) - f(c)| < \varepsilon \text{ for all } x \text{ such that } |x - c| < \delta;$$

that is, δ depends only on ε, and does not, as might be expected, vary from function to function in F.
2. (more generally, of a family, F, of functions from a TOPOLOGICAL SPACE, X, to a METRIC SPACE, at a point, *p*, of X) such that for every $\varepsilon > 0$ there is an open NEIGHBOURHOOD, U, of *p*, such that for all functions f in F,

$$d[f(x), f(p)] < \varepsilon \text{ whenever } x \in U.$$

See CONTINUOUS. Compare UNIFORM CONTINUITY. See also UNIFORM BOUNDEDNESS PRINCIPLE.

equidistant, *adj.* being the same distance from one or more given points, lines, etc. For example, the perpendicular bisector of a line, as in Fig. 144, is the set of points equidistant from its endpoints; a circle is the locus of points equidistant from its centre.

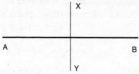

Fig. 144. Any point on XY is **equidistant** from A and B.

equilateral, *adj.* (of a geometrical figure) having all sides of equal length. Any equilateral plane figure is also EQUIANGULAR, and so is REGULAR, as is the hexagon of Fig. 145.

Fig 145. The hexagon is both **equilateral** and equiangular.

equilibrium, *n.* **1.** the state of a system of vectorial quantities at a point in which their RESULTANT is equal to zero.

2. (*Mechanics*) the state in which a given mechanical system remains at rest, which occurs when the total FORCE and the total TORQUE are both zero.

3. (*Physics*) a stable condition in which a system of forces is balanced or distributed in the most efficient manner; that is, a system is in equilibrium at a point at which its POTENTIAL curve is concave up.

4. (*Economics*) see MARKET EQUILIBRIUM, ECONOMY.

equilibrium point, *n.* a point $\mathbf{y}^\varepsilon$ such that, for a given ORDINARY DIFFERENTIAL EQUATION $\mathbf{y}' = f(\mathbf{y})$, $f(\mathbf{y}^\varepsilon) = \mathbf{0}$.

equimeasurable, *adj.* (of a pair of functions, f and g) such that both functions are real-valued and MEASURABLE, and

$$\mu(\{\, f(x) : f(x) > y \,\}) \; = \; \mu(\{\, g(x) : g(x) > y \,\})$$

for all real y; functions that agree almost everywhere are equimeasurable.

equinumerous, *adj.* (of two classes) having the same number of members.

equipollent or **equipotent,** *adj.* **1.** (*Logic*) (of statements or propositions) deducable from one another, EQUIVALENT.

2. (of two classes) having the same CARDINALITY. The *Schröder–Bernstein theorem* establishes that two sets are equipollent if there is an INJECTIVE mapping from each to the other.

equipotent, *adj.* **1.** another word for EQUIPOLLENT.

2. (*Mechanics*) another word for EQUIVALENT (sense 9).

equiprobable, *adj.* (of a number of events) having the same probability. For example, in a fair game of dice with a single die, the occurrence of each face is an equiprobable outcome. The *principle of indifference* is that in the absence of any reason to the contrary, all elementary events should be considered equiprobable.

equivalence, *n.* **1.** the relation that holds between two statements when they are EQUIVALENT, that is, when each implies the other.

2. see EQUIVALENCE RELATION.

3. (*Logic*) also called **biconditional. a.** the binary TRUTH–FUNCTIONAL sentential connective, corresponding to the English *if and only if*, that forms a sentence that is true when either both the component sentences are true or both are false, and is false when only one of them is true. Its TRUTH–TABLE is shown in Fig. 146.

P	Q	P ≡ Q
T	T	T
T	F	F
F	T	F
F	F	T

Fig. 146.
Truth-table for **equivalence**.

b. a sentence that has this as its main connective, usually written: P ≡ Q or P ↔ Q.

c. the relation that holds between two statements when the sentence so formed is true.

See also CONDITIONAL, IMPLICATION.

equivalence class, *n.* a class the elements of which are all those members of the underlying set that are related to one another by an EQUIVALENCE RELATION on it. Since each of the equivalence classes is closed under the equivalence relation, and since two elements belong to the same equivalence class if and only if the relation holds between them, the set of all such classes constitutes a PARTITION of the underlying set. For example, each of the n distinct residue classes modulo n is an equivalence class under congruence mod n; the classes of measurable functions that agree almost everywhere are equivalence classes.

equivalence relation, *n.* a relation that is REFLEXIVE, TRANSITIVE, and SYMMETRIC; it imposes a PARTITION on its domain of definition, so that two elements belong to the same subset if and only if the relation holds between them. For example, since two integers are congruent modulo n if and only if their difference is divisible by n, congruence mod n is reflexive ($a \equiv a$ for all a), transitive (if $a \equiv b$ and $b \equiv c$, then $a \equiv c$), and symmetric ($a \equiv b$ if and only if $b \equiv a$), so that this is an equivalence relation; hence it can be proved that, since every integer has a unique smallest positive remainder when divided by n, and so is congruent mod n to only one of the integers between 1 and n, these congruence classes are both disjoint and exhaustive of all the integers, and so constitute a partition of them.

equivalent, *adj.* **1.** (of two geometric figures) having some particular property in common, such as similar triangles or parallelograms of the same height on the same base.

2. (*Logic*) (of two statements or propositions) related by EQUIVALENCE, each implying the other, being either both true or both false. For example, given the facts of terrestrial anatomy, *This creature has a heart* and *This creature has kidneys* are equivalent, even although the two statements may have different truth values in some other possible world, so that their equivalence is not necessary.

3. (of two equations or inequalities) having the same SOLUTION SET.

4. (of two sets) having the same CARDINALITY.

5. (of two fractions) reducible to the same proper fraction, and so representing the same rational number. For example, $\frac{2}{4}$ and $\frac{3}{6}$ are distinct when regarded formally as fractions, but are equivalent since both represent the rational number $\frac{1}{2}$.

6. (of two matrices A and B) such that there exist NON–SINGULAR matrices C and D with A = CBD.

7. (of two metrics on a given set) generating the same TOPOLOGY, in that the same topology results from using the open balls for either metric as a BASE.

8. (of two IDEALS in an INTEGRAL DOMAIN) related in such a way that there are elements a and b for which $(a)I = (b)J$, where I and J are the given ideals and (a) and (b) are the PRINCIPAL IDEALS generated by a and b respectively.

9. (*Mechanics*) also called **equipollent.** (of two systems of forces) having the same VECTOR SUM and the same TORQUE about the same point.

10. see REPRESENTATION.

equivalent norms, *n.* two norms for which the associated distance functions are EQUIVALENT; that is, $\| \; \|_1$ and $\| \; \|_2$ are equivalent if there are positive constants M and N such that

$$M\| \; \|_1 \; \leq \; \| \; \|_2 \; \leq \; N\| \; \|_1.$$

All norms on a given Euclidean space are equivalent, and so give rise to the same topology.

eradius, *n.* another term for EXRADIUS.

Eratosthenes, see SIEVE OF ERATOSTHENES.

Erdös, Paul (1913–96), one of the most prolific mathematicians of all time, who solved many, and posed more, difficult problems in NUMBER THEORY, GRAPH THEORY, and other areas especially in the application of PROBABILITY within DISCRETE MATHEMATICS. Born in Budapest, he travelled ceaselessly, while co-authoring more than 1500 papers, and is the source of at least as many anecdotes. Mathematicians fancifully define their *Erdös number* as a measure of how close they are to collaboration with him: the 485 who jointly wrote papers with him have an Erdös number of 1, and someone who published jointly with someone with Erdös number n has number $n+1$. This has generated a literature, often pseudonymous, about the properties of the *Erdös collaboration graph* in which points represent mathematicians and vertices represent collaboration.

erect, *vb.* to draw or construct (a line or other geometrical figure) on a given figure, especially perpendicular to it.

ergodic, *adj.* (of a MEASURE–PRESERVING TRANSFORMATION on a measure space) having only trivial invariant subsets. See BIRKHOFF ERGODIC THEOREM.

ergodic hypothesis, *n.* the principle, central to statistical physics, that for a system in STATISTICAL EQUILIBRIUM all accessible STATES have an equal probability of realisation so that the system will pass rapidly through all such states.

ergodic set, *n.* a MINIMAL set of STATES of a MARKOV CHAIN such that the probability of exiting from the set is zero, a minimal STOCHASTICALLY CLOSED set of states.

Erlangen programme or **Erlanger programme,** *n.* (*Geometry*) the influential 19th-century codifying programme for mathematics, enunciated in 1872 at Erlangen by FELIX KLEIN, and based on his famous algebraic definition of geometry as 'the study of those properties of a set that remain invariant when the elements of the set are subjected to the transformations of some transformation group.'

error, *n.* **1.** the difference between some quantity and an approximation to or estimate of it, often expressed as an absolute or relative range, such as ± 5 mm, or ± 5%. See also ABOLUTE ERROR, RELATIVE ERROR.

2. (*Statistics*) see TYPE I ERROR, TYPE II ERROR.

error-correcting code, *n.* (*Information theory*) a mathematical coding system that can recognize and correct a certain proportion of errors in a CODE. Such codes can be constructed using BLOCK DESIGNS so that distinct words of the code are distinguishable in a number of different ways, or are widely separated when regarded as, for example, binary numbers. Compare ERROR-DETECTING CODE. See also HAMMING CODES.

error-detecting code, *n.* (*Information theory*) a mathematical coding system that can recognize but not correct errors in a CODE. See CODING THEORY. Compare ERROR-CORRECTING CODE.

escribed circle, ecircle, or **excircle,** *n.* one of the circles tangent to one side and to the extensions of the other two sides of a triangle, the centre of which is an EXCENTRE and the radii EXRADII, as shown in Fig. 147. Every triangle has three escribed circles.

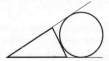

Fig. 147. **Escribed circle**. The circle is one excircle of the triangle.

essential, *adj.* (*Logic*) (of a property) holding of its subject in every POSSIBLE WORLD, so that nothing could be that subject if it lacked that property. So, for example, bipedal locomotion is an essential property of the human species, but not of persons, since there may be creatures in other worlds that we would regard as persons – for example, for the purposes of morality – but that do not walk upright. The intelligibility of such an account, however, is disputed. See *DE RE.*

essential boundedness, *n.* the property of being ESSENTIALLY BOUNDED.

essential domain, *n.* the subset of a given UNIVERSE OF DISCOURSE on which a PARTIAL function, relation, or predicate is defined. This term is useful to distinguish the sense of *domain* in which it is correlative to RANGE from that in which it is correlative to CO–DOMAIN.

essentially bounded, *adj.* (of a MEASURABLE FUNCTION) having the the property that there exists a constant C such that the set $\{x : |f(x)| > C\}$ has zero measure. The infimum of such constants is called the *essential supremum* of $|f|$. For example, if $f(x) = 1/x$ for rational x, and $f(x) = \sin x$ for all other real x, then $f(x)$ is essentially bounded.

essential singularity, *n.* an ISOLATED SINGULARITY of a complex function that is neither a REMOVABLE SINGULARITY nor a POLE. A function f has an isolated singularity at infinity if $f(1/z)$ has one at the origin.

essential supremum, see ESSENTIALLY BOUNDED.

estimate, *vb.* **1.** to calculate an approximate value for an expression.

2. (*Statistics*) to assign a value (a *point estimate*) or a range of values (an *interval estimate*) to a PARAMETER of a population on the basis of SAMPLING STATISTICS. See ESTIMATOR.

3. *n.* any STATISTIC that is intended as an approximation to the correct value of a PARAMETER of a distribution; in particular, a SAMPLE STATISTIC.

estimator, *n.* (*Statistics*) a derived RANDOM VARIABLE that generates ESTIMATES of a PARAMETER of a given distribution, such as $\overline{X}$, the mean of a number of IDENTICALLY DISTRIBUTED random variables, X_i. If $\overline{X}$ is UNBIASED, its observed value, $\overline{x}$, should be close to the EXPECTED VALUE $E(X_i)$. See also SAMPLING STATISTIC.

Euclid, (pronounced *You*-clid) Greek mathematician of 3rd century BC Alexandria, credited with the first axiomatic treatment of geometry in his *Elements*, which also treat proportion and number, including irrationality. He also wrote works on astronomy and on conics (now lost). The *Elements* reached the West in translation from the Arabic, and exerted a profound influence; until very recently, school geometry textbooks were little more than translations of Euclid.

Euclidean, *adj.* of, derived from, or relating to EUCLIDEAN GEOMETRY.

Euclidean algorithm, *n.* another term for EUCLID'S ALGORITHM.

Euclidean construction, *n.* the CONSTRUCTION of a geometric figure using only compasses and a straight edge, the latter being used only to draw lines and not for measurement.

Euclidean distance or **Cartesian distance,** *n.* the standard DISTANCE in EUCLIDEAN SPACE, calculated as the square root of the sum of the squares of the arithmetical differences of the corresponding coordinates of the two points,

$$d(\mathbf{x}, \mathbf{y}) = \sqrt{\sum_{i=1}^{n} (x_i - y_i)^2},$$

where $\mathbf{x} = \langle x_1, x_2, ..., x_n \rangle$, and $\mathbf{y} = \langle y_1, y_2, ..., y_n \rangle$. Thus, in two-dimensional Euclidean space, the distance $|AB|$ between $A = (a_1, a_2)$ and $B = (b_1, b_2)$ is

$$\sqrt{(a_1 - b_1)^2 + (a_2 - b_2)^2}.$$

See also EUCLIDEAN TOPOLOGY.

Euclidean domain or **Euclidean ring,** *n.* an INTEGRAL DOMAIN in which the DIVISION ALGORITHM holds by reason of the existence of a GAUGE or valuation function. A Euclidean domain is a PRINCIPAL IDEAL DOMAIN, and the polynomials over any field are a Euclidean domain with the gauge being the DEGREE of the polynomial. See GAUSSIAN DOMAIN.

Euclidean geometry, *n.* the geometrical system in which EUCLID'S AXIOMS are satisfied; this is in essence the geometry described by EUCLID (although his description was deficient). Euclidean geometry is contrasted with the NON-EUCLIDEAN GEOMETRIES described by Riemann and Lobachevski. In particular the PARALLEL POSTULATE is satisfied, so that exactly one line can be drawn parallel to a given line through any point not on the line.

Euclidean norm, *n.* the norm imposed on a vector by taking the square root of the sum of the squares of its entries. See also EUCLIDEAN TOPOLOGY, FROBENIUS NORM.

Euclidean point space, *n.* (*Mechanics*) the set $\mathbb{R}^n$ of *n*-tuples over the reals (for *n* a positive integer); an *n*-dimensional real AFFINE MANIFOLD.

Euclidean space or **Cartesian space,** *n.* **1.** a finite-dimensional real or complex VECTOR SPACE possessing a SCALAR PRODUCT, so that a EUCLIDEAN DISTANCE may be imposed. Most commonly, one begins with the *n*–fold CARTESIAN PRODUCT of the real or complex FIELDS. This represents the most common and useful abstraction and extension of the mathematical representation of the three-dimensional space of daily experience in terms of CARTESIAN COORDINATES, and is referred to as *Euclidean n-space*, where *n* is the dimension of the space.

2. less commonly, any finite or infinite INNER PRODUCT SPACE.

Euclidean ring, *n.* another term for EUCLIDEAN DOMAIN.

Euclidean topology, *n.* the TOPOLOGY induced on $\mathbb{R}^n$ (the space of *n*-tuples of reals) by taking the METRIC to be the EUCLIDEAN DISTANCE function, and constructing the corresponding neighbourhoods. See also EUCLIDEAN NORM.

Euclidean vector space, *n.* (*Mechanics*) the VECTOR SPACE $\mathbb{R}^n$ of *n*-tuples over the reals (for *n* a positive integer).

Euclid numbers, *n.* the even PERFECT NUMBERS.

Euclid's algorithm or **Euclidean algorithm,** *n.* an iterative method of finding the HIGHEST COMMON FACTOR (HCF) of two integers, polynomials, or elements in a EUCLIDEAN DOMAIN by dividing the larger by the smaller, then the smaller by the remainder of that first division, then remainder of the first division by the remainder of the second, and so on, until the process terminates with remainder zero. For example, to find the HCF of 56 and 12, divide 56 by 12, leaving remainder 8; $12 \div 8$ has remainder 4; now $8 \div 4$ has zero remainder, so that the last divisor, 4, is the required HCF.

Euclid's axioms, *n.* the axioms for EUCLIDEAN GEOMETRY that assert that:

> a straight line may be drawn from any point to any other point;
>
> a finite straight line may be extended continuously in a straight line;
>
> a circle may be described with any centre and any radius;
>
> all right angles are equal to one another; and
>
> if a straight line meets two other straight lines so as to make the sum of the two interior angles on one side of the transversal less than two right angles, the other straight lines, extended indefinitely, will meet on that side of the transversal.

Equivalently, the last axiom may be replaced by PLAYFAIR'S AXIOM.

Eudoxus' axiom or **method of exhaustion,** *n.* the classical resolution to the paradoxes of DICHOTOMY and of the existence of INCOMMENSURABLES, named after the Greek astronomer and mathematician *Eudoxus of Cnidus* (?406 – ?355 BC). In Euclid (proposition X.1) it is given as:

> two unequal magnitudes being set out, if from the greater there be subtracted a magnitude greater than its half, and from that which is

left a magnitude greater than its half, and if this process be repeated continually, there will be left some magnitude that will be less than the lesser magnitude set out.

In Archimedes' hands this became a powerful method of computing volumes and areas (see ARCHIMEDES' METHOD), and was not equalled as a way of describing irrationality until the introduction of DEDEKIND CUTS.

Euler, Leonhard (1707–83), (pronounced *Oi*–ler) Swiss-born mathematician and physicist, who worked mainly in St Petersburg, where he followed the Bernoullis, and in Berlin, at the invitation of Frederick the Great. He was renowned for his ability to perform complex calculations mentally, and continued his work even after he lost his sight. He was one of the most prolific mathematicians of all time, publishing over 500 papers and textbooks on virtually all branches of mathematics (a further 350 have appeared posthumously). His most significant contributions were to analytical geometry, calculus and trigonometry, and thereby to the unification and systematization of all of mathematics.

Euler–Bernoulli law, *n.* (*Mechanics*) the law stating that the BENDING MOMENT of a thin beam is EIκ, where κ is the CURVATURE of the beam, E is the YOUNG'S MODULUS for the material, and I is the MOMENT OF INERTIA of the cross-section about an axis through its CENTRE OF MASS and perpendicular to the plane of the couple.

Euler chain, see EULERIAN CHAIN.

Euler characteristic, *n.* (*Graph theory, algebraic topology*) an invariant of a surface such that for all graphs able to be appropriately embedded in the surface,

$$\text{vertices} + \text{faces} - \text{edges} = e(S).$$

This formula is an extension of EULER'S FORMULA to topological surfaces other than a SPHERE (sense 4). In fact,

$$e(S) = 2 - 2k - j,$$

where k is the number of HANDLES and $j = 0, 1, 2$ is the number of CROSSCAPS of the surface; for the TORUS, $e(S) = 0$.

Euler differential equation, *n.* another name for EULER'S EQUATION.

Euler equation, *n.* an ORDINARY DIFFERENTIAL EQUATION of the form

$$(t + a)^2 y'' + b(t + a) y' + cy = 0,$$

that is solved by writing $t + a = e^\theta$.

Eulerian angles, *n.* (*Mechanics*) the angles θ, ϕ, ψ that determine the rotation of a RIGID BODY about a fixed point O by locating a set of CARTESIAN AXES, OXYZ, fixed in the body in terms of Cartesian axes, Oxyz, fixed in space. The angles θ and ϕ are the SPHERICAL COORDINATES of OZ with respect to Oxyz, and ψ is the angle between the planes OXZ and OzZ.

Eulerian (or **Euler**) **chain** or **trail,** *n.* a CHAIN in a graph that uses each EDGE exactly once.

Eulerian circuit, *n.* a CIRCUIT in a graph that uses each EDGE exactly once.

Eulerian description, *n.* another term for SPATIAL DESCRIPTION.

Eulerian graph, *n.* a GRAPH whose edges form a single CIRCUIT. See EULERIAN CIRCUIT.

Eulerian strain rate, *n.* (*Continuum mechanics*) the symmetric part of VELOCITY GRADIENT; that is, the Eulerian strain rate

$$\Sigma = \tfrac{1}{2}\,(\mathbf{L} + \mathbf{L}^{\mathrm{T}})$$

where **L** is the velocity gradient. Compare BODY SPIN.

Eulerian walk, *n.* another term for TRAIL.

Euler–Lagrange equations, *n.* any of the fundamental NECESSARY CONDITIONS in the CALCULUS OF VARIATIONS. For the simplest problems this requires that the EXTREMAL y_0 that minimizes the functional

$$\int_a^b f(y, y', x)\,\mathrm{d}x$$

should satisfy EULER'S EQUATION at any point where y_0' is smooth. At points where this fails one has also additional *Weierstrass–Erdman corner conditions.*

Euler line, *n.* the line in a triangle on which the ORTHOCENTRE, CENTROID, and CIRCUMCENTRE lie. In the triangle in Fig. 148, X is the point of intersection of the altitudes, Z is the centre of the circumcircle shown, and L, M and N are the midpoints of the sides AB, BC and CA respectively, so that Y is the centroid; thus XYZ is the Euler line.

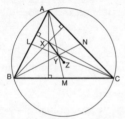

Fig. 148. XYZ is the **Euler line** of triangle ABC.

Euler–Maclaurin summation formula, *n.* another term for the EULER SUMMATION FORMULA.

Euler–Mascheroni constant, *n.* another name for EULER'S CONSTANT.

Euler multiplier, *n.* another term for INTEGRATING FACTOR.

Euler number, *n.* another term for the constant *e*.

Euler phi function or **totient,** *n.* the number theoretic function, written $\varphi(m)$, that counts the number of distinct reduced RESIDUE CLASSES of an integer. φ is MULTIPLICATIVE and so determined by its value on prime powers, for which

$$\varphi(p^{k+1}) = p^k(p-1).$$

An extension of FERMAT'S LITTLE THEOREM due to Euler is that for all integers a relatively prime to m one has the CONGRUENCE

$$a^{\varphi(m)} \equiv 1 \pmod{m}.$$

Euler's circles, *n.* (*Logic*) a diagram in which the terms of CATEGORIAL statements are represented by circles: intersecting circles, as in the first diagram in Fig. 149 opposite, represent classes the intersection of which is not empty; one circle drawn inside another, as in the second diagram, represents the inclusion of one class in the other; and non-intersecting circles, as in the third diagram in the figure, represent disjoint classes. This technique is less sophisticated than VENN DIAGRAMS.

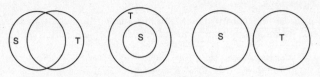

Fig. 149. **Euler's circles** showing the three possible relations of two sets.

Euler's constant or **Euler–Mascheroni constant,** *n*. the constant 0.5772157... (often denoted γ) that is the limit as *n* tends to infinity of the difference between the sum of *n* terms of the HARMONIC SERIES and the NATURAL LOGARITHM of *n*. It is not known whether this number is algebraic or transcendental, nor even whether it is rational or irrational.

Euler's criterion, see LEGENDRE SYMBOL.

Euler's equation, *n*. the DIFFERENTIAL EQUATION

$$\frac{\partial f}{\partial y} = \frac{d}{dx}\left(\frac{\partial f}{\partial y'}\right),$$

the satisfaction of which is required by the EULER–LAGRANGE EQUATIONS, where *f* is a function of *x*, *y*, and *y′*, and *y* is a function of *x* with derivative *y′*.

Euler's equations of motion, *n*. **1.** (*Mechanics*) equations of motion for a RIGID BODY with a fixed point in terms of its PRINCIPAL AXES.

2. (*Continuum mechanics*) the equivalence

$$\mathbf{F} - \frac{1}{\rho}\nabla p = \frac{\partial \mathbf{v}}{\partial t} + \mathbf{v}.\nabla\mathbf{v},$$

where **F** is the applied FORCE, ρ is the DENSITY, *p* is the PRESSURE, *t* is the TIME, and **v** is the VELOCITY at a point. If **F** is CONSERVATIVE, then $\mathbf{F} = -\nabla V$, and $(1/\rho)\nabla p = \nabla P$, so that

$$\frac{\partial \mathbf{v}}{\partial t} - \mathbf{v} \times \text{curl } \mathbf{v} = -\nabla(V + P + \tfrac{1}{2}v^2).$$

Euler's formula, *n*. **1.** (*Graph theory, algebraic topology*) the relation that holds between the numbers of faces, edges and vertices of a three-dimensional polyhedron:

$$\text{vertices} + \text{faces} - \text{edges} = 2.$$

This formula extends to plane GRAPHS as

$$\text{nodes} + \text{regions} - \text{arcs} = 2,$$

where the exterior of the graph is counted as a region. It also extends to graphs on topological surfaces other than a sphere and gives rise to the EULER CHARACTERISTIC of the surface.

2. the identity e^{iz} or $\exp(iz) = \cos z + i \sin z$. See also DE MOIVRE'S FORMULAE.

Euler's laws of motion, *n*. the axiomatic laws of motion for CONTINUUM MECHANICS that state that the force acting on any SUB-BODY is equal to the rate of change of its linear MOMENTUM, and that the TORQUE on any sub-body is equal to the rate of change of its ANGULAR MOMENTUM. Although these are equivalent for finite sets of PARTICLES, their equivalence has not been shown for a continuous body. Compare NEWTON'S LAWS OF MOTION, HAMILTON'S PRINCIPLE OF LEAST ACTION.

Euler (or **Euler–Maclaurin**) **summation formula,** *n.* the asymptotic integration formula for a repeatedly continuously differentiable function on [*a, b*]:

$$\int_a^b f(t)\ dt = \sum_{j=a}^b f(j) - \frac{f(a)+f(b)}{2} - \sum_{k=1}^m B_{2k} \frac{f^{(2k-1)}(b)-f^{(2k-1)}(a)}{2k!} + R_n(x),$$

where B_i are BERNOULLI NUMBERS and the error $R_n(x)$ is less than

$$\frac{4}{(2\pi)^{2n}} \int_1^x \left| f^{2n}(t) \right|\ dt.$$

This is often useful to accelerate convergence of an integral.

Euler trail, see EULERIAN CHAIN.

evaluate, *vb.* to determine the unique member of the RANGE of a FUNCTION corresponding to a given member of its DOMAIN. For example, to evaluate $y = x^2$ at $x = 3$ is to calculate the value of 3^2.

even, *adj.* **1.** (of a number) exactly divisible by two; equal to $2n$ for some integer n. Thus the even numbers form the infinite sequence 2, 4, 6, 8, ... **2.** (of a function) changing neither sign nor absolute value when the sign of the independent variable is changed, so that $f(x) = f(-x)$. The graph of such a function is thus symmetrical about the *y*-axis, as illustrated by the graph of the cosine function shown in Fig. 150.

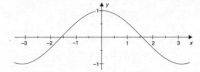

Fig. 150. Cos*x* is an **even** function.

3. (of a PERMUTATION) derivable from the natural ordering by an even number of exchanges of single elements. For example, (3 1 2) is derived from (1 2 3) by exchanging 3 and 1 and then 3 and 2, but (1 3 2) requires an odd number of such pairwise exchanges. See also DIFFERENCE POLYNOMIAL. Compare ODD. See also PARITY.

event, *n.* (*Statistics*) an element of a PROBABILITY SPACE; a possible outcome of a TRIAL. For example, the events that constitute the possible outcomes of a trial involving throwing a pair of dice three times are triples of pairs each member of which is one of the numbers 1 to 6; there are thus $(6^2)^3$ such events in the sample space.

eventually, see NET CONVERGENCE.

evolute, *n.* a curve that describes the locus of the CENTRES OF CURVATURE of another curve (the INVOLUTE) to which its tangents are normal. Fig. 151 opposite shows the evolute, E, of a curve, I, as the ENVELOPE of the NORMALS to the given curve.

evolution, *n.* an algebraic operation in which a root of a number or expression is extracted. Compare INVOLUTION.

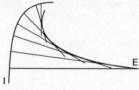

Fig. 151. E is the **evolute** of I.

evolution, *n.* an algebraic operation in which a root of a number or expression is extracted. Compare INVOLUTION.

exa- (symbol **E**), *prefix* denoting a multiple of 10^{18} of the physical units of the SYSTEME INTERNATIONAL.

exact, *adj.* **1.** another word for ACCURATE.

2. (of a DIFFERENTIAL EQUATION) obtained by putting the EXACT DIFFERENTIAL of a function equal to zero, so that when written in the form

$$y'g(x, y) - h(x, y) = 0,$$

the left-hand side is the derivative of some $F(x, y)$. This holds if and only if the coefficients of the differentials in each variable are continuously differentiable and satisfy the INTEGRABILITY CONDITION $g_x = h_y$. The possibility exists that one can usefully obtain exactness by multiplying both g and h by the INTEGRATING FACTOR $m(x, y)$.

3. (more generally, of a DIFFERENTIAL FORM) being the derivative of another form. This forces the form to be CLOSED, and if the region on which it is defined is SIMPLY CONNECTED then this form is also exact (*Poincare's lemma*). Compare CONSERVATIVE VECTOR FIELD.

exact differential, *n.* another term for TOTAL DIFFERENTIAL.

exact divisor, *n.* a FACTOR of a given inteter.

exact line search, see LINE SEARCH METHOD.

exact numerical quantifier, see NUMERICAL QUANTIFIER.

exact sequence, *n.* a sequence of mappings between objects such that the KERNEL of each mapping is the RANGE of the previous one. A *short exact sequence* is a five-term exact sequence of which the initial and terminal objects are trivial. In other words one has $g : X \rightarrow Y$ and $h : Y \rightarrow Z$ such that g is a MONOMORPHISM while h is an EPIMORPHISM whose kernel is the range of g. Compare COMMUTATIVE DIAGRAM.

examination paradox, see UNEXPECTED EXAMINATION PARADOX.

example, *n.* a specific INSTANCE of a general proposition. See also COUNTER-EXAMPLE.

excentre or **ecentre,** *n.* the centre of the ESCRIBED CIRCLE of a triangle.

excess demand, see ECONOMY.

exchange property, see MATROID.

excircle, *n.* another term for ESCRIBED CIRCLE.

excluded middle, *n.* (*Logic*) the principle that every proposition is either true or false, that is, that there is neither a third TRUTH VALUE nor do any statements lack truth value. This is a theorem of CLASSICAL logic but not of INTUITIONIST logic; in the former it is always known for any statement that either it or its negation is true without it necessarily being known which, so that DILEMMA is an acceptable method of proof in classical mathematics. However, in intuitionist logic, the disjunction of a statement and its negation can only be justifiably asserted if one or other of its disjuncts can be justifiably asserted, so that dilemma is not valid. For example, it follows classically from the identity

$$[(\sqrt{2})^{\sqrt{2}}]^{\sqrt{2}} = 2$$

that there exists a pair of irrational numbers α and β for which α^β is rational; the proof proceeds from the fact that we are entitled classically to assert that either $(\sqrt{2})^{\sqrt{2}}$ is rational or it is irrational. Hence, if it is irrational, then the given identity proves the result, but if it is not irrational, then it is rational, and so is itself an instance of the required existential proposition. This argument fails intuitionistically in the absence of a CONSTRUCTIVE determination that $(\sqrt{2})^{\sqrt{2}}$ is irrational, while the usual proof of this by the GELFOND–SCHNEIDER THEOREM is non-constructive.

exclusive, see MUTUALLY EXCLUSIVE. Compare EXHAUSTIVE.

exclusive disjunction, **exclusive or**, or **non-equivalence,** *n.* (*Logic*) **1.** the TRUTH–FUNCTIONAL binary sentential connective that forms a compound sentence that is true whenever one or other but not both of its DISJUNCTS is true; its TRUTH–TABLE is shown in Fig. 152. If no qualification is expressed, disjunction is usually taken to be INCLUSIVE DISJUNCTION, but where exclusive disjunction is explicitly intended, it is sometimes written 'P $\vee$ Q', and often read 'P *aut* Q'; regarded as non-equivalence it is sometimes written 'P $\not\equiv$ Q'. See also DISJUNCTION.

P	Q	P $\vee$ Q
T	T	F
T	F	T
F	T	T
F	F	F

Fig. 152. **Exclusive disjunction.**
Truth-table for exclusive 'or'.

2. the relation that holds between two sentences when the statement so formed is true.

3. a sentence with this as its main connective, such as 'either the treasurer or the secretary will be elected president'.

exhaustive, *adj.* (of a set of sets of elements of some domain) covering the whole domain; having a union equal to the whole domain. For example, the residue classes modulo n constitute an exhaustive COVERING of the integers. Compare MUTUALLY EXCLUSIVE. See also PARTITION, EUDOXUS' AXIOM .

existence, *n.* **1.** in CLASSICAL logic or mathematics, the fact that some putative entity has properties that can be deduced from the theory in question; that is, that some OPEN SENTENCE yields a statement that is true within the theory when the name of that putative entity is substituted for the variable.

2. in INTUITIONIST logic or mathematics, the fact that it is possible to CONSTRUCT some putative entity by a stepwise procedure within the theory.

3. more properly, in either classical or intuitionist logic, the fact that a property has an instance, or that a set has a member. Existence is thus a property not of individuals but of properties or classes; if F is a property, then an *F* exists (in a theory) if $(\exists x)(Fx)$ is true. An individual, *a*, can be said derivatively to exist if $(\exists x)(x = a)$ is true.

existential, *adj.* (*Logic*) **1a.** denoting or relating to a statement, proposition, or formula that asserts the existence of at least one object satisfying a specific condition, or that contains an EXISTENTIAL QUANTIFIER.

b. (*as substantive*) an existential statement or formula.

2. (*as substantive*) an EXISTENTIAL QUANTIFIER.

existential generalization or **existential introduction,** *n.* (*Logic*) the INTRODUCTION RULE for the EXISTENTIAL QUANTIFIER, that permits the inference of $(\exists x)(Fx)$ from any instance *Fa*; for example, from the statement that Montmorency is a unicorn, it can be validly inferred that unicorns exist.

existential instantiation or **existential elimination,** *n.* (*Logic*) the ELIMINATION RULE for the EXISTENTIAL QUANTIFIER, that licenses the inference of a conclusion from an existential premise, when the same conclusion can be derived from an instance without using any other premises about the subject of that instance.

existential quantifier, *n.* (*Logic*) the symbol that indicates that the OPEN SENTENCE that follows is true of at least one member of the relevant UNIVERSE, written $(\exists x)$; $(\exists x)(Fx)$ may represent *something is* (*an*) *F*, *something Fs*, or *there are* (*some*) *Fs*.

exp, *abbrev. and symbol for* the EXPONENTIAL FUNCTION.

expand, *vb.* to express (some quantity or expression) in an extended but equivalent form; thus $(x + y)^2$ can be expanded as

$$x^2 + 2xy + y^2,$$

an expansion that is generalized by the BINOMIAL THEOREM.

expansion, *n.* an expression equivalent to a given expression but in a form appropriate for some specific purpose, especially a sum of terms derived by DISTRIBUTION of all multiplications over additions. For example, the full expansion of $(x + y)(a + b)$ is $ax + ay + bx + by$.

expectation, *n.* (*Statistics*) **1.** another term for EXPECTED VALUE; for example, *expectation of life*.

2. a former term for the numerical PROBABILITY that an event will occur.

expected utility, *n.* (*Statistics*) the WEIGHTED AVERAGE UTILITY of the possible outcomes of a probabilistic situation; the EXPECTED VALUE of the UTILITY function, that is, the sum or integral of the product of the PROBABILITY DISTRIBUTION and the utility function.

expected value or **mathematical expectation,** *n.* (*Statistics*) the sum or integral, over all possible values of a RANDOM VARIABLE, of the product of the value of the variable, or some given function of it, with the probability of that value; this is written $E[f(X)]$. Thus

$$E[f(X)] = \int_a^b f(x)\,\mathrm{p}(x)\,\mathrm{dx},$$

where [*a*, *b*] is the range of values of the variable, and $p(x)$ is its probability function. $E(X)$ is the MEAN of the distribution, and $E[(X - E(X))^2]$ is its VARIANCE. See also MOMENT.

experimental condition, *n.* (*Statistics*) **1.** one of the distinct states of affairs or values of the independent variable for which the dependent variables are measured in order to carry out statistical tests or calculations. For example, a test of a new drug may involve two experimental conditions, in one of which subjects receive the new drug and in the other a placebo.

2. in particular, a condition in which there is some intervention by the experimenter, as contrasted with the CONTROL CONDITION. In this sense, in the example above, only those subjects who receive the drug are subject to the experimental condition.

experimental design, see DESIGN.

explicit, *adj.* (of a function) equating the dependent variable directly with a function of the independent variable, as in $y = f(x)$, so that its values may be directly calculated from those of the independent variables. Compare IMPLICIT.

explicit definition, see DEFINITION.

exponent or **index,** *n.* a number or expression written superscript to another and indicating the POWER to which the latter is to be raised. Positive integral exponents indicate the number of times a term is to be multiplied by itself; for example, $a^3 = a \times a \times a$. The rules for the manipulation of exponents are

$$x^a x^b = x^{a+b}; \quad (x^a)^b = x^{ab}; \quad x^a y^a = (xy)^a,$$

whence all real and complex exponents can be defined; in particular,

$$x^0 = 1; \quad x^{-a} = 1/x^a; \quad x^{1/a} = {}^a\sqrt{x}.$$

Expressions involving real or complex exponents can be expressed in terms of the EXPONENTIAL FUNCTION using the identity

$$a^b = \exp[b(\ln a)],$$

and this is the basis of the use of LOGARITHMS in the evaluation of arithmetic products and quotients. More generally, analogous INDEX LAWS hold in GROUPS.

exponential, *adj.* **1.** (of a function, curve, series or equation) involving, expressible, or describable in terms of the EXPONENTIAL FUNCTION.

2. more generally, involving or describable by an expression containing powers or EXPONENTS, for example x^y.

3. (of any quantity) growing in accordance with a formula expressible in exponential terms. For instance, the national debt grows exponentially.

4. (*as substantive*) an EXPONENTIAL FUNCTION.

exponential decay, *n.* reduction in the value of a quantity over time in which the time required for the quantity to halve its value is constant, whatever that value. This length of time is called the *half-life*, and is a useful measure of the rate of decay. Where $y = Ae^{kt}$, for constants $A > 0$ and k, and t represents time, then when $k < 0$, y exhibits exponential decay. (Compare EXPONENTIAL GROWTH).

exponential distribution, *n.* (*Statistics*) a continuous single parameter distribution used especially when making statements about the lengths of life of certain materials or waiting times between randomly occurring events.

Its PROBABILITY DENSITY FUNCTION is

$$p(x) = \lambda e^{-\lambda x}$$

for positive λ and non-negative x, and it is a special case of the GAMMA DISTRIBUTION.

exponential function (abbrev. **exp**), *n.* the real or complex function defined as the sum of the EXPONENTIAL SERIES,

$$\exp z \;=\; \sum_{n=0}^{\infty} \frac{z^n}{n!} \;=\; 1 + z + \frac{z^2}{2} + \frac{z^3}{6} + \frac{z^4}{24} + \ldots;$$

furthermore, $\exp z = e^z$, where

$$e \;=\; \lim_{n \to \infty} \left(1 + \frac{1}{n}\right)^n \;=\; 1 + 1 + \frac{1}{2} + \frac{1}{6} + \frac{1}{24} + \ldots \;=\; \sum_{n=0}^{\infty} \frac{1}{n!}.$$

This function is the unique solution to the differential equation $y' = y$ with $y(0) = 1$ and so is its own derivative, and is the inverse of the NATURAL LOGARITHMIC FUNCTION, so that

$$\exp(\ln x) \;=\; x \;=\; \ln(\exp x)$$

for all x for which the functions are defined; its graph is shown in Fig. 153. It is the basis of the definitions of the HYPERBOLIC FUNCTIONS, and satisfies EULER'S FORMULA

$$\exp(iy) \;=\; \cos y + i\sin y.$$

For real arguments, $\exp x$ tends to 0 as x tends to $-\infty$.

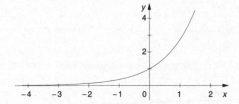

Fig. 153. The graph of the **exponential function,** $\exp x$.

2. any function closely related to the exponential function, and in particular $y = a^x$, for any a.

exponential growth, *n.* increase in the value of a quantity over time in which the rate of change of the quantity is proportional to its value at that time, so that the rate of increase increases with time. Where $y = Ae^{kt}$, for constants $A > 0$ and k, and t represents time, then when $k > 0$, y exhibits exponential growth, and will eventually outgrow any constant power of t. EXPONENTIAL DECAY is negative exponential growth.

exponential matrix, *n.* the matrix

$$e^{At} \;=\; \sum_{j=0}^{\infty} \frac{t^j A^j}{j!},$$

used in the solution of systems of ORDINARY DIFFERENTIAL EQUATIONS.

exponential notation, *n.* another term for SCIENTIFIC NOTATION.

exponential series, *n.* the series

$$\sum_{n=0}^{\infty} \frac{z^n}{n!} = 1 + z + \frac{z^2}{2} + \frac{z^3}{6} + \frac{z^4}{24} + \dots;$$

which converges for any complex number *z* to the EXPONENTIAL FUNCTION exp *z*.

exponential time, see POLYNOMIAL TIME.

exponentiate, *vb.* to raise (a number or quantity) to a POWER.

exportation, *n.* (*Logic*) a rule for separating conjuncts in the antecedent of a conditional, deriving *if P, then if Q then R* from *if P & Q, then R*. Compare IMPORTATION.

express, *vb.* to transform into other equivalent terms. For example, the product $(x + y)(x - y)$ can be expressed as a difference of squares as $x^2 - y^2$.

expression, *n.* any symbol or WELL-FORMED string of symbols of a particular theory. This is the most inclusive term for all the elements of a FORMAL CALCULUS, such as its FORMULAE, names, variables, predicates, relations, functions, sentences and SEQUENTS.

exradius or **eradius,** *n.* a radius of an ESCRIBED CIRCLE of a triangle.

extended, *adj.* (of a CYCLOID, EPICYCLOID or HYPOCYCLOID) described by a point attached to but lying outside, rather than on or within, the circumference of a circle as it rolls without slipping round some other figure; for example, the figure shown in Fig. 154 is an extended cycloid. Compare COMMON, CONTRACTED.

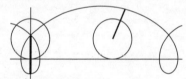

Fig. 154. An **extended** cycloid.

extended plane, *n.* the set of complex numbers together with a POINT AT INFINITY, often symbolized $\mathbb{C}^*$; this is equivalent to the RIEMANN SPHERE. Compare COMPLEX PLANE.

extended precision, see PRECISION.

extended real numbers, *n.* the set $[-\infty, \infty]$ of real numbers, together with the positive and negative INFINITE CARDINALS, often symbolized $\mathbb{R}^*$. Often the field and order operations are partially extended to $[-\infty, \infty]$, so that, for example, $r + \infty = +\infty$ for $r \neq -\infty$; the extended real line is then order equivalent and topologically equivalent to $[0, 1]$. See REAL LINE. See also EXTENDED PLANE, COMPACTIFICATION.

extension, *n.* (*Logic*) **1a.** a MAPPING, the DOMAIN and RANGE of which are supersets of those of the original mapping, and such that the RESTRICTION of the larger mapping to the original domain agrees with the original mapping. For example, the principal complex square root is an extension of the positive square root defined for positive numbers.

b. a function or operator defined on a SUPERSPACE of the domain of a given function and coinciding for arguments for which both are defined. See also HAHN–BANACH THEOREM and TIETZE EXTENSION THEOREM.

2. also called **conservative extension.** a FORMAL THEORY whose PRIMITIVE TERMS, FORMATION RULES, and AXIOMS include those of a given theory, and that contains the given theory in the sense that everything true in the given theory is also true in the extended theory. For example, Zermelo–Frankel set theory is a conservative extension of Peano arithmetic, and first-order predicate calculus of sentential calculus. Compare REDUCT.

3. (*Algebra*) **a.** (of a RING) a ring H, of which a given ring, G, is an IDEAL such that the FACTOR RING H/G is isomorphic to N, where G is extended by N.

b. (of a GROUP) a group H, of which a given group, G, is a NORMAL SUB-GROUP such that the FACTOR GROUP H/G is isomorphic to N, where G is extended by N.

4. (*Logic*) the class of entities to which a given expression correctly applies. For example, the extension of the phrase *satellite of Mars* is the set of which the only members are Phobos and Deimos. Compare INTENSION.

extension field, *n.* (of a smaller FIELD) a field containing the smaller field as a sub-field. See also ALGEBRAIC NUMBER FIELD.

extensional, *adj.* (*Logic*) explicable purely in terms of EXTENSIONS, that is, ignoring differences of meaning that do not affect the extension. For example, since substitution of co-referential terms may change a necessary truth into a contingent one, the truth-value of statements of necessity is not a function of the references of the terms of these statements, and so necessary truth is not an extensional notion. Compare INTENSIONAL. See also EXTENSIVE, EXTENSIONALITY, SUBSTITUTIVITY, TRANSPARENT.

extensionality, *n.* the principle and axiom of set theory stating that sets are definable in terms of their elements alone, however they may have been selected. For example,

$$\{a, b\} = \{b, a\} = \{x : x \text{ is one of the first two letters of the alphabet}\}.$$

extensive, *adj.* (*Logic*) **1.** of or relating to EXTENSION.

2. (of a definition) in terms of the objects to which the term applies rather than a property that they satisfy, for example defining a set as $\{1, 2, 3\}$ rather than $\{x : 1 \leq x \leq 3\}$.

exterior, *n.* the INTERIOR of the COMPLEMENT of a set, or, equivalently, the complement of its CLOSURE.

exterior algebra, see EXTERIOR PRODUCT.

exterior angle, *n.* an angle contained between one side of a polygon and the extension of the adjacent side, such as the angle ABX in Fig. 155.

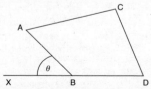

Fig. 155. θ is an **exterior angle** of ABCD.

exterior differential, *n.* the differential of a DIFFERENTIAL FORM, producing a $(k + 1)$-form from a k-form.

exterior multiplication, see MODULE.

exterior penalty function, see PENALTY FUNCTION.

exterior point, *n.* a point lying simultaneously on two tangents to a given CONIC, such as the point E in Fig. 156. Compare INTERIOR POINT.

Fig. 156. E is an **exterior point** of the ellipse.

exterior product, *n.* the unique associative PRODUCT on COVECTORS satisfying

$$\omega \wedge (\zeta + \nu) = (\omega \wedge \zeta) + (\omega \wedge \nu),$$
$$(c\omega) \wedge \zeta = c(\omega \wedge \zeta),$$

and, for $\omega = \alpha_1 \wedge \alpha_2 \wedge ... \wedge \alpha_n$ a product of 1–covectors,

$$\omega(h_1,..., h_n) = \det[\alpha_k h_i].$$

The algebra is then called the *exterior algebra* or the *Grassman algebra*. Compare VECTOR PRODUCT.

external direct product, *n.* **1.** another name for the EXTERNAL DIRECT SUM of groups under multiplication.

2. also called **Cartesian product.** an EXTERNAL DIRECT SUM over infinitely many spaces in which the sum is not required to be finite.

external direct sum, *n.* the CARTESIAN PRODUCT

$$M = M_1 \times M_2 \times ... \times M_n$$

of a finite or infinite set of MODULES over a ring, but with only finitely many non-zero entries in each sequence, where addition and multiplication by a member of the ring, r, are defined as

$$\langle x_1, ..., x_n \rangle + \langle y_1, ..., y_n \rangle = \langle x_1 + y_1, ..., x_n + y_n \rangle$$
$$r\langle x_1, ..., x_n \rangle = \langle rx_1, ..., rx_n \rangle,$$

where each x_i is a member of M_i. The structure M so defined is an R–MODULE and is then denoted

$$M = M_1 \oplus M_2 \oplus ... \oplus M_n.$$

Also, if N_i is the set of *n*-tuples that agree with the members of M_i in the ith position and are zero elsewhere, then the external direct sum of the M_i is isomorphic to the INTERNAL DIRECT SUM of the N_i. Analogous structures may be defined for RINGS and GROUPS. See also EXTERNAL DIRECT PRODUCT.

external division (**of a segment**), *n.* (*Geometry*) the construction of a point E lying beyond the endpoints A and B of a given line segment AB, in such a way that the ratio of the DIRECTED lengths $|AE|$ to $|EB|$ is some given negative number λ; that is, as shown in Fig. 157, the segments AE and EB have opposite sense. Compare INTERNAL DIVISION, INTERNAL AND EXTERNAL DIVISION.

A   B   E   Fig. 157. E **externally divides** AB.

extract, *vb.* **1.** to find the value of (a ROOT).

−*n.* **2.** an uncommon word for a (LOCAL) MAXIMUM or MINIMUM.

extraneous roots, see REDUNDANT.

extrapolate, *vb.* to estimate a value of a function or measurement beyond the values already known, especially by the extension of a curve. For example, if we plot known data points as the dots in Fig. 158, we can extrapolate to the value marked with a cross. Compare INTERPOLATE. See also RICHARDSON EXTRAPOLATION.

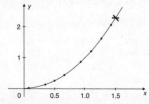

Fig. 158. The value at 1.5 is **extrapolated** from the known points.

extremal, *n.* **1.** the clause in a RECURSIVE DEFINITION that specifies that no items other than those generated by the stated recursive rules fall within the definition. For example, in the PEANO AXIOMS,

> 1 *is an integer;*
> *if n is an integer then so is n + 1;*
> *and nothing else is,*

the third clause is the extremal.

2. a solution arc to a problem in the CALCULUS OF VARIATIONS.

extreme, *n.* **1.** the first or last term of a series or proportion.

2. a MAXIMUM or MINIMUM value of a function.

extreme and mean ratio, *n.* another term for the GOLDEN MEAN.

extreme point, *n.* any point of a CONVEX set that cannot be expressed as the average of distinct points of the set; a corner of a POLYTOPE. Every COMPACT CONVEX subset of EUCLIDEAN SPACE is the CONVEX HULL of its extreme points. This is a special case of the KREIN–MILMAN THEOREM. Compare CARATHEODORY'S THEOREM. See also SIMPLEX METHOD.

extremum, *n.* a point at which a function has either a MAXIMUM or a MINIMUM, which may be either LOCAL or GLOBAL. Any maximization or minimization is an *extremum problem.*

extrinsic, *adj.* (of a property) relating to the space in which the bearer of the property is embedded rather than to its own internal structure; for example, left- and right-hand spirals are extrinsically but not intrinsically distinct. Compare INTRINSIC.

f, 1. *symbol for* an unspecified FUNCTION, as $f(x)$.

2. *abbrev. for* FEMTO-, used in symbols for fractions of the physical units of the SYSTEME INTERNATIONAL.

F, *n.* **1.** the number 15 in HEXADECIMAL notation.

2. a FUNCTION, especially as contrasted with another, f, of which it is an INDEFINITE INTEGRAL, written $F(x) = \int f(x)\,dx$.

3. (*Logic*) an unspecified PREDICATE or property: Fa represents 'a is (or has) F'. The symbol may be used with subscripts to distinguish different predicates and in some circumstances with superscripts to denote the number of arguments it requires: $F_2{}^3 abc$ is a ternary predicate, and the second such designated F.

4. (*Logic*) *symbol for* falsehood (also written **0** or $\perp$). See TRUTH TABLES.

face, *n.* **1.** any of the plane surfaces of a geometrical solid bounded by its edges, such as DCGH in the cube of Fig. 159.

Fig. 159. A cube has six **faces**.

2. relatedly, any subset F of a convex set C that is extreme in the sense that any closed line segment in C with midpoint in F actually lies in F. For example, any subset of a set at which a linear functional achieves its maximum over the given set is a face and is said to be exposed by the functional. A zero-dimensional face is an EXTREME POINT.

3. (*Graph theory*) any of the areas of a CONNECTED PLANE GRAPH, bounded by EDGES and incident with no edges other than those that bound it. The number of such faces is given in terms of the numbers of edges and VERTICES by EULER'S FORMULA.

factor, *n.* **1.** also called **divisor** or **submultiple. a.** any integer or polynomial that exactly divides a given integer or polynomial. For example, 1, 2, 3, 4, 6, and 12 are factors of 12, and its PRIME factors are 2 and 3.

b. any analogous quantity or entity the product of which with another is a given quantity or entity, such as any of a set of cycles of which the composition is a given permutation.

2. (*Statistics, especially in psychology*) a postulated causal influence derived from and used to explain a cluster of responses. See FACTOR ANALYSIS.

3. (*Algebra*) (*as modifier*) see FACTOR SPACE.

factor analysis, *n.* (*Statistics*) any of several techniques for deriving, from a number of given variables, a smaller number of variables that can be used to explain the distribution of the former and are postulated as real causal influences; especially in psychology, the analysis of subjects' responses by identifying co-occurring clusters of responses that may then be explained in terms of a number of independent postulated character traits. *Inverted factor analysis* clusters subjects with similar response patterns in order to construct a set of independent alternative descriptions.

factor group or **quotient group,** *n.* the GROUP, denoted G/H, the members of which are the COSETS of a NORMAL SUBGROUP H in a group G; the product of two cosets is defined as the coset containing the product in the same order of arbitrary representatives of the given cosets. This is an ABELIAN group when the original group is. See also FACTOR SPACE.

factorial, *n.* the function that computes the product of the first *n* natural numbers, written

$$n! = n(n-1)!,$$

and equivalent to the RESTRICTION of the GAMMA FUNCTION to positive integers. For completeness, 0! is defined to be 1. For large *n*,

$$n! \sim n^n e^{-n}.$$

See also STIRLING'S FORMULA.

factorize, *vb.* **1.** to express an integer or polynomial as a PRODUCT of certain of its FACTORS. For example, $x^2 - 1$ can be factorized as $(x+1)(x-1)$.

2. in particular, to give a PRIME factorization or complete factorization.

3. to express any quantity or entity in terms of an analogous set or sequence of elements whose product or COMPOSITION is the given quantity or entity.

factor ring, **quotient ring**, or **residue class ring**, *n.* the RING, denoted R/K, the members of which are the COSETS of an IDEAL K in a ring R; these cosets are called *residue classes,* and are EQUIVALENCE CLASSES the members of which differ by a member of K. Since K is an ideal, the sums and products of elements of the factor ring are unique, and the cosets of a sum or product of ring elements equals the sum or product (in the same order) of their individual cosets. K is the zero of the factor ring, and it has a unit element if R has. See also FACTOR SPACE.

factor space or **quotient space,** *n.* the result of using the structure of a given set (when that is possible) to impose a similar structure on the set of equivalence classes with respect to a given EQUIVALENCE RELATION. For example, the FACTOR GROUP (or quotient group) G/H of a group G by a normal subgroup H is the set of cosets of H in G, and the FACTOR RING (or quotient ring) R/K of a ring R by an ideal K is the set of cosets of K in R; one may similarly define factor topological groups, and, if the equivalence relation corresponds to membership in a vector subspace, one may construct factor vector spaces, factor normed or Banach spaces, and factor Hilbert spaces, where the subspace in these cases lies in the same class. See also THREE SPACE PROPERTY.

factor theorem, *n.* the fundamental result that if $P(x)$ is a POLYNOMIAL over a field and $P(a) = 0$, then $(x-a)$ divides $P(x)$.

faithful, *adj.* (of a REPRESENTATION of a GROUP) having a trivial KERNEL.

faithful module, *n.* a MODULE over a ring for which the ANNIHILATOR

$$A = \{ r \in R : rx = 0 \text{ for all } x \in M \}$$

is zero, where M is the module and R the ring.

fallacy, *n.* an invalid argument or form of argument.

false, *adj.* one of the distinct TRUTH–VALUES of a two-valued logic; a unique ANTI–DESIGNATED truth value.

false position or *regula falsi* (**rule of false position**), *n.* **1.** a method of solving POLYNOMIAL and DIOPHANTINE EQUATIONS by guessing solutions, observing how the value of the function varies, and then using these observations to inform the next guess. For example, to find a solution of $x^2 + y^2 = 100$, we might first guess $x = 5$, $y = 4$, which gives $x^2 + y^2 = 41$, which is too small; next we try doubling each value separately, but $10^2 + 4^2 = 116$, and $5^2 + 8^2 = 89$, which are respectively too large and too small; so next we try intermediate values, such as $x = 8$, $y = 6$, which are indeed a solution.

2. (*Numerical analysis*) more formally, a method of solving POLYNOMIAL and DIOPHANTINE EQUATIONS by successively updating one (in *simple false position*) or two (*double false position*) previous estimates. For example, to use double false position to estimate a root of a UNIMODAL equation known to lie in $[a, b]$, one computes the x-intercept of the line between $(a, f(a))$ and $(b, f(b))$, and uses this value to replace the point at which the value of the function has the same sign. NEWTON'S METHOD uses single false position. See also DICHOTOMY.

family, *n.* **1.** a set of SIMILAR curves or surfaces the equations of which are of the same form and differ from one another only in the values assigned to one or more constants in their general equation. See also PENCIL.

2. a SET, especially a set of subsets of a topological space.

fan, *n.* a SPREAD each node of which has finitely many successors.

Farey sequence, *n.* (*Number theory*) the sequence of all fractions in LOWEST TERMS with denominators not exceeding n, where n is the *order* of the Farey sequence, and listed in increasing order of their sizes. Some authors only consider fractions in the unit interval, and then consider the *Farey table* to be these rows listed one under the other. (Named after the English civil engineer and mathematician, *John Farey* (1766–1826).)

Farkas' lemma, *n.* the result that a linear inequality $\langle f_0, x \rangle \leq 0$ is a consequence of a system

$$\langle f_1, x \rangle \leq 0, \dots, \langle f_n, x \rangle \leq 0,$$

if and only if there are non-negative real numbers $\lambda_1, \dots, \lambda_n$ with

$$\sum_{k=1}^{n} \lambda_k f_k = f_0.$$

The result holds with certain of the inequalities replaced by equalities if the corresponding multipliers are now arbitrary real numbers. This is the key result underpinning the KUHN–TUCKER THEOREM or LINEAR PROGRAMMING DUALITY.

farthest point, *n.* a point not in a given subset of a METRIC SPACE that has a maximal distance from any point in the subset. Compare NEAREST POINT.

Fasbender's theorem, see FERMAT'S PROBLEM.

fast Fourier transform (abbrev. **FFT**), *n.* the name given to various REDUCED COMPLEXITY METHODS of evaluating either direction of the FINITE FOURIER TRANSFORM.

Fatou's lemma, *n.* the theorem that if a sequence $\{ f_n \}$ of non-negative MEASURABLE FUNCTIONS is defined on a measurable set, then

$$\int_E \liminf_{n \to \infty} f_n \, d\mu \;=\; \liminf_{n \to \infty} \int_E f_n \, d\mu \, .$$

(Named after the French analyst, *Pierre Fatou* (1878–1929).) Compare DOMINATED CONVERGENCE THEOREM.

F distribution, *n.* (*Statistics*) a continuous distribution obtained from the ratio of two CHI–SQUARE DISTRIBUTIONS each divided by the number of its DEGREES OF FREEDOM, used to test the equality of the VARIANCES of two NORMALLY DISTRIBUTED variables, and especially the significance of possible causal variables in a REGRESSION. It has probability density function

$$F(\nu, \mu) \;=\; \frac{\nu^{(\frac{\nu}{2})} \, \mu^{(\frac{\mu}{2})}}{B(\frac{\nu}{2}, \frac{\mu}{2})} \cdot \frac{x^{(\frac{\nu}{2} - 1)}}{(\mu + \nu x)^{(\frac{\nu + \mu}{2})}} \, ,$$

where ν and μ are the parameters of the chi-squared distributions.

feasible, *adj.* **1.** (of a CONSTRAINED OPTIMIZATION PROBLEM) possessing CONSISTENT constraints; having a non-empty FEASIBLE SET.
2. (of a point) lying in a prescribed FEASIBLE SET.

feasible directions method, see ZOUTENDIJK'S METHOD.

feasible set, *n.* the set of points satisfying the restrictions of a given CONSTRAINED OPTIMIZATION problem.

Feigenbaum number, *n.* a real number that characterizes the values of the parameters for which the LOGISTIC MAP undergoes PERIOD DOUBLING, defined as the limit of the ratios of the successive differences of the parameters. Similar maps have the same Feigenbaum number. (Named after *Mitch Feigenbaum* (born 1944), American mathematician celebrated for his contribution to the study of CHAOTIC non-linear systems.)

Feit–Thomson theorem, *n.* the theorem of group theory that every finite SIMPLE non-ABELIAN GROUP has even order. This is proved by *REDUCTIO AD ABSURDUM* in some 250 pages; it is one of the longest and most complex proofs in all of mathematics.

Fejer polynomials, see KOROVKIN THEOREMS.

Fejer's condition, see DIRICHLET'S CONDITION.

Fejer's theorem, *n.* the theorem that the ARITHMETIC MEANS of the PARTIAL SUMS of the FOURIER SERIES of any periodic continuous function of the interval $[-\pi, \pi]$ are UNIFORMLY CONVERGENT to the function. Compare KOROVKIN THEOREMS.

femto- (symbol **f**), *prefix* denoting a fraction of 10^{-15} of the physical units of the SYSTEME INTERNATIONAL.

Fenchel conjugate, *n.* (of a CONVEX function on a NORMED SPACE, X) the convex function defined on the DUAL BANACH SPACE X^* by the formula

$$f^*(y) \;=\; \sup \{ \, y(x) - f(x) : x \in X \, \},$$

where $f(x)$ is the given convex function.

Fenchel's duality theorem, *n.* the central convex DUALITY theorem that if

$$f: X \to \ ]-\infty, \infty] \quad \text{and} \quad g: X \to [-\infty, \infty[$$

are respectively a CONVEX and a CONCAVE function, then, if there is a point at which one function is continuous and the other is finite,

$$\inf_x \{f(x) - g(x)\} = \max_{x^*} \{g^*(x^*) - f^*(x^*)\}.$$

Here f^* is the FENCHEL CONJUGATE of f and

$$g^*(x^*) = -(-g)^*(-x^*).$$

This specializes to contain the von Neumann MINIMAX THEOREM and is equivalent to the convex LAGRANGE MULTIPLIER theorem.

Fermat, Pierre de (1601–65), French lawyer and amateur mathematician who is credited with founding modern number theory and (independently of Pascal) the calculus of probability, as well as discovering analytical geometry independently of Descartes. He obtained sophisticated results in the foundations of analytic geometry and differential calculus, but failed to publish them. He claimed to have proved the famous unsolved problem known as FERMAT'S LAST THEOREM.

Fermat number, *n.* a number of the form $F_N = 2^{(2^N)} + 1$. The Fermat number F_{382447} is known to be composite. See FERMAT PRIME.

Fermat prime, *n.* any PRIME NUMBER of the form $2^n + 1$. Fermat conjectured that when n is a power of two, then $2^n + 1$ is always prime. This fails for $n = 32$, and it is not known whether there is an infinitude of Fermat primes. Compare MERSENNE PRIME.

Fermat's last theorem, *n.* the celebrated conjecture in number theory that the equation $x^n + y^n = z^n$, does not have non-trivial integral solutions for x, y, z when n is greater than 2 (when $n = 2$ PYTHAGORAS' THEOREM shows the existence of solutions). Fermat annotated his copy of a translation of Diophantus with 'I have assuredly found an admirable proof of this, but the margin is too narrow to contain it.' This claim is not now given much credence, but the result was known to hold for 'most exponents', that is, to be false on a set of exponents with zero density, and for all exponents below 125 000. A claim to have proved it, in 1988, using ALGEBRAIC GEOMETRY, was proved incorrect, but it is generally accepted that it was finally proved in 1993–94 by the Princeton-based English mathematician Andrew Wiles, using many of the deepest ideas of modern mathematics. See also WARING'S PROBLEM.

Fermat's little theorem, *n.* the result in number theory that for any integer n and prime p that is not its factor, n^{p-1} is CONGRUENT to 1 MODULO p. See EULER PHI FUNCTION.

Fermat's problem or **Steiner's problem,** *n.* the problem, due to Fermat and often considered as the oldest optimization problem with a natural dual formulation, of finding a point in the plane that minimizes the sum of the distances to the vertices of a given triangle. The solution to this problem is contained in *Fasbender's theorem*, which establishes that if this point (the *Torricelli point*) is not a vertex then the minimum sum of these distances is also the maximum altitude of an EQUILATERAL triangle that CIRCUMSCRIBES the given triangle; this triangle has sides perpendicular to the segments joining the Torricelli point to the vertices of the original triangle. See also LOCATION PROBLEM.

Ferrari's method, *n.* a method for solving QUARTIC equations related to CARDANO'S FORMULA.

Ferrer's graph, *n.* the graphical representation of a PARTITION of a positive integer as an array in which each addend is represented by a row of dots.

FFT, *abbrev. for* FAST FOURIER TRANSFORM.

Fibonacci numbers or **Fibonacci sequence,** *n.* the sequence of integers

$$0, 1, 1, 2, 3, 5, 8, 13, 21, 34,...$$

where each number is the sum of the previous two; hence the two-term DIFFERENCE SEQUENCE

$$F_{n+1} = F_n + F_{n-1}$$

with $F_0 = 0$ and $F_1 = 1$. The ratio of successive terms tends to the GOLDEN MEAN as n tends to infinity. With initial values $L_0 = 2$, and $L_1 = 1$, the same recursion generates the *Lucas numbers*. (Named after *Leonardo Fibonacci* (c.1170–1250), known as Leonardo of Pisa, number theorist and algebraist, who introduced the Arabic number system to Europe.)

field, *n.* **1.** a set of entities subject to two binary operations, usually referred to as addition and multiplication, such that the set is a COMMUTATIVE GROUP under the addition, the set excluding the zero element is a commutative group under the multiplication, and the multiplication DISTRIBUTES over the addition; thus the rationals and the reals are fields but the integers are not. See also SKEW FIELD. Compare GROUP, RING, ALGEBRAIC NUMBER FIELD.

2. the set of elements that are either arguments or values of a function, the union of its DOMAIN and RANGE.

3. see VECTOR FIELD, SCALAR FIELD, TENSOR FIELD.

field extension, see EXTENSION FIELD.

field of fractions, *n.* (of an INTEGRAL DOMAIN) the full RING of QUOTIENTS of the domain.

field of integration, *n.* the region over which a definite MULTIPLE INTEGRAL is evaluated.

field of sets, *n.* another term for an ALGEBRA OF SETS.

Fields medal, *n.* the highest award of the International Mathematical Union, given at its quadrennial meetings in recognition of outstanding research (usually by mathematicians under 40 years of age). These medals stem from a bequest by the Canadian analyst *John Charles Fields* (1863–1932), and were first awarded in 1936.

Fiet-Thompson theorem, *n.* the 1963 theorem that the ORDER of a non-abelian SIMPLE finite group is even, or equivalently that every non-Aberlian group of odd order is SOLVABLE.

figurate numbers, *n.* (for each integer n greater than two) the sequence of numbers generated by counting the number of points in the successive members of the sequence of nested regular n-gons, where each figure in the sequence is constructed from the preceding by retaining one vertex in common, extending by one unit each of the sides meeting in the common vertex, and placing a row of unit-spaced points around the preceding member of the sequence, in order to increase by one the number of points per side, as shown in Fig. 160; the first member of each sequence is a single point. Thus the figurate numbers are the TRIANGULAR, SQUARE, PENTAGONAL NUMBERS and so on, and the formula for the kth member of the sequence based on n-gons is then

$$2k + \tfrac{1}{2} nk(k-1) - k^2,$$

where k is often allowed to take negative integral values.

Fig. 160. **Figurate numbers**.

(a) The triangular numbers.   (b) The square numbers.

figure, *n.* **1a.** any arrangement of points, lines, curves or surfaces, constituting a geometrical shape.

b. often, more specifically, such a figure that is CLOSED; in this sense a *plane figure*, such as a circle, encloses an area, and a *solid figure*, such as a sphere, encloses a volume.

2. another word for DIGIT.

3. (*Logic*) one of the four possible arrangements of the three terms of a SYLLOGISM. Compare MOOD.

filter, *n.* (on a set) a family of non-empty subsets of the given set that contains any SUPERSET of a member and such that the family is closed under finite intersection. For example, the *Fréchet filter* on an infinite set is the set of complements of finite sets. A filter is said to converge to a point x if every neighbourhood of x lies in the filter. This leads to a convergence theory essentially equivalent to that of NET CONVERGENCE.

finer, *adj.* (of a TOPOLOGY) strictly containing another topology; the DISCRETE TOPOLOGY is finer than any other.

finitary, *adj.* (of a proof) not involving infinite sets either explicitly or implicitly; for example, a proof, in a denumerable theory, involving an assertion of existence that is not supported by a construction of the entity in question essentially involves quantification over an infinite domain and so is not finitary. See HILBERT'S PROGRAMME. See also CONSTRUCTIVE.

finite, *adj.* **1.** (according to Russell) having a number of elements capable of being put into ONE–TO–ONE CORRESPONDENCE with a bounded initial segment of the natural numbers; able to be counted using a terminating sequence of natural numbers.

2. (equivalently, according to Dedekind) not INFINITE (sense 3).

Compare COUNTABLE, DENUMERABLE.

finite character, *n.* the property of a collection of sets (or a property) that a set is in the collection (or has the property) if and only if this is also true of every non-empty finite subset of the given set.

finite-dimensional, *adj.* (of a VECTOR SPACE) possessing a finite maximal set of LINEARLY INDEPENDENT vectors. For example, the set of ordered pairs of real numbers is finite-dimensional, but the set of real continuous functions on $[0, 1]$ is not.

finite element method, *n.* (*Partial differential equations*) a general numerical method of solving boundary value problems by considering their formulation as VARIATIONAL INEQUALITIES and using DISCRETIZATION, in order to look for an approximate solution that is required to be of a prescribed form, such as a polynomial, on small polygonal subregions or *finite elements*, where the functions are required to meet the given boundary conditions and to be conformable.

finite extension, *n.* a FIELD L that contains a given field K, and is a FINITE–DIMENSIONAL VECTOR SPACE over K.

finite field, see GALOIS FIELD.

finite (or **discrete**) **Fourier transform,** *n.* **1.** the problem of determining the coefficients of the unique polynomial p of degree n that INTERPOLATES given values a_i at w^i for $i = 0, 1, ..., n$ where w is a PRIMITIVE $(n + 1)^{\text{th}}$ ROOT OF UNITY either in the complex FIELD or in a finite field.
2. the equivalent and inverse problem of EVALUATING all of $p(w^i)$ for $i = 0, 1, ..., n$, when the coefficients of p are known. A FAST FOURIER TRANSFORM does this very efficiently by exploiting the roots of unity so that all $n + 1$ evaluations are only marginally more difficult than a single one. This is of great practical importance in many areas exploiting FOURIER ANALYSIS, such as image enhancement.

finite geometry, *n.* a GEOMETRY with only finitely many points and lines, such as a FINITE PROJECTIVE PLANE.

finite group, *n.* a GROUP with finite ORDER.

finite induction, *n.* another word for INDUCTION (sense 1), to distinguish it from TRANSFINITE INDUCTION.

finite intersection property, *n.* (*Topology*) the property of a space that whenever a family of closed subsets is such that any finite subcollection has a non-empty intersection, then so does the entire family. An application of DE MORGAN'S LAWS shows this to be equivalent to the space being COMPACT.

finitely additive measure, see MEASURE.

finitely generated, *adj.* (of an algebraic structure) generated by a finite number of elements. See GENERATE. Compare CYCLIC GROUP.

finite measure, *n.* a MEASURE that assigns a finite value to each measurable set in its measure ring.

finite projective plane, *n.* a square BLOCK DESIGN or CONFIGURATION that consists of $n^2 + n + 1$ points and $n^2 + n + 1$ lines with $n + 1$ points on each line, $n + 1$ lines through each point, each pair of lines meeting at one point, and dually each pair of points lying on one line, where n is the *order* of the finite projective plane. Such planes exist for all prime power orders, and no other order is known to be possible: no plane of order 6 exists, and the case of order 10 is not settled.

finitism, *n.* the doctrine in the philosophy of mathematics that the only entities that may be admitted to mathematics are those that are CONSTRUCTIBLE, and only those propositions may be entertained whose truth can be proved in a finite number of steps. See also FORMALISM, INTUITIONISM. Compare PLATONISM.

first category set, see BAIRE CATEGORY.

first countable, *adj.* (of a topological space) having a countable BASE for the TOPOLOGY at each point of the space, as in any METRIC SPACE. Compare SECOND COUNTABLE.

first derivative, *n.* the DERIVATIVE of a given function, rather than the derivative of any derivative; a derivative of first order. The SECOND DERIVATIVE is the first derivative of the first derivative.

first derivative test, *n.* a test for OPTIMALITY of a CRITICAL POINT of a given function using only the FIRST DERIVATIVE: the critical point c is a LOCAL MINIMUM if in some neighbourhood of c the derivative $f'(x)$ is strictly positive to the left of c and strictly negative to its right; it is a LOCAL MAXIMUM if the derivative is strictly negative to the left and strictly positive to the right of c. For example, in Fig. 161, A is a local maximum and B is a local minimum; the change of derivative from positive at X to negative at Y, and thence to positive at Z, is shown by the tangents at these points. Compare SECOND DERIVATIVE TEST, POINT OF INFLECTION.

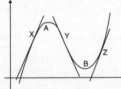

Fig. 161. **First derivative test**. See main entry.

first divided difference sequence, see DIVIDED DIFFERENCE.

first isomorphism theorem, *n.* **1.** also called **homomorphism theorem.** any theorem stating that some specific algebraic structure, G, has the property that if θ is a HOMOMORPHISM, then $G/\ker\theta$, where $\ker\theta$ is the KERNEL of the homomorphism, is ISOMORPHIC to the image of G under the homomorphism.

2. (in usages in which the foregoing is referred to as a homomorphism theorem) a second ISOMORPHISM THEOREM.

first-kind induction, **incomplete induction**, or **special induction**, *n.* INDUCTION in which the inductive step is from the integer n to $n + 1$. Compare COMPLETE INDUCTION.

first-order, *adj.* **1a.** being or relating to the FIRST DERIVATIVE of a function.

b. in particular, (of an ORDINARY DIFFERENTIAL EQUATION) involving the FIRST DERIVATIVE, but no higher order differential coefficients, of the dependent variable with respect to the independent variable.

c. (of a PARTIAL DIFFERENTIAL EQUATION) involving no partial differential coefficient of order greater than 1.

2. (*Logic*) QUANTIFYING only over INDIVIDUALS and not over PREDICATES or CLASSES. *Lower* or first-order *predicate calculus* (LPC) studies the logical properties of such quantification. Compare SECOND-ORDER.

3. see TENSOR.

4. also **of the first order.** having unit ORDER. Compare SECOND-ORDER.

first principles, *n.* **1.** the fundamental assumptions on which a given theory or procedure is based.

2. the AXIOMS of a mathematical or scientific theory.

first species, see SPECIES.

Fisher–Behrens problem, *n.* (*Statistics*) the problem of finding a test for the equality of the MEANS of two NORMALLY DISTRIBUTED populations with different VARIANCES, given a sample from each. This was a central issue of study in the first half of the 20th century; Scheffé devised an exact test that does not use all the information contained in the sample and that is not unique unless the sample sizes are equal and some natural pairing of the samples is used; Behrens, Welsh, and others give approximate solutions using all the information.

Fisher's inequality, *n.* (*Combinatorics*) a result for a BLOCK DESIGN (or configuration), stating that the number of blocks is always greater than or equal to the number of points (VARIETIES).

fish-hook, *n.* (*Logic*) *informal* name for the symbol '⊣', used to represent the relation of ENTAILMENT.

fit, *n.* (*Statistics*) the degree of correspondence between observed and predicted characteristics of a distribution or model. See GOODNESS OF FIT.

fixed point, *n.* **1.** a point that is mapped into itself by a given transformation; for example, 0 and 1 are fixed points of $f(x) = x^2$, since $f(0) = 0$ and $f(1) = 1$.
2. a point that lies in its own image under a given CORRESPONDENCE.
3. (*as modifier*) (of a numerical notation) writing numbers in full with the DECIMAL POINT separating the integral from the fraction part, as contrasted with FLOATING POINT notation in which variable multiples of the base are extracted as factors.

fixed point theorem, *n.* a theorem, such as BROUWER'S or the Banach CONTRACTION PRINCIPLE, giving conditions for a mapping to have a FIXED POINT (sense 1).

fixed set, *n.* a set S such that $T(S) = S$, for a given MAPPING T that may be MULTI-VALUED.

flag, *n.* **1.** (*Computing*) a BOOLEAN variable that indicates the result of a test and so can be used as a condition to switch different parts of a PROGRAM into action.
2. (*Geometry*) a triple consisting of a HALF PLANE, a boundary HALF LINE and its endpoint.

flat, *n.* another term for an AFFINE subspace.

floating point, *n.* (*as modifier*) (of a numerical notation) expressing numbers as multiples of suitable powers of the BASE of the counting system, and so not always employing the decimal point strictly between the integral and fractional parts of the number as is the case in FIXED POINT notation. For example, 123.45 can be written as 12345×10^{-2} or as 1.2345×10^2. When the coefficient of the power of the base is required to be strictly less than the base, this is known as SCIENTIFIC NOTATION.

floor or **greatest integer function,** *n.* (*Computing*) the largest integer not exceeding a given real number. Compare CEILING.

Floquet theorem, *n.* (*Differential equations*) the result that a vector linear differential equation with continuous periodic coefficients, $y' = \mathbf{P}(t)y$, has a FUNDAMENTAL MATRIX of the form

$$Y(t) = \mathbf{Z}(t)e^{Rt}$$

where $\mathbf{Z}$ the same period as $\mathbf{P}$, and R is a matrix of constants.

flow, see NETWORK FLOW.

fluid, *n.* a body of matter that flows when acted upon by any FORCE, however small; glass is thus a fluid. *Fluid mechanics* is the branch of CONTINUUM MECHANICS that specifically studies such bodies. See NEWTONIAN FLUID, SIMPLE FLUID, VISCOUS FLUID, INVISCID FLUID, IDEAL COMPRESSIBLE FLUID.

flux, *n.* (*Continuum mechanics*) a FIELD that represents the transfer of some quantity per unit area; for example, heat flux, energy flux, mass flux, and magnetic flux.

fluxion, *n. largely obsolete term for* the RATE OF CHANGE of a function, derived from Newton's original formulation; his notation $\dot{x}$, with a central superscript dot, for a DERIVATIVE is still used.

focal, *adj.* of, relating to, situated at, or measured from the FOCUS.

focus, *n.* a fixed point on the concave side of a CONIC SECTION, in terms of which, together with its DIRECTRIX and its ECCENTRICITY, the locus of points constituting the conic is defined. Fig. 162 shows both foci (E) of an ellipse, the focus (H) of one branch of an hyperbola, and the focus (P) of a parabola that all share a vertex.

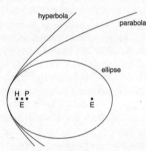

Fig. 162. **Focus**. See main entry.

-fold, *suffix.* indicating the number of elements used in some repetitive process, as in the *n*-fold product of the unit interval.

folium (of Descartes), *n.* a plane curve that intersects with itself at a NODE, on one side of which it forms a loop; the branches on the other side of the node are ASYMPTOTIC to the same line, as shown in Fig. 163. It has standard equation

$$x^3 + y^3 = 3axy,$$

where $x + y + a = 0$ is the equation of the line.

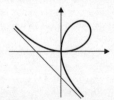

Fig. 163. **Folium**.

follow, *vb.* to be implied (by); to be a logical consequence (of).

force, *n.* a PRIMITIVE TERM in Newtonian mechanics, obeying NEWTON'S LAWS OF MOTION and measured in NEWTONS.

forcing, *n.* a method of constructing INTERPRETATIONS of set theory, introduced by Cohen, and used by him to prove the independence of the AXIOM OF CHOICE and of the CONTINUUM HYPOTHESIS in ZERMELO–FRANKEL SET THEORY.

Ford–Fulkerson algorithm, *n.* a particular labelling algorithm for obtaining an optimal NETWORK FLOW.

forest, *n.* a family of TREES.

forgetful functor, *n.* (*Category theory*) a FUNCTOR obtained by considering a category as another category with simpler objects. For example, the functor from the category of groups to the category of sets that preserves all mappings *forgets* the structure of the groups.

form, see DIFFERENTIAL FORM, LOGICAL FORM, and QUADRATIC FORM.

formal calculus, formal system, or **formal theory,** *n.* (*Logic*) an UNINTERPRETED symbolic system, often including NON–LOGICAL AXIOMS, of which the syntax is precisely defined, and on which a relation of DEDUCIBILITY is defined in purely syntactic terms; a LOGISTIC SYSTEM. Compare FORMAL LANGUAGE.

formal equivalence, *n.* (*Logic*) the relation that holds between two OPEN SENTENCES when their UNIVERSAL CLOSURES are MATERIALLY EQUIVALENT, as between the two sides of a mathematical IDENTITY such as $a + b = b + a$.

formalism, *n.* **1.** the philosophical doctrine that mathematical statements have no extrinsic meaning but that their symbols themselves, regarded as physical objects, exhibit a structure that has useful applications. See FINITISM. Compare LOGICISM, INTUITIONISM, PLATONISM.
2. a FORMAL LANGUAGE, especially one intended as a FORMALIZATION of some fragment of natural language.
3. the mathematical or logical structure of some theory or argument, as distinguished from its subject matter.

formalize, *vb.* (*Logic*) to extract the LOGICAL FORM of an expression; to express in the symbols of some FORMAL SYSTEM.

formal language, *n.* **1.** any language designed for use in situations in which natural language is unsuitable for the required precision, as in FORMAL LOGIC or in COMPUTER PROGRAMS. The symbols and formulae of such a language stand in precisely defined syntactic and semantic relationships.
2. (*Logic*) a FORMAL SYSTEM for which an INTERPRETATION is provided. What distinguishes a formal language from a mere FORMAL CALCULUS is that the SEMANTICS enable us to regard the former as about some subject matter. See also STRUCTURE. Compare LOGISTIC SYSTEM.

formal logic, *n.* **1.** the study of FORMAL SYSTEMS.
2. also called **symbolic logic.** the study of DEDUCTIVE ARGUMENT and of the structure and relations of statements, in which the terms of a FORMAL CALCULUS are used to represent precisely defined categories of expressions. Compare PHILOSOPHICAL LOGIC.
3. any specific formal calculus that can be interpreted as representing natural argument or some species of it.

formally valid, *adj.* another term for VALID (sense 1b).

formal system or **formal theory,** *n.* other terms for FORMAL CALCULUS.

formation rules, *n.* (*Logic*) the set of rules that specify the syntax of a FORMAL CALCULUS; the algorithm that generates all and only the WELL-FORMED FORMULAE (WFFS) of the system.

formula, *n.* **1.** any sequence of symbols of a FORMAL CALCULUS, whether or not complying with the FORMATION RULES of that theory, such as

$$) 5) + 7 \times$$

in elementary arithmetic.

2. such a sequence that is syntactically correct; a sentence or WELL–FORMED FORMULA of any formal theory.

3. a formal expression of some rule or other result, such as Stirling's formula or Frenet formulae.

forward difference, see DIFFERENCE QUOTIENT, DIFFERENCE SEQUENCE.

forward error analysis, *n.* (*Numerical analysis*) the analysis, for a given algorithm, of the error between an exact quantity and its computed approximation. In principle the aim is to determine a measure of the error, that if small, validates the calculation. In practice this will not distinguish ROUNDING ERROR from TRUNCATION ERROR. Compare BACKWARD ERROR ANALYSIS.

foundations of mathematics, *n.* the study of the justification of mathematical rules and axioms. Under the influence of the archetype of Euclid's axiomatic treatment of geometry, and of the unification of apparently disparate branches of mathematics by, for example, Euler, this has often taken the form of a search for a small number of concepts which can be regarded as fundamental in the sense that all others can be derived from them. Whether such derivation is valid is a matter for logical appraisal, but whether it sustains the interpretation placed upon it is a matter for philosophy. Since mathematics makes existential claims, the ontological status of the entities supposedly referred to in mathematical propositions is another area of debate. Particular impetus was given to foundational studies by the discovery of inconsistencies in intuitive mathematical notions, for example RUSSELL'S PARADOX and CANTOR'S PARADOX. See FORMALISM, INTUITIONISM, LOGICISM. See also CONSTRUCTIVISM, PLATONISM, REALISM.

four-colour theorem, *n.* the celebrated result that any planar map can be coloured using at most four colours in such a way that no two adjacent areas are of the same colour. This was conjectured in the 19th century, repeatedly misproven and finally proved in 1976 by a combination of graph theory and sophisticated computing. It follows that it is impossible to add a fifth region to the first diagram of Fig. 164 below in such a way that all five regions share a boundary with one another; an attempt to do so, for example by adding the white region in the second diagram in fact removes one of the original boundaries and so would enable C to be coloured in the same way as B.

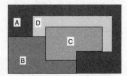 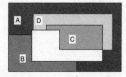

Fig. 164. **Four-colour theorem**. See main entry.

four-current density, see MAXWELL'S LAWS.

four group, see KLEIN FOUR-GROUP.

Fourier, Jean Baptiste Joseph, Baron de (1768–1830), French analyst and physicist, whose study of the conduction of heat had a profound influence on mathematical physics and on the study of real functions. Of humble origin, he became a professor in a military academy, accompanied Napoleon on his Egyptian expedition and was appointed Governor of Lower Egypt; after the French defeat there, he became Prefect of Grenoble and was created a baron. He also published widely on Egyptology, became permanent secretary of the French Academy of Science, and was elected to the Academy of Medicine, the *Académie Française*, and the Royal Society.

Fourier analysis, *n.* the study and application of FOURIER SERIES and related integrals, especially in the study of differential equations, mathematical physics and approximation.

Fourier coefficients, *n.* the coefficients enabling one to express a function formally in terms of its FOURIER SERIES. In standard real form one has

$$a_n = \frac{1}{\pi} \int_0^{2\pi} f(x) \cos(nx) \, dx \qquad (n \geq 0)$$

and

$$b_n = \frac{1}{\pi} \int_0^{2\pi} f(x) \sin(nx) \, dx \qquad (n \geq 1)$$

and in complex terms

$$c_n = \frac{a_n - i b_n}{2} = \frac{1}{2\pi} \int_0^{2\pi} f(x) \exp(-inx) \, dx$$

Fourier series, *n.* an infinite trigonometric series of the form

$$\frac{1}{2} a_0 + \sum_{n=1}^{\infty} [a_n \cos(nx) + b_n \sin(nx)]$$
$$= \tfrac{1}{2} a_0 + a_1 \cos x + b_1 \sin x + a_2 \cos 2x + b_2 \sin 2x + \ldots,$$

where a_0, a_1, b_1, a_2, b_2, etc. are the FOURIER COEFFICIENTS. They are used, especially in mathematical physics, to represent or approximate to any single-valued periodic function by assigning suitable values to the coefficients. A *Fourier sine series* or *Fourier cosine series* is one whose only non-zero terms are the sine or cosine terms respectively. See also DIRICHLET'S CONDITION.

Fourier transform, *n.* **1.** the INTEGRAL TRANSFORM

$$F(y) = \int_{-\infty}^{\infty} f(x) \exp(-iyx) \, dx,$$

that sends a function *f* to another function *F*. The real and imaginary parts of this essentially define the *Fourier cosine* and *Fourier sine transforms*. Under reasonable conditions the Fourier transform is invertible, and its inverse is given by the CAUCHY PRINCIPAL VALUE of

$$f(x) = \frac{1}{2\pi} \int_{-\infty}^{\infty} F(y) \exp(iyx) \, dy.$$

2. the function *F* that is so related to a given function; the image of the given function under *Fourier transformation*.
See also LAPLACE TRANSFORM.

four squares theorem, *n.* (*Number theory*) the theorem, demonstrated by Lagrange, that any positive integer may be expressed as the sum of the squares of four integers. See LAGRANGE'S THEOREM.

fourth harmonic, see HARMONIC POINTS.

fractal or **fractile,** *n.* a set with non-integral HAUSDORFF DIMENSION, such as CANTOR'S TERNARY SET whose dimension is log2/log3 = 0.6309. A *fractal curve* is typified by self-symmetry under magnification, and can be constructed from any regular polygon by replacing each side with the *generator*, and then iterating the process; Fig. 165 shows the first, second, and half of the fourth-generation curves produced from a square by the generator shown in bold. (Common fractal curves are called *snowflakes* or *dragon curves* because of the appearance of the iterates.) Such curves were first constructed by Peano, and were used by Mandelbrot (who defined them as having Hausdorff dimension strictly larger than their TOPOLOGICAL DIMENSION) to study the irregular and fragmentary patterns seen in nature, for example, BROWNIAN MOTION and the distribution of the galaxies. Fractals typically form the boundaries between regions of the complex plane governed by distinct ATTRACTORS of a DYNAMICAL SYSTEM. See also PEANO CURVES, CHAOS.

Fig. 165. Stages in the generator of a **fractal** curve.

fractal dimension, *n.* an extension of the notion of DIMENSION which may have a non-integer value. For example, the KOCH CURVE has dimension ln4/ln3 (approximately 1.26) and the CANTOR SET has dimension ln2/ln3. Intuitively a fractional value between 1 and 2 indicates that the object is 'thicker' than a line with dimension 1, but 'thinner' than an area with dimension 2. Fractal dimension has found many practical applications in the analysis of chaotic or noisy processes (see CHAOS).

fraction, *n.* **1.** a RATIO of two integers, or any number that can be expressed as such a ratio; written *m/n*, where *m* is not a multiple of *n*, and *n* is not zero or one. The rules for addition and multiplication of fractions are

$$\frac{a}{b} + \frac{c}{d} = \frac{ad + cb}{bd}, \qquad \frac{a}{b} \times \frac{c}{d} = \frac{ac}{bd}$$

from which the rules for subtraction, division, and inversion can readily be derived. See DECIMAL FRACTION, VULGAR FRACTION, PROPER FRACTION.

2. any ratio of one quantity or expression (the NUMERATOR) to another non-zero quantity or expression (the DENOMINATOR).

fractional, *adj.* relating to, containing, or constituting FRACTIONS.

fractional linear transformation, *n.* another term for the MÖBIUS TRANSFORMATION.

fractional part, *n.* the difference between the given real number and its INTEGRAL PART. For example, the fractional part of 3.42, written {3.42}, is 0.42, and of −3.42 is 0.58.

frame, *n.* (*Statistics*) an enumeration of a population for the purpose of SAMPLING, especially as the basis of a STRATIFIED SAMPLE.

frame of reference or **frame,** *n.* **1a.** any set of lines, directions, planes, etc., such as the coordinate axes, relative to which the position of a point in a space can be described.

b. any point or set of points that are taken to be fixed and at rest, and relative to which the motion of other objects is measured.

2. the moving TRIHEDRAL formed by the TANGENT, NORMAL and BINORMAL to a curve in three-dimensional space. See FRENET FORMULAE.

3. (*Mechanics*) a particular choice of origin and basis vectors in three-dimensional EUCLIDEAN POINT SPACE and of a fixed initial point of the real line indexing time, to which the observations of a given OBSERVER may be referred. See INERTIAL FRAME OF REFERENCE, ROTATING FRAME OF REFERENCE.

Frattini subgroup, *n.* the SUBGROUP, $\Phi(G)$, of a given group G defined to be the intersection of all maximal subgroups of G; if G has no maximal subgroups, its Frattini subgroup is defined to be G itself. $\Phi(G)$ is a CHARACTERISTIC SUBGROUP.

Fréchet differential, *n.* the function $\delta f(x;)$, derived from a given function f between NORMED SPACES and defined on an open domain, for which

$$\lim_{\|h\| \to 0} \frac{f(x+h) - f(x) - \delta f(x; h)}{\|h\|} = 0.$$

If this limit, $\delta f(x;)$, is continuous and linear in h then the function is said to be *Fréchet differentiable* at x, and the associated linear operator $\delta f(x;)$ is the *Fréchet derivative* of f at x, often written $\nabla f(x)$. Such a derivative is necessarily a linear GATEAUX DERIVATIVE. For example, if f is a real-valued function on Euclidean space and f has continuous partial derivatives, then the Fréchet derivative may be identified with the GRADIENT. (Named after the French analyst, topologist and probability theorist, *Maurice René Fréchet* (1878–1973), who pioneered the study of abstract spaces.)

Fréchet filter, see FILTER.

Fréchet space or **F-space,** *n.* **1.** a complete metrizable LOCALLY CONVEX space. Some authors use the term *F-space* without requiring local convexity.

2. another term for T_1-SPACE. See T–AXIOMS.

Fredholm alternative, *n.* **1.** the ALTERNATIVE THEOREM, holding for a continuous linear operator, A, with closed range, or its representation as a matrix, that either A is SURJECTIVE, or else the adjoint A^* has a non-trivial KERNEL. Hence, either the inhomogeneous equation $Ax = b$ is always solvable or that the homogeneous equation $A^*y = 0$ has a non-zero solution.

2. for a *Fredholm integral equation* of the second kind,

$$y(s) - \lambda \int_a^b K(s, t)\, y(t)\, \mathrm{d}t = f(t),$$

the parallel assertion that, for a continuous KERNEL, either the inhomogeneous equation always has a unique solution, or the homogeneous equation has a non-trivial solution, in which case only certain right-hand side functions lead to solutions and then always to infinitely many (depending on λ). These equations originated in the study of oscillations. See also VOLTERRA INTEGRAL EQUATION.

(Named after the Swedish analyst and physicist, *Erik Ivar Fredholm* (1866–1927).)

free, *adj.* **1.** (of an algebraic structure) consisting of all formal objects satisfying the requisite algebraic conditions without additional relations being imposed; for example, the symmetric group on a set of three elements is not free as a group. See also FREE GROUP, FREE MODULE.

2. (of a TREE) not having a ROOT (or origin).

3. (*Logic*) (of a variable) not BOUND, so that it is at best a place-holder in an OPEN SENTENCE; in some formulations of logic these are interpreted as universally quantified.

4. (of a vector) determined only up to translation; unattached.

5. (of an element of a group) not of finite PERIOD.

free Abelian group, see FREE GROUP.

freedom equation, *n.* a less common name for PARAMETRIC EQUATION.

free elements, (of a module) see TORSION ELEMENTS.

free group, *n.* **1.** a GROUP that has a set of GENERATORS such that the only products of generators and their inverses that equal identity are of the form aa^{-1} or $a^{-1}a$. In a *free Abelian group*, the weaker condition is required that any such product should be reducible by the commutative law to this form; by a general property of Abelian groups, if it has a finite number of generators, then it is free if and only if no element is of finite period. Free groups are fully characterized by their rank, the cardinality of any set of generators, and by the *Schreier–Nielson theorem*, all subgroups of free groups are free and have ranks related by the index of the subgroup. Every group is a homomorphic image of a free group.

2. the set, F_X, of EQUIVALENCE CLASSES, $[u]$, of WORDS in a non-empty set, X, where a pair of elements are equivalent if and only if there is a finite sequence of words starting with one and finishing with the other, such that each is obtained from the preceding by ELEMENTARY REDUCTION; multiplication is defined on F_X by $[u][v] = [uv]$.

free module, *n.* a MODULE that possesses as a BASIS a subset $\{a_1, a_2,\ldots, a_n\}$ in terms of which every non-zero element can be written uniquely in the form $\sum_i u_i a_i$, where the u_i are elements of the RING over which the module is a left module. All vector spaces are free, and an Abelian group is free if and only if it has no element of finite period. Submodules of a free module are not in general free unless the underlying ring is a principal ideal domain. See also TORSION FREE MODULE.

free ultrafilter, see ULTRAFILTER.

Frege, Friedrich Ludwig Gottlob (1848–1925), German mathematician and philosopher, who initiated the 'linguistic turn' in philosophy, and founded the study of mathematical logic. He taught all branches of mathematics at Jena throughout his life, but almost all his publications were logical. His main innovations were the distinction between SENSE and REFERENCE, and a logic of quantification, in which the QUANTIFIERS are treated as properties of properties and for which he devised an idiosyncratic 'concept notation'

(BEGRIFFSSCHRIFT) so elaborate as to discourage understanding; although many of his ideas became familiar through the work of Peano and Russell, it was their notation which became standard. He wrote on the foundations of mathematics, setting out axioms for set theory from which he believed arithmetic could be derived. His bitterness about the indifference and hostility with which his work was received was increased by a sarcastic review by Cantor, who had not even troubled to read the book; then, when the second volume of his development of LOGICISM was already in the press, he received a letter from Bertrand Russell, one of his few admirers, informing him that his axioms were inconsistent (see RUSSELL'S PARADOX). Although Frege did revise his axioms (unsuccessfully), he abandoned the projected third volume, and was too depressed to do any other useful work; his bitterness was compounded by the fate of Germany in the First World War, and his diary reveals, as well as virulent antisemitism, a pathological hatred of Catholics, the French, socialism, and democracy. He did, however, publish three more philosophical papers; in his last years he became convinced of the falsehood of his logicism, but his revised views were never published. Frege is now regarded as a crucial figure in the history of both logic and philosophy, and he and Wittgenstein, whom he greatly influenced, are the source of almost all modern philosophy of language.

French curve, *n.* a stencil or template used for drawing certain curves.

Frenet formulae or **Serret–Frenet formulae,** *n.* the fundamental formulae for a space curve that recapture the unit TANGENT, $\mathbf{T}$, NORMAL, $\mathbf{N}$, and BINORMAL, $\mathbf{B}$, from the CURVATURE κ and TORSION τ of the curve. These are

$$\mathbf{N}' = -\kappa\mathbf{T} + \tau\mathbf{B}, \quad \mathbf{B}' = -\tau\mathbf{N}, \quad \mathbf{T}' = \kappa\mathbf{N},$$

where all derivatives are with respect to arc length. This shows that, up to translation and rotation of the original FRAME, the curvature and torsion characterize a space curve. For this reason the formulae are sometimes described as the *fundamental theorem of space curves*. (Named after the French differential geometer, *Jean Frédéric Frenet* (1816–1900).)

frequency, *n.* **1.** the number of times that an event occurs within some given unit period; the rate of occurrence.

2. the number of times a PERIODIC FUNCTION repeats itself in every unit of the independent variable; the reciprocal of the PERIOD of the function.

3. (*Statistics*) **a. absolute frequency.** the number of individuals in a class, usually the number of occurrences of an event, or individuals with some property; for example, in 100 tosses of a coin, the absolute frequency of heads may be 47.

b. relative frequency. the ratio of the absolute frequency of some phenomenon to the total population under consideration; for example, as the number of tosses of an unbiased coin increases, the relative frequency of heads tends to $\frac{1}{2}$.

frequency distribution, *n.* (*Statistics*) the function of the distribution of a SAMPLE that corresponds to the PROBABILITY DENSITY FUNCTION of the given underlying POPULATION and tends to it as the sample size increases; the set of RELATIVE FREQUENCIES of the SAMPLE POINTS falling within given intervals of the range of the random variable.

frequently, see NET CONVERGENCE.

Fresnel integrals, *n.* the two definite integrals

$$\int_0^\infty \cos(x^2)\,dx = \int_0^\infty \sin(x^2)\,dx = \left[\frac{\pi}{8}\right]^{\frac{1}{2}},$$

used in optical theory.

friction, *n.* (*Mechanics*) a force tangential to two surfaces in contact that is due to the roughness of the materials and is determined by the *coefficient of static friction*, μ, and the *coefficient of kinetic friction*, μ', which both vary only with the materials and units used. If the normal reaction between the surfaces is R and no slipping takes place, the friction cannot exceed μR, but if the surfaces slide over one another, the friction has magnitude μ'R and direction opposing the motion.

Fritz John conditions or **Fritz John theorem,** *n.* a form of the KUHN–TUCKER CONDITIONS that is valid without a CONSTRAINT QUALIFICATION. For a constrained minimization, one adds an additional non-negative multiplier, λ_0, for the objective function, and asserts that not all the multipliers are simultaneously zero. A constraint qualification can be viewed as ensuring that λ_0 is non-zero and so can be scaled to unity. Compare KUHN–TUCKER CONDITIONS.

Frobenius, Ferdinand Georg (1849–1917), German group theorist and analyst, who developed the theory of abstract groups, and made contributions to the theory of differential equations.

Frobenius group, *n.* a GROUP that has a proper subgroup, H, such that, for every x in the relative complement G \ H, the intersection of H with $x^{-1}Hx$ is the identity element.

Frobenius method, *n.* the method of solving ordinary differential equations near a REGULAR SINGULAR POINT, a, by positing a solution of the form

$$(x-a)^\alpha P(x-a)$$

for some power series P and index α that are then found by iteration by substituting the potential solution into the equation. See ORDINARY POINT.

Frobenius norm or **trace norm,** *n.* the norm on matrices that arises by treating a matrix as a vector and using the EUCLIDEAN NORM of that vector:

$$\|A\|^2 = \sum_{i=1}^n \sum_{j=1}^m |a_{ij}|^2.$$

This quantity is also the TRACE of AA^*.

Frobenius theorem, *n.* the theorem that the only finite-dimensional associative DIVISION ALGEBRAS over the reals are the reals, the complex numbers, and the QUATERNIONS.

from above, *adv.* on an interval with a given lower bound. See ABOVE.

from below, *adv.* on an interval with a given upper bound. See BELOW.

from the left, see LEFT–HAND LIMIT.

from the right, see RIGHT–HAND LIMIT.

frontier or **boundary,** *n.* the set of points (*frontier points*) that are members both of the CLOSURE of a given set and of the closure of its complement;

equivalently, the set of points in the closure but not in the INTERIOR of the given set, usually written FrA. For example,

$$Fr((0,1]) = \{0,1\};$$

the frontier of the rationals is the set of all real numbers.

frustum, *n.* **1.** a part of a solid such as a cone or pyramid lying between the base and a plane parallel to the base that intersects with the solid, as for example in Fig. 166.

2. any part of such a solid contained between two parallel planes intersecting with the solid.

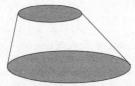

Fig. 166. **Frustum** of a cone.

F-sigma set, *n.* a subset of a topological space expressible as the union of countably many closed sets, usually denoted $\mathbf{F}_\sigma$. In a METRIC space all open sets are of this form, as are the rationals as a subset of the reals. See also G–DELTA, BAIRE CATEGORY.

F-space, *n.* **1.** usually, an alternative term for FRECHET SPACE.

2. in some authors, any complete metrizable space, that is, not requiring local convexity.

Fubini's theorem, *n.* the standard result that legitimates evaluation of MULTIPLE INTEGRALS as ITERATED INTEGRALS and changing the order of integration of an iterated integral. (Named after the Italian analyst, algebraist, and differential geometer, *Guido Fubini* (1879–1943).) See TONELLI'S THEOREM.

full linear group, *n.* another term for the GENERAL LINEAR GROUP.

full measure, *n.* (*as modifier*) (of a set in a MEASURE SPACE) having a set of NULL measure as complement.

full rank, *n.* (*as modifier*) (of a MATRIX) having maximum possible RANK; that is, having rank equal to the number of rows or the number of columns, whichever is less.

function, *n.* a RELATION between two sets that associates a unique element of the second with each element of the first; a MANY–ONE RELATION, written f or $f(x)$; formally, the set of ordered pairs $\langle x, f(x) \rangle$. If we write $y = f(x)$, then y is the VALUE of the function for the ARGUMENT x; if the function is defined between sets S and T with the arguments in S and the values in T, then S is the DOMAIN and T the CODOMAIN of the function. We can write

$$f: S \to T$$

$$f: x \mapsto y.$$

If s is a subset of S, then $f(s)$ is the set of values of $f(x)$ for $x \in s$, and is called the IMAGE of s under the function. The image, $f(S)$, of the domain is the RANGE of the function. Although the terms are usually regarded as synonymous, some authors prefer the term MAPPING, or TRANSFORMATION when dealing with abstract spaces; some use the former terms to indicate that the identity of the function is taken to depend on the specified domain and range as well as on the set of ordered pairs of relata, so that the real-valued square root function is regarded as a different mapping when defined on all reals and when defined on the non-negative reals; 'transformation' is often preferred when the algebraic expression for the value of the function is derived in a uniform way from the expression for the argument. Compare GRAPH, SET–VALUED MAPPING.

functional, *n.* a function whose domain is itself a set of functions, and whose range is another set of functions that may be numerical constants. The term is often reserved for LINEAR FUNCTIONALS.

functional analysis, *n.* the modern abstract study of LINEAR and non-linear FUNCTIONS in terms of the underlying LINEAR SPACES on which the functions are defined, and the DUALS of those spaces. This perspective, growing out of the study of LINEAR OPERATORS and FUNCTIONALS, aims at producing a unifying corpus of results and techniques for linear spaces and linear operators. This is applicable to the study of such diverse areas of mathematics as ALGEBRA, REAL ANALYSIS, NUMERICAL ANALYSIS, CALCULUS OF VARIATIONS, and DIFFERENTIAL EQUATIONS, through the application of general theorems such as the HAHN–BANACH THEOREM, the UNIFORM BOUNDEDNESS PRINCIPLE, the OPEN MAPPING THEOREM, and the RIESZ REPRESENTATION THEOREM.

functional calculus, *n.* **1.** the branch of mathematics that studies the properties of functions and operations upon functions. The BRACHISTOCRONE PROBLEM is an example of a problem of functional calculus.

2. any theory that guarantees that it is possible to mimic familiar constructions on the complex numbers with matrices or operators, for example, with the existence of SQUARE ROOTS of operators.

3. a less common term for PREDICATE CALCULUS.

functional determinant, *n.* the DETERMINANT of the DIFFERENTIAL COEFFICIENTS of *n* functions in *n* variables. See JACOBIAN.

functional equations, *n.* **1.** the branch of mathematics that studies equations in which the variables are functions, and that attempts to establish the properties of functions from the equations that they satisfy.

2. (more precisely and recently) the study of equations of the form $A = 0$, where A is a term containing a finite number of independent variables, a finite number of known functions, and a finite number of unknown functions, that one seeks.

3. (*singular*) an equation of this form; for example,

$$f(x + y) = f(x) + f(y),$$

the condition for a function to be additive.

function space, *n.* a vector space the elements of which are functions, especially continuous or bounded functions; for example, $C[0, 1]$ is the set of functions continuous on the interval $[0, 1]$.

functor, *n.* a function between CATEGORIES that maps objects to objects and MORPHISMS to corresponding morphisms. If a functor F maps the morphisms $M(a, b)$, into the corresponding morphisms $M'[F(a), F(b)]$ of the images of a and b under the functor, that is, if

$$F[M(a, b)] \subset M'[F(a), F(b)],$$

and if

$$F(f \circ g) = F(f) \circ F(g),$$

where $\circ$ is the COMPOSITION of the morphisms, then F is a *covariant functor*; it is a *contravariant functor* if

$$F[M(a, b)] \subset M'[F(b), F(a)],$$

and

$$F(f \circ g) = F(g) \circ F(f).$$

For example, the mapping from G to G/G', the derived subgroup of G, is a functor of the CATEGORY of all groups onto the category of all Abelian groups.

fundamental form, n. **1. first fundamental form.** the QUADRATIC FORM

$$ds^2 = A\, du^2 + 2B\, du\, dv + C\, dv^2,$$

where

$$A = \left(\frac{\partial \mathbf{x}}{\partial u}\right)^2, \qquad B = \left(\frac{\partial \mathbf{x}}{\partial u}\right)\left(\frac{\partial \mathbf{x}}{\partial v}\right), \qquad C = \left(\frac{\partial \mathbf{x}}{\partial v}\right)^2,$$

that determines the metric, and ARC LENGTH, on a given surface.
2. second fundamental form. a further quadratic form,

$$D\, du^2 + 2D'\, du\, dv + D''\, dv^2,$$

where

$$D = \sum_i X_i \frac{\partial^2 x_j}{\partial u^2}, \quad D' = \sum_i X_i \frac{\partial^2 x_i}{\partial u\, \partial v}, \quad D'' = \sum_i X_i \frac{\partial^2 x_j}{\partial v^2},$$

with X_i the direction cosines of the normal to the surface; this enables points of the surface to be classified as flat, elliptic, parabolic, or hyperbolic.
fundamental homotopy group or **groupoid,** see HOMOTOPY.
fundamental matrix, n. the matrix whose columns are the linearly independent FUNDAMENTAL SETS OF SOLUTIONS of a LINEAR HOMOGENEOUS system of ORDINARY DIFFERENTIAL EQUATIONS, or matrix equation, $\mathbf{y}' = A(t)\mathbf{y}$, where $\mathbf{y}$ has dimension n. See also PRINCIPAL SOLUTION MATRIX.
fundamental operations of arithmetic, see OPERATION.
fundamental parallelogram, see PERIODIC FUNCTION.
fundamental sequence, n. another term for CAUCHY SEQUENCE.
fundamental set of solutions, n. any BASIS for the VECTOR SPACE of all solutions of a HOMOGENEOUS system of LINEAR EQUATIONS. Such a treatment is possible since every linear combination of solutions of the system of equations is also a solution. See also FUNDAMENTAL MATRIX.
fundamental system of solutions, n. any set of n LINEARLY INDEPENDENT solutions of a HOMOGENEOUS linear ORDINARY DIFFERENTIAL EQUATION of order n; a set of n solutions is a fundamental system if and only if their WRONSKIAN is non-zero. The general solution of the differential equation is a linear combination of any fundamental set of solutions.

fundamental theorem of algebra, n. the theorem that a complex polynomial of the nth degree has precisely n complex ROOTS, counting MULTIPLICITY, and hence that the complex numbers are ALGEBRAICALLY CLOSED.

fundamental theorem of arithmetic or **unique factorization theorem,** n. the theorem that every positive integer has a unique CANONICAL decomposition as a product of powers of its PRIME FACTORS; that is, if the product

$$\prod_{i=1}^{n} p_i^{k_i} = \prod_{i=1}^{m} p_i^{l_i},$$

where p_i are the successive primes and k_i and l_i their respective exponents, then $n = m$ and $k_i = l_i$ for all i.

fundamental theorem of calculus, n. a theorem stating the relationship between INTEGRATION and DIFFERENTIATION: if the DERIVATIVE, $f(x)$, of $F(x)$ is integrable (in particular if the function is continuously differentiable), so that $F(x)$ is an INDEFINITE INTEGRAL of $f(x)$, then

$$\int_{a}^{b} f(x) \, dx = F(b) - F(a).$$

Conversely, if $F(x)$ is defined to be the integral of $f(x)$ from a to x for all x in $[a, b]$, then f is the derivative of F at every point of the interval at which f is continuous.

fundamental theorem of projective geometry, n. the theorem that three distinct corresponding pairs of points uniquely determine a PROJECTIVITY.

fundamental theorem of space curves, see FRENET FORMULAE.

fuzzy set theory, n. a version of naive set theory in which one allows elements to have *degrees of membership* in a *fuzzy set*; these degrees range from 1, when the element is in the set, to 0 when it is out of the set. The intention is to quantify precisely the intrinsically imprecise.

g

g, (*Mechanics*) *symbol for* the LOCAL GRAVITATIONAL CONSTANT.

G, *abbrev. for* GIGA-, used in symbols for multiples of the physical units of the SYSTEME INTERNATIONAL.

Galois, Évariste (1811–32), French mathematician who made significant contributions to the theory of functions, the theory of equations, and number theory, and whose work became a basis for group theory (a term which he introduced); all this developed from his concern, while still at school, to show the impossibility of the SOLUTION BY RADICALS of the quintic (which had, unknown to Galois, already been shown by Abel), and to describe the general conditions for the solubility of any polynomial equation. Although he had already published some papers, his first submission to the Academy of Sciences, in 1829, was lost by Cauchy, and the second by Fourier. He clashed with the oral examiner for the *École Polytechnique* and was refused a place; after his father committed suicide, he abandoned thoughts of mathematics as a career and enrolled as a trainee teacher, only to be expelled for writing an anti-monarchist article. He was imprisoned twice because of his republican beliefs, and his third submission to the Academy was rejected by Poisson. Galois was killed in a duel, probably provoked by royalist or police agents, at the age of 20, and is generally viewed as one of the two great romantic figures in mathematics (the other being Ramanujan).

Galois correspondence, *n.* a pair of mappings, between completely ordered sets, that are ANTITONE and mutually dominating; thus

$$f: S \rightarrow S'; \quad f': S' \rightarrow S,$$

and for each x in S, x' in S',

$$x \leq f'f(x); \quad x' \leq ff'(x').$$

In GALOIS THEORY such a correspondence, called a DUAL ISOMORPHISM (although it is not an isomorphism), exists between the lattice of intermediate EXTENSION FIELDS and the subgroups of the GALOIS GROUP of a finite normal extension field.

Galois field or **finite field,** *n.* any FIELD that contains only a finite number of elements. The study of such fields was initiated by Galois in 1830.

Galois group, *n.* the GROUP of all AUTOMORPHISMS of the SPLITTING FIELD, K, of a given ALGEBRAIC EQUATION that leave all members of the base field, F, fixed; it is denoted $G(F/K)$. It can be regarded as the group of all PERMUTATIONS of the ROOTS of the equation that leave all relations of the roots invariant. The Galois group of the general equation of degree n is the full SYMMETRIC GROUP.

Galois theory, *n.* the algebraic study of GROUPS of AUTOMORPHISMS of FIELDS in which one associates an EXTENSION FIELD with a given ALGEBRAIC EQUA-

TION. The theory grew out of Galois' highly original study of the solubility of equations, devised in part to prove the impossibility of SOLUTION BY RADICALS of the general QUINTIC, Abel's proof of which was unknown to him. See also CARDANO'S FORMULA.

gambler's ruin, *n.* a RANDOM WALK in which a gambler bets on repeated trials, in each of which he wins with probability strictly between 0 and 1, until he either doubles his initial capital or loses it all; for example, he may start with \$5 and repeatedly bet \$1 until he either has \$10 or goes broke. This gives the TRANSITION PROBABILITIES of a MARKOV CHAIN with two ABSORBING STATES; the probability of ruin varies both with the initial state and with the probability of winning on each trial.

game, *n.* a mathematical representation in GAME THEORY of some situation in which the outcome depends on the choices made by the participants. These include some recreational activities, also commercial, personal and military activities.

game theory or **theory of games,** *n.* (*Statistics, operations research*) the mathematical theory concerning the optimal choice of strategy in situations involving decision-making in competition or a conflict of interest. See GAME. See also MINIMAX THEOREM.

gamma distribution, *n.* (*Statistics*) a continuous two-parameter distribution from which the CHI-SQUARE and EXPONENTIAL DISTRIBUTIONS are derived, written $Ga(\lambda, \nu)$ and defined in terms of the GAMMA FUNCTION as

$$Ga(\lambda, \nu) = \frac{\lambda^\nu \, x^{\nu-1} \, e^{-\lambda x}}{\Gamma(\nu)}.$$

gamma function, *n.* one of the most important SPECIAL FUNCTIONS, which has the property that

$$\Gamma(z+1) = z\Gamma(z),$$

so that for integral n, $\Gamma(n+1) = n!$, the FACTORIAL function, which it thus extends to all real or complex z. It has three standard definitions: the first, due to Euler and to Gauss, defines it as

$$\Gamma(1 + z) = \int_0^\infty x^z \, e^{-x} \, dx,$$

where the real part of z is required to be greater than -1; the second, due to Gauss, is

$$\Gamma(1 + z) = \lim_{n \to \infty} \frac{n! \, n}{(z+1)(z+2) \dots (z+n)},$$

where z is not a negative integer; and the third, due to Weierstrass, is

$$\frac{1}{\Gamma(1 + z)} = e^{\gamma z} \prod_{i=1}^\infty \left[\left(1 + \frac{z}{n}\right) e^{-z/n} \right].$$

The gamma function satisfies

$$\frac{\pi}{\sin(\pi z)} = G(z) \, G(1 - z),$$

for all z whose absolute value is strictly between 0 and 1; thus $\Gamma(1/2) = \sqrt{\pi}$, as required by the NORMAL DISTRIBUTION. See also BETA FUNCTION.

gap series, n. a POWER SERIES with many zero coefficients.

gate, see LOGIC CIRCUIT.

Gateaux derivative, n. the mapping determined by the GATEAUX DIFFEREN-TIAL when this exists in all directions.

Gateaux differential, n. the directional derivative at $\mathbf{x}$ with increment $\mathbf{h}$, of a given function f defined on an open domain, given by

$$\delta f(\mathbf{x}; \mathbf{h}) = \lim_{t \to 0} \frac{f(\mathbf{x} + t\mathbf{h}) - f(\mathbf{x})}{t}.$$

If this limit exists for all $\mathbf{h}$, then the function is said to be *Gateaux differentiable* at $\mathbf{x}$, and if it is linear in $\mathbf{h}$, the mapping

$$T = \delta f(\mathbf{x};\ )$$

is said to be the (*linear Gateaux derivative*) of f at $\mathbf{x}$, written $\nabla f(\mathbf{x})$ and often referred to as the GRADIENT of f. If a mapping of one finite-dimensional vector space into another is a LIPSCHITZ FUNCTION, then any linear Gateaux derivative is automatically a FRECHET DERIVATIVE.

Gateaux smooth, see SMOOTH (sense 2).

gatepost, n. (*Logic, informal*) another word for TURNSTILE.

gauge or **valuation,** n. a mapping g from an INTEGRAL DOMAIN $E \setminus \{0\}$ into the non-negative integers, such that $g(ab) \geq g(a)$ for all a and b in $E \setminus \{0\}$, and such that for b in E and a in $E \setminus \{0\}$ there exist q and r in E for which $b = qa + r$ and either $r = 0$ or $g(r) < g(a)$. For example, the degree of polynomials over a field is a gauge. See also EUCLIDEAN DOMAIN.

gauge function, n. another term for MINKOWSKI FUNCTION.

gauge transformation, n. (*General relativity*) a small COORDINATE CHANGE that transforms one nearly Cartesian coordinate system into another of the same kind.

Gauss, Carl Friedrich (1777–1855), German mathematician and astronomer who is generally regarded as one of the most prolific and influential mathematicians ever. In his doctoral thesis when he was only 22, he developed the concept of complex number and used it to establish the FUNDAMENTAL THEOREM OF ALGEBRA. In 1801, he published *Disquisitiones arithmeticae*, which firmly established NUMBER THEORY as a well-defined branch of mathematics. He was Professor and Director of the observatory at Göttingen from 1807 onward, and was employed by the government to conduct a trigonometric survey of the kingdom of Hanover. He obtained a wide variety of essential results in geometry, algebra, analysis, astronomy, and statistics, as well as contributing to the mathematization of the physics of electricity, magnetism and gravitation.

Gaussian curvature, n. a measure of the curvature at a point of a space surface, given as a ratio of the discriminants of the two FUNDAMENTAL FORMS of the surface, but actually dependent only on the second. The *Gauss–Bonnet theorem* expresses the *integral curvature* (the integral of Gaussian curvature over the surface) as 2π minus the line integral of the GEODESIC CURVATURE with respect to arc length over the boundary of the surface.

Gaussian distribution, n. another name for the NORMAL DISTRIBUTION.

Gaussian domain or **unique factorization domain**, *n.* an INTEGRAL DOMAIN in which every non-zero non-UNIT is uniquely representable as a finite product of irreducible elements, up to permutation. Since the GAUSSIAN INTEGERS are a Gaussian domain, and $5 = (1 + 2i)(1 - 2i)$, 5 is reducible in the Gaussian integers; 3 is irreducible therein, and hence prime, because it is in a EUCLIDEAN DOMAIN. The domain $\mathbb{Q}[x, y]$ is a Gaussian domain, but not a PRINCIPAL IDEAL DOMAIN; for a quadratic number field, a Gaussian domain is a principal ideal domain.

Gaussian elimination or **pivoting**, *n.* the solution of simultaneous equations by the ELEMENTARY OPERATIONS. The term is often reserved for incomplete ELIMINATION, or reduction to a triangular form as opposed to complete JORDAN ELIMINATION. See LU DECOMPOSITION.

Gaussian field, *n.* the FIELD consisting of complex numbers $u + iv$ where u and v are rational.

Gaussian function, *n.* the function $y = \exp(-x^2)$, whose integral from $-\infty$ to ∞ converges to $\sqrt{\pi}$.

Gaussian integer, *n.* a complex number of which the real and imaginary parts are integers; a number of the form $n + im$ where n and m are integers; the algebraic integers in the GAUSSIAN FIELD.

Gaussian plane, *n.* another name for the ARGAND DIAGRAM.

Gaussian reciprocity, *n.* another term for QUADRATIC RECIPROCITY.

Gauss–Jordan elimination, *n.* another term for JORDAN ELIMINATION.

Gauss' lemma, *n.* the result that if a polynomial with integral coefficients factors over the rationals then it factors over the integers.

Gauss–Markov least squares theorem, *n.* the theorem that the LEAST SQUARES estimate of β for the model

$$Y_i = \beta X_i + \varepsilon_i; \quad E[\varepsilon_i] = 0; \quad \text{var}(\varepsilon_i) = \sigma^2; \quad \text{cov}(\varepsilon_i, \varepsilon_j) = 0,$$

has UNIFORM MINIMUM VARIANCE among all unbiased LINEAR ESTIMATES of β. The theorem also holds for MULTIVARIATE distributions.

Gauss–Seidel iteration, *n.* another term for the METHOD OF SUCCESSIVE DISPLACEMENTS.

Gauss' test, *n.* the test for the CONVERGENCE of a POWER SERIES that if

$$\left| \frac{u_n}{u_{n+1}} \right| = 1 - \frac{L}{n} + O\left[\frac{L}{n^{1+\varepsilon}} \right],$$

then $\sum u_n$ converges absolutely if $L > 1$, and diverges, or converges conditionally, if $L \leq 1$.

Gauss' theorem, *n.* another name for the DIVERGENCE THEOREM.

gcd, *abbrev. for* GREATEST COMMON DIVISOR.

gcf, *abbrev. for* GREATEST COMMON FACTOR.

G-delta set, *n.* also written $\mathbf{G}_\delta$. A set expressible as the intersection of countably many open sets; the complement of an F-SIGMA SET.

Gelfand transform, *n.* (*Operator theory*) the mapping that associates a continuous function on the compact space, Δ, of all MAXIMAL IDEALS in the induced weak topology, to an element of a commutative BANACH ALGEBRA, A. This is effected via the formula

$$\hat{x}(h) = h(x)$$

for any complex homomorphism, h, of A (the maximal ideals being identifiable with the homomorphisms). The *Gelfand–Naimark theorem* shows that when A is a B*-ALGEBRA the Gelfand transform is an ISOMETRY of A onto the continuous function space $C(\Delta)$ in uniform norm, and has the property that

$$(\overset{\wedge}{x^*}) = (\overline{\hat{x}}).$$

(Named after *Israil Moiseyevich Gelfand* (1913–) Russian functional analyst, who, without finishing his secondary education, obtained a doctorate for his development of the theory of BANACH ALGEBRAS (commutative normed rings). He has also contributed to special function theory and to the mathematical description of elementary particles and neurophysiology.)

Gelfond–Schneider theorem or **Gelfond's theorem,** *n.* (*Number theory*) the theorem that for complex ALGEBRAIC numbers α and β with α not 0 or 1 and β irrational, α^β is TRANSCENDENTAL. Thus $\sqrt{2}^{\sqrt{2}}$ and $e^\pi = (-1)^{-i}$ are transcendental. (Named after the Russian analyst and number theorist, *Alexander Osipovich Gelfond* (1906–68), who developed much of the basic theory of transcendental numbers.)

general, *adj.* (of a statement, theorem, etc.) not specifying an individual subject but quantifying over a domain. EXISTENTIAL as well as UNIVERSAL statements are general in this sense. Usually, the domain is non-trivial, but in formal contexts this may not be so. See GENERALIZATION.

general induction, *n.* another term for COMPLETE INDUCTION, by contrast with special induction. See INDUCTION.

generalization, *n.* **1a.** a GENERAL statement concerning all the members of some class.
b. the process of inferring such a statement from an INSTANCE.
2. (*Logic*) **a.** the formal derivation of a general statement from a particular one by replacing its subject term with a BOUND VARIABLE and prefixing a QUANTIFIER. In particular, *universal generalization* is the valid inference of a UNIVERSAL statement from a particular one of which the subject is usually an arbitrary individual; *existential generalization* is the valid inference of an EXISTENTIAL statement from a particular statement. For example, *someone is happy* is the existential generalization of *John is happy*.
b. (*Logic*) the statement so inferred. A *universal generalization* ascribes a property to all members of a class; and an *existential generalization* to one or more unspecified members.

generalized continuum hypothesis, see CONTINUUM HYPOTHESIS.

generalized coordinates, *n.* (for a set of particles with a finite number, m, of degrees of freedom) a set of variables, often denoted $q_1 \ldots q_m$; the minimum number of COORDINATES necessary to describe the motion of the set. See LAGRANGIAN.

generalized delta function, see KRONECKER DELTA.

generalized eigenvalue problem, *n.* the problem of finding scalars λ and vectors $\mathbf{x}$ solving $A\mathbf{x} = \lambda B\mathbf{x}$ where A and B are given matrices or linear operators, and B is POSITIVE DEFINITE. In the *classical eigenvalue problem*, B is the identity matrix or operator. See LATENT ROOT.

generalized function, see DISTRIBUTION.

generalized inverse, *n.* another term for the PSEUDO-INVERSE of a matrix.

generalized maximum likelihood ratio test statistic, *n.* the ratio of the MAXIMUM LIKELIHOOD of drawing a given sample under a given hypothesis to the maximum likelihood of drawing that sample under the hypothesis that the given hypothesis is false. Some authors use the term for the inverse of this ratio.

generalized mean-value theorem, *n.* another term for CAUCHY'S MEAN–VALUE THEOREM. See also MEAN–VALUE THEOREM.

generalized nilpotent, *adj.* an element x in a BANACH ALGEBRA, such that

$$\lim_{n \to \infty} \|x^n\|^{\frac{1}{n}} = 0.$$

The set of such elements is called the *radical* of the algebra.

generalized polynomial, *n.* a function of the form

$$\sum_{i=1}^{n} c_i \, p_i$$

for fixed continuous functions p_i and arbitrary scalars. See UNICITY.

generalized ratio test, see RATIO TEST.

general linear group or **full linear group,** *n.* the GROUP, often written $GL(V)$, of all LINEAR TRANSFORMATIONS of a finite-dimensional VECTOR SPACE, V, that are INVERTIBLE; equivalently, where V is over a field F, the group of all NON-SINGULAR $n \times n$ matrices over F, written $GL(n, F)$. Compare SPECIAL LINEAR GROUP.

general solution, *n.* **1.** a relation between the variables of an ORDINARY DIFFERENTIAL EQUATION that satisfies the equation but contains distinct arbitrary constants of the same number as the order of the equation. Compare SINGULAR SOLUTION, PARTICULAR INTEGRAL.
2. a solution of a PARTIAL DIFFERENTIAL EQUATION of order n that contains the maximum number of independent arbitrary functions; this number may be less than n.

generate, *vb.* **1.** to provide a precise criterion for membership in a set, in the form of an algorithm whose application recursively yields all and only the members of the set. For example, the formation rules of a language generate all and only its well-formed expressions; the basis elements in a vector space generate the space.
2. (of a subset of a structure such as a ring, group, or module) to enable all elements of the group to be constructed by recursive application of the operations defined on the structure to the members of the subset; thus the structure is contained in the closure of the set of its *generators* under these operations. For example, the set of all transpositions generates the symmetric group and the set of all 3-cycles generates the alternating group. A *finitely-generated* structure is the closure of a finite set of *generators*. Clearly, any BASIS of a vector space generates the space.

generating function, *n.* **1.** (for a sequence) a formal POWER SERIES of which the coefficients are the given sequence. This often allows one to study the sequence by analytic techniques, and is used both in combinatorics and in analysis. For example, the generating function of the Fibonacci numbers is

$$\frac{x + x^2}{1 - (x + x^2)}.$$

2. see LAMBERT SERIES.

generating set, n. a set whose elements GENERATE a given algebraic structure.

generator, n. **1.** also called **generatrix.** a point, line, or plane whose motion subject to certain constraints sweeps out a given geometric figure, as shown for the CYLINDER in Fig. 167; an ELEMENT of a SURFACE OF REVOLUTION.

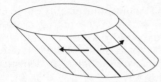

Fig. 167. The **generator** of a cylinder.

2. one of a set of elements that together GENERATE a group.

3. an element of a CYCLIC GROUP the powers of which exhaust the group.

4. see RULED SURFACE.

generatrix, n. another term for GENERATOR (sense 1).

generic, *adj.* **1.** applicable or referring to an entire class of objects; general.

2. (of a set in a complete metric space), expressible as a DENSE G–DELTA SET, and so (in the sense of BAIRE CATEGORY) a RESIDUAL SET.

genus, n. **1.** a measure of the CONNECTEDNESS of a CLOSED SURFACE, equal to $1 - K/4\pi$, for K the integral GAUSSIAN CURVATURE. See also EULER'S FORMULA.

2. (of an ALGEBRAIC plane curve) the difference between the maximum number of DOUBLE POINTS a curve of the given DEGREE may possess, and the actual number of the given curve.

3. (for a topological surface) a pair (p, q), where p is the number of HANDLES and q is the number of CROSS–CAPS of the surface.

4. a class of non-equivalent primitive binary QUADRATIC FORMS with given discriminant, each form representing the same integers.

5. the least natural number m such that an ENTIRE FUNCTION has a *Weierstrass product expansion*

$$z^n \exp[g(z)] \prod_n \left(1 - \frac{z}{a_n}\right) \exp w\left(\frac{z}{a_n}\right),$$

where g is entire,

$$w(z) = \frac{z}{1} + \frac{z^2}{2} + \frac{z^3}{3} + \ldots + \frac{z^m}{m},$$

and n ranges over the natural numbers. If no such expression exists the genus is infinite. Compare ORDER.

geod, *abbrev. for* GEODESIC.

geodesic or **geodetic** (abbrev. **geod**), *adj.* **1.** relating to or involving the geometry of curved surfaces.

2. (*as substantive*) also called **geodesic curve.** the shortest curve between two points on a curved surface lying wholly on the surface. For example, in

Fig. 168, the bold curve is a geodesic joining the points A and B on the surface of a hemisphere; it is an arc of a GREAT CIRCLE of the sphere.

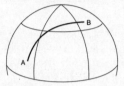

Fig. 168. AB is a **geodesic** on the surface of a sphere.

geodesic curvature, *n.* (of a space curve at a point) the quantity whose magnitude is the CURVATURE of the projection of the curve onto the TANGENT PLANE at the given point, and whose sign is positive if the PRINCIPAL NORMAL to the projected curve is equal to the VECTOR PRODUCT of the unit normal to the surface and the tangent to the curve at the point. This depends on the surface that the curve is viewed as lying in.

geom, *abbrev. for* GEOMETRY, GEOMETRIC.

geometer or **geometrician,** *n.* one who studies GEOMETRY.

geometric or **geometrical** (abbrev. **geom**), *adj.* **1.** of, relating to, using the methods of, or according with the principles of GEOMETRY.

2. representing, consisting of, formed by, or characterized by abstract mathematical points, lines, surfaces, etc. rather than actual physical shapes. A *geometric figure* is usually taken to be a diagram, but mathematics is concerned only with its abstract and not with its physical properties, so that diagrams are not themselves geometric figures properly so called, but merely representations of them.

geometric distribution, *n.* (*Statistics*) the distribution of the number, x, of BERNOULLI TRIALS required to obtain a first success; where the probability of success in each trial is p, the probability that $x = r$ is

$$p(1-p)^{r-1},$$

with mean $1/p$.

geometric form of the Hahn–Banach theorem, *n.* another name for the SEPARATION THEOREM OF MAZUR.

geometric mean, *n.* the nth root of the product of a sequence or set of n quantities, numbers or expressions; for example, the geometric mean of 3 and 4 is $\sqrt{(3 \times 4)} = 2\sqrt{3}$. The geometric mean of a set of numbers is always less than their ARITHMETIC MEAN except when all the numbers are equal, when identity holds.

geometric progression, *n.* a sequence of numbers whose successive members differ by a constant multiplier. For example, 1, 2, 4, 8, 16,.... In general the nth term of a progression, of which the first member is a and each member is r times the preceding, is ar^{n-1}. Compare ARITHMETIC PROGRESSION.

geometric series, *n.* a series whose terms form a GEOMETRIC PROGRESSION, such as

$$1 + \frac{1}{2} + \frac{1}{4} + \frac{1}{8} + \ldots$$

This converges only if the absolute ratio of successive terms is less than 1.

The sum of finite initial segments of the series is

$$a + ar + ar^2 + \dots + ar^{n-1} = \frac{a(r^n - 1)}{r - 1},$$

whence, if the infinite series converges, its sum is $a/(1-r)$.

geometrize, *vb.* to use or apply geometric methods to; to represent in geometric form.

geometry, *n.* **1.** the elementary study of the properties and relations of CON-STRUCTIBLE plane figures.

2. the study of the geometric properties of objects, as defined in 1872 by Klein's ERLANGEN PROGRAM.

3. some specific mathematical system or axiomatization of these properties and relations, such as EUCLIDEAN GEOMETRY, RIEMANNIAN GEOMETRY.

4. see PROJECTIVE GEOMETRY, FINITE GEOMETRY, DIFFERENTIAL GEOMETRY.

Gergonne point, *n.* the point of intersection of the CEVIANS through the points of tangency of the INCIRCLE of a triangle. (Named after the French projective geometer, *Joseph Diaz Gergonne* (1771–1859), who shares credit with Poncelet for formulating the principle of DUALITY.)

Gerschgoren circle theorem, *n.* (*Matrix theory*) the result that all the EIGEN-VALUES of a matrix $\{a_{ij}\}$ lie within the circles centred at a_{ii} and with radii

$$R_i = \sum_{i = j} | a_{ij} |.$$

In addition, the union of any k of these circles, if disjoint from the remaining circles, contains exactly k characteristic values counting multiplicity.

Gibbs phenomenon, *n.* the necessary behaviour of partial sums of a FOURIER SERIES near a jump discontinuity of a function of bounded variation; since convergence is not uniform the curves of the partial sums tend to approximate vertical segments longer than the jump, by a proportion of precisely

$$\frac{2}{\pi} \int_0^\pi \frac{\sin x}{x} \ dx = 1.17898 \dots.$$

(Named after the American theoretical physicist and chemist *Josiah Willard Gibbs* (1839–1903), who developed VECTOR ANALYSIS and statistical MECHAN-ICS. He was initially trained as an engineer.)

giga- (symbol **G**), *prefix* denoting a multiple of 10^9 of the physical units of the SYSTEME INTERNATIONAL.

GIMPS, *acronym for* Great Internet Mersenne Prime Search, a research project in which more than 200 000 individuals worldwide cooperate to run a search algorithm in spare background capacity on their personal computers. In December 2001, it discovered the 39th known MERSENNE PRIME, a 4-million digit number.

given, *adj.* **1.** known or determined independently.

2. stipulated or assumed for the purposes of a specific construction or proof; for example, an epsilon-delta proof of continuity typically begins 'Given e > 0'.

3. (*as substantive*) an axiom or assumption of a proof.

glb, *abbrev. for* GREATEST LOWER BOUND.

global or **in the large** (*im grossen*), *adj.* (of a mathematical relation, property, etc.) holding for all values of the variables without restriction. For example, Fig. 169 shows the graph of $y = x^4 - 5x^2 + x + 4$, which has a global minimum at $x = -1.64$, but not at $x = 1.52$, where the minimum is only LOCAL.

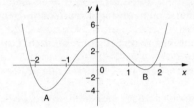

Fig. 169. A **global** minimum (A) and a local minimum (B).

gnomon, *n.* the geometric figure remaining after a parallelogram has been removed from one corner of a larger similar parallelogram, such as that in Fig. 170.

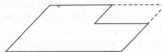

Fig. 170. **Gnomon**.

gnomonic projection, *n.* the projection of a sphere from its centre onto any tangent plane; for example, Fig. 171 shows the gnomonic projection of a circle on the surface of a sphere onto an ellipse on a tangent plane. Compare STEREOGRAPHIC PROJECTION.

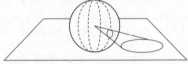

Fig. 171. The **gnomonic projection** of a circle.

Gödel, Kurt (1906–78), Czech-born American logician who proved a number of fundamental metamathematical results that bear his name; in the course of these proofs he developed the theory of RECURSIVE FUNCTIONS, and, since he thereby showed the unattainability of the aims of HILBERT'S PROGRAMME and (on some interpretations) of LOGICISM, he brought about a complete reassessment of the FOUNDATIONS OF MATHEMATICS. He also proved both the AXIOM OF CHOICE and the CONTINUUM HYPOTHESIS to be consistent with the standard axioms of set theory.

Gödel number, *n.* a number uniquely associated with every formula of a formal theory: if the primitive symbols of the calculus are each associated with a unique number, then any sequence of symbols can be uniquely represented by a product of powers of the prime numbers in order, with the nth prime having as exponent the number associated with the nth symbol. This device is extended in the proof of GÖDEL'S THEOREM by assigning to a proof a product of powers of primes in the same way, where the exponents of successive primes are the Gödel numbers of the formulae of each line of the proof in turn. Gödel then showed that valid deducibility could be represented as a number-theoretic property of the number assigned to a proof. Gödel numbers are also used to identify TURING MACHINES.

Gödel's completeness theorem, *n.* the result that a theory is CONSISTENT if and only if it has a MODEL.

Gödel's proof or **Gödel's theorem,** *n.* (*Logic*) the crucial result that, in an AXIOMATIC FORMAL CALCULUS of the complexity of NUMBER THEORY (PEANO ARITHMETIC), it is impossible to prove CONSISTENCY without using methods from outside the system. Gödel showed this by proving that validity corresponds to a property of GÖDEL NUMBERS, describing the construction of the Gödel number corresponding to an assertion that the formula with that number is not provable, and then proving that were arithmetic COMPLETE, that statement would have that property. It follows from his theorem that HILBERT'S PROGRAMME of devising a decision algorithm for all of mathematics is unattainable, and that the logicist doctrine of the deducibility of all of mathematics from the axioms of logic is false.

Gödel statement, *n.* a statement that asserts its own unprovability, especially that used in GÖDEL'S PROOF and given in terms of its GÖDEL NUMBER.

Goldbach's conjecture, *n.* (*Number theory*) the conjecture that every even number (greater than or equal to 6) can be written as a sum of two odd primes. (Named after the Prussian-born number theorist and analyst, *Christian Goldbach* (1690–1764), who became Professor of Mathematics at, and the historian of, the Russian Imperial Academy. He was also the tutor of Peter the Great, and was a member of the Tsar's Foreign Ministry. He corresponded with Euler and the Bernoullis and contributed to the theory of differential equations and infinite series. He also conjectured that all odd numbers are the sum of three odd primes; *Vinogradov's theorem* shows this true of all except possibly finitely many odd numbers.)

golden mean, **golden section**, or **extreme and mean ratio,** *n.* the proportion of the division of a line so that the smaller is to the larger as the larger is to the whole, or of the sides of a rectangle so that the ratio of their difference to the smaller equals that of the smaller to the larger, supposed in classical aesthetic theory to be uniquely pleasing to the eye. This yields

$$G = \frac{\sqrt{5} - 1}{2} = 0.618\ 033\ 988 \ldots$$

of which the inverse is $1.618\ 033\ 988\ldots = G + 1$, which is also sometimes referred to as the golden ratio. It is a consequence of the definition that if one draws a rectangle with sides in the golden ratio (a *golden rectangle*), and then removes from it a square, the rectangle that remains has the same proportions as the original. If this process is repeated as shown in Fig. 172, then the successive points of division lie on a LOGARITHMIC SPIRAL.

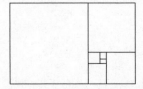

Fig. 172. A sequence of **golden rectangles**.

The golden mean is also the limit both of the CONTINUED FRACTION

$$\cfrac{1}{1+\cfrac{1}{1+\cfrac{1}{1+\cfrac{1}{1+\cdots}}}},$$

and of the ratio of successive terms of the FIBONACCI NUMBERS.

-gon, *suffix denoting* a POLYGON; for example, an *enneagon* and a *chilliagon* are polygons with nine and a thousand sides respectively.

Goodman's paradox, *n.* (*Logic, philosophy*) the paradox of INDUCTION that past experience provides identically strong evidence for incompatible predictions. If *grue* is defined as the property of being green until a specified future time and blue thereafter, everything that is or has been green is also grue, but while its past greenness gives rise to a prediction that it will remain green, its past grueness provides exactly the same grounds for predicting that it will remain grue – which after the specified time will be the incompatible colour blue. (Named after the American philosopher of language and science *Nelson Goodman* (1906–).) See also HEMPEL'S PARADOX.

goodness of fit, *n.* (*Statistics*) the extent to which observed sample values of a variable approximate to values derived from a theoretical distribution, often measured by a CHI-SQUARED TEST.

googol, *n.* the number represented by 1 followed by a hundred zeros, 10^{100}. (This exceeds the number of atoms in the universe, which is only of the order of 10^{85}.)

googolplex, *n.* the number represented in the decimal system by 1 followed by a GOOGOL of zeros; the googolth power of 10, $10^{(10^{100})}$.

grad, *abbrev. and symbol for* GRADIENT (sense 3).

gradient, *n.* **1a.** the slope of a line measured as the ratio of its vertical change to its horizontal change. Thus the gradient of the line joining points (x_1, y_1) and (x_2, y_2) in the Cartesian plane, as shown in Fig. 173a below, is

$$m = \frac{y_2 - y_1}{x_2 - x_1},$$

and the *gradient form* of the equation of a line through the point (x_1, y_1) is

$$\frac{y - y_1}{x - x_1} = m,$$

that is,

$$y = mx + (y_1 - mx_1);$$

where the equation of a line is given as $y = mx + c$, m is the gradient of the line, and c its intercept with the y-axis.

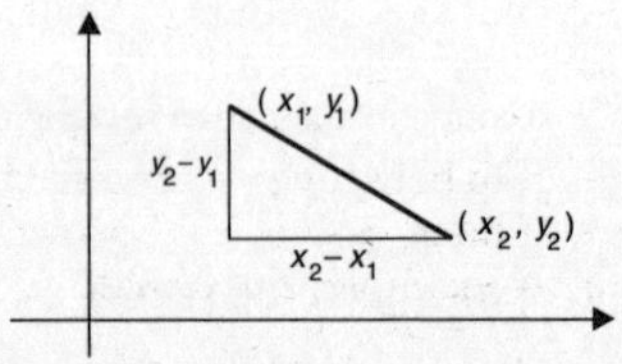

Fig. 173a. The **gradient** of a line.

b. more generally, the slope of a curve at a point measured as the slope of the tangent at that point; the gradient of the curve for $x = a$ is the INSTANTANEOUS RATE OF CHANGE in the value of the function. This is given by the limit, as x approaches a, of the ratio of the change, Δy, in the dependent variable to the change, Δx, in the independent variable, as shown in Fig. 173b. See DERIVATIVE.

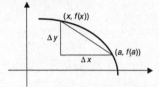

Fig. 173b. The **gradient** of a curve.

3. the vector whose components parallel to each coordinate axis are the partial derivatives of a given function with respect to the variable mapped on that axis, and whose direction is that in which the derivative of the function has its maximum value; the vector

$$\left(\frac{\partial \mathbf{F}}{\partial x}, \frac{\partial \mathbf{F}}{\partial y}, \frac{\partial \mathbf{F}}{\partial z} \right).$$

Often it is necessary that the partials be continuous, in which case the gradient is identifiable with the derivative of the vector function. The gradient is written grad F or ∇F. See also FRECHET DERIVATIVE, CURL, DIVERGENCE.

4. more generally, for a CARTESIAN TENSOR

$$\mathrm{T}_{ijk} \; \mathbf{e}_i \otimes \mathbf{e}_j \otimes \mathbf{e}_k \otimes \dots$$

the quantity

$$\frac{\partial}{\partial x_p} \left(\mathrm{T}_{ijk} \right) \mathbf{e}_i \otimes \mathbf{e}_j \otimes \mathbf{e}_k \otimes \dots \otimes \mathbf{e}_p.$$

gradient method, $n.$ another name for STEEPEST DESCENT.

Gram determinant, $n.$ the DETERMINANT of a GRAM MATRIX.

Gram matrix, $n.$ the self-adjoint matrix G whose i,j^{th} entry is the inner product $\langle y_i, y_j \rangle$ of the ith and the jth elements of a given finite sequence of vectors, $\{y_i\}$ in an INNER PRODUCT space. For example, the Gram matrix of $(1, 2)$ and $(1, -1)$ is

$$\begin{bmatrix} (1, 2) \; (1, 2) & (1, 2) \; (1, -1) \\ (1, 2) \; (1, -1) & (1, -1) \; (1, -1) \end{bmatrix} = \begin{bmatrix} 5 & -1 \\ -1 & 2 \end{bmatrix}.$$

(Named after *Jörgen Pedersen Gram* (1859–1916), Danish number theorist and analyst.)

Gram–Schmidt process, $n.$ an iterative method of converting any LINEARLY INDEPENDENT family of vectors in an INNER PRODUCT space into an ORTHONORMAL system.

graph, $n.$ **1.** a drawing showing the relationship between certain sets of quantities or numbers by means of a series of lines, points, etc., plotted with respect to a set of coordinate axes. See, for example, BAR GRAPH.

2. a drawing showing a functional relationship between two or more variables by means of a curve, surface, etc. containing all and only those points the coordinates of which satisfy the given relation. For example, Fig. 174 shows part of the graph of $y = x \sin(x+1) - 1$, since the coordinates of every point that lies on the curve satisfy this relation.

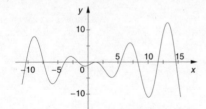

Fig. 174. The **graph** of a function.

3a. the set of points $\langle x, y \rangle$ where $y = f(x)$, as contrasted with the FUNCTION f.
b. the set of points $\langle x, y \rangle$ where $y \in f(x)$ as contrasted with the SET–VALUED FUNCTION f.
4. (*Graph theory*) a set of points (VERTICES) and line segments (EDGES) that connect some of these vertices, used both in the study of TOPOLOGY and in COMBINATORICS and the construction of COMBINATORIAL ALGORITHMS. See also TREE.

graphic, *adj.* using, pertaining to, or determined by GRAPHS, for example, a graphic representation of a function is a graph in sense 2.

graphical solution, *n.* a method of solving two SIMULTANEOUS EQUATIONS by plotting the GRAPH of each equation and identifying the points of intersection of the curves. Since the coordinates of these points satisfy both equations, they constitute a solution of the equations.

graph paper, *n.* paper printed with intersecting lines, for drawing graphs, diagrams, etc. The lines are usually equally spaced horizontally and vertically, but this is not the case of LOG PAPER.

graph theory, *n.* the branch of mathematics concerned with the study and application of planar graphs and their generalizations.

Grassmann algebra, see EXTERIOR PRODUCT.

gravitational constant, see LOCAL GRAVITATIONAL CONSTANT, UNIVERSAL GRAVITATIONAL CONSTANT.

gravity, *n.* (*Mechanics*) the force on bodies towards one another due to their MASS. According to *Newton's law* the gravitational force between two particles with masses m_1 and m_2, with distance r between them, is $\gamma m_1 m_2 r^{-2}$, where γ is the UNIVERSAL GRAVITATIONAL CONSTANT. A solid uniform sphere acts under gravity like a particle of the same mass at its centre.

great circle, *n.* **1.** a circular section of a sphere; a circle drawn on the surface of a sphere that has the same radius by a diametral plane, as the sphere. The MINOR ARC of the great circle through two points on the surface of a sphere is the shortest line between them lying wholly on the surface. For example, Fig. 175 opposite shows the great circle through A and B; the bold arc is the shortest distance between them on the surface. Compare SMALL CIRCLE. See also GEODESIC.

Fig. 175. **Great circle**.

2. such a line on the surface of the earth, or its representation on a map. The shortest route between two points on the earth's surface lies on a great circle.

greatest, *adj.* (of an element of an ordering) uniquely MAXIMAL, that is, greater than every other member of the set; so, for example, the set of subsets of a given set ordered under inclusion has the given set itself as a greatest element.

greatest common factor or **greatest common divisor** (abbrev. **gcf**, **gcd**), *n.* another name for HIGHEST COMMON FACTOR.

greatest integer function, *n.* another name for FLOOR.

greatest lower bound, *n.* another term for INFIMUM.

greedy algorithm, *n.* any of a class of combinatorial algorithms, so called because they attempt as much improvement as possible at each iteration. Compare MYOPIC ALGORITHM.

Green-elastic or **hyperelastic,** *adj.* (*Continuum mechanics*) (of a BODY) such that there is a scalar function of the DEFORMATION GRADIENT whose rate of change is the STRESS-POWER per unit volume. Although all ELASTIC bodies in nature have been found to be Green-elastic, in principle a Green-elastic body is elastic but not ELASTIC.

Green's functions, *n.* a set of integrating KERNELS used in solving non-homogeneous differential equations with boundary conditions, and involving the FUNDAMENTAL SOLUTIONS to the corresponding homogeneous differential equations. (Named after *George Green* (1793–1841), a self-taught miller's son and baker from Nottingham, who was elected to a Cambridge fellowship after graduating aged 43. He had by then already made valuable contributions to vector analysis, pioneered the mathematical description of electricity and magnetism, studied *n*–dimensional spaces, and published studies of a number of problems of applied mathematics.)

Green's theorem, *n.* **1.** the identity

$$\iint (f\nabla g - g\nabla f) \cdot \mathbf{n}\, \mathrm{dS} = \iiint (f\nabla^2 g - g\nabla^2 f)\, \mathrm{dV},$$

where *f* and *g* are SMOOTH functions, and S and V are respectively the surface and volume of a closed surface of which **n** is the unit NORMAL.

2. a special case of STOKE'S THEOREM, stating that

$$\int_{\partial\Omega} (f\,\mathrm{dx} + g\,\mathrm{dy}) = \iint_{\Omega} \frac{\partial f}{\partial x} - \frac{\partial g}{\partial y}\ \mathrm{dA},$$

when f and g are continuously differentiable on a planar region containing a set Ω with reasonable boundary. The case where $f = -y$, $g = x$ gives a useful formula for the area of Ω. Various corresponding *Green's identities* exist involving the LAPLACIAN.

243

Gregory's series, *n.* the MACLAURIN SERIES for arctan:

$$\arctan x \;=\; x \;-\; \frac{x^3}{3} + \frac{x^5}{5} - \dots ,$$

the special case of which with $x = 1$ sums to $\pi/4$ and is also called *Leibniz' series.* (Named after the Aberdeen-born astronomer and algebraist *James Gregory* (1638–75), who studied in Padua and held chairs at St Andrews and Edinburgh, although the series seems to have been known earlier in India. He distinguished convergent and divergent series, anticipated the study of power series and the differential and integral calculus, and gave a proof of the FUNDAMENTAL THEOREM OF CALCULUS.)

Grelling's paradox or **heterological paradox,** *n.* the semantic paradox, discovered by the German mathematician *Kurt Grelling* in 1908, resulting from defining 'heterological' as a description of words that do not describe themselves, so that, for example, 'short' is not heterological since it is a short word, but 'long' is heterological. It then follows that 'heterological' is heterological if and only if it is not. Although superficially of the same form as RUSSELL'S PARADOX, and solvable by Russell's theory of TYPES, such paradoxes are generally regarded as having a different origin, and as being avoided by distinguishing between OBJECT LANGUAGE and METALANGUAGE. See SELF-REFERENCE.

group, *n.* a set that is closed under an ASSOCIATIVE binary operation with respect to which there exists a unique IDENTITY ELEMENT within the set and every element has an INVERSE within the set; for example, the integers are a group under addition, but not under multiplication. Compare RING, FIELD.

group homorphism, see HOMOMORPHISM.

groupoid, *n.* **1.** a set together with a binary operation under which it is closed. The positive real line with the binary operation given by

$$a * b \;=\; \sqrt{(a^2 + b)}$$

produces a groupoid that is not a SEMI-GROUP.
2. a CATEGORY in which every arrow is invertible. In this sense a groupoid is like a group with multiplication only partially defined. See HOMOTOPY.

group representation, *n.* see REPRESENTATION.

group ring, *n.* the set of all formal sums $\sum_x \alpha_x x$, where x ranges over a multiplicative group, and the α_x, of which all but finitely many are zero, are elements of a field (usually taken to be the complex numbers). The group ring of the group G over a field is usually denoted R_G, and multiplication and addition are defined by

$$\sum_{x \in G} \alpha_x x \;+\; \sum_{x \in G} \beta_x x \;=\; \sum_{x \in G} (\alpha_x + \beta_x) x$$

and

$$\left(\sum_{x \in G} \alpha_x x \right)\left(\sum_{x \in G} \beta_x x \right) \;=\; \sum_{x \in G} \left(\sum_{st = x} \alpha_s \beta_t \right) x .$$

grue, see GOODMAN'S PARADOX.

guard digits, *n.* (*Numerical analysis*) digits added to the PRECISION of a calculation so as to allow for ROUNDING ERROR, cancellation error, or other numerical error.

h

h, *abbrev. for* HECTO-, used in symbols for multiples of the physical units of the SYSTEME INTERNATIONAL.

ℍ, *symbol for* the set of QUATERNIONS.

Haar condition, *n.* the condition, for continuous functions $g_1,\ldots, g_n$ on a set, that the determinant $\det[g_i(x_j)]$ does not vanish for any subset of distinct points $x_1,\ldots, x_n$. This holds for 1, $x,\ldots, x^n$ on any interval. See UNICITY, VANDERMONDE DETERMINANT. (Named after the Hungarian analyst *Alfred Haar* (1885–1933).)

Haar measure, *n.* a non-zero MEASURE, μ, on a SIGMA–RING, S, generated by compact subsets of a TOPOLOGICAL GROUP that is LOCALLY COMPACT, such that the measure is either *left-invariant* ($\mu(xA) = \mu(A)$ for all x in G and A in S) or *right-invariant* ($\mu(Ax) = \mu(A)$ for all x in G and A in S). In the commutative case these coincide, and on a compact group any left-invariant measure is also right-invariant, and vice versa. Such a measure is unique up to a multiplicative constant. For example, LEBESGUE MEASURE is such a measure.

Hadamard, Jacques Salomon (1865 – 1963)**,** French algebraist, analyst, number theorist and mathematical physicist, who proved the PRIME NUMBER THEOREM, made important contributions to the study of functions of complex variables, and developed FUNCTIONAL ANALYSIS.

Hadamard configuration, see HADAMARD MATRIX.

Hadamard design, *n.* a BLOCK DESIGN with $4\lambda + 3$ points for some integral λ, and with $2\lambda + 1$ blocks, any two distinct points of this design being contained in exactly λ blocks; that is, a $(4\lambda + 3, 2\lambda + 1, \lambda)$ design. A Hadamard design is a SYMMETRIC DESIGN.

Hadamard inequality, *n.* **1.** the inequality that the DETERMINANT of a POSITIVE DEFINITE matrix is no greater than the product of the DIAGONAL entries of the matrix.

2. the inequality that the DETERMINANT of a SQUARE matrix is no greater than the product of the EUCLIDEAN NORMS of each row in the matrix.

Hadamard matrix, *n.* an $n \times n$ matrix with all entries equal to ± 1, and with an inverse that is its TRANSPOSE divided by n, where n must be divisible by 4. These matrices give rise to a class of symmetric BLOCK DESIGNS called *Hadamard configurations*.

Hadamard product, *n.* the MATRIX of the entry by entry products of two given matrices of the same dimensions; that is, where C is the Hadamard product of A and B, $c_{ij} = a_{ij} b_{ij}$.

Hadamard three-circle theorem, see THREE–CIRCLE THEOREM.

Hahn–Banach theorem, *n.* (*Functional analysis*) the EXTENSION theorem proving that a LINEAR functional defined on a subspace of a vector space and DOMINATED thereon by a SUBLINEAR function defined on the entire space has a linear extension still dominated by the sublinear function. The *geometric form of the Hahn–Banach theorem* is known as the SEPARATION THEOREM OF MAZUR.

Hahn decomposition, *n.* a partition (P, N) of a SIGMA–RING, X, equipped with a SIGNED MEASURE, μ, such that $P \cap N = \emptyset$, $P \cup N = X$, and, for all A in the sigma-ring on X, $\mu(A \cap P) \geq 0$, and $\mu(A \cap N) \leq 0$.

half-angle formula, *n.* any formula giving the value of a trigonometric or elliptic function at half a given argument, such as

$$\sin \frac{x}{2} = \pm \left(\frac{1 - \cos x}{2} \right)^{1/2}, \quad \cos \frac{x}{2} = \pm \left(\frac{1 + \cos x}{2} \right)^{1/2}.$$

Compare DOUBLE–ANGLE FORMULA.

half-closed, see HALF–OPEN.

half-life, *n.* the time required for a quantity undergoing EXPONENTIAL DECAY to halve its value.

half-line, *n.* any proper connected unbounded subset of a LINE in a Cartesian space; a translate of a RAY.

half-open or **half-closed,** *adj.* (of an INTERVAL) including one endpoint but excluding the other; of the form [*a, b*) or (*a, b*].

half-plane, *n.* **1a.** any subset of the two-dimensional CARTESIAN SPACE bounded by a line, possibly containing a half-line in the boundary line.
b. more often, the set of points (*x, y*) where

$$c \leq ax + by \quad \text{or} \quad c \geq ax + by$$

(*closed half-planes*) or corresponding strict inequalities (*open half-planes*), especially such a subset when the line is parallel to an axis so that the set is defined by $x \geq a$, $x \leq a$, $y \geq b$, or $y \leq b$. The shaded area in Fig. 176 shows the half-plane $x > b - (b/a)y$.

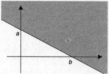

Fig. 176. The line is the boundary between two **half-planes**.

2. any subset of the complex numbers subject to such a condition on their real and imaginary parts. See ARGAND DIAGRAM.

half-space, *n.* all points on one side of a HYPERPLANE in a Euclidean or more general space; it is *open* if the hyperplane is excluded, and *closed* if it is included, but it may be neither. A half-space in one dimension is a HALF-LINE; in two dimensions it is a HALF-PLANE. Two *complementary* half-spaces PARTITION the space.

Halley's method, *n.* a method of solving an equation in one variable, $f(x) = 0$ by the iteration

$$x_{n+1} = x_n - 2 \frac{f(x_n) f'(x_n)}{f(x_n) f''(x_n) - 2f'(x)^2}$$

where x_0 is a first approximation to the root, and f' and f'' are the first two derivatives of f. This is equivalent to a special case of the iteration

$$x_{n+1} = x_n - (N{+}1)\frac{(1/f(x_n))^{(N)}}{(1/f(x_n))^{(N+1)}}$$

for $N=1$. When $N=0$ this reduces to NEWTON'S METHOD, which converges more slowly, but requires less computation per iteration.

Hall subgroup, n. a SUBGROUP of a finite group that has the property that $|H|$ and the number of left cosets of H in G, $|G:H|$, are relatively prime.

halt, vb. (of a TURING MACHINE) to reach a designated final state.

halting problem, n. the UNSOLVABLE PROBLEM of determining whether a TURING MACHINE will HALT when presented with a given input string.

Hamburger moment problem, see MOMENT PROBLEM.

Hamel basis, n. **1.** a BASIS for a VECTOR SPACE, especially when the basis vectors are ORTHOGONAL.

2. in particular, the non-countable basis for the REAL NUMBERS regarded as a vector space over the RATIONALS, in terms of which every non-zero real can be written uniquely as

$$\sum_{i=1}^{n} x_i\, b_i,$$

where the x_i are non-zero rationals and the b_i are elements of the basis.

Hamel dimension, see DIMENSION.

Hamilton, Sir William Rowan (1805–65), Irish algebraist, astronomer, and physicist, who discovered QUATERNIONS. He was elected Astronomer Royal for Ireland and appointed to a Chair at Trinity College while still an undergraduate, and later became President of the Royal Irish Academy.

Hamiltonian, n. a function H such that a given PARTIAL DIFFERENTIAL EQUATION of first order can be rewritten as

$$\partial u/\partial t = -H(t, x_1, ..., x_n, p_1, ..., p_n),$$

where the variables are all functions of the parameter t. This is a *Hamilton–Jacobi* type differential equation. The Hamiltonian exists for any equation

$$F(x_0, x_1, ..., x_n, u, p_0, ..., p_n) = 0,$$

where $p_k = \partial u/\partial x_k$, that does not depend explicitly on u. The *Hamiltonian canonical form* is then

$$\frac{dx_k}{dt} = \frac{\partial H}{\partial p_k}, \quad \frac{dp_k}{dt} = -\frac{\partial H}{\partial x_k}.$$

Such systems occur in classical mechanics, in CONTROL THEORY and elsewhere. See also PONTRYAGIN'S MAXIMUM PRINCIPLE.

Hamiltonian chain or **path,** n. a CHAIN in a graph that uses each VERTEX exactly once.

Hamiltonian circuit, n. a CIRCUIT in a graph that uses each VERTEX exactly once.

Hamiltonian function, n. (*Mechanics*) the sum

$$-\mathrm{L} + \sum \dot{q}\, \frac{\partial \mathrm{L}}{\partial \dot{q}},$$

where L is the LAGRANGIAN, and q ranges over the GENERALIZED COORDINATES.

247

Hamiltonian graph, *n.* a graph containing a HAMILTONIAN CIRCUIT.

Hamiltonian walk, *n.* a PATH in a graph.

Hamilton's equations of motion, *n.* (*Mechanics*) the equations

$$\dot{q} = \frac{\partial H}{\partial p}, \quad \dot{p} = \frac{-\partial H}{\partial q},$$

where H is the HAMILTONIAN FUNCTION,

$$p = \frac{\partial L}{\partial \dot{q}},$$

and q ranges over the GENERALIZED COORDINATES; they are equivalent to LAGRANGE'S EQUATIONS.

Hamilton's principle of least action, *n.* (*Mechanics*) a variant of Newton's second law for a discrete set of PARTICLES, and EULER'S LAWS OF MOTION for a GREEN–ELASTIC BODY under the influence of CONSERVATIVE forces; this states that a set of particles under the influence of conservative forces will move in the period t_0 to t in such a way as to minimize the ACTION from t_0 to t of the given set. See NEWTON'S LAWS OF MOTION.

Hamming codes, *n.* (*Information theory*) a class of efficient ERROR–CORRECTING CODES, usually BINARY CODES, that allow the recipient to diagnose and correct a certain number of transmission errors per word, on the presumption that the chance of a very high proportion of errors is negligible.

ham sandwich theorem, *n.* a colourful name for the theorem that given three volumes in Euclidean three-space, there is at least one plane which simultaneously bisects the three volumes. As a consequence, any sandwich may be cut in such a way that the two pieces have equal amounts of filling and equal amounts of each slice of bread.

handedness, see ENANTIOMORPHIC.

handle, *n.* (*Topology*) a piece of a surface constructed by identifying two disjoint disks on another surface as the ends of a cylinder. This produces a TORUS from a SPHERE, as shown in Fig. 177, and a coffee cup from a soup bowl. See EULER'S FORMULA.

Fig. 177. A solid with one **handle**, topologically equivalent to a torus.

hangman paradox, *n.* see UNEXPECTED EXAMINATION PARADOX.

Hankel matrix, *n.* a MATRIX whose entries are related by $a_{ij} = b_{i+j}$, so that they are constant on lines parallel to the OFF DIAGONAL. Compare TOEPLITZ MATRIX. (Named after the German analyst and geometer *Hermann Hankel* (1839–73).)

Hardy–Weinberg ratio, *n.* the ratio of the frequencies of the three possible combinations obtained in a random mating of a population, the individuals of which each have a pair of genes either of which can only be one of two exhaustive alternatives. Where characteristic A occurs with probability p, and therefore B has probability $1 - p$, then the ratio of the pairs of genes in the first generation offspring is

$$AA : AB : BB = p^2 : 2p(1-p) : (1-p)^2.$$

This, and its corollary, the *Hardy–Weinberg law*, which describes the conditions for genetic equilibrium and states that the proportions of dominant and recessive genes tend to remain constant in a randomly mating population unless there are outside influences, were discovered independently by Wilhelm Weinberg, a German doctor, and by the Cambridge mathematician *Godfrey Harold Hardy* (1877–1947); the latter also solved many problems of number theory, and made substantial contributions to the theories of summation, Fourier series, and special functions; after he retired he published *A Mathematician's Apology*, seeking to explain the importance of the subject to a wider public.

harmonic, *adj.* **1.** able to be expressed in terms of SINE and COSINE functions. **2.** of or relating to numbers of which the reciprocals form an ARITHMETIC PROGRESSION. **3.** (of a function) having two-dimensional LAPLACIAN equal to zero; occurring as the real or imaginary part of an ANALYTIC function. **4.** (*Projective geometry*) having CROSS RATIO equal to −1. See HARMONIC RATIO. **5.** (*as substantive*) a component of a periodic quantity, such as a musical tone, with a frequency that is an integral multiple of the frequency of the given oscillation. The *first harmonic* is the fundamental frequency itself; the *second harmonic* has twice the fundamental frequency (and in music is called the first overtone); the *third harmonic* has three times the fundamental frequency (the second overtone), etc. However, in non-technical musical usage, the *first harmonic* is identified with the first overtone, etc. Fig. 178 shows one cycle of a fundamental frequency with its first three overtones superimposed upon it.

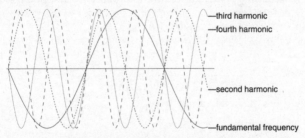

Fig. 178. The first four **harmonics** of a given frequency.

harmonic analysis, *n.* **1.** the representation of a periodic function by means of the summation and integration of simple trigonometric functions; *harmonic synthesis* reconstitutes the function from these components. See also FOURIER SERIES.
2. the study of such representations and their properties.

harmonic conjugate or **fourth harmonic,** *n.* the fourth point, collinear with a given pair of points, A and B, and another collinear point, X, that can be found by constructing the COMPLETE QUADRANGLE corresponding to a trapezoid ABCD (shown by the bold lines in Fig. 179), for which X is the point of intersection of the extensions of AB and CD; then if Y is the sixth point of the complete quadrangle, and the diagonals of the trapezoid intersect at Z, then the intersection, H, of YZ with the line XAB is the fourth harmonic of X with respect to A and B. These four collinear points are then an *harmonic set* as their CROSS RATIO is a HARMONIC RATIO in which A, B and X, H are *conjugate pairs*; iterating the process leads to an *harmonic net of rationality*. Dually, a COMPLETE QUADRILATERAL produces an *harmonic pencil* of four lines. See also POLE AND POLAR.

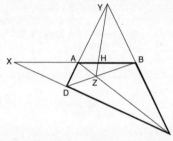

Fig. 179. **Harmonic conjugate.**
See main entry.

harmonic mean, *n.* the reciprocal of the ARITHMETIC MEAN of the reciprocals of a given set of positive numbers; it is always less than or equal to the GEOMETRIC MEAN of the same set of numbers. For example, the arithmetic mean of $\frac{1}{2}$, $\frac{1}{3}$, and $\frac{1}{4}$ is $\frac{13}{36}$; so the harmonic mean of 2, 3, and 4 is $\frac{36}{13}$.

harmonic net of rationality, see HARMONIC CONJUGATE.

harmonic pencil, see HARMONIC CONJUGATE.

harmonic points, *n.* **1.** the points of INTERNAL AND EXTERNAL DIVISION of a line segment in the HARMONIC RATIO. See also APOLLONIUS' CIRCLE.

2. fourth harmonic point. another term for HARMONIC CONJUGATE.

harmonic progression, *n.* a sequence of numbers of which the reciprocals form an ARITHMETIC PROGRESSION, such as 1, $\frac{1}{2}$, $\frac{1}{3}$, $\frac{1}{4}$, ...

harmonic ratio, *n.* (*Projective geometry*) a CROSS RATIO of four points (HARMONIC POINTS) that is equal to –1; that is, such that the directed ratio

$$(A, B; D, E) = \frac{AC . BD}{AD . BC} = -1.$$

harmonic series, *n.* **1.** the infinite series

$$1 + \frac{1}{2} + \frac{1}{3} + \frac{1}{4} + ...,$$

of which the partial sums diverge at the same rate as log *n*. See also EULER'S CONSTANT.

2. any series the ABSOLUTE VALUES of whose terms form an HARMONIC PROGRESSION, such as

$$1 - \frac{1}{4} + \frac{1}{7} + \frac{1}{10} + ...$$

Compare GEOMETRIC SERIES, LOGARITHMIC SERIES.

harmonic set, see HARMONIC CONJUGATE.

harmonic synthesis, see HARMONIC ANALYSIS.

Hasse's algorithm, see THREE–X PROBLEM.

Hausdorff dimension, *n.* (of a set S in a finite-dimensional space) the unique positive EXTENDED REAL NUMBER D for which S has finite *d*-dimensional HAUSDORFF MEASURE for $d <$ D and infinite measure for $d >$ D. (Named after the German analyst and topologist *Felix Hausdorff* (1868–1942). Compare TOPOLOGICAL DIMENSION. See also FRACTAL.

Hausdorff distance, *n.* the distance between two sets A and B in a METRIC SPACE, defined as the supremum of the DISTANCES from points in A to the set B and from points in B to the set A. When restricted to compact convex subsets of Euclidean space, this distance becomes a complete metric, the *Hausdorff metric,* and provides a useful tool in the study of many economic and ISOPERIMETRIC PROBLEMS.

Hausdorff maximality theorem, *n.* the theorem that every non-empty PAR-TIALLY ORDERED SET contains a maximal TOTALLY ORDERED subset. See also ZORN'S LEMMA.

Hausdorff measure, *n.* the MEASURE of a set, S, in a finite-dimensional space with respect to a positive function *h* induced by

$$\mu_k(S) = \lim_{r \to 0+} \inf \{ \Sigma\, h(r_n) : C(r) \},$$

where $C(r)$ represents any finite COVER of S by BALLS of radius less than *r*. In the most standard case, one uses

$$h(r) = \gamma(d)\, r^d.$$

Here $\gamma(d)$ is defined in terms of the GAMMA FUNCTION as

$$\gamma(d) = \frac{\Gamma(\tfrac{1}{2})^d}{\Gamma(1 + \tfrac{d}{2})},$$

which for integer *d* is the volume of a *d*-dimensional UNIT BALL. The meas-ure is then called *d*-dimensional or *Hausdorff d-measure.* The Hausdorff *d*-measure of a subset of Euclidean *n*-space is less than or equal to *n*, and is no smaller than its TOPOLOGICAL DIMENSION. See FRACTAL.

Hausdorff metric, see HAUSDORFF DISTANCE.

Hausdorff moment problem, see MOMENT PROBLEM.

Hausdorff space, *n.* a TOPOLOGICAL SPACE in which every pair of distinct points have a pair of disjoint open NEIGHBOURHOODS. See also SEPARATION AXIOM, T–AXIOMS, COMPACTIFICATION.

haversine, *n.* half the value of the VERSED SINE of an angle.

hazard rate function, *n.* the probability of failure of a component, or the death of an organism, within the next unit of time, Δt, given survival to time *t*, defined as

$$\Delta t \cdot h(t) = P[\, X \le (t + \Delta t) : X > t\,].$$

hcf, *abbrev. for* HIGHEST COMMON FACTOR.

heat equation or **diffusion equation,** *n.* the partial differential equation

$$\nabla^2 u = c^2\, \frac{\partial u}{\partial t},$$

where ∇^2 is the LAPLACIAN in one, two, or three dimensions, generally solved by using FOURIER SERIES.

heating, *n.* (*Continuum mechanics*) the sum of integrals

$$\int_R \rho \, r \, dv \; + \; \int_{\partial R} h \, da,$$

where the first integral is over the volume, and the second over the surface area, of the current CONFIGURATION, **R**, of a given SUB–BODY with DENSITY ρ, r the heat supply, and h the heat FLUX.

hect-, *prefix denoting* six. For example, a *hectad* is a set of six elements, and a *hectic* equation has degree six.

hecto- (symbol **h**), *prefix denoting* a multiple of 10^2 of the physical units of the SYSTEME INTERNATIONAL.

-hedron, *suffix denoting* a POLYHEDRON; for example, an *enneahedron*, and an *eikosihedron* are polyhedra with nine and twenty sides respectively.

height, *n.* **1.** the length of the ALTITUDE of a polygon in a given orientation. **2.** the maximum of the absolute values of the coefficients of a given polynomial or LINEAR FORM. Compare LENGTH (sense 5).

Heine–Borel covering theorem, *n.* the theorem that a subset of EUCLIDEAN SPACE is closed and bounded if and only if it is COMPACT. Compare BOLZANO–WEIERSTRASS THEOREM.

Heine's theorem, *n.* the result that if M and N are METRIC SPACES, A is a COMPACT subset of M, and f is a continuous function from A to N, then f is UNIFORMLY CONTINUOUS on A. (Named after the German analyst *Heinrich Heine* (1821–81).)

helicoid, *n.* any solid or surface shaped like a screw thread.

helix, *n.* a curve that lies on the surface of a cylinder or cone at a constant angle to line segments that sweep out the surface. The *circular helix* has equations

$$x = a\cos t, \quad y = a\sin t, \quad z = bt.$$

See also SPIRAL.

Helly's theorem, *n.* the result that if a finite family of bounded closed subsets of an n-dimensional vector space is such that every subcollection with at least $n+1$ members has a common point, then the entire family has a common point. The example of the sides of a triangle shows that the subcollections must have $n+1$ members and that n are insufficient. See also CARATHEODORY'S THEOREM, RADON'S THEOREM. (Named after the Austrian analyst, geometer, topologist, and physicist *Eduard Helly* (1884–1943).)

hemi-, *prefix denoting* half.

hemicycle, *n.* a semicircle or semicircular structure.

hemisphere, *n.* half of a sphere, bounded by a plane containing its centre.

hemispheroid, *n.* one half of a spheroid.

Hempel's paradox or **confirmation paradox,** *n.* the paradox of INDUCTION that shows that logically equivalent statements are not equivalent for the purposes of confirmation by experience: we ordinarily regard every sighting of a black raven as supporting the hypothesis that all ravens are black, which is equivalent to the proposition that all non-black things are non-ravens, so that whatever tends to confirm one equally tends to confirm the other. Further, both statements have the same universal affirmative form, so that each should be equally supported by instances that instantiate both its subject and predicate; hence observations of non-black non-ravens, such as white handkerchiefs, should tend to confirm the second statement, and so also the first, yet this is patently absurd. The paradox may be resolved by regarding such statements as involving RESTRICTED QUANTIFIERS: it is not asserted of everything whatever that if it is a raven it is black, but only of ravens that all of them are. (Named after the German-born American positivist philosopher of science *Carl Gustav Hempel* (1905–).) See also GOODMAN'S PARADOX.

hendeca-, *prefix denoting* eleven; for example, a *hendecagon* and a *hendecahedron* are respectively an eleven-sided plane figure and an eleven-faced solid.

hepta-, *prefix denoting* seven. For example, a *heptagon* and a *heptahedron* are respectively a seven-sided plane figure and a seven-faced solid; a *heptangular* figure has seven angles.

heptad, *n.* a group or sequence of seven.

hereditary, *adj.* **1.** (of a set) containing all those elements that have a given relation to any element of the set; closed under that relation. For example, the positive integers are hereditary with respect to ≥ (greater than or equal to) on the integers.

2. (of a property) transferred by a given relation, so that if x has the property and Rxy, then y also has the property. For example, evenness is hereditary with respect to addition of 2 but not 3.

3. (of a topological property) possessed by any subspace of a topological space that has the property.

Hermite, Charles (1822–1901), influential French analyst, algebraist, and number theorist, who developed the theory of functions, used ELLIPTIC FUNCTIONS to solve the general quintic in one variable, and proved that e is TRANSCENDENTAL. Although he had already done original mathematical work, he found examinations difficult and it took him six years to obtain his first degree.

Hermite interpolation, see LAGRANGE INTERPOLATION FORMULA.

Hermite's polynomials, *n.* the polynomials $H_n(x)$, given by the GENERATING FUNCTION

$$e^{2tx-t^2} = \sum_{n=0}^{\infty} \frac{H_n(x)\,t^n}{n!} \;;$$

the polynomial $H_n(x)$ is a solution of the differential equation

$$y'' - 2xy' + 2ny = 0$$

for the integer n.

Hermitian, *n.* **1.** a matrix over the complex numbers, whose TRANSPOSE is equal to the matrix of the COMPLEX CONJUGATES of its entries, so that it is its own HERMITIAN CONJUGATE. For example, every real symmetric matrix is Hermitian.

2. an operator equal to its own HERMITIAN CONJUGATE.

Hermitian conjugate, another name for the ADJOINT of a matrix or operator.

Hermitian vector space, *n.* another term for UNITARY SPACE.

Heron or **Hero of Alexandria,** first-century Greek mathematician, physicist, and inventor, whose major geometric work was lost until 1896. He also invented a jet-propelled rotary steam engine, and a method of calculating square roots similar to that used by modern computers.

Heronian mean, *n.* the NEO–PYTHAGOREAN MEAN given by

$$\tfrac{1}{3} [a + \sqrt{ab} + b].$$

Heron's formula or **Hero's formula,** *n.* the formula for the area of a triangle in terms of the lengths of its sides:

$$A = \sqrt{[s(s-a)(s-b)(s-c)]}$$

where *a*, *b* and *c* are the sides opposite the respective vertices, and *s* is the semi-perimeter $\tfrac{1}{2}(a + b + c)$.

Heron's method or **Hero's method,** *n.* NEWTON'S METHOD for determining the SQUARE ROOT of a number.

Hessenberg form, *n.* a matrix that either has only zeros more than one row above the main diagonal, as in Fig. 180, (a *lower Hessenberg form*), or has only zeros more than one row below the main diagonal (an *upper Hessenberg form*). Compare TRIANGULAR MATRIX.

$$\begin{bmatrix} 4 & 1 & 0 & 0 \\ 8 & 5 & 2 & 0 \\ 2 & 9 & 6 & 3 \\ 4 & 3 & 1 & 7 \end{bmatrix}$$

Fig. 180. A lower **Hessenberg form**.

Hessian, *n.* **1.** also called **Hessian matrix.** the matrix of which the entries are the second PARTIAL DERIVATIVES of a given function; for example, the Hessian of $f(x, y) = x^2 - y^2$ is

$$\begin{bmatrix} 2 & 0 \\ 0 & -2 \end{bmatrix}.$$

The analogue of the SECOND DERIVATIVE TEST for functions of more than one variable uses the Hessian to identify locally optimal values of the function: the Hessian is positive definite at a local minimum, negative definite at a local maximum, and indefinite at a saddle point; if it is singular, the test is indeterminate.

2. the DETERMINANT of the Hessian matrix.

(Named after the German differential geometer *Ludwig Otto Hesse* (1811–1874).)

heterological paradox, *n.* another name for GRELLING'S PARADOX of SELF-REFERENCE.

heteroscedastic, *adj.* (*Statistics*) **1.** (of a number of distributions) having different VARIANCES.

2. (of a BIVARIATE or MULTIVARIATE distribution) not having any variable with constant variance for all values of the other or others.

3. (of a random variable in a multivariate distribution) having different variances for different values of the other variables.

Compare HOMOSCEDASTIC.

heuristic, *adj.* using or obtained by informal methods or reasoning from experience, often since no precise algorithm is known or is relevant. Compare MECHANICAL.

hex, *abbrev. for* HEXADECIMAL notation.

hexa-, *prefix denoting* six; for example, a *hexagon* has six sides, and a *hexangular* figure has six vertices.

hexad, *n.* a set or sequence of six.

hexadecimal, *adj.* **1.** denoting or using PLACE VALUE NOTATION to base 16, usually written using the digits 0 to 9 and the letters A to F; for example, the hexadecimal 2B7E represents

$$(2 \times 16^3) + (11 \times 16^2) + (7 \times 16) + 14 = 11134$$

in decimal notation, and is sometimes written $2B7E_{16}$. Hexadecimal notation is useful in computing since each hexadecimal digit is equivalent to a BYTE of four BITS.

2. (*as substantive*) a number in hexadecimal notation.

hexafoil, see MULTIFOIL.

hexagon, *n.* a plane figure with six sides, such as that shown in Fig. 181. A REGULAR hexagon can be constructed by marking off on the circumference of a circle CHORDS equal in length to its radius.

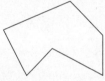

Fig. 181. A **hexagon** with a re-entrant angle.

hexagram, *n.* a star-shaped figure formed by extending the sides of a regular hexagon to meet at six points, or by superimposing two equilateral triangles one of which is inverted; drawing the diagonals of the hexagon yields twelve congruent equilateral triangles, as shown in Fig. 182 in which the original hexagon is shown bold. See also PASCAL'S MYSTIC HEXAGRAM THEOREM.

Fig. 182. **Hexagram**.

hexahedron, *n.* a solid figure with six plane faces; a regular hexahedron is a cube.

higher arithmetic, *n.* another name for NUMBER THEORY.

higher derivative, see SECOND DERIVATIVE.

higher homotopy group, see HOMOTOPY.

higher mathematics, *n.* mathematics of greater abstraction than the traditional school syllabus of arithmetic, algebra, geometry and trigonometry, including ANALYSIS, TOPOLOGY, LINEAR ALGEBRA, NUMBER THEORY, etc.

higher partial derivative, see PARTIAL DERIVATIVE.

highest common factor, **greatest common factor**, or **greatest common divisor** (abbrev. **hcf**, **gcf**, **gcd**), *n.* an integer *d* that exactly DIVIDES (sense 2) two given integers *a* and *b*, and is such that if *c* divides *a* and *b*, then *c* divides *d*; this definition extends to finite sets of integers and to INTEGRAL DOMAINS. For example, the highest common factor of 12, 60 and 84 is 12. See EUCLID'S ALGORITHM.

high precision, see PRECISION.

Hilbert, David (1862–1943), German mathematician, Professor at Göttingen from 1895 until his death, who was made a Fellow of the Royal Society in 1928. He is best known for his influential work in the foundations of geometry and mathematics in general; HILBERT'S PROGRAMME was the motivation for the development of COMPUTABILITY THEORY. His collection of 23 problems, now referred to as HILBERT'S PROBLEMS, profoundly influenced the course of 20th-century mathematics; these problems, many of which still unsolved, are listed in Appendix 3. His other achievements included HILBERT'S BASIS THEOREM of RING theory, and his investigations of HILBERT SPACE theory and number theory.

Hilbert cube, *n.* the compact subset of the HILBERT SPACE of SQUARE SUMMABLE sequences

$$\{ a_n \}_{n=1}^{\infty}$$

with $|a_n| \leq 1/n$, or similar ORDER–INTERVALS.

Hilbert matrix, *n.* the finite or infinite HANKEL MATRIX with entries

$$\{ a_{n,m} \}_{n,m=1}^{\infty} = \left\{ \frac{1}{n+m-1} \right\}_{n,m=1}^{\infty}.$$

The infinite matrix induces a Hilbert space operator on the SQUARE SUMMABLE sequences with OPERATOR NORM equal π.

Hilbert norm, see MEAN SQUARE.

Hilbert's basis theorem, *n.* the result that if R is a left (or right) NOETHERIAN RING, then the POLYNOMIAL RING $R[X_1,..., X_n]$ is a left (or right) Noetherian ring.

Hilbert space, *n.* a real or complex LINEAR SPACE on which an INNER PRODUCT is defined, and in which all CAUCHY SEQUENCES converge to a limit. Thus, it is a BANACH SPACE the norm on which is induced by an inner product. The most common realizations are the L_2 spaces. See L_p SPACE. Compare UNITARY SPACE, INNER PRODUCT SPACE.

Hilbert's paradox or **infinite hotel paradox,** *n.* the paradox stated by Hilbert that a hotel with an infinite number of rooms may be full, and yet able to accommodate another guest; this can be done by moving the existing guest in each room *n* to room *n* + 1, thereby leaving room 1 free for the latecomer. If an infinite number of additional guests arrive, inconvenience to the earlier arrivals is minimized if each of them moves from room *n* to room 2*n*, thereby leaving all the odd-numbered rooms vacant for the new arrivals. This is a direct consequence of Cantor's account of infinite CARDINALITY. See also TRISTRAM SHANDY PARADOX.

Hilbert's problems, *n.* a collection of 23 problems that remained unsolved when Hilbert listed them in 1901, and that have occupied much attention since then. These include the RIEMANN HYPOTHESIS, FERMAT'S LAST THEOREM, the GELFOND–SCHNEIDER THEOREM, and the CONTINUUM HYPOTHESIS. Many remain unsolved. A full list of the problems, with an indication of current knowledge, appears in Appendix 3.

Hilbert's programme, *n.* the motivating problem of METAMATHEMATICS, proposed by Hilbert in 1920 in support of his FORMALISM in the FOUNDATIONS OF MATHEMATICS, of formalizing all of mathematics and showing by purely syntactic means that finitary methods could never lead to contradiction. This is equivalent to finding a decision algorithm for all of mathematics, and although it was shown unattainable by GÖDEL'S PROOF in 1931, the project nonetheless led to the development of PROOF THEORY and COMPUTABILITY THEORY. See also TURING MACHINE.

Hilbert transform, *n.* **1.** the TRANSFORM

$$g(x) = \frac{1}{\pi} \int_{0}^{\infty} \frac{f(x+t) - f(x-t)}{t} \, dt.$$

2. also called **Hilbert singular integral.** the transform

$$g(x) = \frac{1}{2\pi} \int_{0}^{2\pi} f(t) \cotan \frac{t-x}{2} \, dt,$$

used in the theory of FOURIER TRANSFORMS.

histogram, *n.* (*Statistics*) a figure that represents a FREQUENCY DISTRIBUTION, consisting of contiguous rectangles with widths proportional to the size of the respective CLASS INTERVALS and areas proportional to the RELATIVE FREQUENCIES of the phenomenon in question in each interval, as shown in Fig. 183. A BAR CHART is a histogram in which all the class intervals are of equal width or that represents a DISCRETE RANDOM VARIABLE. See also STEM–AND–LEAF DIAGRAM.

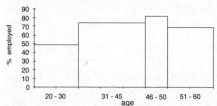

Fig. 183. **Histogram** showing percentage employed by age group.

Hodge conjecture, *n.* (*Geometry*) the conjecture (one of the MILLENNIUM PRIZE PROBLEMS) that for certain spaces called projective algebraic varieties, when approximating the shape of a given object by using simple geometric building blocks, the pieces called Hodge cycles are actually (rational linear) combinations of geometric pieces called algebraic cycles. This arises against from generalising the geometrical tool of approximating the shape of complex objects by combining simple building blocks of increasing dimension, using pieces with no clear geometric interpretation. (Named after *William Hodge* (1903–75), Scottish mathematician, whose principal interest was the relationship between geometry, analysis and topology.)

Hölder condition, see HÖLDER–CONTINUOUS.

Hölder-continuous, *adj.* (of a function between NORMED SPACES) satisfying the *Hölder condition* of order *p*, that for some positive $p \le 1$,

$$\| f(x) - f(y) \| \le k \| x - y \|^p$$

for all *x* and *y* in some given set. This condition is also known as the Lipschitz condition of order *p*; see LIPSCHITZ FUNCTION.

Hölder means, *n.* the multidimensional MEANS defined by

$$H_p(a_1, \ldots, a_n) = \left[\frac{1}{n} \sum_{i=1}^{n} (a_i)^p \right]^{\frac{1}{p}}$$

for an extended real number *p* and positive $a_1, \ldots, a_n$. Thus H_1 is the ARITHMETIC MEAN, and H_2 the ROOT MEAN SQUARE.

Hölder's inequality *n.* **1.** the integral inequality that holds for measurable functions on a set S:

$$\int_s |fg| \le \left[\int_s |f|^p \right]^{\frac{1}{p}} \left[\int_s |g|^q \right]^{\frac{1}{q}},$$

where $1 \le p, q \le \infty$ and $1/p + 1/q = 1$, and all the integrals are taken with respect to the same measure.
   2. the specialization of the previous inequality to sequences:

$$\sum |f_i g_i| \le \left[\sum |f_i|^p \right]^{\frac{1}{p}} \left[\sum |g_i|^q \right]^{\frac{1}{q}}.$$

See L_p SPACE.

holomorphic, *adj.* (of a complex function) another word for ANALYTIC; more properly, satisfying the CAUCHY–RIEMANN equations in a region.

homeomorphic, *adj.* related by a HOMEOMORPHISM.

homeomorphism, *n.* a ONE–TO–ONE CORRESPONDENCE that is CONTINUOUS in both directions between the points of two geometric figures or between two topological spaces. It is an equivalence relation and preserves topological properties; if it also preserves distances it is an ISOMETRY. See also DIFFEOMORPHISM.

homogeneous, *adj.* **1.** (of a polynomial) having all its terms of the same DEGREE. For example, $x^2 + 2xy + y^2$ is homogeneous in the second degree.
   2a. (of an equation) consisting of an equality between a homogeneous function and zero.
   b. (of a system of linear equations) having the form $A\mathbf{x} = \mathbf{0}$. where $\mathbf{x}$ is the vector of variables, $\mathbf{0}$ is the zero vector, and A is the matrix of coefficients.
   3. (of an ORDINARY DIFFERENTIAL EQUATION) able to be as expressed as an expression of the form $y^{(n)} = f(x, y)$; it is homogeneous of degree γ if

$$f(\lambda x, \lambda y) = \lambda^\gamma f(x, y).$$

Ordinary differential equations of any order that are homogeneous of degree zero can be solved by writing them in terms of $v = y/x$.
   4. (of a function on a vector space) such that

$$f(t x_1, \ldots, t x_n) = t f(x_1, \ldots, x_n)$$

for every non-zero scalar t. More generally, if

$$f(tx_1,\ldots, tx_n) = t^d f(x_1,\ldots, x_n)$$

for every non-zero scalar t then f is homogeneous of degree d; if this holds for positive t, then f is said to be *positively homogeneous* of degree d.

homogeneous coordinates or **projective coordinates,** n. (*Geometry*) **1.** the representation of a point with CARTESIAN COORDINATES (x, y) as (x', y', t) where $x' = tx$ and $y' = ty$; clearly this representation is not unique, and any triple (tx, ty, t), for non-zero t, will represent the same point. For a given plane, this is equivalent to selecting a point not in the plane, and representing any point in it by the line joining it to the reference point; this then allows any point of the plane to be represented in terms of a given set of basis points as a linear combination whose coefficients are the homogeneous coordinates of the point with reference to those basis points. This enables problems to be solved in the AUGMENTED EUCLIDEAN plane that are insoluble in the Cartesian plane. An analogous representation is possible in spaces of other dimensionality. Clearly, given homogeneous coordinates (x, y, z) with z non–zero, the *inhomogeneous coordinates* $(x/z, y/z)$ of points in the ordinary plane can be recovered.

2. the representation of points in a two-dimensional ALGEBRAIC GEOMETRY by equivalence classes of proportional triples from the ground field, and similarly for higher dimensions. The coordinates are determined by the choice of a TRIANGLE OF REFERENCE and a UNIT POINT.

homologous, *adj.* playing the same role in distinct but related figures or functions, such as the corresponding points of a figure and a projection of it; for example, the triangles ABC and XYZ in Fig. 184 are homologous with respect to the projection of one plane onto the other.

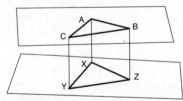

Fig. 184. **Homologous** triangles in skew planes.

homology group, n. one of a class of groups constructed in algebraic topology and based upon FACTOR GROUPS of fixed-dimensional cyclic chains of SIMPLICES of SIMPLICIAL COMPLEXES. Here the factor is taken over the chains that are *homologous to zero* in the sense that they arise as the boundaries of chains of one higher dimension. Homology groups are useful in classifying topological spaces in the same sense that HOMOTOPY or CONNECTIVITY properties are.

homomorphic image, n. the algebra that is the RANGE of a HOMOMORPHISM.

homomorphism, n. a mapping from one algebraic structure to another under which the structural properties of its domain are preserved in its range in the sense that if $*$ is the operation on the domain, and $\circ$ is the operation on the range, then

$$\theta(x*y) = \theta(x) \circ \theta(y).$$

In particular, a *group homomorphism* is a mapping θ such that both domain and range are GROUPS, and

$$\theta(xy) = \theta(x)\,\theta(y)$$

for all x and y in the domain; a *ring homomorphism* is a mapping θ from one RING to another such that

$$\theta(x + y) = \theta(x) + \theta(y) \quad \text{and} \quad \theta(xy) = \theta(x)\,\theta(y)$$

for all x and y in the domain; a *module homomorphism* is a mapping such that

$$\theta(x + y) = \theta(x) + \theta(y) \quad \text{and} \quad \theta(rx) = r\theta(x)$$

for all x and y in the R-module and r in the ring R (where, if R is a FIELD, then θ is a linear mapping). In group theory, homomorphisms are taken to be SURJECTIVE unless otherwise stated to the contrary. The *natural epimorphism* or *natural homomorphism* is the homomorphism ν from G to the FACTOR GROUP G/K, that is defined for rings and modules by $\nu(x) = x + K$ and for groups by $\nu(x) = xK$. See also ISOMORPHISM, EPIMORPHISM, MONOMORPHISM, MORPHISM, ENDOMORPHISM.

homomorphism theorem, *n.* another name for FIRST ISOMORPHISM THEOREM.

homoscedastic, *adj.* (*Statistics*) **1.** (of a number distributions) having equal VARIANCES.

2. (of a BIVARIATE or MULTIVARIATE distribution) having one variable of which the variance is constant for all values of the other or others.

3. (of a random variable in a multivariate distribution) having constant variance for all values of the other variables.

Compare HETEROSCEDASTIC.

homothetic, *adj.* (*Euclidean geometry*) (of a pair of figures) similar, with corresponding sides parallel.

homothety, *n.* a linear transformation that involves no rotation, the composition of a TRANSLATION and a central DILATION. It is of the form

$$x' = kx, \quad y' = ky,$$

and is a *stretching* if $k > 1$, and a *shrinking* if $0 < k < 1$. See also SIMILITUDE.

homotopy, *n.* a continuous DEFORMATION of one function or curve into another. A homotopy between a function and a constant is called *null* or *inessential.* The *fundamental homotopy group* at a point x of a topological space S, denoted $\prod_1(S, x)$, is composed of equivalence classes of closed paths through x, with juxtaposition of paths as multiplication. *Higher homotopy groups* are constructed analogously using generalized paths mapping HYPERCUBES into the space. The *fundamental homotopy groupoid* of a topological space S, denoted $\prod_1(S)$, is composed of equivalence classes of paths from x to y, with juxtaposition of paths as multiplication, where possible.

Hooke's law, *n.* (*Mechanics*) the law stating that the TENSION in a stretched ELASTIC body is kx, where x is the extension and k is a constant. If the body is a spring, k is the *stiffness*; if the body is a cable, then $k = \lambda/a$, where a is the natural length and λ is the *modulus of elasticity*; and if the body is a thin beam then $k = EA$, where E is YOUNG'S MODULUS and A is the area of the cross-section.

Horner's method, *n.* an iterative method for finding real roots of algebraic equations by identifying the largest integer less than some root, and changing the variable by subtracting this integer, so that the root of the new equation lies between 0 and 1; this process is then repeated to identify the interval of next smallest order in which the root lies, and the root to any required degree of accuracy is the sum of the amounts subtracted in the successive changes of variable. (Named after the English algebraist, *William George Horner* (1786–1837), whose main career was as a schoolmaster.)

Horner's rule, *n.* an efficient method of evaluating a polynomial by writing

$$p(x) = ((\dots(a_n x + a_{n-1})x + a_{n-2})x + \dots + a_1)x + a_0,$$

and so nesting the multiplications. When the coefficients are small integers, this greatly reduces the computational work required.

hotel paradox, see HILBERT'S PARADOX.

hull, *n.* another word for SPAN.

Hungarian method, *n.* a method due to Kuhn for solving ASSIGNMENT PROBLEMS, anticipating later PRIMAL–DUAL METHODS.

Hurwitz' theorem, *n.* the result that, given an irrational number ζ, there are infinitely many distinct rationals h/k with

$$\left| \zeta - \frac{h}{k} \right| < \frac{1}{\sqrt{5}\, k^2},$$

the constant $\sqrt{5}$ being best possible. Compare THUE–SIEGEL–ROTH THEOREM.

Huygen's formula, *n.* the estimate that the length of an arc of a circle approximately equals eight-thirds of the chord subtending half of the given arc less one-third of the chord subtending the entire arc. (Named after the Dutch astronomer, physicist and mathematician, *Christian Huygens* (1629–95), whose early work on analysis contributed to the discovery of calculus.)

hydrodynamics, *n.* the branch of MECHANICS that studies fluids in motion.

hydrostatic, *adj.* **1.** (of a second-order CARTESIAN TENSOR) equal to the product of a SCALAR FIELD and the identity tensor, such as a tensor representing the pressure at a point in a fluid at rest.

2. (of a stress) such that the tangential component of the STRESS VECTOR is zero, and the normal component is independent of the unit outward normal for all possible surfaces. See also PRESSURE. Compare SHEAR STRESS.

hydrostatics, *n.* the branch of MECHANICS that studies fluids at rest, founded by Archimedes.

hyp., *abbrev. for* HYPOTENUSE, HYPOTHESIS.

hyper-, *prefix denoting* some entity of more than three dimensions, such as HYPERSPACE, HYPERCUBE.

hyperbola, *n.* a CONIC SECTION with ECCENTRICITY greater than 1, formed by a plane that cuts both bases of a cone; it consists of two branches ASYMPTOTIC to two intersecting fixed lines, and has two FOCI. When symmetrical about the coordinate axes it has equation:

$$\frac{x^2}{a^2} - \frac{y^2}{b^2} = 1,$$

where the TRANSVERSE AXIS coincides with the x-axis, the CONJUGATE AXIS lies on the y-axis, $2a$ is the distance between the two intersections with the x-axis, and $b = \sqrt{(e^2 - 1)}$, where e is the eccentricity. Its parametric equations are

$$x = a\sec\theta, \quad y = b\tan\theta$$

and, as shown in Fig. 185, the general hyperbola has asymptotes $y = \pm(b/a)x$.

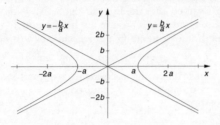

Fig. 185. The general **hyperbola** and its asymptotes.

hyperbolic, *adj.* **1.** of or relating to HYPERBOLAE or HYPERBOLIC FUNCTIONS.
2. (of a second-order PARTIAL DIFFERENTIAL EQUATION) having DISCRIMINANT $b^2 - 4ac$ positive, where

$$au_{xx} + bu_{xy} + cu_{yy} + du_x + eu_y + fu = h$$

is the general form of a second-order partial differential equation.

hyperbolic cylinder, *n.* a CYLINDER in which the fixed curve is a HYPERBOLA and the GENERATORS are perpendicular to the plane of the hyperbola. It is a QUADRIC, and in a suitable coordinate system has equation

$$x^2/a^2 - y^2/b^2 = 1.$$

hyperbolic function, *n.* any of a group of functions of an angle originally defined in terms of the TRIGONOMETRIC FUNCTIONS or in terms of EXPONENTIAL FUNCTIONS. The primary hyperbolic functions are SINH (*hyperbolic sine*), defined as

$$\sinh z = -i\sin iz = \frac{e^z - e^{-z}}{2},$$

and COSH (*hyperbolic cosine*), defined as

$$\cosh z = \cos iz = \frac{e^z + e^{-z}}{2}$$

for complex z; TANH (*hyperbolic tangent*) is defined as the ratio of sinh to cosh. The reciprocals of these three functions are respectively COSECH (*hyperbolic cosecant*), SECH (*hyperbolic secant*), and COTH (*hyperbolic cotangent*). Their inverses are ARC–SINH, ARC–COSH, ARC–TANH, etc. The hyperbolic functions satisfy the identity

$$\sinh^2\alpha - \cosh^2\alpha = 1,$$

so that the point $(\cosh\alpha, \sinh\alpha)$ lies on the rectangular hyperbola

$$x^2 - y^2 = 1,$$

as shown in Fig. 186; this is analogous with the trigonometric identity $\sin^2\theta + \cos^2\theta = 1$, where the point $(\cos\theta, \sin\theta)$ lies on the unit circle. In the hyperbolic case, the parameter α is twice the area enclosed by the x-axis, the arc of the unit hyperbola between the vertex and ($\cosh\alpha$, $\sinh\alpha$), and the line joining this point to the origin, as shown in the figure, whence

$$\cosh^{-1}x = \sinh^{-1}y = \alpha.$$

In general, the hyperbolic functions satisfy all the trigonometric identities except for changes in the sign of second-degree terms in sinh.

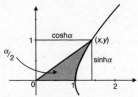

Fig. 186. **Hyperbolic function**.
($\cosh\alpha$, $\sinh\alpha$) describes a hyperbola.

hyperbolic geometry, *n.* another name for LOBACHEVSKIAN GEOMETRY.

hyperbolic paraboloid, see PARABOLOID.

hyperbolic spiral, *n.* a SPIRAL whose RADIUS VECTOR has length inversely proportional to its angle with the polar axis, so that its equation in POLAR COORDINATES is $r\theta = k$. It has an asymptote at $y = k$, as illustrated in Fig. 187.

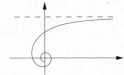

Fig. 187. **Hyperbolic spiral.**

hyperboloid of one sheet, *n.* a geometric surface of which the sections parallel to the three coordinate axes are HYPERBOLAE or ELLIPSES. The standard equation of a hyperboloid of one sheet is

$$\frac{x^2}{a^2} + \frac{y^2}{b^2} - \frac{z^2}{c^2} = 1,$$

where *a*, *b* and *c* are constants, and the axes of symmetry are taken as the coordinate axes; it is a RULED SURFACE, and in this case the cross-sections parallel to the *x*–*y* plane are elliptical, and those parallel to the other two coordinate planes are hyperbolic, as shown in Fig. 188a. When $a = b$, the *x*–*y* sections are circular, and this is the *hyperboloid of revolution of one sheet* obtained by revolving the hyperbola

$$\frac{x^2}{a^2} - \frac{z^2}{c^2} = 1,$$

around the *z*-axis.

Fig. 188. (a) **Hyperboloid of one sheet**.
(b) **Hyperboloid of two sheets**.

(a)   (b)

hyperboloid of two sheets, *n.* a geometric surface that consists of two parts separated by a finite distance, and the cross-sections of which parallel to two of the coordinate axes are HYPERBOLAE, and those parallel to the third axis are ELLIPSES, except for the interval in which there is no intersection; an example is shown in Fig. 188b above. The standard equation of a hyperboloid of two sheets is

$$\frac{x^2}{a^2} - \frac{y^2}{b^2} - \frac{z^2}{c^2} = 1,$$

where a, b and c are constants, and the axes of symmetry are taken as the coordinate axes. When $b = c$, the y–z sections are circular, and this is the *hyperboloid of revolution of two sheets* obtained by revolving the hyperbola

$$\frac{x^2}{a^2} - \frac{y^2}{b^2} = 1,$$

around the x-axis.

hypercomplex numbers, *n.* various generalizations of the complex numbers such as the QUATERNIONS; the original problem was to find such a DIVISION ALGEBRA (of which the only examples are the quaternions and the Cayley algebra) but the subject developed into a study of more general finite-dimensional algebras.

hypercube, *n.* a figure, usually in a Euclidean space of four or more dimensions, with all its sides equal and all its angles right; the n-fold CARTESIAN PRODUCT of an interval.

hyperelastic, *adj.* another term for GREEN–ELASTIC.

hypergeometric differential equation, *n.* the DIFFERENTIAL EQUATION

$$x(x-1)y'' + [(a+b+1)x - c]y' + aby = 0,$$

whose solution for c not a positive integer, and $|x| < 1$ can be expressed as

$$y = k_1 F(a, b; c; x) + k_2 x^{1-c} F(a-c+1, b-c+1; 2-c; x),$$

where $F(a, b; c; x)$ is the HYPERGEOMETRIC FUNCTION.

hypergeometric distribution, *n.* (*Statistics*) the DISTRIBUTION of the number of elements with a given property in a sample of size n chosen from a population of size N of which D have the property. The distribution has parameters D, N, and n; and

$$P(X = d) = \frac{\binom{D}{d}\binom{N-D}{n-d}}{\binom{N}{n}}.$$

hypergeometric function, *n.* one of a class of functions that can be represented as POWER SERIES of which the coefficients are appropriately defined products and quotients of FACTORIALS and GAMMA FUNCTION values. Almost all the important functions of mathematical physics can be given hypergeometric reformulations, and are written $F(a, b; c; x)$. These arise by FROBENIUS METHOD as analytic solutions of the HYPERGEOMETRIC DIFFERENTIAL EQUATION, and generate the HYPERGEOMETRIC SERIES. See also ELEMENTARY FUNCTION, SPECIAL FUNCTION.

hypergeometric series, *n.* a POWER SERIES of which the coefficients are products and quotients of FACTORIALS and GAMMA FUNCTIONS. If $F(a, b; c; x)$ is the HYPERGEOMETRIC FUNCTION,

$$F(a, b; c; x) = \sum_{n=0}^{\infty} \frac{(a)_n (b)_n}{(c)_n \, n!} z^n$$

is valid at least for z in the unit disk, where $(a)_n$ is the POCHAMMER SYMBOL (or rising factorial) defined as a ratio of GAMMA FUNCTIONS.

hyperplane, *n.* a translate of the NULL SPACE of any linear functional; a three-dimensional space in four dimensions, or more generally an $(n-1)$-space in n dimensions.

hyper-real numbers, *n.* a rigorous formulation of INFINITESIMAL real numbers in NONSTANDARD ANALYSIS, based in essence on the COMPACTNESS THEOREM of MODEL THEORY.

hyperspace, *n.* **1.** any space of more than three DIMENSIONS.

2. often in science fiction a space–time continuum of four dimensions in which time travel is possible.

hypocycloid, *n.* a curve described by a point on the circumference of a circle as it rolls around the inside of another fixed coplanar circle. The form depends on the ratio of the radii of the circles. The ASTROID or star curve, as shown in Fig. 189, is a *common hypocycloid* with equation

$$x^{2/3} + y^{2/3} = r^{2/3}.$$

The second figure is an *extended hypocycloid*; in each case, P_1 is the initial position of the point on the radius of the smaller circle, and P_2 is another position as it rolls around the inside of the larger circle. Compare EPICYCLOID, CYCLOID.

Fig. 189. Common and extended **hypocycloid**.

hypotenuse, *n.* the side opposite the right angle in a right-angled triangle, as shown in Fig. 190.

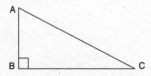

Fig. 190. AC is the **hypotenuse** of triangle ABC.

hypothesis, *n.* **1.** an assumption used in an argument without being endorsed; a supposition.

2. an unproved theory; a conjecture.

hypothesis testing, *n.* (*Statistics*) the theory, methods, and practice of testing one hypothesis concerning the parameters of the distribution of a population (the NULL HYPOTHESIS) against another (the ALTERNATIVE HYPOTHESIS); consideration of a suitable TEST STATISTIC enables the null hypothesis to be rejected at a given SIGNIFICANCE LEVEL only if the test statistic lies in its CRITICAL REGION for that significance level. Compare STATISTICAL INFERENCE.

hypothetical, *adj.* (*Logic*) another term for CONDITIONAL.

hypothetical syllogism, *n.* an argument of the form

> *if P then Q*
> *if Q then R*
> so *if P then R.*

i

i or **I,** *symbol for* **1.** the number 1 in ROMAN NUMERALS.

    2. the IDENTITY FUNCTION, for which $I(x) = x$ for all x.

    3. (nonstandardly) the IDENTITY ELEMENT of a GROUP; the symbol e is more common.

i, (conventionally printed in bold type) **1.** a UNIT VECTOR, usually one in the direction of the x-axis. Compare **j**, **k**. See DIFFERENTIAL OPERATOR.

    2. a unit QUATERNION.

i, *symbol for* the IMAGINARY NUMBER $\sqrt{-1}$, the square root of -1.

I, *symbol for* the IDENTITY matrix; the n by n unit matrix is denoted I_n when it is necessary to indicate the rank of the matrix.

𝕀 or *ℐ*, *symbol for* the INTEGERS; see ℤ.

icosa-, *prefix denoting* 20; for example, an *icosagon* has 20 sides.

icosahedron, *n.* a solid with 20 faces. A *regular icosahedron* has faces that are equilateral triangles.

icosidodecahedron, *n.* one of the ARCHIMEDIAN SOLIDS, with 20 triangular faces and 12 pentagonal faces. It can be obtained from either an ICOSA-HEDRON or a DODECAHEDRON by cutting off the corners in such a way that the new vertices lie at the midpoints of the original edges.

ideal, *n.* a SUBRING of a RING that is closed under subtraction and under multiplication by any ring element whatever. In the non-commutative case, one distinguishes *left ideals* and *right ideals*. In the absence of qualification as a left or right ideal, the term may be taken to refer to a *two-sided ideal*, that is, one closed under multiplication both on the left and on the right; this is occasionally called a *normal subring*. The multiples of any fixed integer form an ideal in the ring of integers. An ideal in a ring R is an R-MODULE.

ideal element, *n.* any element added to a mathematical theory in order to eliminate special cases. For example, adding the ideal element $i = \sqrt{-1}$ to the reals allows all algebraic equations to be solved, and the IMPROPER POINTS of the projective plane permit the assertion, without excluding parallel lines, that every pair of lines intersect.

ideal fluid, *n.* (*Continuum mechanics*) a FLUID that is neither VISCOUS nor COMPRESSIBLE.

ideal point, *n.* an IDEAL ELEMENT, especially a POINT AT INFINITY.

idempotent, *adj.* **1.** (of a matrix, function or ring element) having the property that it is equal to its own square. For example, the matrix

$$\begin{bmatrix} 2 & 2 \\ -1 & -1 \end{bmatrix}$$

is idempotent.

2. (of an operation) having the property that every element of its domain is idempotent with respect to it; for example, set-theoretic intersection and union are idempotent, since $S \cup S = S = S \cap S$. A RING of which every member is idempotent is called a BOOLEAN RING.

identical, *adj.* **1.** also called **numerically identical.** (of one entity usually referred to in two different ways) being one and the same individual or set of individuals. For example, the classes $\{x : x \text{ is even}\}$ and $\{x : x+1 \text{ is odd}\}$ are identical; congruent triangles have numerically identical properties but are not themselves numerically identical. If A is identical with B, written $A = B$, then the set $\{A, B\}$ has only one member.

2. also called **qualitatively identical.** (usually of two numerically distinct individuals) exactly alike, having the same properties; congruent triangles are identical in this sense, but since they are distinct triangles, they are not numerically identical.

3. identical relative to some relation. See RELATIVE IDENTITY.

identical equation, *n.* an IDENTITY (sense 3).

identically distributed, *adj.* (of a set of RANDOM VARIABLES) having the same CUMULATIVE DISTRIBUTION FUNCTION.

identical proposition, *n.* any necessary truth, but especially a CATEGORIAL identity, such as *whatever is triangular has three sides.* See IDENTITY (sense 3).

identity, *n.* **1.** also called **numerical identity.** the property or fact of being the same individual. For example, one speaks of the identity of $a^{1/2}$ and $\sqrt{a}$.

2. (*Logic*) the RELATION that trivially holds between every entity and itself, formally defined as the set of ordered pairs $\langle x, x \rangle$ for all x in the underlying domain.

3. a universally true equation, one that is not to be solved for the value of its variables that makes it true, but that is true for all values of its variables. For example, in

$$(x + y)(x - y) = x^2 - y^2,$$

or trigonometric identities such as

$$\sin^2\theta + \cos^2\theta = 1.$$

Identities are sometimes written using the ' $\equiv$ ' sign.

4. see IDENTITY ELEMENT.

5a. also called **qualitative identity.** the property of being exactly alike, or the relation that holds between such entities. For example, the identity of congruent triangles can be shown by superimposition.

b. see RELATIVE IDENTITY.

6. (*Logic*) an assertion that the relation of identity holds, such as 'e^x is $\exp x$'; any statement of which the operator of widest scope is '='.

identity element, identity, or **neutral element,** *n.* (for a set endowed with a binary operation) a member of the set such that the result of applying the operation to that element and any member of the set is equal to the latter member. A *multiplicative identity* is a UNITY; an *additive identity* is often called a ZERO. For example, the identity for multiplication in ordinary arithmetic is 1, since $x \times 1 = x = 1 \times x$. In some mathematical theories it is necessary to distinguish between a LEFT identity, l, such that $l \times x = x$ for all x, and a

RIGHT identity, r, such that $x \times r = x$ for all x; when an identity element does exist, it is both the unique left and the unique right identity for the theory. See also INVERSE.

identity function, *n.* a function whose value for any given argument is equal to that argument; $x + 0$ is thus an identity function on the real numbers.

identity matrix, *n.* a DIAGONAL MATRIX in which every diagonal entry is 1 and the other elements are all 0; the n by n unit matrix is denoted I_n when it is necessary to indicate its rank.

identity of indiscernibles, see LEIBNIZ'S LAW.

identity theorem or **uniqueness theorem,** *n.* the theorem asserting that if f and g are complex functions ANALYTIC on a region G, and $f(x) = g(x)$ for all z in any set containing a CLUSTER POINT in G, then $f(x) = g(x)$ for all z in G.

if, *conj.* **1.** the usual conjunction in a CONDITIONAL sentence. In ordinary usage some relevance of antecedent to consequent is expected.

2. (*Logic*) the usual rendition of the sentential connective in MATERIAL CONDITIONALS as a TRUTH FUNCTION in the sentential calculus. Since the truth of the compound sentence depends only on the TRUTH–VALUES of the components, 'if $2 + 2 = 4$, then gold is a metal' must be assigned the same truth–value as 'if $2 + 2 = 4$, then $2 + 3 = 5$', namely *true.*

See MATERIAL IMPLICATION. Compare ONLY IF.

iff, *conj. abbrev. for* **if and only if,** indicating that the two sentences connected are NECESSARY and SUFFICIENT CONDITIONS for one another's truth. Usually this is used for EQUIVALENCE in the METALANGUAGE rather than as the BICONDITIONAL in the OBJECT LANGUAGE.

iid, (*Statistics*) *abbrev. for* INDEPENDENT IDENTICALLY DISTRIBUTED.

ill-conditioned, *adj.* (*Numerical analysis*) **1.** (of a problem) having a large CONDITION NUMBER.

2. (of a computation) numerically unstable. While it is possible to have an ill-conditioned IMPLEMENTATION for a WELL–CONDITIONED problem, the converse is implausible.

-illion, *suffix denoting* a number of millions, the multiple being indicated by the prefix. In North American and French usage, each term in the sequence *million, billion, trillion, quadrillion, quintillion,* etc. is a factor of 1000 greater than the preceding; in UK and German usage, each is a factor of a million greater. Thus in the USA, the nth member of the series is $1000^{(n+1)}$, that is, $10^{3(n+1)}$, while in the UK it is $1\,000\,000^n$, that is, 10^{6n}; however, usage is tending to standardize on the former convention.

ill-posed problem, see WELL–POSED PROBLEM.

im, *abbrev. and symbol for* the IMAGINARY PART of a number; that is, where a and b are real, $\text{im}(a + bi) = b$; for example, $\text{im}(2 + 3i) = 3$.

image, *n.* **1.** (of a point or number) the value of a FUNCTION corresponding to the given value of the independent variable, or the point into which the given point is mapped. For example, the image of $x = 2$ under the function $y = x^2$ is $y = 4$.

2. (of a set) the set of values taken by a given function or SET–VALUED function for arguments in that set, written $f(S)$ where S is the given set;

$$f(S) = \{y : y = f(x),\ x \in S\}.$$

3. (of a function) the set of all values of the function; its RANGE as distinct from its CODOMAIN.

imaginary, *adj.* involving only IMAGINARY NUMBERS; having REAL PART equal to zero. The usage *purely imaginary* emphasises the contrast with other COMPLEX numbers. See also IMAGINARY PART.

imaginary axis, *n.* the *y*-axis of an ARGAND DIAGRAM, along which the IMAGINARY PART of the complex number to be represented is measured, as shown in Fig. 191.

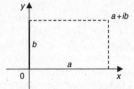

Fig. 191. The *y*-axis is the **imaginary axis**.

imaginary circle, *n.* the set of (imaginary) points that satisfy an equation of the form $(x - a)^2 + (y - b)^2 = -r^2$.

imaginary number, *n.* any number of the form ki, where $i = \sqrt{-1}$ and k is any real number, usually excluding zero except where the continuity of the IMAGINARY AXIS requires its inclusion. Often called *purely imaginary numbers* to avoid confusion with the COMPLEX NUMBERS.

imaginary part, *n.* the coefficient of i in any complex number, function or expression; the imaginary part of a complex number $z = a + ib$ is the real number b and is denoted Imz, im$z$, or im(z).

im grossen, see GLOBAL.

im kleinen, see LOCAL.

immediate inference, *n.* any form of argument that derives a conclusion from a single premise. In the traditional theory of the SYLLOGISM, the derivation from a categorial statement of its CONVERSE, OBVERSE, or CONTRAPOSITIVE, are, when valid, examples of immediate inferences.

implementation, *n.* (of an algorithm) an explicit computer program that executes a given mathematical algorithm. Different implementations of the same algorithm may behave quite differently.

implication, see MATERIAL IMPLICATION. See also STRICT IMPLICATION.

implicit, *adj.* (of a function) not expressing the value of the dependent variable directly (explicitly) as a function of the independent variable, but rather stating a relationship which the variables must jointly satisfy. The general form of such a function is

$$F(x, y) = 0, \quad F(x, y, z) = 0, \text{ etc.}$$

For example, $y^2 + xy + x^2 = 0$ cannot be universally expressed in the form $y = f(x)$. Compare EXPLICIT.

implicit definition, see DEFINITION.

implicit differentiation, *n.* (of a function) the computation of the derivative or partial derivative of an IMPLICIT function without explicitly determining the function. For example, given $F(x, y) = 0$, with two real variables, the CHAIN RULE applied to $F(x, g(x)) = 0$ gives

$$F_x + F_y g'(x) = 0,$$

and hence $g'(x)$ must be $-F_x/F_y$.

implicit function theorem, *n.* a theorem that gives conditions under which an EXPLICIT equation equivalent to a given IMPLICIT function exists locally. For example, suppose that $F(x, y) = 0$ represents n equations in $m + n$ unknowns and has a solution a (n variables) and b (m variables). Near a and b, one may find a unique function $y = g(x)$ with $b = g(a)$ whenever F is continuously differentiable and the matrix of partial derivatives with respect to b variables is FULL RANK. Then $F(x, g(x)) = 0$, and IMPLICIT DIFFERENTIATION allows the computation of the partial derivatives of g at a.

imply, *vb.* to enable a conclusion to be validly inferred.

importation, *n.* (*Logic*) a rule for conjoining the antecedents of a set of repeated conditionals as in

if P then if Q then R.
so *if P and Q then R.*

Compare EXPORTATION.

impossible set, *n.* another term for a NIKODYM SET.

impossibility theorem, *n.* any theorem asserting the impossibility of some much sought after, often intuitively plausible, result; for example, ARROW'S IMPOSSIBILITY THEOREM, Godel's INCOMPLETENESS THEOREM, or the impossibility of the SOLUTION BY RADICALS of the QUINTIC.

impredicative definition, *n.* (*Logic*) a definition in terms that require quantification over a range that includes that which is to be defined, such as *having all the properties of a great general* where one of the properties so ascribed must be that property itself. See TYPE.

improper fraction, *n.* a FRACTION in which the numerator has a greater ABSOLUTE VALUE than the denominator, such as $\frac{7}{3}$, or is a polynomial of higher degree, such as

$$\frac{x^2 + 3}{x + 1}$$

improper integral, *n.* a DEFINITE INTEGRAL with one or both limits of integration infinite, or having an INTEGRAND that becomes infinite between the limits of integration. The improper integral

$$\int_a^\infty f(x)\,dx$$

is said to CONVERGE if

$$\lim_{L \to \infty} \int_a^L f(x)\,dx$$

exists, and to DIVERGE otherwise. Similarly, if the integrand has a singularity at an endpoint of the interval of integration, then the integral is defined as the limit of the proper integrals as the limit of integration tends to that endpoint from within the interval; if the singularity is in the interior of the interval, then the integral is the sum of the two improper integrals on the subintervals above and below the singularity.

improper point, *n.* an IDEAL ELEMENT of the AUGMENTED EUCLIDEAN GEOMETRY; a point at which parallel lines meet.

impulse, *n.* (*Mechanics*) **1.** the integral of a force **F** between times t_1 to t_2,

$$\int_{t_1}^{t_2} \mathbf{F}\, dt.$$

If the force is acting on a set of particles, P, then the impulse is the total change of linear momentum of P.

2. a force applied to a mechanical system for a short time, such as a hammer blow.

imputation, *n.* (*Game theory*) an efficient, feasible, and individually rational PAYOFF in a game involving coalitions. See also EFFICIENT POINT.

inaccessible cardinal, *n.* a CARDINAL σ with the property that $\sigma > \aleph_0$ (ALEPH-NULL) and such that $\sigma > \tau$ implies $\sigma > 2^\tau$, while the union of every subset of σ with cardinality less than σ also has cardinality less than σ. The existence of such cardinals is stipulated by the *axiom of inaccessibility*, which is independent of ZERMELO–FRANKEL SET THEORY together with the AXIOM OF CHOICE.

incentre, *n.* the centre of the INCIRCLE of a figure (where one exists), especially a triangle. See Fig. 192 below. Compare CIRCUMCENTRE.

incidence, *n.* the partial coincidence of two geometric figures; for example, a line through a point in the interior of a closed curve is incident to the curve at at least two points.

incidence matrix, *n.* (*Graph theory*) a matrix of which the rows and columns correspond to vertices and edges of a graph, with ij^{th} entry 1 if vertex i lies on edge j and 0 otherwise. More generally, one counts the number of times an edge is INCIDENT to a vertex. A similar matrix method may be used to describe a DIGRAPH or other structure. Compare ADJACENCY MATRIX.

incident, *adj.* lying on or passing through. Regarding a line or plane as a set of points, a point and a line are incident if the former belongs to the latter. Thus in the axioms of geometry one may say that any two distinct points are incident with just one line, and any two lines are incident with at least one point.

incircle, *n.* a circle INSCRIBED in a triangle so that each of the sides of the triangle is a tangent, of which the centre is the INCENTRE, and the radius is the *inradius*. Fig. 192 shows the construction of the incircle: the centre is the point of intersection of the bisectors of the angles of the given triangle, and the radius is the perpendicular from that point to any side. Compare CIRCUMCIRCLE.

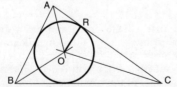

Fig. 192. The **incircle** of triangle ABC. O is the incentre, and OR the inradius.

inclination or **elevation,** *n.* **1.** the angle between a straight line (or its extension) and the positive direction of the *x*-axis, measured by convention in the anticlockwise direction from the axis; thus θ is the angle of inclination of the bold line in Fig. 193. Compare DECLINATION.

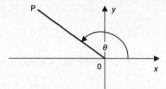

Fig. 193. θ is the **inclination** (sense 1) of OP to Ox.

2. the smaller angle between a line and its projection onto a plane; for example, in Fig. 194 the angle YXZ is the angle of inclination of XY to the plane CDEF.

3. the smaller DIHEDRAL angle between two planes; since XY in Fig. 194 is perpendicular to the edge CD, angle YXZ is also the inclination of the plane ABCD to the plane of CDEF.

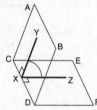

Fig. 194. θ is the **inclination** (senses 2 and 3) of XY and ABCD to CDEF.

inclined plane, *n.* a non-horizontal plane that makes an angle to the horizontal of less than a right angle.

inclusion, *n.* the relation between two sets that obtains when all the members of the one are members of the other, usually written S ⊂ T for *strict inclusion*, that is, if there exists an element *a* of T that is not a member of S, and S ⊆ T if the inclusion is *weak*, that is, if the two sets may also be identical; however, some authors use S ⊂ T for weak inclusion, and S ⊊ T for the strong relation. See also SUBSET.

inclusion–exclusion principle, *n.* the elementary but powerful counting principle, with application to probability theory, that the number of elements in a finite set that possess at least one out of *k* possible properties is equal to the number possessing exactly one of the properties, less the number possessing exactly two properties, plus the number possessing exactly three properties, and so on, up to the number possessing all *k*. Thus the number of elements in A ∪ B ∪ C is

$$|A| + |B| + |C| - |A \cap B| - |A \cap C| - |B \cap C| + |A \cap B \cap C|$$

where $|X|$ is the CARDINALITY of X.

inclusive, *adj.* (of a mathematical property defined on intervals or sets) such that, if it holds for a given interval or set, it also holds for all its subintervals or subsets.

inclusive disjunction or **inclusive or,** *n.* (*Logic*) **1.** the binary TRUTH–FUNC-TIONAL sentential connective that forms a compound sentence that is true whenever one or both of its DISJUNCTS is true; its TRUTH–TABLE is shown in

Fig. 195. Where P and Q are the disjuncts, the disjunction is usually written P ∨ Q, and often read 'P *vel* Q'. If no qualification is stated, disjunction is usually taken to be inclusive.

P	Q	P ∨ Q
T	T	T
T	F	T
F	T	T
F	F	F

Fig. 195. **Inclusive disjunction**. Truth table for inclusive 'or'.

2. the relation that holds between two sentences when the statement so formed is true.

3. a sentence with this as its main connective.

Compare EXCLUSIVE DISJUNCTION.

incommensurable, *adj.* (of two quantities) not being in rational proportion to one another, not COMMENSURABLE; for example, as was discovered by Pythagoras, 2 and √2 are incommensurable.

incomparable, *adj.* not able to be compared using a given relation. In a PARTIAL ORDER some pairs of terms are incomparable; for example, in the Cartesian PRODUCT ORDER, ⟨1,1⟩ > ⟨0,0⟩, but ⟨0,1⟩ and ⟨1,0⟩ are incomparable. Compare INDIFFERENT.

incompatible, *adj.* (of a set of propositions, statements, equations, etc.) not capable of all being true at the same time or under the same conditions; INCONSISTENT.

incomplete, *adj.* **1.** (*Logic*) (of a formal theory) not so constructed that the addition to the AXIOMS of a proposition that is not a theorem renders the whole theory INCONSISTENT; thus the theory does not contain as many theorems as it may without inconsistency.

2. (of an expression) meaningless or unanalyzable by itself; requiring some specific type of context in order to constitute a meaningful expression. For example, Russell's theory of descriptions holds that DEFINITE DESCRIPTIONS are incomplete, being capable of analysis only as the subjects of statements and not by themselves.

3. not COMPLETE in order, or as a METRIC SPACE. The vector space of polynomials endowed with CHEBYSHEV NORM form an incomplete normed space. The rationals are incomplete in order and topology, but can be given an equivalent complete metric.

incomplete elliptic integral, see ELLIPTIC INTEGRAL.

incomplete induction, *n.* another term for FIRST–KIND INDUCTION, by contrast with COMPLETE INDUCTION. See INDUCTION.

incompressibility, *n.* (*Continuum mechanics*) the property of a FLUID of having constant DENSITY, that is, its density remains unaltered under changes in PRESSURE.

inconsistency, *n.* (*Logic*) **1.** a SELF–CONTRADICTORY proposition.

2. the property of being INCONSISTENT.

inconsistent, *adj.* **1.** (of a set of equations) not simultaneously SATISFIABLE; having the property that there are no values of the variables for which all of the set are true. For example, the equations $x + 2y = 5$ and $x + 2y = 6$ are inconsistent.

2. (of a set of propositions or statements) incapable of all being true at once, INCOMPATIBLE.

3. (*Logic*) **a.** (of a set of WELL–FORMED FORMULAE of a formal THEORY) such that an explicit contradiction may be derived from them by the derivation rules of the theory.

b. (*Logic*) (of a formal theory) having an explicit contradiction as a theorem, or more generally having an ATOMIC statement as a theorem, whence excluding nothing from its theorems.

increasing, *adj.* (of a function of a single variable) having the property that $f(x) \geq f(y)$ for all pairs of arguments such that $x > y$. This property may be LOCAL or GLOBAL. For example, the graph shown in Fig. 196 is increasing for $0 \leq x \leq 2$, but it is only locally increasing, as its value falls for $x < 0$ and $x > 2$. If $f(x) > f(y)$ for all $x > y$ the function is *strictly increasing*. The symbol $\uparrow$ is sometimes used for either the strict or weak property. See also MONOTONE, ISOTONE, ANTITONE.

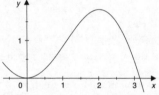

Fig. 196. A locally **increasing** function on]0,2[.

increment, *n.* a small but finite change in the value of a variable or function, written δx or Δx. A negative increment is sometimes called a *decrement*. See also DELTA, EPSILON–DELTA NOTATION.

indefinite, *adj.* (of a QUADRATIC FORM or matrix) neither POSITIVE (SEMI-) DEFINITE nor negative (semi-) definite.

indefinite integral, *n.* any function whose DERIVATIVE is a given function; it is usually written as a function of the given function as

$$\int f(x) \, dx,$$

where $f(x)$ is the given function. For example, x^2, $x^2 + 5$, $x^2 - 3$, and any function of the form $x^2 + c$, where c is a constant, are indefinite integrals of $y = 2x$. The schema, here $x^2 + c$, is usually given as the general form of an indefinite integral, but c (referred to as the *constant of integration*) is often omitted; the common indefinite integrals listed in Appendix 2 are given in this form. See INTEGRAL. Compare DEFINITE INTEGRAL. See also ANTIDIFFERENTIATE, FUNDAMENTAL THEOREM OF CALCULUS.

in-degree, *n.* (of a graph or network) see DEGREE.

independent, *adj.* **1.** (of a system of linear equations or vectors) not LINEARLY DEPENDENT.

2. (*Statistics*) **a.** (of two or more random variables) distributed so that the value of one variable will have no effect on that taken by the others. Thus the probability of each random variable taking each of some sequence of specified values equals the product of their separate probabilities of taking these values. More formally and generally, for random variables $X_1,...,X_n$ on a given probability space with values in a real or complex EUCLIDEAN SPACE it is required that

$$P\,[\,X_1 \in B_1, \, ... \, , X_n \in B_n\,]\;=\;\prod_{k=1}^{n}\,P\,[X_k \in B_k\,]$$

for arbitrary BOREL SETS $B_1,..., B_n$.

b. (of two or more events) such that the occurrence of one does not affect the probability of any of the others. Consequently, the probability of any set of independent events occurring equals the product of their individual probabilities. Compare STATISTICAL DEPENDENCE.

3. (*Logic*) (of a set of statements, propositions, or formulae) **a.** not validly derivable from one another or from any set of the others, so that if all are axioms of some theory, none can be dispensed with without loss.

b. more generally, not logically related, so that in no case can the truth value of one be inferred from that of the others.

4. (of a set of MEASURABLE functions, $\{f_\gamma\}_{\gamma \in \Gamma}$, from a measure set X with $\mu(X) = 1$ to $\mathbb{R}^n$) such that

$$\mu\left[\,\bigcap_{k=1}^{n}\,f_{\gamma_i}^{-1}(A_i)\,\right]\;=\;\prod_{i=1}^{k}\,\mu\left[\,f_{\gamma_i}^{-1}(A_i)\,\right]\,,$$

for every finite collection, $A_1,..., A_k$, of Borel sets in $\mathbb{R}^n$ and any collection $\gamma_1, ..., \gamma_k$ of indices.

independent variable, *n.* **1.** a VARIABLE in a mathematical equation or statement, whose value determines that of the DEPENDENT VARIABLE; in $y = f(x)$, x is the independent variable. See also ARGUMENT.

2. (*Statistics*) also called **predictor**. the VARIABLE that defines a set of distinct EXPERIMENTAL CONDITIONS, or that an experimenter deliberately manipulates in order to observe its relationship with some other quantity.

indeterminate, *n.* **1a.** a VARIABLE or PLACE–MARKER in an expression.

b. a purely formal, uninterpreted, symbol. Sometimes a specific contrast is intended with variables for which certain identities may hold by virtue of the structural properties of a given domain; for example, in the integers modulo a prime p, the element $x^p = x$ for all x, permitting cancellation; however, regarded as an indeterminate, x^p is a distinct entity from x. See also COMMUTING INDETERMINATE.

2. an expression that, if evaluated naively, has no defined value, but that may be capable of evaluation by an alternative method. For example,

$$\lim_{x \to 0}\;\frac{x^2}{x}$$

is an indeterminate of form $\raisebox{0.3ex}{$0$}\!/\!\raisebox{-0.3ex}{$0$}$, but it can be evaluated after cancellation of x, and similarly for

$$\lim_{x \to 0}\;\frac{\tan\,(nx)}{\tan\,(mx)}\,.$$

See also L'HÔPITAL'S RULE.

3. (of a set of simultaneous equations) having an infinite SOLUTION SET.

index, *n.* **1.** another word for EXPONENT.

2a. one of a specified set of symbols, usually written subscript to the right of a variable, used to distinguish each in turn of a set or sequence of otherwise identical symbols such as variables, as $\langle x_1, x_2, x_3 \rangle$; the index of x_3 is 3. See INDEX SET.

b. *vb.* (of a set of symbols) to distinguish the elements of a set or sequence onto which they are mapped by associating each with a distinct element of an index set. To make this explicit, the sequence $\langle x_1, x_2, x_3 \rangle$ can be written $\langle x_i \rangle_i$ for $i \in \{1, 2, 3\}$.

3. a number or expression written within a RADICAL, and indicating which ROOT is to be extracted, such as $\sqrt[3]{8} = 2$.

4. (of a subgroup H of a finite group G) the number of left COSETS of the subgroup in the group, written $[G:H]$. When E is the unit subgroup, $[G:E]$ is the ORDER of G. LAGRANGE'S THEOREM shows that

$$[G:H]\ [H:E]\ =\ [G:E]$$

so that the order of H divides that of G.

5. (of a point) another word for WINDING NUMBER.

index laws, *n.* the rules that govern the manipulation of POWERS of elements in a SEMIGROUP. The laws are

$$x^m x^n\ =\ x^{m+n}; \quad (x^m)^n\ =\ x^{mn}$$

In a GROUP these can be extended by

$$x^{-m}\ =\ (x^m)^{-1}; \quad x^0\ =\ e$$

where e is the IDENTITY ELEMENT of the group. See also EXPONENT.

index number, *n.* (*Statistics*) a measure of the change, relative to some specific BASE PERIOD, in some variable, such as the price, volume, or value of a commodity, the gross national product, or the general level of prices. The value of the variable in the base period is conventionally given the index number 100, and the index number of any other period is in proportion with it; thus an index number of 250 indicates that the value of the variable is two and a half times that of the base period.

index set, *n.* a set whose elements are used to INDEX those of some other set, such as Λ in $\bigcup_{\lambda \in \Lambda} A_\lambda$, the union of all the sets A_λ, one for each λ in Λ.

indicator function, *n.* the extended real-valued function that takes value zero on a given set and $+\infty$ off the set. It is denoted i_c or δ_c, and is convex precisely if the set C is. Compare CHARACTERISTIC FUNCTION. See also SUPPORT FUNCTION.

indicial equation, *n.* the equation that determines the index used in FROBENIUS METHOD for solving second-order regular differential equations.

indifference, *n.* **1a.** the fact that some given ORDERING is INDIFFERENT between the members of some set.

b. the relation that holds between such elements with respect to such an ordering.

2. principle of indifference. the principle that in the absence of any reason to the contrary, each possible outcome of an experiment is to be treated as EQUIPROBABLE. See MATHEMATICAL PROBABILITY.

indifference curve, *n.* a set of points for which a given UTILITY FUNCTION maintains constant value, and which are therefore INDIFFERENT to one another with respect to that function. For example, an isotherm, which passes through all points on a map where the temperature is equal, is an indifference curve for any utility function that depends solely upon temperature; a contour map likewise shows a set of indifference curves with respect to height. Similarly, one speaks of *indifference surfaces* in more than two dimensions. The term is due to Edgeworth; see EDGEWORTH BOX.

indifference sets, *n.* the EQUIVALENCE CLASSES of points that are pairwise preferred to one another under a given ORDERING; that is, the sets of points between which the WEAK ordering is INDIFFERENT. These are all singletons exactly when the order is antisymmetric.

indifference surface, see INDIFFERENCE CURVE.

indifferent, *adj.* **1a.** (of an ORDERING) giving no precedence to any element of a given set in the sense that it orders each of these elements above the other or others; that is, the ordering holds between any pair of members of this set, but is symmetric upon it. For example, in the set of combinations of k elements chosen from n, cardinality is an indifferent ordering, since all elements have cardinality k.
b. (of elements of an ordering) mutually preferred one to the other. This is not to be confused with being INCOMPARABLE, when neither can be said to be preferred to the other; the set of such elements is an INDIFFERENCE SET.
2. (*Statistics*) not showing or having preferences; not rendering any of the possible outcomes more probable than the others. See also UNBIASED.

indirect proof, *n.* a common mathematical term for *REDUCTIO AD ABSURDUM*; proof of a conclusion by showing its negation to be self-contradictory or to contradict known axioms. Compare DIRECT PROOF.

indirect proportion, *n.* another term for INVERSE PROPORTION.

indirect variation, *n.* another term for INVERSE PROPORTION.

indiscernibility of identicals, see LEIBNIZ'S LAW.

indiscernible, *adj.* qualitatively identical. See IDENTITY.

indiscrete topology, *n.* the TOPOLOGY on a given space whose only open sets are the whole space and the empty set.

individual, *n.* (*Logic*) **1.** a unique object as opposed to a property or class. See also PARTICULAR.
2. any element in the domain of a given theory.

indivisible, *adj.* (usually of integers or polynomials) **1.** not able to be exactly divided; PRIME.
2. not exactly divisible by a given number or quantity; RELATIVELY PRIME. For example, 8 is indivisible by 3.

induced measure, *n.* the MEASURE induced by the OUTER MEASURE, μ^*, on the μ^*-measurable sets.

induced topology or **relative topology,** *n.* the TOPOLOGY defined on a subset of the underlying set of a given topological space by taking its open sets to be intersections of the open sets of the given topology with the subset. For example, the open sets of the induced topology on a line in the Euclidean plane are the segments of the line in the interior of the open disks.

induction, *n.* **1.** also called **mathematical induction** or **finite induction.**
a. also called **first-kind**, **incomplete**, or **special induction.** a method of proving a proposition that all integers have a certain property *P* by proving

   (i) the BASE clause, $P(1)$; and
   (ii) the RECURSION clause: if $P(n)$ for an integer *n*, then $P(n+1)$.

For any finite *m*, $P(m)$ now follows by a finite number of steps of MODUS PONENS; the universal generalization is a consequence of PEANO'S AXIOMS. For example, one may prove by induction that $\sum_n$, the sum of the first *n* natural numbers is equal to $\frac{1}{2}n(n+1)$ by first observing that the base clause is obviously true for $n = 1$. The recursion clause requires us to show that if the hypothesis is true of *n* it is true of $n+1$:

$$\sum_{n+1} = 1 + 2 + \ldots + n + (n+1) = \sum_n + (n+1),$$

which *ex hypothesi* is $\frac{1}{2}n(n+1) + (n+1)$; extracting the common factor yields

$$(n+1)\left(\frac{n}{2}+1\right) = \frac{(n+1)(n+2)}{2}$$

as required, so that the result is established for all *n*.
b. also called **complete**, **general**, or **second-kind induction.** inductive argument in which the inductive step is from all integers less than *n* to all integers less than $n+1$; that is, the recursion clause concerns proper subsets, rather than members, of the set in question.
Compare TRANSFINITE INDUCTION.
2. the application of recursive rules. See RECURSION.
3. (*Logic*) a process of reasoning in which a general conclusion is drawn from a set of particular premises, often drawn from experience or from experimental evidence. The conclusion goes beyond the information contained in the premises and does not follow necessarily from them. Thus an inductive argument may be highly probable, yet lead from true premises to a false conclusion; for example, large numbers of sightings at widely varying times and places provide very strong grounds for the falsehood that all swans are white. Compare DEDUCTION. See also GOODMAN'S PARADOX, HEMPEL'S PARADOX.

inductive, *adj.* of, relating to, or using INDUCTION. For example, *inductive proof* or *inductive argument* is argument by induction.

inductive definition, *n.* another name for RECURSIVE DEFINITION.

inductive order, *n.* (*Set theory*) an ORDER in which every non-empty subset has at least one minimal element. Compare WELL–ORDERED.

inductive step, *n.* another name for the RECURSION CLAUSE of INDUCTION, or for an application of that clause.

inelastic, *adj.* (of a function) having ELASTICITY less than unity. In economics, demand for a good is said to be inelastic if an increase in the price results in a decrease in revenue. Compare ELASTIC.

inequality, *n.* **1.** the relationship between two numbers, quantities, etc. that holds when they are COMPARABLE but not EQUAL, so that they are related by a STRICT ORDERING.

2. a statement that such a relation holds; the negation of an EQUALITY or EQUATION, written $a \neq b$.

3. any of the specific relationships:

$a < b$ (a is less than b),

$a \leq b$ (a is less than or equal to b),

$a > b$ (a is greater than b), or

$a \geq b$ (a is greater than or equal to b).

inertia, *n.* (of a HERMITIAN MATRIX) the ordered triple of which the entries are the number of positive, negative, and zero CHARACTERISTIC ROOTS of the matrix. Compare SIGNATURE.

inertial frame of reference, *n.* (*Mechanics*) a FRAME OF REFERENCE in which the rates of change of position of the origin and basis vectors are constant.

inertia tensor, *n.* (*Continuum mechanics*) (for a set of axes) the TENSOR represented by the SYMMETRIC MATRIX the entries of whose PRINCIPAL DIAGONAL are the MOMENTS OF INERTIA of the RIGID BODY, and whose other elements are the negative of the PRODUCTS OF INERTIA for the respective coordinate planes.

inessential, *adj.* (depending upon context) not central, not difficult, not important, removable, etc.

inf, *abbrev. and symbol for* INFIMUM.

infeasible, *adj.* **a.** (of a CONSTRAINED OPTIMIZATION problem) subject to INCONSISTENT constraints; having an empty FEASIBLE SET.

b. (of a point) not lying in a prescribed FEASIBLE SET.

inference, *n.* **1.** any process or mode of reasoning from premises to a conclusion. In this sense one says 'his inference was invalid', since the inference is an argument rather than a statement and so cannot be said to be true or false.

2. the conclusion inferred. In this sense one says 'his inference was false', since the inference is a statement, not an argument, and so can be true or false but not valid or invalid.

See also INDUCTION, DEDUCTION.

inferential statistics, *n.* another name for STATISTICAL INFERENCE.

inferior limit, *n.* another name for LIMIT INFERIOR.

infimal, *n.* constituting or relating to an INFIMUM.

infimal convolution, see CONVOLUTION.

infimum or **greatest lower bound** (abbrev. **inf, glb**), *n.* the unique largest member of the set of LOWER BOUNDS for some given set, equal to its MINIMUM if the given set has a least member. The infimum τ of a set T may be defined as satisfying the two conditions, that $\tau \leq t$ for all t in T, and that for all $t > \tau$ there is a $t' < t$ in T. For example, the geometric sequence $\frac{1}{2}$, $\frac{1}{4}$, $\frac{1}{8}$, ... has as a lower bound every real number less than or equal to zero; it has no least member, and so no minimum but its infimum is 0. Compare SUPREMUM.

infinite, *n.* **1.** (of a number or quantity) not FINITE; having a size or absolute value that is greater than any natural number. See TEND TO INFINITY.

2. (of a set or sequence) having an unbounded number of terms; not able to be counted by a terminating sequence of natural numbers, that is, not able to be put into one-to-one correspondence with a bounded initial segment of them. This sense differs from the preceding in that the real numbers of the interval $[a, b]$ are an infinite set, even although it is not unbounded.

3. (of a set) able to be put into one-to-one correspondence with a proper subset of itself. See also CANTOR'S DIAGONAL THEOREM.

4. (of a GROUP) having infinite ORDER.

5. (of an integral) having INFINITY as one or both limits of integration. See IMPROPER INTEGRAL.

infinite descent, *n.* a method of proof by contradiction, introduced by Fermat and used in number theory, that may be used to prove that a result known to be false for some, normally small, positive integer p is false for all positive integers, or to prove that a result that is assumed to be true for all integers $n \geq n_0 > 0$ and false for all $n < n_0$ is in fact false for all positive integers. In particular, if a result is assumed to be true for some positive integer m, and it can be shown that it would follow for $m - k$, and so by iteration for $m - 2k$, $m - 3k$, and so forth, it then remains to show that for any m there exists a positive integer r of the form $r = m - ik$, for integral i, such that $r = p$ or $r < n_0$, contradicting the assumption.

infinite hotel paradox, *n.* another name for HILBERT'S PARADOX.

infinite-order, *adj.* (of an element a of a group with identity e) such that there is no finite integer n such that $a^n = e$. See ORDER (sense 6).

infinite product, *n.* the PRODUCT of an INFINITE SEQUENCE of terms, usually indexed by, or expressed as a function on, the natural numbers, and often written in the form

$$\prod_{i=1}^{\infty} a_i,$$

or, more briefly, $\prod_i a_i$. An infinite product of non-zero complex numbers is said to CONVERGE if, as n tends to infinity, the partial product

$$\prod_{i=1}^{n} a_i$$

converges to a non-zero limit, and DIVERGES TO ZERO if that limit exists but is zero. For example,

$$\prod_{i=1}^{\infty} (1 + z^{2^i}) = \frac{1}{1-z}$$

for $|z| < 1$. If a sequence has a finite number of zero terms, one determines its convergence by considering convergence of the non-zero tail; the value of the product will, however, be zero in both the convergent and divergent cases. These conventions allow one to convert products securely into series by taking logarithms. A series $\{a_i\}$ is absolutely convergent exactly if

$$\prod_{i=0}^{\infty} (1 + |a_i|) < \infty.$$

See also WALLIS PRODUCT.

infinite regress, *n.* a putative explanation or construction in terms of something that itself requires a similar explanation or construction; such an explanation is therefore vacuous. However, it should not be confused with an infinite progression of elements, each of which is constructed from those already constructed. For example, PEANO'S AXIOMS for arithmetic do not constitute a regress since each number is generated from the preceding, and the first is given by an axiom; thus anything can be shown to be a number by tracing back a finite number of steps to that axiom. However, it is not possible to dispense with that axiom, since without it the process of establishing that anything is an integer would be regressive, and would never terminate. See also VICIOUS CIRCLE.

infinite sequence, *n.* a SEQUENCE the members of which are indexed by the NATURAL NUMBERS, $\mathbb{N}$; the range of a function of $\mathbb{N}$. Thus the sequence is equinumerous with, and ordered by, $\mathbb{N}$.

infinite series, *n.* the sum of an INFINITE SEQUENCE of terms, usually indexed by, or expressed as a function on, the NATURAL NUMBERS, $\mathbb{N}$, and often written in the form

$$\sum_{i=0}^{\infty} a_i \, ,$$

or, more briefly, $\sum_i a_i$. If the sequence of PARTIAL SUMS tends to a limit as the number of terms increases, the series is said to CONVERGE; for example,

$$\sum_{i=0}^{\infty} \frac{1}{2^n} = 1 + \frac{1}{2} + \frac{1}{4} + \frac{1}{8} + \dots$$

converges to 2.

infinitesimal, *adj.* **1a.** (usually of an increment) approaching zero as a limit, arbitrarily small. Informally, infinitely small.

b. (*as substantive*) an infinitesimal increment or quantity. In early treatments of the CALCULUS, a DERIVATIVE was treated as a ratio of infinitesimals and an INTEGRAL as a sum of products of infinitesimals. These had to be non-zero for the ratio to be well-defined and the products to be non-zero, but zero for the rate of change so derived to be instantaneous and the upper and lower sums to be equal. This paradoxical conception was later largely abandoned in favour of the EPSILON–DELTA treatment of limits, but see HYPER–REAL NUMBERS.

2. in NON–STANDARD ANALYSIS, having zero STANDARD PART. See ARCHIMEDEAN PROPERTY.

infinitesimal calculus, *n.* another name for CALCULUS.

infinity, *n.* **1.** a value greater than any computable value, such as the index of the limit of a non-terminating sequence of values; this value is denoted ∞. For example, the sequence $f(0) = 1$, $f(1) = 2$, $f(2) = 4$,..., of which the nth member is 2^n, is said to tend to infinity as n tends to infinity (written $n \to \infty$). In general, this usage is to be understood in terms of the LIMITS of sequences of values, and is defined formally in terms of EPSILON–DELTA NOTATION; for example, to say that the quotient of 1 by 0 is infinity is to be interpreted as asserting that $1/x$ increases without bound as x tends to zero, that is, that for any positive ε, however small, there is an N such that $1/x > N$ for all $|x| < \varepsilon$.

2. the supposed value of an undefined expression regarded as the limit of a sequence of analogous expressions, such as the POINT AT INFINITY at which parallel lines are said finally to meet, or the sum of certain series that do not converge. These are put on a proper footing by the addition to the relevant theory of IDEAL ELEMENTS or by COMPACTIFICATION.

3. an imprecise term for an ALEPH, an infinite CARDINAL.

infix notation, *n.* **1.** the usual notation for binary operators in which they are written between their arguments; for example, P ∨ Q represents disjunction, and *x* + *y* addition. This requires bracketing to avoid ambiguity in complex expressions such as '2 + 3 × 5'; in general, if one uses a BINARY TREE diagram with the operators at the nodes, as in Fig. 197, to represent the structure of an expression, the infix representation is obtained by reading the nodes from left to right; more properly, one starts at the top and reads in order at each node its left branch, the node itself, and its right branch, enclosing the whole in brackets, and iterating where necessary. Compare POLISH NOTATION, REVERSE POLISH NOTATION.

Fig. 197. **Infix notation.**
Tree diagram for (P ∨ Q) & (R → S).

2. a notation for BINARY RELATIONS in which the symbol for the relation is written between its arguments; thus *x has R to y* is written *xRy*. This is less common for most relations than PREFIX NOTATION, but identity is always written infix as *x* = *y*. See also POSTFIX NOTATION.

inflection or **inflexion,** *n.* a change of CURVATURE at a point from positive to negative or vice versa. See POINT OF INFLECTION.

information, *n.* **1.** a mathematical abstraction of the content of any meaningful statement or data, enabling the study of the most efficient way of recording or transmitting them. A BIT is the unit of information, recording no more than the presence or absence of a single feature. See INFORMATION THEORY.

2. also called **uncertainty.** formally, a real-valued function of events in a PROBABILITY SPACE that depends only on the probability of events, and is such that events of probability one have zero uncertainty, that uncertainty increases as probability drops, and that the uncertainty of the simultaneous occurence of two independent events is the sum of their individual uncertainties. The only measurable functions meeting these requirements are of the form

$$I(E) = -c \log(P(E))$$

for positive constants c. Thus for a measurable partition ξ of the space, one has the corresponding *information function*

$$I(\xi) = -c \sum_{E \in \xi} \log(P(E)) \chi_E,$$

the expected value of which is the ENTROPY of the partition. Usually, c is chosen to yield a logarithm to base two, with information measured in BITS.

information theory, *n.* a collection of mathematical theories, based on PROBABILITY theory, that are concerned with methods of coding, decoding, storing, and retrieving INFORMATION, and with the likelihood of a given degree of accuracy in the transmission of a MESSAGE through a CHANNEL that is subject to probabilities of failure that are described by a PROBABILITY LAW.

inherent round-off, see ROUND–OFF ERROR.

inhomogeneous, *adj.* not HOMOGENEOUS, especially of a system of linear equations (often differential equations) or inequalities. See also FREDHOLM ALTERNATIVE.

inhomogeneous coordinates, see HOMOGENEOUS COORDINATES.

initialize, *vb.* to set variables or parameters at the commencement of an algorithm. For example, when computing the determinant of a matrix by PARTIAL PIVOTING, one initializes the value of the determinant to unity, and then iteratively UPDATES that value during the computation.

initial condition, *n.* a BOUNDARY CONDITION involving a differential equation at the beginning of the relevant time period, such as the initial velocity and acceleration of a particle of which the position vector is sought.

initial line, *n.* the single axis for POLAR COORDINATES.

initial segment, *n.* a finite subsequence consisting of successive terms of an infinite sequence beginning with the first term; that is all elements of the sequence with indices less than a given one. Similarly, an initial segment of an ordered set is the set of all elements less than (or less than or equal to) a given one in terms of that ordering.

injection, *n.* an INJECTIVE mapping.

injective, *adj.* **1.** (of a function, mapping, etc.) associating two sets in such a way that different members of the domain are paired with different members of the CODOMAIN, although not all of the latter need be members of the specified RANGE; thus, in Fig. 198, T is the codomain and T^+ is the range, so that the mapping from S to T^+ (as well as that from S to T) is injective. In some usages, an injective map is called one-to-one, but this term may cause confusion, as it is also used for a bijection. For example, the mapping associating oldest sons to fathers is injective even when the codomain is all men; $f(x) = x^2$ is not injective on the reals since $f(x) = f(-x)$, but it is injective on the positive real numbers. Compare BIJECTIVE, SURJECTIVE. See also MONOMORPHISM.

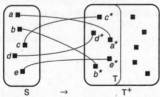

Fig. 198. An **injective** mapping from S to T^+.

2. (of a left R-MODULE Q) having the property that whenever there is a left R-module A with a submodule B such that there is a HOMOMORPHISM f of B into Q, then f can be extended to a homomorphism g of A into Q.

Inn, *abbrev. and symbol for* the set of INNER AUTOMORPHISMS of a given group.

inner automorphism (abbrev. **Inn**), *n.* a group (or ring) AUTOMORPHISM induced by conjugation; for an element *a* in the group or ring the mapping

$$r \rightarrow a^{-1}ra.$$

The set of all inner automorphisms of a group form a normal subgroup of the automorphism group that is isomorphic to the factor group of the given group by its centre.

inner Jordan measure or **inner Jordan content**, *n.* the supremum of the volumes of finite collections of disjoint HYPERCUBES lying inside the set; thus, a type of internal QUADRATURE.

inner measure, *n.* **1.** the supremum of the measures of LEBESGUE MEASURABLE subsets of a set in EUCLIDEAN SPACE.

2. see INNER JORDAN MEASURE.

Compare OUTER MEASURE.

inner product, *n.* **1a.** the product of two vectors defined in a given INNER PRODUCT SPACE.

b. in particular, in a real EUCLIDEAN SPACE, the SCALAR PRODUCT that is the sum, $\sum x_i y_i$, of the products of the corresponding entries of the vectors $\mathbf{x} = \langle x_i \rangle$ and $\mathbf{y} = \langle y_i \rangle$.

2. any analogous operation, especially the multiplication of the ith row of one matrix by the jth column of another CONFORMABLE matrix to yield the $(i, j)^{\text{th}}$ entry of their product.

3. (for TENSORS) see METRIC TENSOR.

inner product space, *n.* **1.** a complex vector space endowed with an INNER PRODUCT; a SESQUILINEAR SEMIDEFINITE FORM that satisfies:

$$\langle x, y \rangle = \overline{\langle y, x \rangle}$$
$$\langle cx + y, z \rangle = c\langle x, z \rangle + \langle y, z \rangle$$
$$\langle x, cy + z \rangle = c\langle x, y \rangle + \langle x, z \rangle$$

$\langle x, x \rangle \geq 0$, with equality only for $x = 0$.

for all x, y, z in the space and complex scalars c. In such a space, a norm is always introduced by the formula

$$\|x\| = \langle x, x \rangle.$$

2. a real vector space endowed with an INNER PRODUCT; a BILINEAR SEMIDEFINITE FORM that satisfies

$$\langle x, y \rangle = \langle y, x \rangle$$
$$\langle rx + y, z \rangle = r\langle x, z \rangle + \langle y, z \rangle$$
$$\langle x, ry + z \rangle = r\langle x, y \rangle + \langle x, z \rangle$$

$\langle x, x \rangle \geq 0$, with equality only for $x = 0$.

for all x, y, z in the space and real scalars r.

Compare HILBERT SPACE, UNITARY SPACE.

input–output model or **Leontief model**, *n.* (*Mathematical economics*) a matrix model of economic production due to Leontief, in which one commences with *n* goods together with a *technology matrix* A whose i,j^{th} entry specifies the number of units of good i necessary to produce one unit of good j. The quantity

$$y = (I - A)x$$

then represents the *output* of the economy corresponding to *input x*. Thus one can determine whether and how a given net output is feasible by solving the matrix system.

input set, *n.* (*Information theory*) the set of signals from which the sender selects one to represent a MESSAGE.

inradius, see INCIRCLE.

ins, *abbrev. and symbol for* INSIDE.

inscribe, *vb.* to construct (a geometric figure) inside another figure so that the two have points in common but do not intersect. An inscribed polygon has every vertex lying on the given figure, so that the square in Fig. 199 is not inscribed in the regular hexagon. See also INCIRCLE. Compare CIRCUMSCRIBE.

Fig. 199. A square **inscribed** in a circle inscribed in a hexagon.

inside (abbrev. **ins**), *n.* the set of all points, denoted ins Γ, not lying on a given CONTOUR Γ, for which the WINDING NUMBER is not zero. Compare OUTSIDE.

insoluble, *adj.* another word for UNSOLVABLE.

insolvable, *adj.* another word for UNSOLVABLE.

instance, *n.* (*Logic*) **1.** an expression that is derived from another given expression by INSTANTIATION.

2. a SUBSTITUTION INSTANCE.

instantaneous, *adj.* **1.** (of any property of a function of time) occurring at or associated with a given instant, or as a limit as a time interval approaches zero. For example, instantaneous velocity is the derivative of displacement or position with respect to time.

2. (more generally) any analogous property of a function at a unique value of its independent variable. See also DERIVATIVE.

3. (of interest) see COMPOUND INTEREST.

instantiation, *n.* (*Logic*) **1.** the process of deriving a particular statement from a general one by eliminating a QUANTIFIER and replacing the BOUND VARIABLE with a name or other referring expression. This may, however, not be a valid syntactic ELIMINATION RULE. For example, *existential instantiation* is the derivation of an instance, *Fa*, from the existentially quantified statement, $(\exists x)Fx$, which is not a valid inference, although it does yield a *typical instance* of the existential.

2. the result of such a process, whether valid or not. Thus, *Fa* is an instantiation of $(\exists x)Fx$.

instantiation rule, *n.* (*Logic*) any syntactic ELIMINATION RULE specifying the conditions for the validity of the derivation of a particular statement from a general one by eliminating a specified QUANTIFIER and replacing the BOUND VARIABLE with a name or other referring expression. In particular, *Janet is rational* can be validly inferred by UNIVERSAL INSTANTIATION from *all women are rational*.

int, 1. *abbrev. and symbol for* INTERIOR.

2. the symbol for the INTEGRAL PART or FLOOR of a real number.

integer, *n.* **1.** also called **signed number** or **directed number**. a number that may be expressed as the sum or difference of two natural numbers; a member of the set $\{\ldots, -3, -2, -1, 0, 1, 2, 3, \ldots\}$, usually denoted $\mathbb{Z}$. The integers are the closure of the natural numbers under subtraction, and are identified with the rational numbers with denominator 1.

2. an ALGEBRAIC INTEGER. The usual integers are then referred to as *rational integers* if a distinction is necessary.

integer lattice, *n.* a subset of EUCLIDEAN *n*-SPACE that is closed under addition and subtraction; most usually, such a set constructed as all integral combinations of *n* linearly independent points or generators. The celebrated *Minkowski theorem* asserts that any symmetric CONVEX BODY of volume greater than $2^n d(\Lambda)$ contains a non-zero member of the lattice Λ, where $d(\Lambda)$ is the determinant of the matrix of which the rows are the coefficients of the generators.

integer part, see INTEGRAL PART.

integer programming, *n.* the extension of LINEAR PROGRAMMING in which some variables (in *mixed integer programming*) or all variables (in *pure integer programming*) are restricted to the integers; most computing problems are expressible in this form.

integrable, *adj.* **1.** (of a function) possessing a finite integral in the sense of LEBESGUE or RIEMANN INTEGRATION, or otherwise.

2. (of a differential equation or form) being the derivative of another form; EXACT.

integrability condition, *n.* (of a differential equation or form) a condition guaranteeing that it is EXACT; thus, in a simply connected region, it suffices that the form be CLOSED.

integral, *n.* **1.** (for a given function, $f(x)$) the limit, evaluated by the INTEGRAL CALCULUS, of the sum of the rectangular ELEMENTS $f(x)\delta x$, where δx is a subinterval of a partition of an interval of values of the independent variable, and the limit is taken as the number of subintervals tends to infinity and the length of each tends to zero. This is sometimes expressed in terms of an infinite sum of INFINITESIMAL quantities. The area, lightly shaded in Fig. 200, between the curve $y = f(x)$ and the *x*-axis, and between $x = a$ and $x = b$ is the limit of the sum of such rectangular elements of area, of which

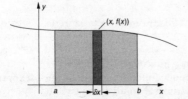

Fig. 200. The **integral** from *a* to *b* is the limit of the sum of elements $f(x)\delta x$.

one is represented by the darker shaded rectangle in the Figure; the bases of these rectangles form a partition of $[a, b]$. This is the DEFINITE INTEGRAL of $f(x)$ from a to b, and is written

$$\int_a^b f(x) \, dx.$$

An INDEFINITE INTEGRAL or ANTIDERIVATIVE, written

$$\int f(x) \, dx,$$

is any other function of x whose DERIVATIVE is $f(x)$, and is unique up to a constant; for example, the indefinite integral of ax^n is

$$\frac{ax^{n+1}}{n+1} + c,$$

where c is a constant; however, this *constant of integration* is often omitted, and this practice has been followed in the list of common indefinite integrals in Appendix 2. Integrals are formally defined in terms of UPPER and LOWER DARBOUX SUMS; and the definite and indefinite integrals are related by the FUNDAMENTAL THEOREM OF CALCULUS. These concepts can be extended by the ITERATED INTEGRAL to MULTIPLE INTEGRATION. See also RIEMANN INTEGRAL, LEBESGUE INTEGRAL.

2a. the symbolic representation of a definite or indefinite integral.

b. the symbol $\int$.

3. a solution of a DIFFERENTIAL EQUATION.

4. *adj.* of or pertaining to INTEGERS. An *integral polynomial* is a polynomial with rational integer coefficients.

integral calculus, *n.* the branch of CALCULUS concerned with the evaluation of INTEGRALS and their application to the determination of areas, volumes, etc., and to the solution of DIFFERENTIAL EQUATIONS. Compare DIFFEREN-TIAL CALCULUS.

integral convolution, see CONVOLUTION.

integral curvature, see GAUSSIAN CURVATURE.

integral domain, *n.* **1.** (*Number theory*) a non-zero COMMUTATIVE RING with a (multiplicative) IDENTITY in which the (additive) ZERO has no ZERO DIVI-SORS; for example the integers, but not the integers mod m unless m is prime. A ring is an integral domain if and only if $ax = ay$ implies $x = y$, and if an integral domain is finite it is a FIELD.

2. (*Algebra*) a non-zero commutative ring with no zero divisors, whether or not it possesses a multiplicative identity.

Compare DIVISION RING, EUCLIDEAN DOMAIN. See CANCELLATION LAW.

integral equation, *n.* a FUNCTIONAL EQUATION involving integrals; similarly an *integro-differential equation* involves both integrals and derivatives. See also VOLTERRA EQUATION.

integral part or **integer part,** *n.* (of a real number) the largest integer not greater than the given number; thus the integral part of 3.42, written [3.42] or int 3.42, is 3, and of −3.42 is −4. Compare FRACTIONAL PART.

integral polynomial, *n.* a POLYNOMIAL with integer coefficients.

integral rational, *adj.* (of an expression, function, equation, etc.) express-ible as a ratio of POLYNOMIALS with integral coefficients.

integral reduction formulae, see REDUCTION FORMULAE.

integral test, *n.* a test for the CONVERGENCE of an infinite series $\sum_n f(n)$, where f is continuous, non-negative, and MONOTONE decreasing on $[1, \infty)$, by virtue of the fact that the series converges if and only if the IMPROPER INTEGRAL

$$\int_1^\infty f(x)$$

converges. For example,

$$\sum_{n=1}^\infty \frac{n}{n^2 + 1}$$

diverges, since

$$\int_1^\infty \frac{x}{x^2 + 1} \, dx = \lim_{b \to \infty} \frac{1}{2} \ln \frac{b^2 + 1}{2},$$

which tends to infinity.

integral transform, *n.* an operator such as the FOURIER TRANSFORM, LAPLACE TRANSFORM, and MELLIN TRANSFORM that is defined in terms of some integral KERNEL;

$$\int_A k(x, y) f(x) \, dy$$

is an integral transform of f where A is a range fixed for that transform, and k is the kernel. For example, the Fourier transform has kernel

$$\frac{1}{2\pi} \exp ixy$$

and range $[0, \infty]$. These transforms are often used in the solution of partial DIFFERENTIAL EQUATIONS.

integrand, *n.* the function that is integrated in an INTEGRAL.

integrate, *vb.* to find the INTEGRAL of a function. See also ANTIDIFFERENTIATE.

integrating factor or **Euler multiplier,** *n.* a function $m(x, y)$ by which a DIFFERENTIAL EQUATION of the form

$$y'f(x, y) - g(x, y) = 0$$

can be multiplied so that the resulting differential equation is EXACT.

integration, *n.* **1.** the operation by which an INTEGRAL is calculated.

2. the study of integration and integrals. See also FUNDAMENTAL THEOREM OF CALCULUS, RIEMANN INTEGRAL.

integration by parts, *n.* the INTEGRATION of a product of two differentiable functions, by the rule

$$\int F(x) \, G'(x) \, dx = F(x) \, G(x) - \int F'(x) \, G(x) \, dx$$

where $F'(x)$ and $G'(x)$ are the first derivatives of $F(x)$ and $G(x)$. For DEFINITE INTEGRALS, the corresponding formula is

$$\int_a^b F(x) \, G'(x) \, dx = \left[F(x) \, G(x) \right]_a^b - \int_a^b F'(x) \, G(x) \, dx,$$

where, for any function ϕ,

$$\left[\phi(x) \right]_a^b = \phi(b) - \phi(a).$$

For example,

$$\int x \sin x \, dx = x(-\cos x) - \int \frac{d}{dx}(x). \, (-\cos x) = -x \cos x + \sin x.$$

integro-differential equation, *n.* a FUNCTIONAL EQUATION involving both integrals and derivatives.

intended interpretation, *n.* (for a FORMAL CALCULUS) a mathematical or other theory that is an INTERPRETATION of the given calculus when the latter was devised in order to display the formal properties of the theory. For example, the intended interpretation of first-order predicate calculus is a fragment of natural language.

intension, *n.* (*Logic*) the set of characteristics or properties by which the REFERENT or referents of a given expression is determined; the sense of an expression that determines its reference in every POSSIBLE WORLD, as opposed to its actual reference. For example, the intension of *prime number* may be *having no non-trivial integral factors*, whereas its EXTENSION would be the set { 2, 3, 5, 7, 11, … }.

intensional, *adj.* (*Logic*) incapable of explication solely in terms of the set of objects to which the given concept is applicable, and so requiring, for example, explanation in terms of meaning or understanding. Compare EXTENSIONAL. See also OPAQUE.

intercept, *n.* **1.** a point at which two figures intersect.

2. the point at which a given figure intersects with a specified coordinate axis, or the value of that coordinate at that point. If a straight line has intercepts with the axes at $(a, 0)$ and $(0, b)$, as in Fig. 201, then the *intercept form* of its equation is $x/a + y/b = 1$, that is $bx + ay = ab$.

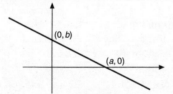

Fig. 201. The **intercepts** (sense 2) of the line $ay + bx = ab$ with the axes.

3. a line segment lying between two points of intersection of the line with a given figure.

interest, see COMPOUND INTEREST.

interior (abbrev. **int**), *n.* **1.** the set of points lying strictly between the end points of an interval.

2. the set of all INTERIOR POINTS of a given set in a TOPOLOGICAL SPACE; the largest OPEN subset of the set, equal to the union of all the open sets contained in the given set. The interior of A is denoted IntA or A°. Compare CLOSURE.

3. (of a simple closed curve) the bounded region determined by the curve, as ensured by the JORDAN CURVE THEOREM.

interior angle, *n.* **1.** any angle formed by two adjacent sides of a polygon and lying in its interior, as all the marked angles in Fig. 202 opposite. In a REFLEX polygon such as that shown, at least one interior angle will be the larger of the angles formed by a pair of sides.

Fig. 202. All the **interior angles** (sense 1) are indicated.

2. (with respect to a TRANSVERSAL of a pair of lines) either of the pairs of angles that lie on the same side of the transversal and between the other two lines, such as the marked angles in Fig. 203.

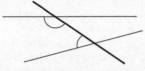

Fig. 203. **Interior angles** (sense 2) with respect to a transversal.

interior penalty function, see PENALTY FUNCTION.

interior point, *n.* **1.** (*Topology*) a point of a given set contained within an OPEN subset of it. For example, 0.5 is an interior point of the real interval [0, 1], whereas 0 is not.

2. (of a conic in Euclidean geometry) a point not on the conic through which no tangent to the conic runs. Compare EXTERIOR POINT.

intermediate value theorem, *n.* another name for BOLZANO'S THEOREM.

internal and external division (in fixed proportion), *n.* (*Geometry*) the construction of points D and E such that D divides a given line segment AB internally (see INTERNAL DIVISION) and E divides AB externally (see EXTERNAL DIVISION) in the same proportion though with opposite signs. Thus the ratio of the DIRECTED lengths

$$\frac{|AD|}{|DB|} = -\frac{|AE|}{|EB|}$$

is the same positive number λ, which uniquely determines the points D and E; in addition, determination of either D or E uniquely fixes the other. So, for example, as in Fig. 204, if D divides AB internally in the ratio 2:1, then E divides it externally in the ratio 6:3. See also HARMONIC POINTS, MEAN AND EXTREME PROPORTION.

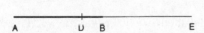

Fig. 204. D **internally divides** AB, and E externally divides it, in the same proportion.

internal direct product, *n.* the INTERNAL DIRECT SUM of groups under multiplication.

internal direct sum, *n.* **1.** a sum of a finite or infinite number of IDEALS of a RING, NORMAL SUBGROUPS of a group under addition, or submodules of a MODULE over a ring, such that the ideals or submodules $J_1 \ldots J_n$ satisfy the condition that, for all $0 \le i \le n$,

$$J_i \cap \sum_{\substack{j=1 \\ i \ne j}}^{n} J_j = \{0\}.$$

The sum is then written

$$J_1 \oplus J_2 \oplus \ldots \oplus J_n.$$

See also INTERNAL DIRECT PRODUCT, EXTERNAL DIRECT SUM.

internal division (**of a segment**), *n.* (*Euclidean geometry*) the construction of a point D between the endpoints A and B of the given line segment AB, in such a way that the ratio of the DIRECTED lengths is positive, that is

$$\frac{|AD|}{|DB|} = \lambda > 0.$$

For example, in Fig. 204, D divides AB internally in the ratio 2:1. Compare EXTERNAL DIVISION, INTERNAL AND EXTERNAL DIVISION.

internal energy, *n.* (*Continuum mechanics*) the energy inherent in a body other than the KINETIC ENERGY; formally, a MEASURE on the subsets of a BODY that is ABSOLUTELY CONTINUOUS with respect to MASS. See INTERNAL ENERGY DENSITY.

internal energy density, *n.* (*Continuum mechanics*) the SCALAR FIELD, $\phi(x, t)$, unique almost everywhere, such that the INTERNAL ENERGY of a SUB–BODY of a BODY with DENSITY ρ is given by the integral

$$\int \phi(x, t)\, \rho \; dv$$

over the volume of the CONFIGURATION of the sub-body at time t.

interpolate, *vb.* **1.** to estimate a value of a function between the values already known. Compare EXTRAPOLATE.

2. to approximate a function by a simpler one with given *interpolating* values or derivative values, for example, by SPLINE FITTING or LAGRANGE INTERPOLATION.

interpreted, *adj.* (of a FORMAL CALCULUS) endowed with an INTERPRETATION.

interpretation, *n.* (*Logic*) **1.** an allocation of significance to the terms of a purely FORMAL SYSTEM; an assignment of RANGES to the BOUND VARIABLES, denotations to the individual constants, and extensions to the predicate constants of a formal calculus, so that the closed well-formed formulae of the calculus have a truth-value in the interpretation.

2. a function from the formal calculus to POSSIBLE WORLDS that yields such an assignment.

See also STRUCTURE.

interquartile range, *n.* (*Statistics*) the difference between the first and third QUARTILES, that is, between the value of a variable below which lie 25% of the population, and that below which lie 75%; a measure of the spread of a distribution. See also PERCENTILE.

intersect, *vb.* **1.** (of two geometric figures) to have points in common.

2. (of two sets) to have a non-empty INTERSECTION; not to be DISJOINT.

intersection, *n.* **1.** (*Geometry*) a point or set of points common to two or more figures.

2. also called (*archaic*) **product.** (*Set theory*) **a.** the set of elements that are members of two or more given sets, written $S \cap T$ or $\bigcap_i S_i$, often read as *cap.* In the VENN DIAGRAM of Fig. 205 opposite, the sets S and T are represented by the regions shaded respectively vertically and horizontally; their intersection is the region shaded in both directions.

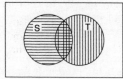

Fig. 205. The **intersection** S ∩ T is cross-hatched.

b. the binary operation that forms such a set from two given sets.

c. more generally, the intersection over any COLLECTION of subsets

$$C = \{ C_\alpha : \alpha \in A \}$$

of a given set X is the set of which the elements lie in each member of the collection. This is denoted

$$\bigcap_{\alpha \in A} C_\alpha$$

or ∩C. If the collection of subsets is empty, ∩∅ = X while ∪∅ = ∅; so, to avoid the seeming paradox that the intersection is not contained in the union it is sometimes useful to replace the universe X by ∪C before computing the intersection denoted ∩*. This only changes the intersection over an empty collection and ensures that ∩*C is always a subset of ∪C.

intersection graph, *n.* (of a family of sets) a GRAPH in which a pair of vertices are joined by an edge if and only if the intersection of the sets represented by those vertices is non-empty.

interval, *n.* **1.** the set containing all real numbers or points between two given real numbers or points. A *closed interval*, [a, b], includes the endpoints and so is a set of the form $\{x : a \leq x \leq b\}$, where *a* and *b* are the ENDPOINTS, while an *open interval*, (a, b), does not, and so is of the form $\{x : a < x < b\}$. On the real line, the *half-open* (or *half-closed*) interval from *a* to *b* is written [a, b) or (a, b], where square brackets indicate the inclusion of the endpoint (corresponding to a weak inequality in the set notation), and round parentheses its exclusion (corresponding to a strong inequality); reversed brackets,]a, b[, are also used with the latter meaning. Unbounded intervals are also written in this notation; for example [a, ∞) is the unbounded interval $x \geq a$, which is regarded as closed, while (a, ∞) is the open interval $x > a$; the real line, ℝ = (∞, −∞), is both open and closed. **2. order interval.** more generally, a subset of a POSET that contains all elements BETWEEN any two elements of the subset.

interval analysis, *n.* (*Numerical analysis*) an analysis of the error in a numerical calculation, in which an INTERVAL is estimated within which the true value of the computation lies.

interval estimate, *n.* another term for CONFIDENCE INTERVAL.

interval graph, *n.* an INTERSECTION GRAPH based on a family of INTERVALS of a partial order, especially the real line.

interval of convergence, see CIRCLE OF CONVERGENCE.

interval of existence, *n.* the real interval, I, on which a function is a solution of a system of ORDINARY DIFFERENTIAL EQUATIONS for all values in the interval. See also SOLUTION CURVE.

293

interval scale, *n.* (*Statistics*) a scale of measurement of data in accordance with which the differences between the values can be quantified in absolute but not relative terms and for which any zero is merely arbitrary. For example, dates are measured on an interval scale since differences can be measured in years, but no sense can be given to a ratio of dates. Compare ORDINAL SCALE, RATIO SCALE, NOMINAL SCALE.

interview problem or **secretary problem,** *n.* (*informal*) the sampling problem in which a succession of applicants is being randomly screened for a job. The interviewer must pick the chosen candidate as soon as he or she is seen, or else lose that candidate irrevocably, and the question becomes that of determining how many to sample. Asymptotically, the best strategy is to interview a proportion of $1/e$ of the applicants and then to select the next candidate better than any yet seen.

in the large, *adj.* another term for GLOBAL.

in the small, *adj.* another term for LOCAL.

into, *prep.* **1.** divided into, in expressions such as '5 into 2 won't go'.
2. less commonly, multiplied by; for example, in elementary arithmetic '$(x-a)(x+a) = (x^2-a^2)$' is sometimes read as '$(x-a)$ into $(x+a)$ is (x^2-a^2)'.
3. (*also as adj.*) (of a function) having an IMAGE contained within (a given set). For example, the function $y = x^2$ maps the integers into the non-negative integers; in some usages, a one-to-one mapping is both into and ONTO. See INJECTIVE.

intransitive, *adj.* (of a relation) having the property that if it holds between one element and a second, and between that second and a third, then it must fail to hold between the first and third. For example, the relation *mother of* is intransitive, as no-one's mother can be his or her mother's mother, or the successor function, since no integer can be its own successor. An intransitive relation is never TRANSITIVE for any set of arguments whatever. Compare NON–TRANSITIVE.

intrinsic, *adj.* (of a property) relating only to the bearer of the property and not to the space in which it is embedded; thus a left-hand and a right-hand glove are intrinsically distinct but extrinsically homeomorphic. Being COMPACT is an intrinsic property of a subset of a topological space, while being an OPEN SET is not. Compare EXTRINSIC.

introduction rule, *n.* (*Logic*) any syntactic RULE OF INFERENCE stating the conditions under which a formula or statement containing a specified operator may be validly derived from others that may not contain it. For example, *conjunction introduction* is the rule in sentential calculus that permits the inference of the conjunction A & B from the two separate premises A and B; *universal introduction* is the rule in predicate calculus that permits (under certain conditions) the inference of the universally quantified statement $(x)Fx$ from an arbitrary instance Fa. Compare ELIMINATION RULE.

intuitionism, *n.* **1.** the philosophical doctrine that mathematics cannot intelligibly comprehend the properties of most infinite sets, and that only what can be shown to be provable by FINITARY METHODS can be justifiably asserted. For example, it follows classically from the identity

$$\left((\sqrt{2})^{\sqrt{2}}\right)^{\sqrt{2}} = 2$$

that there exists a pair of irrational numbers α and β for which α^β is rational: classically we are entitled to assert that either $(\sqrt{2})^{\sqrt{2}}$ is rational or it is irrational; if it is the latter, then the given identity proves the result, and if it is not, it is rational and so is itself an instance of the required existential proposition. This classical argument fails intuitionistically without a CONSTRUCTIVE determination that $(\sqrt{2})^{\sqrt{2}}$ is irrational (which it is by the GELFOND–SCHNEIDER THEOREM).

2. the reconstruction of mathematics in accordance with this doctrine. See FORMALISM, LOGICISM, FINITISM.

intuitionist or **intuitionistic,** *adj.* (of a logical or mathematical system) constructed in accordance with the principles of INTUITIONISM, so that, for instance, the law of EXCLUDED MIDDLE does not hold since we may have no justification for asserting either a given statement or its negation. The principle of DOUBLE NEGATION, DILEMMA, and the equivalence of *something is not-F* with *not everything is F* also fail in intuitionist systems. For example, consider the infinite sequence of which all the elements are 0 except for a 1 for the element of which the index corresponds to the first completed instance of the sequence of digits '0123456789' in the decimal expansion of π; this sequence does not occur in the first 2^{25} digits of π, and until it is found to occur or, less plausibly, is proven (constructively) not to, the sequence is not well-formed intuitionistically, but clearly converges to zero classically. Compare CLASSICAL.

invalid, *adj.* (of an argument) not VALID; having a conclusion that does not follow from the premises, i.e. that may be false when the premises are all true. Consequently, one shows an argument to be invalid either by constructing a COUNTEREXAMPLE or else by devising a POSSIBLE WORLD in which the premises are true but the conclusion is false.

invariable, *n.* a CONSTANT.

invariance of domain theorem, *n.* the result that if U is an open set in Euclidean *n*-space, and if another set E in Euclidean *n*-space is its homeomorphic image, then E is open. It is a consequence of this result that the dimension *n* is a topological invariant. See also OPEN MAPPING THEOREM.

invariant, *n.* **1.** an entity, property, relationship, etc. that is unaltered by a particular transformation of coordinates. For example, Euclidean distance is invariant under rotation. An *invariant subgroup* for the INNER AUTOMORPHISMS of a finite group is a NORMAL SUBGROUP. An *invariant subset* S of a mapping or multifunction *T* is such that *T*(S) lies in S. A *translation invariant metric* on a group or vector space assigns the same distance to a pair of points and any translate of the pair.

2. a quantity or set of quantities that characterize the relevant properties of an object within a given class.

3. (*Fractal theory*) (of a subset, E, of Euclidean *n*-space) such that, for CONTRACTIONS $\psi_1, \ldots, \psi_m$,

$$E = \bigcup_{i=1}^{m} \psi_i(E);$$

for example, the Cantor ternary set is invariant under

$$\psi_1 = \frac{x}{3}, \quad \psi_2 = \frac{2x+1}{3}.$$

invariant subspace conjecture, *n.* the conjecture that every continuous linear operator on an infinite-dimensional HILBERT SPACE possesses a non-trivial INVARIANT proper LINEAR subspace. This clearly holds for matrices in more than one dimension as is shown by consideration of the span of an eigenvector. It has recently been shown that the analogous conjecture is false in the Banach space of absolutely summable sequences.

inverse, *adj.* **1a.** (of an element) related to a given element of a set on which an operation is defined in such a way that its product with the given element under that operation is the IDENTITY element, *e.* See also INVERSE MATRIX.

b. (*as substantive*) an inverse element. For example, the inverse of a function under composition is its INVERSE FUNCTION; the *additive inverse* of x is $-x$, and the *multiplicative inverse* (or RECIPROCAL) of x is $1/x$. For non-commutative non-associative operations, an element, x, may have distinct LEFT and RIGHT inverses, x_l and x_r respectively, such that $x_l x = e = x x_r$.

2. (of a relationship) relating two quantities in such a way that an increase in one corresponds to a decrease in the other and vice versa; thus speed is in INVERSE PROPORTION to the time taken to cover a given distance. Compare DIRECT.

3. (*as substantive*) another term for RECIPROCAL.

inverse correlation, see CORRELATION.

inverse function, *n.* a function, usually written f^{-1}, whose DOMAIN and RANGE are respectively the range and domain of a given function, f, and under which the image, y, of an element, x, is the element of which x was the image under the given function, that is $f^{-1}(x) = y$ if and only if $f(y) = x$; the function whose composition with the given function is the IDENTITY FUNCTION. In order that the inverse should have a unique value for each argument, and so be properly a function, the given function must be INJECTIVE. For example, the extraction of positive square roots, $\sqrt{x}$, is the inverse of squaring, x^2, since $y = x^2$ if and only if $x = \sqrt{y}$, and

$$\sqrt{(x^2)} = (\sqrt{x})^2 = x;$$

however, without the restriction to positive values, the square root function on the domain of real numbers does not have an inverse. See also LEFT INVERSE, RIGHT INVERSE.

inverse function theorem, *n.* the special case of the IMPLICIT FUNCTION THEOREM stating that a continuously differentiable function on Euclidean space possesses an INVERSE FUNCTION in a neighbourhood of a point at which it has a NON–SINGULAR JACOBIAN. See also LIUSTERNIK'S THEOREM.

inverse image, *n.* another term for COUNTERIMAGE.

inverse image set, *n.* **1. weak** or **lower inverse image set.** the set of which each element has an image under a given CORRESPONDENCE F with a non-zero intersection with a given set B; the set

$$\{ x : F(x) \cap B \neq \varnothing \},$$

denoted $F^-(B)$ or $F^w(B)$.

2. strong or **upper inverse image set.** the set of which each element has an image under a given CORRESPONDENCE F that is contained in a given set B; the set

$$\{\, x : F(x) \subset B \,\},$$

denoted by either $F^+(B)$ or $F^s(B)$.

If the given correspondence is ONE-TO-ONE, then both inverse image sets may be identified with the COUNTERIMAGE.

inverse matrix, *n.* the matrix INVERSE of a given matrix with respect to matrix multiplication. Such a matrix, denoted A^{-1}, exists precisely when A is NON-SINGULAR, and it is then computable by GAUSSIAN ELIMINATION (in practice), or (in principle) by the ADJOINT formula

$$A^{-1} = \frac{\text{adj}(A)}{\det(A)} \, ;$$

that is, the transpose of the matrix of the quotients of its cofactors by its determinant. See also CRAMER'S RULE.

inverse proportion, **inverse variation**, **reciprocal variation**, or **indirect variation,** *n.* the relationship that holds between two variable quantities when an increase by a certain multiple in one brings about or is associated with a decrease by the same factor in the other. If quantities *a* and *b* are in indirect proportion then $a_1/a_2 = b_2/b_1$. Compare DIRECT PROPORTION.

inverse variation, *n.* another term for INVERSE PROPORTION.

inversion, *n.* (*Euclidean geometry*) a transformation under which the image of each point, P, on a half-line through the origin, O, of a given circle or norm sphere with radius *r*, is the point, Q, on the same half-line such that $|OP|\,|OQ| = r^2$. If one adjoins an IDEAL POINT at infinity, this become a bijection of the extended *inversive plane* $P°$ that preserves circles.

inversive, *adj.* pertaining to INVERSION and the inversive plane.

inverted factor analysis, (*Statistics*) see FACTOR ANALYSIS.

invertible, *adj.* **1.** possessing an INVERSE.

2. (of an IDEAL, X, in a ring, R) such that in an OVER-RING of R, if

$$A = \{\, s \in S : \ sX \subseteq R \,\}, \quad B = \{\, s \in S : Xs \subseteq R \,\}$$

then $AX = XB = R$.

inviscid fluid, *n.* (*Continuum mechanics*) a fluid that is not a VISCOUS FLUID.

involute, *n.* any of a family of curves that is the locus of the free end of a thread that is held taut as it is unwound from a given curve, the EVOLUTE; the thread is thus always tangential to the given curve, as shown in Fig. 206, and its NORMALS are tangents to the evolute. The members of this family of curves are normal TRAJECTORIES of the tangents of the given curve.

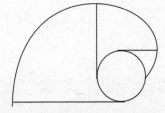

Fig. 206. The **involute** of a circle.

involution, *n.* **1.** an operator of which the square is the identity operator; if $f(f(x)) = x$ for all x, then f is an involution. See also B*–ALGEBRA.

2. the algebraic operation of raising some expression to a specified power. Compare EVOLUTION.

involve, *vb.* to raise to a specified power.

irrational, *adj.* **1.** not expressible as a ratio of integers.

2. (of an equation, etc.) containing one or more variables in irreducible RADICAL form or raised to a fractional power, such as $\sqrt{(x^2 + 1)} = x^{5/3}$.

3. (*as substantive*) an IRRATIONAL NUMBER.

irrational number, *n.* any REAL or COMPLEX NUMBER that cannot be expressed as the ratio of two integers, such as π, e, $\log_2 3$ or $\sqrt{2}$. The imaginary number i is irrational, but is a GAUSSIAN INTEGER. An irrational is usually defined as a limit of a sequence of rational numbers or as a partition of the rationals into those greater and those less than it. See DEDEKIND CUT.

irreducible, *adj.* **1.** (of a polynomial over a field) unable to be factorized into polynomials of lower degree over the same field; for example, $(x^2 + 1)$ is irreducible over the reals, but can be factorized as $(x + i)(x - i)$ over the complex numbers. Such a polynomial is also PRIME.

2. (of a non-zero element in an INTEGRAL DOMAIN) not expressible as a product of two non-units: if $a = bc$ then either b or c is a UNIT. All other non-zero elements are either units or are *reducible*. The number 5 is irreducible in the integers but since $5 = (2 + i)(2 - i)$, it is reducible as a GAUSSIAN INTEGER. By a theorem of Fermat, this is also true of any prime that is congruent to 1 MODULO 4, being expressible as a sum of two integral squares. Compare PRIME.

3. (of a RADICAL) unable to be expressed as a rational expression; for example, $\sqrt{(x + 1)}$ is irreducible.

irreflexive, *adj.* (of a relation) not holding between any member of its domain and itself. For example, *is distinct from* is an irreflexive relation, since nothing can be distinct from itself. An irreflexive relation is never REFLEXIVE on any subdomain whatever. Compare NON-REFLEXIVE.

irrotational, *adj.* **1.** (of a vector field) having zero CURL.

2. (*Continuum mechanics*) (of a MOTION) having zero VORTICITY. In this case the VELOCITY is given by the GRADIENT of a SCALAR FIELD. An irrotational motion is equivalent to a POTENTIAL FLOW.

isochoric motion, *n.* (*Continuum mechanics*) a MOTION in which volume is preserved, and so the determinant of the DEFORMATION GRADIENT is identically one.

isocline, see DIRECTION FIELD.

isodiametric, *adj.* having all DIAMETERS of equal length.

isogon, *n.* an equiangular polygon.

isogonal or **isogonic,** *adj.* having, making, or involving equal angles. An *isogonal transformation* in the plane preserves angles and has the form

$$x' = a_1 x + b_1 y + c_1, \quad y' = a_2 x + b_2 y + c_2$$

where either $a_1 = b_2$ and $a_2 = -b_1$, or else $a_2 = b_1$ and $a_1 = -b_2$.

isolated ordinal, *n.* (*Set theory*) an ORDINAL with an immediate PREDECESSOR, and so not a LIMIT ORDINAL.

isolate, *vb.* (of a ROOT of an equation) to determine an interval within which it is the only root.

isolated point, *n.* a point that is not a CLUSTER POINT of a given set; a point that has a PUNCTURED NEIGHBOURHOOD that does not intersect the given set. Any countable subset of a metric space has an isolated point. See PERFECT SET.

isolated singularity, *n.* (of a complex function) a point in a PUNCTURED NEIGHBOURHOOD of which the function is ANALYTIC and at which the function is not continuous. The singularity is either REMOVABLE, if

$$\lim_{z \to a} (z - a)f(z) = 0,$$

or is ESSENTIAL or is a POLE.

isometric, *adj.* (of a method of projecting a figure in three dimensions) having the three axes equally inclined and all lines drawn to scale. For example, *isometric graph paper* has three axes and is ruled in equilateral triangles, as shown in Fig. 207 overleaf.

isometry, *n.* an AUTOMORPHISM or HOMEOMORPHISM that preserves distance.

isomorphic, *adj.* **1.** related by an ISOMORPHISM.

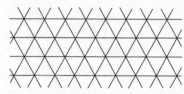

Fig. 207. **Isometric** graph paper.

2. (of MODELS) related by a ONE-TO-ONE CORRESPONDENCE that preserve relations, functions, and constants. Isomorphic models have the same CARDINALITY. Compare ELEMENTARILY EQUIVALENT.

3. (of NORMAL SERIES of a group) having normal factors that are isomorphic under a suitable permutation.

isomorphism, *n.* a ONE–TO–ONE CORRESPONDENCE between the elements of two or more sets that preserves the structural properties of the domain; a BIJECTIVE HOMOMORPHISM. For example, the bijective mapping from the integers to those rationals that are of the form $n/1$ retains the order of the elements, and the sum or product of the images of two elements equals the image of their sum or product; the logarithmic function is an isomorphism between the reals under addition and the positive reals under multiplication, since it is a bijection under which $x = yz$ if and only if $\ln x = \ln y + \ln z$. See also AUTOMORPHISM, DUAL ISOMORPHISM. Compare EPIMORPHISM, MONOMORPHISM.

isomorphism theorems, *n.* a set of results concerning the FACTOR SPACES, where they exist, of sets on which a particular algebraic structure is imposed, such as groups, rings or modules. For groups, the *first isomorphism theorem* states that $S/\ker\theta$ is isomorphic to the image of the HOMOMORPHISM θ, where $\ker\theta$ is the kernel of θ; the *second isomorphism theorem* states that $(G/N)/(K/N)$ is isomorphic to (G/K), where K and N are NORMAL SUBGROUPS of G with N contained in K; and the *third isomorphism theorem* states that $(AB)/B$ is isomorphic to $A/(A \cap B)$. Analogous results hold for rings and modules with respect to ideals and submodules respectively, in which case the third isomorphism theorem states that $(A + B)/B$ is isomorphic to $A/(A \cap B)$. The first isomorphism theorem is also known as the *homomorphism theorem*, in which case the second and third may be referred to as the first and second, but there is no agreed convention. Other isomorphism theorems are SCHREIER'S REFINEMENT THEOREM and the JORDAN–HÖLDER THEOREM.

isoperimetric figure, see ISOPERIMETRIC PROBLEMS.

isoperimetric problems, *n.* extensions of the classical VARIATIONAL problem of finding which of all plane figures with equal perimeter (*isoperimetric figures*) has greatest area. See also DIDO'S PROBLEM.

isosceles, *adj.* **1.** (of a triangle) having two sides of equal length, and the angles opposite those sides equal, as the triangle ADE in Fig. 208 below.

2. (of a trapezium) having the two non-parallel sides of equal length, such as the quadrilateral ABCD in Fig. 208.

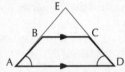

Fig. 208. An **isosceles** triangle (ADE) and an isosceles trapezium (ABCD).

isotone, *adj.* (of a function between ordered spaces) order-preserving; MONOTONE INCREASING. Compare ANTITONE.

isotropic, *adj.* (of a tensor) having components that remain unchanged under an arbitrary change of basis.

iterated integral, *n.* a repeated integral of the form that arises when evaluating a multiple integral by FUBINI'S THEOREM;

$$\int_a^b dx \int_c^d dy \int_e^f f(x, y, z) \, dz,$$

in which one first integrates with respect to z, then y, then x, treating the remaining variables as parameters. Compare MULTIPLE INTEGRATION.

iterated kernel, see KERNEL.

iterated map, *n.* a MAP T of a space X to itself with associated iteration $x_{n+1} = T(x_n)$, for $n = 0, 1, 2, \ldots$ See LOGISTIC MAP, MANDELBROT SET.

iterated series, *n.* a double or MULTIPLE SERIES of the form

$$\sum_{n=0}^{\infty} \sum_{m=0}^{\infty} a_{n,m}.$$

iteration, *n.* **1.** the repeated application of a mathematical procedure, where each step is applied to the output of the preceeding. See, for example, JACOBI ITERATION, NEWTON-RAPHSON METHODS, SIMPLEX METHODS.

2. each successive step of such an iterative process. See also RECURSION.

iterative, *adj.* involving or characterized by ITERATION; another word for RECURSIVE.

j

j, (conventionally printed in bold type) **1.** a UNIT VECTOR, usually one in the direction of the *y*-axis. Compare **i**, **k**. See DIFFERENTIAL OPERATOR.
2. a unit QUATERNION.

j, (*Engineering and physics*) *symbol for* the IMAGINARY NUMBER, $\sqrt{-1}$.

J, (*Mechanics*) *symbol for* JOULE.

Jackson–Bernstein theorems, *n.* a set of theorems concerning CHEBYSHEV APPROXIMATION that describe the best possible convergence rates for classes of functions that have given smoothness properties, and, conversely, deduce smoothness properties from the rates. For example, if *f* is a LIPSCHITZ FUNCTION satisfying the Lipschitz condition of order α, the error in best approximation by a trigonometric polynomial of degree *n* is at worst $O(n^{-\alpha})$, and if this rate holds for $0 < \alpha < 1$, the function has Lipschitz order α. Correspondingly, a function is entire if and only if the error tends to zero better than geometrically.

Jacobi, Karl Gustav Jacob (1804–51), German mathematician who, independently of Abel, made dramatic progress in the theory of ELLIPTIC FUNCTIONS, number theory (in which his work was admired by Gauss), differential determinants, and various branches of analysis, geometry, and mechanics. He was first an Extraordinary and then an Ordinary Professor at the University of Königsberg (1827–42), and later at Berlin, and died of smallpox in 1851.

Jacobian or **Jacobian determinant,** *n.* a function derived from a set of *n* simutaneous equations in *n* variables, of which the value at any point is the $n \times n$ determinant of the JACOBIAN MATRIX of partial derivatives of those equations evaluated at that point. If

$$u_j = f_j(x_1, x_2, \ldots, x_n),$$

this is generally written

$$\frac{\partial(f_1, f_2, \ldots, f_n)}{\partial(x_1, x_2, \ldots, x_n)} = \frac{\partial(u_1, u_2, \ldots, u_n)}{\partial(x_1, x_2, \ldots, x_n)}.$$

If the Jacobian is non-zero, the equations have a non-trivial solution.

Jacobian elliptic functions, *n.* a class of ELLIPTIC FUNCTIONS which arise by inversion of ELLIPTIC INTEGRALS, and of which the three most basic are *sn*, *cn*, and *dn*. For MODULUS *k*, the function $y = sn(z) = sn(z, k)$ is obtained by inversion of

$$z = \int_0^y \left(1 - t^2\right)^{-\frac{1}{2}} \left(1 - k^2 t^2\right)^{-\frac{1}{2}} dt,$$

and the other two functions are defined by

$$sn^2z + cn^2z = 1$$
$$k^2sn^2z + dn^2z = 1$$
$$cn(0) = dn(0) = 1.$$

Then sn is doubly periodic with periods $4K(k)$ and $2iK(k')$. Here K is the COMPLETE ELLIPTIC INTEGRAL of the first kind and k' is the complementary MODULUS. These functions may be rigorously defined in terms of THETA FUNCTIONS.

Jacobian matrix, *n.* (of m functions of n variables) the $m \times n$ matrix of which the ith row is the vector of the PARTIAL DERIVATIVES of the ith function, evaluated at the point in question. For example, the Jacobian matrix of

$$f_1(x, y) = x^2 + xy + y^2, \quad f_2(x, y) = x^2y^2$$

at $(1, 2)$ is

$$\begin{bmatrix} 2x + y & x + 2y \\ 2xy^2 & 2y^2x \end{bmatrix}(1, 2) = \begin{bmatrix} 4 & 5 \\ 8 & 4 \end{bmatrix}.$$

Jacobi equation, *n.* a DIFFERENTAL EQUATION of the form

$$(a_1 + b_1x + c_1y)(x\,dy - y\,dx) - (a_2 + b_2x + c_2y)\,dy + (a_3 + b_3x + c_3y)\,dx = 0,$$

closely related to the BERNOULLI EQUATION.

Jacobi iteration, *n.* the iterative method of solving a linear matrix system, $\mathbf{Ax} = \mathbf{b}$, by repeatedly solving

$$\mathbf{x}_{NEW(i)} = \frac{1}{a_{ii}} \left[\mathbf{b}_i - \sum_{i \neq j} a_{ij} \mathbf{x}_{OLD(j)} \right]$$

This assumes $a_{ii} \neq 0$, and will not always converge even then. Compare METHOD OF SUCCESSIVE DISPLACEMENTS.

Jacobi's method, *n.* a method for solving a first-order PARTIAL DIFFERENTIAL EQUATION of the form

$$F\left(x_1, x_2, \dots, x_n, \frac{\partial z}{\partial x_1}, \frac{\partial z}{\partial x_2}, \dots, \frac{\partial z}{\partial x_n}\right) = 0 ,$$

in which the dependent variable z does not occur explicitly; it is an elaboration of CHARPIT'S METHOD for equations with more than two independent variables.

Jacobi symbol, see LEGENDRE SYMBOL.

Jacobson radical, see RADICAL.

James' theorem, *n.* the result that a subset of a BANACH SPACE is COMPACT in the WEAK TOPOLOGY if and only if it is weakly closed and every continuous linear functional achieves its supremum on the set. In particular, a Banach space is REFLEXIVE if and only if every continuous linear functional supports the unit ball at some point.

jamming, *n.* (*Numerical analysis*) the behaviour of a minimization algorithm whose CONVERGENCE is not GLOBAL, and which may therefore converge to a point that is not even a STATIONARY POINT of the function in question. See also ZIGZAGGING.

Jensen inequality, *n.* **1.** the inequality

$$f(\sum \lambda_i a_i) \leq \sum \lambda_i f(a_i)$$

whenever $\sum \lambda_i = 1$ and $\lambda_i \geq 0$, that is satisfied by all convex combinations of points in the domain of a convex function, and is equivalent to the convexity of the function f.
2. various corresponding integral inequalities, often phrased in terms of PROBABILITY.

(Named after the Danish algebraist, analyst, and engineer, *Johan Ludvig William Valdemar Jensen* (1859–1925), pioneer of convex function theory.)

join, *n.* the binary operator the value of which is the SUPREMUM of a pair of elements in a LATTICE; given a pair of elements x and y of the lattice, their join, written $x \vee y$, is the element m such that $m \geq x$, $m \geq y$, and there is no $n < m$ that has the same relations to x and y. Compare MEET.

joint density function, *n.* (*Statistics*) a function of two or more RANDOM VARIABLES from which can be obtained a single PROBABILITY that all the variables will jointly take specified values or fall in specified intervals. For example, given random variables X and Y on a space with probability P,

$$P[(X,Y) \in B] = \int_B f(x, y) \, dx \, dy$$

defines a joint probability on $\mathbb{R}^2$ with density f, with respect to LEBESGUE MEASURE; and for all BOREL SETS on the line

$$P(X \in B) = \int_B f(x) \, dx$$

where $f(x) = \int f(x, y) \, dy$.

joint variation, *n.* a variation in which the values of one variable depend upon those of several other variables, as in $z = kxy$, for some constant k.

Jordan algebra, *n.* a non-associative ALGEBRA, exemplified by $n \times n$ matrices with the product of A and B defined as AB + BA. (Named after the French analyst, topologist, group theorist and algebraist, *Marie-Ennemond Camille Jordan* (1838–1922), whose work on permutation groups and the theory of equations directed attention to the significance of the work of Galois.)

Jordan block or **Jordan factor,** *n.* a square matrix of the form

$$J(\lambda) = S + \lambda I$$

where λ is a scalar and S is SUPER DIAGONAL with entries equal to one. Note that a one-dimensional block may have any field entry.

Jordan content, *n.* another term for CONTENT.

Jordan contour or **Jordan curve,** *n.* another term for SIMPLE CLOSED CURVE.

Jordan curve theorem, *n.* the fundamental theorem, important in complex analysis, that states that any SIMPLE CLOSED CURVE (*closed Jordan arc*) has an inside and an outside, and thus that the plane is divisible into two disjoint regions each with the curve as its boundary. More generally, the *Jordan–Brouwer separation theorem* proves that any topological $(n-1)$-sphere separates n-dimensional Euclidean space into two parts.

Jordan decomposition, *n.* the expression of of a SIGNED MEASURE as the difference of two non-negative measures. Often these two measures are required to be mutually SINGULAR, in which case they uniquely define the *positive* and *negative parts* of the signed measure.

Jordan elimination or **Gauss–Jordan elimination,** *n.* a variant of GAUSSIAN ELIMINATION in which the elimination is *completed*; that is, it is continued until the ECHELON REDUCED matrix is replaced by an identity matrix (in the square case).

Jordan factor, *n.* another term for JORDAN BLOCK.

Jordan–Holder theorem, *n.* the theorem that any two COMPOSITION SERIES of a finite group are isomorphic. See also ISOMORPHISM THEOREM, SCHRIER REFINEMENT THEOREM.

Jordan inner measure or **Jordan inner content,** see INNER JORDAN MEASURE.

Jordan measure, *n.* another term for CONTENT.

Jordan normal form, *n.* (of a matrix) a matrix SIMILAR to a given one and expressible in BLOCK DIAGONAL MATRIX form as

$$\text{diag}[\, J(\lambda_1),\dots,J(\lambda_k),\dots,J(\lambda_n)\,]$$

where each $J(\lambda_k)$ is a JORDAN BLOCK. This representation is unique up to permutation of the blocks. If the scalar field is not algebraically closed a more complicated form results. Over the reals it is necessary to introduce factors corresponding to (real) irreducible quadratic factors of the characteristic polynomial.

Jordan outer measure or **Jordan outer content,** see OUTER MEASURE.

Jordan product, *n.* the symmetric product $\frac{1}{2}[AB + BA]$ of two matrices or operators A and B. Compare LIE PRODUCT.

joule (symbol **J**), *n.* (*Mechanics*) the standard unit of WORK, equal to the work done by a FORCE of one NEWTON when its point of application moves a distance of one METRE; its units, expressed in terms of the fundamental units of the SYSTEME INTERNATIONAL, namely kilograms (kg), metres (m), and seconds (s), are $\text{kg}\,\text{m}^2/\text{s}^2$.

Jourdain's paradox, *n.* the version of the LIAR PARADOX formulated by the French mathematician Jourdain in 1913: on one side of a card is written 'The statement on the other side of this card is true', and on the other side is written 'The statement on the other side of this card is false'; if the former is true, so is the latter which states that the former is false, whence the latter is false so that the former is true. It should be noted that there is no paradox if both statements read 'The statement on the other side is false'; it then simply follows that one of these statements is false. As with other semantic paradoxes, a strict distinction between OBJECT LANGUAGE and METALANGUAGE, or a THEORY OF TYPES blocks the construction of the paradox.

Julia set, *n.* the boundary between two sets of points in the complex plane determined by the behaviour of a given function f that maps the plane to itself. Starting with an initial value $z = a$, the sequence obtained by iteration of the function,

$$a,\ f(a),\ f(f(a)),\ f(f(f(a))),\ \dots,$$

may either stay BOUNDED, or take arbitrarily large values. For a given function, this property divides the complex plane into two regions, and the Julia set associated with that function is the boundary between the two regions. The most studied family of functions is $f(z) = z^2 - c$, where c is a complex parameter. The Julia set varies in the complex plane, and is CON-NECTED for some values of as *c*, and disconnected for others; the best-known geometric FRACTAL, the *Mandelbrot set,* is the boundary between these two sets of values of the parameter *c*. See also ATTRACTOR.

jump or **saltus,** *n.* the absolute value of the differences between the left- and right-hand limits of a given function (usually a function of bounded variation) at an interior point of its domain:

$$|f(x+) - f(x-)|.$$

At an endpoint, the appropriate limit is compared to the function value; that is, the jump at the endpoints of $]a, b[$ is respectively

$$|f(a+) - f(a)| \quad \text{and} \quad |f(b) - f(b-)|.$$

jump discontinuity, *n.* a point at which a function (usually a function of bounded variation) is discontinuous by virtue of a JUMP; for example,

$$f(x) = \begin{cases} 1 & x < 2 \\ 2 & x \geq 2, \end{cases}$$

of which the graph is shown in Fig. 209, has a jump discontinuity at $x = 2$.

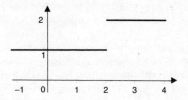

Fig. 209. **Jump discontinuity**.

k

k, 1. *abbrev. for* KILO-, used in symbols for multiples of the physical units of the SYSTEME INTERNATIONAL.

2. (conventionally printed in bold type) **a.** a UNIT VECTOR, usually one in the direction of the *z*-axis. Compare **i**, **j**. See DIFFERENTIAL OPERATOR.

b. a unit QUATERNION.

K, *symbol for* the SPECIAL FUNCTION, called the COMPLETE ELLIPTIC INTEGRAL OF THE FIRST KIND, for which the formula is

$$K(k) = \int_0^{\pi/2} (1 - k^2 \sin^2 \theta)^{-\frac{1}{2}} \, d\theta;$$

this gives the period of a pendulum.

Kakeya's problem, *n.* the insoluble problem of finding the plane set of minimum area that permits a unit line to be reversed by being moved continuously in the set. (Named after the Japanese analyst and geometer, *Soichi Kakeya.*)

Kakutani fixed point theorem, *n.* the theorem that every CORRESPONDENCE Γ that maps a compact convex subset C of a locally convex space into itself, with a closed GRAPH and convex non-empty images (that is, $\Gamma(x)$ is a non-empty convex subset of C for all $x \in C$), has a FIXED POINT, $x \in \Gamma(x)$. This extends BROUWER'S THEOREM.

Kantorovitch inequality, *n.* the inequality, valid for any POSITIVE DEFINITE MATRIX P, and any non-zero vector x, that

$$\frac{\langle x, x \rangle \langle x, x \rangle}{\langle x, \mathrm{P}x \rangle \langle x, \mathrm{P}^{-1}x \rangle} \geq \frac{4mM}{(m+M)^2},$$

where m and M are respectively the smallest and the largest EIGENVALUES of P. This is useful in estimating convergence rates for DESCENT METHODS.

Karmarkar method, *n.* a POLYNOMIAL TIME ALGORITHM for LINEAR PROGRAMMING that is based on projective transformations and is of the nature of interior PENALTY FUNCTION METHODS. This appears to compete well with algorithms based on SIMPLEX METHOD for certain classes of programs. Compare KHACHIYAN ALGORITHM.

Karush–Kuhn–Tucker theorem, see KUHN–TUCKER CONDITIONS.

Katetov's interpolation theorem, *n.* the result that if a lower SEMICONTINUOUS real-valued function f, majorizes an upper semicontinuous real-valued function g, both with domain a NORMAL TOPOLOGICAL SPACE, then there is a continuous function, h, such that

$$f(x) \geq h(x) \geq g(x).$$

This theorem has TIETZE'S EXTENSION THEOREM as an immediate corollary.

Kelvin's circulation theorem, *n.* (*Continuum mechanics*) the theorem that the CIRCULATION of a frictionless fluid is invariant in time in the presence of CONSERVATIVE FORCES. The theorem also holds for a fluid of uniform density that is not COMPRESSIBLE.

Kepler's laws of planetary motion, *n.* (*Mechanics*) the laws stating that each planet travels in an ELLIPSE with the sun at one FOCUS, that the RADIUS VECTOR from the sun to the planet sweeps out equal areas in equal times, and that the squares of the PERIODIC times of the planets are proportional to the cubes of the SEMI–AXES of the elliptic orbit. These laws, with a slight correction to the last, are deducible from Newton's law of GRAVITY.

kernel, *n.* **1.** the set of elements of the domain of a MAPPING that have the IDENTITY ELEMENT of the range as their image. The kernel of a HOMO-MORPHISM from a GROUP to another is a NORMAL SUBGROUP; the kernel of a ring homomorphism is the counterimage of zero and is an IDEAL. See ISOMORPHISM THEOREM. See also NULL SET.

2a. a function the product of which with a given function is integrated to produce an INTEGRAL TRANSFORM of the given function, thus

$$g(t) = \int_a^b K(s, t) \, f(s) \, dt$$

has kernel K.

b. iterated kernels. the sequence of kernels defined by $K_0 = K$ and

$$K_{n+1}(s, t) = \int_a^b K(s, r) \, K_n(r, t) \, dr.$$

These arise in the solution of integral equations, the sum

$$- \sum_{n=0}^{\infty} \lambda^n K_{n+1}(s, t)$$

being called a *resolvent kernel.* See VOLTERRA'S INTEGRAL EQUATION, FOURIER TRANSFORM, LAPLACE TRANSFORM.

3. (for a regular BOREL MEASURE) another term for SUPPORT.

ket, see ANGLE BRACKETS.

kg, (*Mechanics*) *symbol for* KILOGRAM.

Khachiyan algorithm, *n.* the first POLYNOMIAL TIME ALGORITHM for LINEAR PRO-GRAMMING based on the ELLIPSOIDAL METHOD. This is primarily of theoretical interest, since while implementations of the SIMPLEX METHOD may exhibit exponential convergence in pathological cases, in practice it performs very well. Compare KARMARKAR METHOD.

kilo- (symbol **k**), *prefix denoting* a multiple of 10^3 of the physical units of the SYSTEME INTERNATIONAL.

kilogram (symbol **kg**), *n.* (*Mechanics*) the standard unit of MASS; one of the fundamental units of the SYSTEME INTERNATIONAL. It is currently defined as the mass of a certain cylinder of platinum and iridium kept at Sèvres in France, but is likely to be redefined in terms of the mass of atoms of some element.

kilowatt (symbol **kW**), *n.* (*Mechanics*) the POWER of 1000 WATTS.

kinematics, *n.* the study of the motion of bodies without reference to mass or force. Compare DYNAMICS.

kinematic viscosity, *n.* (*Continuum mechanics*) the ratio of the VISCOSITY to the DENSITY of a BODY.

kinetic energy, *n.* **1.** (*Mechanics*) the quantity $\frac{1}{2}\, m\mathbf{v}.\mathbf{v}$ for a PARTICLE of MASS *m* and VELOCITY **v**, and the sum of the individual kinetic energies for a set of particles. For a RIGID BODY the kinetic energy is

$$\tfrac{1}{2}mv^2 + \tfrac{1}{2}A\omega_1^2 + \tfrac{1}{2}B\omega_2^2 + \tfrac{1}{2}C\omega_3^2,$$

where *m* is the MASS, *v* is the speed of the CENTRE OF MASS, A, B, C are the PRINCIPAL MOMENTS OF INERTIA, and ω_1, ω_2, ω_3 are the ANGULAR VELOCITIES about the PRINCIPAL AXES.

2. (*Continuum mechanics*) more generally, the integral

$$\frac{1}{2}\int \rho(\mathbf{v}.\mathbf{v})\ d\mathrm{v},$$

over the volume of the current CONFIGURATION of a SUB–BODY where **v** is the velocity of a point with DENSITY ρ.

kinetic friction, see FRICTION.

kinetics, *n.* **1.** another name for DYNAMICS.

   2. the branch of MECHANICS, including both dynamics and KINEMATICS, concerned with the study of bodies in motion.

   3. the branch of dynamics that excludes the study of bodies at rest.

kite, *n.* a CONVEX QUADRILATERAL with two pairs of equal adjacent sides, as in Fig. 210. Compare DELTOID.

Fig. 210. **Kite**.

kittygory, see CATEGORY.

Klein, Christian Felix (1849–1925), German mathematician who introduced the ERLANGEN PROGRAMME. He attempted to classify and unify geometry by means of a general group-theoretic definition (*Kleinian geometry*) and was influential in the study of ELLIPTIC FUNCTIONS. He was Professor at Erlangen from 1872 to 1875, then between 1880 and 1913 was Professor at Munich, Leipzig, and Göttingen, and played a leading role in the scientific and mathematical community of his time. He also wrote about mathematics for the general public, and founded a mathematical encyclopædia, which he continued to supervise until his death.

Klein bottle, *n.* a closed surface with only one side and no interior; if it is cut in half lengthwise, two MÖBIUS STRIPS result. It cannot be realized in three–dimensional space, but a model can be formed by inserting the narrower end of an open tapered tube through the wider end and then sealing the ends to one another, as shown in Fig. 211.

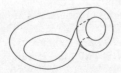

Fig. 211. **Klein bottle**. See main entry.

Klein four-group, n. the smallest non-cyclic group; it consists of four PERMUTATIONS

$$e = (1), \quad a = (12)(34), \quad b = (13)(24), \quad c = (14)(23),$$

and is also realizable as the matrix group

$$\pm \begin{bmatrix} 0 & 1 \\ 1 & 0 \end{bmatrix}, \quad \pm \begin{bmatrix} 1 & 0 \\ 0 & 1 \end{bmatrix}.$$

knapsack problem, n. the INTEGER PROGRAMMING problem of maximizing

$$\sum_{k=1}^{n} c_k x_k,$$

subject to

$$\sum_{k=1}^{n} w_k x_k \leq K,$$

where the variables x_i are non-negative integers. The problem can be viewed as trying to maximize the value of goods packed into a knapsack. This is an NP–COMPLETE PROBLEM.

Knaster fixed point theorem, see TARSKI FIXED POINT THEOREM.

knot, n. a closed space curve, formed by looping and interlacing a piece of string and joining the ends together; a set of points in Euclidean three-space topologically equivalent to a circle. Any two knots are thus topologically equivalent, but it may not be possible to transform one into the other by continuous DEFORMATION, that is, without breaking the string.

knot theory, n. the branch of geometry that studies the possible types of KNOTS and their deformability properties, first studied by Gauss and his student Listing.

knot polynomial, n. a polynomial associated with a KNOT, such that if two knots are equivalent, then their knot polynomials are equal.

Koch curve, n. the FRACTAL curve of HAUSDORFF DIMENSION $\ln 4 / \ln 3$, whose generator is formed by erecting an equilateral triangle on the middle third of a straight line; the curve is then defined as the limit, Γ_ω, of the succession of recursively defined curves, where Γ_{n+1} is constructed from Γ_n by replacing each line segment with the generator.

Kolmogorov–Smirnoff test, n. (*Statistics*) the GOODNESS OF FIT test of which the test STATISTIC is

$$\max_{i=1\dots n} \left| \frac{i}{n} - F(X_i) \right|,$$

where F is the hypothesized distribution, and the X_i are in increasing order. This test statistic is independent of F when the hypothesis is true; tables for the statistic for different values of n at various significance levels are available.

Kolmogorov space, n. another name for T_0-SPACE (see T–AXIOMS).

Kolmogorov's three series theorem, n. (*Probability*) the result that a series of INDEPENDENT RANDOM VARIABLES, $\{ f_n \}$, converges pointwise (almost everywhere) if and only if three numerical series converge. When the variables are uniformly bounded, this reduces to verifying the convergence of the sums of the EXPECTATIONS, $E(f_n)$, and of the VARIANCES, $\sigma^2(f_n)$. (Named after the Russian probabilist, topologist, and analyst, *Andrei Nikolaevich Kolmogorov* (1903–87), who in 1933 established the set theoretic foundations of probability theory.)

Königsberg bridge problem, *n.* the problem of determining whether the seven bridges over the River Pregel at Königsberg in Prussia could be traversed once and only once, starting at any point in the town and returning to the same point (an EULERIAN CIRCUIT of the town); the positions of the bridges are shown schematically in Fig. 212. This was a foundational problem in GRAPH THEORY, and it was shown to be impossible by Euler's demonstration that an Eulerian circuit exists in a connected graph if and only if each vertex has even degree.

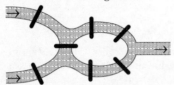

Fig. 212. **Königsberg bridge problem.** See main entry.

König's theorem, *n.* the result that the number of edges in a maximum MATCHING of a BIPARTITE GRAPH equals the number of vertices in a minimal *covering* of the graph, that is, in a subset K of vertices such that each edge has a vertex in K.

Korovkin theorems, *n.* a class of theorems regarding UNIFORM APPROXIMATION of which the most basic is that if a sequence $\{L_n\}$ of non-negative linear operators on $C[a,b]$ is such that $L_n(x^k)$ converges uniformly to x^k for $k = 0, 1, 2$, then $L_n(f)$ converges uniformly to f for all continuous functions. The BERNSTEIN POLYNOMIALS are such a sequence, which establishes the WEIERSTRASS APPROXIMATION THEOREM. A second basic result is that if a sequence $\{L_n\}$ of non-negative linear operators on the periodic function space $C[-\pi, \pi]$ is such that $L_n(f)$ converges uniformly to f for 1, sin, and cos, then $L_n(f)$ converges uniformly to f for all continuous functions. The *Fejer polynomials* defined by

$$L_n(f) = \frac{1}{n} \sum_{k=0}^{n-1} S_n(f),$$

which compute the CESARO SUMS of the partial FOURIER SUMS for f, satisfy the hypotheses. Compare DIRICHLET'S CONDITION.

Krasnoselskii theorem, see STAR.

Krein–Milman theorem, *n.* the theorem stating that a COMPACT CONVEX subset of a locally convex space is the closed convex HULL of its EXTREME POINTS.

Kronecker, Leopold (1823–91), German algebraist, number theorist, and philosopher of mathematics, who was captivated by number theory while at school. He earned his doctorate while managing his family business until he retired, aged 30, to pursue mathematics full-time. In his heated correspondence with Weierstrass, he was the first critic of non-constructive existence proofs in classical analysis, sought the reconstruction of all of mathematics in terms of the positive integers.

Kronecker delta, *n.* the function, usually denoted δ_{ij}, of two variables that takes the value 1 when $i = j$ and is zero otherwise; thus the $m \times m$ identity matrix may be written as $[\,\delta_{ij}\,]_m$. The *generalized delta function* has k subscripts and k superscripts, and is zero unless the subscripts are a PERMUTATION of the superscripts, in which case it is the SIGNATURE of the permutation, and has value 1 for an even permutation and -1 for an odd permutation; when $k = 3$, it is sometimes written ε_{ijk}.

Kronecker's lemma, *n.* the result that if

$$\sum_{n=1}^{\infty} \frac{a_n}{n}$$

converges, then

$$\frac{1}{N} \sum_{n=1}^{N} a_n$$

tends to zero as N tends to infinity.

Kuhn–Tucker conditions or **Karush–Kuhn–Tucker theorem,** *n.* (*Optimization*) a result extending the LAGRANGE METHOD OF MULTIPLIERS to constraints defined by equalities and inequalities. It ensures that, given an appropriate CONSTRAINT QUALIFICATION, any point x_0 that minimizes $f(x)$ subject to

$$g_1(x) \le 0,\ldots, g_n(x) \le 0,\ \ g_{1+n}(x) = 0,\ldots, g_{m+n}(x) = 0$$

will occur at a STATIONARY POINT of the *Lagrangian*

$$\mathrm{L}(x, \lambda) = f(x) + \sum \lambda_i g_i$$

where the summation is over the BINDING CONSTRAINTS with $g_k(x_0) = 0$, and where the MULTIPLIERS corresponding to inequality constraints are non-negative. This holds in particular if there are no equality constraints and the inequality constraints satisfy SLATER'S CONDITION. Compare FRITZ JOHN CONDITION.

kurtosis, *n.* (*Statistics*) a measure of the concentration of a distribution about its mean, especially

$$\mathrm{B}_2 = \frac{m_4}{(m_2)^2},$$

where m_2 and m_4 are, respectively, the second and fourth MOMENTS of the distribution about the mean. A NORMAL DISTRIBUTION or any other with $B_2 = 3$ is called *mesokurtic*; if $B_2 < 3$ it is *platykurtic*, and if $B_2 > 3$ it is *leptokurtic*. Compare SKEWNESS.

kW, (*Mechanics*) *symbol for* KILOWATT.

L or l, *symbol for* 50 in ROMAN NUMERALS.

label, *vb.* to associate some entity with each NODE of a TREE in order to distinguish it from all others. For example, as shown in Fig. 213, a rooted binary tree can be labelled with binary numbers of which the digits represent the direction taken at each successive node on the route from the root to the associated vertex.

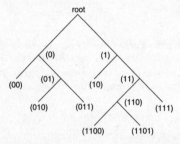

Fig. 213. A binary tree **labelled** by the route to each node.

labelled tree, *n.* a TREE in which some entity LABELS each NODE .

lacunary series, *n.* a series in which the terms with non-zero coefficients are widely separated; in particular, a POWER SERIES in which there is a subsequence of the powers $\{\lambda_n\}$ corresponding to non-zero coefficients that satisfies

$$\lambda_{(n_k + 1)} > (1 + \varepsilon)\lambda_{n_k}$$

for some $\varepsilon > 0$, so that infinitely often there is a geometric jump in the size of the power.

lacunary value, *n.* a value not taken by an ANALYTIC FUNCTION in its domain of definition. By PICARD'S THEOREM, a non-constant ENTIRE function has at most one lacunary value.

Lagrange, Joseph Louis, Comte de (1736–1813), Italian-born French mathematician and physicist who, in 1755, became Professor of Geometry at the Royal Artillery School at Turin, where he founded the Academy of Sciences. He later succeeded Euler as Director of Mathematics at the Berlin Academy of Science, returning to France after the death of Frederick the Great. His treatment of ISOPERIMETRIC PROBLEMS laid a foundation for the calculus of variations, and he made significant contributions to many other branches of mathematics, including probability theory, number theory, the theory of equations, and the foundations of group theory.

Lagrange form of the remainder, *n.* the expression of the REMAINDER, or the error, of a TAYLOR SERIES; that is, the difference between a function and its nth Taylor polynomial. For example, $R_n(f, a)$ is the difference

$$f(a + h) - \left\{ f(a) + \frac{f'(a)}{1!} h + \ldots + \frac{f^{(n)}(a)}{n!} h^n \right\}$$

expressed as

$$R_n(f, a) = \frac{f^{(n+1)}(c)}{(n+1)!} h^{n+1}$$

for some c between a and $a + h$. Correspondingly, the *Cauchy form of the remainder* is given by

$$R_n(f, a) = \frac{f^{(n+1)}(a + th)}{n!} h^{n+1}(1 - t)^n,$$

where t is strictly between 0 and 1.

Lagrange interpolation formula, n. the formula for the unique polynomial of degree at most n that INTERPOLATES a given real function, $f(x)$, at $n + 1$ distinct points, $x_0 \ldots x_n$:

$$P_n(x) = \sum_{k=0}^{n} L_k(x) f(x_k),$$

where

$$L_k(x) = \prod_{\substack{j=0 \\ j \neq k}}^{n} \frac{x - x_j}{x_k - x_j}.$$

The *Hermite interpolation* produces the unique polynomial of degree at most $2n - 1$ that takes prescribed function and first derivative values at n distinct points.

Lagrange method of multipliers, n. the method of reducing a CONSTRAINED OPTIMIZATION PROBLEM to an unconstrained problem by the addition of LAGRANGE MULTIPLIERS. The method is valid if the GRADIENTS of the constraints are LINEARLY INDEPENDENT at x_0. If, in addition, the OBJECTIVE FUNCTION is CONVEX and the constraints are AFFINE, then any such stationary point is optimal. See also KUHN–TUCKER CONDITION.

Lagrange multipliers, n. a sequence of real numbers λ_i, such that a point x_0, that minimizes $f(x)$ subject to $g_1(x) = 0, \ldots, g_m(x) = 0$ will be a STATIONARY POINT of the *Lagrangian*

$$L(x, \lambda) = f(x) + \sum_{i=1}^{m} \lambda_i g_i(x).$$

Lagrange problem, n. the special case of a control theory BOLZA PROBLEM in which there is no extra term depending on the endpoints.

Lagrange's equations, n. (*Mechanics*) the equations

$$\frac{d}{dt} \left\{ \frac{\partial L}{\partial \dot{q}_i} \right\} = \frac{\partial L}{\partial q_i} \quad [1 \leq i \leq m]$$

for a set of PARTICLES with LAGRANGIAN L, that are described by the GENERALIZED COORDINATES $q_1, \ldots, q_m$, where m is the number of DEGREES OF FREEDOM of the set. These equations are a consequence of HAMILTON'S PRINCIPLE OF LEAST ACTION.

Lagrange's identity, n. (*Differential equations*) the identity

$$vLu - uL^*v = \langle k(v), P. k(u) \rangle',$$

where $\langle \, , \, \rangle$ is the inner product,

$$Lu = \sum_{k=0}^{n} p_{n-k} \, u^{(n-k)}$$

and

$$L^*v = \sum_{k=0}^{n} (-1)^k \, [p_{n-k} \, u]^{(k)}$$

are respectively a linear differential operator and its scalar ADJOINT, u' and $u^{(k)}$ indicate differentiation, $k(u)$ is the vector of derivatives $(u,...,u^{(n-1)})$, and the CONCOMITANT MATRIX, P, has coefficients $p_{i,j}$ defined by

$$p_{i,j} = \sum_{h=1}^{n-j+1} (-1)^{h-i} \binom{h-1}{i-1} \, [p_{n-h+1}]^{(h-i)}$$

for $i \leq n-j+1$, and is zero otherwise. If one obtains a solution w to the adjoint equation $L^*v = 0$, Lagrange's identity allows one to use this solution as an INTEGRATING FACTOR in the sense that any solution of $Lu = 0$ must satisfy

$$\langle k(w), \text{P.} \, k(u) \rangle = c$$

for an arbitrary constant c, and this equation is of lower order.

Lagrange's linear equation, n. the partial DIFFERENTIAL EQUATION of the form

$$\sum_{i=1}^{n} P_i(x_1, x_2, ..., x_n)\frac{\partial z}{\partial x_i} = R(x_1, x_2, ..., x_n),$$

where P_i and R are differentiable functions. If the equation is integrable, its general solution is $\phi(u_1, u_2, ..., u_n) = 0$, where ϕ is an arbitrary function, and the u_i are independent solutions of the SIMULTANEOUS DIFFERENTIAL EQUATIONS

$$\frac{dx_1}{P_1} = \frac{dx_2}{P_2} = \, ... \, = \frac{dx_n}{P_n}.$$

A Lagrange's linear equation may also have a SPECIAL INTEGRAL.

Lagrange's theorem, n. **1.** (*Number theory*) the theorem that every positive integer is expressible as the sum of four SQUARE NUMBERS. Jacobi subsequently gave an exact formula for $\mathbf{r}_4(k)$, the number of such representations. Three squares suffice unless n is of the form $4^n(8k+7)$. For example, $7 = 4 + 1 + 1 + 1$; $12 = 4 + 4 + 4$. See also WARING'S PROBLEM.

2. (*Group theory*) the theorem that the ORDER of every finite GROUP G is equal to the product of the order of any subgroup H of G and the number of cosets of H in G (the index of H in G), that is,

$$|G| = |H| \, . \, |G:H|.$$

Lagrangian, n. **1.** (*Mechanics*) the difference between the KINETIC ENERGY and the POTENTIAL ENERGY of a set of PARTICLES.
2. see LAGRANGE MULTIPLIERS.

Lagrangian description, n. another term for MATERIAL DESCRIPTION.

lambda or **lambda operator,** n. the ABSTRACTION operator, λ, that is defined by the rules of LAMBDA ABSTRACTION and LAMBDA CONVERSION, and is studied by LAMBDA CALCULUS.

lambda abstraction, *n.* the operation, in LAMBDA CALCULUS, that forms an expression denoting a function (or, equivalently, a class or predicate) from any expression whatever, by prefixing to it the ABSTRACTION operator, λ, and a BOUND VARIABLE; the notatifupλx[...] can be most generally read as 'is an x such that ...'. For example, lambda abstraction applied to a binary relation such as x *is to the left of* y (here written Lxy), yields

$\lambda x[Lxy]$, the predicate ... *is to the left of* y;

$\lambda y[Lxy]$, the predicate x *is to the left of* ...; and

$\lambda xy[Lxy]$, the relation ... *is to the left of* ...

lambda calculus, *n.* (*Logic*) the study of ABSTRACTION in terms of the operator LAMBDA (λ). This formulation is equivalent to COMBINATORIAL LOGIC, and was originated by the American computer pioneer Alonzo Church in 1940; it was important in the development of computer languages.

lambda conversion, *n.* the rule,

$$\lambda x[Fx](a) = F(a)$$

for the APPLICATION of the LAMBDA operator, λ; that is, the application to a of the abstracted predicate ... *is an* x *such that* Fx, is equivalent to Fa. More generally,

$$\lambda x[... x ...](a) = ... a ... ;$$

in particular, $\lambda x[x](a) = a$, and $\lambda x[y](a) = y$.

Lambert series, *n.* (*Number theory*) a series of the form

$$F(x) = \sum_{n=1}^{\infty} f(n) \frac{x^n}{1 - x^n} ,$$

where $F(x)$ is said to be the *generating function* of $f(n)$.

lamina, *n.* a surface with infinitesimal, or uniform positive but negligible, thickness.

language, *n.* an INTERPRETED system; a formal calculus endowed with a SEMANTIC interpretation. See FORMAL LANGUAGE.

Laplace, Pierre Simon, Marquis de (1749–1827), French analyst, probability theorist, and physicist, who was originally educated by the charity of neighbours, and is often regarded as the greatest exponent of celestial mechanics since Newton, proving that planetary perturbations do not disturb but maintain the stability of the solar system. He also proved that respiration is a form of combustion. His standing survived the fall of successive regimes: he was for six weeks Napoleon's Minister of the Interior, later became Chancellor of the Senate, was ennobled both under the Empire and by Louis XVIII, and was elected President of the *Académie Française*.

Laplace equation, *n.* the partial differential equation

$$\frac{\partial^2 U}{\partial x^2} + \frac{\partial^2 U}{\partial y^2} + \frac{\partial^2 U}{\partial z^2} = 0,$$

that involves the LAPLACIAN, or its analogue in n dimensions, that is satisfied by electromagnetic, gravitational, and other potentials. See also DIRICHLET'S PROBLEM.

Laplace transform, *n.* the INTEGRAL TRANSFORM

$$g(y) = \int\limits_0^\infty e^{-xy} f(x)\, dx,$$

which may be viewed as a form of the FOURIER TRANSFORM and is used in the solution of differential equations. See also KERNEL.

Laplacian or **Laplace operator,** *n.* the DIFFERENTIAL OPERATOR

$$\frac{\partial^2}{\partial x^2} + \frac{\partial^2}{\partial y^2} + \frac{\partial^2}{\partial z^2}$$

or its analogue in *n* dimensions; the sum of the second partial derivatives of a function. The Laplacian of *f* is written $\nabla^2 f$ (referred to as *nabla squared*, DEL squared, or GRAD squared) or Δf. See also DIRICHLET'S PROBLEM.

large, *adj.* see GLOBAL.

latent root, **characteristic root**, **characteristic value**, or **eigenvalue**, *n.* **1.** (for a matrix, A) a root, λ, of the CHARACTERISTIC EQUATION

$$\det(A - tI) = 0,$$

of the matrix.

2. more generally, for a linear operator A, a solution λ, of the equation

$$AX = \lambda X \quad (X \neq 0).$$

latent vector, *n.* another term for EIGENVECTOR.

lateral face, *n.* any side of a POLYHEDRON other than a base.

Latin square, *n.* (*Statistics*) one of a set of square arrays of *n* rows and *n* columns, especially as used in statistics and studied in COMBINATORIAL ANALYSIS, built up from *n* different symbols in such a way that each symbol occurs exactly once in each row and column. See also OFFICER PROBLEM.

a	*b*	*c*	*d*	
c	*d*	*b*	*a*	
b	*a*	*d*	*c*	Fig. 214. **Latin square**.
d	*c*	*a*	*b*	

lattice, *n.* **1.** an algebra endowed with two binary operations, denoted $\wedge$ and $\vee$, often called MEET and JOIN, that are SYMMETRICAL and ASSOCIATIVE, and for which

$$x \wedge x = x = x \vee x$$

and

$$x \wedge (x \vee y) = x = x \vee (x \wedge y).$$

For example, the SUPREMUM and the INFIMUM of a pair of functions define a lattice, as do the INTERSECTION and the UNION of subsets of a set. See UPPER BOUND, LOWER BOUND. See also BOOLEAN ALGEBRA, PARTIAL ORDERING, INTEGER LATTICE.

2. a partially ordered set in which each pair of elements has a SUPREMUM and an INFIMUM.

latus rectum, *n.* (*pl.* **latera recta**) a chord that passes through the FOCUS, and is perpendicular to the major axis, of a CONIC. Fig. 215 opposite shows one of the latera recta of an ellipse and a hyperbola, and the latus rectum of a parabola, all of which share a vertex, where E, H, and P are foci of the respective conics.

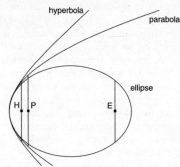

Fig. 215. **Latus rectum**.
See main entry.

Laurent expansion or **Laurent series,** *n.* (for a function that is analytic in a punctured disk or annulus) the expression of the given function as a doubly infinite POWER SERIES:

$$f(z) = \sum_{n=-\infty}^{\infty} a_n (z - a)^n.$$

For example, the Laurent expansion of

$$f(z) = \frac{e^z}{(z + 1)^2}$$

in $\{ z : 0 < |z + 1| < \infty \}$ is

$$f(z) = \frac{e}{(z + 1)^2} + \frac{e}{(z + 1)} + \sum_{k=0}^{\infty} \frac{e (z + 1)^k}{(k + 2)!}.$$

The function has a REMOVABLE SINGULARITY at a if all the negative coefficients (that is, for negative n) are zero, it has a POLE at a if only finitely many negative coefficients are non-zero, and it has an ESSENTIAL SINGULARITY at a otherwise. In the first case the series coincides with the TAYLOR SERIES. (Named after the French analyst, *Matthieu Hermann Laurent* (1841–1908).)

law, *n.* a general principle, rule or theorem, often restricted to a particular theory, such as a commutative law, or the law of large numbers

law of averages, *n.* (*informal*) the supposed expectation that a possible event is bound to occur regularly with a FREQUENCY approximating to its PROBABILITY; thus, after five heads in a row the 'law of averages' would make tails the better bet. This is in fact false. Compare LAW OF LARGE NUMBERS.

law of large numbers, *n.* (*Statistics*) the fundamental statistical result that the average of a sequence of n independent identically distributed random variables tends to their common mean as n tends to infinity, whence the relative FREQUENCY of the occurrence of an event in n independent repetitions of an experiment tends to its PROBABILITY as n increases without limit. This is often mis-stated as the supposed LAW OF AVERAGES. See also STRONG LAW OF LARGE NUMBERS, WEAK LAW OF LARGE NUMBERS.

laws of motion, see EQUATIONS OF MOTION.

lawyer paradox, *n.* the paradox, traditionally ascribed to the Greek philosopher Protagoras and of which the earliest known source is in Cicero, concerning a law instructor who agrees to waive his fees for an impecunious student in return for the student's undertaking to pay him if and only if he wins his first case; however, after qualifying, the student takes up another career, and some time later is sued by his tutor for his fees. The tutor reasons that he cannot lose: if he wins the case, the student will be bound by the ruling to pay; while if he loses the case, the student wins and so is obliged by their agreement to pay. Conversely, the student reasons that he cannot lose: if he wins the case, the ruling will be that he need not pay; while if he loses the case, he is exempted by their agreement from paying. The paradox is resolved by observing that, in fact, the agreement is subtly incoherent in that the quantifiers in the agreement permit an instantiation that introduces SELF–REFERENCE. See also LIAR PARADOX.

lcd, *abbrev. for* LEAST (or lowest) COMMON DENOMINATOR.

lcm, *abbrev. for* LEAST (or lowest) COMMON MULTIPLE.

leading coefficient, *n.* the coefficient of the term of highest degree in a polynomial in a single variable. For example, in

$$5x^3 - 3x^2 + 2x - 1,$$

the leading coefficient is 5.

leading diagonal, *n.* another term for the MAIN DIAGONAL of a matrix.

least, *adj.* (of an element in an ordering) uniquely MINIMAL; less than every other member of the set. For example, the set of subsets of a given set, ordered by inclusion, has the empty set as least element.

least (or **lowest**) **common denominator** (abbrev. **lcd**), *n.* the smallest integer, or polynomial of lowest degree, that is exactly divisible by all the DENOMINATORS of a given set of fractions, and that can therefore be used as denominator of a set of fractions equal to each of the given fractions; for example, $\frac{1}{2}$, $\frac{1}{3}$, and $\frac{1}{5}$, have lcd 30, and can be expressed as $\frac{15}{30}$, $\frac{10}{30}$, and $\frac{6}{30}$, respectively.

least (or **lowest**) **common multiple** (abbrev. **lcm**), *n.* the smallest number or quantity that is exactly divisible by all of a given set of numbers or quantities; for example, the lcm of 3, 5 and 10 is 30.

least integer function, *n.* another term for CEILING.

least residue, see RESIDUE CLASS.

least squares, *n.* a method of extrapolating from observed values of a relationship to a functional relationship and especially to a curve that best fits the known data, by seeking the solution for which the sum of the squares of the differences between the observed and the theoretical values is least.

least squares theorem, see GAUSS–MARKOV LEAST SQUARES THEOREM.

least upper bound (abbrev. **lub**), *n.* another term for SUPREMUM.

Lebesgue, Henri-Léon (1875–1941), French mathematician who generalized the RIEMANN INTEGRAL to the LEBESGUE INTEGRAL by incorporating the concepts of JORDAN MEASURE and BOREL MEASURE in that of LEBESGUE MEASURE. He established fundamental theorems of Lebesgue measure theory, applied LEBESGUE INTEGRATION to the study of FOURIER SERIES, and worked on the development of abstract measure theory. He was elected to the French Academy of Sciences and the Royal Society of London.

Lebesgue covering, *n.* a COVERING of a set in Euclidean *n*-space by countable families of BOXES.

Lebesgue decomposition of a measure, see SINGULAR (sense 4).

Lebesgue integral, *n.* the integral of a MEASURABLE function, *f*, over a subset, E, of a MEASURE SPACE with respect to its MEASURE, μ, written

$$\int_E f \, d\mu;$$

for example, with respect to Lebesgue measure, the integral of any measurable function, *f*, over the rationals, $\mathbb{Q}$,

$$\int_{\mathbb{Q}} f \, d\mu = 0.$$

A Lebesgue integral can be constructed by the device of taking the limit of integrals of SIMPLE FUNCTIONS approximating the function, and is equal to the RIEMANN INTEGRAL

$$\int_E f(x) \, dx$$

if E is a bounded interval on which *f* is bounded, with discontinuities that form a set of measure 0.

Lebesgue integration, *n.* a generalization of INTEGRATION to MEASURABLE functions over subsets of a MEASURE SPACE with respect to its MEASURE, often LEBESGUE MEASURE. See also LEBESGUE INTEGRAL.

Lebesgue measurable, see MEASURABLE.

Lebesgue measure, *n.* (on the line or in Euclidean space) the MEASURE obtained by restricting Lebesgue OUTER MEASURE to the SIGMA–ALGEBRA of MEASURABLE subsets. The one-dimensional Lebesgue measure of an interval (a, b) is $b - a$.

Lebesgue outer measure, see OUTER MEASURE.

Lebesgue's theorem, *n.* the theorem that if *f* is a LEBESGUE INTEGRABLE function on the reals, then

$$\lim_{h \to 0} \int_0^h \left| f(x + t) - f(x) \right| dt$$

is zero almost everywhere.

Leech lattice, *n.* the 24-dimensional INTEGER LATTICE corresponding to the STEINER TRIPLE SYSTEM $S(5,8,24)$ that contains 759 sets. Let K be the group generated by all OCTADS of the base set B = {1, 2, ..., 24} under the operation of SYMMETRIC DIFFERENCE; it has 2^{12} elements, consisting of the octads, their complements, 2576 dodecads, the empty set, and B. Let $C \in K$; the lattice then consists of $\bigcup_{m,C} C(m)$, where $C(m)$ is the set of all 24-dimensional integer vectors the sum of which is congruent to 0 modulo 4, and such that

$$x_i \equiv 0 \quad (\text{mod } 4) \quad \text{if } i \in C$$
$$x_i \equiv 2 \quad (\text{mod } 4) \quad \text{if } i \notin C.$$

left, *adj.* (of an operator in a non-COMMUTATIVE theory) acting on the left: I_l is a left IDENTITY if $I_l x = x$ for all x; l_x is a left INVERSE of x if $l_x x = I_l$. Compare RIGHT.

left-handed, *adj.* (of a coordinate system) having the orientation corresponding to a LEFT–HANDED TRIHEDRAL, as in Fig. 216.

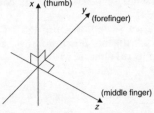

x ▲ (thumb)

y ➚ (forefinger)

(middle finger)

z

Fig. 216. **Left-handed trihedral**.
A left-handed coordinate system.

left-handed trihedral, *n.* a configuration of three directed lines that do not lie in the same plane and that have a negative TRIPLE PRODUCT. It is referred to as left-handed because the thumb and first two fingers of the left hand are in this orientation, as shown in Fig. 216 above: if the thumb is placed in the positive direction of the first line, then the angle between the other two fingers is less than π. The alternative possibility defines a RIGHT–HANDED TRIHEDRAL.

left-hand limit, *n.* the ONE–SIDED LIMIT of a function defined on an interval FROM BELOW or *from the left*; the limit where x is restricted to values less than a, written

$$\lim_{x \to a-} f(x) = f(a-).$$

Compare RIGHT–HAND LIMIT.

left-invariant, see HAAR MEASURE.

leg, *n.* either side that forms the right angle of a RIGHT–ANGLED TRIANGLE.

Legendre, Adrien-Marie (1752–1833), French mathematician who established many important results, especially in number theory and ELLIPTIC INTEGRALS, conjectured the PRIME NUMBER THEOREM and the law of QUADRATIC RECIPROCITY, and published an elementary geometry textbook. He also published works on the orbits of comets and geodesy, and was appointed to a number of official positions.

Legendre polynomials, *n.* the COMPLETE set of ORTHOGONAL polynomials, $P_i(x)$, on $[-1, 1]$, and defined by $P_0 = 1$, and

$$P_n(x) = \frac{1}{2^n n!} \frac{d^n}{dx^n} (x^2 - 1)^n.$$

The nth polynomial, for which this is *Rodrigues' formula*, solves LEGENDRE'S DIFFERENTIAL EQUATION with $p = n$.

Legendre's differential equation, *n.* the differential equation

$$(1 - x^2)y'' - 2xy' + p(p + 1)y = 0,$$

whose solution, when p is a natural number, is the pth LEGENDRE POLYNOMIAL. It is closely related to the HYPERGEOMETRIC DIFFERENTIAL EQUATION.

Legendre's identity, *n.* the identity relating COMPLETE ELLIPTIC INTEGRALS: for any $0 < k < 1$

$$K(k)E[\sqrt{(1-k^2)}] + E(k)K[\sqrt{(1-k^2)}] - K(k)K[\sqrt{(1-k^2)}] = \pi/2.$$

Legendre symbol, *n.* (*Number theory*) the symbol $(a\,|\,p)$ or $(a\,/\,p)$ defined for prime p and for a relatively prime to p: $(a\,|\,p)$ is 1 if a is a QUADRATIC RESIDUE (mod p) and is -1 otherwise. By *Euler's criterion,*

$$(a\,|\,p) \equiv a^{\left[\frac{p-1}{2}\right]} \qquad (\text{mod } p)$$

when p is an odd prime, while

$$(2\,|\,p) \equiv (-1)^{\left[\frac{p^2-1}{8}\right]} \qquad (\text{mod } p)$$

The Legendre symbol is MULTIPLICATIVE, and is extended multiplicatively to the case where p is a product of odd primes p_i (not necessarily distinct):

$$(a\,|\,p) = (a\,|\,p_1)\ldots(a\,|\,p_n).$$

In this case, the symbol is called the *Jacobi symbol*. There is a further extension due to Kronecker.

Lehmann–Scheffé theorem, *n.* the result that if $T(x)$ is a COMPLETE SUFFICIENT STATISTIC, and $S(x)$ is an UNBIASED ESTIMATE of θ, then the CONDITIONAL EXPECTATION of $S(x)$ given $T(x)$ is a UNIFORM MINIMUM VARIANCE unbiased estimate for θ.

Leibniz, Gottfried Wilhelm (1646–1716), German philosopher, logician, mathematician and scientist. His initial academic training was in the law, and he left his home town of Leipzig for ever when he was denied a doctorate of law there on the grounds that he was only 20; however, he was immediately awarded the degree at Nürnberg, where he was also offered a Chair. Refusing this offer, he instead entered the service of the Elector of Mainz, for whom he travelled widely; he was befriended by all the leading intellectuals of his age, and at this time developed the differential and integral CALCULUS independently of Newton, perfected Pascal's calculating machine, and laid the foundations of DYNAMICS. He served the Duke of Hanover as everything from a schools inspector to a mining engineer (in which capacity he postulated the molten origin of the Earth), but he also continued his researches, describing the BINARY SYSTEM and laying the foundations of TOPOLOGY. He postulated a perfect universal language of thought, in which logical relations would be transparent. He extended his duties as Historian to the House of Brunswick from genealogy to comparative linguistics and geology in pursuit of a universal history, and in this capacity he also influenced the succession of George I to Queen Anne of Britain.

Leibniz's alternating series test, see ALTERNATING SERIES TEST.

Leibniz's law, *n.* (*Logic*) **1.** the principle of the *identity of indiscernibles*, stating that two expressions satisfy exactly the same predicates if and only if they both refer to the same individual. This is sometimes written

$$(\forall F)(Fa \leftrightarrow Fb) \equiv a = b,$$

where it should be noted that the UNIVERSAL QUANTIFIER ranges over predicates and so is SECOND–ORDER; this is thus not a well-formed formula of LOWER PREDICATE CALCULUS.
2. the weaker principle of the *indiscernibility of identicals*, that if $a = b$ then whatever is true of a is true of b.

Leibniz' series, *n.* another name for GREGORY'S SERIES for π.

Leibniz' theorem, *n.* the theorem that the nth derivative, $[uv]^{(n)}$, of the product of two functions u and v is the SERIES with BINOMIAL COEFFICIENTS

$$\sum_{i=0}^{n} \binom{n}{i} u^{(i)} v^{(n-1)}.$$

lemma, *n.* a subsidiary result proved in order to simplify the proof of a required theorem.

lemniscate or **lemniscate of Bernoulli,** *n.* a closed plane curve consisting of two symmetrical loops meeting at a node, as shown in Fig. 217 below, with equation

$$(x^2 + y^2)^2 = a^2(x^2 - y^2),$$

where a is the greatest distance from the curve to the origin; its polar equation is $r^2 = a^2 \cos 2\theta$, where the node is taken as the origin. The lemniscate arises as the locus of the foot of the perpendicular from the origin to the tangents of a RECTANGULAR HYPERBOLA, and is also the locus of the vertex A of a triangle ABC when the side BC is fixed, and A is permitted to move subject to the condition that

$$4|AB| \cdot |AC| = |BC|^2.$$

Gauss was led to the theory of ELLIPTIC FUNCTIONS by his determination of the arc length of the lemniscate.

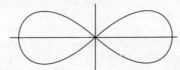

Fig. 217. **Lemniscate**.

length, *n.* **1.** (*Euclidean geometry*) a measure of the extent of a line segment that is zero only when its endpoints coincide and is otherwise positive, and that is unaffected by changing the orientation of the line. In a Cartesian coordinate system, the length of the straight line segment AB, where A is the point (a_1, a_2) and B is (b_1, b_2), is

$$\sqrt{(a_1 - b_1)^2 + (a_2 - b_2)^2},$$

as is made clear by Fig. 218a, in which the sides of the right-angled triangle are parallel to the axes; in an n-dimensional space, this generalizes to

$$\sqrt{\sum_{i=1}^{n} (a_i - b_i)^2}.$$

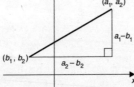

Fig. 218a. **Length.** (sense 1). See main entry.

2. arc-length. a measure of an arc of a curve, equal to the length of the straight line segment obtained by deforming it into that shape without dilatation; if the line is the graph of the function $y = f(x)$ between a and b, and $f'(x)$ is continuous, then its length is the integral

$$\int_a^b \sqrt{1 + \left(\frac{dy}{dx}\right)^2}\ dx.$$

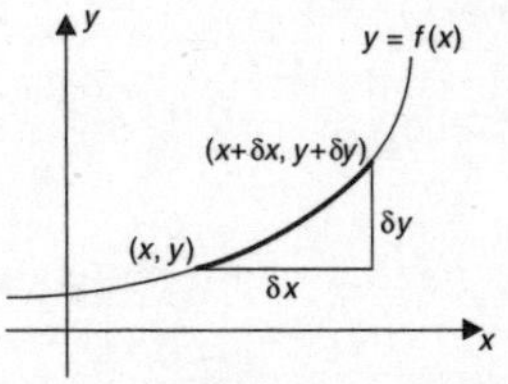

Fig. 218b. **Length**. (sense 2).
See main entry.

As is clear from Fig. 218b, the length of an element of the curve approximates to

$$\sqrt{dx^2 + dy^2} = \sqrt{1 + \left(\frac{dy}{dx}\right)^2}\ dx;$$

the limit of the sum of such elements as their number tends to infinity and each dx tends to zero is thus the length of the curve. For example, the arc length of a semi-circle of unit radius is

$$\int_{-1}^1 \sqrt{1 + \frac{x^2}{1 - x^2}}\ dx = \int_{-1}^1 \sqrt{\frac{1}{1 - x^2}}\ dx = \pi.$$

3. a non-negative number associated with a VECTOR and indicating its magnitude without regard to direction, written $|\mathbf{v}|$, or sometimes v. If $\mathbf{v}$ has ORTHOGONAL components $(x, 0, 0)$, $(0, y, 0)$ and $(0, 0, z)$ in the directions of the respective coordinate axes, then

$$v = |\mathbf{v}| = \sqrt{x^2 + y^2 + z^2}.$$

Two vectors $\overrightarrow{AB}$ and $\overrightarrow{BA}$ with resultant zero have the same length. Compare MEASURE, NORM, METRIC, DISTANCE.

4. the NORM in a NORMED SPACE, **X**, of functions defined by

$$\|f\|_X = \sup_{x \in X} \|f(x)\|_X$$

for a function f into a compact normed space.

5. (of a polynomial) the sum of the absolute values of the coefficients. Compare HEIGHT.

6. (of a CYCLE) the number of elements moved by the given permutation. A transposition has length two.

length-preserving transformation, *n.* a linear transformation, A, such that

$$\|A\mathbf{x}\| = \|\mathbf{x}\|$$

for all vectors $\mathbf{x}$ in a NORMED VECTOR SPACE.

Leontief model, *n.* another name for INPUT–OUTPUT MODEL.

leptokurtic, *adj.* (*Statistics*) (of a distribution) having KURTOSIS, B_2, greater than 3; that is, more heavily concentrated about the mean than a NORMAL DISTRIBUTION. Compare PLATYKURTIC, MESOKURTIC.

letter-box principle, see PIGEON–HOLE PRINCIPLE.

letter problem, see MONTMORT MATCHING PROBLEM.

level curve or **level surface,** *n.* the points of constant value for a given scalar function; $f(x, y) = c$ typically gives rise to a level curve or CONTOUR LINE, while $f(x, y, z) = c$ typically gives rise to a level surface.

level set, *n.* (for a given real-valued function f) any *lower level set* of the form

$$L(r) = \{ x : f(x) \leq r \}$$

or

$$L(r) = \{ x : f(x) < r \},$$

or *upper level set* of the form

$$U(r) = \{ x : f(x) \geq r \}$$

or

$$U(r) = \{ x : f(x) > r \}.$$

See also SEMICONTINUOUS.

lexical order or **lexicographic order,** *n.* the arrangement of a set of items in accordance with a RECURSIVE ALGORITHM, such as the entries in a dictionary, whose order depends on their first letter unless those are identical, in which case it is the second that decides, and so on. The *reverse-lexical order* orders by last letters; this is particularly useful for ordering finite words constructed from an infinite alphabet, as in an infinite-dimensional vector space. When polynomials are ordered in this way, a polynomial is greater than zero if its highest-order term has a positive coefficient, and the order is a TOTAL ORDERING.

l'Hôpital's rule or **l'Hospital's rule,** *n.* a rule permitting the evaluation of the limit of an INDETERMINATE quotient of functions as the quotient of the limits of their derivatives. For example,

$$\lim_{x \to 0} \frac{\sin x}{x}$$

is an indeterminate of the form $^0/_0$, but it can be evaluated as

$$\lim_{x \to 0} \frac{\cos x}{1} = 1.$$

(Named after the French analyst and geometer, *Guillaume François Antoine de l'Hôpital, Marquis de St Mesme* (1661–1704), who was the author of the first textbook on differential calculus, but is believed to have bought the rights to this rule from its discoverer.)

Liapunov (or **Lyapunov**) **convexity theorem,** *n.* the theorem that, given a finite number of signed FINITE MEASURES $\{ \mu_1, ..., \mu_n \}$, the associated *vector measure* μ with values in n-space defined by

$$\mu(E) = (\mu_1(E), ..., \mu_n(E))$$

for all E in the measure algebra M, has a compact range for each measurable set E. That is,

$$R_\mu(E) = \{ \mu(F) : F, E \in M, F \subseteq E \}$$

is compact, and is convex if each measure is NON–ATOMIC.

Liapunov (or **Lyapunov**) **function,** *n.* a function *V* constructed in order to establish that a point (for example, zero) is STABLE for an AUTONOMOUS differential equation system $y' = f(y)$ with *f* continuous. In one form, one requires that locally the function *V* has continuous partial derivatives, is strictly positive except at zero, and has the derivative along a trajectory

$$\frac{dV(y(t))}{dt}$$

negative for all solutions. One may conclude that zero is a Liapunov-stable point of the system. This is called *Liapunov's second* or *direct method*. If, in addition, the derivative along a trajectory is strictly negative away from the origin, then the origin is actually asymptotically stable.

Liapunov (or **Lyapunov**) **stability,** see STABLE.

liar paradox, *n.* (*Logic*) the paradox of SELF-REFERENCE generated by consideration of the truth value of the statement *this statement is false*, which, if true, is false, and, if false, is true; it is attributed to Epimenides the Cretan in the form *all Cretans are liars*.

Lie brackets, see LIE PRODUCT.

Lie commutator, *n.* another name for LIE PRODUCT.

Lie group, *n.* a TOPOLOGICAL GROUP that can be given an ANALYTIC STRUCTURE such that the group operation and inversion are analytic. Thus, the coordinates of the product of two elements are analytic functions of the coordinates of the elements. See MANIFOLD. (Named after the Norwegian group theorist, analyst, and geometer, *Marius Sophus Lie* (1842–99), who collaborated with Felix Klein and succeeded him as Professor in Leipzig. He developed the theory of invariance and made substantial contributions to the theory of differential equations.)

Lie product or **Lie commutator,** *n.* the binary operation that sends elements *a* and *b* to the *Lie bracket* [*a*, *b*] which is defined as $ab - ba$.

likelihood, *n.* (*Statistics*) the probability of a given sample being randomly drawn, regarded as a function of the parameters of the population. A *likelihood ratio* is any ratio of likelihoods. See LIKELIHOOD RATIO TEST. See also MAXIMIZED LIKELIHOOD.

likelihood ratio test, *n.* the statistical TEST in which the NULL HYPOTHESIS is rejected for small values of $\lambda = p_0/p_1$, where each p_i is the maximum of the probabilities $P[\mathbf{X} \mid \theta]$, where θ ranges over the possibilities permitted by the respective hypotheses, and $\mathbf{X}$ is a vector of observations. This is an invaluable HYPOTHESIS TESTING technique. See NEYMAN–PEARSON LEMMA. See also GENERALIZED MAXIMUM LIKELIHOOD RATIO TEST STATISTIC.

like terms, *n.* terms of an expression which are the same except for the numerical coefficients. Thus the sum of two polynomials is the sum of like terms; for example, the sum of $3x - 4y$ and $2x + 6y$ is $(3x + 2x) + (-4y + 6y) = 5x + 2y$.

limacon of Pascal, *n.* the locus of a point, P, on a line which revolves around a point, C, on a fixed circle, that that the point remains at a fixed distance from the intersection, X, of the line with the circle; Fig. 219 overleaf shows a number of positions of P. It has standard equation $r = a\cos\theta + b$, and for equal *a* and *b* this gives a CARDIOID.

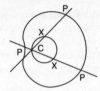

Fig. 219. **Limacon of Pascal**.

lim inf, *abbrev. for* LIMIT INFERIOR.

limit, *n.* **1.** a value that is approached increasingly closely by a function $f(x)$ or a sequence a_n, as the independent variable, x or n, increases without restriction. If for any small $\varepsilon > 0$ there is a large integer N such that

$$|f(x) - k| < \varepsilon, \text{ for all } x > N,$$

k is said to be the *limit of $f(x)$ as x tends to infinity*, written

$$\lim_{x \to \infty} f(x) = k.$$

2a. a value that a function $f(x)$ approaches increasingly closely, as the independent variable, x, approaches a stated value, a. If for any $\varepsilon > 0$ there is a δ such that

$$|f(x) - k| < \varepsilon, \text{ for all } x \text{ such that } |x - a| < \delta,$$

then k is the *limit of $f(x)$ as x tends to a*, written

$$\lim_{x \to a} f(x) = k.$$

If $f(x)$ is continuous at a,

$$\lim_{x \to a} f(x) = f(a).$$

This formalism replaces the NON–STANDARD definition of limits in terms of INFINITESIMALS: that $f(x)$ is at most infinitesimally distant from $f(a)$ for x infinitesimally distant from a.

b. a point at which the preceding condition holds either for values less than x, in which case it is a *limit on the left* (a LEFT–HAND LIMIT), or for values greater than x, in which case it is a *limit on the right* (a RIGHT–HAND LIMIT). If a function has both a left-hand limit and a right-hand limit and they are equal, then it has a limit at the point.

3a. in a METRIC SPACE, a value, k, such that, for any $\varepsilon > 0$, there is a $\delta > 0$ such that

$$d(f(x), k) < \varepsilon, \text{ for all } x \text{ such that } d(x, a) < \delta,$$

that is, as x tends to a. This is analogous to the preceding, with metrics replacing absolute value.

b. similarly, in a TOPOLOGICAL SPACE, a value, k, such that, for any NEIGHBOURHOOD $N(k)$ of k, there is a neighbourhood $N(a)$ of a such that

$$f(x) \in N(k), \text{ for all } x \in N(a),$$

that is, as x tends to a.

4. a point such that a sequence or NET in a topological space is *eventually* in every NEIGHBOURHOOD of the limiting point; that is, x_α tends to x if, for each neighbourhood V of x, there exists β in the DIRECTED SET A such that x_α belongs to V for $\alpha \geq \beta$.

5. (*Measure theory*) a MEASURABLE function, f, such that a given sequence $\{f_n\}$, of measurable functions is CONVERGENT IN MEASURE to f; that is, for every $\varepsilon > 0$, there is an N such that

$$\mu(\{ x : |f_n(x) - f(x)| > \varepsilon \}) < \varepsilon$$

for all $n > N$.

See TEND TO, EPSILON–DELTA NOTATION, CONVERGE, CONTINUOUS, SUM (sense 2). See also LIMIT INFERIOR, LIMIT SUPERIOR.

limit inferior or **lower limit** (abbrev. **lim inf**), *n*. **1.** the LIMIT as n tends to infinity of the INFIMA of the subsequences of the elements beyond the nth element of a given real sequence:

$$\liminf_{n \to \infty} a_n = \lim_{n \to \infty} [\inf \{a_m : m \ge n\}].$$

This produces the smallest CLUSTER POINT of the sequence, and may equal negative infinity. The sequence has a limit if and only if the limit inferior and LIMIT SUPERIOR are identical, and in that case the limit is their common value.

2. the set of points that are in all but finitely many members of a given sequence of sets $\{A_n\}$:

$$\liminf_{n \to \infty} A_n = \bigcup_{m=1}^{\infty} \left\{ \bigcap_{n \ge m} A_n \right\}.$$

The sets are said to have a limit if the limit inferior and limit superior agree, and the limit is then this common set. For example, the limit of a nested decreasing sequence of sets is the intersection of the collection. There are topological analogues of these concepts.

limit of integration, *n*. either of the endpoints of an interval over which a DEFINITE INTEGRAL is to be evaluated. In the integral

$$\int_a^b f(x) \, dx,$$

a is the *lower limit*, and b the *upper limit of integration*.

limit ordinal, *n*. (*Set theory*) a non-zero ORDINAL with no immediate PREDECESSOR. For example, the ordinal of the natural numbers ω is a limit ordinal. Compare ISOLATED ORDINAL.

limit point, *n*. another term for CLUSTER POINT.

limit superior or **upper limit** (abbrev. **lim sup**), *n*. **1.** the LIMIT as n tends to infinity of the SUPREMA of the subsequences of the elements beyond the nth element of a given real sequence:

$$\limsup_{n \to \infty} a_n = \lim_{n \to \infty} [\sup \{a_m : m \ge n\}].$$

This produces the largest CLUSTER POINT of the sequence, and may equal positive infinity. The sequence has a limit if and only if the limit superior and the LIMIT INFERIOR are identical, and in that case the limit is their common value.

2. the set of points that are in infinitely many members of a given sequence of sets $\{A_n\}$:

$$\limsup_{n \to \infty} A_n = \bigcap_{m=1}^{\infty} \left\{ \bigcup_{n \ge m} A_n \right\}.$$

The sets are said to have a limit if the limit superior and limit inferior agree, and the limit is then this common set. There are topological analogues of these concepts.

lim sup, *abbrev. for* LIMIT SUPERIOR.

Lindelöf space, *n.* (*Topology*) a TOPOLOGICAL SPACE in which every COVER by open sets contains a countable subcollection that still covers the space. *Lindelöf's theorem* states that every SECOND COUNTABLE space is Lindelöf. (Named after *Ernst Leonard Lindelöf* (1870–1946), Finnish topologist and analyst.) Compare COMPACT.

Lindemann theorem, *n.* (*Number theory*) the result that if $\alpha_1,...,\alpha_n$ are algebraic and distinct, while $\beta_1,...,\beta_n$ are algebraic and not all zero, then

$$\beta_1 \exp(\alpha_1) + ... + \beta_n \exp(\alpha_n) \neq 0.$$

In particular, either α or $\exp\alpha$ is TRANSCENDENTAL when α is non-zero; thus $\log\alpha$ and $\exp\alpha$ are transcendental for algebraic arguments, α. Since $\exp(2\pi i) = 1$, the transcendentality of π follows. A recent theorem due to Baker shows that any non-vanishing algebraic linear combination of logarithms of algebraic numbers is transcendental. (Named after the German analyst and geometer, *Carl Louis Ferdinand von Lindemann* (1852–1939).) Compare GELFOND–SCHNEIDER THEOREM.

line, *n.* **1.** an undefined primitive concept of EUCLIDEAN GEOMETRY; in CARTE-SIAN GEOMETRY, a straight one-dimensional geometrical figure of infinite length and no thickness; any pair of points uniquely determine a line, of which the segment between the given points is the shortest path between them. In two dimensions, the line joining (x_1, y_1) and (x_2, y_2) has the equation $y = mx + c$, where m is the SLOPE or GRADIENT,

$$m = \frac{y_1 - y_2}{x_1 - x_2},$$

and c is the intercept with the y-axis of a Cartesian coordinate system, as shown by the bold line in Fig. 220. Any other point (x, y) is COLLINEAR with these two points if

$$\frac{y - y_2}{x - x_2} = \frac{y_1 - y_2}{x_1 - x_2} = m,$$

so that the line is the infinite set of all such points. The *slope-intercept equation* of the line is in the form $y = mx + c$, where c is the y-intercept; and the *point–slope equation* is of the form

$$y - y_0 = m(x - x_0).$$

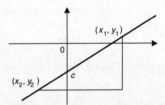

Fig. 220. **Line**. See main entry.

In three or more dimensions, the direction of the line is given by its DIREC-TION COSINES; in this case, for a point to lie on the line determined by two given points this equation must hold for each ratio of pairs of corresponding coordinates. The points on a line are uncountably infinite, isomorphic with the real numbers, and constitute a CONTINUUM.

2. a LINE SEGMENT.

3. an undefined element of an axiomatic geometry (along with POINT). A line is often viewed as a distinguished set of points, satisfying given incidence relations. See also PROJECTIVE PLANE.

4. a one-dimensional subspace of a vector space, or an algebraic geometry; the AFFINE HULL of two distinct points.

lineal element, see DIRECTION FIELD.

linear, *adj.* **1.** of or relating to the first degree; having no variable raised to any power consisting only of constant multiples of the variables. For example, a LINEAR COMBINATION of x, y, and z has the form $ax + by + cz$, where a, b, and c are numerical coefficients; a LINEAR DIFFERENTIAL EQUATION is a linear combination of derivatives of order less than some finite n.

2. proportional; constituting or based upon a LINEAR SCALE.

linear algebra, *n.* the branch of mathematics concerned with LINEAR EQUATIONS, MATRICES, DETERMINANTS, VECTOR SPACES, etc.

linear combination, *n.* a sum of the respective products of the elements of some set with constant coefficients. (It is sometimes required that not all the constants are zero.) For example, a linear combination of vectors $\mathbf{u}$, $\mathbf{v}$ and $\mathbf{w}$ is any sum of the form

$$a\mathbf{u} + b\mathbf{v} + c\mathbf{w},$$

where a, b, and c are scalars.

linear congruence, *n.* a CONGRUENCE of the form $ax \equiv b \pmod{m}$, or, more generally, of the form

$$f(x) \equiv 0 \pmod{m},$$

where f is a LINEAR FUNCTION of several variables. Compare QUADRATIC CONGRUENCE.

linear convergence, *n.* convergence of a sequence subject to the condition that

$$\limsup_{n \to \infty} \frac{|x_{n+1} - x|}{|x_n - x|} = L$$

is strictly between 0 and 1 so that when the limit exists the error is effectively reduced by some fixed proportion. If the convergence is worse or better than linear it is said to be *sub-linear* or *super-linear* respectively. See RATE OF CONVERGENCE.

linear dependence, *n.* the property of being LINEARLY DEPENDENT.

linear differential equation, *n.* **1.** an ORDINARY DIFFERENTIAL EQUATION that does not contain any products (including powers greater than 1) of the derivatives and the dependent variable. A first-order linear ordinary differential equation has the form

$$y' + a(x)y = b(x),$$

which is HOMOGENEOUS if $b(x) = 0$. There is an algebraic technique for the solution of homogeneous linear equations, which is always effective if all the coefficients in the equation are constants and the ORDER is less than 5. A second-order linear differential equation can be reduced to a first-order linear equation if the differential operator can be factorized into linear factors, or if a PARTICULAR SOLUTION of the homogeneous equation is known. If the COMPLEMENTARY FUNCTION is known, the general solution can be found by VARIATION OF PARAMETERS. See EULER EQUATION, FROBENIUS METHOD.

2. a PARTIAL DIFFERENTIAL EQUATION that does not contain any products (including powers greater than 1) of partial derivatives and the dependent variable. A COMPLETE SOLUTION of such an equation can be found as the sum of a COMPLEMENTARY FUNCTION, which is a complete solution of the HOMOGENEOUS equation, and a PARTICULAR INTEGRAL. There is an algebraic technique, similar to the one for ordinary differential equations, for finding a complete solution of an equation of the form

$$\sum_{i=0}^{n} a_i \frac{\partial^n z}{\partial x^i \partial y^{n-i}} = f(x, y),$$

where x and y are independent variables, a_i is a constant, and $f(x, y)$ is a differentiable function. See LAGRANGE'S LINEAR EQUATION.

linear eccentricity, see ECCENTRICITY.

linear equation, *n.* **1.** a POLYNOMIAL equation of the first degree, of the form $y = ax + b$.

2. any equation of the form $Ax = b$ where A is a matrix or a linear operator. See FREDHOLM ALTERNATIVE.

linear estimate, *n.* (*Statistics*) an ESTIMATE that is a LINEAR COMBINATION of observations.

linear fractional transformation, *n.* another name for the MÖBIUS TRANSFORMATION.

linear functional, *n.* a LINEAR mapping, *f*, from a VECTOR SPACE into its base field, that satisfies

$$f(\alpha\mathbf{u} + \beta\mathbf{v}) = \alpha f(\mathbf{u}) + \beta f(\mathbf{v}),$$

where α and β are scalars, and **u** and **v** are vectors. The set of all (continuous) linear functionals endowed with pointwise operations comprises the algebraic (or continuous) *dual vector space.*

linear independence, *n.* the property of not being LINEARLY DEPENDENT.

linearly accessible set, *n.* another name for NIKODYM SET.

linearly dependent, *adj.* such that there is a LINEAR COMBINATION of the given elements, with not all coefficients zero, that equals zero. For example, **u**, **v** and **w** are linearly dependent vectors if there exist scalars *a*, *b* and *c*, not all zero, such that

$$a\mathbf{u} + b\mathbf{v} + c\mathbf{w} = 0.$$

Elements are said to be *K-linearly dependent* if there is a set of such constants that are elements of some given K; for example, the vectors $(1, \pi)$ and (π, π^2) are $\mathbb{R}$-linearly dependent, but not $\mathbb{Q}$–linearly dependent (where $\mathbb{R}$ is the set of real numbers and $\mathbb{Q}$ is the set of rationals), since one of the required coefficients is a multiple of the irrational number π. See also BASIS.

linearly independent, *adj.* not LINEARLY DEPENDENT.

linear mapping, *n.* a MAPPING, θ, between VECTOR SPACES, such that

$$\theta(x+y) = \theta(x) + \theta(y) \text{ and } \theta(\lambda x) = \lambda\theta(x),$$

where λ is a field element. A linear mapping is a HOMOMORPHISM of vector spaces, and if the domain and range are finite-dimensional it may be represented by a matrix.

linear operator, *n.* an OPERATOR between vector spaces that preserves addition and scalar multiplication, often written $L(X,Y)$, where X and Y are the vector spaces. Finite-dimensional linear operators are identifiable with MATRICES.

linear order or **linear ordering,** see ORDERING.

linear programming, *n.* the study of OPTIMIZATION problems that can be solved by seeking the maximum or minimum values of a linear function of non-negative variables subject to constraints expressed as linear equalities or inequalities. This is of considerable practical and theoretical importance in operations research and in economics. See DUALITY THEORY OF LINEAR PROGRAMMING. Compare INPUT–OUTPUT MODELS, SIMPLEX METHOD, INTEGER PROGRAMMING.

linear regression, *n.* (*Statistics*) a REGRESSION that is LINEAR in the unknown parameters, whatever the order of the known parameters. For example,

$$E(X) = \alpha + \beta t + \gamma t^2$$

is linear.

linear scale, *n.* a scale, distances along which are proportional to the quantities they represent, as shown in Fig. 221. Compare LOGARITHMIC SCALE.

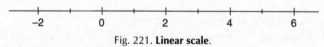

Fig. 221. **Linear scale**.

linear space, *n.* a former term for VECTOR SPACE, especially one consisting of ordered *n*–tuples of real or complex numbers.

linear span, *n.* the smallest subspace containing a given set in a vector space; the set of all LINEAR COMBINATIONS of the vectors in the given set.

linear topological space, *n.* another term for TOPOLOGICAL VECTOR SPACE.

linear transformation, *n.* a LINEAR MAPPING from a vector space into another, especially itself.

line at infinity, *n.* the set of IMPROPER POINTS added to the EUCLIDEAN PLANE to form the real AUGMENTED EUCLIDEAN GEOMETRY for the PLANE; the set of IDEAL POINTS at which parallel lines are held to meet.

line integral, *n.* another term for CURVILINEAR INTEGRAL.

line of flow, *n.* (*Continuum mechanics*) another term for STREAMLINE.

line search method, *n.* (*Numerical analysis*) one of a number of techniques for determining the minimum or maximum of a function on a line or line segment. This is an important component of most DESCENT METHODS. *Exact line search* produces the exact optimum and is usually only feasible with QUADRATIC functions. *Approximate line search* settles for an estimate of the optimum; popular methods include DICHOTOMOUS LINE SEARCH and *Armijo's method.*

line segment, *n.* the portion of a LINE between two of the points on it and, in EUCLIDEAN GEOMETRY, constituting the shortest distance between them on the plane. An open line segment of finite length is homeomorphic (for example, under a stretching transformation) with the whole real line.

Liouville function, *n.* the number theoretic function

$$\lambda(n) = (-1)^{e(n)},$$

defined for an integer *n*, where $e(n)$ is the number of prime factors of *n*, counting repetition. (Named after the French number theorist, differential geometer, and analyst, *Joseph Liouville* (1809–82), who proved the existence of transcendental numbers.)

Liouville numbers, *n.* an uncountable set of numbers that are TRANSCENDENTAL by virtue of being too well approximated by rationals. Precisely, a number *x* such that, for each integer *n*, there is a rational number p/q (with $q > 1$) with

$$\left| x - \frac{p}{q} \right| < \frac{1}{q^n}.$$

The set is second BAIRE CATEGORY and MEASURE zero. The infinite *Liouville series* $\sum x^{n!}$ produces a Liouville number for all rational *x* in $]0, 1[$.

Liouville's theorem, *n.* **1.** (*Complex analysis*) the result that a BOUNDED ENTIRE function is constant. This leads to MORERA'S THEOREM, and to an analytic proof of the FUNDAMENTAL THEOREM OF ALGEBRA.

2. (*Statistical physics*) a classical theorem that postulates that the number of STATES accessible to a system is proportional to the accessible volume of PHASE SPACE.

Lipschitz condition, see LIPSCHITZ FUNCTION.

Lipschitz function, *n.* a function between NORMED SPACES with the property that the distance between function values is bounded by a constant multiple of the distance between the arguments. If the function satisfies the *Lipschitz condition* that

$$\|f(x) - f(y)\| \leq k \|x - y\|$$

for all *x* and *y* in a set *A* or at a point x_0, then *f* is *k-Lipschitz* on *A* or at x_0. For example, $f(x) = x^2$ is 2–Lipschitz on $(-1, 1)$, since

$$\left| x^2 - y^2 \right| = \left| x + y \right| \left| x - y \right| \leq 2 \left| x - y \right|.$$

When $k = 1$, the function is NON–EXPANSIVE, and is a CONTRACTION if $k < 1$. The *Rademacher theorem* proves that every finite-dimensional Lipschitz function is differentiable almost everywhere. More generally, if a function satisfies the *Lipschitz condition of order p* (also known as the *Hölder condition*) that

$$\|f(x) - f(y)\| \leq k \|x - y\|^p$$

for some $0 < p \leq 1$, for x and y in a set A, then f is said to be *Hölder-continuous* on A. (Named after the German analyst, algebraist, number theorist and physicist, *Rudolph Otto Sigismund Lipschitz* (1832–1903).)

Lissajous figures or **Bowditch curves,** *n.* the curves, of particular importance in electronics, produced by the intersection of two SINUSOIDS with perpendicular axes; they may differ in phase, as in the first row of Fig. 222, and in frequency as in the second row.

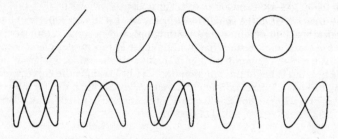

Fig. 222. **Lissajous figures**.

literal, *adj.* **1.** (of an expression) using letters to denote constants, coefficients, etc. For example, in $ax^2 + b$, x is a variable, but a and b are literal constants. Compare NUMERICAL. See also ALGEBRA.

2. (*as substantive*) (*Logic*) an expression consisting of a single ATOMIC symbol or its negation, especially a CONSTANT as contrasted with a VARIABLE.

lituus, *n.* a trumpet-shaped curve that is the locus of points such that the square of the length of their radius vector is inversely proportional to the angle between the x-axis and the radius vector; its polar equation is $r^2 = a/\theta$. As shown in Fig. 223, it is asymptotic to the axis, and winds round the origin without ever reaching it.

Fig. 223. **Lituus**.

Liusternik's theorem, *n.* the infinite-dimensional INVERSE FUNCTION THEOREM that establishes that if a mapping, T, between BANACH SPACES that is CONTINUOUSLY DIFFERENTIABLE in the sense of the FRECHET DIFFERENTIAL, has a SURJECTIVE derivative at a point a, then there is a neighbourhood U of $T(a)$ and a constant $K > 0$ such that, for each element $y \in U$, there is a solution to $y = T(x)$ with

$$\|x - a\| \leq K \|y - T(a)\|.$$

See INVERSE FUNCTION THEOREM.

live, *vb.* (of a MEASURE) to be CONCENTRATED on (some set); that is, if there is a set A such that for all measurable E, $\mu(E) = \mu(A \cap E)$, then μ is said to live on A.

ln, *abbrev. and symbol for* NATURAL LOGARITHM; $\ln(e^x) = x$.

load, *n.* the CONTACT FORCE density of a body.

Lobachevskian geometry or **hyperbolic geometry,** *n.* a NON–EUCLIDEAN GEO-
METRY in which a line has at least two parallels through a given point; it has
a model in the interior of a circle, with lines represented by chords; it can
also be modelled as the geometry of lines on a saddle, whence the name
hyperbolic. (Named after the Russian geometer, *Nikolai Ivanovitch Lobachevski* (1793–1856).) Compare ELLIPTIC GEOMETRY.

local, **relative**, or **in the small** (*im kleinen*), *adj.* (of some property of a topological space or of a function) holding only within some NEIGHBOURHOOD
of the given point rather than for all values. For example, a *local maximum*
is a value of the function greater than any adjacent value, but it may not be
the greatest value of the function over its whole range; the curve shown in
Fig. 224 has a local maximum at A, but y is greater than this for x less than
−2.5 or greater than 2. Compare GLOBAL. See also LOCALLY COMPACT,
LOCALLY CONNECTED, LOCALLY EUCLIDEAN.

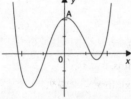

Fig. 224. A is a **local** maximum.

local gravitational constant, *n.* the acceleration of any particle close to sea
level at any place on earth, and due to the Earth's GRAVITY alone. This is
denoted g and is equal to γMR^{-2}, where γ is the UNIVERSAL GRAVITATIONAL
CONSTANT, M is the MASS of the Earth, and R is the radius of the Earth at
that point. The value of g is 9.8321 ms^{-2} at the North Pole, 9.7801 ms^{-2} at
the equator, and 9.8119 ms^{-2} at London.

locally compact, *adj.* (of a TOPOLOGICAL SPACE) having the property that
every point has a NEIGHBOURHOOD that is a subset of a COMPACT set. Euclidean space is locally compact, as is every compact space.

local coordinate system, *n.* (*Differential topology*) another term for CHART.

locally connected, *adj.* (of a TOPOLOGICAL SPACE) having the property that,
for every point p and every NEIGHBOURHOOD U of p, there is a CONNECTED
OPEN set V that is a subset of U such that $p \in V$. A space is locally connected
if and only if the COMPONENTS of all open subsets are open.

locally convex, *adj.* (of a TOPOLOGICAL VECTOR SPACE) having NEIGHBOUR-
HOOD BASE consisting of CONVEX sets.

locally Euclidean, *adj.* (of a TOPOLOGICAL SPACE) having the property that
every point p has a NEIGHBOURHOOD that is HOMEOMORPHIC to an open
subset of a fixed EUCLIDEAN SPACE, of which the dimension is associated to
the space. A locally Euclidean space can be given the structure of a LIE
GROUP.

locally finite, *adj.* (of a COLLECTION of subsets of a TOPOLOGICAL SPACE) having the property that every point p has a NEIGHBOURHOOD that only finitely
many of the members of the collection intersect.

local uniform convexity, see UNIFORM CONVEXITY.

location problems, *n.* various generalizations of FERMAT'S PROBLEM, originating with Alfred Weber in 1909, in which one seeks to locate a point in space in such a way as to minimize a weighted sum of distances (in some metric) from a given set of points. For example, these problems arise when trying to place central facilities to serve a number of communities.

locus, *n.* a set of points that satisfy or are determined by some specified condition. For example, the locus of points that are equidistant from two given points is the perpendicular bisector of the line segment joining the given points. See also PENCIL.

log, *abbrev. and symbol for* LOGARITHM. $\log_{10} x$ is the COMMON LOGARITHM of x, and $\log_e x$ is its NATURAL LOGARITHM, often written $\ln x$; $\log x$ without the base explicitly indicated usually denotes the natural logarithm in mathematical texts and the common logarithm in scientific texts, unless context dictates otherwise.

logarithm, *n.* the power to which a BASE must be raised to yield a given number, usually abbreviated $\log x$ or $\log_b x$, where b is the base. COMMON LOGARITHMS have base 10, and NATURAL LOGARITHMS, written $\ln x$, have base e. $\log_b x$ is the inverse of b^x so that if $b^x = y$ then $\log_b y = x$ and

$$\log_b b^x = x = b^{(\log_b x)}$$

so, in particular, for the natural logarithms,

$$\log_e e^x = \ln \exp x = x = \exp \ln x = e^{(\log_e x)}.$$

Thus base change is effected by

$$\log_a x = \frac{\log_b x}{\log_b a}.$$

Logarithms are used to simplify multiplication, division and exponentiation since if $a = b \times c$, $\log a = \log b + \log c$, and

$$x^n = \text{antilog}(n \log x).$$

logarithmic, *adj.* **1.** of, relating to, using, or containing LOGARITHMS; for example, a LOGARITHMIC FUNCTION.

2. consisting of, or using, points or lines, the distances of which from a fixed point or line are proportional to the logarithms of numbers, such as a LOGARITHMIC SCALE.

logarithmic convexity, *n.* the CONVEXITY of the LOGARITHM of a given function. For example, $\exp(x + y)$ is logarithmically convex.

logarithmic coordinate system, *n.* a CARTESIAN coordinate system in which the axes are marked with LOGARITHMIC SCALES.

logarithmic derivative, *n.* the derivative of the LOGARITHM of a given function; thus, by the CHAIN RULE,

$$\frac{f'(x)}{f(x)}$$

is the logarithmic derivative of f at x. For example, the digamma function is the logarithmic derivative of the gamma function.

logarithmic differentiation, *n.* differentiation after taking LOGARITHMS of both sides of an identity. This is particularly helpful with differentiation of product expressions.

logarithmic function, *n.* **1.** the function $\log_e x$ or $\ln x$ that is defined for positive x as the inverse of the EXPONENTIAL FUNCTION or as the definite integral from 1 to x of $1/t$ with respect to t. Its value for any argument is its NATURAL LOGARITHM, so that

$$\ln \exp x = 1 = \exp \ln x;$$

its derivative is $1/x$, and it is ASYMPTOTIC to the y-axis with y tending to $-\infty$ as x tends to 0 from above, as shown in Fig. 225. The logarithm is extended into the complex plane by

$$\log z = \log |z| + i \arg z$$

which is a MULTIVALUED FUNCTION; its PRINCIPAL PART is constructed by taking the PRINCIPAL VALUE of the argument. This provides an ANALYTIC CONTINUATION of the logarithm to the CUT PLANE $\mathbb{C} \setminus [-\infty, 0]$.

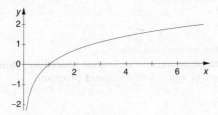

Fig. 225. **Logarithmic function**.

2. any function containing the logarithmic function or the logarithm of a function to any base.

logarithmic scale, *n.* a scale distances along which are proportional to the LOGARITHMS of the marked indices. For example, if 10 is marked one inch along the scale, each successive inch will be marked 100, 1000, etc. In general, if the first unit on the scale represents b, the nth will represent b^n, so that a measured quantity of k will be represented as $\log_b k$ units along the scale; for example, Fig. 226 above shows a scale to base 2. See also LOG PAPER. Compare LINEAR SCALE.

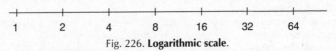

Fig. 226. **Logarithmic scale**.

logarithmic series, *n.* the alternating series

$$1 - \tfrac{1}{2} + \tfrac{1}{3} - \tfrac{1}{4} + \dots.$$

which converges to $\ln 2$, or more generally, the series

$$x - \frac{x^2}{2} + \frac{x^3}{3} - \dots + (-1)^{n-1} \frac{x^n}{n} = \ln(1+x),$$

valid inside the unit disk and at 1. Compare HARMONIC SERIES.

logarithmic spiral, **logistic spiral**, or **equiangular spiral,** *n.* a SPIRAL, such as that in Fig. 227, in which the angle between the RADIUS VECTOR and the polar axis is proportional to the logarithm of the length of the vector, so that its POLAR EQUATION is $\log r = a\theta$.

Fig. 227. **Logarithmic spiral**.

logic, *n.* **1.** the analysis, without regard to meaning or context, of the patterns of reasoning by which conclusions are validly derived from sets of premises. See also LOGICAL FORM, FORMAL LOGIC, DEDUCTION, VALID.

2. any particular FORMAL CALCULUS or FORMAL LANGUAGE in which are defined AXIOMS and RULES OF INFERENCE.

3a. the system and principles of reasoning used in a particular field of study; for example, the logic of quantum theory.

b. (*informal*) a particular method of argument or reasoning. In this sense one says 'his logic led him to the wrong conclusion'.

4. (*informal*) the process of reasoning or any analogous procedure; thus one speaks of checking the logic of a proof or computer program.

logical axioms, *n.* axioms valid in every STRUCTURE for a given formal language; the axioms for the LOGICAL CONSTANTS of the theory.

logical consequence, *n.* the relation that obtains between the CONCLUSION and the PREMISES of an argument when it is formally VALID; the relation that holds between one statement and a set of statements when it is impossible for the former to be false in any MODEL in which the latter are all true.

logical constant, *n.* one of the CONNECTIVES of a given system of FORMAL LOGIC, especially those of the SENTENTIAL CALCULUS, namely *not, and, or,* and *if ... then ...,* or their formal equivalents, NEGATION, CONJUNCTION, DISJUNCTION, and IMPLICATION.

logical form, *n.* **1.** the syntactic structure that may be shared by different expressions, as abstracted from their content and articulated by the LOGICAL CONSTANTS of a particular LOGICAL SYSTEM.

2. in particular, the least detailed structure of an argument by virtue of which it can be shown to be logically VALID. Thus

> *John is tall and thin.*
> so *John is tall*

has the same logical form as

> *London is large and dirty.*
> so *London is large,*

namely

> *P and Q*
> so *P*

in SENTENTIAL CALCULUS. Although these can also be formalized as

> *Fa and Ga*
> so *Fa*

in the PREDICATE CALCULUS, the latter is not in this sense the logical form of the given arguments, as their validity is demonstrable at sentential level without recourse to analysis of the predicate structure.

3. by contrast, the most detailed structure at a particular level of analysis of a statement, this level of detail being required in order to anticipate all logical relationships that this statement may have with any others. For example,

> *if London is big, noisy and dirty, it is noisy and dirty*

is said to have logical form in sentential calculus

> *if (P and Q and R), then (Q and R)*

even though we would say that the logical form, in the preceding sense, of

> *London is big, noisy and dirty.*
> so *it is noisy and dirty*

is

> *P and Q*; so *Q*

This disparity arises because we may wish to infer, say, *London is noisy* from the same sentence, and wish its logical form to allow for this too. We could assign this sentence a more detailed form in sentential calculus, but there is no unique most detailed logical form, incorporating, for example, modal and tense logic analyses, as that would require assurance that there would be no future logical discoveries.

logically equivalent, *adj.* having the same TRUTH-VALUE in all circumstances. In sentential logic, therefore, logically equivalent compound sentences have the same TRUTH-TABLE.

logically possible, *adj.* capable of being described without self-contradiction.

logical product, *n.* another term for CONJUNCTION or INTERSECTION.

logical sum, *n.* another term for DISJUNCTION or UNION.

logical truth, *n.* **1.** another term for TAUTOLOGY.

2. the property of being true under any possible state of affairs.

logic circuit or **logic gate,** *n.* an electronic circuit used in digital computers to perform a single BOOLEAN operation on the values of two or more input signals. There are six standard logic circuits corresponding to different LOGICAL CONSTANTS; for example, the AND gate gives an output of 1 when and only when both its inputs are 1. These are the basic building blocks of computers; for example, a single-digit binary adder consists of an OR gate together with an AND gate to provide the carry digit to the next element.

logicism, *n.* the philosophical theory that all of mathematics can be deduced from logic (including set theory). Frege's attempt to achieve this was far advanced when Russell discovered that his axioms permitted the derivation of what is now known as RUSSELL'S PARADOX; the various subsequent attempts to rehabilitate logicism, including Russell's, gave rise to increasingly unnatural sets of axioms, and the initiative in the FOUNDATIONS OF MATHEMATICS was grasped instead by FORMALISM and INTUITIONISM.

logistic, *n.* **1.** the curve with equation

$$y = \frac{k}{1 + e^{a + bx}}$$

where *b* is less than zero. This has a horizontal asymptote *y* = *k* at infinity which is approached from below, and has one intermediate inflection point; for example, Fig. 228 shows the graph of $y = 3/(1 + \exp(1 - 2x))$. The logistic curve is frequently used to model growth in biological populations for which saturation occurs.

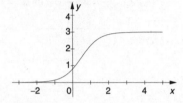

Fig. 228. **Logistic**.

2. SEXAGESIMAL arithmetic, using a place-value base of 60.

logistic map, *n.* an ITERATED MAP of of the form $x \rightarrow ax(1-x)$ the interval [0,1], for $0 < a < 4$. For $0 < a < 1$, the point 0 is an attracting fixed point. For $1 < a < 3$, the point 0 is a repelling fixed point and $1 - 1/a$ is an attracting fixed point. For $3 < a < 1 + \sqrt{6}$, there are two repelling periodic points of period 2. For increasing values of *a* the periodic orbits bifurcate (see BIFURCATION) into orbits of period 2, 4, 8, ... See also FEIGENBAUM NUMBER.

logistic spiral, *n.* another term for LOGARITHMIC SPIRAL.

logistic system, *n.* (*Logic*) an UNINTERPRETED FORMAL CALCULUS containing a set of AXIOMS and a set of RULES OF INFERENCE, especially a system of FORMAL LOGIC. Compare FORMAL LANGUAGE.

log–normal distribution, *n.* the DISTRIBUTION of a random variable, X, when log X is a RANDOM VARIABLE with a NORMAL DISTRIBUTION.

log paper, *n.* graph paper, one of the axes of which is a LOGARITHMIC SCALE, as shown in Fig. 229; *double log paper* has logarithmic scales on both axes.

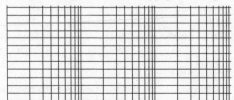

Fig. 229. **Log paper**.

log tables, *n.* tables listing the values of LOGARITHMS, especially COMMON LOGARITHMS, of numbers, usually to four significant digits.

long division, *n.* **1.** an algorithm for division by a number of more than a single digit that proceeds by subtracting from the initial segment of the dividend the largest multiple of the divisor less than that initial segment;

this process is repeated for the successive remainders augmented by the next digit of the dividend. For example, to divide 379 by 16, first subtract 2×16 from 37; this leaves a remainder of 5 to which the final 9 is appended giving 59, which is next divided by 16, giving 3 with remainder 11; the quotient is then the sequence of partial quotients, 23, and the remainder 11. The process can then be continued, as shown in the usual format in the first example in Fig. 230, to calculate any number of decimal places.

2. a similar method of dividing one polynomial by another, as in the second example in Fig. 230.

$$
\begin{array}{r}
23.6\ldots \\
16\,\overline{\big)\,379.0\ldots} \\
32 \\
\overline{59} \\
48 \\
\overline{110} \\
96 \\
\overline{140}
\end{array}
\qquad
\begin{array}{r}
x\;-\;\;4\;+5x^{-1} \\
x^{2}+2x\,\overline{\big)\,x^{3}-\,2x^{2}-\,3x\,+\,10} \\
\underline{x^{3}\,+\,2x^{2}} \\
-\,4x^{2}\,-\,3x \\
\underline{-\,4x^{2}\,-\,8x} \\
5x\,+\,10 \\
\underline{5x\,+\,10}
\end{array}
$$

Fig. 230. **Long division**. See main entry.

long radius, *n.* the line, or the distance, between the CENTRE and a VERTEX of a regular polygon. Compare SHORT RADIUS.

Fig. 231. A **long radius** of a square.

loop, *n.* **1.** an EDGE of a GRAPH that joins a vertex to itself.

2. a non-ASSOCIATIVE algebra with one binary operation. It is a QUASI-GROUP with an IDENTITY (which is necessarily unique).

3a. a RECURSIVE procedure that does not terminate, especially through an error in programming a computer.

b. *vb.* to repeat without terminating, such as when a program loops.

Lorenz attractor, *n.* the ATTRACTOR of the LORENZ EQUATIONS.

Lorenz equations, *n.* the system of ORDINARY DIFFERENTIAL EQUATIONS

$$x' = 3(y - x), \quad y' = -xz + 26.5x - y, \quad z' = xy - z,$$

which exhibits remarkable chaotic behaviour.

Lorentz force equation, *n.* (*Electromagnetism*) a single vector equation that describes the force acting on a moving charge due to the presence of an ELECTRIC FIELD and a MAGNETIC FIELD. The equation is

$$\mathbf{F} = q(\mathbf{E} + \mathbf{v} \times \mathbf{B}),$$

where $\mathbf{F}$ is the force, q is the quantity of charge, $\mathbf{E}$ is the electric field, $\mathbf{B}$ is the magnetic field, and $\mathbf{v}$ is the velocity of the charge.

Lorenz group, *n.* the GROUP of TRANSFORMATIONS on four-dimensional EUCLIDEAN SPACE that leaves invariant the QUADRATIC FORM

$$x_0^2 - x_1^2 - x_2^2 - x_3^2 = 0.$$

The fundamental postulate of the Theory of Special Relativity is that space–time is a differential MANIFOLD endowed with a Lorenz group structure.

Löwenheim–Skolem theorem, *n.* (*Logic*) the crucial result in MODEL THEORY that if a COUNTABLE THEORY has a MODEL, then it has a countable model, and indeed a model of every cardinality greater than or equal to $\aleph_0$. For example, it shows that there exist non-standard models of arithmetic: since the theory is countable, by the Löwenheim–Skolem theorem, it has an uncountable model, which is clearly non-standard. Compare COMPACTNESS THEOREM.

lower bound, *n.* a value less than or equal to all of a set of given values. For example, in the LATTICE of subsets of $\{1, 2, 3, 4, 5\}$ ordered by set inclusion, the greatest lower bound of $\{1, 2, 3\}$ and $\{2, 3, 4\}$ is $\{2, 3\}$. Compare UPPER BOUND. See also INFIMUM, MINIMUM.

lower Darboux integral, *n.* another term for LOWER INTEGRAL.

lower Darboux sum, *n.* another term for LOWER SUM.

lower Hessenberg form, see HESSENBERG FORM.

lower integral or **lower Darboux integral,** *n.* the limit, as the lengths of its subintervals tend to zero, of the LOWER SUMS of a function on that interval; if this exists and is equal to the UPPER INTEGRAL then the function is RIEMANN INTEGRABLE.

lower inverse image set, see INVERSE IMAGE SET.

lower level set, see LEVEL SET.

lower limit, *n.* **1.** the smaller of the two LIMITS OF INTEGRATION of an interval over which a definite integral is taken.

2. another term for the LIMIT INFERIOR of a sequence.

Compare UPPER LIMIT.

lower predicate calculus (abbrev. **LPC**), or **first-order predicate calculus,** *n.* a formalization of PREDICATE CALCULUS in which quantification is only over individuals and not over classes or predicates.

lower semicontinuous, see SEMICONTINUOUS.

lower sum or **lower Darboux sum,** *n.* the weighted sum of the products of the INFIMAL values of a function on a succession of subintervals of a given interval with the lengths of the subinterval; whence, the area, shown shaded in Fig. 232, between the *x*-axis and the STEP FUNCTION of which the value in

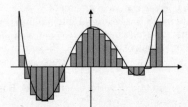

Fig. 232. **Lower sum**. See main entry.

341

each subinterval is the infimum of the given function on that subinterval. The limit of this sum as the lengths of the subintervals tend to zero is the LOWER INTEGRAL of the function. Compare UPPER SUM. See RIEMANN INTEGRAL.

lower-triangular, *adj.* (of a square matrix) having all entries above the main diagonal equal to zero, such as

$$\begin{bmatrix} 1 & 0 & 0 & 0 \\ 2 & 3 & 0 & 0 \\ 4 & 5 & 6 & 0 \\ 7 & 8 & 9 & 1 \end{bmatrix}.$$

Compare UPPER TRIANGULAR, HESSENBERG FORM.

lowest common denominator, *n.* another name for LEAST COMMON DENOMINATOR.

lowest common multiple, *n.* another name for LEAST COMMON MULTIPLE.

lowest terms, *n.* the unique expression of a RATIONAL NUMBER as a ratio of two RELATIVELY PRIME integers.

loxodrome or **loxodromic spiral,** *n.* any curve on a SURFACE OF REVOLUTION that cuts all MERIDIANS at a constant angle, especially on a sphere.

lozenge, *n.* another word for RHOMBUS.

LPC, *abbrev. for* LOWER PREDICATE CALCULUS.

l_p space, *n.* the BANACH SPACE, denoted $l_p(S)$, consisting of all functions that map a set S into the real or complex scalars and have

$$\|f\|_p = \left\{ \sum_S |f(x)|^p \right\}^{\frac{1}{p}} < \infty,$$

for $1 \le p \le \infty$ (the CHEBYSHEV NORM being intended for $p = \infty$); this quantity defines the l_p *norm*. In the most standard case, S is the set of natural numbers, giving the l_p *sequence spaces.* For $1 \le p < \infty$, the DUAL NORMED SPACE to $l_p(S)$ can be identified with $l_q(S)$ where $1/p + 1/q = 1$. Compare L_p SPACE. See also HÖLDER'S INEQUALITY, HILBERT SPACE.

L_p space, *n.* the BANACH SPACE, denoted $L_p(S, \mu)$, of all MEASURABLE functions that map a MEASURE SPACE S into the real or complex scalars, and have

$$\|f\|_p = \left\{ \int_S |f(s)|^p \right\}^{\frac{1}{p}} < \infty,$$

for $1 \le p \le \infty$ (the essential CHEBYSHEV NORM being intended for $p = \infty$); this quantity defines the L_p *norm*. The Lebesgue integrable functions correspond to members of L_1, and L_2 is the set of all Lebesgue SQUARE–INTEGRABLE functions on an interval. Strictly, $L_p(S, \mu)$ consists of the EQUIVALENCE CLASSES of functions that agree everywhere except on a set of measure zero. In the most standard case, S is a bounded interval and μ is LEBESGUE MEASURE, giving rise to the *Lebesgue integral spaces.* For a sigma finite measure μ and $1 \le p < \infty$, the DUAL NORMED SPACE to $L_p(S, \mu)$ can be identified with $L_q(S, \mu)$ where $1/p + 1/q = 1$. Compare l_p SPACE. See also HÖLDER'S INEQUALITY, HILBERT SPACE.

L series, see DIRICHLET SERIES.

lub, *abbrev for* LEAST UPPER BOUND.

Lucas numbers, *n.* the sequence of integers

$$2, 1, 3, 4, 7, 11, 18, 29, ...$$

derived from the same DIFFERENCE EQUATION as the FIBONACCI NUMBERS, but using different initial values. (Named after the French mathematician *Edouard Lucas* (1842–91).)

LU decomposition, *n.* (of a matrix) the factorization of a NON-SINGULAR matrix A as the product LU, where L and U are respectively LOWER and UPPER TRIANGULAR matrices. More generally, for any square matrix A one may write A as the product LPU with P a PERMUTATION MATRIX.

lune, *n.* **1.** a section of the surface of a sphere enclosed between two semi-circles that intersect at diametrically opposite points on the sphere.

2. a crescent-shaped figure formed on a plane surface by the intersection of the arcs of two circles, such as the shaded section of Fig. 233.

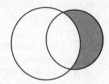

Fig. 233. **Lune.**

Lusin's theorem or **Luzin's theorem,** *n.* the result that an almost everywhere finite MEASURABLE function on EUCLIDEAN SPACE is continuous except on a set of arbitrarily small measure: given such a function f and d > 0, there exists a continuous function g such that $f(x) = g(x)$ except on a set of measure less than d. (Named after the Russian analyst, topologist, and logician, *Nikolai Nikolaevitch Lusin* (1883–1950).)

m

m, 1. (*Mechanics*) *symbol for* METRE.

2. *abbrev. for* MILLI-, used in symbols for fractions of the physical units of the SYSTEME INTERNATIONAL.

M, 1. *symbol for* the number 1000 in ROMAN NUMERALS.

2. *abbrev. for* MEGA-, used in symbols for multiples of the physical units of the SYSTEME INTERNATIONAL.

Maclaurin, Colin (1698–1746), Scottish mathematician and physicist, who developed Newton's work in both subjects. He entered Glasgow University at age 11, was appointed Professor of Mathematics at Aberdeen at 19, was elected to a Fellowship of the Royal Society at 21, moved to the Chair at Edinburgh on Newton's recommendation at 27, and in 1740 shared a prize offered by the French Academy of Science for an essay on tides with Euler and Daniel Bernoulli; he also wrote a defence of Newton's theory of FLUXIONS against the philosopher Berkeley. During the 1745 rebellion he was active in preparing the defences of Edinburgh against Bonnie Prince Charlie, and had to flee to England when the city fell to the rebels; although he returned after their swift defeat, his health never recovered and he died the following year.

Maclaurin series, *n.* power series for certain functions obtained by the application of MACLAURIN'S THEOREM; a TAYLOR SERIES expanded around zero, such as

$$\sin x = x - \frac{x^3}{3!} + \frac{x^5}{5!} - \frac{x^7}{7!} + \ldots$$

$$\cos x = 1 - \frac{x^2}{2!} + \frac{x^4}{4!} - \frac{x^6}{6!} + \ldots$$

$$\exp x = 1 + \frac{x}{1} + \frac{x^2}{2!} + \frac{x^3}{3!} + \ldots$$

$$\ln(x+1) = x - \frac{x^2}{2} + \frac{x^3}{3} - \frac{x^4}{4} + \ldots \qquad (|x| < 1).$$

Maclaurin's formula or **Maclaurin's theorem,** *n.* the theorem of mathematical analysis stating that if a real-valued function, *f,* is infinitely differentiable in an open neighbourhood of the origin, then $f(x)$ can be approximated locally as the sum of $f(0)$ and an initial segment of the series of terms of the form

$$f_n(x) = \frac{1}{n!} f^{(n)}(0) \, x^n,$$

where $f^{(n)}(x)$ is the nth derivative of $f(x)$. This is a special case of TAYLOR'S THEOREM, and the function can be expanded on this neighbourhood as the sum of this power series as n tends to infinity, provided the remainder term in Taylor's theorem tends to zero. For example, to order six

$$\sin x = x - \frac{x^3}{6} + \frac{x^5}{120} - \cdots$$

For complex functions, if f is HOLOMORPHIC for all z such that $|z| < r$, then

$$f(z) = \sum_{n=0}^{\infty} \frac{1}{n!} f^{(n)}(0)\, z^n$$

for all z in the disc.

macrostate, see STATE.

magic square, n. a square array of rows of integers arranged so that the sum of the integers in each column, row, and diagonal is the same. For example, Fig. 234 shows a square of *order* four that is found in Durer's 1514 engraving 'Melancholia'. Such squares date back to both Chinese and Western antiquity. A rule due to de La Loubere (c.1670) gives squares of all odd orders.

16	3	2	13
5	10	11	8
9	6	7	12
4	15	14	1

Fig. 234. **Magic square.**

magnetic field, n. (*Electromagnetism*) a VECTOR FIELD created by a moving charge, or CURRENT, that describes the interactions of that current with others in its neighbourhood.

magnitude, n. **1.** a number assigned to a quantity usually as a multiple of a given unit of that quantity, and thus enabling comparisons to be made on basis of the ratio of any two such quantities.

2. a non-negative scalar associated with a quantity, such as its ABSOLUTE VALUE or LENGTH. For example, x and $-x$ have the same magnitude but opposite POLARITY. Compare MEASURE.

main connective, n. (*Logic*) the CONNECTIVE that has widest SCOPE in a given expression. For example, in 'not everyone is either rich or poor' it is *not*, while in 'either everyone is rich or everyone is poor' it is *or*.

main diagonal or **leading diagonal,** n. the diagonal from top left to bottom right of a square matrix or determinant, or the entries on that diagonal, that is, the entries a_{ii}.

major, *adj.* (*Geometry*) (of an ARC, SECTOR, or SEGMENT of a circle) being the larger of two such figures determined by the same points on the circumference. In Fig. 235, the arc AXB is the *major arc*, the region OAXB is the *major segment*, and the region ABX is the *major sector* determined by the points A and B. Compare MINOR.

Fig. 235. The **major** arc, major sector and major segment include the point X.

major axis, *n.* the longer axis of an ellipse, or the longest of an ellipsoid. The bold axis, AC, in Fig. 236 is the major axis, and BD the MINOR AXIS.

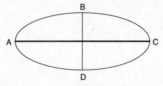

Fig. 236. AC is the **major axis** of the ellipse.

majorize, *vb.* to dominate with respect to some measurement scale, especially with respect to a PARTIAL ORDERING.

major premise, *n.* (*Logic*) the premise of a SYLLOGISM that contains the predicate of the conclusion.

major term, *n.* (*Logic*) the predicate of the conclusion of a SYLLOGISM.

Mandelbrot set, *n.* the set of complex numbers, c, for which the JULIA SET of $f(z) = z^2 - c$, is connected. The Mandelbrot set is an extremely complex object, but lies entirely within a distance of 2 from the origin, and is the best-known geometric FRACTAL. The boundary between the Mandelbrot set and its complement is often called the *Mandelbrot separator*. (Named after Polish-born, French-educated, and US-based mathematician *Benoit Mandelbrot* (born 1924), who also held chairs of engineering at Yale, economics at Harvard, and physiology at the Einstein College of Medicine.)

manifold, *n.* **1.** a collection of objects of a set. For example, an *affine manifold* is merely an AFFINE subset of a vector space.

2. (*Differential topology*) a TOPOLOGICAL SPACE each point of which has a neighbourhood HOMEOMORPHIC to the interior of a sphere in a Euclidean space of fixed dimension. When this holds, the neighbourhood of a point, together with the function mapping it into $\mathbb{R}^n$, is a *chart* or (*local*) *coordinate system*, and a collection of charts that cover the manifold is called an *atlas*. Formally, M is an *n*-dimensional manifold if there is a locally finite open cover, $\{U_\lambda : \lambda \in \Lambda\}$, of M such that, for each λ, there is a map ϕ_λ that maps U_λ homeomorphically onto an open subset of $\mathbb{R}^n$; the pair $(\phi_\lambda, U_\lambda)$ is then a chart, and the set

$$\Phi = \{(\phi_\lambda, U_\lambda) : \lambda \in \Lambda\}$$

is an atlas for M. The pair (M, Φ) is a $C^{(r)}$–manifold if Φ is a $C^{(r)}$ DIFFERENTIAL STRUCTURE. Compare ANALYTIC STRUCTURE.

Mann–Whitney test, *n.* (*Statistics*) a statistical test of the difference between the MEDIANS of the distributions of data collected in two experimental conditions applied to unmatched groups of subjects by comparing the distributions of the RANKS of the scores.

mantissa, *n.* the fractional part of a COMMON LOGARITHM, representing the digits of the given number but not its ORDER OF MAGNITUDE. For example, the mantissa of log45 and the mantissa of log4.5 are both 0.6532. Compare CHARACTERISTIC.

many–one, *adj.* (of a mapping or function) capable of associating the same element of the RANGE of a function with more than one member of the DOMAIN, or holding between more than one first argument and the same second argument of a relation, as represented by the diagram of Fig. 237.

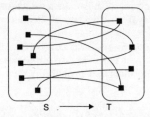

Fig. 237. A **many–one** mapping from S to T.

many-valued logic, *n.* **1.** a logical system in which the TRUTH VALUES that a statement may have are not restricted to simple truth and falsehood. If the DESIGNATED values are interpreted as species of truth, and the antidesignated values as species of falsehood, a TRUTH–VALUE GAP may still remain.
2. the study of such systems.

map, *n.* **1.** another word for MAPPING (although this conflicts with the usual English meaning).
2. the IMAGE of a given element or set under a mapping.

mapping or **map,** *n.* a FUNCTION or TRANSFORMATION. The language of mappings is often preferred for functions between abstract spaces. For some authors, a mapping is an ordered pair made up of a function together with a given CODOMAIN (a prescribed set containing the RANGE of the function) while a function remains a set of ordered pairs. In this sense, the mapping

$$f: \mathbb{I}^+ \to \mathbb{R} : n \mapsto \frac{1}{n}$$

of the positive integers into the rationals, taking each to its reciprocal, is different from the mapping

$$f: \mathbb{I}^+ \to \,]0, 1] : n \mapsto \frac{1}{n},$$

although both are the set of ordered pairs $\langle n, 1/n \rangle$. Where there is no qualification stated, there is generally no implication that the mapping is either INJECTIVE or SURJECTIVE; but in some non-technical usage a mapping is taken to be ONE-TO-ONE unless stated to the contrary. Discrete mappings are often represented by diagrams such as that in Fig. 237 above. See also DOMAIN.

marginal distribution, *n.* the PROBABILITY DISTRIBUTION FUNCTION of a component of a RANDOM VECTOR. For example, if $\mathbf{X} = (X_1, X_2)$ is a two-dimensional continuous random vector with probability distribution function $P_X(x_1, x_2)$, then the marginal probability distribution function of X_1 is

$$P_{X_1}(x_1) = \int_R P_X(x_1, x_2)\, dx_2.$$

marginal expectation, *n.* (*Statistics*) the EXPECTATION of a given component of a RANDOM VECTOR. This is related to the CONDITIONAL EXPECTATION by

$$E[E(X_1 \mid X_2)] = E(X_1),$$

where $\mathbf{X} = (X_1, X_2)$ is the random vector.

marginal probability, *n.* (*Statistics*) in a MULTIVARIATE distribution, the probability of one variable, or function of several of these variables, taking a specific value or falling in a specific interval, irrespective of the values of the other variables.

marginal probability measure, *n.* (*Statistics*) a MEASURE constructed from an OUTER MEASURE on a product space by restriction to one of the two factors: if μ^* is an outer measure on $X \times Y$, the marginal probability measure is a measure α that satisfies $\alpha^*(A) = \mu^*(A \times Y)$.

market equilibrium, see ECONOMY.

Markov (or **Markoff**) **chain** or **process,** *n.* (*Statistics*) a sequence of events, usually called *states*, the probability of each of which is dependent only on the event immediately preceding it. (Named after the Russian probabilist, topologist, number theorist, and algebraist, *Andrei Andreiëvich Markov* (1856–1922), who developed the theory of STOCHASTIC PROCESSES.) See also TRANSITION MATRIX.

marksman, *n.* (*Statistics*) an ESTIMATE that has smaller VARIANCE than any other estimate of the same parameter; formally, an estimate $M(X)$ associated with another estimate $S(X)$, such that

$$M(X) = E[S(X) \mid T(X)]$$

where $T(X)$ is any COMPLETE SUFFICIENT STATISTIC.

marriage theorem, *n.* (*informal*) the result that a PERFECT MATCHING exists for a BIPARTITE GRAPH of which all the vertices have the same non-zero degree. The name arises since the application of the result to a community in which each female knows exactly k males and each male exactly k females (where 'knowing' is taken to be a symmetric relation) shows that it is possible to pair all men and women exactly.

martingale, *n.* (*Statistics*) **1.** a sequence of RANDOM VARIABLES $\{x_n\}$ such that the conditional expectation of each x_{n+1}, given $x_1, x_2, ..., x_n$, is x_n.
2. in particular, a betting system in which one doubles the stakes after each loss and halves them after each win.

Mascheroni's constant, *n.* another term for EULER'S CONSTANT.

mass, *n.* (*Continuum mechanics*) a primitive concept in mechanics; it is assumed to be additive for disjoint bodies and, in Newtonian dynamics, to be constant for a given set of PARTICLES that may either constitute a BODY or be discrete. It is standardly measured in KILOGRAMS, and in practice the mass of a body is found by measuring its WEIGHT. Formally, mass is a MEASURE, and if the set of particles is a body it is required that the measure $m(\chi^{-1}(\cdot))$, where χ is a CONFIGURATION, should be ABSOLUTELY CONTINUOUS with respect to LEBESGUE MEASURE.

mass gap, see YANG–MILLS THEORY.

matched-pairs design, *n.* (*Statistics*) an experimental design concerned with measuring the differences in the values of the dependent variables for matched pairs of subjects. These are chosen to eliminate individual differences and are then respectively subjected to the CONTROL CONDITION and the EXPERIMENTAL CONDITION. Compare BETWEEN–SUBJECTS DESIGN, WITHIN–SUBJECTS DESIGN.

matching, *n.* (*Graph theory, Combinatorics*) a subset of the edges of a graph such that no two edges share the same vertex. Vertices matched by an edge are called *mates*. A matching is *perfect* if it is *complete* in the sense that all vertices are *mated*. A combinatorial *matching problem* seeks to find a matching of a given graph that maximizes the total value (extrinsically decided) of edges used. See also ASSIGNMENT PROBLEM, MARRIAGE THEOREM, TRANSPORTATION PROBLEM.

mate, see MATCHING.

material conditional, *n.* a statement whose main connective is MATERIAL IMPLICATION; the term is used to distinguish the TRUTH–FUNCTIONAL conditional statements, familiar in logic, mathematics and computing, from other context-sensitive forms of conditional statement common in ordinary language, such as those of STRICT IMPLICATION.

material derivative, *n.* (*Continuum mechanics*) the PARTIAL DERIVATIVE

$$\frac{\partial}{\partial t} \, \mathrm{T}(\mathbf{X}, \, t) \, \Big|_{\mathbf{X}} ,$$

of a FIELD, $\mathrm{T}(\mathbf{X}, \, t)$, defined on the REFERENCE CONFIGURATION of a BODY. This is equivalent to the DIFFERENTIAL OPERATOR

$$\frac{D}{Dt} \, = \, \frac{\partial}{\partial t} \, + \, \mathbf{u} \, . \, \nabla ,$$

when T is regarded as a field over the CURRENT CONFIGURATION.

material description or **Lagrangian description,** *n.* the description of physical phenomena associated with the deformation of a body in terms of FIELDS that are defined over the reference, rather than the current, CONFIGURATION. Compare SPATIAL DESCRIPTION.

material implication or **implication,** *n.* (*Logic*) **1.** the binary TRUTH–FUNCTIONAL sentential connective that assigns the value *false* to a compound sentence only when its first argument (the ANTECEDENT) is true and the second (the CONSEQUENT) is false, and the value *true* otherwise, without consideration of relevance, etc.; its TRUTH–TABLE is shown in Fig. 238. The closest approximation in natural language is *if ... then ...* . See MATERIAL CONDITIONAL.

P	Q	$P \rightarrow Q$
T	T	T
T	F	F
F	T	T
F	F	T

Fig. 238. Truth table for **material implication**.

2. a compound sentence formed with this connective, written '$P \rightarrow Q$' or '$P \supset Q$', where P is the antecedent and Q the consequent; a CONDITIONAL statement.

3. the relationship that holds between an ordered pair of sentences when there are no circumstances in which the first is true and the second false.

4. paradoxes of material implication. a number of inferential patterns that follow immediately from the definition: a falsehood materially implies any statement whatever, and anything whatever materially implies a truth. Compare STRICT IMPLICATION.

math or **maths,** *abbrev. for* MATHEMATICS or MATHEMATICAL.

mathematical, *adj.* of, used in, pertaining to, using, or characterized by MATHEMATICS or its methods and principles, and in particular its precision.

mathematical expectation, *adj.* (*Statistics*) another name for EXPECTED VALUE.

mathematical induction, *n.* another name for INDUCTION (sense 1). See also PEANO'S AXIOMS.

mathematical logic, *n.* FORMAL LOGIC, especially that branch concerned with the FOUNDATIONS OF MATHEMATICS.

mathematical model, *n.* a fragment of a mathematical or formal theory that reflects some aspect of a particular physical, social, economic, technological, natural or other phenomenon or process and enables predictions to be made about its behaviour. This may require simplification of the problem, and the predictions may therefore only be approximate. For example, it is possible to construct a computer model of the national economy in order to test the probable results of certain changes in government policy. See INPUT–OUTPUT MODEL.

mathematical probability or **classical probability,** *n.* (*Statistics*) **1.** the probability of an event consisting of n out of m possible equally likely occurrences, defined to be n/m. See also INDIFFERENT.

2. the study of such probabilities.

mathematical programming, *n.* the theory and application of the OPTIMIZATION of functions, often subject to CONSTRAINTS given in terms of functions. The most basic problem studies optimization of a real-valued function over a prescribed set. This involves establishing the existence of optimal points, characterizing such points, and finding algorithms for their computation. See SIMPLEX METHOD, PENALTY METHODS, QUADRATIC PROGRAMMING PROBLEM, DUALITY THEORY OF LINEAR PROGRAMMING, LAGRANGE METHOD OF MULTIPLIERS.

mathematics, *n.* **1.** a group of related subjects, including ALGEBRA, GEOMETRY, TRIGONOMETRY and CALCULUS, concerned with the study of number, quantity, shape, and space, and their inter-relationships, applications, generalizations and abstractions. It is not too much of a parody of the history of mathematics to see it as developing from the arithmetical description of a few common-place concerns, albeit that the results obtained at this early stage, by, for example, Babylonian and Chinese astronomers, were both sophisticated and accurate. As this process of arithmetization expanded, more general conceptions of NUMBER were developed, together with the crucial insight that these results had general validity, and represented functional relationships; the lack of suitable notations for VARIABLES and FUNCTIONS, however, hampered the development of mathematics for a millennium. Although it was recognized at an early stage that justification requires rigorous PROOF which must ultimately rest upon unexceptionable AXIOMS (until the limitations of the axiomatic method were proven in the present century), and for two millennia Euclid's treatment of geometry was regarded as the paradigm not only of mathematics but of science in general, the present concept of rigour in proof and definition, as exemplified by EPSILON–DELTA NOTATION, largely originates with Cauchy and others in the 19th century. The search for certainty also gave rise, as in Descartes' ANALYTIC GEOMETRY, to a search for unification of superficially distinct branches of the subject, which in turn led to an increasing level of abstraction as mathematicians directed their attention, in ABSTRACT ALGEBRA, to the properties of structures and operators as such. The question of the indispensability of such accepted axioms as the parallel postulate of EUCLIDEAN GEOMETRY requires consideration of the INDEPENDENCE of the axioms, while the discovery of the PARADOXES of the infinite demands consideration of their CONSISTENCY. This new METAMATHEMATICS was given a further boost by the development of computing and the need to describe the properties of its ALGORITHMS; this brings logic itself under scrutiny, so that the historic themes of axiomatization and generalization converge at the interface with philosophy and logic in the FOUNDATIONS OF MATHEMATICS.

2. mathematical operations and processes involved in the solution of a problem or the study of some scientific field.

See also HIGHER MATHEMATICS, APPLIED MATHEMATICS.

Mathieu's differential equation, *n.* the differential equation, originating in the study of vibrations in the two-dimensional WAVE EQUATION, of the form

$$\frac{d^2 y}{dt^2} + (a + 16q \cos 2t) y = 0.$$

For *a* appropriately related to *q*, periodic solutions (with period 2π) exist; of these, the even and odd solutions are called *Mathieu functions*.

matric, matrical or **matricial,** *adj.* of, or pertaining to, a MATRIX. A *matric polynomial* is constructed by replacing each power of the indeterminate in a polynomial by the same power of a given matrix. The CAYLEY–HAMILTON THEOREM ensures that a matrix is always a zero of the matric polynomial arising from the characteristic polynomial of the given matrix.

matrix, *n.* a rectangular array of elements, usually themselves members of a FIELD and referred to as SCALARS, set out in rows and columns; matrices are used to facilitate the solution of problems such as the transformation of coordinates, and are usually indicated by enclosing the entire array within brackets. An *m*-by-*n* (written $m \times n$) matrix has *m* rows and *n* columns, and the entry denoted a_{ij} lies at the intersection of the ith row and the jth column; for example,

$$\begin{bmatrix} a_{11} & a_{12} & a_{13} & a_{14} \\ a_{21} & a_{22} & a_{23} & a_{24} \\ a_{31} & a_{32} & a_{33} & a_{34} \end{bmatrix}$$

is a 3-by-4 matrix. A matrix with the same number of rows and columns is a *square matrix* and a square matrix with all entries zero except those at the intersection of the pairs of rows and columns with the same number is a *diagonal matrix*; an $n \times n$ diagonal matrix with all non-zero entries equal to 1 is the *identity matrix* of order *n*. Addition is defined for matrices with the same dimensions: the elements of the sum are equal to the sums of the corresponding elements of the addends. The product of matrices is defined only if they are CONFORMABLE, that is, if the number of columns of the first equals the number of rows of the second, in which case the ij^{th} entry of the product is the INNER PRODUCT of the ith row of the first with the jth column of the second; that is, if A is the $m \times n$ matrix $[a_{ij}]$ and B is the $n \times p$ matrix $[b_{ij}]$, then AB is the $m \times p$ matrix of which the ik^{th} entry is

$$\sum_{j=1}^{n} a_{ij}\, b_{jk}.$$

It is provable that both addition and multiplication of matrices are associative, and multiplication distributes over addition, but while addition is commutative, multiplication is not. Subtraction and division are defined in the usual way, and a matrix is only invertible if it is NON–SINGULAR, that is, if its DETERMINANT is non-zero. A *matrix equation* in which the product of an $m \times n$ MATRIX OF COEFFICIENTS and a vector of n variables is set equal to a vector of m constants is equivalent to a system of m SIMULTANEOUS EQUATIONS in n unknowns. See also ELEMENTARY MATRIX OPERATIONS.

matrix of coefficients, *n.* the $m \times n$ matrix A in the set of linear equations $A\mathbf{x} = \mathbf{b}$, where $\mathbf{x}$ is a vector of n variables, and $\mathbf{b}$ is a vector of m constants; this is equivalent to a system of m SIMULTANEOUS EQUATIONS in n unknowns.

matroid, *n.* (*Combinatorics*) a collection of subsets of a set such that any subset of a member of the collection belongs to the collection, and such that if the sets $\{a_1,\ldots, a_k\}$ and $\{b_1,\ldots, b_{k+1}\}$ are in the collection, then so is the set $\{a_1,\ldots, a_k, b_i\}$ for some $i \le k + 1$. Matroids are thus characterized by this *exchange property*. Given any graph, the set of all sets of edges containing no CYCLE is a matroid.

max, *abbrev. and symbol for* MAXIMUM or MAXIMAL.

max-flow min-cut theorem, *n.* the important duality theorem, due to Ford and Fulkerson, that the value of a maximum value NETWORK FLOW is equal to the total capacity of a minimum NETWORK CUT. See also FORD–FULKERSON ALGORITHM.

maximal, *adj.* **1.** (of an element of an ORDERING or LATTICE) having no element greater than it; being the GREATEST element of a CHAIN. A maximal element need not be the unique greatest element unless the ordering is TOTAL; for example, the set of proper subsets of a given set ordered by inclusion has no greatest element, but every set formed by removing a single member from the given set is maximal. See MAXIMUM. Compare MINIMAL.
2. (of an ORTHOGONAL or ORTHONORMAL sequence) such that whenever the INNER PRODUCTS of any given element with the sequence elements are zero, so is the given element. Unless the inner product space is complete, not every maximal orthonormal sequence need be a Schauder BASIS, but clearly every basis is maximal.

maximal domain, see DOMAIN.

maximal ideal, *n.* a proper IDEAL that is maximal with respect to inclusion.

maximin, *adj.* (*Game theory*) being or relating to a strategy or value that maximizes the minimum value of a function; for example, the maximin of the family $\{f_n\}$ is a function f_i such that $\min f_i > \min f_j$ for all $j \ne i$. Compare MINIMAX THEOREM.

maximizing, see PAYOFF.

maximize, *vb.* to find or yield the MAXIMUM value of (a function).

maximized likelihood, *n.* the probability of randomly drawing a given sample from a population, maximized over the possible values of the population parameters.

maximum (abbrev. **max**), *n.* **1.** the largest element of a set S, usually denoted maxS; for example, the negative numbers have no maximum, but the non-positive numbers have maximum 0, although both have 0 as SUPREMUM. See MAXIMAL.

2. the highest value of a function, usually denoted max*f*; it is a GLOBAL maximum if this condition is satisfied in comparison with all other values of the function. A LOCAL maximum is a value greater than any other in a neighbourhood of its argument, and, in the case of a real differentiable function on an open set, is identified by having zero derivative and negative second derivative since the tangent to the curve changes from rising to falling at this point. In Fig. 239, the left-hand maximum is global and the other local. See FIRST DERIVATIVE TEST and SECOND DERIVATIVE TEST.

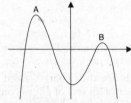

Fig. 239. A is a global and B a local **maximum**.

maximum condition, *n.* the condition on a MODULE that every non-empty set of submodules has a MAXIMAL member. See also NOETHERIAN MODULE, ASCENDING CHAIN CONDITION. Compare MINIMUM CONDITION.

maximum likelihood, *n.* (*Statistics*) **1.** the non-BAYESIAN rule that, given an experimental observation, one should utilize as POINT ESTIMATES of parameters of a distribution, those values that give the highest CONDITIONAL PROBABILITY to that observation, irrespective of the *a priori* probability assigned to the parameters.

2. (*as modifier*) (of a test, method, estimator, etc.) concerning or in terms of MAXIMIZED LIKELIHOOD. See GENERALIZED MAXIMUM LIKELIHOOD RATIO TEST STATISTIC.

maximum modulus theorem, *n.* the theorem of complex analysis that if a function is analytic and attains its maximum modulus in some open region, it is constant. As a result, if the region D is bounded and the modulus of the function f is continuous on the closure of D, then $|f(z)|$ achieves its maximum on the boundary of D. If the function vanishes nowhere in the domain, the minimum modulus is also achieved on the boundary.

maximum value theorem, *n.* the theorem of mathematical analysis, due to Weierstrass, that a real-valued continuous function on a CLOSED BOUNDED INTERVAL (or, more generally, any COMPACT set) achieves its SUPREMUM (and INFIMUM); that is, it has a supremum and takes that value for some argument, and so, *a fortiori*, has a bounded range. For example, on the closed interval $[0, 1]$, the function $y = x^3 - x^2 + 1$ achieves its maximum at $x = 1$, but on the open interval $]0, 1[$ it does not, as for all $x < 1$ there is an $x < x' < 1$ for which $f(x') > f(x)$.

353

maximum principle or **minimum principle,** see PONTRYAGIN'S MAXIMUM PRINCIPLE.

Maxwell, James Clerk (1831–79) Scottish physicist whose first paper was published by the Royal Society of Edinburgh while he was still at school. He became the first Cavendish Professor of Physics at Cambridge, and wrote on a wide range of aspects of electromagnetic radiation, for which he published the first complete account, as well as mechanics, the kinetic theory of gases, and astronomy.

Maxwell's equations (**of electromagnetism**), *n.* a set of four equations which summarize the laws of electricity and magnetism. These equations relate the ELECTRIC FIELD and MAGNETIC FIELD vectors **E** and **B** to their sources, electric charge, CURRENT, and changing fields. Maxwell's equations are, in differential form:

$$\nabla . \mathbf{E} = \frac{\rho}{\varepsilon_0} \qquad \nabla \times \mathbf{E} = -\frac{\partial \mathbf{B}}{\partial t}$$

$$\nabla . \mathbf{B} = 0 \qquad \nabla \times \mathbf{B} = \mu_0 \varepsilon_0 \frac{\partial \mathbf{E}}{\partial t} + \mu_0 \mathbf{j},$$

where ρ is the density of charge, ε_0 and μ_0 are fundamental constants (called, respectively, the permeability and the permittivity of free space), and **j** is the CURRENT density. In principle all problems in electromagnetism may be solved using Maxwell's equations. It may be easily shown that Maxwell's equations imply that the electric and magnetic field vectors obey a WAVE EQUATION which describes ELECTROMAGNETIC RADIATION propagating through a vacuum with the velocity of light. Although essentially a classical result, Maxwell's equations are unchanged in form when reformulated in special relativity. See also ELECTROMAGNETIC POTENTIALS.

Mayer problem, *n.* the special case of BOLZA'S PROBLEM in control theory, of which the integrand is zero.

Mazur separation theorem, see SEPARATION THEOREM OF MAZUR.

meagre, see BAIRE CATEGORY.

mean, *n.* **1.** another word for AVERAGE. See ARITHMETIC MEAN. Compare GEOMETRIC MEAN, ARITHMETIC–GEOMETRIC MEAN, HARMONIC MEAN.

2. (*Statistics*) a PARAMETER or STATISTIC, written $\mu(\text{X})$ for the population mean and $\overline{\text{X}}$ for a sample mean. The population mean is obtained by summing or integrating the product of each possible value of a variable with its probability, the sum or integral being taken over the whole range of the variable; the sample mean is obtained by summing the observed values in the sample and dividing by their number. Compare MEDIAN, MODE. See also EXPECTED VALUE.

3. the INTEGRAL of a continuous function, *f*, over an interval $]a, b[$, divided by the length of the interval,

$$\frac{1}{b-a} \int_a^b f(x) \; \mathrm{d}x.$$

Since the integral is the area between the graph of *f* and the axis between *a* and *b*, the mean of the function is the height of the rectangle on the same base and with the same area, and so, in an obvious sense, is the average value of *f* on this interval.

4. the second and third terms in a PROPORTION; for example, the terms b and c in the proportion $a : b = c : d$.

5. any function of two or more positive variables of which the value always lies between the minimum and the maximum values of the arguments, such as the HÖLDER MEANS. See also NEOPYTHAGOREAN MEANS.

6. see CONVERGENT IN MEAN.

mean and extreme proportion, *n.* the relationship, written $a : b :: c : d$, between four numbers or quantities when $ad = bc$.

mean curvature or **mean normal curvature,** *n.* (at a point of a surface) the sum of the PRINCIPAL CURVATURES.

mean deviation, *n.* (*Statistics*) **1.** the difference between an observed value of a variable and its MEAN.

2. also called **mean deviation from the mean**, **average deviation.** a measure of the dispersion of a distribution, derived by computing the mean of the absolute values of the differences between the observed values of the variable and their mean. Compare STANDARD DEVIATION.

mean (or **weak**) **ergodic theorem,** *n.* a weaker form of the BIRKHOFF ERGODIC THEOREM, due to von Neumann, in which one only obtains functions CONVERGENT IN MEAN square. See ERGODIC.

mean error, see PROBABLE ERROR.

mean normal curvature, see MEAN CURVATURE.

mean proportional, *n.* another term for GEOMETRIC MEAN, especially of two numbers.

mean square, *adj.* of or pertaining to the sum or integral of the squares of the absolute values of a given series or integral. The *mean square norm* or HILBERT NORM of a function f on an interval $]a, b[$ is given by

$$\int_a^b |f(t)|^2 \, dt.$$

See also CONVERGENT IN MEAN, L_p SPACE.

mean-value theorem, *n.* **1.** an elementary result in mathematical analysis, due to Lagrange, that states that, if a real function, f, is continuous on a closed interval $[a, b]$ and differentiable on the open interval $]a, b[$, then there is a point in the open interval at which the first derivative of the function equals

$$\frac{f(b) - f(a)}{b - a} \; ;$$

thus there is a point on any arc of the graph of the function at which the tangent is parallel to the chord joining the end points of the arc. The *generalized mean-value theorem*, known as CAUCHY'S MEAN–VALUE THEOREM, extends this to show that given two such functions, f and g, one can solve

$$f'(c)\,[\,g(b) - g(a)\,] \;=\; g'(c)\,[\,f(b) - f(a)\,]$$

for some c in $[a, b]$.

2. any of the corresponding results for integrals. The *first mean-value theorem for integrals* asserts that the average value of the definite integral of a continuous function over an interval is attained. The *generalized mean-value* *theorem* extends this to show that, when *f* is continuous and *g* is non-negative and integrable, then

$$\int_a^b f(t)\, g(t)\, \mathrm{dt} = f(c) \int_a^b g(t)\, \mathrm{dt}$$

for some *c* in $]a, b[$. The *second mean-value theorem* states that when *g* is integrable and *f* is monotone, then

$$\int_a^b f(t)\, g(t)\, \mathrm{dt} = f(a) \int_a^c g(t)\, \mathrm{dt} + f(b) \int_c^b g(t)\, \mathrm{dt}$$

for some *c* in $[a, b]$. When *f* is non-negative and non-increasing there is a further specialization, *Bonnet's mean-value theorem*:

$$\int_a^b f(t)\, g(t)\, \mathrm{dt} = f(a) \int_a^x g(t)\, \mathrm{dt}$$

for some *x* in $[a, b]$.

measurability-preserving transformation, *n.* a one-to-one transformation between MEASURE SPACES such that both the given map and its inverse are MEASURABLE.

measurable, *adj.* **1a.** (of a set) belonging to the SIGMA–ALGEBRA in a given MEASURABLE SPACE.

b. (of a set E with respect to a given OUTER MEASURE μ^*) having the property that for every subset A of the space,

$$\mu^*(A) = \mu^*(A \cap E) + \mu^*(A \setminus E).$$

2. (of a function or transformation between MEASURE ALGEBRAS) such that the inverse image of a measurable set in the range space is measurable in the domain space. Thus a real-valued function is *Borel measurable* if the inverse image of every open (or Borel) set is Borel measurable, and a real-valued function is *Lebesgue measurable* if the inverse image of every open (or Borel) set is Lebesgue measurable. When the range of *f* is the extended real numbers, $f^{-1}(\pm\infty)$ is also required to be measurable. Intuitionists hold that all sets are Lebesgue measurable since the construction of non-measurable sets depends upon the axiom of choice, and it has been shown that the assumption that all sets are Lebesgue measurable is consistent with the remaining axioms of set theory.

measurable cover, *n.* (for a set) a collection of measurable sets whose union contains the given set.

measurable space, *n.* a set together with a SIGMA–ALGEBRA on the set.

measurable kernel, *n.* a set K contained in a given set E, such that every measurable subset of E \ K has measure zero.

measure, *n.* **1.** a signed scalar associated with a vector and indicating both its MAGNITUDE and its SENSE but not its ORIENTATION. Two vectors $\overrightarrow{AB}$ and $\overrightarrow{BA}$ with resultant zero have measures of opposite polarity. See also NORM. Compare LENGTH, METRIC.

2. a non-negative EXTENDED real-valued function defined on the subsets of a set that is ADDITIVE, or, more commonly, COUNTABLY ADDITIVE, for disjoint subsets and is zero for the empty set. The former is often called a *finitely additive measure*. If the function is permitted to take either sign, it is a *signed measure*. See also MEASURE SPACE, LEBESGUE MEASURE.

measure algebra or **field of sets,** *n.* a Boolean SIGMA–ALGEBRA endowed with a measure.

measure preserving transformation, *n.* a one-to-one transformation between MEASURE SPACES that is MEASURABILITY PRESERVING and that preserves the MEASURE of sets. More generally, a MEASURABLE transformation such that the measure of every measurable set and of its inverse image agree.

measure ring, *n.* a Boolean SIGMA RING endowed with a measure.

measure space, *n.* a MEASURABLE SPACE endowed with a non-negative MEASURE. Any non–empty set, the collection of all its subsets, and the counting measure (the CARDINALITY of these subsets) define a measure space. It is often desired that the measure be positive except on the zero element, and this may be achieved by identifying sets with null symmetric difference; the sets of finite measure in the resulting ring form a metric space in which the metric is given by the measure of the symmetric difference of two sets.

measure theory, *n.* the study of MEASURABLE sets and functions, introduced by Lebesgue in order to generalize the RIEMANN INTEGRAL.

mechanical or **mechanistic,** *adj.* (of a procedure) not requiring interpretation; able to be applied by a suitably programmed computer. Compare HEURISTIC.

mechanics, *n.* the application of mathematical methods to the study of the equilibrium and motion of bodies in a particular frame of reference, including STATICS, DYNAMICS, and KINEMATICS.

medial triangle, *n.* another term for MEDIAN TRIANGLE.

median, *n.* **1.** (*Statistics*) the middle value of a FREQUENCY DISTRIBUTION, such that the probabilities of the variable taking a value below and above it are equal. In a DISCRETE distribution the median is the middle term, or, if there are an even number, the average of the two middle terms, when the values are written out in ascending order. For example, the median of both the sets of values {1, 7, 31} and {2, 5, 9, 16} is 7. Compare MEAN, MODE.

2. a straight line joining one vertex of a triangle to the midpoint of the opposite side. All three such lines coincide in the CENTROID, as shown in the first diagram of Fig. 240. See also MEDIAN TRIANGLE.

3. a straight line joining the midpoints of the non-parallel sides of a trapezium, and parallel to the two parallel sides, as shown in the second diagram in Fig. 240.

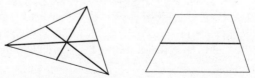

Fig. 240. **Medians** of a triangle and a trapezium.

median triangle or **medial triangle,** *n.* (of a given triangle) the triangle whose vertices are the midpoints of the sides of the original triangle. In

Fig. 241 the triangle XYZ is the median triangle of ABC; as is obvious, it is similar to, and congruent with, the other three triangles formed within the original triangle. See MIDPOINT THEOREM.

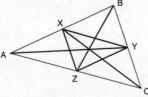

Fig. 241. The bold triangle XYZ is the **median triangle** of ABC.

mediator, *n.* the perpendicular bisector of a line segment, or, more generally, any AXIS OF SYMMETRY.

meet, 1. *n.* the binary LATTICE operator whose value is the INFIMUM of a pair of elements of a lattice; given a pair of elements *x* and *y* of the lattice, their meet, written $x \wedge y$, is the element *j* such that $j \leq x, j \leq y$, and there is no $k > j$ that has the same relations to *x* and *y*. Compare JOIN.

2. *vb.* to intersect, as of two sets or lines.

mega- (abbrev. **M**), *prefix denoting* a multiple of 10^6 of the physical units of the SYSTEME INTERNATIONAL.

Mellin transform or **Mellin inversion formula,** *n.* an INTEGRAL TRANSFORM that is defined by

$$g(s) = \int_0^\infty x^{s-1} f(x) \, dx,$$

and that is invertible under reasonable conditions. The Mellin transform of the negative exponential is the GAMMA FUNCTION. (Named after the Finnish analyst and physicist, *Robert Hjalmar Mellin* (1854–1933).) Compare FOURIER TRANSFORM.

member or **element,** *n.* any individual object belonging to a set or logical class.

menage problem, the problem of determining how many seating ARRANGE-MENTS exist at a circular dining table for *k* couples in such a way that *n* of the couples sit together and men and women alternate; this problem is in general extremely complex. With 10 couples, none of whom sit in adjacent seats, there are 3 191 834 419 200 distinct possibilities.

Menelaus' theorem, *n.* (*Geometry*) the theorem that, for any triangle, points X, Y, and Z respectively on the sides BC, CA, and AB (extended if necessary) are collinear if and only if the product of the DIRECTED RATIOS

$$\frac{BX}{XC} \frac{CY}{YA} \frac{AZ}{ZB} = -1.$$

For example, in Fig. 242, Y and Z internally divide AC and AB respectively in the ratio 3:1 and 1:2, so that X divides BC externally in the ratio 6:1.

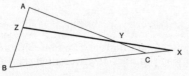

Fig. 242. **Menelaus' theorem.** See main entry.

(Named after *Menelaus of Alexandria*, first century AD Greek mathematician, whose most important, and only extant, work was on spherical geometry, for which he gave an axiomatic treatment analogous to Euclid's for plane geometry; he introduced spherical triangles, and thus made an important contribution to classical astronomy.) See also CEVA'S THEOREM.

mensuration, *n.* the study of the measurement of geometric magnitudes such as length.

mereology, *n.* the formal study of the logical properties of the relation of whole and part.

meromorphic, *adj.* (of a complex function in a domain) ANALYTIC except at POLES.

meromorphism, *n.* **1.** (*Analysis*) a complex function all of whose singularities are POLES; it is thus ANALYTIC elsewhere in the given domain.

2. (*Algebra*) a MONOMORPHISM from a GROUP into itself.

meridian, meridian curve, or **meridian section,** *n.* a plane section of a SURFACE OF REVOLUTION, such as a paraboloid, that contains the axis of revolution. Compare PARALLEL SECTION.

Mersenne, Marin (1588–1648), French monk, theologian, philosopher, and number theorist, whose wide travels enabled him to act as a channel of communication amongst such European academics as Descartes and Galileo, whom he championed, and Fermat, Pascal, and Huygens. He also inspired the invention of the pendulum clock.

Mersenne number or **Mersenne prime,** *n.* any PRIME NUMBER of the form $2^n - 1$, where n must itself be prime. Thus 3, 7, 31, 127 are Mersenne primes. Not all such numbers are primes, and it is unknown whether there are infinitely many of them; the largest known is $2^{13\,466\,917} - 1$, a number with 4 053 946 digits, which was discovered in December 2001 by the GIMPS project. Compare FERMAT PRIME. See also PERFECT NUMBER.

Merten's theorem, *n.* the theorem that the product of the values of two complex series, one of which is ABSOLUTELY CONVERGENT, is the value of the series of which the coefficients are the CAUCHY PRODUCT of the terms of the given series; that is,

$$(\textstyle\sum_n a_n)\ (\textstyle\sum_n a_n)\ =\ \textstyle\sum_n \{ \sum_{j+k=\,n} a_j b_k \}.$$

If both series are absolutely convergent, then so is the Cauchy product.

mesh-fineness or **norm,** *n.* (of a PARTITION of an interval) the SUPREMUM of the set of distances between successive members of the partition.

mesokurtic, *adj.* (*Statistics*) (of a distribution) having KURTOSIS $B_2 = 3$, and thus concentrated around its mean to the same extent as a NORMAL DISTRIBUTION. Compare LEPTOKURTIC, PLATYKURTIC.

message, *n.* a discrete piece of information, whose likelihood of accurate transmission through a CHANNEL with a given PROBABILITY LAW is the subject of study by INFORMATION THEORY.

metalanguage, *n.* the language in which another language or system of symbols, the OBJECT LANGUAGE, is described. For example, when one says that 'P & Q' is true if and only if P and Q are both true, one is defining the connective '&' in the object language (in this case, sentential calculus) in terms of the conjunction of the metalanguage (in this case English). See also FORMAL LANGUAGE, NATURAL LANGUAGE.

metamathematics, *n.* the logical study and analysis of the reasoning, principles, and rules that govern the use and combination of mathematical symbols, numbers, etc., including questions of both SYNTAX and SEMANTICS such as those studied by COMBINATORIAL LOGIC, and whether axiom systems are INDEPENDENT, CONSISTENT, or COMPLETE. In so far as INTUITIONIST LOGIC regards itself as descriptive of the best practice of mathematical proof, it is a metamathematical enterprise.

metavariable, *n.* (*Logic*) a variable in the METALANGUAGE that can take expressions of the OBJECT LANGUAGE as values. So, for example, in the rule of sentential calculus that $A \& B \vdash A$, 'A' is a metavariable that can be replaced by any well-formed formula to yield a valid sequent of the calculus.

method of exhaustion, *n.* another name for EUDOXUS' AXIOM.

method of false position, see FALSE POSITION.

method of linear interpolation, *n.* another name for the SECANT METHOD for finding a zero of a real function.

method of partial fractions, see PARTIAL FRACTIONS.

method of successive displacements or **Gauss–Seidel iteration,** *n.* the variant of the JACOBI ITERATION, in which one immediately uses the new information. This usually out-performs the Jacobi method. Explicitly, this yields the iterative method of solving a linear matrix system $A\mathbf{x} = \mathbf{b}$ by repeatedly solving for $\mathbf{x}_{NEW}$ from

$$\mathbf{x}_{NEW(i)} = \frac{1}{a_{ii}} \left[\mathbf{b}_i - \sum_{i<j} a_{ij} \mathbf{x}_{OLD(j)} - \sum_{i>j} a_{ij} \mathbf{x}_{NEW(j)} \right].$$

This assumes that $a_{ii} \neq 0$, and will not always converge even then.

metre (symbol **m**), *n.* the standard unit of LENGTH; one of the fundamental units of the SYSTEME INTERNATIONAL. It was once defined in terms of the length of a bar of platinum kept under constant conditions in Paris, but is now defined by stipulating that the speed of light is 299 792 458 metres per second.

metric, *n.* **1.** a non-negative SYMMETRIC binary function defined for a given set, often denoted $d(x, y)$, $\delta(x, y)$ or $\rho(x, y)$ and referred to as DISTANCE, that satisfies the TRIANGLE INEQUALITY

$$\delta(x, y) + \delta(y, z) \geq \delta(x, z),$$

and is zero only if $x = y$. More generally, a *pseudo-metric* allows $\delta(x, y) = 0$ for $x \neq y$; a *quasi-metric* or *skew-metric* need not be symmetric, while a *semi-metric* may fail the triangle inequality. For example, the ordinary distance function in the plane is a metric, since $|AB| = |BA|$, the triangle inequality follows from Pythagoras' theorem, and $|AB| = 0$ only if A and B coincide. Another metric is

$$\sup_{0 \leq x \leq 1} |f(x) - g(x)|$$

on the continuous real functions on the interval $[0, 1]$. See also METRIC SPACE. Compare NORM, MEASURE.

2. of, using, or pertaining to the METRIC SYSTEM.

metrical, *adj.* of or relating to measurement.

metrication, *n.* the conversion (of an instrument, measuring system, etc.) to the METRIC SYSTEM.

metric density, *n.* the METRIC OUTER DENSITY of a MEASURABLE set. A point with unit density is called a *point of density* for the set, and a point with zero density is called a *point of dispersion* for the set. Almost all points of any set are points of density of the set. The set is measurable if and only if almost all points in its complement are points of dispersion of the set.

metric density theorem, *n.* the theorem that if E is a MEASURABLE subset of the real numbers, and λ is a LEBESGUE MEASURE, then

$$c(x) = \lim_{a \to 0} \frac{\lambda(E \cap (x - a, x + a))}{2a}$$

exists and agrees with the CHARACTERISTIC FUNCTION, $\chi_E(x)$, of E, almost everywhere. More generally, if λ^n is a Lebesgue measure in $\mathbb{R}^n$, then

$$\lim_{\varepsilon \to 0} \frac{\lambda(E \cap B_\varepsilon(x))}{2\varepsilon}$$

exists and converges almost everywhere to $\chi_E(x)$.

metric outer density, *n.* for a given set, A, in Euclidean space with respect to Lebesgue OUTER MEASURE, μ^*, at a point x, the limit, if it exists, as $\mu^*(I_x)$ tends to zero, of the ratio

$$\mu^*(A \cap I_x)/\mu^*(I_x),$$

over all closed intervals I_x containing x. A point with unit outer density is called a *point of density* for A, and a point with zero outer density is called a *point of dispersion* for A. If the set in question is measurable, one speaks simply of METRIC DENSITY.

metric projection, see PROJECTION.

metric space, *n.* a set endowed with a METRIC; this induces a TOPOLOGY on the set in which Ω is open if and only if for all x in Ω, there is a positive ε such that the OPEN BALL $B_\varepsilon(x)$ is contained in Ω.

metric system, *n.* any physical measuring system of which the units and subunits are related by multiples of ten, and of which the METRE is the unit of length. For example, metric units in the SYSTEME INTERNATIONAL (SI) are based on the metre, kilogram, and second (the *mks* system); a previous system based on the centimetre, gram and second (the *cgs* system) is now superseded.

metric tensor, *n.* a TENSOR, **g**, that is SYMMETRIC and of type $(0, 2)$, such that the MATRIX $[\mathbf{g}_{ab}]$ is NON–SINGULAR. The *contravariant metric tensor* defined by **g** is the tensor with components $\mathbf{g}^{ab}$ such that $\mathbf{g}^{ab}\mathbf{g}_{bc} = \delta^a_c$ (that is, it is the MATRIX INVERSE of $\mathbf{g}_{ab}$). The metric tensor and the contravariant metric tensor provide T and T* respectively with an INNER PRODUCT.

metrizable, *adj.* (of a TOPOLOGY) compatible with a METRIC. *Urysohn's metrization theorem* asserts the metrizability of any REGULAR, HAUSDORFF, SECOND COUNTABLE space. For example, the set of real numbers under the discrete topology is metrizable, but under the ZARISKI TOPOLOGY it is not. See also COMPACTUM.

Michael's continuous selection theorem, see SELECTION.

micro- (symbol μ), *prefix denoting* a fraction of 10^{-6} of the physical units of the SYSTEME INTERNATIONAL.

microstate, see STATE.

middle term, *n.* (*Logic*) the term of a SYLLOGISM that occurs in both premises but not the conclusion.

midline, *n.* the MEDIAN of a trapezoid.

midpoint, *n.* the point on a line segment that is equidistant from its endpoints.

midpoint theorem, *n.* the theorem that the line joining the midpoints of two sides of a triangle is parallel to the third side and half its length. See MEDIAL TRIANGLE.

Millennium Prize Problems, *n.* a collection of seven problems, mostly classical but some modern, published by the Clay Mathematics Institute of Cambridge, Massachusetts (CMI) to celebrate mathematics in the new millennium and to recapture the impact of HILBERT'S PROBLEMS. All have resisted solution and a one million dollar prize is offered for the solution to each. A full list of the problems is given in Appendix 4.

milli- (symbol **m**), *prefix denoting* a fraction of 10^{-3} of the physical units of the SYSTEME INTERNATIONAL.

milliard, *n.* (*in UK usage*) a thousand million; a BILLION in North American usage.

min, *abbrev. and symbol for* MINIMUM or MINIMAL.

minimal, *adj.* (of an element of an ordering or LATTICE) having no element smaller than it; being the least element of a CHAIN. A minimal element may not be the unique LEAST element unless it is a TOTAL ORDERING. For example, the set of non-empty subsets of a given set ordered by inclusion has no least element, but every singleton is minimal. See MINIMUM. Compare MAXIMAL.

minimal ideal, *n.* a proper non-zero IDEAL that is minimal with respect to inclusion.

minimal surface, *n.* a surface of which the MEAN CURVATURE vanishes identically. A smooth surface minimizing the area spanned by a given contour is minimal in this sense.

minimax, *adj.* (*Game theory*) being or relating to a strategy or value that minimizes the maximum value of a function; for example, the minimax of the family $\{f_n\}$ is a function f_i such that $\max f_i < \max f_j$ for all $j \neq i$. See also MINIMAX THEOREM.

minimax strategy, *n.* (*Game theory*) an optimal MIXED STRATEGY.

minimax theorem, *n.* (*Game theory*) a theorem justifying the exchange of order in taking the minimum and maximum of a SADDLE FUNCTION:

$$\min_X \max_Y F(x, y) = \max_Y \min_X F(x, y).$$

This number, if it exists, is called the *value* of the associated TWO–PERSON GAME. The *Sion minimax theorem* shows that this minimax exists when X and Y are compact, and $F(x, \cdot)$ is lower semicontinuous and quasi-convex, while $F(\cdot, y)$ is upper semicontinuous and quasi-concave. The most celebrated case is *von Neumann's minimax theorem*, in which X and Y are poly-hedra and F is bilinear, corresponding to a PAYOFF matrix.

minimize, *vb.* to find, or yield, the MINIMUM value of a given function.

minimizing, see PAYOFF.

minimum (abbrev. **min**), *n.* **1.** the least element of a set, usually denoted min S. For example, the positive numbers have no minimum, but the non-negative numbers have minimum 0, although both have 0 as INFIMUM. See MINIMAL.

2. the lowest value of a function, usually denoted min *f*; it is a GLOBAL minimum if this condition is satisfied with respect to all other values of the function. A LOCAL minimum is a value less than any other in a neighbourhood of its argument, and is identified in the real differentiable setting by having zero first derivative and positive second derivative, since the tangent to the curve changes from falling to rising at this point. In Fig. 243 the left-hand minimum is global, and the other is local. See FIRST and SECOND DERIVATIVE TESTS.

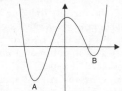

Fig. 243. A graph with a global **minimum** at A and a local **minimum** at B.

minimum condition, *n.* a condition on a MODULE that every non-empty set of submodules has a MINIMAL member. See also ARTINIAN MODULE, DESCENDING CHAIN CONDITION. Compare MAXIMUM CONDITION.

minimum polynomial, *n.* the polynomial of least degree, unique up to multiplication by a scalar, that is associated with a given matrix, transformation, or ALGEBRAIC element over a field, such that a polynomial function of the given matrix is zero; any other such polynomial is a multiple of the minimum polynomial.

Minkowski function or **gauge function,** *n.* a non-negative, positively HOMOGENEOUS, CONVEX function with value zero at zero. Given a convex set C, an associated gauge is constructed by setting

$$g_C(x) = \inf\{t > 0 : x \in tC\}.$$

An everywhere finite (symmetric) gauge is a NORM, and there is a duality between gauge and SUPPORT functions. (Named after the Russian-born, Swiss-German number theorist, algebraist, analyst and geometer, *Hermann Minkowski* (1864–1909), who developed the theory of four-dimensional space-time that laid the mathematical foundation for relativity theory.)

Minkowski's inequality, *n.* the theorem that the EUCLIDEAN NORM satisfies the TRIANGLE INEQUALITY and so is actually a NORM.

Minkowski theorem, *n.* the seminal result in the geometry of numbers that a symmetric convex body in the plane must contain a non zero lattice point in its interior if its area exceeds 4. See also INTEGER LATTICE.

Minkowski world or **Minkowskian space–time,** *n.* a four-dimensional model of physical space and time used in relativity theory; formally, a space in which three coordinates specify the position of a point in space, and the fourth represents the time at which an event occurs at that point.

minor, 1. *adj.* (*Geometry*) (of an ARC, SECTOR, or SEGMENT of a circle) being the smaller of two such figures determined by the same points on the circumference. In Fig. 244, the arc AYB is the *minor arc*, the region OAYB is the *minor segment*, and the region ABY is the *minor sector* determined by the points A and B. Compare MAJOR.

Fig. 244. The **minor** arc, **minor** sector, and **minor** segment include the point Y.

2. *n.* **a.** the DETERMINANT of a square submatrix of a given matrix. A *signed minor* is a COFACTOR.

b. also called **complementary minor.** (of an element of a matrix or determinant) the determinant of the submatrix obtained by deleting the row and column containing the given element from the given matrix or determinant; this has order one less than the given matrix or determinant. More generally, the complementary minor of a given minor or submatrix is the determinant of the submatrix obtained by deleting all the rows and columns in which elements of the given minor or submatrix occur.

minor axis, *n.* the shorter axis of an ellipse, or the shortest of an ellipsoid. The bold axis, BD, in Fig. 245 is the minor axis, and AC the MAJOR AXIS.

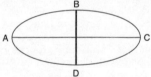

Fig. 245. The bold line is the **minor axis** of the ellipse.

minor premise, *n.* (*Logic*) the premise of a SYLLOGISM that contains the subject of the conclusion.

minor term, *n.* (*Logic*) the subject of the conclusion of a SYLLOGISM.

minuend, *n.* the number from which another, the SUBTRAHEND, is to be subtracted.

minus, *prep.* **1.** reduced by the SUBTRACTION of.

2. more generally, operated upon by any analogous operation, such as the RELATIVE COMPLEMENT A \ B.

3. *adj.* (*prenominal*) **a.** (of a specified number) less than zero, having a negative value; thus −4 is read as 'minus four'.

b. involving or indicating subtraction or negativity. Minus numbers are written with the MINUS SIGN prefixed.

minus sign, *n.* the symbol '−', indicating subtraction or a negative quantity.

minute or **minute of arc,** *n.* a 60th part of a DEGREE of arc. See also SECOND.

mixed, *adj.* **1.** (of a number or polynomial expression) having both an integral and rational part; for example,

$$5\tfrac{1}{3}, \quad \text{or} \quad 2x^2 + 4x + \frac{2}{3x}.$$

2. (of a partial derivative) involving derivatives with respect to more than one variable; for example,

$$\frac{\partial^3 f}{\partial x\ \partial y\ \partial z}$$

is a mixed derivative.

mixed congruential method, *n.* the specific CONGRUENTIAL METHOD used by a RANDOM NUMBER GENERATOR to produce the sequence

$$n_{i+1} \equiv [an_i + c] \pmod{m},$$

where a and c are non-negative integers, m is a large number relative to the WORD size of the specific computer, and n_0 is the SEED. A typical set of requirements then imposed is that c is odd with

$$c/m \sim 0.2113,$$

$$\frac{m}{\sqrt{10}} < a < m - \sqrt{m},$$

$$a \equiv \pm 3 \pmod 8,$$

and that the binary representation of a should have no obvious pattern.

mixed partial derivative, see PARTIAL DERIVATIVE.

mixed strategy, *n.* in a TWO–PERSON ZERO–SUM GAME, a probabilistic combination of a player's alternatives or *pure strategies*. The MINIMAX THEOREM of von Neumann can be viewed as establishing the existence of optimal mixed strategies or *minimax strategies*.

mixed surd, see SURD.

mixed type boundary conditions, *n.* BOUNDARY CONDITIONS for a PARTIAL DIF- FERENTIAL EQUATION stated in the form

$$g\,\frac{\partial u}{\partial \mathbf{n}} + u = f$$

on the boundary, where g is some function, possibly constant, and

$$\frac{\partial u}{\partial \mathbf{n}} = \nabla u \cdot \mathbf{n}$$

is the normal derivative, for u the dependent variable.

mks, *abbrev. and symbol for* the METRIC SYSTEM that uses the metre, kilogram, and second as units.

M matrix, *n.* a matrix M of the form $M = \lambda I - P$, where P is a matrix with POSITIVE entries for $\lambda \geq \sigma(P)$, the SPECTRAL RADIUS of P.

Möbius, August Ferdinand (1790–1868), German geometer, topologist, number theorist, and statistician, who spent most of his professional life as Professor of Astronomy in Leipzig where he erected the university observatory. Although he is best known for the MÖBIUS STRIP, this discovery was in fact only found amongst his papers after his death.

Möbius function, *n.* the MULTIPLICATIVE arithmetic function $\mu(r)$ defined by $\mu(r) = 1$ for $r = 1$; $\mu(r) = -1$ for prime r; $\mu(r) = (-1)^r$, where r is the number of prime factors of n, provided none of them is repeated; and $\mu(r) = 0$ for numbers whose prime decomposition includes repeated primes. Hence, for example,

$$\mu(30) = \mu(3).\mu(2).\mu(5) = -1,$$

and

$$\mu(32) = \mu(2^5) = 0.$$

See also MÖBIUS INVERSION FORMULA.

Möbius inversion formula, *n.* the formula whereby, given any arithmetic function *f*, one considers the related arithmetic function

$$F(n) = \sum_{d|n} f(d)$$

where the summation is over the divisors *d* of *n*; then one recovers *f* as

$$f(n) = \sum_{d|n} F(d) \, \mu\left(\frac{n}{d}\right),$$

where μ is the MÖBIUS FUNCTION, and *f* is multiplicative exactly when *F* is.

Möbius strip, *n.* a one-sided continuous surface, formed by twisting a rectangular strip of material through 180° and joining the ends, as shown in Fig. 246. The shortest closed path round the Möbius strip is of length 2*l*, where *l* is the length of the original unjoined strip; thus, if the surface is painted, starting at an arbitrary point and continuing until one returns to the starting point, and the strip of paper is then cut, it is found to be coloured on both sides. See also KLEIN BOTTLE.

Fig. 246. **Möbius strip.**

Möbius transformation or **fractional linear transformation,** *n.* a rational invertible transformation of the complex plane,

$$w = \frac{az + b}{cz + d},$$

with $ad \neq bc$. These are CONFORMAL mappings.

mod, *abbrev. and symbol for* MODULUS or MODULO.

modal, *adj.* **1.** (*Statistics*) of, relating to, or constituting a MODE.

2. (*Logic*) of or relating to a MODALITY.

modal interval, *n.* (*Statistics*) one of a set of CLASS INTERVALS that has the highest FREQUENCY. See also MODE.

modality, *n.* (*Logic*) **1.** the property of a statement of being classified under one of the concepts studied by MODAL LOGIC, and especially possibility and necessity.

2. any of the modal qualifiers themselves, or the operators representing them. For example, in Lewis' strongest system (S5), all iterated modalities are reducible to either possibility or necessity, and so these are the only two distinct modalities.

modal logic, *n.* **1.** the logical study of such concepts as possibility, necessity, contingency, etc., and of FORMAL SYSTEMS of which the intended interpretation includes such concepts.

2. the study of analogous families of concepts, such as moral, epistemological and psychological concepts, and of systems intended to represent them. See ALETHIC LOGIC, DEONTIC LOGIC, EPISTEMIC LOGIC, DOXASTIC LOGIC.

3. any FORMAL SYSTEM capable of being interpreted as a MODEL for the behaviour of such concepts.

mode, *n.* **1.** (*Statistics*) that one of a range of values that has the highest FREQUENCY. Compare MEAN, MEDIAN.

2. (*Logic*) another word for MOOD.

model, *n.* **1a.** See MATHEMATICAL MODEL.

b. *vb.* to abstract such a mathematical description of some process.

2. formally, a theory in which a given sentence or set of sentences is true. For example, Peano's axioms have a model in arithmetic (their intended interpretation) but they also have non-standard models that are not isomorphic to arithmetic.

3. (*Logic*) an INTERPRETATION of a FORMAL CALCULUS under which the theorems derivable in that system are assigned the value *true*, that is are mapped onto truths.

model theory, *n.* the branch of logic that studies the properties of MODELS; the SEMANTIC study of formal systems. Model theory is concerned with the concepts of truth, satisfaction, and validity, which are defined extrinsically for a formal system, as contrasted with PROOF THEORY, which is concerned only with the study of the intrinsic property of syntactic deducibility.

modular arithmetic, *n.* any system of arithmetic, denoted $\mathbb{Z}_n$, with a given finite MODULUS, n, in which two numbers are EQUIVALENT when they differ by an integral multiple of the modulus; this is sometimes referred to as *clock arithmetic* by analogy with arithmetic on a clock face (which has modulus 12). It is a RING, and if n is PRIME it is a FIELD. Modular arithmetic can be regarded either as operating with RESIDUE CLASSES related by identity, or with integers related by CONGRUENCE.

modular equation, *n.* an identity of the form $f(x) = f(x^n)$ for integral n, where n is referred to as the *order* of the modular equation. For example,

$$f(x) = \frac{2\sqrt{f(x^2)}}{1 + f(x^2)}$$

is a second-order modular equation. *Modular functions*, which satisfy these equations, were studied by RAMANUJAN and yielded approximations to many digits of π.

modular field, *n.* a FIELD of finite non-zero CHARACTERISTIC n, the smallest positive integer such that the n-fold sum of the multiplicative unit of the given field is zero; it can be shown that n must be prime. Any finite field is modular; for example, in $\mathbb{Z}_p$ (the integers modulo some prime p) the multiplicative unit is 1, and the sum of p such units is zero (mod p), so that the integers mod p are a field with characteristic p. If no such integer exists the field is said interchangeably to have characteristic 0 or ∞.

modular form, see MODULAR FUNCTION.

modular function, *n.* **1.** a function that is MEROMORPHIC in the upper half of the complex plane and that is AUTOMORPHIC for the MODULAR GROUP or one of its subgroups. There are more general *modular forms* that are merely required to be automorphic forms of dimension $-2m$ with respect to the modular group

$$F\left(\frac{az+b}{cz+d}\right) = (cz+d)^{-2m}f(z),$$

for some positive half-integer m.

2. a function that satisfies a MODULAR EQUATION.

modular group, *n.* the GROUP consisting of all MÖBIUS TRANSFORMATIONS with integral coefficients and determinant equal to unity.

modular representation, *n.* a REPRESENTATION over a FIELD of prime CHARACTERISTIC.

module, *n.* a COMMUTATIVE GROUP M endowed with an *exterior multiplication* (either on the left or right) that is associative and distributive, and multiplies group elements by elements of a RING R (called *scalars*) to produce group elements; then M is a *module over R*, or an *R-module*. If, in addition, R is a UNITARY ring, then M is said to be a *unitary R-module* if the product $i * g = g$, where i is the identity element of the ring, and g is any element of the group. Every commutative group may be viewed as a module over the integers. A vector space is a module in which R is a FIELD. Every ring R may be viewed as an R-module over itself, and an IDEAL in R is an R-module.

module homomorphism, see HOMOMORPHISM.

modulo *n* or **mod** *n*, *adj.* (of a relationship) holding in MODULAR ARITHMETIC to the specified MODULUS. For example, $15 + 9 \equiv 3 \pmod 7$. The relation of CONGRUENCE mod n is basic to the study of divisibility in number theory, and is an equivalence relation under which the equivalence classes are the RESIDUE CLASSES mod n.

modulus (abbrev. **mod**), *n.* **1.** also called **absolute value.** a positive real number that is a measure of the magnitude of a COMPLEX NUMBER, and is equal to the square root of the sum of the squares of the REAL and IMAGINARY PARTS of the given number. Thus the absolute value of $x + iy$ is

$$|x + iy| = \sqrt{x^2 + y^2},$$

and the absolute value of $4 + 3i$, for example, is $\sqrt{(4^2 + 3^2)} = 5$. This is equal to the length of its POSITION VECTOR in the ARGAND DIAGRAM, as shown in Fig. 247. Compare ARGUMENT.

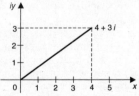

Fig. 247. The **modulus** (sense 1) of $4 + 3i$ is 5.

2. the BASE of a system of MODULAR ARITHMETIC; an integer that can be divided exactly into the difference between two given integers, so that they are CONGRUENT MODULO that divisor. For example, 7 is a modulus of 11 and 25. **3.** the number by which a LOGARITHM to one BASE has to be multiplied to find the logarithm of the same number to a different base.

4. the variable, usually denoted k, of the complete ELLIPTIC INTEGRALS and ELLIPTIC FUNCTIONS. The quantity $\sqrt{(1 - k^2)}$ is the *complementary modulus*.

5. (of continuity, convexity, smoothness, etc.) one of a number of measures of the given property. The standard *modulus of continuity* of a real-valued continuous function on a metric space is defined by

$$\omega(\delta) = \sup\{\, |f(x) - f(y)| \,:\, d(x,y) \leq \delta \,\}$$

and f is UNIFORMLY CONTINUOUS exactly when

$$\lim_{\delta \to 0} \omega(\delta) = \omega(0).$$

modulus of elasticity, see HOOKE'S LAW.

modus ponens, modus ponendo ponens, or **rule of detachment,** *n.* (*Logic*) the principle or rule of inference that whenever a CONDITIONAL statement and its ANTECEDENT are both given to be true, then its CONSEQUENT may be validly inferred. For example, given

if it's Tuesday this must be Belgium.

and

it is Tuesday.

we can validly DETACH the consequent of the conditional to infer

this must be Belgium.

modus tollens or *modus tollendo tollens,* *n.* (*Logic*) the principle or rule of inference that whenever a CONDITIONAL statement and the negation of its CONSEQUENT are both given to be true, the negation of its ANTECEDENT may be validly inferred. For example, given

if it's Tuesday this must be Belgium.

and

this is not Belgium.

we can infer

it can't be Tuesday.

molecular, *adj.* (*Logic*) (of a sentence, formula, etc.) compound; not ATOMIC but capable of analysis into atomic formulae of the appropriate kind; a conjunction or implication is thus a molecular statement.

moment, *n.* (*Statistics*) **1.** also called **moment about the origin.** the EXPECTATION of a specified power of the deviations from zero of all the values of a RANDOM VARIABLE. The power of the deviations is the *order* of the moment; thus, the *second moment* is the expectation of the squares of the deviations.
2. moment about the mean or **central moment.** analogously, the expected value of a given power of the deviations of a random variable from its MEAN.

moment generating function, *n.* (*Statistics*) the expected value of $\exp(t^T X)$, usually written $M_X(t)$, where X is a given n-dimensional RANDOM VECTOR. The moment generating function does not always exist, but when it does, it is related to the CHARACTERISTIC FUNCTION C_X by

$$M_X(t) = C_X(-it).$$

It has the useful property that the rth MOMENT about the origin of a single random variable X is $r! M_X^{(r)}(0)$, where $M_X^{(r)}(0)$ is the rth derivative of the moment generating function evaluated at zero; analogous formulae exist for moments of random vectors.

moment of a force, *n.* (*Mechanics*) another term for TORQUE.

moment of inertia, *n.* (*Continuum mechanics*) (about a fixed axis) the integral

$$\int p^2 \rho \, dv,$$

over the volume of a BODY with density ρ, where p is the perpendicular distance of its points from the axis.

moment of momentum, *n.* (*Mechanics*) another term for ANGULAR MOMENTUM.

moment of relative momentum, *n.* (*Mechanics*) another term for RELATIVE ANGULAR MOMENTUM.

moment problem, *n.* the problem of whether a given sequence of real or complex numbers is a MOMENT SEQUENCE for some MEASURE or METRIC DENSITY. The *Stieltjes moment problem* seeks a measure on $[0, \infty]$; the *Hamburger moment problem* seeks a measure on $[-\infty, \infty]$; and the *Hausdorff moment problem* seeks a measure on a finite interval.

moment sequence, *n.* the sequence of integrals, with respect to a given measure, of either x^n (the nth *moment*) or of $\exp(in\theta)$ (a *trigonometric moment sequence*). See also MOMENT PROBLEM.

momentum, *n.* **1.** the product of the mass and velocity of a body. For a continuous mass distribution dm over a body, B, with velocity, **v**, this produces

$$p = \int_B \mathbf{v} \, dm.$$

2. (*Continuum mechanics*) more generally, the integral

$$\mathbf{M}(\mathbf{R}_t) = \int \rho \, \mathbf{v} \, dv,$$

over the volume of the CONFIGURATION, $\mathbf{R}_t$, of a SUB–BODY, **R**, at time t, where **v** is the velocity of the points of **R**, and ρ is the density.

momentum space, *n.* (*Statistical physics*) a space of s dimensions representing a system with s DEGREES OF FREEDOM in which the rectangular coordinates specify the MOMENTUM of the points within the system.

monad, *n.* (*Non-standard analysis*) the set of HYPER–REAL NUMBERS whose difference from a given real number is infinitesimal, written

$$\mu(a) = \{ b \in \mathbb{R}^* : a - b \text{ is infinitesimal} \}.$$

monadic, unary, or **singulary,** *adj.* (of an operator, predicate, etc.) having only a single argument place. Negation, inversion and set complementation are monadic operators. A monadic relation is a (one-place) predicate.

Monge's methods, *n.* methods involving TOTAL DIFFERENTIAL EQUATIONS for solving PARTIAL DIFFERENTIAL EQUATIONS of the forms

$$R \frac{\partial^2 z}{\partial x^2} + S \frac{\partial^2 z}{\partial x \partial y} + T \frac{\partial^2 z}{\partial y^2} = V$$

and

$$R \frac{\partial^2 z}{\partial x^2} + S \frac{\partial^2 z}{\partial x \partial y} + T \frac{\partial^2 z}{\partial y^2} + U \left[\frac{\partial^2 z}{\partial x^2} \frac{\partial^2 z}{\partial y^2} - \left(\frac{\partial^2 z}{\partial y^2} \right)^2 \right] = V,$$

where x and y are independent variables and R, S, T, U, V are differentiable functions of $x, y, z, \partial z/\partial x$, and $\partial z/\partial y$.

monic, *adj.* (of an ARROW, $h : a \to b$, in a CATEGORY) such that for parallel arrows, $f_1, f_2 : d \to a$, if $h \circ f_1 = h \circ f_2$ then $f_1 = f_2$.

monic polynomial, *n.* a polynomial with leading coefficient +1.

monodromy theorem, *n.* **1.** any of a number of fundamental theorems of COMPLEX ANALYSIS, asserting the uniqueness of ANALYTIC CONTINUATION: if a function of a complex variable is analytic in a disc inside a simply connected domain, and if the function can be analytically continued along all polygonal arcs in the domain, then the function extends to a single-valued analytic function on the entire domain.

2. monodromy theorem of Darboux. in particular the theorem that if a function of a complex variable, analytic in a domain bounded by a simple closed curve, is continuous in the closed region and is one-to-one on the boundary curve, then the function is one-to-one throughout the region.

monoid, *n.* **1.** a SEMIGROUP with an IDENTITY.

2. a GROUPOID, usually commutative.

monomial, *n.* an expression consisting of a single term, such as $5ax$.

monomorphism, *n.* an INJECTIVE HOMOMORPHISM. Compare EPIMORPHISM, ISOMORPHISM.

monotone or **monotonic,** *adj.* (of a sequence or function) consistently INCREASING (ISOTONE) or DECREASING (ANTITONE) in value, so that either

$$f(x_1) > f(x_2) \quad \text{for all } x_1 > x_2,$$

or

$$f(x_1) < f(x_2) \quad \text{for all } x_1 > x_2.$$

These may be called *strictly monotone* functions to distinguish them from those satisfying either

$$f(x_1) \geq f(x_2) \quad \text{for all } x_1 > x_2,$$

or

$$f(x_1) \leq f(x_2) \quad \text{for all } x_1 > x_2,$$

which are *weakly monotone*; however, the symbols $\uparrow$ for an increasing and $\downarrow$ for a decreasing function are sometimes used without distinction. All these properties may be either LOCAL or GLOBAL. For example, the curve in Fig. 248 is locally monotone increasing for $x < 1.4$ and for $x > 2.6$, and locally monotone decreasing for $1.4 < x < 2.6$, and is clearly not globally monotone.

Fig. 248. A piecewise **monotone** function.

monotone convergence theorem, *n.* the result that if $\{f_n\}$ is a pointwise-increasing sequence of non-negative measurable functions defined on a measurable set E, then

$$\lim_{x \to \infty} \int_E f_n \, d\mu = \int_E f \, d\mu,$$

where f denotes the pointwise limit of the sequence (which may be either finite or infinite).

monotone multifunction, see MONOTONE RELATION.

monotone relation, *n.* a subset of the product of a BANACH SPACE and its DUAL, with the property that if (x_1, y_1) and (x_2, y_2) lie in the subset, then

$$\langle (y_1 - y_2), (x_1 - x_2) \rangle \geq 0.$$

If $y \in T(x)$, then T defines a *monotone multifunction*. Even for a function in two dimensions, having a monotone graph and being ISOTONE are distinct properties .

monster group, *n.* the largest irregular finite simple group, of which the order is approximately 8×10^{53} (exactly

$$2^{46} \times 3^{20} \times 5^9 \times 7^6 \times 11^2 \times 13^3 \times 17 \times 19 \times 23 \times 29 \times 31 \times 41 \times 47 \times 59 \times 71, \text{ or}$$

808 017 424 794 512 875 886 459 904 961 710 757 005 754 368 000 000 000).

Monte Carlo method, *n.* (*Statistics*) a heuristic mathematical technique for evaluation or estimation of intractable problems by probabilistic simulation and sampling. For example, in *Monte Carlo integration*, the integral

$$I = \int_a^b f(x) \, dx$$

may be approximated by

$$\hat{I} = \frac{b-a}{n} \sum_{i=1}^n f(x_i),$$

where the x_i are independent observations from a UNIFORM DISTRIBUTION on the interval (a, b); this is because the expected value $E(\hat{I}) = I$, and so the precision of the estimate increases with n.

Montmort matching problem or **letter problem,** *n.* the problem of determining the probability that in a random pairing of two naturally paired sets (such as two packs of cards, or a set of addressed letters and envelopes) there is at least one correct matching. This probability tends to $1 - 1/e$ as the number of items increases without bound, and the problem is equivalent to the INTERVIEW PROBLEM.

mood, *n.* (*Logic*) **1.** also called **mode.** one of the concepts or operators studied or formalized by MODAL LOGIC.

2. one of the 64 possible combinations of the three categorial statements of a SYLLOGISM. Compare FIGURE.

Moore–Osgood theorem, *n.* a theorem guaranteeing that the order of taking ITERATED LIMITS may be changed. For example, suppose $f: X \times Y \to Z$ is a mapping between PSEUDO–METRIC SPACES; if

$$\lim_{x \to a} f(x, y) = f(a, y) \quad \text{UNIFORMLY for } y \neq b,$$

and

$$\lim_{y \to b} f(x, y) = f(x, b) \quad \text{POINTWISE for } x \neq a,$$

then

$$\lim_{x \to a} \lim_{y \to b} f(x, y) = \lim_{y \to b} \lim_{x \to a} f(x, y).$$

The most common application has Y as the set of extended natural numbers with b as ∞.

Moore–Penrose inverse, see PSEUDO–INVERSE.

Moore–Smith convergence, *n.* another term for NET CONVERGENCE.

Moore–Smith sequence, *n.* another term for NET.

Moore space, *n.* a TOPOLOGICAL SPACE in which there exists a sequence $\{G_n\}$ satisfying the conditions that each G_i is a family of open sets whose union is the whole space; that for each i, G_{i+1} is contained in G_i; and that if x, y are distinct members of an open set R, then there is an n for which x is a member of a set U whose closure is in G_n but which does not contain y. (Named after the American topologist, *Robert Lee Moore* (1882–1974).)

Mordell's conjecture, *n.* the conjecture that algebraic equations with GENUS greater than 1 have only a finite number of rational solutions. It was eventually proved by G. Faltings in 1983. (Named after Louis Mordell (1888–1972), an American number theorist.)

Morera's theorem, *n.* (*Complex analysis*) the consequence of CAUCHY'S INTEGRAL FORMULA that a function that is continuous on a region and of which the CONTOUR INTEGRAL over all triangles is zero, must be analytic.

morphism, *n.* **1.** (*Category theory*) another term for ARROW.
2. see HOMOMORPHISM, ISOMORPHISM, EPIMORPHISM, MONOMORPHISM, HOMEOMORPHISM.

Moscow papyrus, *n.* one of the two principal sources of knowledge of ancient Egyptian mathematics, along with the RHIND PAPYRUS, dated around 1850 BC.

motion, *n.* (*Continuum mechanics*) any change of the position of a body; formally, a one-parameter family of CONFIGURATIONS, χ_t, of a BODY **B**, with time t as the parameter. Thus for X a point of **B**, the position of X at time t is given by $\chi_t(X)$, and so the motion may be identified with the mapping $\chi(X, t) = \chi_t(X)$ from **B** $\times$ $\mathbb{R}$ into three-dimensional EUCLIDEAN POINT SPACE.

Motzkin's theorem, *n.* the result that if two disjoint finite sets, S and T, of points in the plane do not both lie on the same line, then there is either a line passing through at least two points of S and missing T entirely, or a line passing through at least two points of T and missing S entirely.

mountain pass lemma, *n.* a theorem giving conditions for a potentially unbounded differentiable function to have a CRITICAL POINT: if

$$f(0) < \inf\{f(x) : \|x\| = 1\} > f(a)$$

for some a with $\|a\| > 1$, there will be a critical point b with

$$f(b) \geq \inf\{f(x) : \|x\| = 1\}$$

if f satisfies some 'growth condition' such as that $f(x)$ tends to infinity with x. Geometrically, this critical point is located in the 'mountain pass'.

moving average, *n.* (*Statistics*) a sequence derived from a given sequence of values by starting with each element of the original sequence in turn and taking the averages of a given number of successive elements of the original sequence; this device is often used in TIME SERIES to even out short-term fluctuations and so to make a trend clearer. Note that the sequence of n-term moving averages must be $n-1$ terms shorter than the given sequence; for example, the 3-term moving average of 4, 6, 8, 7, 9, 8 is

$$\text{ave}(4, 6, 8), \text{ave}(6, 8, 7), \text{ave}(8, 7, 9), \text{ave}(7, 9, 8) = 6, 7, 8, 8.$$

M test, see WEIERSTRASS M TEST.

mu-function, *n.* (*Number theory*) the MÖBIUS FUNCTION, $\mu(n)$.

multi–, *prefix denoting* many; for example, a *multiangular* figure has many angles; a *multinomial* is a sum of more than one term.

multiant, *n.* a generalized form of DETERMINANT for arrays of more than two dimensions.

multicollinearity, *n.* (*Statistics*) the condition occurring when two or more of the independent variables in a REGRESSION EQUATION are very highly CORRELATED.

multicriteria optimization or **multiobjective optimization,** *n.* the study of OPTIMIZATION PROBLEMS with more than one objective criterion to satisfy. An example of such a problem is attempting in an industry to minimize environmental damage while maximizing profit. Typically, there is no maximum and one must settle for an EFFICIENT POINT or use additional external criteria.

multifoil, *n.* a symmetric plane figure constructed by placing congruent arcs of a circle around a regular polygon so that the ends of the arcs bisect the sides. A *trefoil* has three sides, a *quatrefoil*, as shown in Fig. 249, has four, and a *hexafoil* has six sides.

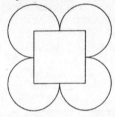

Fig. 249. A **multifoil** of four leaves.

multifunction, *n.* another word for SET–VALUED FUNCTION, especially in complex analysis.

multilinear function or **multilinear form,** *n.* a function of several VECTOR–SPACE variables that is linear in each variable separately, especially in finite dimensions. The *degree* is then the number of such variables. Such a form is *alternating* if all underlying vector spaces are identical and if interchanging two variables merely changes the sign of the expression.

multinomial, *n.* **1.** an algebraic expression with more than one term. In some usages, the term is used for an expression with 'many' terms, as opposed to a BINOMIAL, TRINOMIAL, etc.

2. another term for POLYNOMIAL.

multinomial coefficient, *n.* the coefficient

$$\binom{n}{n_1 \ldots n_m} = \frac{n!}{n_1! \ldots n_m!} ,$$

where the n_i are non-negative integers whose sum is n, corresponding to the number of ways of choosing n_i objects of type i without regard to order, in such a way that the total number of objects chosen is n.

multinomial distribution, *n.* the distribution of a *multinomial random variable*, **X**, representing the number of times each of m possible outcomes occurs in n independent repetitions of an experiment. If p_i is the probability of the ith outcome, then its probability function is

$$P(\mathbf{X} = \mathbf{x}) = \frac{n!}{n_1! \dots n_m!} \, p_1^{n_1} \dots p_m^{n_m}.$$

If $\mathbf{X}$ is distributed in this way we write $\mathbf{X} \sim \mathrm{Mu}(n, \mathbf{p})$, where $\mathbf{p}$ is the vector of probabilities p_i. This distribution represents the probability of selecting n_i objects from the ith of m categories in such a way that the total number selected, irrespective of order, is n, when, for each i, each object in the ith category has probability p_i of being selected. The marginal distributions are themselves multinomial, and in particular

$$P(X_i = n_i) = {}^{n}C_{n_i}(1 - p_i)^{n-n_i} \, p_i^{n_i},$$

which is the BINOMIAL DISTRIBUTION, $\mathrm{Bi}(n, p_i)$.

multinomial random variable, see MULTINOMIAL DISTRIBUTION.

multinomial theorem, n. the generalization of the BINOMIAL THEOREM to n variables:

$$(x_1 + x_2 + \dots x_m)^n = \sum \frac{n!}{n_1! \, n_2! \dots n_m!} \, x_1^{n_1} \, x_2^{n_2} \, \dots x_m^{n_m},$$

where the sum is over all non-negative integers summing to n.

multinormal distribution, see MULTIVARIATE NORMAL DISTRIBUTION.

multi–objective optimization, n. another term for MULTICRITERIA OPTIMIZATION.

multiple, n. any number or polynomial that is the PRODUCT of a given number or polynomial and an integral multiplier; for example, $3\frac{1}{2}$ is a multiple of $1\frac{3}{4}$, or $x^2 - y^2$ of $x + y$.

multiple integral or **repeated integral,** n. an expression requiring MULTIPLE INTEGRATION.

multiple integration or **repeated integration,** n. the INTEGRATION of a function of two or more variables;

$$\int_{a_1}^{b_1} \dots \int_{a_k}^{b_k} f(x_1, x_2, \dots x_k) \, \mathrm{d}x_1 \, \mathrm{d}x_2 \, \mathrm{d}x_k$$

is a *definite multiple integral* over the region of k-space defined by $a_i \le x_i \le b_i$ for each i; without the limits of integration it is an *indefinite multiple integral*. Compare ITERATED INTEGRAL.

multiple precision, see PRECISION.

multiple regression, n. a REGRESSION function that gives the conditional expectation of a random variable in terms of more than one other random variable.

multiple root or **repeated root,** n. one of a number of equal roots of the same polynomial or equation. This occurs when the polynomial has $(x - a)^n$ as a factor, for n greater than 1; the largest such n is the *multiplicity* or *order* of the root a. A root of order 1 is a *simple root*. Equivalently, it is a root of the polynomial equation that is also a root of one or more derivatives of the polynomial.

multiple sequence, *n.* a SEQUENCE indexed over a number of INDEX SETS, such as

$$\{ x_{i,j,k} : 0 \le i < n, \ 1 < j \le m, \ 0 < k \le p \}.$$

multiple series, *n.* a SERIES indexed over a number of INDEX SETS; the sum of all the elements of a MULTIPLE SEQUENCE, which in the finite case is independent of order. See PRINGSHEIM CONVERGENCE.

multiple-valued function, *n.* another term for SET–VALUED function, especially in complex analysis.

multiplicand, *n.* a number to be multiplied by another, the MULTIPLIER.

multiplication, *n.* **1.** a binary arithmetical operation defined initially for positive integers in terms of repeated ADDITION, by which the PRODUCT of two quantities is calculated, usually written $a \times b$, $a.b$, or ab. To multiply a by integral b is to add a to itself b times; multiplication by rational numbers can then be defined to take advantage of the ASSOCIATIVE and COMMUTATIVE properties of multiplication, and the fact that DIVISION is its inverse, by virtue of which, for example, multiplication by $\frac{3}{2}$ is performed by first multiplying by 3 and then dividing the product by 2.

2. any analogous binary operation. See PRODUCT.

3. (for SUBGROUPS) the binary operation that forms the *product* HK of two subgroups H and K of a given group, where

$$HK = \{ hk : h \in H, k \in K \}$$

is not in general a subgroup unless either H or K is contained in the NORMALIZER of the other.

4. (for IDEALS) the binary operation that forms the *product*

$$LK = \left\{ \sum_{j=1}^{n} l_j k_j : l_j \in L, \ k_j \in K \right\}$$

of two ideals L and K of a ring R; LK is then also an ideal of R, and is the intersection of the given ideals.

multiplication sign, *n.* the symbol ' $\times$ ' placed between numbers to be multiplied, as in $3 \times 4 = 12$.

multiplication table, *n.* a table setting out the results of multiplying every pair of a range of numbers, or elements of a group, ring or other algebraic structure. In any place-value notation, only a finite number (equal to the base) of distinct multiplication tables are needed to enable any product to be calculated.

multiplicative, *adj.* **1.** (of a function) DISTRIBUTING over MULTIPLICATION, so that $f(xy) = f(x) f(y)$. See also HOMOMORPHISM.

2. (of an arithmetic function) multiplicative only for relatively prime arguments, so that $f(xy) = f(x) f(y)$ if x and y have no common divisors. An arithmetic function that is multiplicative in the previous sense is then called *totally multiplicative.*

multiplicative group, *n.* **1.** in a FIELD, the GROUP obtained by just considering the non-zero elements of the group with respect to its operation of multiplication.

2. generally, any group in which the product of elements a and b is written ab and the group identity is denoted 1.

multiplicative identity, *n.* an IDENTITY ELEMENT under a multiplicative operation; a UNITY.

multiplicative inverse, see INVERSE.

multiplicity, see MULTIPLE ROOT.

multiplier, *n.* **1.** the number by which another number, the MULTIPLICAND, is multiplied.

2. see LAGRANGE MULTIPLIERS.

multiplier methods, *n.* a class of optimization methods that use variants of the LAGRANGE MULTIPLIERS in combination with PENALTY FUNCTION METHODS to solve constrained optimization problems.

multiply, *vb.* to combine two numbers or quantities by MULTIPLICATION; to calculate their PRODUCT.

multiply out, *vb.* to expand an expression by applying the DISTRIBUTIVE LAW to all the multiplications over any other operations, for example, expressing $(x + a)^2$ as $x^2 + 2ax + a^2$.

multivalued function, *n.* another term for SET–VALUED function, especially in complex analysis.

multivariate, *adj.* (*Statistics*) (of a distribution) involving a number of distinct, though not necessarily independent, random variables.

multivariate normal distribution, *n.* the multidimensional extension of the NORMAL DISTRIBUTION; the joint distribution of any n normal random variables has probability density function

$$\frac{\exp\left[-\frac{1}{2}(x - \mu)' \sum^{-1}(x - \mu)\right]}{\sqrt{(2\pi)^n |\sum|}},$$

where μ is the vector of means, and the VARIANCE–COVARIANCE MATRIX, $\sum$, is positive definite.

Müntz' theorem, see WEIERSTRASS APPROXIMATION THEOREM.

mutatis mutandis, *adv.* (*Latin*) the necessary changes having been made; similarly to the previous case, but with some minor and obvious adjustment to the new case. The term is often used when some work but no imagination is demanded of the reader.

mutually exclusive, *adj.* (of a pair of properties) each precluding the other; for example, membership of DISJOINT classes. Compare EXHAUSTIVE. See PARTITION.

myopic algorithm, *n.* any of a class of algorithms that are so called because they look only at very local information at each iteration, as with STEEPEST DESCENT, and often pay a price for this short sightedness. Compare GREEDY ALGORITHM.

n, *abbrev. for* NANO–, used in symbols for fractions of the physical units of the SYSTEME INTERNATIONAL.

n-, *prefix indicating* an unspecified but specific finite number of variables, dimensions, arguments, terms, etc. For example, an *n-gon* is a polygon with *n* sides, an *n-tuple* is an ordered set with *n* members, an *n-fold* or *n-times* process is one iterated *n* times, and *n-dimensional* space is space with *n* dimensions.

N, (*Mechanics*) *symbol for* NEWTON.

$\mathbb{N}$ or $\mathcal{N}$, *symbol for* the set of NATURAL NUMBERS or positive INTEGERS. Compare $\mathbb{R}, \mathbb{Q}, \mathbb{Z}$.

nabla squared, *n.* the LAPLACIAN, written ∇^2. See also DIFFERENTIAL OPERATOR.

naive set theory, *n.* the presentation of the theory of SETS in the style of an informal mathematical theory, regarding it as a body of given knowledge rather than the consequences of a set of UNINTERPRETED axioms. The naive theory seeks a formalization of the intuitive conception of the properties of sets, by postulating a minimal set of independent non-logical axioms that are intended to enable theorems to be derived that codify this conception. This was the approach of the early set theorists such as Cantor and Russell, but their discovery of many paradoxes (including those that bear their names) led to the realization that the intuitive notion of a set could not be rigorously sustained, and that instead, a more modest approach was necessary, in the formal manner of AXIOMATIC SET THEORY.

name, *n.* a word or symbol that has or purports to have a REFERENCE, but not in virtue of the literal meaning of the expression. 'The Holy Roman Empire' is a name rather than a DESCRIPTION, since it refers to that political entity, irrespective of whether it is holy, Roman, or an empire. Names are primitive terms in the PREDICATE CALCULUS; in some presentations, atomic sentences are primitive and predicates are obtained by deleting a name, while in others, predicates are primitive and an atomic sentence consists of the concatenation of a predicate and a name. An atomic sentence is true if and only if the bearer of the name (its reference) satisfies the predicate, and general statements are obtained by replacing the name with a VARIABLE and prefixing a quantifier.

nano- (symbol **n**), *prefix for* a fraction of 10^{-9} of the physical units of the SYSTEME INTERNATIONAL.

Nansen's formula, *n.* (*Continuum mechanics*) the identity,

$$\text{da} = \{\ \mathbf{F}^{-T}\det\mathbf{F}\,\text{dA}\ \},$$

where da and dA are elements of area in the current and reference CONFIGURATION respectively, and **F** is the DEFORMATION GRADIENT.

Napier, John (1550–1617), Scottish theologian and amateur mathematician, who invented the notation of the decimal point and NAPERIAN LOGARITHMS, and contributed to the theory of spherical triangles, for the solution of which he devised *Napier's rules* and *Napier's analogies*, a set of trigonometric identities. In the guise of scholarship, he published a virulent attack on the Roman Catholic church, with an admonitory dedication to James VI of Scotland, who hoped to succeed Elizabeth I on the English throne and was contemplating an alliance with Catholic Spain; he also designed new weapons for use in defending Protestant Scotland against these 'enemies of God'.

Napierian logarithm, *n.* **1.** the original logarithm due to John Napier, actually equal to

$$10^7\log_{1/e}\left(\frac{x}{10^7}\right).$$

2. now often another name for the NATURAL LOGARITHM, as opposed to the Briggsian or COMMON LOGARITHM.

Napier's bones, *n.* a set of graduated rods once used as an aid for multiplication and division; a primitive form of SLIDE RULE.

nappe, *n.* either of the two parts into which a CONE is divided by the vertex, as shown in Fig. 250.

Fig. 250. Two **nappes** of a cone.

natural base, see NATURAL LOGARITHM.

natural boundary condition, see SPLINE–FITTING.

natural cubic spline, see SPLINE–FITTING.

natural deduction, *n.* a system of FORMAL LOGIC that has no AXIOMS but proceeds by the recursive application of a set of RULES OF INFERENCE to assumptions that are made for the purposes of a particular deduction. Such a system uses SEQUENTS to record which assumptions are operative at any stage; the sequent $\Delta \vdash B$ states that B is conditional upon the set of assumptions Δ. Strictly, the rules of a natural deduction system are for the derivation of sequents from sequents rather than for the derivation of individual WELL–FORMED FORMULAE. Viewed in this way, the rule permitting assumptions merely allows any instance of the schema $A \vdash A$ to be founded upon without further justification. Some rules permit assumptions to be discharged, and a theorem is a sequent from which all the assumptions have been discharged, so that it is unconditionally true. Compare AXIOMATIC SYSTEM.

natural density, (of a sequence of non-negative integers) see SCHNIRELMANN DENSITY.

natural epimorphism or **natural homomorphism,** *n.* an EPIMORPHISM from a GROUP G to its FACTOR GROUP, G/N, where N is a NORMAL SUBGROUP of G, given by mapping an element to its left or right COSET, xN or Nx respectively. Analogous epimorphisms exist for rings and modules with respect to ideals and submodules respectively.

natural language, *n.* an ordinary language as spoken or written, contrasted with a symbolic FORMAL LANGUAGE.

natural logarithm, *n.* a LOGARITHM whose BASE is the *natural base, e,* usually written $\ln x$ or $\log_e x$; the inverse of the EXPONENTIAL FUNCTION, having the property that if $\ln x = y$ then $\exp y = x$; the derivative of $\ln x$ is $1/x$. See also NAPIERIAN LOGARITHM. Compare COMMON LOGARITHM.

natural number, *n.* one of the counting numbers; a number that can represent the CARDINALITY of a finite set of objects, usually identified with the positive INTEGERS 1, 2, 3, 4, ... There is some disagreement about whether 0 should be included, as is usual for the WHOLE NUMBERS. The natural numbers are often represented as $\mathbb{N}$. See also PEANO ARITHMETIC.

natural transformation, *n.* a mapping between two FUNCTORS that preserves structure; a CANONICAL transformation. A finite-dimensional vector space has a natural identification with its second dual but not with its first dual.

naught, *n.* variant spelling (*especially in USA*) of NOUGHT.

Navier–Stokes equation, *n.* (*Continuum mechanics*) the identity, for a SIMPLE NEWTONIAN VISCOUS FLUID with density ρ,

$$\rho\mathbf{a} = \rho\mathbf{b} - \nabla p + \mu\nabla^2\mathbf{v} + \tfrac{1}{3}\mu\nabla(\nabla\cdot\mathbf{v}),$$

where $\mathbf{a}$ is the ACCELERATION, $\mathbf{b}$ the BODY FORCE DENSITY, $\mathbf{v}$ the VELOCITY, p the PRESSURE function, and μ the VISCOSITY. The solution to these equations, which are ubiquitous in modern computational science, is not fully understood, and one of the MILLENNIUM PRIZE PROBLEMS is to develop a sophisticated theory which would aid understanding of complex phenomena such as the wake of a boat or the turbulence following a jet.

nbd, *abbrev. for* NEIGHBOURHOOD.

nearest point, *n.* a point not in a given subset of a metric space that has minimal DISTANCE from the points of the subset; a BEST APPROXIMATION. Such a point exists whenever the subset is compact, but it is usually not unique. Compare FARTHEST POINT.

necessary, *adj.* (*Logic*) **1.** (of a statement, formula etc.) true under all INTERPRETATIONS or in all possible circumstances.
2. (of an inference) VALID, having a conclusion that is true whenever the premises are true.
3. (of a property) ESSENTIAL, so that its subject could not lose it and continue to be the entity it is.

necessary condition, *n.* **1.** something ENTAILED by the truth of some statement or the obtaining of some state of affairs, that is required to be true as a precondition for the latter to be true; so if the necessary condition is false, then what it was a condition for must also be false. If P is a necessary condition for Q, Q IMPLIES P, and this relation is often expressed as 'Q only if P'. Although a necessary condition may also be a SUFFICIENT CONDITION, this is not in general true; for example, it is a necessary condition for a series to converge that the successive terms tend to zero, but this is not sufficient, as the harmonic series exemplifies. However, if P is a necessary condition of Q, then Q is a sufficient condition for P; so for example, to show that the successive terms tend to zero, it is sufficient to know that the series converges.

2. (*Optimization theory*) a necessary condition for an OPTIMUM point that one hopes is easy to verify, such as determination of a stationary point in unconstrained optimization, or of a Kuhn–Tucker point in constrained optimization, and that in the presence of an additional SUFFICIENT CONDITION guarantees optimality. See KUHN–TUCKER CONDITIONS.

necessity, *n.* (*Logic*) **1.** the property of being NECESSARY or of being a NECESSARY CONDITION.

2. a statement asserting that some property is ESSENTIAL or that some statement is necessarily true.

3. the operator in MODAL LOGIC that indicates that the expression it takes as argument is true in all POSSIBLE WORLDS, usually written □ or **L**.

Necker cube, *n.* a figure that appears to represent a three-dimensional solid, as shown in the second diagram of Fig. 251; in fact, however, it is impossible to construct such a solid. (Named after the Swiss mathematician and physicist, *Louis Necker* (1730–1804).) See also PENROSE TRIANGLE.

Fig. 251. A real and a **Necker cube**.

needle problem, *n.* (*Probability*) the problem, due to Buffon, of determining the probability of a needle landing on one of a family of parallel lines when dropped at random on a plane; when the length of the needle is l, and the lines are a units (where $a > l$) apart, the probability, appropriately formalized, is $2l/\pi a$. This provides a MONTE–CARLO METHOD, although a poor one, of computing π.

negation, *n.* (*Logic*) **1.** the monadic TRUTH–FUNCTIONAL sentential operator that forms one sentence from another and corresponds to the English *not*. Its TRUTH–TABLE is given in Fig. 252.

P	–P
T	F
F	T

Fig. 252. Truth table for **negation**.

2. a sentence so formed. It is usually written –P, ~P, ¬ P, or $\overline{P}$, where P is the given sentence, and is true when the latter is false, and false when the latter is true.

negative, *adj.* **1.** (of a set of values or quantity) less than zero; for example, the *negative integers* are defined as the result of subtracting natural numbers from zero; a *negative acceleration* is a deceleration or retardation.

2. (*prenominal*) a less common word for MINUS (sense 3), denoting a negative quantity, as in 'negative three'.

3a. measured in a direction opposite to that regarded as POSITIVE; having the same MAGNITUDE but an opposite SENSE to an equivalent positive quantity. See LENGTH.

b. in particular, (of an angle) measured in a clockwise direction, especially from the positive direction of the *x*-axis of a coordinate system.

4. (*Logic*) **a.** (of a categorial proposition) denying the satisfaction by the subject of the predicate, as in *some men are irrational*, or *no pigs have wings*.

b. (of an expression) containing a privative term or a NEGATION sign.

c. (*as substantive*) a negative statement.

5. another term for NEGATIVE SEMIDEFINITE.

negative binomial distribution, *n.* the distribution of a DISCRETE RANDOM VARIABLE with PROBABILITY DISTRIBUTION FUNCTION,

$$P(X = k) = \binom{k-1}{r-1} p^r (1-p)^{k-r},$$

where p is the probability of a success, and k is the number of BERNOULLI TRIALS required to obtain r successes. This distribution is useful in the modelling of accidents.

negative correlation, see CORRELATION.

negative definite, see NEGATIVE SEMIDEFINITE.

negatively dependent, see STATISTICALLY DEPENDENT.

negative semidefinite or **negative,** *adj.* (of a matrix or a self-adjoint operator on HILBERT SPACE) having $\langle Ax, x \rangle \leq 0$ for all x. If the scalar field is complex, the requirement that A be self-adjoint is redundant. The operator is *negative definite* if $\langle Ax, x \rangle < 0$ for all $x \neq 0$, for which it suffices to check the strict negativity of the leading PRINCIPAL MINORS, obtained by deleting all but the first n rows and columns. Compare POSITIVE SEMIDEFINITE.

neighbourhood (abbrev. **nbd**), *n.* **1.** also called ε**-neighbourhood.** (in a EUCLIDEAN or METRIC SPACE) the OPEN SET of all points whose DISTANCE from a given point, a, is strictly less than a specified value; the set of points

$$\{ x : d(x, a) < \varepsilon \},$$

written $N(\varepsilon, a)$. An *open* ε-*neighbourhood* is also called an OPEN BALL or open SPHERE. In this notation, a function is said to have a limit as x tends to a if there is a p such that

for every $\varepsilon > 0$, there exists a $\delta > 0$ such that
$f(x) \in N(\varepsilon, p)$ for all x such that $x \in N(\delta, a)$.

Compare EPSILON–DELTA NOTATION.

2a. more generally, any set in a TOPOLOGICAL SPACE that contains an open set that contains the given point; in a Euclidean or metric space, this is any set that contains an ε-neighbourhood, so that every ball is an open neighbourhood, but an open neighbourhood is not necessarily a ball. Some authors avoid this usage and restrict the unqualified term to open neighbourhoods. A *closed* ε-*neighbourhood* of a point a in a metric space is defined by

$$\{ x : d(x, a) \leq \varepsilon \}.$$

b. punctured neighbourhood. derivatively, a neighbourhood of a point from which the given point itself has been deleted; that is, a punctured ε–neighbourhood of a, written $N'(\varepsilon, a)$, is $N(\varepsilon, a) \setminus \{a\}$.

3. (of infinity) a neighbourhood of an ideal point added in a COMPACTIFICATION. Thus $]-r, \infty]$ is an open neighbourhood of $+\infty$ on the real line.

neighbourhood base, *n.* a collection of NEIGHBOURHOODS that form a BASE for a TOPOLOGY, so that every open set of the topology can be expressed as a union of some of these neighbourhoods.

neo-Pythagorean means, *adj.* one of ten MEANS defined by the later Pythagorean school in terms of proportions, of which the first three are the *Pythagorean means*, corresponding to the ARITHMETIC, GEOMETRIC, and *subcontrary* or HARMONIC MEANS. The fourth mean is the *counter-harmonic mean*,

$$\frac{a^2 + b^2}{a + b}.$$

nested, *adj.* (of a sequence of sets or intervals) such that each is a subset or subinterval of the preceding. In Euclidean space, if nested intervals are non-empty, bounded, and CLOSED, there must be at least one point common to all. This fact is called the *nested interval theorem*. See CANTOR INTERSECTION THEOREM.

net, *n.* **1.** a diagram of a hollow solid consisting of the plane shapes of the faces so arranged that the diagram could be folded to form the solid. Fig. 253 shows nets of a cube and a triangular pyramid.

Fig. 253. **Nets** of a cube and a pyramid.

2. also called **Moore–Smith sequence.** a generalization of the concept of a SEQUENCE to permit talk of convergence in non-metrizable topological spaces: a net of a set S is a mapping from a DIRECTED SET D into S. See NET CONVERGENCE.

3. see EPSILON NET.

net convergence or **Moore–Smith convergence,** *n.* (*Topology*) the property of a NET of a set S with respect to a DIRECTED SET D, that a sequence $\{x_d\}$, where x_d is the element of S associated with d in D, is *eventually* in every neighbourhood V of some x, in the sense that there exists d such that x_e lies in V for all e in D with $e \geq d$; the net is then said to *converge* to x. The net $\{x_d\}$ is *frequently* in every neighbourhood V of x if, for each d in D, there exists e in D with $e \geq d$, such that x_e lies in V; the set of such e is then COFINAL in D. Then x is a CLUSTER POINT of a set A if and only if there is a net in A that converges to x. See FILTER.

network, *n.* a directed GRAPH together with a *source* (a vertex or *node* with no entering edges or *arcs*), a *terminal* or *sink* (a vertex with no exiting arcs), and a *capacity* or bound on each arc.

network cut, *n.* a subset consisting of all arcs in a NETWORK (with one source and one sink) that originate in a given set of vertices, S, containing the source and not the sink and that terminate in the relative complement of S. See also MAX–FLOW MIN–CUT THEOREM.

network flow, *n.* an assignment to each arc in a NETWORK of some non-negative quantity less than the capacity of the arc, in such a way that the total amount entering and exiting each intermediate node balances. In many combinatorial optimization problems, one wishes to maximize the *value* of the network, which is defined as the total flow reaching the terminal node. This is applicable, for example, to routing telephone calls or flights between two cities through a choice of a number of fixed intermediate cities.

Neumann, John von (1903–57), Hungarian-born American mathematician, who taught at Berlin and Hamburg, and then entered the US in 1930, becoming a member of the Institute for Advanced Study at Princeton in 1933. He is best known for the founding of GAME THEORY, but his enormously wide range of work included mathematical economics, computer science, quantum theory, set theory, group theory, operational calculus, probability, mathematical logic and the foundations of mathematics.

Neumann condition, *n.* a BOUNDARY CONDITION for a PARTIAL DIFFERENTIAL EQUATION, where the normal derivative, $\partial u / \partial \mathbf{n}$, of u, defined as $\nabla u . \mathbf{n}$, is given at each point.

Neumann function, *n.* (*Partial differential equations*) the function

$$Y_\nu(x) = \frac{\cos \nu \pi J_\nu(x) - J_{-\nu}(x)}{\sin \nu \pi},$$

where $J_\nu(x)$ is the BESSEL FUNCTION, and ν is the *order* of the Neumann function.

Neumann-type boundary conditions, *n.* see DIRICHLET'S PROBLEM.

neutral element, *n.* another word for IDENTITY ELEMENT, especially a multiplicative identity or UNITY.

new maths, *n.* **1.** the approach to mathematical education in which the principles of set theory are introduced at an elementary level as the foundations of arithmetic.

2. consequently, a general term for set theory and associated topics in any elementary mathematics syllabus.

newton (abbrev. **N**), *n.* the standard unit of FORCE, equal to the force required to accelerate a MASS of one KILOGRAM at one METRE per SECOND per second.

Newton, Sir Isaac (1643–1727), English physicist, astronomer, and mathematician, who is widely recognized as the greatest and most influential scientist of all time. He obtained his degree after a very undistinguished career at Cambridge, but he then largely invented DIFFERENTIAL CALCULUS and extensively explored its applications, as well as developing his theories of colour and planetary motion while the University was closed during the Great Plague. He also made major contributions to algebra, analytic geometry, and the theory of equations, and is probably best known for his law of gravity and laws of motion, although these were only published at the insistence of the astronomer Halley. As well as being Lucasian Professor at Cambridge, he represented the University in Parliament and defended it against James II; he was Master of the Mint and succeeded Pepys as President of the Royal Society.

Newton–Cotes formulae, *n.* a class of numerical QUADRATURE methods extending the TRAPEZOIDAL RULE and SIMPSON'S RULE.

Newtonian fluid, *n.* (*Continuum mechanics*) a BODY in which the DEVIATORIC part of the STRESS TENSOR depends linearly on the VELOCITY GRADIENT.

Newtonian viscous fluid, *n.* (*Continuum mechanics*) a NEWTONIAN FLUID that is a VISCOUS FLUID.

Newton–Raphson method, *n.* a method of minimizing a twice-differentiable function, by the application of NEWTON'S METHOD to the gradient of a function as a means of approximating a CRITICAL POINT. This involves computing with the inverse of the HESSIAN matrix and is generally avoided in practice by the use of QUASI–NEWTON METHODS. Compare STEEPEST DESCENT.

Newton's identities, *n.* the formulae expressing the powers of the sums of the roots of a polynomial in terms of the coefficients of the polynomial. If

$$p(x) = x^n + a_1 x^{n-1} + a_2 x^{n-2} + \ldots + a_n$$

and

$$s_k = r_1^{\,k} + \ldots + r_n^{\,k},$$

where $r_1, \ldots, r_k$ are the roots of the polynomial, then for $k < n$

$$s_k + s_{k-1}a_1 + s_{k-2}a_2 + \ldots + s_1 a_{k-1} + k a_k = 0;$$

while for $k \geq n$

$$s_k + s_{k-1}a_1 + s_{k-2}a_2 + \ldots + s_{k-n} a_n = 0.$$

This may be viewed as expressing each s_k in terms of the elementary SYMMETRIC FUNCTIONS.

Newton's law, see GRAVITY.

Newton's laws of motion, *n.* (*Mechanics*) the fundamental laws describing the behaviour of PARTICLES. These state:

Newton's first law: a particle that is at rest (or in motion) will remain at rest (or in motion) until acted upon by an external force;

Newton's second law: the rate of change of linear MOMENTUM of a particle is equal to the total applied force; and

Newton's third law: every action has an equal and opposite reaction.

See HAMILTON'S PRINCIPLE OF LEAST ACTION, EULER'S EQUATIONS OF MOTION.

Newton's method, *n.* the iterative formula for approximately solving an equation $f(x) = a$ by repeatedly computing in one dimension

$$x_{\text{NEW}} = x_{\text{OLD}} - \frac{f(x_{\text{OLD}}) - a}{f'(x_{\text{OLD}})}.$$

This corresponds to approximating the function by its tangent, and is guaranteed to exhibit a QUADRATIC RATE OF CONVERGENCE if the initial estimate is sufficiently good. For computing the square root of A one solves $x^2 = A$ by this method, which reduces to *Heron's method:*

$$x_{\text{NEW}} = \frac{1}{2}\left[x_{\text{OLD}} + \frac{A}{x_{\text{OLD}}} \right].$$

In several dimensions one uses

$$x_{\text{NEW}} = x_{\text{OLD}} - G^{-1}[f(x_{\text{OLD}}) - a],$$

where G is the matrix of partial derivatives of f evaluated at x_{OLD}.

Neyman–Pearson lemma, *n.* (*Statistics*) the theorem that of all TESTS of a given hypothesis of the same SIGNIFICANCE LEVEL, the LIKELIHOOD RATIO TEST has maximal POWER.

***n*-gon,** *n.* a regular POLYGON with *n* SIDES.

nibble, *n.* (*informal*) half a BYTE; a byte of four BITS.

Nikodym set, **impossible set**, or **linearly accessible set,** *n.* a subset of the unit square with unit area, such that for every point in the set there is a line that intersects the set only in this point. This was constructed by Nikodym in 1927, and in 1952 Davis found such a set with uncountably many lines through each point.

nilpotent, *adj.* (of a matrix, function, or ring element) having the property that some integral power of the given quantity is zero. For example, the matrix

$$\begin{bmatrix} 0 & 1 \\ 0 & 0 \end{bmatrix}$$

is nilpotent since

$$\begin{bmatrix} 0 & 1 \\ 0 & 0 \end{bmatrix}\begin{bmatrix} 0 & 1 \\ 0 & 0 \end{bmatrix} = \begin{bmatrix} 0 & 0 \\ 0 & 0 \end{bmatrix}.$$

nilpotent group, *n.* a GROUP G for which the ascending chain of groups

$$1 = Z_0 \subseteq Z_1 \subseteq \ldots \subseteq Z_n$$

with Z_{k+1}/Z_k equal to the CENTRE of G/Z_k terminates finitely with $G = Z_n$.

nine-point circle, *n.* the circle, discovered by Poncelet, upon which lie the feet of the three altitudes to any triangle, the midpoints of the three sides, and the midpoints of the line segments between the ORTHOCENTRE and the three vertices of the triangle. The *nine-point centre* of this circle lies on the EULER LINE midway between the orthocentre and the CIRCUMCENTRE. In Fig. 254, L, M, and N are the midpoints of the sides of the triangle ABC; X, Y, and Z are the feet of the altitudes and P the orthocentre of the triangle; the circle LXMYNZ bisects AP, BP, and CP.

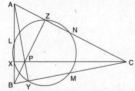

Fig. 254. **Nine-point circle**.
See main entry.

nines complement, see CASTING OUT NINES.

node, *n.* **1.** a point at which two or more branches of a graph meet. Compare CRUNODE, SPINODE, TACNODE.

2. another word for a VERTEX of a TREE, NETWORK, or DIGRAPH.

Noether, Amelie Emmy (1882–1935)**,** German-born abstract algebraist, who was educated at Erlangen. She was championed by Hilbert and Klein against objections to the appointment of a woman to a post at Göttingen, where she developed the theory of IDEALS and a general theory of non-commutative algebra. In 1933 she was dismissed along with all other Jewish academics, and took refuge from the Nazis at Princeton.

Noetherian module, *n.* a MODULE that satisfies the ASCENDING CHAIN CONDITION so that every strictly increasing chain of SUBMODULES is finite. This is equivalent to satisfying the MAXIMUM CONDITION that every non-empty set of submodules has a maximal member. Even if the module is not UNITARY, every submodule, including the module itself, is FINITELY GENERATED. For example, the set $\mathbb{Z}$ of integers form a Noetherian $\mathbb{Z}$-module, but the rationals do not. See ARTINIAN MODULE.

Noetherian radical, see RADICAL.

Noetherian ring, *n.* a RING in which every strictly ascending (increasing) chain of (left or right) IDEALS is finite; a ring that, when considered as a (left or right) R-module, is a NOETHERIAN MODULE. It is said to be *left* or *right Noetherian* according as this holds for left or right ideals. The integers form a Noetherian ring which is not an ARTINIAN RING.

noise, *n.* another term for RANDOM ERROR, especially in information theory and engineering.

nominal data, *n.* (*Statistics*) data classifiable by a NOMINAL SCALE.

nominal scale, *n.* (*Statistics*) a DISCRETE classification of data; one in which data are neither measured nor ordered, but subjects are merely allocated to distinct categories. For example, a record of students' course choices constitutes nominal data that could be correlated with, say, school results. Compare ORDINAL SCALE, INTERVAL SCALE, RATIO SCALE.

non-, *prefix indicating* the negation of some property; for example, a *non-commutative* operator is one that is not commutative.

nona- or **ennea-,** *prefix denoting* nine; for example, a *nonagon* is a nine-sided polygon.

non-atomic, *adj.* (of a MEASURE or MEASURE RING) such that no member of the measure space is an ATOM. Some authors use *continuous* in a similar sense. Compare ATOMIC.

non-constructive, *adj.* (of a proof or definition) not CONSTRUCTIVE, not enabling the conclusion to be proved or the object to be constructed in a finite number of steps, since, for example, it involves quantifying over an infinite domain, as does the AXIOM OF CHOICE. See also EXCLUDED MIDDLE.

non-deterministic polynomial time algorithm, *n.* another term for NP–DECISION PROBLEM.

non-denumerable, *adj.* infinite and not able to be put in ONE–TO–ONE CORRESPONDENCE with the integers; having CARDINALITY greater than the integers. This differs from being not DENUMERABLE for authors who distinguish denumerable and countable. See also CONTINUUM HYPOTHESIS.

non-dominated, *adj.* (of a point in a partial ordering) MINIMAL or MAXIMAL.

non-elementary, *adj.* (of a proof) using methods of COMPLEX ANALYSIS; more informally, using any advanced techniques. An elementary proof may be very difficult, and a non-elementary proof easy.

non-empty, *adj.* (of a set) having MEMBERS; not identical with the EMPTY SET.

non-empty word, *n.* (*Group theory*) a formal expression of the form

$$x_1^{\varepsilon_1}, x_2^{\varepsilon_2}, \dots x_n^{\varepsilon_n},$$

where the x_i are members of a given non-empty set X, all the $\varepsilon_i = \pm 1$, and the *length* of the word is the positive integer n. With the product of two words defined by

$$\left[x_i^{\varepsilon_i}\right]\left[y_j^{\delta_j}\right] = x_1^{\varepsilon_1}, \dots x_n^{\varepsilon_n}, y_1^{\delta_1}, \dots y_m^{\delta_m},$$

the non-empty words form a SEMIGROUP, and with multiplication extended to the EMPTY WORD, the set of all words is a MONOID.

non-equivalence, *n.* (*Logic*) another name for EXCLUSIVE DISJUNCTION; the relation that holds between two sentences or propositions when one is EQUIVALENT to the negation of the other, or the operator which forms an assertion of such a relation between two given expressions.

non-Euclidean geometry, *n.* the study of geometrical systems in which the PARALLEL POSTULATE of EUCLIDEAN GEOMETRY is replaced, resulting in fundamental changes in the properties of the space. If there are at least two parallels to a given line through the same point, the result is LOBACHEV-SKIAN GEOMETRY; if there are no parallels, it is ELLIPTIC GEOMETRY.

non-expansive mapping, *n.* a LIPSCHITZ FUNCTION on a metric space, of which the Lipschitz constant does not exceed 1.

non-logical axioms, *n.* axioms valid only in a given STRUCTURE for a THEORY. Compare LOGICAL AXIOM.

non-measurable, *adj.* (of a set) not belonging to a given MEASURE ALGEBRA; in particular, not LEBESGUE MEASURABLE. For example, if, for α in the interval $[0, 1]$, the equivalence classes C_α are defined by

$$C_\alpha = \{x: \ x - \alpha \in \mathbb{Q}\},$$

and N is then formed by choosing one element from each C_α, then N is not Lebesgue measurable. However, since this depends upon the axiom of choice, intuitionists deny the existence of this set, and hold that all sets are Lebesgue measurable. Compare MEASURABLE. See also METRIC DENSITY.

non-negative, *adj.* (of a number or quantity) not NEGATIVE; either positive or zero.

non-parametric statistics, *n.* the branch of statistics that studies data measurable on an ORDINAL or NOMINAL SCALE. See MANN–WHITNEY TEST, WILCOXON TEST.

non-principal ultrafilter, see ULTRAFILTER.

non-reflexive, *adj.* **1.** (*Logic*) (of a relation) neither REFLEXIVE nor IRRE-FLEXIVE; holding between some members of its domain and themselves, but failing to hold between other members and themselves, and thus having both reflexive and irreflexive restrictions; for example ... *is confident in* ..., which is reflexive on its restriction to self-confident people.

2. (of a BANACH SPACE) not REFLEXIVE; that is, not canonically identifiable with its second dual Banach space; possessing a UNIT BALL that is not compact in the WEAK TOPOLOGY.

non-residue, see RESIDUE (sense 2).

nonsense correlation, *n.* (*Statistics*) a CORRELATION supported by data but having no basis in reality, such as between incidence of the common cold and ownership of televisions.

non sequitur, *n.* (*Latin*: it does not follow) (*Logic*) **1.** an INVALID argument; one in which the conclusion does not follow from the premises.

2. the conclusion of such an argument; a statement which does not follow from what was previously said.

non-singular, *adj.* (of a LINEAR TRANSFORMATION or MATRIX) possessing an inverse. A finite-dimensional non-singular linear transformation corresponds to a matrix with non-vanishing DETERMINANT.

non-standard, *adj.* **1.** (of a mathematical system) ELEMENTARILY EQUIVALENT, but not ISOMORPHIC, to the usual MODEL for a given set of AXIOMS; for example, the non-standard real numbers have infinitesimal elements.

2. (of an element of the non-standard real numbers) not corresponding to an element of the ordinary REAL NUMBERS; for example, every infinitesimal and every infinite number is non-standard.

non-standard analysis, *n.* a formalization of analysis that allows rigorously for the existence of INFINITESIMALS.

non-standard real number, *n.* an element of a NON–STANDARD MODEL of the axioms of the REAL NUMBERS; finite non-standard real numbers correspond to the sum of an ordinary real number (its *standard part*) and an INFINITESIMAL (its *non-standard part*).

non-symmetric, *adj.* (*Logic*) (of a relation) not SYMMETRIC, nor ASYMMETRIC, nor ANTISYMMETRIC; holding between some pairs and failing to hold for some other pairs of members, x and y, of its domain when it holds between y and x, and thus having both symmetric and asymmetric restrictions. For example, ... *is a brother of* ..., is symmetric on its restriction to males.

non-transitive, *adj.* (*Logic*) (of a relation) neither TRANSITIVE nor INTRANSITIVE; holding between some pairs and failing to hold between other pairs of members, x and z, of its domain when it is given to hold both between x and y and between y and z. For example ... *is a half-brother of* ... is non-transitive, since the half-brother of my half-brother may be myself, my full brother, another half-brother, or no relation of mine.

non-trivial, *adj.* **1.** (of a substructure of a given algebraic structure) neither EMPTY nor a TRIVIAL pathological example; sometimes it is also required that the substructure be PROPER. For example, a non-trivial subgroup of a group does not consist only of the identity of the group, nor a non-trivial ring of its zero.

2. (of a solution of an equation) not zero; for example, $\sqrt{3}$ is the non-trivial solution to $x^3 = 3x$.

3. not obvious.

non-zero ring, *n.* a ring that is not the ZERO RING.

norm, *n.* **1.** the LENGTH of a vector expressed as the square root of the sum of the squares of its ORTHOGONAL COMPONENTS.

2a. a non-negative real-valued function defined on the members of a vector-space, satisfying the conditions that

$$\| -x \| = \| x \|,$$

$$\| tx \| = | t | \cdot \| x \| \text{ for scalar } t,$$

and the triangle inequality

$$\| x + y \| \leq \| x \| + \| y \|$$

where $\| x \|$ is the norm of x.

b. (*as a prefix*) with respect to a given norm or in the topology induced by a given norm. For example, a *norm-convergent* or *norm-bounded* function is convergent or bounded with respect to some norm; a *norm-compact* function is compact with respect to the norm topology; and a *norm-one* element has unit norm.

3. an everywhere finite MINKOWSKI FUNCTION or GAUGE.

4. (of an ALGEBRAIC NUMBER) the product of all the CONJUGATES to the given number; for example, the norm of an algebraic number of the form $a + b\sqrt{d}$ is $a^2 - db^2$ for any integers a, b, and d.

5. another word for MESH–FINENESS.

6. (*Statistics*) another term for the MODE of a distribution.

normable, *adj.* (of a TOPOLOGICAL LINEAR SPACE) compatible with a NORM; such that there exists a norm for which the topology imposed by the norm is the given topology.

normal, *adj.* **1a.** perpendicular to a given line or plane, or to the tangent to a curve or surface at the point of contact. For example, in Fig. 255, the bold line PN is normal to the curve at the point P, where its tangent is the line PT.

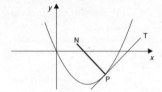

Fig. 255. NP is **normal** (sense 1) to the curve.

b. (*as substantive*) a line drawn normal to another line, plane, curve or surface. The normal to a surface expressed as $F(x, y, z) = 0$, is given by the GRADIENT VECTOR $\nabla F(x, y, z)$.

2. see PRINCIPAL NORMAL.

3. see NORMAL TOPOLOGICAL SPACE, NORMAL SUBGROUP.

4. (of a set of functions) PRECOMPACT with respect to the topology induced by the CHEBYSHEV NORM.

5. another term for NORMALLY DISTRIBUTED.

6. (of a property) usual, desirable, regular.

normal closure, *n.* the smallest NORMAL SUBGROUP of a group, G, that contains a given non-empty subset, X, denoted X^G.

normal cone, see POLAR SET.

normal curvature, *n.* (of a surface at a point) the CURVATURE of a NORMAL SECTION of the surface at the given point. A positive sign is taken if the principal normal to the section points in the same direction as the normal to the surface, and a negative sign is taken otherwise.

normal curve, *n.* (*Statistics*) a symmetrical *bell-shaped curve* representing the PROBABILITY DENSITY FUNCTION of a NORMAL DISTRIBUTION. The area of a vertical section of the curve represents the probability that the random variable lies between the values that delimit the section, and these probabilities can be discovered from statistical tables. In Fig. 256 opposite, the vertical line is the MEAN (which coincides with the MEDIAN and MODE), and the units on the baseline are STANDARD DEVIATIONS.

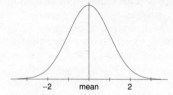

Fig. 256. **Normal curve**.

normal derivative, *n.* (of a function defined on a curve or surface) the DIRECTIONAL DERIVATIVE in the direction of the outward pointing NORMAL at a given point of the curve or surface

$$\frac{\partial h(\mathbf{x})}{\partial n} = \nabla h(\mathbf{x}) \cdot \mathbf{n}.$$

normal distribution or **Gaussian distribution,** *n.* (*Statistics*) a distribution that is CONTINUOUS and is SYMMETRICAL with MEAN, MEDIAN and MODE coincident. It is an easy distribution to handle computationally, and is the most commonly treated, primarily because many quantitative measurements appear to be approximately normally distributed; this is in part a consequence of the CENTRAL LIMIT THEOREM. The graph of the PROBABILITY DENSITY FUNCTION of a normal distribution is a NORMAL CURVE and is given by

$$\frac{\exp\dfrac{-(x-\mu)^2}{2\sigma^2}}{\sigma\sqrt{2\pi}},$$

where μ is the mean and σ^2 the VARIANCE. See also MULTIVARIATE NORMAL DISTRIBUTION.

normal equations, *n.* the equations that describe the best LEAST SQUARE approximation for the distance of a vector from a set of n vectors. The scalars minimizing

$$\| x - a_1 x_1 - a_2 x_2 - \dots - a_n x_n \|$$

will satisfy

$$\langle x, a_1 x_1 + a_2 x_2 + \dots + a_n x_n \rangle = 0,$$

which can be expressed in terms of the GRAM MATRIX of the vectors x_1, $x_2, \dots, x_n$.

normal extension field, *n.* an EXTENSION FIELD such that the group of automorphisms leaving the base field fixed (a *relative automorphism*) is such that no other element is left fixed by the entire set of automorphisms; some usages also require that the DEGREE of the extension is finite. Alternatively, if any irreducible polynomial has one root in the extension, it has all its roots in the extension.

normal factor, *n.* any of the FACTOR GROUPS of members of a NORMAL SERIES for a group by the next member of the series.

normal family, *n.* **1.** another term for a set of functions that are PRECOMPACT, especially in the topology of uniform convergence on compact subsets of a given set in Euclidean space.

2. a family of complex functions that are ANALYTIC on a common DOMAIN and of which any infinite sequence of members contains a subsequence converging uniformly on compact sub-regions either to an analytic limit or identically to infinity. See ASCOLI'S THEOREM.

normal form, *n.* a canonical form to which a given structure or object may be reduced, especially of a matrix or in logic. See JORDAN NORMAL FORM, CONJUNCTIVE NORMAL FORM, DISJUNCTIVE NORMAL FORM, PRENEX NORMAL FORM, SKOLEM NORMAL FORM.

normalize, *vb.* to put in some NORMAL FORM, especially to divide a non-zero vector by its NORM, to divide a polynomial by its leading coefficient, or to change the variable in a probability distribution so that its mean becomes zero and its variance unity.

normalizer, *n.* (with respect to a given subset of a GROUP) the set of elements, g, of the group, such that $g^{-1}Hg = H$, where H is the given subset. Where G is the underlying group, the normalizer of the subset H is denoted $N_G(H)$. Compare CENTRALIZER.

normally distributed or **normal,** *adj.* having a NORMAL DISTRIBUTION.

normal matrix, *n.* a matrix A that commutes with its ADJOINT; that is, such that $AA^* = A^*A$. The JORDAN NORMAL FORM of a normal matrix is a diagonal matrix. See also SPECTRAL THEOREM.

normal number, *n.* a real number whose expansion to a given base is such that each block of digits of fixed length tends to occur equally often as the number of digits in the expansion tends to infinity. Thus, in base 10, one out of every 100 two-digit blocks should on average be 37 and one in ten single-digit blocks should be 7. It is known that almost all numbers are *absolutely normal* – that is, are normal in every base – but no naturally occurring irrational numbers are provably so.

normal operator, *n.* an operator that commutes with its ADJOINT; that is, such that $AA^* = A^*A$. See also SPECTRAL THEOREM.

normal section, *n.* a SECTION of a surface by a plane containing both a normal to the surface, and a prescribed tangent at a given point.

normal series, *n.* a finite sequence of subgroups $\{G_k\}$ of a given GROUP, G, beginning with the group itself, and terminating in the TRIVIAL SUBGROUP, such that each G_{k+1} is a NORMAL SUBGROUP of G_k. Some authors call this simply a *series* and reserve the term 'normal series' for a series in which each subgroup G_k is normal in G. If no two members of the sequence are identical, the normal series is said to be *without repetitions,* and the FACTOR GROUPS G_k/G_{k+1} are the *normal factors* of the series. See also COMPOSITION SERIES, PRINCIPAL SERIES.

normal stress, *n.* that part of a STRESS VECTOR, $t(n)$, that acts along the unit outward normal, n, to the surface of a fluid; it is equal to

$$[\,t(n)\,.\,n\,]\,n.$$

See also COMPRESSIVE NORMAL STRESS, TENSILE NORMAL STRESS.

normal subgroup or **invariant subgroup,** *n.* a SUBGROUP that is left invariant NER AUTOMORPHISMS of the group, or equivalently, has identical left- and right-COSETS in the group. This occurs as the kernel of some homomorphism of the given group into another group. This latter definition is sensible also for MONOIDS. For example, the centre of a group is a normal subgroup.

normal subring, *n.* a two-sided IDEAL.

normal topological space, *n.* a TOPOLOGICAL SPACE in which, for every pair of disjoint closed sets, there is a pair of disjoint open sets in which they are respectively contained. If this follows when the sets are merely disjoint from each other's closures, the space is said to be *completely normal*. See also URYSOHN'S LEMMA, and TIETZE'S EXTENSION THEOREM.

normed ring, *n.* another term for BANACH ALGEBRA.

normed space, *n.* a VECTOR SPACE endowed with a NORM.

normed vector ring, *n.* another term for BANACH ALGEBRA.

not, *adv.* the usual English expression for logical NEGATION. To avoid ambiguity about the scope of the operator, it is usual in formal contexts to read sentential negation as *it is not the case that...* .

nought or **naught,** *n.* another word for ZERO.

nowhere dense set, *n.* (*Topology*) a set whose CLOSURE has empty INTERIOR. See also BAIRE CATEGORY.

nowhere differentiable function, *n.* a CONTINUOUS FUNCTION on the real line that is not differentiable at even a single point. An early example was constructed by Weierstrass.

NP, *n.* the class of NP–DECISION PROBLEMS; see P VERSUS NP.

NP-complete problem, *n.* a DECISION PROBLEM, such as the SATISFIABILITY PROBLEM, for which a POLYNOMIAL TIME ALGORITHM can exist if and only if P EQUALS NP.

NP-decision problem or **non-deterministic polynomial time algorithm,** *n.* a computational problem to which there is a yes or no answer, such as the SATISFIABILITY PROBLEM, for which a POLYNOMIAL TIME ALGORITHM exists to verify any guessed solution.

NP-hard problem, *n.* a DECISION PROBLEM, whether or not in NP, for which a polynomial time reduction to an NP–COMPLETE PROBLEM exists.

***n*-space,** *n.* a SPACE with *n* DIMENSIONS.

***n*th** or ***n*th,** *adj.* having an unspecified ORDINAL NUMBER, often either the final member of a finite sequence of elements $\langle a_1,..., a_n \rangle$, or an arbitrary member of an infinite sequence. For example, the nth member of the sequence $\langle 0, k, 2kx, 3kx^2, ... \rangle$ is of the form knx^{n-1}; the positive n^{th} root of a positive number x is $x^{1/n}$.

***n*-tuple,** *n.* an ORDERED SET of *n* elements.

nucleus, *n.* (of an integral operator or equation) another word for KERNEL. See VOLTERRA'S INTEGRAL EQUATION.

null, *adj.* **1.** empty, equal to zero, or having measure zero.
　　2. inessential; of no significance.

null graph, *n.* the GRAPH a representation of which is shown in Fig. 257.

Fig. 257. **Null graph.**

null hypothesis, *n.* (*Statistics*) **1.** the residual hypothesis that cannot be rejected unless the TEST STATISTIC used in the HYPOTHESIS TESTING problem lies in the CRITICAL REGION for a given SIGNIFICANCE LEVEL.

2. in particular, especially in psychology, the hypothesis that certain observed data are a merely random occurrence.

nullity, *n.* the DIMENSION of the KERNEL or NULL SET of a matrix or operator.

null measure or **zero measure,** *n.* a MEASURE of a subset of *n*-dimensional Euclidean space for which, for every ε > 0, there is a COVERING of the subset by rectangles whose volume is less than ε.

null sequence, *n.* a CAUCHY SEQUENCE the limit of which is zero.

null set, *n.* **1.** the SET with no elements; the EMPTY set, written ∅.

2a. the set of arguments of a function for which the value of the function is zero. See also KERNEL.

b. (*Measure theory*) in particular, for a given MEASURE, a set of measure zero. For example, the CANTOR SET and all countable sets on the line are null sets for LEBESGUE MEASURE.

null space, *n.* the NULL SET of a linear operator.

null vector, *n.* the zero vector.

number, *n.* **1.** also called **natural number.** one of a unique sequence of elements used for counting a collection of individuals, as when one says that the number of Jesus's disciples is twelve. See also CARDINAL NUMBER.

2. any quantity derived by extension of these. CLOSURE under subtraction gives the WHOLE NUMBERS (INTEGERS); under division, the RATIONAL NUMBERS; and under extraction of roots, the COMPLEX NUMBERS. These can be arranged in a hierarchical classification: every number is a complex number; a complex number is the sum of a REAL NUMBER and an IMAGINARY NUMBER, the latter itself equal to the product of a real number with *i* (the square root of −1); a real number is either a RATIONAL NUMBER or an IRRATIONAL NUMBER; a rational number may be either an integer or a FRACTION, while an irrational number may be either an ALGEBRAIC NUMBER (as are all rational numbers) or a TRANSCENDENTAL NUMBER.

3. the concept of CARDINALITY in abstraction from its applications. In this sense, number can be defined without reference to counting.

4. see ORDINAL NUMBER.

5. the symbol used to represent a number, a NUMERAL.

number field, *n.* another term for ALGEBRAIC NUMBER FIELD.

number line or **real line,** *n.* an infinite line on which points are taken to represent the REAL NUMBERS by their distance from a fixed origin; the axis of a one-dimensional coordinate system.

number theory or **higher arithmetic,** *n.* the mathematical theory that studies the properties and relations of INTEGERS, and their extensions, both algebraic and analytic. This includes the study of divisibility, primality and factorization properties, PARTITION properties, IRRATIONAL and TRANSCENDENTAL numbers, ALGEBRAIC INDEPENDENCE, rates of approximation, the representation of numbers as, for example, sums of squares, and the integer solution of polynomials in several unknowns.

numerable, *adj.* another word for DENUMERABLE.

numeral, *n.* a symbol representing a number, especially a single digit. See
ARABIC NUMERAL, ROMAN NUMERAL.

numerator, *n.* the DIVIDEND in a VULGAR FRACTION or ratio. The numerators
of $\frac{7}{8}$ and $x/(x-1)$ are 7 and x respectively. Compare DENOMINATOR.

numerical, *adj.* **1.** containing or using constants, coefficients, terms or elements represented by numbers; for example, $3x^2 + 4y = 2$ is a numerical
equation. Compare LITERAL.

2. another word for ABSOLUTE. For example, the *numerical value* of a signed
number is its absolute value.

numerical analysis, *n.* the branch of mathematics concerned with the study
of computation and of its accuracy, stability, and often its implementation
on a computer. One central concern is the determination of appropriate
numerical MODELS for applied problems. Another is the construction and
analysis of robust and efficient ALGORITHMS for problems such as those of
numerical integration and differentiation, solution of differential equations, and COMBINATORIAL and CONSTRAINED OPTIMIZATION problems. With
the rapid development of high-speed, relatively cheap computing devices
with large memories, the relative advantage and feasibility of different
methods is in considerable flux.

numerical eccentricity, *n.* the constant $\varepsilon = e/a$ for a family of similar CONICS,
where e is the (linear) ECCENTRICITY, and a is the major semi-axis of the
conic.

numerical identity, see IDENTITY.

numerical quadrature, see QUADRATURE.

numerical quantifier, *n.* (*Logic*) **1.** any of the sequence of QUANTIFIERS,
$(\exists_n x)$, read as *there are at least n Fs*. The first member, $(\exists_1 x)$, is defined by

$$(\exists_1 x)Fx \equiv (\exists x)(Fx)$$

to be equivalent to the EXISTENTIAL QUANTIFIER, $(\exists x)$, and the sequence is
defined recursively by

$$(\exists_{n+1} x)Fx \equiv (\exists_n x)(Fx \ \& \ (\exists y)(Fy \ \& \ x \neq y));$$

that is, *there are at least n + 1 Fs* is equivalent to *there are at least n Fs different
from some F*.

2. exact numerical quantifier. a restriction on the above, definable in terms
of it as

$$(\mathbf{n} x)Fx \equiv (\exists_n x)Fx \ \& \ -(\exists_{n+1} x)Fx.$$

Alternatively, $(\mathbf{1} x)Fx$ can be defined as the UNIQUE QUANTIFIER,

$$(\exists! x)Fx \equiv (\exists x)(Fx \ \& \ (\forall y)(Fy \rightarrow x = y)),$$

whence by recursion

$$(\mathbf{n+1} x)Fx \equiv (\mathbf{n} x)(Fx \ \& \ (\mathbf{1} y)(Fy \ \& \ x \neq y)).$$

numerical range, *n.* (for a matrix or for a continuous linear operator T on a
HILBERT SPACE) the convex set of values in the complex plane traced out by

$$\langle Tx, x \rangle \ \text{for} \ \|x\| \leq 1.$$

See RAYLEIGH QUOTIENT.

O

0, *symbol for* falsehood, especially in computing, and sometimes in TRUTH-TABLES.

***O* and *o* notation,** see ORDER NOTATION.

object, see CATEGORY.

objective, *adj.* (*Continuum mechanics*) (of a TENSOR FIELD defined on a BODY **B**) invariant under RIGID BODY MOTIONS of **B**.

objective function, *n.* the function of which one seeks the OPTIMUM in an optimization problem, especially one with CONSTRAINTS. For example, in many problems, an economist views a cost function or a profit function as an objective function.

object language, *n.* the language described by another language, the META-LANGUAGE.

oblate spheroid, *n.* a SURFACE OF REVOLUTION swept out by an ellipse rotating about its MINOR AXIS; for example, the earth is an oblate spheroid. Compare PROLATE SPHEROID.

oblique, *n.* **1.** (of lines, planes, etc.) neither perpendicular nor parallel.

2. (of a geometric figure, especially a triangle) not containing a right angle.

oblique angle, *n.* any angle that is neither a right angle nor any multiple of a right angle.

oblique coordinate system, *n.* a COORDINATE SYSTEM in which the axes are not at right angles. Compare CARTESIAN COORDINATES.

oblong, *adj.* **1.** RECTANGULAR.

2. *n.* another word for RECTANGLE.

observer, *n.* (*Mechanics*) a formalization of the notion of a person who observes events and records their positions and times; that is, a BIJECTION that assigns to each event in the physical world a position in three-dimensional EUCLIDEAN POINT SPACE and a time indexed by the real line.

obtuse, *adj.* **1.** (of an angle) strictly greater than a RIGHT ANGLE but strictly less than a STRAIGHT ANGLE; angle A in Fig. 258 is obtuse.

2. (of a triangle) having one interior angle greater than a right angle, as the triangle in Fig. 258.

Compare ACUTE.

Fig. 258. An **obtuse** triangle; angle A is obtuse.

obverse, *n.* (*Logic*) a categorial statement derived from a given statement by changing its predicate from positive to negative or vice versa, and negating the whole statement. For example, the obverse of *all cats are mammals* is *no cats are non-mammals*.

obvious, *adj.* (of a step in a PROOF) following directly, though not necessarily cognitively obvious. It is alleged that when G H Hardy was once challenged about a point in a proof, he spent 40 minutes scribbling complex calculations, then responded "Yes, I was right, it is obvious."

oct-, *prefix denoting* eight. For example, an *octangle* is a figure with eight angles; an *octillion* is $(1\,000\,000)^8$ in the UK, and $1000 \times (1000)^8$ in the USA.

octad or **ogdoad,** *n.* a set or sequence of eight elements.

octagon, *n.* a POLYGON with eight sides.

octahedron, *n.* a POLYHEDRON with eight plane faces; in a REGULAR octahedron the faces are equilateral triangles.

octal, *adj.* (of a number or notation) written in or pertaining to PLACE VALUE NOTATION with BASE 8, often used in computing; for example, 371.24_8 represents the number

$$(3 \times 8^2) + (7 \times 8^1) + (1 \times 8^0) + (2 \times 8^{-1}) + (4 \times 8^{-2}) = 249.25_{10}.$$

octant, *n.* any of the eight trihedral sections into which the three-space is divided by the Cartesian coordinate axes. The first octant is that bounded by the positive directions of the three axes; the second, third, and fourth are counted anti-clockwise above the x–y plane, and the fifth to eighth are respectively below the first to fourth. See also ORTHANT, QUADRANT.

odd, *adj.* **1.** (of an integer) not exactly divisible by 2; divisible by 2 with a remainder of 1; of the form $2n + 1$ for some integer n. See also ODD PART.
2. (of a function) changing sign but not absolute value when the sign of the independent variable is changed, so that $f(x) = -f(-x)$; for example, $\sin x$ and $\tan x$ are odd functions. Such a function, as illustrated in Fig. 259, is symmetrical about the origin.

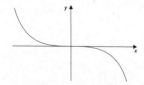

Fig. 259. The graph of an **odd** function.

3. (of a PERMUTATION) derived from the natural order by an odd number of TRANSPOSITIONS; thus, the permutation $(1, 3, 2)$ is odd but $(3, 1, 2)$ is EVEN, since it can be achieved by exchanging 1 and 3 and then exchanging 1 and 2. See also DIFFERENCE POLYNOMIAL.
See also PARITY. Compare EVEN.

odd part, *n.* the largest ODD factor of a given natural number, often written $\text{odd}(n)$. For example, the odd part of 52 is 13, and $\text{odd}(16) = 1$.

ode, *abbrev. for* ORDINARY DIFFERENTIAL EQUATION.

off diagonal, *n.* in a square matrix or array, the DIAGONAL between the lower left and upper right-hand corners.

officer problem, *n.* the problem, due to Euler, of placing 36 officers of six ranks and six regiments in a LATIN SQUARE.

ogdoad, *n.* another term for OCTAD.

ogive, *n.* (*Statistics*) a graph of a CUMULATIVE FREQUENCY DISTRIBUTION.

oh notation, see ORDER NOTATION.

omega, *n.* the smallest infinite ORDINAL; the ordinal of the natural order of the natural numbers, written ω.

omega inconsistency, *n.* (*Philosophy*) the apparent paradox that occurs when the PRINCIPLE OF INDUCTION fails; that is, when it is not possible to infer from the fact that each element of a domain has a property that all of the elements have it. It is so called since the paradigm case is that of the finite ORDINAL NUMBERS, each of which has a finite successor while they clearly do not all have a finite successor, since the set of finite ordinals is the smallest infinite ordinal, OMEGA. A philosophical example, due to Russell, is that it is part of the concept of desire that one wants each of one's desires to be satisfied, but amongst those desires is the apparently inconsistent desire to face new challenges, that is, to leave some of one's desires unsatisfied; thus one can satisfy any of one's desires but not all of them. This paradox seems best resolved by observing that the scope of the universal quantifier has changed.

one, *n.* **1.** the smallest NATURAL NUMBER; the first non-null CARDINAL NUMBER; the second smallest NATURAL NUMBER in the construction of the ORDINALS; the number of members of a set that has all of its members identical; UNITY.

2. the IDENTITY ELEMENT under multiplication in a RING.

one-many, *adj.* (of a function or mapping) capable of associating a single member of the domain with more than one member of the range of a SET–VALUED FUNCTION; holding between the same first argument and more than one second argument of a binary relation, as illustrated by the diagram of Fig. 260. For example, $\sin^{-1}x$ and $\sqrt{x}$ are one-many functions on the positive real numbers; ... *is the mother of* ... is a one-many relation, since a mother may have many children but every child has exactly one mother.

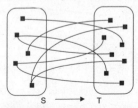

Fig. 260. A **one-many** mapping from S to T.

one-one, *adj.* another term for ONE-TO-ONE.

one-point compactification or **Alexandroff compactification,** *n.* a COMPACTIFICATION that adds a single point, designated ∞, to a HAUSDORFF SPACE, and adds complements of closed subsets of the original space as neighbourhoods of this point. The REAL NUMBERS can be compactified in this manner.

one-sided, *adj.* **1.** (of a surface) having the property that any two points can be joined without crossing an edge. See MÖBIUS STRIP, KLEIN BOTTLE.

2. (of the LIMIT of a function of a single real variable at a point, p) the limit as computed when the function is viewed as restricted to the half-line $]-\infty, p[$ or to the half-line $]p, \infty[$. The function has a DISCONTINUITY at p if the one-sided limits do not exist or are not equal, and is continuous at p if they are finite and equal, in which case that value is the *two-sided* limit at p. For example, the one-sided limit of $x/|x|$ at zero from above is +1 and the one-sided limit from below is –1. See also LEFT-HAND LIMIT, RIGHT-HAND LIMIT.

one-tailed, *adj.* (*Statistics*) (of a SIGNIFICANCE TEST) concerned with the hypothesis that an observed value of a TEST STATISTIC differs significantly from a given value, where error is only relevant in one direction. For example, in testing whether weighing scales give short measure, a consumer does not regard overweight goods as a significant error. Compare TWO-TAILED.

one-to-one or **one-one** (**1–1**), *adj.* **1.** (of two sets of individuals) having elements that are able to be paired with one another without remainder; equinumerous. For example, the natural numbers *n* correspond one-to-one with the points $\langle n, n \rangle$ on the real plane.

2. also called **bijective** or **into and onto.** (of a mapping) associating a unique member of the codomain with every member of the domain of a function, or a unique first argument with each second argument of a binary relation, and vice versa, as illustrated by the diagram in Fig. 261. For example, there is a 1–1 correspondence between the members of a rugby team and the natural numbers 1 to 15; $x \rightarrow 2x$ is a 1–1 mapping from the integers to the even integers. These terms are often reserved for the case when the CODOMAIN and RANGE coincide. Compare SURJECTIVE, INJECTIVE.

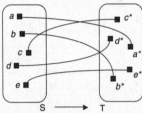

Fig. 261. A **one-to-one** correspondence between S and T.

only if, *conj. indicating* a NECESSARY CONDITION, the converse of IF, so that *P only if Q* is equivalent to *if P, Q.*

***O* notation,** see ORDER NOTATION.

onto, *prep.* **1.** (of a mapping) to the whole of the given set; having the whole CODOMAIN as its RANGE.

2. (*as adj.*) another term for SURJECTIVE.

ontology, *n.* the (set of) entities whose existence is implicit in any given theory; the set over which its QUANTIFIERS can range. For example, FIRST-ORDER PREDICATE CALCULUS has an ontology of INDIVIDUALS.

opacity, *n.* the property of being an OPAQUE context.

opaque, *adj.* (of a MODAL context) not permitting variables within the scope of the MODAL operator to be bound by QUANTIFIERS of wider scope. For example, from

John believes someone robbed him

one cannot infer

There is someone whom John believes robbed him,

since he may have no idea who the culprit might be. Similarly,

He ought to do something

is not equivalent to

There is something he ought to do.

Likewise, the principle of SUBSTITUTIVITY fails in opaque contexts. For example, even though √9 = 3,

Adam believes that 3 is odd

may be true, and

Adam believes that √9 is odd

false. Opacity is responsible for the difference between *DE RE* and *DE DICTO* modality. Compare TRANSPARENT.

open, see OPEN SET.

open ball, *n.* (*Topology*) an ε-NEIGHBOURHOOD or OPEN SET in a METRIC SPACE.

open cover or **open covering,** *n.* a COVER of a set using OPEN subsets.

open disc, *n.* an OPEN BALL, especially in the CARTESIAN PLANE $\mathbb{R}^2$.

open interval, *n.* a set of real numbers lying between but excluding its endpoints, written $]x, y[$ or (x, y), where x and y are the endpoints; the HALF-OPEN INTERVALS $]x, y]$ and $[x, y[$ are not OPEN SETS as they do not contain neighbourhoods of one endpoint. All the foregoing are BOUNDED, with INFIMUM x and SUPREMUM y, but only $[x, y[$ has a MINIMUM and only $]x, y]$ has a MAXIMUM; $]x, \infty[$ is open and unbounded, and $\mathbb{R}$ is also an open interval. See also INTERVAL. Compare CLOSED INTERVAL.

open mapping, *n.* a function that sends open sets in the domain space to open sets in the range space. Compare CLOSED MAPPING.

open mapping theorem, *n.* a theorem giving conditions for a mapping to be an OPEN MAPPING. The *open mapping theorem for linear operators* asserts that a continuous surjective linear mapping between BANACH SPACES is open. The *open mapping theorem for analytic functions* asserts that an analytic function that is non-constant on a given domain is open thereon. See also INVARIANCE OF DOMAIN THEOREM.

open set, *n.* **1.** (in a METRIC SPACE) a set of points around each of which there is an OPEN BALL lying within the set. See also OPEN INTERVAL.
2. a member of a TOPOLOGY; one of the family of subsets of a topological space that constitute the topology.
3. a set of which the COMPLEMENT is a CLOSED SET.

open sentence, **sentential function**, or **propositional function,** *n.* (*Logic*) a well-formed expression containing a FREE occurrence of a variable that can be replaced by a name to yield a sentence; for example, *x is wise*, or *x gave y to z*. In a formulation of the PREDICATE CALCULUS in which names and atomic sentences are primitive, an open sentence is obtained from a well-formed sentence by replacing a name uniformly with a variable. An open sentence does not have a truth-value, but a sequence will *satisfy* it if the replacement of its variables by the respective elements of the sequence yields a true sentence; if each of the variables is bound by a quantifier, the resulting CLOSED SENTENCE also has a truth-value.

open theory, *n.* (*Logic*) a THEORY that contains only OPEN SENTENCES.

operand, *n.* a quantity or entity upon which a mathematical OPERATION is performed; an argument of an operation.

operate, *vb.* to perform an OPERATION on; to take as an argument. For example, negation and complementation operate respectively upon disjunctions and set unions according to DE MORGAN'S LAWS.

operation, *n.* **1.** any procedure, such as addition, multiplication, set union, conjunction, etc., that generates a unique value according to mechanistic rules from one or more given numbers or values as arguments. The *fundamental operations of arithmetic* are taken to be addition, subtraction, multiplication, and division; extraction of square roots is often added to this list.
2. a function determined by such a procedure.

operator, *n.* **1a.** any symbol used to indicate an OPERATION, such as the integration operator ∫, and the differential operator Δ.
b. the function determined by such an operation.
2. a MAPPING, such as a LINEAR OPERATOR.

operator norm, *n.* the NORM placed on the vector space, B[X, Y], of continuous LINEAR OPERATORS between two normed spaces X and Y, by the formula

$$\| T \| = \sup \{ \| T(x) \| : \| x \| \le 1 \}.$$

With this norm, B[X, Y] becomes a normed space that is COMPLETE precisely when Y is.

ophelmal, *adj.* an obscure word for OPTIMAL.

opposite, *adj.* (*as substantive*) (*Euclidean geometry*) the side of a right-angled triangle that is not an arm of a given angle; for example, in Fig. 262, the opposite side to θ is AB.

Fig. 262. AB is **opposite** θ.

opposite angles, *n.* the pair of angles formed between the opposite directions of two intersecting lines, and sharing a vertex but having no arm in common, such as θ and ϕ in Fig. 263.

Fig. 263. ϕ and θ are **opposite angles**.

opposite ring, *n.* a REVERSE ring; the RING constructed from a given noncommutative ring by using REVERSE multiplication.

optimal, *adj.* pertaining to, having, or being a value that is an OPTIMUM; a generic term for MAXIMAL, MINIMAL, SUPREMAL, or INFIMAL.

optimal assignment problem, see ASSIGNMENT PROBLEM.

optimal control, *n.* another name for CONTROL THEORY.

optimization, *n.* the determination of OPTIMAL values of a function, often subject to CONSTRAINTS.

optimization theory, *n.* the branch of mathematics that is concerned with the analysis and solution of problems in MATHEMATICAL PROGRAMMING such as those of INTEGER PROGRAMMING, CONTROL THEORY, or the CALCULUS OF VARIATIONS.

optimize, *vb.* to seek or find the OPTIMAL value of a function, often subject to certain given constraints. See also MATHEMATICAL PROGRAMMING.

optimum, *n.* a generic term for the MAXIMUM, MINIMUM, SUPREMUM, and INFIMUM of a set or function.

or, *n.* the usual English expression for DISJUNCTION. See INCLUSIVE OR, EXCLUSIVE OR.

orbit, *n.* **1.** (of a mapping, *f*, at a point, *x*) the sequence of points generated by repeated COMPOSITION of *f* at *x*; that is,

$$\{ f^{[n]}(x) : \ n = 0, 1, 2, ... \},$$

where $f^{[n]}(x) = f \circ f \circ ... \circ f$ (*n* times) and f^0 is the identity function; if this set is finite, the orbit is said to be *closed*. The closure of the orbit will, under reasonable conditions, contain a FIXED POINT of *f*. The term is especially used for PERMUTATIONS. See also DYNAMICAL SYSTEM.

2. (*Group theory*) the set of products, under an ACTION of a group G on a non-empty set, of all elements of the group with a given element of the set; that is,

$$x^G = \operatorname{orb}_G(x) = \{ gx : g \in G \}.$$

The distinct orbits of a set form a partition of the set.

3. the TRAJECTORY of an ORDINARY DIFFERENTIAL EQUATION.

orbit stabilizer theorem, *n.* the theorem that, for a group G that ACTS on a non-empty set X, the cardinality of the ORBIT of an element of X is the INDEX of the STABILIZER of that element in G.

order, *n.* **1.** the number of times a given function must be differentiated to obtain a given DERIVATIVE. For instance, f''', the third derivative of the function *f*, is of order 3.

2. the order of the highest order derivative in a DIFFERENTIAL EQUATION. For example,

$$\frac{\partial^2 y}{\partial x^2} + \left(\frac{\partial y}{\partial x}\right)^3 = 0$$

is an equation of second order. Compare DEGREE.

3. the position of elements in a sequence. For example, in an ordered set, the order of elements makes a difference, and $\langle a, b \rangle = \langle b, a \rangle$ if and only if $a = b$.

4. the number of rows or columns in a square matrix or determinant.

5. the number of elements of a finite GROUP or SET; its CARDINALITY.

6. also called **period.** (for an element *a* in a GROUP) the smallest number of times the given element has to be multiplied by itself to yield the IDENTITY element of the group; the least positive integer *n* such that $a^n = e$, written $|a| = n$. If $\langle a \rangle$ is the set generated by *a*, then $|a| = |\langle a \rangle|$. If no such integer exists the element has *infinite order*.

7. the MULTIPLICITY of a ZERO or a POLE.

8. the number of poles, counting multiplicity, in any fundamental parallelogram of a doubly PERIODIC FUNCTION, such as an ELLIPTIC FUNCTION.

9. (for an ENTIRE function) the quantity

$$\lambda = \limsup_{r \to \infty} \frac{\log \log m(r)}{\log r},$$

where m(*r*) is the maximum modulus of the given entire function on the disc of radius *r*. The order and GENUS, γ, of an entire function satisfy $\gamma \leq \lambda \leq \gamma + 1$. See THREE–CIRCLE THEOREM.

10. an alternative term for ORDERING.

11. see ORDER OF MAGNITUDE, order of SYMMETRY, or ORDER OF CONVERGENCE.

12. of the order of. a. having approximately the magnitude of.

b. (of a function) approximating to a constant multiple of another function for large values of the argument. Similarly, one speaks of functions as being *of lower order* or *of higher order* than a given function.

c. (of a function) having another function as an ASYMPTOTE as their arguments tend to infinity; having 1 as the limit of their quotient.
See ORDER NOTATION.

order-complete, see COMPLETE.

ordered arrangement, *n.* another term for PERMUTATION.

ordered geometry, *n.* an abstract geometry, more primitive than Euclidean geometry, in which the primitive elements are points and the relation of intermediacy or betweenness.

ordered pair, *n.* an ORDERED SET of two elements.

ordered set, *n.* **1.** a SEQUENCE of elements that is distinguished both by the identity and by the order of those elements, so that $\langle a, b \rangle$ is not identical with $\langle b, a \rangle$ unless $a = b$. The members of an ordered set are often enclosed by ANGLE BRACKETS.

2. a set endowed with an ORDERING; an alternative and frequently preferred term for a PARTIALLY ORDERED SET.

ordered structure, *n.* a structure (group, field, vector space, etc.) endowed with an ORDERING that preserves the underlying operations. Thus an *ordered field* is a field together with an order with the property that, for $a < b$ and any c, $a + c < b + c$, and, for $a < b$ and any $0 < c$, $ac < bc$.

ordered vector space, *n.* a VECTOR SPACE endowed with a PARTIAL ORDERING that respects addition and positive multiplication. See ORDERED STRUCTURE.

order ideal, *n.* the IDEAL of a COMMUTATIVE RING with identity defined as the set of elements, r, of the ring such that $rm = 0$, where m is a given member of an R-module. The order ideal is denoted $O(m)$. See also ANNIHILATOR.

ordering, *n.* (*Logic*) any of a number of categories of RELATIONS that permit at least some members of their domain to be placed in order. A *linear ordering* (or *simple ordering*) is REFLEXIVE, ANTISYMMETRIC, TRANSITIVE and CONNECTED (COMPLETE), and thus enables every member to be ordered relative to every other; for example, *less than or equal to* on the integers. A *partial ordering* is reflexive, antisymmetric and transitive, and thus generates CHAINS of comparable elements; members of distinct chains may be incomparable, as with set inclusion. Either of these orderings is *strong* or *strict* if it is ASYMMETRIC instead of reflexive and antisymmetric, such as *strictly less than*, or proper set inclusion. An ordering is a *well-ordering* if every non-empty subset has a least member under the ordering, i.e. a unique member that has the given relation to all other members of the subset. A *pre-ordering* or *quasi-ordering* is reflexive and transitive. The *reverse ordering* sets a less than (or equal to) b exactly when b is less than (or equal to) a. There is much variation in the usage of these terms. See also LATTICE, PARTIAL ORDER, TREE.

order interval, *n.* **1.** a subset, I, of an ORDERED SET that contains all elements lying between elements of the subset, so that if $a \in$ I, $b \in$ I and $a \le c \le b$ then $c \in$ I.

2. more particularly, a subset of an ordered set that is of the form

$$\{c : a \leq c \leq b\}.$$

order notation, *n.* a notation for indicating comparisons between the elements of one sequence and those of another, or between two functions on a set, and useful for replacing inequalities by equalities. Given sequences $\{a_n\}$ and $\{b_n\}$, if there is a constant M such that

$$|a_n| \leq M|b_n| \text{ for all } n,$$

then we write $a_n = O(b_n)$, and read this as 'a_n is big Oh of b_n', and similarly for two functions on a set. It is often useful to identify cases where two sequences are O of each other, in which case we write $a_n = \Theta(b_n)$, read 'a_n is *theta* of b_n'. If

$$\lim_{n \to \infty} \left| \frac{a_n}{b_n} \right| = 0,$$

then a_n is said to be $o(b_n)$ as $n \to \infty$, read as 'a_n is little oh of b_n'. Compare ASYMPTOTIC.

order of convergence, *n.* another term for RATE OF CONVERGENCE.

order of magnitude or **order,** *n.* the approximate size of something, especially in powers of 10.

order of symmetry, see SYMMETRY.

order-preserving, *adj.* (of a mapping between ORDERED SETS) ISOTONE.

order statistic, *n.* the ordering $X_{(1)} \leq \ldots \leq X_{(n)}$ of a given set of RANDOM VARIABLES, $X_1, \ldots, X_n$, usually INDEPENDENT and IDENTICALLY DISTRIBUTED, in which case, if they have PROBABILITY DISTRIBUTION FUNCTION P_X, then the joint probability distribution function of the order statistic is given by

$$P[x_{(1)}, \ldots, x_{(n)}] = n! \prod_{j=1}^{n} P_X[x_{(j)}].$$

ordinal, *n.* a set of which every member is also a subset (a *transitive set*) that contains only transitive elements. This can be used to generate the TRANSFINITE SEQUENCE,

$$\emptyset, \ \{\emptyset\}, \ \{\emptyset, \{\emptyset\}\}, \ \{\emptyset, \{\emptyset\}, \{\emptyset, \{\emptyset\}\}\}, \ \ldots;$$

the members of this sequence can be regarded as the canonical members of the sequence of ORDINAL NUMBERS, and so can be abbreviated 0, 1, 2, 3, ... Compare CARDINAL NUMBER.

ordinally similar, *adj.* (of two RELATIONS) such that there is a ONE–TO–ONE CORRESPONDENCE between their domains that preserves order under the given relations.

ordinal number, *n.* a measure of a set that takes account of the order as well as the number of its elements; defined to be the set of all WELL–ORDERED sequences that are ORDINALLY SIMILAR. Compare CARDINAL NUMBER.

ordinal scale, *n.* (*Statistics*) a scale on which data are shown simply in accordance with some order, in the absence of appropriate units of measurement. For example, a squash ladder is an ordinal scale since one can say only that one competitor is better than another, but not by how much. Compare INTERVAL SCALE, RATIO SCALE, NOMINAL SCALE.

ordinary differential equation (abbrev. **ode**), *n.* a DIFFERENTIAL EQUATION containing no PARTIAL DERIVATIVES.

ordinary point, *n.* **1.** a non-ISOLATED point of a curve at which the curve has a smooth tangent and does not cross itself; a point that is not a SINGULAR POINT.

2. a point *a* of a second-order DIFFERENTIAL EQUATION

$$y'' + P(x)y' + Q(x)y = 0,$$

for which both *P* and *Q* are analytic around *a*. See FROBENIUS METHOD.

ordinary representation, *n.* a REPRESENTATION over the complex numbers.

ordinate, *n.* the vertical or *y*-COORDINATE of a point in a two-dimensional system of CARTESIAN COORDINATES, equal to the distance of the point from the *x*-axis measured parallel to the *y*-axis; for example, in Fig. 264 the ordinate of P is 3. Compare ABSCISSA.

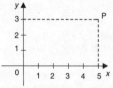

Fig. 264. The **ordinate** of P is 3.

orientable, *adj.* (of a surface or manifold) able to be given an ORIENTATION in the sense that a consistent direction can be given to points of the surface by TRIANGULATION. Equivalently, the surface contains no MÖBIUS STRIP, so that one cannot move a small oriented circle on the surface in such a way that it returns to its initial position with orientation reversed. A surface, such as a Möbius strip or a KLEIN BOTTLE, that cannot be given such an orientation is *non-orientable*.

orientation, *n.* the position of a geometric figure relative to a coordinate system; especially, the SENSE of a directed line. Thus a simple closed curve in the plane may be given either a clockwise or an anticlockwise orientation. In three dimensions, a set of coordinate axes may form a LEFT– or RIGHT–HANDED TRIHEDRON, and orientation is reversed in a mirror, as illustrated in Fig. 265. See ENANTIOMORPHIC. See also DIRECTION NUMBERS.

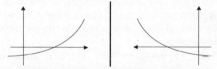

Fig. 265. Reflection reverses **orientation**.

origin, *n.* **1.** the point of intersection of all the axes of a COORDINATE SYSTEM; especially in CARTESIAN COORDINATES, in which the respective coordinates are measured parallel to the axes. The origin is thus the point of which the coordinates are all zero.

2. the ROOT of a TREE.

orthant, *n.* any of the eight regions into which three-space is divided by the coordinate planes; especially the *positive orthant* consisting of all vectors with non-negative coordinates. Similarly in *n*-dimensional space, the positive orthant consists of all vectors with non-negative coordinates. See also OCTANT, QUADRANT.

orthocentre, *n.* the point of intersection of all three ALTITUDES of a triangle; for example, the point X in Fig. 266.

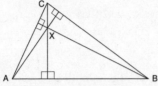

Fig. 266. X is the **orthocentre** of ABC.

orthogonal or **orthographic,** *adj.* **1.** perpendicular, normal; sometimes written ⊥.

2. (of a set of elements of some algebraic structure) having the property that the product (as defined for the given structure) of any pair of distinct elements is zero. If, in addition, the product of any element with itself is unity, then the elements are ORTHONORMAL.

3. (of a pair of LATIN SQUARES) such that each pair of corresponding symbols occurs exactly once.

orthogonal complement, *n.* (of a VECTOR or set of vectors) the subspace of vectors ORTHOGONAL to the given vector or set. For example, in a Euclidean vector space, the orthogonal complement of $\mathbf{x} = (1, 0, 0)$ has as a basis $\mathbf{y} = (0, 1, 0)$ and $\mathbf{z} = (0, 0, 1)$; equivalently, in three-dimensional Euclidean geometry, the y–z plane is the orthogonal complement of the x-axis.

orthogonal functions, *n.* a set of functions $f_1, ..., f_n$ (that may be infinite), of which every pair of distinct functions satisfies the identity

$$\int_a^b f_i(x) \, f_j(x) \, \mathrm{d}x \; = \; 0,$$

on some given range of integration $]a, b[$, or for a more general set and MEASURE. The LEGENDRE POLYNOMIALS are orthogonal on $]-1, 1[$. Compare ORTHONORMAL FUNCTIONS.

orthogonal group, *n.* the GROUP of $n \times n$ ORTHOGONAL MATRICES over the reals, written $\mathrm{O}(n)$. See also SPECIAL ORTHOGONAL GROUP.

orthogonal matrix, *n.* a MATRIX that is the inverse of its TRANSPOSE, so that any pair of distinct rows or distinct columns are ORTHOGONAL VECTORS. Compare UNITARY MATRIX.

orthogonal projection, *n.* the PROJECTION of a figure on a line, plane, etc. in such a way that the line joining the corresponding elements is perpendicular to the line, plane, etc. The orthogonal projection of a vector on a line, such as one of the coordinate axes, is the component of the vector in the direction of that line; for example in Fig. 267, XY is the orthogonal projection of AB on CD.

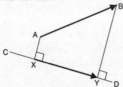

Fig. 267. XY is the **orthogonal projection** of AB on CD.

orthogonal sequence, *n.* a SEQUENCE of pair-wise ORTHOGONAL VECTORS, often functions, such as the LEGENDRE POLYNOMIALS. Compare ORTHONORMAL SEQUENCE.

orthogonal vectors, *n.* a set of VECTORS of which the SCALAR PRODUCT (or INNER PRODUCT) of any pair of distinct vectors is zero.

orthographic, *adj.* another word for ORTHOGONAL.

orthonormal, *adj.* (of a set of elements of some algebraic structure) having the property that the product (as defined for the given structure) of any pair of distinct elements is zero, and the product of any element with itself is unity. Compare ORTHOGONAL.

orthonormal functions, *n.* a set of functions $f_1, ..., f_n$ (that may be infinite) that satisfy the identity, on some given range of integration $]a, b[$,

$$\int_a^b w(x)\, f_i(x)\, f_j(x)\, \mathrm{dx} = \begin{cases} 0 & \text{if } i \neq j \\ 1 & \text{if } i = j \end{cases},$$

where $w(x)$ is a *weight* function; the integration may also be over a more general set.

orthonormal matrix, *n.* an uncommon term for ORTHOGONAL MATRIX.

orthonormal sequence, *n.* an ORTHOGONAL SEQUENCE of VECTORS, often functions, all of UNIT NORM. The standard basis in Euclidean space is orthonormal.

Osborne's rule, *n.* the rule that states that TRIGONOMETRIC IDENTITIES can be transformed into the corresponding identities for HYPERBOLIC FUNCTIONS by multiplying out fully, replacing all trigonometric functions with their hyperbolic analogues, and then changing the signs of any terms involving the product of two sinhs; for example, given

$$\cos(x - y) = \cos x \cos y + \sin x \sin y,$$

Osborne's rule permits us to infer

$$\cosh(x - y) = \cosh x \cosh y - \sinh x \sinh y.$$

osc, *abbrev. and symbol for* OSCILLATION.

oscillate, *vb.* (of a function, sequence, or series) not to tend either to a finite limit or to infinity, or else to do so in such a way that every NEIGHBOURHOOD of the limit contains both values greater than and values less than it. For example, the sequence $\langle 0, 1, 0, 1, 0, ... \rangle$ and the series

$$1 - \tfrac{1}{2} + \tfrac{1}{4} - \tfrac{1}{8} + ...$$

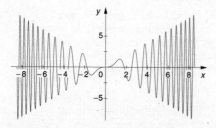

Fig. 268. A non–periodic **oscillating** function.

oscillate; the function $x \sin x^2$ oscillates, but is not periodic, as can be seen from the section of its graph shown in Fig. 268 above. Compare CONVERGE, DIVERGE.

oscillation (abbrev. **osc**), *n.* a measure of the spread of a bounded set; the SUPREMAL difference between pairs of elements, defined as the difference between the SUPREMUM and INFIMUM of the set. The oscillation of a function on an interval is also known as its *saltus*. See also BOUNDED VARIATION.

osculate, *vb.* (of two curves) to meet at a point at which they have a TANGENT in common, as shown in Fig. 269. A circle that osculates with another curve at a point is thus not necessarily its OSCULATING CIRCLE at that point; if the verb is used in this latter sense it is required that the curves have the same CURVATURE as well as a common tangent at the common point.

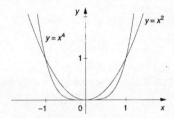

Fig. 269. $y = x^2$ and $y = x^4$ **osculate** at the origin.

osculating circle, *n.* an older term for CIRCLE OF CURVATURE.

osculating plane, *n.* the plane spanned by the unit TANGENT and unit PRINCIPAL NORMAL vectors to a space curve at a given point; the curvature of the curve is evaluated in this plane. The unit BINORMAL is the normal vector to the osculating plane.

osculation or **tacnode,** *n.* a point at which two branches of a curve OSCULATE; that is, have a common tangent, when both branches extend on both sides of the point . For example, $y = x^2$ and $y = x^4$ have a point of osculation at the origin, as shown in the graph in Fig. 269 above. In some usages the stronger condition that the curves also have the same CURVATURE at the common point is demanded. Compare CUSP.

out-degree (of a graph or network), see DEGREE.

outer automorphism, *n.* any AUTOMORPHISM except an INNER AUTOMORPHISM.

outer Caratheodory measure, see OUTER MEASURE.

outer Jordan measure, see OUTER MEASURE.

outer Lebesgue measure, see OUTER MEASURE.

outer measure, *n.* **1.** also called **Caratheodory outer measure.** a SET FUNCTION constructed preliminary to, and sharing many of the properties of, a MEASURE. Explicitly, an outer measure, μ^*, on a set S is an ISOTONE, COUNTABLY ADDITIVE, extended-real-valued SET FUNCTION defined for all subsets of the set and assigning value zero to the empty set; that is

$$\mu^*(\varnothing) = 0,$$

$$\mu^*(E) \le \mu^*(F) \quad \text{if F contains E,}$$

and

$$\mu^*(\bigcup_{n=1}^{\infty} E_n) \leq \sum_{n=1}^{\infty} \mu^*(E_n).$$

2a. Lebesgue outer measure. the particular outer measure of a set in Euclidean n-space computed by taking the infimum of the sum of the volumes (*content*) of any *Lebesgue covering* of the set by countable families of open finite order intervals (*boxes*):

$$\mu^*(E) = \inf\{\sum |I_n| : E \text{ lies in } \bigcup I_n\}.$$

The restriction of this function to the LEBESGUE MEASURABLE subsets defines LEBESGUE MEASURE in n-space.

b. Jordan outer measure. an analogous measure defined using only finite covers.

Compare INNER MEASURE.

outlier, *n.* (*Statistics*) in a sample, a point widely separated from the main cluster of points. See SCATTER DIAGRAM.

output set, *n.* (*Information theory*) the set of signals of which the receiver can observe one at a time. Compare INPUT SET.

output variable, *n.* another term for STATE VARIABLE.

outside, *n.* (for a CONTOUR in the Euclidean plane) the set denoted outΓ, defined as the set of points not lying on the curve for which the WINDING NUMBER is zero. Compare INSIDE.

ovals of Cassini, *n.* (*Geometry*) the locus of the vertex of a triangle when the product of the sides adjacent to the vertex is held constant and the opposite side is held fixed. When the constant is one quarter of the length of the fixed side, this produces a LEMNISCATE; for a smaller ratio, two distinct ovals are obtained; Fig. 270 shows the figure obtained for larger values.

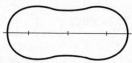

Fig. 270. **Ovals of Cassini.**

over, *prep.* with respect to; especially where one algebraic structure is defined in terms of another; for example, if F is a field with respect to which the vector space V is defined, then V is said to be a vector space over F.

overdetermined, *adj.* (of a system of equations, usually linear) involving more equations than variables. Compare UNDERDETERMINED.

overlap, *n.* (*Differential topology*) the map

$$\phi\psi^{-1} : \psi(U \cap V) \to \phi(U \cap V),$$

where (ϕ, U) and (ψ, V) are CHARTS. It is said to be a $C^{(r)}$-*overlap* if this map is $C^{(r)}$; that is, if it is r-times continuously differentiable.

over-ring, *n.* a RING, such as a QUOTIENT RING, in which a given ring can be embedded.

p, 1. the usual symbol for an unspecified PRIME NUMBER, used also as a prefix. See P-ELEMENT, P-GROUP, P-SUBGROUP.

2. *abbrev. for* PICO-, used in symbols for fractions of the physical units of the SYSTEME INTERNATIONAL.

3. (*Logic*) see P (sense 2).

P, 1. *abbrev. for* PETA-, used in symbols for multiples of the physical units of the SYSTEME INTERNATIONAL.

2. also written **Pr.** *standard notation for* a PROBABILITY MEASURE.

3. (*Logic*) also written **p.** the usual symbol for an unspecified sentence or proposition.

4. *n.* the class of DECISION PROBLEMS for which there are POLYNOMIAL TIME ALGORITHMS.

Pa, (*Mechanics*) *symbol for* PASCAL.

pad, *vb.* (*Computing*) to expand a vector, matrix, or other quantity in order to bring it to a desired length, dimension, or the like. This is usually arranged by adding zeros, or empty elements. So, for example, padding with zeros permits us to assume that a matrix is square of size 2^n.

Padé approximation, *n.* the rational function analogue of the TAYLOR SERIES. Precisely, a *Padé approximation of order* (m, n) to a function f at a point a is a rational function P_n/Q_m such that for x near a

$$\left| \frac{f(x) - P_n(x)}{Q_m(x)} \right| = O\left((x - a)^\nu\right),$$

with ν as large as possible. Here P_n and Q_m are polynomials of degrees m and n respectively in $(x - a)$. An $(n + m + 1)$-times continuously differentiable function always has $\nu > n$, and typically $\nu \geq n + m + 1$.

pair, *n.* **1.** a set with two members, written $\{a, b\}$.

2. an ORDERED SET with two members, written $\langle a, b \rangle$.

3. *vb.* to associate in pairs.

paired-sample problem, *n.* (*Statistics*) any problem requiring a test to be applied to two related samples. Compare TWO-SAMPLE PROBLEM.

paired vector spaces, *n.* two VECTOR SPACES X and Y over a given field that are endowed with a bilinear mapping, $\langle , \rangle$, from X × Y into the scalar field. Often the space Y is already a DUAL vector space for X, and $\langle x, y \rangle = y(x)$.

pairwise, *adj.* taken two at a time; for example, Cartesian coordinate axes are pairwise perpendicular.

Pappus' theorem, 1. (*Geometry*) the theorem of PROJECTIVE GEOMETRY that states that if six vertices of a hexagon lie alternately on two lines then the three pairs of opposite sides (extended where necessary) meet in collinear points. This produces a self-dual CONFIGURATION of nine lines and nine points with three points on each line and three lines through each point, the *Pappian plane*. This is shown in Fig. 271, where ACE and BDF are the given lines, on which the vertices ABCDEF of the bold-drawn hexagon lie alternately; the extensions of the pairs of sides AB and DE, BC and EF, and CD and FA intersect at X, Y and Z respectively, and these points are found to lie on the dotted line. Compare DESARGUE'S THEOREM, FINITE GEOMETRY.

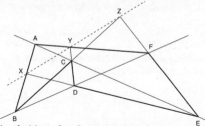

Fig. 271. **Pappus' theorem** (sense 1). See main entry.

2. (*Analysis*) **a.** the theorem that the area of a surface of revolution, formed by revolving a curve in the plane around a line that does not intersect the curve, is equal to the product of the length of the curve and the circumference of the circle described by the CENTROID of the curve. This follows from the fact that the area of the whole surface is the sum of the surface areas of cylindrical ELEMENTS and the mean radius of these cylindrical elements tends to the distance of the centroid of the curve from the axis as their mesh-fineness increases. Fig. 272 shows the surface of revolution of $y = f(x)$ from $x = a$ to b around the x-axis; the darker shading shows an element of this surface.

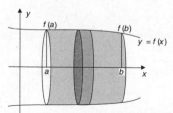

Fig. 272. **Pappus' theorem** (sense 2). See main entry.

b. the theorem that the volume of a solid of revolution, formed by revolving a planar region around a line in the plane not cutting the region, is equal to the product of the area of the region and the circumference of the circle described by the centroid of the region. If the region is regarded as the entire area between a curve and a given line as in Fig. 272 above, the solid is the sum of cylindrical elements, and the result follows from the fact that the volume of a cylinder, $\pi r^2 h$, is identical with

$$rh \times 2\pi(r/2),$$

that is, the product of the area between the axis and a generator and the circumference of the circle of half the radius of the base, so that the average such radius is the ordinate of the centroid of the region.

(Named after *Pappus of Alexandria* (fl. c. 300 – 320 AD), who compiled an historical and critical summary of the most important results of Greek mathematics.)

par. *abbrev. for* PARALLEL.

parabola, *n.* a CONIC SECTION formed by the intersection of a cone with a plane parallel with the GENERATOR; a conic with unit ECCENTRICITY, with canonical equation $y^2 = 4ax$, where $2a$ is the distance between FOCUS and DIRECTRIX, when the curve is symmetrical about the x-axis and has its vertex at the origin; Fig. 273 shows a parabola with $a = 1$. The parametric equations of a parabola are

$$x = at^2, \quad y = 2at.$$

A parabola is the projection of an arc of a circle onto an oblique plane through the chord joining the endpoints of the arc.

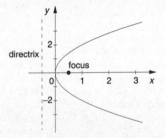

Fig. 273. A **parabola** showing its focus and directrix.

parabolic, *adj.* **1.** of, relating to, or shaped like a PARABOLA or PARABOLOID. **2.** (of a second order PARTIAL DIFFERENTIAL EQUATION) having DISCRIMINANT $b^2 - 4ac$ equal to zero, where

$$au_{xx} + bu_{xy} + cu_{yy} + du_x + eu_y + fu = h$$

is the general form of a second-order partial differential equation.

parabolic spiral. *n.* a SPIRAL in which the length of the RADIUS VECTOR is proportional to the square root of its angle with the polar axis, so that its equation in POLAR COORDINATES is $r^2 = k\theta$; as shown in Fig. 274, it is symmetrical about the origin.

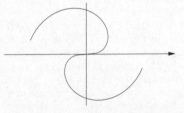

Fig. 274. **Parabolic spiral**.

paraboloid, *n.* a three-dimensional surface or solid that has parabolic sections parallel to two coordinate planes, and either elliptical or hyperbolic sections parallel to the third coordinate plane; these are *elliptical paraboloids* and *hyperbolic paraboloids* respectively, as shown in Fig. 275 opposite. Its equation when its axes of symmetry coincide with the coordinate axes is

$$\frac{x^2}{a^2} \pm \frac{y^2}{b^2} = 2cz.$$

Fig. 275. Elliptic and hyperbolic **paraboloids**.

paracompact, *adj.* (of a topological space) such that every open covering admits a LOCALLY FINITE refinement; this holds, for example, in compact or metrizable spaces. See also PARTITION OF UNITY.

paradox, *n.* an apparently absurd or self-contradictory statement for which there is *prima facie* support, or an explicit contradiction derived from apparently unexceptionable premises such as the axioms of set theory. Some, such as RUSSELL'S PARADOX and CANTOR'S PARADOX, required revision of the intuitive conception from which the paradox was derivable; others, such as GRELLING'S PARADOX, depend upon the strict inadmissibility of the description of the paradox; while yet others, such as the paradoxes of MATERIAL IMPLICATION or the SKOLEM PARADOX, merely draw attention to some feature of the formal theory which is contrary to intuition.

parallel (abbrev. **par**), *adj.* **1a.** (of a pair or set of lines in Euclidean geometry) never meeting or intersecting however far extended, sometimes written A ∥ B.

b. (of a set of curves) remaining a constant distance apart.

c. (of a set of HYPERPLANES) being TRANSLATES one of the other.

2. (of a pair of lines in AFFINE or AUGMENTED EUCLIDEAN GEOMETRY) meeting at a POINT AT INFINITY.

3. having the same direction; representable by vectors that are scalar multiples of one another, such as parallel forces. More generally, if **x** and **y** are vectors in an INNER PRODUCT SPACE, then **x** and **y** are parallel if and only if

$$\| \langle x, y \rangle \| = \| x \| \| y \| = \sqrt{\langle x, x \rangle \langle y, y \rangle}.$$

– (*as substantive*) **4.** a line parallel to a given line.

5. another term for PARALLEL SECTION.

parallelepiped or **parallelopiped,** *n.* a solid, the six faces of which are all PARALLELOGRAMS, such as that shown in Fig. 276.

Fig. 276. **Parallelepiped**.

parallelogram, *n.* (*Euclidean geometry*) a QUADRILATERAL with opposite sides parallel, and hence equal in length. An equilateral parallelogram is a RHOMBUS, an equiangular one is a RECTANGLE, and one that is both equilateral and equiangular is a SQUARE. Compare TRAPEZIUM.

parallelogram law, *n.* **1.** the norm identity valid in an INNER PRODUCT SPACE:

$$\| x + y \|^2 + \| x - y \|^2 = 2 \| x \|^2 + 2 \| y \|^2,$$

for all vectors **x** and **y**. This extends the PARALLELOGRAM RULE to an arbitrary inner product space, and holds only if the norm is induced by an inner product:

$$\langle x, x \rangle = \| x^2 \|.$$

See also POLARIZATION IDENTITY.

2. another term for PARALLELOGRAM RULE.

parallelogram of periods, see PERIODIC FUNCTION.

parallelogram rule or **parallelogram law,** *n.* a rule for finding the RESULTANT of two VECTORS by constructing a PARALLELOGRAM, each pair of parallel sides of which represents the magnitude to scale, and the direction and sense of the given vectors; its diagonal then represents the direction and magnitude of the resultant. A *parallelogram of forces* is a diagram in which the combined effect of two forces acting on the same body is determined using this rule. For example, if the two forces are represented by the vectors $\overrightarrow{OA}$ and $\overrightarrow{OB}$ in Fig. 277, then their resultant, to the same scale, is $\overrightarrow{OR}$, where R completes the parallelogram.

Fig. 277. **Parallelogram rule**.
A parallelogram of forces.

parallelopiped, *n.* variant spelling of PARALLELEPIPED.

parallel postulate, *n.* the axiom of EUCLIDEAN GEOMETRY stating that if two straight lines are cut by a third, the two will meet on the side of the third on which the sum of the interior angles is less than two right angles; equivalently, PLAYFAIR'S AXIOM states that through a given point only one line can be drawn parallel to a given line. This was regarded as self-evident until the 19th century, when NON–EUCLIDEAN GEOMETRIES were devised in which all the remaining axioms of Euclidean geometry were retained, but this was false. Since these other geometries are CONSISTENT, the parallel postulate must be INDEPENDENT of the remaining Euclidean axioms.

parallel section or **parallel,** *n.* a section of a SURFACE OF REVOLUTION, such as a paraboloid, that is perpendicular to the axis of revolution. Compare MERIDIAN.

parameter, *n.* **1.** an arbitrary constant whose value affects the specific nature but not the formal properties of a mathematical expression, such as the arbitrary constants *a* and *b* in $ax^2 + bx + c = 0$.

2. a variable that is not being considered and that may, for present purposes, be regarded as a constant, such as *y* in the PARTIAL DERIVATIVE

$$\frac{\partial}{\partial x} f(x, y).$$

3. one of a number of auxiliary variables in terms of which all the variables in an IMPLICIT functional relationship can be explicitly expressed. See PARAMETRIC EQUATIONS.

4. (*Statistics*) a characteristic of the distribution of a POPULATION, such as its mean, variance, or moments about the mean, as distinct from the characteristics of a SAMPLE drawn from the population. Compare STATISTIC.

5. (of a CONIC) the length of the LATUS RECTUM, in terms of which the given conic can be described. For the ellipse and the hyperbola, it is $2p = 2b^2/a$, where a and b are the semi-axes; for the parabola $y^2 = 2px$, it is $2p$.

parametric equations or **freedom equations,** *n.* a set of equations expressing a number of quantities as EXPLICIT functions of the same set of independent variables (the PARAMETERS), and equivalent to some direct functional relationship between these quantities. For example, the circle $x^2 + y^2 = r^2$ has parametric equations

$$x = r\cos\theta, \quad y = r\sin\theta$$

in terms of the parameter θ.

parametric statistics, *n.* (*Statistics*) the branch of statistics concerned with data measurable on INTERVAL or RATIO SCALES, so that arithmetic operations are applicable to them, enabling PARAMETERS such as the MEAN of the distribution to be defined.

parametrization, *n.* the description of a curve, surface, or other object by PARAMETRIC EQUATIONS.

parentheses, *n.* another term for round BRACKETS.

Pareto optimal point or **Pareto efficient point,** *n.* (*Optimization*) another term for an EFFICIENT point. (Named after the Italian economist and sociologist, *Vilfredo Pareto* (1848–1923), who was educated as a mathematician and physicist, worked initially as an engineer, and is best known for his application of mathematics to economics.)

parity, *n.* (of an integer) being EVEN or ODD; that is, divisible or indivisible by 2. Thus, n and $n + 2$ have the same parity, and if n has odd parity, then $n + 1$ has even parity.

Parseval's theorem, *n.* the theorem of FOURIER ANALYSIS stating that BESSEL'S INEQUALITY holds as equality for a SQUARE INTEGRABLE function. The term is also applied to extensions of this to any COMPLETE ORTHONORMAL SEQUENCE in a HILBERT SPACE. (Named after the French mathematician, *Marc Antoine Parseval de Chênes* (1755–1836), who was forced to flee France after he published poems critical of Napoleon.)

partial, *adj.* **1.** (of an operation) concerned with only one of the independent variables of its argument.

2. (of a mapping, relation, or predicate) defined only on part of the UNIVERSE of discourse, this part being called the *domain* of the mapping, or, to avoid confusion with the domain of discourse or the set on which the mapping is defined albeit not for all elements, its *essential domain*.

3. (*as substantive*) a PARTIAL DERIVATIVE.

partial derivative or **partial differential coefficient,** *n.* the DERIVATIVE of a function of two or more variables with respect to one of these variables, the others being regarded as constant; written

$$\frac{\partial f}{\partial x} \quad \text{or} \quad f_x.$$

Higher partial derivatives arise as partial derivatives of partial derivatives. If different variables are used in the repeated process, a *mixed partial derivative* arises. Thus,

$$\frac{\partial^2 f(x, y, z)}{\partial x\, \partial y} = f_{yx}(x, y, z) = f_{2,1}(x, y, z)$$

is a mixed second partial derivative, which is the result of differentiating the partial derivative $\partial f/\partial y$ with respect to x. In general, the *order* of the derivative is the number of times the given function is differentiated. The ambiguity about the sequence in which the derivatives are taken is removed when the partial derivatives of that order are continuous around the point; then the order is irrelevant, and, for example, $f_{xxy} = f_{xyx}$. Compare TOTAL DERIVATIVE.

partial differential coefficient, *n.* another term for PARTIAL DERIVATIVE.

partial differential equation (abbrev. **pde**), see DIFFERENTIAL EQUATION.

partial fraction, *n.* one of a set of FRACTIONS into which a more complex fraction can be resolved; for example,

$$\frac{x^2 - x}{x^3 - x^2 + x - 1} = \frac{1}{x^2 + 1} - \frac{1}{x - 1}.$$

The *method of partial fractions* refers in particular to the representation of a proper rational function over the reals as a finite sum of terms, each of which is a proper rational function of which the denominator is a power of a linear or irreducible quadratic polynomial. Thus, integration of rational functions reduces to integration of these terms.

partially ordered set, *n.* a set endowed with a PARTIAL ORDER.

partial order or **partial ordering,** *n.* a RELATION that is REFLEXIVE, ANTISYMMETRIC and TRANSITIVE, but not necessarily CONNECTED; this imposes a LATTICE on a set. A partial ordering generates CHAINS of elements, but, unless it is a TOTAL ORDERING, there will be pairs of elements in the set between which the relation does not hold in either order; for example, in the TREE diagram of Fig. 278, where $x \le y$ if and only if there is a path from the origin to y via x, clearly it is neither the case that $A \le B$, nor $B \le A$. In particular, although each finite chain will have both a MAXIMAL and a MINIMAL element, these may not be unique. For example, the subsets of the integers have a partial ordering under set INCLUSION; in this case there is a unique minimal element, the empty set, that is contained in all members of the domain, and a unique maximal element, the set of integers itself, that contains all elements of the domain, but not all pairs of elements are related, as neither of $\{1, 2, 3\}$ and $\{2, 3, 4\}$ contains the other. If the domain is restricted to non-empty proper subsets, and the ordering is strict inclusion, there are infinitely many maximal and minimal elements. See also POSET, ORDERING, ZORN'S LEMMA.

Fig. 278. **Partial ordering**.
See main entry.

partial pivoting, *n.* the pivoting strategy in which the PIVOT ELEMENT is chosen to be an allowable element of maximum magnitude. This tends greatly to improve numerical performance of an elimination method.

partial product, *n.* the product of an initial segment of an infinite product. For example, the sequence $\langle a_1, a_2, a_3, \dots \rangle$ has a product if and only if the sequence of partial products $\langle a_1, a_1a_2, a_1a_2a_3, \dots \rangle$ has a limit that is non-zero.

partial recursive function, *n.* a RECURSIVE PARTIAL FUNCTION.

partial sum, *n.* the sum of an initial segment of an infinite series. For example, the series $a_1 + a_2 + a_3 + \dots$ has a sum if and only if the sequence of partial sums $\langle a_1, a_1 + a_2, a_1 + a_2 + a_3, \dots \rangle$ has a limit.

particle, *n.* (*Mechanics*) a body consisting of a MASS located at a point in space; a body can be regarded as approximately a particle if it is small compared with the extent of its motion, or can be treated exactly as a particle at its CENTRE OF MASS when its disposition about its centre of mass is not required. See also CONTINUUM MECHANICS.

particular, *adj.* of, concerning, being, or denoting an INDIVIDUAL. Compare UNIVERSAL.

particular integral, *n.* a function that satisfies a given DIFFERENTIAL EQUATION, especially one that also satisfies certain given INITIAL CONDITIONS or BOUNDARY CONDITIONS.

particular solution, *n.* a relation between the variables of a DIFFERENTIAL EQUATION that satisfies the given equation, especially one functionally given, in which all constants of integration are evaluated in order to satisfy some given conditions. Particular solutions are used to reduce the problem of solving linear equations to that of solving homogeneous linear equations. Compare GENERAL SOLUTION.

partition, *n.* **1a.** a set of DISJOINT and EXHAUSTIVE subclasses of a given class that divide it in such a way that each member of the given class is a member of exactly one such subclass. Such a division is possible if and only if there is an EQUIVALENCE RELATION that relates two elements of the underlying class if and only if they are members of the same subclass. For example, since two integers are congruent modulo n if and only if their difference is divisible by n, congruence mod n is reflexive ($a \equiv a$ for all a), transitive (if $a \equiv b$ and $b \equiv c$, then $a \equiv c$), and symmetric ($a \equiv b$ if and only if $b \equiv a$), so that this is an equivalence relation; hence it can be proved that, since every integer has a unique smallest positive remainder when divided by n, and so is congruent mod n to only one of the integers between 0 and $n-1$, these congruence classes are both disjoint and exhaustive of all the integers, and so constitute a partition of them. See also COVERING.
b. such a division of a class into a set of subsets.
2. a division of a given positive integer into positive integral parts. For example,

$$5 = 4 + 1 = 3 + 2 = 3 + 1 + 1$$
$$= 2 + 2 + 1 = 2 + 1 + 1 + 1 = 1 + 1 + 1 + 1 + 1$$

exhibits the seven possible partitions of 5. See also PARTITION FUNCTION.
3. a division of a given matrix into CONFORMABLE submatrices.
4. a finite sequence of points $\{x_k\}$ of a given interval $[a, b]$ such that

$$a = x_1 < x_2 < \dots < x_n = b.$$

A partition of an interval therefore yields a finite number of subintervals that are pairwise disjoint.

partition function, n. **1.** the function, defined on the positive integers, that counts the number of distinct PARTITIONS (sense 2) of a given argument; for example, $p(4) = 5$ while $p(5) = 7$; in general, $p(n)$ is ASYMPTOTICALLY equal to

$$p(n) \sim \frac{\exp\left[\pi\sqrt{\frac{2n}{3}}\right]}{4n\sqrt{3}}.$$

2. the summation, of importance in statistical physics, given by

$$\sum \exp \frac{-E_i}{kT},$$

where the summation is over all possible STATES of the system, E_i is the energy of the ith STATE, k is a constant that relates the mean KINETIC ENERGY and the ABSOLUTE TEMPERATURE, T, of the system.

partition of unity, n. (*Topology*) a family of non-negative continuous functions on a topological space that sums to unity and is *locally finite* in that all but a finite number of them vanish on some neighbourhood of each point. The partition is *subordinate* to a given covering if each function is zero outside some member of the covering. Such a subordinate partition exists for each open covering whenever the space is PARACOMPACT, and so if the space is METRIZABLE, REGULAR LINDELOF, or COMPACT.

pascal, n. (*Mechanics*) the unit of PRESSURE, TENSION, or STRESS in the SYSTEME INTERNATIONAL, defined as one NEWTON per square METRE. (Named after the French mathematician, *Blaise Pascal* (1623–62), physicist, and philosopher, who is credited with inventing the first calculating machine, and who, independently of Fermat, laid the foundations of probability theory.)

Pascal's mystic hexagram theorem, n. (*Projective geometry*) the theorem that if a hexagon is inscribed in a CONIC, then the points of intersection of pairs of opposite sides are collinear. This has a dual in BRIANCHON'S THEOREM.

Pascal's triangle, n. the triangular array of integers, with 1 at the apex, in which each number is the sum of the two numbers above it in the preceding row; an initial segment is shown in Fig. 279. The nth line of the triangle is the sequence of coefficients of $x^k a^{n-k}$ in the expansion of the binomial $(x + a)^n$. See also BINOMIAL THEOREM, COMBINATION.

```
            1
          1   1
        1   2   1
      1   3   3   1          Fig. 279. Pascal's triangle.
    1   4   6   4   1
  1   5  10  10   5   1
```

Pascal's wager, n. a probabilistic philosophical argument that it is in one's rational self-interest to act as if God exists, since the infinite punishments of Hell, provided they have a positive probability, however small, outweigh any countervailing temporal advantage; once this is taken into account, the UTILITY of the religious life exceeds that of the sybaritic.

path, *n.* **1.** (*Graph theory*) also called **Hamiltonian walk.** a WALK in which each VERTEX, except possibly the first, occurs only once; it is a *closed path* (or *cycle*) if its endpoint is identical with its initial point. Compare TRAIL.

2. (in a TREE) a MONOTONE sequence of edges, of which the first member is the ROOT of the tree.

3. (*Topology*) the mapping inducing an ARC; a path is a continuous mapping from the closed interval [0, 1] such that the images of the endpoints are the two given points. For example, $x = \cos\pi t$ and $y = \sin\pi t$ define a path onto the part of the unit circle in the upper half-plane. See also PATH-CONNECTED.

path-connected or **pathwise connected,** *adj.* (of a TOPOLOGICAL SPACE) having the property that, for any two points, a PATH joining them can be found in the space. In the usage in which an ARC is required to be a HOMEOMORPHIC image of the unit interval, pathwise connectedness is slightly less arduous than being ARC–CONNECTED; often, however, they are not distinguished. A path-connected space is CONNECTED, but a connected set is not necessarily path-connected; for example,

$$\{y = \sin(1/x) : x \in (\mathbb{R} \setminus \{0\})\} \ \cup \ [-1, 1]$$

is connected but not path-connected.

pathological, *adj.* (of a mathematical entity) satisfying the conditions of a theory or theorem but contrary to intuition as to the general nature of the objects concerned, and therefore regarded as bizarre or defective. For example, a function that is everywhere continuous but nowhere differentiable is regarded as pathological. However, what is counted as pathological may change with the development of the relevant theory.

payoff, *n.* (*Game theory*) the positive or negative amount accruing to each player after all players have chosen strategies. In a two-person ZERO–SUM GAME, this is determined by the i,j th entry of the *payoff matrix* $[\ a_{ij}\ ]$; when the first or *maximizing* player chooses strategy i, and the second or *minimizing* player chooses strategy j, the first player receives a_{ij} and the second receives $-a_{ij}$.

pde, *abbrev. for* PARTIAL DIFFERENTIAL EQUATION.

pdf, (*Statistics*) *abbrev. for* PROBABILITY DENSITY FUNCTION.

pe, (*Statistics*) *abbrev. for* PROBABLE ERROR.

Peano, Guiseppe (1858–1932), Italian analyst, and founder of symbolic logic. He is best known for his work in the FOUNDATIONS OF MATHEMATICS where he attempted to derive all of mathematics from fundamental principles, using the notation which he invented and which was later adopted as standard. He also published two innovatory works on the theory of functions, and invented the artificial language, Interlingua.

Peano arithmetic, *n.* the theory of the NATURAL NUMBERS as defined by PEANO'S AXIOMS.

Peano curve, *n.* a continuous curve that passes through every point of the unit square, or more generally, a class of similar iteratively defined curves that trace out FRACTALS. Fig. 280 overleaf shows the first, second, and third iterations of the curve produced by replacing the sides of a square by the generator shown in bold, and then iterating this process.

Fig. 280. **Peano curve**. See main entry.

Peano's axioms, *n.* a set of axioms, first stated by Dedekind, that yield a system isomorphic to the natural numbers by defining a first member, and a unique SUCCESSOR to each member, by excluding loops, and by providing for mathematical induction. See PRINCIPLE OF INDUCTION, ORDINAL.

Peano space, *n.* a METRIC CONTINUUM that is LOCALLY CONNECTED. A Peano space is ARC–WISE CONNECTED.

Pearson's correlation coefficient or **Pearson's product moment correlation coefficient,** *n.* (*Statistics*) a statistic that measures the linear relationship between two variables in a sample, and is used as an ESTIMATE of the CORRELATION, ρ, in the whole population. (Named after the English mathematician, *Karl Pearson* (1857–1936), who was a pioneer of statistics, and the inventor of the CHI–SQUARED TEST. He was also a lawyer, a professor of mechanics, a philosopher, a writer of fiction, and a professor of eugenics whose racist views were partly responsible for the misappropriation of eugenics by Nazis and American racists.)

Peaucellier's cell, *n.* a mechanical device for tracing the INVERSIVE image of a locus of points; that is, for drawing circular arcs of any radius including the infinite. (Named after the French engineer and army officer, *Charles-Nicolas Peaucellier* (1832–1913).)

pedal curve, *n.* the locus of the foot of the perpendicular from a fixed point to a variable tangent on a given curve.

pedal triangle, *n.* **1.** (of a point with respect to a given triangle) the triangle formed by the feet of the perpendiculars drawn from the point to the sides of a given triangle (or their extensions). For example, in Fig. 281, XYZ is the pedal triangle of ABC with respect to the point P. See also SIMSON LINE.

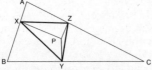

Fig. 281. XYZ is the **pedal triangle** (sense 1) of P in ABC.

2. the triangle whose vertices are the feet of the altitudes of a given triangle; the altitudes of a triangle bisect the angles of its pedal triangle. In Fig. 282, XYZ is the pedal triangle of the obtuse triangle ABC so that two of its vertices lie on extensions of the sides of the original triangle.

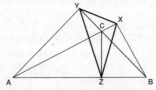

Fig. 282. XYZ is the **pedal triangle** (sense 2) of ABC.

p-element, *n.* (of a GROUP) an element of order p^α for p a prime and α a positive integer. See also CAUCHY'S LEMMA.

Pell's equation, *n.* (*Number theory*) the Diophantine equation of the form

$$x^2 - Dy^2 = \pm N,$$

where D and N are natural numbers usually with $N = 1$ and with D square-free. (Mistakenly ascribed to *John Pell* (1610–85), an English algebraist and astronomer; it should properly be ascribed to Fermat.)

penalty function, *n.* **1. exterior penalty function.** (for a set S) a non-negative continuous function P that is zero on S and strictly positive off S. If the set is expressed as

$$S = \{\, x : g_1(x) \le 0,\ldots, g_n(x) \le 0,\ \ h_{n+1}(x) = 0,\ldots, h_{n+m}(x) = 0 \,\},$$

then the associated penalty function is usually taken as

$$P(x) = \sum_{i \le n} \left[\max\{0, g_i(x)\} \right]^p + \sum_{i > n} h_i(x)^p,$$

for some $p \ge 1$.

2. interior penalty function or **barrier function.** (for a set S with non-empty interior) a non-negative function B that is continuous over the interior of S and approaches infinity as the boundary is approached from within the set. If the set is expressed as

$$S = \{\, x : g_1(x) \le 0,\ldots, g_n(x) \le 0 \,\},$$

then the associated penalty function is often taken as

$$B(x) = - \sum_{i \le n} \log \left| g_i(x) \right|,$$

or

$$B(x) = - \sum_{i \le n} \left| g_i(x) \right|^p,$$

for some negative p.

See PENALTY FUNCTION METHODS.

penalty function methods, *n.* **1.** the class of optimization methods that seek to solve a CONSTRAINED OPTIMIZATION PROBLEM by instead solving a sequence of unconstrained problems constructed by adding internal or external PENALTY FUNCTIONS to the objective. Penalty function methods have considerable computational drawbacks but some conceptual advantages. For example, the problem of minimizing $f(x)$, subject to a vector constraint of the form $h(x) = 0$, can be replaced by the sequence of unconstrained problems

$$\text{minimize } f(x) + K \| h(x) \|^2,$$

in EUCLIDEAN NORM. Under reasonable conditions as K increases to infinity, solutions to these problems converge to a solution of the original problem, since for large K one pays a large penalty for unfeasibility. This simple quadratic penalty function is sometimes called a *Courant penalty function.*

2. exact penalty function methods. penalty function methods where it is possible to obtain the solution to the original problem by solving the unconstrained problem for some fixed K.

pencil, *n.* **1.** a coplanar family of lines or rays that pass through a common vertex, or a family of parallel lines; an example of the former is shown in Fig. 283. Compare BUNDLE, LOCUS.

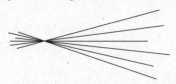

Fig. 283. A **pencil** of coincident lines.

2. more generally, a family of geometric figures with a common property, such as a pencil of circles that all intersect at the same pair of points; these are in fact lines in an abstract geometry.

Penrose triangle, *n.* a figure that appears to represent a three-dimensional triangular solid, but which is in fact impossible to construct. Fig. 284 shows a real solid triangle with the top surface shaded; it is impossible consistently to shade the Penrose triangle, as in fact each vertex is a perspective drawing of a right angle. (Named after the British mathematician, *Roger Penrose,* and his geneticist father, who also designed the PENROSE IMPOSSIBLE STAIRCASE of Fig. 285) See also NECKER CUBE.

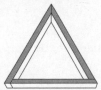

Fig. 284. A real and a **Penrose triangle**.

Penrose impossible staircase, *n.* a paradoxical drawing that appears to show a continuous four-sided staircase in which each side separately appears to be a perspective drawing of rising steps, as shown in Fig. 285. This forms the basis of many of M.C. Escher's drawings.

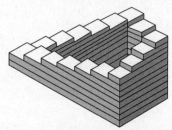

Fig. 285. **Penrose impossible staircase**.

penta-, *prefix denoting* five. For example, a *pentangle* is a figure with five angles; a *pentahedron* a solid with five faces.

pentacle, *n.* another word for PENTAGRAM.

pentad, *n.* a set or sequence of five.

pentagon, *n.* a five-sided polygon.

pentagonal number, *n.* a FIGURATE NUMBER of the form $n(3n \pm 1)$.

pentagram, **pentangle**, or **pentacle**, *n*. a star-shaped figure formed by extending all the sides of a regular PENTAGON to meet in pairs, as shown in Fig. 286.

Fig. 286. **Pentagram**.

pentahedron, *n*. a solid figure with five plane faces; a pyramid with a rectangular base.

pentangle, *n*. another word for PENTAGRAM.

P equals NP, *n*. the conjecture, generally disbelieved, that every NP–DECISION PROBLEM possesses a POLYNOMIAL TIME ALGORITHM. See P VERSUS NP.

per cent, *adv*. (*Latin*) in every hundred; expressing a proportion as a fraction of which the denominator is 100. For example, 5% is 5/100; a 5% solution is one in which the active ingredient constitutes 5 parts out of every 100 of solution. Compare PER MIL.

percentage, *n*. a proportion, ratio or rate expressed with a denominator of 100. Compare PERMILLAGE.

percentile or **centile**, *n*. (*Statistics*) one of the 99 values of a random variable that divide its distribution in such a way that an integral percentage of the population lie below that value. For example, the 90th percentile is the value of a variable such that 90% of the relevant population are below that value. See QUARTILE, INTER–QUARTILE RANGE.

perfect, *adj*. (*prenominal*) precisely factorizable into equal integral or polynomial ROOTS. For example, 36 and $x^2 + 2x + 1$ are perfect squares, 27 is a perfect cube, and 32 is a perfect fifth power.

perfect matching, *n*. a MATCHING in which all VERTICES are matched by EDGES.

perfect number, *n*. a number the sum of whose distinct integral divisors, including 1 but excluding itself, is that number itself. For example, 6 is a perfect number since $1 + 2 + 3 = 6$. Euclid proved that $2^{n-1}(2^n - 1)$ is an even perfect number when $2^n - 1$ is a MERSENNE PRIME; these are now called *Euclid numbers*, and Euler proved that all even perfect numbers are of this form for some positive prime number *n*. Thus 6, 28, and 496 are perfect and correspond to values of 3, 7, and 31 for *n* in the formula. Whether there are infinitely many even perfect numbers or any odd perfect numbers remain unsolved problems. See also SIGMA FUNCTION. Compare ABUNDANT NUMBER, AMICABLE NUMBERS, DEFICIENT NUMBER.

perfect set, *n*. (*Topology*) a set that equals its DERIVED SET; a set that is closed and dense in itself. See ISOLATED POINT.

perfect square or **square number**, *n*. an integer that is the square of another integer, such as 1, 4, 9, 16, ... See also FIGURATE NUMBER.

perigon, *n*. another term for ROUND ANGLE.

perimeter, *n*. **1.** the curve enclosing a region of a surface.
2. the length of such a curve.

period, *n*. **1.** an interval, generally the smallest, after which a PERIODIC FUNCTION takes the same values; a constant *k* such that $f(x) = f(x + k)$ for all *x*.

For example, since $\sin\theta = \sin(\theta + 2n\pi)$, $2n\pi$ is a period of $\sin\theta$ for all integers n; its *principal period*, as shown in Fig. 287, is 2π.

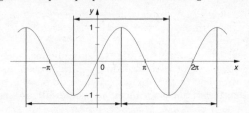

Fig. 287. The **period** of a periodic function.

2. (of an element of a GROUP) another term for ORDER (sense 6).

3. (of a point with respect to a function f) the smallest positive integer n such that the n-fold COMPOSITION $f^n(x) = x$.

period doubling, *n.* the change in the periodic ORBITS of a PARAMETRIZED family of transformations, in which a periodic orbit of period n BIFURCATES becoming a pair of orbits of period $2n$. See also FEIGENBAUM NUMBER.

periodic, *adj.* regularly repeating; for example, a periodic continued fraction or decimal expansion.

periodic function, *n.* **1.** a function with values that are repeated for all integral multiples of a constant increment of the independent variable. For example, as shown in Fig. 287 above,

$$\sin\theta = \sin(\theta + 2\pi) = \sin(\theta + 4\pi), \dots.$$

2. doubly periodic function. a complex function with two non-zero minimal periods, ω_1 and ω_2, that are non-collinear in the sense that one is not a real multiple of the other. All periods are of the form $n\omega_1 + m\omega_2$ for integers n and m, so that

$$f(z + n\omega_1 + m\omega_2) = f(z).$$

Any parallelogram with corners

$$z, \ z + \omega_1, \ z + \omega_2, \ z + \omega_1 + \omega_2$$

is called a *fundamental parallelogram* or a *parallelogram of periods* for this function. See also ELLIPTIC FUNCTION.

permanent, *n.* the sum of all products of elements of a given square matrix, where each product contains exactly one element from each column and row. The permanent of a matrix is thus constructed in the same way as the DETERMINANT, except that the signs of the products do not alternate. See VAN DER WAERDEN'S CONJECTURE.

per mil or ***per mill,*** *adv.* in every thousand; expressing a proportion as a fraction of which the denominator is 1000; sometimes written $\%_{00}$. Compare PER CENT.

permillage, *n.* a proportion, ratio or rate, expressed with denominator 1000. Compare PERCENTAGE.

permutable, *adj.* another word for COMMUTATIVE.

permutation or **ordered arrangement,** *n.* **1.** an ordered arrangement of a specified number of objects selected from a set. The number of distinct permutations of r objects from n is

$$\frac{n!}{(n-r)!},$$

usually written $_nP_r$ or ^nP_r. For example, there are six distinct permutations of two objects selected out of three: $\langle 1,2\rangle$, $\langle 1,3\rangle$, $\langle 2,1\rangle$, $\langle 2,3\rangle$, $\langle 3,1\rangle$, $\langle 3,2\rangle$. Compare COMBINATION.

2. any rearrangement of all the elements of a finite sequence, such as $(1, 3, 2)$ and $(3, 1, 2)$. It is *odd* or *even* according as the number of exchanges of position yielding it from the original order is odd or even. It is a *cyclic permutation* if it merely advances all the elements a fixed number of places; that is, if it is a CYCLE of maximal LENGTH. A *transposition* is a cycle of degree two, and all permutations factor as products of transpositions. See also SIGNATURE.

3. any BIJECTION of a set to itself, where the set may be finite or infinite.

permutation group or **substitution group,** *n.* a GROUP of PERMUTATIONS, where multiplication is defined as successive permutation. If the group is finite, this corresponds isomorphically to a group of PERMUTATION MATRICES. The group of all permutations of n objects has $n!$ elements and determines the full SYMMETRIC GROUP, and the group of all EVEN permutations of n objects is called the ALTERNATING GROUP; these are GENERATED by the transpositions and 3-cycles respectively. Up to isomorphism, all finite groups may be realized as subgroups of the full permutation group. If one identifies permutations with BIJECTIONS of the underlying set, this remains true of all groups. For an infinite set, there is a group of all permutations (the *complete symmetric group*), and there is a symmetric group of permutations that move only a finite number of symbols, which contains the alternating group.

permutation matrix, *n.* a square matrix with one unit element in each row and column and with all other entries zero. This corresponds to precisely one permutation of order n; where the permutation takes the ith element to the $\sigma(i)^{\text{th}}$, the matrix is zero except in the $[i, \sigma(i)]^{\text{th}}$ positions, where it is unity.

permutation representation, *n.* a GROUP of permutations of a set that is ISOMORPHIC to a given group; the permuted set may be the given group itself. See CAYLEY REPRESENTATION THEOREM.

permute, *vb.* to reorder a sequence of elements.

perpendicular, *adj.* (*Euclidean geometry*) **1.** at or forming a right angle; sometimes written $\perp$.

2. (*as substantive*) a line drawn perpendicular to another, or to a plane, etc. See also NORMAL, ORTHOGONAL.

perpendicular distance, *n.* (*Euclidean geometry*) the distance from a point to a line measured by the length of a perpendicular to the given line passing through the point. It is therefore the shortest distance between the given point and a point on the line.

Perron–Frobenius theorem, *n.* any of various results generalizing the theorem that a matrix, A, of strictly positive coefficients has a positive EIGENVECTOR that is unique up to constant multiples; this has an EIGENVALUE equal to the SPECTRAL RADIUS of A and all other eigenvalues have smaller modulus. The adjoint A* has a positive eigenvector with the same LATENT ROOT.

perspective, *adj.* (*Projective geometry*) (of two plane figures) such that their points can be put in one-to-one correspondence so that pairs of corresponding points lie on concurrent lines (or dually, corresponding lines meet in collinear points). Thus, two sets are perspective from a point (the *centre of perspectivity*) if pairs of corresponding points are joined by lines through the centre; for example, in Fig. 288 the irregular pentagons KLMNO and ABCDE are perspective from P. Two sets are perspective from a line (the *axis of perspectivity*) if pairs of corresponding lines meet on the axis. A theorem of Desargues' shows that these two notions are equivalent. A *perspective transformation* is a PERSPECTIVITY.

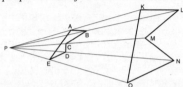

Fig. 288. The two bold figures are **perspective** with respect to P.

perspectivity, *n.* (*Projective geometry*) a transformation under which two figures are PERSPECTIVE, as in Fig. 288. A perspectivity may be from a pencil of lines to a range of points, from one range of points to another, or from one pencil of lines to another; the fundamental theorem is that any PROJECTIVITY is a product of at most three perspectivities.

perturbation, *n.* **1.** (of an equation or of an optimization problem) a change (usually slight) in the values of some of the underlying parameters, made to obtain the desired solution or to study the stability of a given solution.

2. (*Mechanics*) a small DISPLACEMENT in the orbit of a PARTICLE.

peta- (symbol **P**), *prefix* denoting a multiple of 10^{15} of a unit of the SYSTEME INTERNATIONAL.

p-**group,** *n.* a GROUP in which the order of every element is a power of p, where p is prime; a finite p-group has order p^{α} for some natural number α.

phase or **argument,** *n.* an angle θ = ph$z$ such that

$$r(\cos\theta + i\sin\theta)$$

is equal to a given complex number $z = x + iy$. The *principal phase* is that value of θ in the half-open interval $(-\pi, \pi]$ radians. See also AMPLITUDE. Compare MODULUS.

phase space, *n.* **1.** (*Statistical physics*) a space of $2s$ dimensions representing a system with s DEGREES OF FREEDOM. It is the union of the COORDINATE SPACE and MOMENTUM SPACE of the system in which the rectangular coordinates represent the position and momentum of the points of the system.

2. (*Differential equations*) the space with coordinates $y, \dot{y}, \ddot{y}$, etc., the successive derivatives of the dependent variable.

phi, see EULER PHI FUNCTION.

philosophical logic, *n.* the branch of philosophy that studies the relationship between FORMAL LOGIC and ordinary language, and especially the extent to which the former can be held accurately to represent the latter, or, equivalently, to which the latter is an adequate model for the former.

pi, *n.* **1.** a TRANSCENDENTAL NUMBER, π, that is the ratio of the circumference of any circle to its diameter, approximately equal to $3.141\,592\,653\,589\,79\ldots$. Its expansion has now been calculated using two different algorithms, whose results agree to 1.24×10^{12} decimal places (December 2002). The sequence 0123456789 has been found in the decimal expansion of π beginning at the 17 387 594 880th digit after the decimal point, and in consequence the status of many famous INTUITIONISTIC examples has changed. It has also been shown that binary digits of π can be calculated without knowing the preceding digits; it is thus known, for example, that, starting with the 999 999 999 999 999th bit, the binary expansion continues:

$$10\ 01\ 10\ 00\ 10\ 00\ 01\ 01\ 10\ 10\ 11\ 00\ 00\ 01\ 10\ 10\ 01\ 11\ 00\ 10\ 11 \ldots$$

The number of RADIANS in a complete circle is 2π, so that in CIRCULAR MEASURE, $\pi/2$ radians is a right angle. It may be defined as twice the least positive zero of $\cos x$, and can be calculated in principle in many ways such as, for example, from accelerated forms of arc-tangent identities such as GREGORY'S SERIES for $\pi/4$, or from consideration of MODULAR EQUATIONS. The notation π apparently originates from the letter p for 'periphery'.

2. the symbol, $\prod$, denoting a product: the product of the elements $x_a, \ldots, x_b$ is written

$$\prod_{i=a}^{b} x_i.$$

Compare SIGMA.

3. the symbol, π, denoting a permutation.

4. the arithmetic function, $\pi(n)$, denoting the number of primes not exceeding n. See also PRIME NUMBER THEOREM.

Picard's method, *n.* the iterative solution of an ordinary differential equation, essentially by application of the BANACH CONTRACTION mapping method to the equivalent integral equation. (Named after the French analyst, group theorist, and mechanist, *Charles Émile Picard* (1856–1941), who became Permanent Secretary of the mathematical section of the French Academy of Science.)

Picard's theorems, *n.* **1. Picard's first theorem.** the result that a nonconstant entire function can miss at most one finite complex value from its range; that is, can have at most one LACUNARY VALUE.

2. Picard's second theorem. an extension of the foregoing, stating that, in every neighbourhood of an ESSENTIAL ISOLATED SINGULARITY, an analytic function takes all finite values, except possibly for one; $\sin(1/z)$ is an example. See CASORATI–WEIERSTRASS THEOREM.

Pick's theorem, *n.* the result that the area of a simple lattice polygon, P, is $I(P) + \frac{1}{2} B(P) - 1$ where $I(P)$ is the number of lattice points inside P and $B(P)$ is the number of lattice points on the boundary of P including the vertices.

pico- (symbol **p**), *prefix* denoting fractions of 10^{-12} of the physical units of the SYSTEME INTERNATIONAL.

pid, *abbrev for* PRINCIPAL IDEAL DOMAIN.

piecewise or **sectionally,** *adv.* behaving in a certain manner or possessing a given property (e.g. continuity, monotonicity, differentiability) except at finitely many points where certain compatibility conditions are often demanded. For example, a function is piecewise smooth on an interval if it is

427

continuous thereon and is continuously differentiable except at finitely many points where the derivative may have jump discontinuities.

pie chart, *n.* (*Statistics*) a circular diagram divided into sectors of which the areas are proportional to the magnitudes of the quantities represented, as shown in Fig. 289.

Fig. 289. **Pie chart**.

pigeon-hole principle, **Dirichlet's principle**, **drawer principle**, or **letter-box principle,** *n.* the fundamental enumerative principle that if a set of *n* objects is partitioned into fewer than *n* subsets, then at least one subset has at least two members.

pitch, *n.* (of a HELIX) the displacement parallel to the axis of a point as it makes one revolution about the axis.

pivotal function, *n.* a function of data, derived from an experiment and the parameter of interest, regarded as a RANDOM VARIABLE with a probability distribution independent of any unknown parameters.

pivot element, see PIVOTING.

pivoting, *n.* (in LINEAR PROGRAMMING) another name for GAUSSIAN ELIMINATION. The element on which a given elimination is based being called a *pivot element*.

place marker, *n.* a DUMMY VARIABLE or INDETERMINATE that serves merely to articulate a pattern; for example, of terms of an identity, or of component sentences of an argument.

place value, *n.* the particular power of the base of a counting system that is represented by a particular position in a PLACE–VALUE NOTATION; for example, units, tens, hundreds, etc. in the decimal system, so that in 5374 the place-value of the 3 is 100.

place-value notation or **positional notation,** *n.* an arithmetical notation that represents numbers as a sequence of digits in such a way that the successive digits represent multiples of successive powers of the BASE. For example, in decimal notation the term '34.5' represents

$$(5 \times 10^{-1}) + (4 \times 10^{0}) + (3 \times 10^{1}) = 0.5 + 4 + 30;$$

in binary notation, '1011' represents

$$(1 \times 2^{0}) + (1 \times 2^{1}) + (0 \times 2^{2}) + (1 \times 2^{3}) = 1 + 2 + 0 + 8 = 11.$$

plagiarism, *n.* a labour-saving but lawyer-enriching form of research, relatively rare in mathematics, but ascribed, solely on grounds of scansion, by then Harvard mathematician Tom Lehrer in his 1953 song to the 19th-century Russian geometer Nikolai Ivanovich Lobachevski. The best known such allegation in mathematics was that Leibniz had stolen Newton's ideas for the infinitesimal CALCULUS – a charge exacerbated by the fact that it was Leibniz's presentation and notation which was widely adopted. In fact, however, there was no foul play, and this was simply a matter of two people working on the same problems reaching the same solution entirely independently of one another. Lehrer, however, imagined an acolyte of his fictional Lobachevski copying every chapter from a different source and even the index from an old Vladivostok telephone directory – and invited performers "to heed its basic precept and plagiarise the author's version". "Plagiarise! Plagiarise, let no-one else's work evade your eyes," he sang, "Only be sure always to call it please 'research'."

planar, *adj.* **1.** of, relating to, or lying on a PLANE.

2. (of a GRAPH) having the property of being isomorphic to a PLANE GRAPH.

planar point, see UMBILICAL POINT.

Plancherel theorem, *n.* the result that the FOURIER TRANSFORM, viewed as an operator on $L_1(\mathbb{R}^n) \cap L_2(\mathbb{R}^n)$, extends uniquely to a linear ISOMETRY of $L_2(\mathbb{R}^n)$ onto $L_2(\mathbb{R}^n)$.

plane, *n.* **1.** a flat surface; a geometrical figure that has the property that the line joining any two of its points lies completely on its surface. The equation of a plane in CARTESIAN three-dimensional space is $ax + by + cz = d$, where the VECTOR (a, b, c) is the NORMAL to the plane.

2. any subgeometry of dimension 2 of an algebraic geometry.

plane angle, *n.* an angle between two intersecting lines.

plane figure, see FIGURE.

plane geometry, *n.* the study of the properties of and relations between figures all drawn on the same PLANE.

plane graph, *n.* a GRAPH drawn in a plane and having edges that meet only at vertices.

plane of symmetry, *n.* a plane with respect to which a figure in three dimensions is SYMMETRIC.

planimeter, *n.* a mechanical integrating device that measures the area of an irregular plane figure, such as the area under a curve, by moving a point attached to an arm round the perimeter of the figure.

planimetry, *n.* the measurement of plane areas.

Plateau's problem, *n.* the problem of determining the MINIMAL SURFACE with a given twisted curve as its boundary. These are often solved empirically by means of soap-film experiments.

Plato, *n.* Greek philosopher (c. 428–348 BC) who advocated a realist account of mathematical entities and truth (see PLATONISM). His name is also linked with the five PLATONIC SOLIDS. Plato's insistence that mathematics should be an essential part of the education of the Guardians of his ideal Republic did much to establish the high status of mathematics in Western civilization. See QUADRIVIUM.

Platonic solid, *n.* any one of the five REGULAR (sense 2) POLYHEDRA, namely the cube, the regular tetrahedron, the regular octahedron, the regular dodecahedron, and the regular icosahedron, which were once imbued with great mystical significance. Johannes Kepler (1571–1630) was led to his discovery of the laws of planetary motion and his defence of Copernican astronomy by circumscribing or inscribing the orbits of the five other known planets around the Platonic solids, with an inscribed or circumscribed circle representing the orbit of the Earth; the results agree with observation, allowing for eccentricity, to within approximately 5%. Compare ARCHIMEDEAN SOLID.

Platonism, *n.* the philosophical theory that mathematical objects have a real existence in advance of and independently of human knowledge of them and of any physical instantiation of them, and therefore that mathematical truth does not consist in, but is the aim of, the construction of proofs. See also REALISM. Compare CONSTRUCTIVISM, INTUITIONISM, FORMALISM.

platykurtic, *adj.* (*Statistics*) (of a distribution) having KURTOSIS B_2 less than 3; less heavily concentrated about the mean than a NORMAL DISTRIBUTION. Compare LEPTOKURTIC, MESOKURTIC.

Playfair's axiom, *n.* the alternative formulation of the PARALLEL POSTULATE, the fifth of EUCLID'S AXIOMS, as the assertion that through a point not on a given line there is one and only one line parallel to the given line.

plot, *vb.* **1.** to locate or mark (points) on a graph relative to a coordinate system.

2. to draw (a curve) through these points.

plurality, *n.* **1.** a number greater that one.

2. the qualification of a sentence by such expressions as 'many', 'most', 'few', 'a few', etc. Whence *plural logics* study such concepts by analogy with the standard QUANTIFIERS of PREDICATE CALCULUS. See also RESTRICTED QUANTIFIER.

plus, *prep.* **1.** increased by the ADDITION of; for example, *four plus two* is written '4 + 2'.

2. more generally, operated upon by any defined ADDITION, such as the SYMMETRIC DIFFERENCE; for example, the first operator in both $a + (-b)$ and $A \oplus (A \cup B)$ would be read as 'plus'.

3. *adj.* (*prenominal*) **a.** (of a specified number) greater than zero, having a positive value; thus +4 is read as 'plus four'.

b. involving or indicating addition; the plus sign is used to indicate positive increments.

plus sign, *n.* **1.** the symbol '+', indicating ADDITION or any analogous operation, such as the SYMMETRIC DIFFERENCE of sets, or the DIRECT SUM of groups.

2. the symbol '+', indicating a POSITIVE quantity.

Pochhammer symbol or **rising factorial,** *n.* the symbol $(a)_n$ defined as

$$(a)_n = a\,(a+1)...(a+n-1) = \frac{\Gamma(a+n)}{\Gamma(a)}.$$

Poincaré, Jules Henri (1854–1912), prolific French mathematician and physicist who was Professor of Mathematics and Science at the *Université de Paris*, and made major contributions to virtually all branches of mathematics. He originated the study of AUTOMORPHIC FUNCTIONS, was a pioneer of topology, an astronomer, a probability theorist, a philosopher, and a member of the *Académie Française*, and became President of the French Academy of Science.

Poincaré conjecture, *n.* the conjecture that in four-dimensional space, a SIMPLY CONNECTED three-dimensional COMPACT manifold is topologically equivalent to a three-sphere. The corresponding result for two-dimensional surfaces in three-dimensional space is in essence that a solid piece of rubber with no holes can be stretched and deformed into a sphere, whereas a body with holes, such as a TORUS, can not. The question was

posed by Poincaré but has resisted solution and is one of the MILLENNIUM
PRIZE PROBLEMS. Its four-dimensional analogue has recently been shown
false by Michael Freedman, and in April 2002, Martin Dunwoody of Southampton University announced a proof of the three-dimensional case, although this remains to be confirmed.

Poincaré's lemma, *n.* the result that every closed DIFFERENTIAL FORM defined
on a SIMPLY CONNECTED region is EXACT. Compare CONSERVATIVE VECTOR
FIELD.

point, *n.* **1.** a basic element (along with LINE) of an AXIOMATIC geometry;
informally, a geometrical element having no dimensions; in Cartesian
space, an element that can be located by a single *n*-tuple of coordinates.
See also PROJECTIVE PLANE.

2. one element of a line or curve distinguished by the value of the independent variable, such as a point of inflection.

3. an element of a TOPOLOGICAL SPACE or a VECTOR SPACE.

4. (*Combinatorics*) a VARIETY of a BLOCK DESIGN.

point at infinity, *n.* **1.** an IDEAL ELEMENT in AFFINE geometry. See DESARGUE'S
THEOREM.

2. the point added in the one-point COMPACTIFICATION of the complex
plane. The extended plane may then be identified with a sphere of which
it is the conformal image under STEREOGRAPHIC PROJECTION. The point at
infinity then corresponds to the POLE of the projection.

pointed, see WEDGE.

point estimate, *n.* (*Statistics*) a specific value that is an ESTIMATE of a PARA-
METER of a population on the basis of SAMPLING STATISTICS. Compare CON-
FIDENCE INTERVAL.

point evaluation, *n.* a LINEAR FUNCTIONAL that assigns to each member of a
space of functions its value at a given point, so that $\delta_t(f) = f(t)$.

point mass, *n.* a MEASURE the support of which is singleton.

point measure, *n.* a MEASURE, μ, for which there is a point, p, such that for
every measurable set E, $\mu(E) = 1$ if p is in E, and $\mu(0) = 0$ otherwise; that is,
$\mu(E) = \chi_E(p)$, the characteristic function of E.

point of contact, *n.* another term for TANGENCY POINT.

point of density, see METRIC DENSITY.

point of dispersion, see METRIC DENSITY.

point of inflection, *n.* a point on a curve at which it crosses its tangent, and
CONCAVITY changes from up to down or vice versa; its SECOND DERIVATIVE is
zero and changes sign at the point. For example, Fig. 290 shows the tangents at the two points of inflection of one period of a sinusoid.

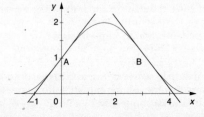

Fig. 290. A and B are
points of inflection.

point process, *n.* (*Probability*) a sequence of EVENTS, usually in time, where the interval between any pair of consecutive events follows a common PROBABILITY DISTRIBUTION; for example, emissions from a radioactive source.

point-set topology, *n.* another name for TOPOLOGY (sense 1).

point-slope equation, see LINE.

point spectrum, see SPECTRUM.

point-to-set mapping, *n.* another name for SET-VALUED FUNCTION.

pointwise convergent, *adj.* (of a sequence of functions) CONVERGENT in the sense that, for each value of the argument, the sequence of function values converges, so that a pointwise convergent sequence is not necessarily UNIFORMLY CONVERGENT. Compare CONVERGENT IN MEAN, CONVERGENT IN MEASURE. See also DOMINATED CONVERGENCE.

pointwise ergodic theorem, *n.* another name for the BIRKHOFF ERGODIC THEOREM. Compare MEAN ERGODIC THEOREM.

Poisson, Siméon Denis (1781–1840), French analyst, applied mathematician and probabilist, who was intended for a medical career, but studied under and was befriended by Laplace and Lagrange. Although at one stage a professor of astronomy, his main work was on the mathematical theory of electricity and magnetism. He wrote the standard work on mechanics, and his gravitational equations are now used in interpreting satellite orbits; his approximation to the binomial distribution is used to model traffic flow and radioactive decay.

Poisson's differential equation, *n.* the differential equation

$$\nabla^2 v = -u$$

where ∇^2 is the three-dimensional LAPLACIAN.

Poisson distribution, *n.* (*Statistics*) a distribution, written $\text{Po}(\lambda)$, representing the number of events occurring randomly in a fixed time at an average rate of λ. The corresponding PROBABILITY DENSITY FUNCTION is given by

$$P(k) = \frac{\lambda^k e^{-\lambda}}{k!},$$

for $k = 0, 1, 2, \ldots$. For large n and small p, with $np = \lambda$, the BINOMIAL DISTRIBUTION $\text{Bi}(n, p)$ approximates to $\text{Po}(\lambda)$.

Poisson's integral, *n.* the integral $U^*(r, \theta)$ defined as

$$\frac{1}{2\pi} \int_0^{2\pi} \frac{U(\phi)\left[a^2 - r^2\right]}{a^2 - 2ar\cos(\theta - \phi) + r^2} \, d\phi,$$

where $U(\phi)$ is continuous on the boundary of a disk of radius a in the complex plane, and the coefficient of $U(\phi)$ in the integrand is the *Poisson kernel*. This gives a HARMONIC extension of U to the interior of the disk, the extension being continuous on the closed disk, and so solves the DIRICHLET PROBLEM on the disk. Similar formulae exist for other boundary conditions.

polar, see POLE AND POLAR.

polar angle, *n.* the angle, measured anticlockwise, between the POLAR AXIS of a system of POLAR COORDINATES and the line joining the origin to any given point, for example, the angle θ in Fig. 291 opposite.

polar axis, *n.* the fixed line in a system of POLAR COORDINATES from which the POLAR ANGLE is measured anticlockwise; for example, the line OX in Fig. 291.

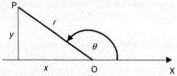

Fig. 291. **Polar coordinates.**
P is the point (r, θ).

polar cone, see POLAR SET.

polar coordinates, *n.* **1.** a system of coordinates, generally written (r, θ) or $[r, \theta]$, in which points in a plane are located by means of the length, r, of the line joining the point to the origin or *pole*, and the angle, θ, swept out by that line from a single axis.

2. the coordinates of a point expressed in such a system. As is apparent from Fig. 291, the polar coordinates of the point with CARTESIAN COORDINATES (x, y) are

$$(r, \theta) = (\sqrt{x^2 + y^2}, \arctan(y/x)),$$

so that in the ARGAND DIAGRAM, this point represents the complex number $x + iy$, where r and θ are respectively the MODULUS and AMPLITUDE (or *anomaly* or *azimuth*) of the complex number; conversely, the Cartesian coordinates of (r, θ) are

$$(x, y) = (r\sin\theta, r\cos\theta).$$

Compare SPHERICAL COORDINATES.

polar decomposition, *n.* the representation of an invertible matrix with complex entries as the composition of a POSITIVE SEMIDEFINITE MATRIX and a UNITARY matrix, or the similar representation of a normal bounded linear operator on a complex Hilbert space.

polar decomposition theorem, *n.* the result that any invertible matrix over the complex numbers may be written as the product of a UNITARY matrix and a POSITIVE DEFINITE HERMITIAN matrix.

polar equation, *n.* an equation in POLAR COORDINATES.

polarity, *n.* **1.** positivity or negativity. For example, $y = x^2$ has two roots of equal magnitude and opposite polarity.

2. the relation between POLE AND POLAR

polarization identity, *n.* **1.** the identity, for a complex INNER PRODUCT SPACE,

$$4\langle x, y \rangle = \|x + y\|^2 - \|x - y\|^2 + i\|x + iy\|^2 - i\|x - iy\|^2 ,$$

that reconstructs the INNER PRODUCT from the NORM.

2. the corresponding identity for a real INNER PRODUCT SPACE:

$$4\langle x, y \rangle = \|x + y\|^2 - \|x - y\|^2.$$

Compare PARALLELOGRAM LAW.

polar set, *n.* the set of vectors, denoted S^0 or S°, such that $\langle x, s \rangle \leq 1$ for all s in a given set S in real HILBERT SPACE; it is closed, convex, and contains zero. When S is a vector subspace, S° coincides with the ORTHOGONAL COMPLEMENT, and when S is a cone, S° coincides with the *polar cone* or *normal cone* of vectors such that $\langle x, s \rangle \leq 0$ for all s in S. The *bi-polar set*, denoted S^{00} or S^{oo}, consists of all vectors that are polar to the polar of S, and coincides with the closed CONVEX HULL of S and zero. These definitions have extensions to normed spaces and paired vector spaces. In a complex vector space, **V,** the (absolute) polar set consists of all vectors such that $|\langle x, s \rangle| \leq 1$ for all s in **V.**

pole, *n.* **1.** an ISOLATED SINGULARITY of an ANALYTIC FUNCTION that is neither REMOVABLE nor ESSENTIAL. In this case, there exists a function g that is analytic and non-zero around the point a; that is, such that for the given function f, one can write

$$f(z) = g(z)(z - a)^{-m},$$

for some non–negative integer m that is the *order* of the pole. The pole is *simple* when $m = 1$.

2. the origin of a system of POLAR COORDINATES.

3. in the STEREOGRAPHIC PROJECTION of a sphere, the point that generates the image of a variable point as the intersection with a plane of the line joining the pole to the variable point.

4. see POLE AND POLAR.

pole and polar, *n.* (of a CONIC) a point (the *pole* of the line) and the line (the *polar* of the point) that is the locus of the point of intersection of the tangents to a given conic at the two points at which a SECANT through the pole cuts the conic (these are the HARMONIC CONJUGATES of the pole with respect to the secant). Analytically, the equation of the polar is obtained by replacing the coordinates of the contact point, in the equation of a general tangent to the conic, by the coordinates of the given pole. When the point lies exterior to the conic so that two tangents can be drawn from it to the conic, the polar is the secant through the corresponding contact points. Fig. 292 shows the two cases, where the pole P is outside an ellipse and when it is inside it; in each case the line XY is the polar of P. See also CONJUGATE (sense 5).

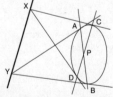

Fig. 292. **Pole and polar.** See main entry.

Polish notation or **prefix notation,** *n.* a logical notation that dispenses with the need for brackets by writing the operators (especially LOGICAL CONSTANTS) before their arguments. For example, *P or Q* ($P \vee Q$) is written A*pq*, and *if P then Q* ($P \rightarrow Q$) is written C*pq*.

$$(P \rightarrow (Q \vee R)) \rightarrow ((P \rightarrow R) \vee S)$$

is thus written unambiguously as

$$CC p A q r A C p r s.$$

If one uses a binary tree diagram such as Fig. 293 opposite, with the operators at the nodes, to represent the structure of an expression, the Polish representation is obtained by reading from the top down and from left to right where otherwise indifferent, so that, at each node one reads in order the node itself, the left branch, and the right branch, iterating where necessary; in Fig. 293 one thus reads from the top down, reading the bold branch at each node before returning to the node and reading the other branch. See also REVERSE POLISH NOTATION. Compare INFIX NOTATION.

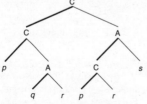

Fig. 293. **Polish notation**.
See main entry.

Polish space, *n.* the HOMEOMORPHIC image of a COMPLETE SEPARABLE METRIC SPACE. See also SOUSLIN SET.

polyadic, *adj.* (of a relation, operation, etc.) having several argument places; for example, ... *moves* ... *from* ... *to* ..., which might be represented as

$$M\,p\,o\,x_1\,y_1\,z_1\,t_1\,x_2\,y_2\,z_2\,t_2,$$

where *p* is a person, *o* is an object, and each quadruple $\langle x_i, y_i, z_i, t_i \rangle$ the coordinates of a place and time.

polygon, *n.* a closed plane figure bounded by three or more straight line segments that terminate in pairs at the same number of vertices, and do not intersect other than at their vertices; in Fig. 294, only the first figure is a polygon. The sum of the interior angles is $(n-2) \times 180°$ where *n* is the number of sides; the sum of exterior angles is always 360°. A *regular polygon* has all its sides and all its angles equal. Specific polygons have names that indicate the number of sides, such as TRIANGLE, QUADRILATERAL, PENTAGON, HEXAGON, etc. A polygon is *concave*, as the first example in Fig. 294, if it is not CONVEX.

Fig. 294. Only the first figure is a **polygon**.

polygon of forces, *n.* (*Mechanics*) a POLYGON whose sides are proportional to, and in the same direction as, a set of forces acting at a point; the diagram is a polygon (that is, is closed) if and only if the forces are in equilibrium.

polyhedral, *adj.* of, pertaining to or consisting of a POLYHEDRON.

polyhedral angle, *n.* a figure formed by the intersection at a common vertex of three or more planes, such as the faces of a POLYHEDRON.

polyhedron, *n.* **1.** a solid figure, or its surface, that is bounded by four or more POLYGONAL faces in such a way that pairs of faces meet along edges, and three or more edges meet at each vertex. There are only five *regular* *polyhedra*, the PLATONIC SOLIDS, all the faces of which are identical regular polygons making equal angles with each other, so that all the edges are of equal length. Specific polyhedra are named according to their number of faces, such as TETRAHEDRON, PENTAHEDRON, HEXAHEDRON, etc.; Fig. 295 shows an irregular hexahedron, with one base shaded.

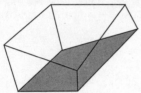

Fig. 295. **Polyhedron.**

2. convex polyhedron. in a EUCLIDEAN GEOMETRY of three dimensions, the intersection of finitely many closed half-spaces, and a *convex polytope*; the convex hull of finitely many points. Polytopes in this case coincide with bounded polyhedra.

polylogarithm, *n.* the function

$$\mathrm{Li}_n(z) = \sum_{k=1}^{\infty} \frac{z^k}{k^n},$$

defined for integral $n \geq 2$ and z in the unit disk; Li_2 is called the *dilogarithm* and Li_3 is called the *trilogarithm*.

polynomial, *adj.* **1.** (of an expression) containing two or more terms, such as $4x^3 - 5xy + 3y^2 - 7$.

2. (*as substantive*) **a.** a mathematical expression consisting of a sum of terms each of which is a product of a constant and one or more variables or INDETERMINATES raised to a non-negative integral power. If there is only a single variable, x, the general form is given by

$$a_0 x^n + a_1 x^{n-1} + a_2 x^{n-2} + \ldots + a_{n-1} x + a_n,$$

where the a_i are real numbers. There are only a finite number of distinct polynomials in a variable over a finite field, but there are infinitely many distinct polynomials in an indeterminate since distinct powers of the indeterminate are treated as distinct formal entities even if their values are identical.

b. (in some usages) also called **multinomial.** any mathematical expression consisting of a sum of a number of terms.

3. (of an algorithm) such that, with respect to a suitable measure, the effort required to execute the given algorithm is a polynomial in an appropriate measure of the size of the input data; the effort is often measured in time. See POLYNOMIAL TIME ALGORITHM, NP–DECISION PROBLEM.

polynomial domain, *n.* a EUCLIDEAN DOMAIN of POLYNOMIALS over a FIELD, where the VALUATION function $v(p(x))$ is the degree of the polynomial $p(x)$.

polynomial ring, *n.* **1.** the RING, denoted $R[X]$, of formal POLYNOMIALS in X over a ring R, where X is a COMMUTING INDETERMINATE; that is, the ring of expressions of the form

$$\sum_{j=0}^{n} a_j X^j,$$

where a_j is an element of R and n is a natural number. If

$$\sum_{j=0}^{n} a_j X^j \text{ and } \sum_{j=0}^{n} b_j X^j$$

are members of the polynomial ring R[X], then their sum is

$$\sum_{j=0}^{n} (a_j + b_j) X^j,$$

and their product is

$$\sum_{j=0}^{2n} \left(\sum_{i=0}^{j} a_j b_{j-i} \right) X^j.$$

The definition can be extended inductively:

$$R[X_1, \ldots, X_n] = (R[X_1, \ldots, X_{n-1}])[X_n],$$

where the X_i are distinct commuting indeterminates. Further, if R is a NOETHERIAN RING with identity, then $R[X_1, \ldots, X_n]$ is Noetherian; this is the *Hilbert basis theorem*. In particular, if K is a FIELD, then $K[X_1, \ldots, X_n]$ is a GAUSSIAN DOMAIN.

2. the ring, denoted R(X), of *rational polynomials* over X, that is, of ratios of elements of R[X].

polynomial time, *n.* a measure of the COMPLEXITY of an ALGORITHM. An algorithm is said to be a *polynomial time algorithm*, or to run in polynomial time, if it solves a given computational problem in a time that is a POLYNOMIAL in the size of the input, that is if the number of elementary operations required to complete its computation can be expressed as such a polynomial function. Algorithms which do not run in polynomial time are said to run in EXPONENTIAL TIME. See MAX–FLOW MIN–CUT ALGORITHM.

polyproperty, see THREE–SPACE PROPERTY.

polytope, *n.* the analogue of a polyhedron in n-dimensional space. See also POLYHEDRON (sense 2).

Poncelet, Jean Victor (1788–1867)**,** French pioneer of PROJECTIVE GEOMETRY, whose principal career was as a military engineer. He served under Napoleon in his march on Moscow, when he was left for dead; later he greatly improved the efficiency of turbines and water-wheels while Professor of Mechanics at Metz. He discovered the PRINCIPLE OF DUALITY and used POINTS AT INFINITY in order to increase the generality of geometric results such as DESARGUES' THEOREM.

Pontryagin's maximum (or **minimum**) **principle,** *n.* (*Control theory*) one of many extensions of the principle that, at the optimal solution of a control problem, a HAMILTONIAN system is solvable, and the corresponding Hamiltonian is maximized. For the simplest fixed end-point problem of minimizing

$$\int_a^b F(t, x(t), u(t))$$

subject to

$$x'(t) = \phi(t, x(t), u(t))$$

with STATE CONSTRAINTS $x(a) = c(a)$ and $x(b) = c(b)$ and CONTROLS required to satisfy $u(t) \in U$ for almost all $a \le t \le b$, and for a suitable subset of

Euclidean space, the HAMILTONIAN is

$$H(t, x, p, u, \lambda) = \langle p, \phi(t, x, u) \rangle - \lambda F(t, x, u);$$

the ADJOINT variables or CONJUGATE VARIABLES p must satisfy the ADJOINT EQUATION

$$-p'(t) = \frac{\partial H(t, x, p, u, \lambda)}{\partial x},$$

the inequality

$$H(t, x(t), p(t), u(t), \lambda) \geq H(t, x(t), p(t), w, \lambda)$$

for all w in U, and the TRANSVERSALITY CONDITONS

$$p(a) \perp x(a) \text{ and } p(b) \perp x(b).$$

In addition $\lambda \geq 0$ holds and either p or λ is non-zero.

population or **universe,** *n.* (*Statistics*) the entire underlying set of individuals from which SAMPLES are drawn.

porism, *n.* a type of mathematical proposition discussed by Euclid, but no longer identifiable; it is thought to have affirmed the possibility of finding conditions under which a problem is indeterminate or has innumerable solutions.

poset, *acronym for* PARTIALLY ORDERED SET. If every pair of elements has a GREATEST LOWER BOUND and a LEAST UPPER BOUND under this ordering, the poset is a LATTICE.

positional notation or **positional system,** *n.* another term for PLACE–VALUE NOTATION.

position vector, *n.* a vector of which the components are the coordinates of a given point; the directed line from the origin to that point. Thus, in Fig. 296, $\overrightarrow{OP}$ is the position vector of P. See also RADIUS VECTOR.

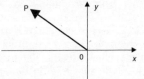

Fig. 296. $\overrightarrow{OP}$ is the **position vector** of P.

positive, *adj.* **1.** having a value greater than zero.

2a. measured in a direction opposite to that considered to be NEGATIVE.

b. having the same magnitude but opposite SENSE to an equivalent negative quantity.

3. (of a direction) moving from lower to higher values of a scale such as a coordinate axis.

4. (of an angle) measured in an anticlockwise direction.

5a. (of a self-adjoint operator or matrix) POSITIVE SEMI–DEFINITE. See also POSITIVE DEFINITE.

b. (of an operator or matrix between ordered vector spaces) ISOTONE. In this sense, a matrix is often called positive if it has non-negative entries.

6. (*Logic*) (of a statement, proposition, etc.) not NEGATIVE. See AFFIRMATIVE.

7. see POSITIVE SET.

positive correlation, see CORRELATION.

positive definite, *adj.* (for A, a matrix or self–adjoint operator on HILBERT SPACE) having $\langle Ax, x \rangle > 0$ for all $x \neq 0$. See POSITIVE SEMIDEFINITE.

positively dependent, see STATISTICALLY DEPENDENT.

positively homogeneous, *adj.* (of a function) HOMOGENEOUS, but only for positive scalars:

$$f(\lambda x) = \lambda f(x) \text{ for every scalar } \lambda > 0.$$

If

$$f(\lambda x) = \lambda^p f(x)$$

for some $p > 0$, then f is positively homogeneous of degree p.

positive orthant, *n.* the ORTHANT in which all coordinates are positive.

positive semidefinite or **positive,** *adj.* (of a matrix or self-adjoint operator on HILBERT SPACE) having $\langle Ax, x \rangle \geq 0$ for all x. If the scalar field is complex, the requirement that A be self-adjoint is redundant. The operator is *positive definite* if $\langle Ax, x \rangle > 0$ for all $x \neq 0$, for which it suffices to check the strict positivity of the leading PRINCIPAL MINORS, obtained by deleting all save the first n rows and columns. The analogous definitions are made for NEGATIVE SEMIDEFINITE and negative definite operators, and also for QUADRATIC FORMS.

positive set, *n.* a set, P, of elements of an ordered FIELD that is closed under addition and multiplication, and has the property that for every non-zero member, x, of the field, either x or $-x$ is in P.

possible, *adj.* **1.** able to exist or be true; not entailing any contradiction.

2. (*Logic*) (of a statement, formula, etc.) capable of being true under at least some interpretation, in some POSSIBLE WORLD, or in some circumstances. That P is possibly true is usually written 'MP' or '◊P'.

possible world, *n.* (*Logic*) a semantic device in MODAL LOGIC that formalizes the notion of what the world might have been like; that is, a complete world-description. A statement is *necessary* (necessarily true) if and only if it is true in every possible world, and *possible* (possibly true) if and only if it is true in at least one.

posterior, **posterior probability**, or *a posteriori* **probability,** *n.* (*Statistics*) the probability assigned to some parameter, or to an event, on the basis of its observed frequency in a sample, calculated from a PRIOR PROBABILITY by BAYES' THEOREM. See also EMPIRICAL PROBABILITY.

postfix notation, *n.* **1.** another term for REVERSE POLISH NOTATION.

2. an uncommon notation for relations that is used by computers in their internal representation, in which the symbol for the relation is analogously written after both its arguments, as in xyR. Compare INFIX NOTATION, PREFIX NOTATION.

post-multiplication, *n.* (by a matrix or ring element) multiplication on the right rather than on the left (*pre-multiplication*), this distinction being important in any non-commutative setting.

postulate, *n.* an AXIOM of a specific theory, especially one of the five axioms laid down by Euclid for plane geometry and contrasted by him with 'common notions' or axioms of complete generality.

potency, *n.* another word for POWER (sense 3).

potential energy, *n.* (*Mechanics*) (at a point P in a CONSERVATIVE VECTOR FIELD of force) the WORK done in moving from some reference point, R, to P along any curve joining R to P.

potential flow, *n.* (*Continuum mechanics*) a MOTION of a BODY in which the VELOCITY is given by the GRADIENT of a SCALAR FIELD. This is equivalent to an IRROTATIONAL MOTION.

potential function, *n.* an HARMONIC function; any twice continuously differentiable function satisfying LAPLACE'S EQUATION in some region of three-space.

potential theory, *n.* the study of POTENTIAL FUNCTIONS.

power, *n.* **1.** the number of times a number or expression is multiplied by itself, for which the symbol is an INDEX. For example, a^3 is the third power of a. See also EXPONENT.

2. (of a point with respect to a circle) the quantity $d^2 - R^2$ where R is the radius of the given circle and d is the distance from the given point to the centre of the circle.

3. also called **potency.** the CARDINALITY of a set.

4. (*Statistics*) the probability of rejecting the NULL HYPOTHESIS in a statistical test when it is in fact false; the power of a test of a given null clearly depends on the particular ALTERNATIVE HYPOTHESIS against which it is being tested.

5. (*Mechanics*) the WORK rate of a force with a moving point of application, given by dL/dt, where L is the work done, and t is the TIME. The standard unit of power is the WATT.

6. (*Continuum mechanics*) the sum of integrals

$$P(\mathbf{R}_t) = \int_{\mathbf{R}} \rho\, \mathbf{b} \cdot \mathbf{v}\, dv + \int_{\partial \mathbf{R}} \mathbf{t} \cdot \mathbf{v}\, da,$$

respectively over the volume and the surface area of the CONFIGURATION at time t of a SUB–BODY, $\mathbf{R}$, of which ρ is the DENSITY, $\mathbf{b}$ is the BODY FORCE DENSITY, $\mathbf{v}$ is the velocity, and $\mathbf{t}$ is the CONTACT FORCE density.

power residue, see RESIDUE (sense 2).

power rule, *n.* the specialization of the CHAIN RULE to the case of an arbitrary power of a given function. If $f(x) = g(x)^r$, then

$$\frac{df(x)}{dx} = \left[rg(x)^{r-1} \right] \frac{dg(x)}{dx}.$$

power series, *n.* a series with terms that contain ascending positive integral powers of a variable around a given point a (its *centre*) and have coefficients that do not depend on the variable; thus it has the general form

$$a_0 + a_1(x - a) + a_2(x - a)^2 + a_3(x - a)^3 + \ldots.$$

For example,

$$1 + x + \frac{x^2}{2!} + \frac{x^3}{3!} + \ldots$$

is a power series for e^x. A power series with real coefficients is said to be real or complex according as both x and a are real or complex. See TAYLOR SERIES, LAURENT EXPANSION.

power set, *n.* a set of which the elements are all the subsets of a given set, written $P(S)$ or 2^S, and having strictly more members than the given set. See CANTOR'S DIAGONAL THEOREM.

pp, *abbrev. for **presque partout** (French),* (of a measure) almost everywhere. See ALMOST ALL.

***p*-primary module,** *n.* a MODULE over an INTEGRAL DOMAIN in which for each element of the module there is a positive α such that $p^\alpha x = 0$, where p is a PRIME of the integral domain.

Pr or **P,** *standard notation for* a PROBABILITY MEASURE.

precedence, *n.* the order in which a nested sequence of OPERATIONS are to be evaluated; for example, in $[(2 + 3) \times 5]$, the addition has precedence over the multiplication. The BRACKETS are also said to have precedence over one another. The greater the precedence of an operator, the narrower its SCOPE.

precision, *n.* (*Numerical analysis*) the accuracy with which a given calculation is performed. On most computers *single precision* is somewhere between 9 and 16 decimal digits; *double precision* refers to calculation of twice that length, and *multiple, high,* or *extended precision* to calculations involving greater accuracy.

precompact, *adj.* (of a set A in a LOCALLY CONVEX SPACE) such that, for any neighbourhood U of zero, there is a finite set F for which A is a subset of F + U. In a metrizable setting this agrees with the definition of a TOTALLY BOUNDED set, and the two terms are often used interchangeably. In a precompact metric space, X, for all $\varepsilon > 0$ there are a finite number of points $x_1, ..., x_n$ such that

$$X \subset \bigcup_{i=1}^{n} B_\varepsilon(x_i).$$

precondition, *vb.* (*Numerical analysis*) to rescale or otherwise adjust a matrix or other quantity so as to improve its CONDITION NUMBER prior to beginning a computation. For example, one may precondition a matrix by transforming it in order to increase the number of near unit eigenvalues.

predecessor, *n.* (of an ORDINAL) a number having the given number as SUCCESSOR. Thus the infinite ordinal, ω, has no predecessor but is the predecessor of the number $\omega + 1$.

predicate, *n.* (*Logic*) **1a.** an expression that ascribes a property to some thing or things (its *subjects*); a predicate with more than one subject is a RELATION.

b. a property, characteristic or attribute that may be affirmed or denied of something. The categorial statement *all men are mortal* relates two predicates, ... *is a man* and ... *is mortal.* See SYLLOGISM.

c. the term of a categorial proposition that is affirmed or denied of its subject; in the same example, *all men* is the subject and *is mortal* is the predicate in this sense.

2a. formally, in some treatments of the PREDICATE CALCULUS, a term that is derived from an ATOMIC SENTENCE by the deletion of a NAME, these being the primitive terms of the system; in other treatments, names and predicates are primitive, and an atomic sentence is defined to be the result of combining them by replacing each variable in the predicate by a name. Predicates are usually written in functional notation, as, for example, $F(x)$ and $R(x, y)$, and yield well-formed sentences when an appropriate sequence of referring expressions replaces the variables in order, or when all the variables are bound by QUANTIFIERS. A predicate cannot itself be true or false, but it is sometimes said to be true if its universal closure is true; that is, if it holds for every element in the relevant domain. It is *satisfied* by, or is true of, a sequence of referring expressions if the uniform replacement of each of its variables by the elements of the sequence in order yields a true sentence.

b. consequently, in SEMANTICS, a function from individuals or sequences to truth-values, the truth set of the function being the EXTENSION of the predicate. In this context it is sometimes a useful device, due to Tarski, to treat sentences as zero-place predicates.

predicate calculus or **functional calculus,** *n.* the system of SYMBOLIC LOGIC concerned not only with representing the logical relations between sentences or propositions as wholes, but also with considering their internal structure in terms of subject and PREDICATE. The primitive terms are individual NAMES, predicates, and VARIABLES that may be bound by QUANTIFIERS. If quantification is restricted to individuals, it is *lower* (or *first order*) *predicate calculus* (LPC), and is consistent, complete, but not decidable. See also LOGICAL FORM. Compare SENTENTIAL CALCULUS.

predicative, *adj.* (*Logic*) (of a definition) given in terms that do not require quantification over entities of the same type as that which is thereby defined. Part of Russell's solution to the paradoxes of self-reference, as contained in his theory of TYPES, was to require all definitions to be predicative. Compare IMPREDICATIVE DEFINITION.

predicted variable, *n.* (*Statistics*) a modern term for DEPENDENT VARIABLE.

predictor, *n.* (*Statistics*) a modern term for INDEPENDENT VARIABLE.

predual space, *n.* a normed space of which some given space is the DUAL BANACH SPACE. The space of sequences convergent to zero, endowed with the supremum norm, is a predual for $l_1(\mathbb{N})$. See l_p- SPACE.

preference or **preference order,** *n.* a PARTIAL ORDER or other similar relation, especially in an economic or related setting.

prefix notation, *n.* **1.** another term for POLISH NOTATION.

2. the usual notation for relations in which the symbol for the relation is written before its arguments, as in *Rxy*. Compare INFIX NOTATION, POSTFIX NOTATION.

pre-Hilbert space, *n.* an incomplete INNER PRODUCT SPACE.

pre-image, *n.* another word for COUNTERIMAGE.

premise or **premiss,** *n.* a statement from which a conclusion is drawn in a particular ARGUMENT. It may, in particular, be an AXIOM of the relevant theory, or merely an assumption taken to be true for the purposes of discovering its consequences.

pre-multiplication, *n.* (by a matrix or ring element) multiplication on the left rather than on the right (*post-multiplication*), this distinction being important in any non-commutative setting.

prenex normal form, *n.* (*Logic*) a formula of PREDICATE CALCULUS of the form

$$(Qx_1)(Qx_2)\ldots(Qx_n)B$$

where each (Qx_i) is an existential or universal QUANTIFIER, the variables are distinct, and B is an OPEN SENTENCE. Each formula may be shown equivalent to one in prenex form by use of the PRENEX OPERATIONS. One of the values of this equivalence is that the DECISION PROBLEM is soluble for some classes of prenex expressions.

prenex operation, *n.* (*Logic*) any of the operations by which any well-formed formula of PREDICATE CALCULUS can be transformed into equivalent formulae in PRENEX NORMAL FORM; for example,

$$-(\exists x)Fx \equiv (\forall x)-Fx; \quad -(\forall x)Fx \equiv (\exists x)-Fx;$$
$$((\exists x)Fx \to P) \equiv (\forall x)(Fx \to P).$$

pre-ordering, see ORDERING.

presentation, *n.* (*Group theory*) a set, X, of GENERATORS together with a set, R, of RELATIONS, such that the group generated by X subject to the relations of R is ISOMORPHIC to a given group; a presentation of a group is usually denoted ⟨X, R⟩. For example,

$$\langle\, a,\, b;\ \ a^2 = b^n = (ab)^2 = 1 \,\rangle$$

is a presentation of the dihedral group of degree *n*, for $n \geq 3$. Formally, for R a subset of the FREE GROUP on X, ⟨X, R⟩ is a presentation of G if and only if G is isomorphic to the free group on X factored by the NORMAL CLOSURE of R in the free group.

pressure, *n.* (*Continuum mechanics*) the FORCE per unit area; formally, a scalar FIELD, *p*, for which the STRESS VECTOR, $\mathbf{t}(\mathbf{n})$, is equal to $-p\mathbf{n}$ in a HYDROSTATIC stress, where **n** is the unit outward normal to the given surface.

presuppose, *vb.* to require a condition to be satisfied as a precondition for a statement to be either true or false, or for a speech act to be felicitous; for example, the question *have you stopped beating your wife?* presupposes that the person addressed both has a wife and used to beat her.

pre-theoretical or **pretheoretic,** *adj.* informal or intuitive; in advance of, or contrasted with, formalization as a THEORY.

prima facie, *adv.* (*Latin*) at first sight; usually used to distinguish an intuitive, pre-theoretical conception from a formal or theoretical one; for example, many things that seem *prima facie* paradoxical turn out merely to be badly expressed.

primal–dual methods, *n.* algorithms for solving CONSTRAINED OPTIMIZATION problems that explicitly use both *primal* information (information about the original problem) and *dual* information (information concerning the Lagrangian and associated multipliers). The SIMPLEX METHOD may be viewed in such terms.

primality test, *n.* a method of determining whether or not a given number is prime, without factorising it. See AKS PRIMALITY TEST.

primal linear program, see DUALITY THEORY OF LINEAR PROGRAMMING.

prime, *adj.* **1a.** (of an integer) having no FACTORS except itself and UNITY.
b. (*as substantive*) see PRIME NUMBER.
2. prime to. not divisible by (another number, polynomial, etc.) See RELATIVELY PRIME, IRREDUCIBLE.
3. more generally, of a non-zero, non-unit element *p* of an INTEGRAL DOMAIN, such that, whenever *p* divides *ab*, then *p* divides *a* or *b*. Any such element is IRREDUCIBLE.
4. (of a RING) such that the product of a pair of IDEALS is zero only if one of the ideals is itself zero. Any simple ring is prime. Compare SEMIPRIME.

5. *n.* a small superscript symbol used to distinguish entities otherwise designated by the same letter such as a change of variables from $\langle x, y, z \rangle$ to $\langle x', y', z' \rangle$, or $f'(x)$, the derivative of the function $f(x)$. x' is read 'x-prime'.

prime factorization, *n.* the decomposition of an integer into a product of PRIME FACTORS.

prime ideal, *n.* an IDEAL I with the property that if $ab \in$ I then either $a \in$ I or $b \in$ I. A non-zero non-unit element is PRIME precisely when the ideal it generates is.

prime number or **prime,** *n.* **1a.** a natural number, other than 1, divisible by no integers other than unity and itself, such as 2, 3, 5, 7, 11, 13, 17, 19, 23, ... The proof that there are infinitely many prime numbers was known to the Greeks, but the largest known (December 2001) is $2^{13\ 466\ 917} - 1$, a 14-million digit MERSENNE NUMBER, discovered by the GIMPS project in 2001.
b. more generally, any integer having this property, such as –2, –3, –5, ...
2. analogously, in any number FIELD, an non-unit element divisible only by the multiplicative unity and itself.
Compare COMPOSITE NUMBER.

prime number theorem, *n.* the celebrated theorem that $\pi(x)$, the number of primes less than x, is ASYMPTOTIC to $x/\ln x$. This was first stated by Gauss in 1792 on the basis of much numerical computation, but it was not proved until Hadamard and de la Vallée–Poussin did so independently in 1896; it was one of the major motivations for the development of analytic number theory. Equivalently, the nth prime is asymptotic to $n \ln n$, as n tends to infinity. More generally, Landau proved in 1900 that if $\pi_k(x)$ is the number of integers not exceeding x that have k distinct prime factors, then

$$\pi_k(x) \sim \frac{x}{\ln x} \frac{(\ln(\ln x))^{k-1}}{(k-1)!}.$$

prime subfield, *n.* the unique SUBFIELD that is contained in every subfield of a FIELD. It is isomorphic either to the field of rational numbers or to the field of integers modulo some prime p.

primitive, *adj.* **1.** (of a term, statement etc.) given in the initial specification of the theory. For example, the axioms are the primitive statements of any theory. See also UNDEFINED ELEMENT.
2. *n.* another word for the ANTIDERIVATIVE of a function. Thus $\log x$ is a primitive for $1/x$.

primitive term, *n.* any of the expressions explicitly stated to be a term of a THEORY, such as names, predicates, and variables in the predicate calculus. Although the primitive terms refer to the UNDEFINED ELEMENTS of a theory, they are implicitly defined by its axioms.

primitive polynomial, *n.* a polynomial over an INTEGRAL DOMAIN with identity, such that the greatest common divisor of its coefficients is the identity.

primitive root *n.* a power RESIDUE of a natural number n such that the powers of the residue generate the entire REDUCED RESIDUE SYSTEM; equivalently, an element with order equal to the TOTIENT of n, $\phi(n)$. An odd prime p has $\phi(p-1)$ primitive roots.

primitive root of unity, *n.* a complex number z such that $z^n = 1$ but $z^k \neq 1$ for every $k < n$. So is z is a primitive root of unity of order n, it is not a root of unity of any lower order. The primitive nth roots of unity are of the form $\exp(2\pi ri/n)$. They are the zeros of the CYCLOTOMIC polynomial $n(x)$.

principal, *adj.* conventionally chosen as the PRINCIPAL VALUE of some function or operation, such as a principal PHASE or PRINCIPAL DIAGONAL.

principal axes, *n.* **1.** (*Euclidean geometry*) the axes of a CONIC or QUADRIC surface with respect to which the figure has an equation that is the sum of multiples of squares.

2. (*Mechanics*) the coordinate axes with respect to which a RIGID BODY has a diagonal INERTIA TENSOR, or, equivalently, the PRODUCTS OF INERTIA are zero.

principal curvatures, *n.* (at a point of a surface) the NORMAL CURVATURES in the PRINCIPAL DIRECTIONS of the surface at the point. The reciprocals of these two numbers are the *principal radii of normal curvature*.

principal diagonal, *n.* the MAIN DIAGONAL of a matrix.

principal directions, *n.* (at a point of a surface) the directions in which the RADIUS OF NORMAL CURVATURE is maximized or minimized.

principal domain, *n.* another term for PRINCIPAL IDEAL DOMAIN.

principal ideal, *n.* an IDEAL in a ring that is generated by a single element. A principal ideal integral domain is called a PRINCIPAL IDEAL DOMAIN.

principal ideal domain (abbrev. **pid**) or **principal domain,** *n.* an INTEGRAL DOMAIN all of whose ideals are PRINCIPAL IDEALS; then prime and irreducible elements coincide, and the domain is a unique factorization domain. There are exactly nine imaginary quadratic fields $Q(\sqrt{d})$ whose subring of algebraic integers yield principal ideal domains, those with $-d = 1, 2, 3, 7, 11, 19, 43, 67,$ or 163.

principal ideal ring, *n.* **1.** a RING in which every ideal is a PRINCIPAL IDEAL.

2. loosely, a PRINCIPAL IDEAL DOMAIN.

principal invariants, *n.* (for a matrix) the coefficients in the CHARACTERISTIC POLYNOMIAL of the given matrix.

principal minor, *n.* the determinant of a PRINCIPAL SUBMATRIX of a given square matrix. See also COFACTOR.

principal moment of inertia, *n.* (*Mechanics*) a MOMENT OF INERTIA of a RIGID BODY with respect to PRINCIPAL AXES.

principal normal, *n.* (to a curve at a point in Euclidean space) the unit derivative of the unit TANGENT VECTOR (with respect to arc length) at the given point, denoted **N**. See also FRENET FORMULAE.

principal part, *n.* **1.** the RESTRICTION of a MULTI–VALUED FUNCTION to arguments for which it takes its PRINCIPAL VALUE.

2. any of the sides and interior angles of a triangle.

principal phase, see PHASE.

principal radii of normal curvature, see PRINCIPAL CURVATURES.

principal series or **chief series,** *n.* a NORMAL SERIES $(G_0, ..., G_n)$ such that for all k between 0 and $n-1$ inclusive, G_{k+1} is maximal in the set of normal subgroups of G properly contained in G_k.

principal solution matrix, *n.* a FUNDAMENTAL MATRIX, Y, for which $Y(t_0)$ is the identity matrix for a given argument t_0.

principal stretches, *n.* (*Continuum mechanics*) the EIGENVALUES of the positive definite symmetric tensor, **U**, such that, for **F** the DEFORMATION GRADI-ENT of the motion of a given body, there is an orthogonal tensor **R** such that $\mathbf{F} = \mathbf{RU}$.

principal stretchings, *n.* (*Continuum mechanics*) the LATENT ROOTS of the EULERIAN STRAIN RATE.

principal submatrix, *n.* a submatrix of a given square matrix, obtained by deleting similarly indexed rows and columns. See also COFACTOR.

principal ultrafilter, see ULTRAFILTER.

principal value, *n.* a conventionally chosen, unique member of a set of values all of which are functionally related to the same argument, as with the complex LOGARITHM, the ANTIDERIVATIVE, or the inverse of a PERIODIC FUNCTION. For example, $\sin n\pi = 0$ for all integral n, but the principal value of $\arcsin 0$ is taken to be 0; the principal value of $\sqrt{4}$ is 2 although -2 is also a root. The function that has as its values the principal values of a many-valued function is conventionally indicated by writing it with a capital letter; thus, Cotan^{-1} is the principal value of the inverse cotangent, and, especially in complex functions, Ln is the principal value of the natural logarithm.

principle of angular momentum, *n.* (*Mechanics*) the principle that for any mechanical system, the TORQUE is equal to the rate of change of the ANGU-LAR MOMENTUM.

principle of duality, *n.* the DUALITY of lines and points in PROJECTIVE GEO-METRY. See DESARGUES' THEOREM.

principle of indifference, see INDIFFERENCE.

principle of induction or **principle of mathematical induction** *n.* the prin-ciple that if a given integer, n, has a certain property, and if it can be proven that if any integer, m, no smaller than n has the property, so has $m + 1$, then all integers greater than n have it. See INDUCTION (sense 2a).

Pringsheim convergence, *n.* the CONVERGENCE of a MULTIPLE SERIES

$$\sum_{\substack{1 \le k_1 \le n_1 \\ \cdots \\ 1 \le k_m \le n_m}} a_{k_1 \ldots k_n}$$

in the sense that the PARTIAL SUMS approach a fixed limit as the minimum of the separate upper limits of summations tends to infinity; that is, as the least index used, irrespective of position, tends to infinity. For example, in the three-dimensional case, for $\sum_{ijk} a_{ijk}$ to converge in this sense requires that the limit

$$\lim_{\min (n,m,p) \to \infty} \sum_{i=1}^{n} \sum_{j=1}^{m} \sum_{k=1}^{p} a_{ijk}$$

exists. (Named after the German analyst, *Alfred Pringsheim* (1850–1941).)

prior probability, **prior**, or *a priori* **probability,** *n.* (*Statistics*) the probability assigned to a parameter or an event in advance of any empirical evidence, often subjectively or on the assumption of the principle of INDIFFERENCE. Compare POSTERIOR PROBABILITY.

prism, *n.* a POLYHEDRON with two parallel and congruent polygonal BASES, so that all parallel cross-sections are also congruent with the bases, and there-fore all sides are parallelograms; Fig. 297 opposite shows an irregular pentagonal prism with a base shaded. Its volume is the product of the area of its base and the perpendicular distance between the planes of the bases. Compare PRISMOID, PRISMATOID.

Fig. 297. A **prism** with one of the bases shaded.

prismatoid, *n.* a POLYHEDRON all of whose vertices lie in one or other of two parallel planes; all lateral faces are therefore either triangular or quadrilateral. In Fig. 298, the parallel bases are shaded dark and only the light shaded side is triangular; the other faces are all skew quadrilaterals. Compare PRISM, PRISMOID.

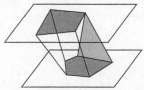

Fig. 298. A **prismatoid**: the parallel sides have dark shading.

prismoid, *n.* a PRISMATOID with planar sides and an equal number of vertices in each of the parallel planes. The lateral sides must therefore be either trapezia or parallelograms, as shown in Fig. 299, where the non-congruent irregular pentagonal bases are shaded. Compare PRISM.

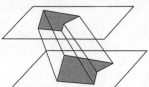

Fig. 299. A **prismoid** with the parallel sides shaded.

prismoidal formula, *n.* **1.** see SIMPSON'S RULE.
2. the formula for the volume of a PRISMATOID given by

$$\tfrac{1}{6}\,h\,(B_1 + B_2 + 4M),$$

where h is the height, B_1 and B_2 are the areas of the bases, and M is the area of the cross-section equidistant from the bases.

prisoner's dilemma, *n.* (*Game theory*) any generalization of the classic situation in which two prisoners are separated and each is told that if he turns Queen's evidence but the other does not confess, he will be released and receive a reward; if neither confesses, they will both have to be released without further penalty, but if both confess, both will be convicted of a charge carrying an intermediate penalty. This leads to a seeming paradox in that each player has a dominant strategy, that if used simultaneously, is worse for both than if each selects his dominated strategy. The arms race is similarly viewed: mutual disarmament is the dominated outcome that both parties would prefer to armed parity or 'mutually assured destruction'.

probability, *n.* **1.** (*Statistics*) a measure or estimate of the degree of confidence one may have in the occurrence of an event, measured on a scale from 0 (impossibility) to 1 (certainty). It may be defined to be the proportion of favourable outcomes to the total number of possible outcomes if these are indifferent (MATHEMATICAL PROBABILITY), or the proportion observed in a sample (EMPIRICAL PROBABILITY), or the limit of this as the sample size tends to infinity (RELATIVE FREQUENCY); measures of *subjective probability* are also used.

2. also called **axiomatic probability.** the formal study of random or chance events, usually in terms of PROBABILITY MEASURE, independent RANDOM VARIABLES, and their generalizations. This arises in part from the study of PERMUTATIONS and COMBINATIONS, and its results are applied to the analysis of empirical data by the construction of statistical tests, and are further developed in GAME THEORY and INFORMATION THEORY. Compare STATISTICS. See also CONDITIONAL PROBABILITY, BAYES' THEOREM.

3. another word for PROBABILITY MEASURE.

probability density function (abbrev. **pdf**), *n.* (*Statistics*) a function representing the relative distribution of the FREQUENCY of a continuous RANDOM VARIABLE from which parameters such as its mean and variance can be derived, and having the property that its integral from *a* to *b* is the probability that the variable lies in this interval. Its graph is the limiting case of a HISTOGRAM as the amount of data increases and the class intervals decrease in size. The corresponding function of two or more random variables is their JOINT DENSITY FUNCTION. Compare CUMULATIVE DISTRIBUTION FUNCTION, FREQUENCY DISTRIBUTION.

probability distribution function or **probability function,** *n.* a function that constitutes a PROBABILITY MEASURE.

probability function, *n.* (*Statistics*) **1.** also called **probability mass function.** the function of which the values are the probabilities of the distinct outcomes of a DISCRETE RANDOM VARIABLE.

2. another name for PROBABILITY DISTRIBUTION FUNCTION.

probability integral transform, *n.* the transformation, $F_x(X)$, of a RANDOM VARIABLE, X, with an invertible cumulative distribution function F_x, to a UNIFORM DISTRIBUTION on $[0, 1]$.

probability law, *n.* (*Information theory*) the part of the definition of a CHANNEL that specifies for each member of the INPUT SET, the probabilities of each member of the OUTPUT SET being received, when that member of the input set is transmitted.

probability mass function, *n.* another name for PROBABILITY FUNCTION.

probability measure, *n.* a MAPPING, P, of a SAMPLE SPACE, X, into the interval $[0, 1]$, such that P maps the whole space to 1, and is additive for the (possibly infinite) union of disjoint sets that lie in the SIGMA–ALGEBRA of EVENTS; P is then a PROBABILITY DENSITY FUNCTION. See also PROBABILITY.

probability space, *n.* (*Statistics*) a totally finite MEASURE SPACE together with an associated PROBABILITY MEASURE that assigns unit measure to the entire space; the probability space is then (X, Σ, P), where X is the SAMPLE SPACE, Σ is the SIGMA–ALGEBRA of EVENTS, and P is the probability measure.

probable error (abbrev. **pe**), *n.* (*Statistics*) the deviation from the true value within which an observation is likely to lie with probability 50%. In a NORMAL DISTRIBUTION this is equal to 0.674σ, where σ is the STANDARD DEVIATION (in this context also referred to as the *mean error*).

process, see STOCHASTIC PROCESS.

problem, *n.* a mathematical question; a statement requiring proof or a method of solution.

produce, *vb.* to extend (a line segment).

product, *n.* **1.** the result of MULTIPLICATION of two or more numbers, quantities, etc. The *continued product* of a sequence of terms $x_k, \ldots, x_m$ is written

$$\prod_{i=k}^{m} x_i, \text{ or } \prod_{i \in I} x_i$$

where the index set I is more general. See also INFINITE PRODUCT.

2. set product. another name for INTERSECTION.

3. logical product. another name for CONJUNCTION or INTERSECTION.

4. the result of any analogous binary or *n*-ary operation, or the expression for such a result, such as the VECTOR PRODUCT.

5. the SCALAR multiple of a VECTOR in VECTOR ANALYSIS or in a VECTOR SPACE.

6. see CARTESIAN PRODUCT, DIRECT PRODUCT.

production function, *n.* (*Optimization, Mathematical economics*) an OBJECTIVE FUNCTION giving the level of production obtainable from a given level of resource input. See COBB–DOUGLAS FUNCTION.

product measure, *n.* the MEASURE μ, placed on the finite CARTESIAN PRODUCT of MEASURE SPACES (M_i, μ_i), by defining

$$\mu\left(\prod_{i=1}^{n} S_i\right) = \prod_{i=1}^{n} \mu_i(S_i)$$

for all products of sets S_i measurable in the coordinate spaces. The measure then extends uniquely to the sigma-algebra generated by sets of the form $\prod_i S_i$.

product of inertia, *n.* (*Mechanics*) (for a set of particles) the sum of the terms

$$\sum_i m_i p_i q_i,$$

where m_i is the MASS of the ith particle, and p_i and q_i are its directed distances from the two planes with which the product is associated. For a continuous body, the product of inertia is

$$\int pq\rho \, dV,$$

where ρ is the DENSITY at the point, and V is the volume of the body.

product order, *n.* the ORDERING placed on the CARTESIAN PRODUCT of ORDERED SETS by defining

$$\{x_i\} \leq \{y_i\} \text{ if } x_i \leq_i y_i \text{ for each index order.}$$

product rule, *n.* **1.** (*Calculus*) the rule for differentiating the PRODUCT of two DIFFERENTIABLE FUNCTIONS:

$$\frac{d(fg)}{dx} = f\frac{dg}{dx} + g\frac{df}{dx}.$$

Compare QUOTIENT RULE.

product topology, *n.* the TOPOLOGY placed on the (possibly infinite) product of TOPOLOGICAL SPACES X_i, by taking as BASES all sets that are finite CARTESIAN PRODUCTS of open sets of each X_i in turn; that is, the basis elements are

$$\{ U_1 \times U_2 \times \ldots \times U_i \times \ldots \}$$

for any U_i in the topology on X_i, with all but finitely many $U_i = X_i$. The topology on Euclidean *n*-space viewed as an *n*-fold product of the real line agrees with the Euclidean topology.

program, *n.* a sequence of instructions (usually called *statements*) that constitute an algorithm that determines how a computer carries out some task. A *mathematical program* requests the solution of an OPTIMIZATION problem.

progression, *n.* a SEQUENCE of terms, especially one whose successive pairs of members have a constant relation. See ARITHMETIC PROGRESSION, GEOMETRIC PROGRESSION, HARMONIC PROGRESSION.

project, *vb.* to draw a PROJECTION of.

projection, *n.* **1.** the image of one figure or object upon another under a MAPPING. For example, the intersection with a plane of a pencil of rays passing through the boundary of a given figure is a PERSPECTIVE projection; Fig. 300 shows a projection of an irregular figure in plane P to plane Q.

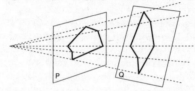

Fig. 300. The **projection** of a figure on a non-parallel plane.

2. an IDEMPOTENT LINEAR MAPPING from a linear space to itself.

3. metric projection. a mapping from a METRIC SPACE onto a set that associates each point in the metric space with its NEAREST POINT in the set. This is an IDEMPOTENT mapping that coincides in HILBERT SPACE with the ORTHOGONAL PROJECTION in the case of a closed vector subspace.

projection methods, *n.* methods of solving CONSTRAINED OPTIMIZATION problems by combining DESCENT METHODS for unconstrained problems with an appropriate PROJECTION onto the feasible set to maintain feasibility at each iteration. For example, when attempting to minimize a function over the orthant, if line search produces a point *x* with some negative coordinates, one might replace these coordinates by zeros. This replaces *x* by its positive part, which is equivalent to orthogonal projection onto the orthant.

projective, *adj.* (of a left R-MODULE P) having the property that given a HOMOMORPHISM, *f*, of P into the left R-module B, and *g* a homomorphism of the left R-module A onto B, then there is a homomorphism *h* of P into A such that *f* equals the composite of *g* and *h* (written *gh*).

projective coordinates, *n.* another name for HOMOGENEOUS COORDINATES.

projective geometry, *n.* **1.** the branch of mathematics concerned with the study of properties of geometrical figures that are invariant under PROJECTION. See also DESARGUES' THEOREM.

2. an abstract geometry concerning points, lines, and the incidence of points and lines, but with no metrical properties.

projective plane, *n.* an abstract geometry of two dimensions that is defined either as an ALGEBRAIC GEOMETRY whose group of transformations is a GENERAL LINEAR GROUP, or by a set of axioms, with POINT and LINE as primitive terms, that includes the *propositions of incidence*, namely that there is exactly one line through any two distinct points, and that there is at least one common point on any two distinct lines. The geometry can be constructed as a generalization of AUGMENTED EUCLIDEAN GEOMETRY. See also DUALITY.

projectivity or **projective transformation,** *n.* the composition of an arbitrary number of PERSPECTIVITIES in a PROJECTIVE GEOMETRY. See CROSS RATIO.

prolate spheroid, *n.* a SURFACE OF REVOLUTION swept out by an ellipse rotating about its MAJOR AXIS. Compare OBLATE SPHEROID.

prolate trochoid, see TROCHOID.

proof, *n.* a sequence of statements, each of which is either validly derived from those preceding it, or is an axiom or assumption, and the final member of which, the *conclusion*, is the statement of which the truth is thereby established. A *direct proof* proceeds linearly from premises to conclusion; an *indirect proof* (also called *REDUCTIO AD ABSURDUM*) assumes the falsehood of the desired conclusion and shows that to be impossible. See also INDUCTION, DEDUCTION, VALID.

proof by contradiction, *n.* an INDIRECT PROOF of the falsehood of an hypothesis by deriving a contradiction from it.

proof by contraposition, *n.* proof of a statement $P \rightarrow Q$ by first proving its CONTRAPOSITIVE $-Q \rightarrow -P$.

proof theory, *n.* the branch of logic that studies the SYNTACTIC properties of FORMAL THEORIES, especially the syntactic characterization of DEDUCTIONS as VALID; this is in direct contrast with MODEL THEORY, which studies SEMANTIC properties.

propagated round-off, see ROUND–OFF ERROR.

proper, *adj.* **1.** (of a relation, etc.) distinguished from a weaker relation by excluding the case where the relata are identical. For example, every set is a subset of itself, but a *proper subset* must exclude at least one member of the containing set. See also STRICT.

2. (*Algebra*) (of a subfield, subgroup, subring, etc.) being a PROPER SUBSET of the given field, group, ring, etc.

proper class, *n.* a CLASS that cannot be a member of other classes.

proper divisor, *n.* an integral DIVISOR of an integer which is not equal to the given integer itself. For example, 1, 2, 3, and 4 are the only proper divisors of 12; 12 is an integral divisor but not a proper divisor.

proper factor, *n.* a FACTOR of a given number other than the number itself; unity is a proper factor. A *perfect number* is one for which the the the sum of the proper factors equals the number itself.

proper fraction, *n.* another term for SIMPLE FRACTION.

proper mapping, *n.* (*Topology*) a CONTINUOUS mapping such that the inverse image of a COMPACT set is compact.

proper point, *n.* a real point in an AUGMENTED EUCLIDEAN GEOMETRY that corresponds to a point of the Euclidean plane, and at which non-parallel lines intersect.

proper subset, *n.* a SUBSET that is strictly contained in, and so excludes some elements of, a given set.

proportion, *n.* **1.** a LINEAR relationship between two variable quantities or their inverses; corresponding elements of two sets that are in proportion are in a constant ratio. For example, according to the gas laws, pressure is directly proportional to temperature but inversely proportional to volume. **2.** a relationship between four numbers or quantities in which the ratio of the first pair equals the ratio of the second pair; written

$$a : b = c : d, \text{ or } a : b :: c : d.$$

proportional, *adj.* **1.** involving or related by a PROPORTION.

2. (*as substantive*) a term of a proportion, especially an unknown; in

$$a : b = c : x,$$

x is the fourth proportional. See also MEAN (sense 4).

proposition, *n.* (*Logic*) **1.** the content of a sentence that affirms or denies something and is capable of being true or false; a statement. Thus my statement, 'I am warm,' and your statement to me, 'You are warm,' express the same proposition; conversely, although we both utter the same words, 'I am warm,' we express different propositions since our assertions are about different subjects.

2. the meaning of such a sentence. In this sense, 'I am warm' always expresses the same proposition, whoever may utter it. Compare STATEMENT.

propositional calculus, *n.* another term for SENTENTIAL CALCULUS.

propositional function, *n.* another term for OPEN SENTENCE.

propositions of incidence, *n.* (*Geometry*) the propositions that there is exactly one line through any two distinct points, and that there is at least one common point on any two distinct lines. See PROJECTIVE PLANE.

protractor, *n.* an instrument used for measuring angles; usually a flat, transparent plastic semicircle, marked in degrees round the circumference.

prove, *vb.* to provide a PROOF of a proposition; the latter is then a THEOREM of the system within which it is proven.

proximal or **proximinal,** *adj.* nearest. A proximal point in a set is a NEAREST POINT.

Prüfer *p*-group, *n.* for a prime number p, the set of all complex numbers z for which there exists a non-negative integer k such that

$$z^{p^k} = 1;$$

the Prüfer *p*-group is denoted $Z(p^\infty)$. (Named after the German group theorist, geometer, and analyst, *Heinz Prüfer* (1896–1934).)

Prüfer substitution, *n.* the first-order system

$$R(x)\,y' = z, \quad z' = -Q(x)\,y,$$
$$y = A(x)\sin\theta(x), \quad z = A(x)\cos\theta(x),$$

equivalent to the STURM–LIOUVILLE EQUATION

$$[R(x)\,y']' + Q(x)\,y = 0,$$

and used in its solution.

pseudo-inverse or **generalized inverse,** *n.* any of several generalizations of the INVERSE of a matrix or of a bounded linear operator with closed range on HILBERT SPACE, often denoted $A^\dagger$ for a given matrix or operator A; a linear operator that coincides with the inverse for an invertible operator. Usually it is required that

$$AA^\dagger A = A \text{ and } A^\dagger AA^\dagger = A^\dagger.$$

This defines a *semi-inverse.* One of the most frequently used inverses is the *Moore–Penrose inverse,* which is the unique solution to

$$AXA = A \text{ and } XAX = X,$$

with AX and XA self–adjoint. It can be obtained as the BEST APPROXIMATION to A in the Hilbert norm, in the sense that $A^\dagger y$ is the unique element z of least norm such that

$$\| A(z) - y \| = \min_x \| A(x) - y \|.$$

pseudo-metric, see METRIC.

pseudo-prime, *n.* **1.** also called **Carmichael number.** an integer q such that, for all integers a, one has

$$a^q \equiv a \pmod{q},$$

as holds for all primes, by FERMAT'S LITTLE THEOREM. The only composite pseudo-prime less than 1000 is 561, so that testing a number for pseudo-primality appears to give very strong evidence that it is prime.

2. *a*-**pseudo-prime.** an integer that satisfies the above condition for the particular integer a.

pseudo-random, *adj.* (of a sequence of numbers) generated by the software RANDOM NUMBER GENERATOR of a computer or calculator. Such numbers are not truly RANDOM since their pattern recurs cyclically and they are often required, for statistical purposes, to be reproducible.

pseudo-sphere, *n.* the surface of revolution of a TRACTRIX about its asymptote. This produces a surface, such as that shown in Fig. 301, of which the TOTAL CURVATURE at each point is constant and negative (a *pseudo-spherical surface*). Lines on this surface do not conform to the parallel postulate, so that the possibility of constructing the surface provides a relative consistency proof for non-Euclidean geometry.

Fig. 301. **Pseudo-sphere**.

psi function, *n.* another name for the DIGAMMA FUNCTION.

p-**subgroup,** *n.* a SUBGROUP that is a P-GROUP. See also SYLOW SUBGROUP.

Ptolemy's theorem, *n.* the result attributed to Claudius Ptolemy of Alexandria around 150 AD that in a CYCLIC QUADRILATERAL the product of the (lengths of the) diagonals equals the sums of the products of the pairs of opposite sides.

punctual, *adj.* consisting of or confined to a single point. See INSTANTANEOUS.

punctured neighbourhood or **deleted neighbourhood,** *n.* a NEIGHBOURHOOD of a point from which the point itself is removed, usually written with a prime to indicate deletion, as $N'(\delta, a)$ in a METRIC SPACE, or more generally $U'(a)$ in a TOPOLOGICAL SPACE.

pure geometry, *n.* another name for SYNTHETIC GEOMETRY.

purely atomic, *adj.* (of a MEASURE) such that there is a countable set in the SUPPORT of the measure with null complement. The measure is thus a countable combination of point masses, each based on an ATOM. Some authors use *purely discontinuous* in a similar sense. See ATOM.

purely discontinuous, *adj.* a less common term for PURELY ATOMIC.

purely imaginary, see IMAGINARY.

pure mathematics, *n.* the study of mathematical systems and structures in the abstract, as opposed to study of, or motivated by, their applications. The term is usually used in university syllabuses for ANALYSIS, ALGEBRA, GEOMETRY, and subjects derived from them, but the distinction from APPLIED MATHEMATICS is conventional. See MATHEMATICS.

pure set, *n.* a set that has no URELEMENTS.

pure strategy, see MIXED STRATEGY.

pure surd, see SURD.

p-value, *n.* (*Statistics*) the probability that a given TEST STATISTIC takes either the observed value or one that is less likely under the NULL HYPOTHESIS. If fixed in advance, this is the SIGNIFICANCE LEVEL of the test.

P versus NP problem, *n.* (*Computer science*) the problem of determining whether questions exist whose answer, if known, can be quickly checked (for example by computer), but which require a much longer time to solve if the answer is not known. A prosaic example is the difference between being asked at a party whether one knows an identified individual, and being asked if one knows anyone in the room; the former can be answered instaneously while the latter requires a laborious tour of the room. There certainly seem to be many such mathematical questions: for example whether a given large integer is the product of two primes, but it may just be that we have not yet devised a suitable method for solving them quickly. Stephen Cook formulated the P versus NP problem in 1971, and it has been included in the MILLENNIUM PRIZE PROBLEMS. In August 2002 Agrawal, Kayal and Saxena published the first deterministic POLYNOMIAL TIME algorithm for testing primality of an integer. It is based on a new extension of FERMAT'S LITTLE THEOREM in which the integers a^p and a are replaced by polynomials $(x - a)^p$ and $(x^p - a)$. The method is simple and while not yet practical has already been dramatically improved and highlights how much we still have to learn.

pyramid, *n.* a POLYHEDRON with one polygonal face (the *base*) and all other

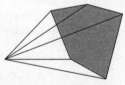

Fig. 302. **Pyramid.**

faces (*lateral faces*) triangular with a common vertex; the solid of Fig. 302 is an irregular pentagonal pyramid, of which the shaded side is the base.

Pythagoras of Samos (c. 380–300 BC), Greek philosopher, mathematician and mystic, who founded a school of thought that influenced Plato. He and his followers believed that 'All is number', recognizing the mathematical nature of music, and extending this through the 'music of the spheres' to a mystical numerical harmony of the universe.

Pythagoras' theorem, *n.* the theorem in EUCLIDEAN GEOMETRY that, in a right-angled triangle, the square of the length of the HYPOTENUSE is equal to the sum of the squares of the lengths of the other two sides; Fig. 303 represents this result geometrically. The lengths of the sides of a right-angled triangle, in integer multiples of an appropriate unit, thus form a PYTHAGOREAN TRIPLE of integers.

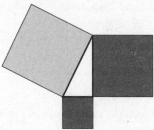

Fig. 303. **Pythagoras' theorem**. The sum of the areas of the dark squares equals that of the light square.

Pythagorean means, *n.* the ARITHMETIC, GEOMETRIC, and HARMONIC MEANS. See also NEO–PYTHAGOREAN MEANS.

Pythagorean numbers, *n.* another term for PYTHAGOREAN TRIPLE.

Pythagorean triple or **Pythagorean numbers,** *n.* any set of three integers satisfying the DIOPHANTINE EQUATION

$$x^2 + y^2 = z^2.$$

As proved by Euclid and Diophantus, these have the form

$$x = a^2 - b^2, \quad y = 2ab, \quad z = a^2 + b^2,$$

for any integral a and b. FERMAT'S LAST THEOREM asserts that there are no analogous triples for higher powers, and WARING'S PROBLEM seeks to generalize further.

q

$\mathbb{Q}$ or $\mathcal{Q}$, *symbol for* the set of RATIONAL NUMBERS. Compare $\mathbb{N}$, $\mathbb{R}$, $\mathbb{Z}$.

q-binomial, *adj.* pertaining to the *q-binomial theorem*; that is, the identity, valid for any y and $0 < q < 1$, that

$$\sum_{m=0}^{n} y^m q^{\left[\frac{m(m+1)}{2}\right]} \begin{bmatrix} n \\ m \end{bmatrix} = (1 + yq)\left(1 + yq^2\right)\dots(1 + yq^m),$$

where the *q-binomial coefficient* is defined for $0 < m < n$ by

$$\begin{bmatrix} n \\ m \end{bmatrix} = \begin{bmatrix} n \\ m \end{bmatrix}_q = \frac{(1 - q^n)(1 - q^{n-1})\dots(1 - q^{n-m+1})}{(1 - q)(1 - q^2)\dots(1 - q^m)},$$

and

$$\begin{bmatrix} n \\ 0 \end{bmatrix} = \begin{bmatrix} n \\ n \end{bmatrix} = 1.$$

The q-binomial coefficient tends to the BINOMIAL COEFFICIENT as q increases to unity, and satisfies

$$\begin{bmatrix} n \\ m \end{bmatrix} = \begin{bmatrix} n-1 \\ m \end{bmatrix} + q^{n-m}\begin{bmatrix} n-1 \\ m-1 \end{bmatrix}.$$

This is an example of a *q-expansion* or *q-series*. See also TRIPLE–PRODUCT IDENTITY OF JACOBI, ROGERS–RAMANUJAN IDENTITIES.

q-bit, alternative spelling of QUBIT.

QED, *abbrev. for* **quod erat demonstrandum**. (*Latin*) that which was to be proved. This is often written at the end of a proof to indicate that its conclusion has been reached.

QEF, *abbrev. for* **quod erat faciendum**. (*Latin*) that which was to be done. This is often written at the end of a construction to indicate that whatever was required has been achieved.

q-expansion, see Q-BINOMIAL.

QR algorithm, *n.* a stable and reasonably efficient numerical method of solving EIGENVALUE problems, which begins by finding a matrix in upper HESSENBERG FORM that is SIMILAR to the given one. One then recursively computes a sequence of orthogonal upper triangular (QR) decompositions that produces a diagonal matrix with the desired eigenvalues.

q-series, see Q-BINOMIAL.

quadrangle, *n.* (*Euclidean geometry*) **1.** a plane figure consisting of four points, each joined by at least two lines to two other points. The line segments between the vertices may intersect (so that the figure is not a POLYGON). A quadrangle is *convex* if both diagonals lie inside, RE–ENTRANT if one lies outside and *crossed* if both lie outside; these are shown in Fig. 304 with the light lines showing the diagonals. Compare QUADRILATERAL. See also DUAL.

Fig. 304. Convex, re-entrant and crossed **quadrangles**.

2. a COMPLETE QUADRANGLE, consisting of a total of six lines and all their points of intersection, including three DIAGONAL POINTS.

quadrant, *n.* **1a.** a quarter of the circumference of a circle.

b. the area within a circle enclosed between two perpendicular radii and the circumference.

2. any of the four sections into which the coordinate plane is divided by the coordinate axes, counted counter-clockwise, as in Fig. 305. Compare OCTANT, ORTHANT.

Fig. 305. The **quadrants** of the coordinate plane.

quadrate, *n.* any cube, square, cuboid or rectangle.

quadratfrei, adj. (*German*) another term for SQUARE–FREE.

quadratic, *adj.* **1a.** of the second degree.

b. (of an expression, function, or equation) containing one or more terms in which the variable is raised to the second power, but no variables are raised to a higher positive power.

2. (*as substantive*) a QUADRATIC EQUATION

quadratic congruence, *n.* a CONGRUENCE of the form

$$f(x) \equiv 0 \pmod{m},$$

where f is a QUADRATIC POLYNOMIAL with integer coefficients. Compare QUADRATIC RESIDUE, LINEAR CONGRUENCE.

quadratic convergence, see RATE OF CONVERGENCE.

quadratic equation or **quadratic,** *n.* an equation of the form

$$ax^2 + bx + c = 0,$$

of which the roots are in general given by the *quadratic formula*

$$x = \frac{-b \pm \sqrt{b^2 - 4ac}}{2a},$$

in which $b^2 - 4ac$ is the DISCRIMINANT. See also SOLUTION BY RADICALS.

quadratic form or **quadratic,** *n.* an expression of the form $\mathbf{X}^T\mathbf{A}\mathbf{X}$, where $\mathbf{A}$ is a square matrix and $\mathbf{X}$ is a column of variables. This is the generalization of expressions of the form ax^2, $ax^2 + by^2 + 2cxy$, etc. See also MULTILINEAR FUNCTION.

quadratic programming problem, *n.* (*Optimization*) a problem in MATHE-MATICAL PROGRAMMING in which the OBJECTIVE FUNCTION is a QUADRATIC and the CONSTRAINTS are LINEAR.

quadratic reciprocity or **Gaussian reciprocity,** *n.* the important number-theoretic result that if p and q are distinct odd prime numbers, p is a QUADRATIC RESIDUE of q if and only if q is a quadratic residue of p, unless both are CONGRUENT to 3 modulo 4, in which case the opposite holds; in that one case, p is a quadratic residue of q if and only if q is not a quadratic residue of p.

quadratic residue, *n.* a number CONGRUENT to a perfect square MODULO a given modulus; a is a quadratic residue (mod n) if and only if

$$x^2 \equiv a \pmod{n}$$

is soluble for integral x. When n is prime, this holds precisely when the LEGENDRE SYMBOL, $(a \mid n)$, is unity.

quadratics, *n.* the branch of algebra that studies QUADRATIC EQUATIONS.

quadrature, *n.* **1.** the construction of a square with the same area as a given figure or surface, especially a circle. See also CONSTRUCTIBLE.

2. the calculation of planar areas by repeated use of quadrature in the previous sense.

3. numerical quadrature. the evaluation of a definite integral by a formula involving weighted sums of function values at given points. The rule is said to have *order n* if it is exact for polynomials of that degree. The TRAPEZOIDAL RULE has order 1 while SIMPSON'S RULE has order 3.

quadric, *adj.* **1a.** of the second degree.

b. characterized by an algebraic equation of the second degree, usually in two or three variables.

2. (*as substantive*) a quadric curve, surface, or function.

quadrilateral, *adj.* **1.** having or formed by four sides.

2a. (*as substantive*) more rarely called **tetragon.** a four-sided polygon; a plane figure bounded by four straight line segments, each joining two vertices and not intersecting elsewhere between the vertices. A QUADRANGLE is thus not a quadrilateral unless it is a POLYGON.

b. a COMPLETE QUADRILATERAL that consists of four infinite lines, all six of their points of intersection, and three diagonals on which opposite pairs of these points lie. See also DUAL.

quadrinomial, *n.* an algebraic expression containing four terms.

quadrivium, *n.* the higher division of the ancient and medieval educational curriculum, comprising arithmetic, geometry, astronomy, and music. (from Latin, meaning "four ways", where four roads meet.) Compare TRIVIUM.

qualitative identity, see IDENTITY, RELATIVE IDENTITY.

quality control, *n.* (*Statistics*) the application of statistical sampling techniques to the maintenance of the quality of a product.

quantic, *n.* a homogeneous function of two or more variables in a form with rational and integral coefficients, especially a QUADRATIC FORM; for example, $x^2 + 3xyz + y^2$ is a ternary quadric quantic.

quantifier, *n.* (*Logic*) **1.** a symbol of PREDICATE CALCULUS that contains a variable and indicates the generality of the OPEN SENTENCE in which that variable occurs. In particular, the EXISTENTIAL QUANTIFIER is written $(\exists x)$, as in $(\exists x)Fx$, and is rendered *there exists an F, something Fs, Fs exist,* or *some-*

thing is an F. The UNIVERSAL QUANTIFIER is written (x) by logicians and $(\forall x)$ by mathematicians, as in $(x)Fx$ or $(\forall x)Fx$, and is rendered *everything Fs* or *everything is an F.* In a semantic INTERPRETATION of quantified formulae, it is necessary to specify a RANGE of quantification. For example,

$$(\forall x)(x^2 \geq 0) \quad x \in \mathbb{R}$$

is read 'for all real x, $x^2 \geq 0$'; this would be written by logicians as

$$(x)((x \in \mathbb{R}) \rightarrow (x^2 \geq 0)).$$

2. any analogous symbol in an extended logic, such as $(Mx)Fx$ for *most x are F*, or RESTRICTED QUANTIFIERS such as $(x : Fx)Gx$ for *all Fs are Gs*. See also NUMERICAL QUANTIFIER.

quantify, *vb.* **1.** to discover a numerical value or to express a quantity.

2. (*Logic*) to use a quantifier to BIND a variable.

quantity, *n.* **1.** an entity that has a magnitude or numerical value.

2. a numerical expression; a variable that ranges over numbers.

3. (*Logic*) the property of being either universal or particular.

quantize, *vb.* to restrict a quantity or variable to values that are integral multiples of some unit.

quantum computer, *n.* a radical, still largely theoretical, alternative to the VON NEUMANN ARCHECTECTURE of present DIGITAL COMPUTERS. First posited by Richard Feynman in 1982, and detailed by Deutsch (1985), it relies on exploiting quantum mechanical properties of quantization, entanglement, and especially quantum interference. *Interference* means that outcomes of quantum processes depend on all possible histories of a process, and makes quantum computers exponentially more powerful than classical ones. *Entanglement* means that spatially separated, non-interacting, systems with prior interaction may still have locally inaccessible information in common. This is what makes quantum encryption possible. The field exploded after 1994 when Peter Shor discovered a quantum algorithm for efficient FACTORIZATION of very large numbers. Such methods would render current CODING THEORY and encryption techniques, such as RSA, ineffective.

quartic, *adj.* **1a.** polynomial of the fourth degree.

b. characterized by an algebraic equation of the fourth degree, usually in two or three variables.

See also BIQUADRATIC, BIQUADRATE.

2. (*as substantive*) a quartic curve, surface, or function. See CARDANO'S FORMULA.

quartile, *n.* (*Statistics*) any of the three values of a variable that divide its distribution into four intervals with equal probability; the 25th, 50th, or 75th PERCENTILES. The 25th and 75th percentiles are respectively the *lower quartile* and the *upper quartile*. See INTER–QUARTILE RANGE.

quasi-, *adj.* sharing some important property with a given class of objects, such as *quasi-analytic* functions, and *quasi-linear equations*.

quasi-concave, *adj.* (of a function) having CONVEX upper LEVEL SETS; that is, all sets of the form $\{x : f(x) \geq r\}$ are convex for all r. Compare CONCAVE, QUASI-CONVEX.

quasi-convex, *adj.* (of a function) having CONVEX lower LEVEL SETS; that is, all sets of the form $\{x : f(x) \leq r\}$ are convex for all r, as holds for any convex function. Compare QUASI–CONCAVE.

quasi-group, *n.* a GROUPOID in which every element has both a unique left INVERSE and a unique right inverse, which need not be equal unless the associative law holds. If a quasi-group is commutative, each element has at least one inverse, but it need not be unique.

quasi-linear, *adj.* (of an nth order ORDINARY DIFFERENTIAL EQUATION) such that it can be written in the form

$$y^{(n)} = f(x, y', \ldots, y^{(n-1)}).$$

Quasi-linear ordinary differential equations of any order can be written as a system of first-order ordinary differential equations.

quasi-linear equation, see DIFFERENTIAL EQUATION.

quasi-metric, see METRIC.

quasi-Newton method, *n.* (*Numerical analysis*) any of a class of methods for solving non-linear equations or UNCONSTRAINED OPTIMIZATION problems, based on NEWTON'S METHOD but using rough estimates of the HESSIANS or JACOBIANS involved in order to reduce computational cost. These estimates are often recursively estimated and are refined during the calculation. In *quasi-Newton minimization* procedures, the approximate Hessian H is usually given a RANK one or rank two update, and a LINE SEARCH METHOD is then used in the direction of $-H\nabla f(x)$ to produce the next estimate.

quasi-ordering, see ORDERING.

quasi-quotation, *n.* a device, written $\ulcorner \ldots \urcorner$, whereby it is possible to refer to generalized expressions, some parts of which are variables and others are operators that appear in their own right. For example, it is not possible to state the truth conditions for a conjunction in the form

'P and Q' if and only if P is true and Q is true,

in a way that is consistent with the fundamental principle, with which any notation must comply, that variables be uniformly replaced by their instances. Here, if the variables stand for unquoted sentences, the second clause is ill-formed, while if they represent quoted sentences, the conjunction is ill-formed. Instead we require that the variables represent quoted expressions, but CONCATENATE P, the symbol '&', and Q, in such a way that we delete the final quotation mark in the replacement of the first variable together with that preceding the ampersand, and the initial quotation mark in the replacement of the second variable together with that following the ampersand. The quasi-quotation $\ulcorner P \& Q \urcorner$ is then defined to be this concatenation,

$$P \wedge \text{'\&'} \wedge Q,$$

and is the required quotation of the conjunction.

quasi-tautology, *n.* (*Logic*) a TAUTOLOGICAL CONSEQUENCE of instances of identity and equality axioms in a THEORY.

quaternary, *adj.* having four variables.

quaternion, n. a generalized complex number, x, consisting of four components such that

$$x = x_0 + x_1\mathbf{i} + x_2\mathbf{j} + x_3\mathbf{k},$$

where the x_i are real numbers and $\mathbf{i}$, $\mathbf{j}$, $\mathbf{k}$ are HYPERCOMPLEX NUMBERS for which

$$\mathbf{i}^2 = \mathbf{j}^2 = \mathbf{k}^2 = \mathbf{ijk} = -1.$$

The set of all quaternions is denoted by $\mathbb{H}$, and forms a DIVISION RING (a SKEW FIELD). The *quaternion group* is the set

$$\{\pm 1, \pm\mathbf{i}, \pm\mathbf{j}, \pm\mathbf{k}\}$$

with respect to the multiplication of the unit quaternions.

quatrefoil, see MULTIFOIL.

qubit or **q-bit,** n. the QUANTUM COMPUTER analogue of a computer BIT. It models the fact that atoms can be prepared in a SUPERPOSITION of two different electronic states. Likewise, a quantum two-state system – a quantum bit or qubit – can be prepared in a superposition of its two logical states 0 and 1. Thus, one qubit can simultaneously be both 0 and 1.

queue, see QUEUEING THEORY.

queueing theory, n. the mathematical study of waiting lines or *queues*. A typical model considers customers entering a queue, often when the time between successive customers is exponentially distributed, in which case the number of customers per unit time is distributed as a POISSON DISTRIBUTION. The queue has a finite or infinite bound, and there is some mechanism for serving customers (often first come, first served) in a number of parallel service channels. The service time for each customer is also described probabilistically, often by a GAMMA DISTRIBUTION. The theory then attempts to determine the expected queue lengths, waiting times, etc. in a steady state of the system.

quinary, *adj.* pertaining to the PLACE VALUE NOTATION with BASE five.

quindecagon, n. a polygon with 15 sides.

quintic, *adj.* of or relating to the fifth degree. A *quintic equation* is not in general amenable to SOLUTION BY RADICALS.

quod erat demonstrandum, see QED.

quod erat faciendum, see QEF.

quotient, n. **1a.** the result when one number, quantity, or expression (the DIVIDEND) is DIVIDED by another (the DIVISOR).

b. the RATIO of two numbers or quantities.

2. the INTEGRAL PART of such a result. See also REMAINDER.

3. (*as modifier*) (of an algebraic structure) see FACTOR SPACE.

quotient group, n. another name for FACTOR GROUP.

quotient ring, n. **1.** another term for FACTOR RING.

2. an OVER–RING, Q, of a given ring, R, with some REGULAR elements, that consists of elements of the form ab^{-1} (in a *right quotient ring*) or $b^{-1}a$ (in a *left quotient ring*) where a is any element and b is a regular element of R, and the inverses of all regular elements of R are in Q. Any semi-prime right Noetherian ring has a semi-simple right quotient ring, and any integral domain has a quotient ring that is a field, the FIELD OF FRACTIONS.

quotient rule, *n.* (*Calculus*) the rule for differentiating the QUOTIENT of two DIFFERENTIABLE functions:

$$\frac{d\left(\frac{f}{g}\right)}{dx} = \frac{g\frac{df}{dx} - f\frac{dg}{dx}}{g^2}.$$

Compare PRODUCT RULE.

quotient space, *n.* another name for FACTOR SPACE.

quotient topology, *n.* (on a set, Y, with respect to a topological space, X, and a mapping, *f*, from X onto Y) the finest TOPOLOGY with respect to which *f* is CONTINUOUS.

r, 1. *abbrev. for* RADIUS.

2. (*Number theory*) **a.** the function, $r(n)$, that counts the number of representations of a natural number, n, as a sum of two integral squares, counting representations as different even when they only differ in order or sign. This is equal to $4[d_1(n) - d_3(n)]$, where $d_i(n)$ is the sum of the divisors of n that are congruent to i modulo 4, and is MULTIPLICATIVE. For example, $r(5) = 8$; in general $r(p) = 8$ for any prime that is congruent to 1 modulo 4, a result first proved by Fermat. See DIVISOR FUNCTION.
b. the function, $r_k(n)$, that counts the number of representations of a natural number, n, as a sum of k integral squares, counting representations as different even when they only differ in order or sign. LAGRANGE'S THEOREM shows that $r_4(n)$ is never zero.

R, *symbol for* an unspecified RELATION, usually written either in PREFIX NOTATION as Rxy or (for binary relations) in INFIX NOTATION as xRy, meaning x has relation R to y. Sometimes superscripts are used to indicate the number of ARGUMENTS; in this notation, $R_i^n(x_1,\ldots, x_n)$ is an n-place relation, R_i, and $R^n(x_1,\ldots, x_m)$ is not WELL-FORMED unless $n = m$.

$\mathbb{R}$ or $\mathfrak{R}$, *symbol for* the set of REAL NUMBERS. Compare $\mathbb{N}$, $\mathbb{Q}$, $\mathbb{Z}$.

Raabe's test, *n.* a test for the CONVERGENCE of a series: if $a_n > 0$, and

$$\frac{a_{n+1}}{a_n} < 1 - \frac{A}{n}$$

for some constant A greater than 1, and sufficiently large n, then $\sum a_n$ converges; for example, the HYPERGEOMETRIC SERIES converges by this test.

racecourse paradox, *n.* another name for the ACHILLES PARADOX of Zeno.

rad, *abbrev. for* RADIAN, RADIUS, RADIX.

Rademacher functions, *n.* the family of functions $\{f_n\}$ defined on the unit interval by the formula

$$f_x(x) = (-1)^{i+1} \quad \text{if} \quad \frac{i-1}{2^n} \le x < \frac{i}{2^n},$$

for each positive integer n. Alternatively, $f_n(x)$ can be defined to be the sign of $\sin(2^n \pi x)$, for a non-negative integer n. (Named after the German-born analyst and number theorist, *Hans Adolph Rademacher* (1892–1969), who was persecuted by the Nazis and took refuge in the USA.)

Rademacher theorem, see LIPSCHITZ FUNCTION.

radial, *adj.* (of a line) extending from a point, in some some way analogously with a radius of a circle. For example, *radially related* figures are SIMILAR, being images of one another under a HOMOTHETY.

radial component, *n.* (*Mechanics*) the COMPONENT of a VECTOR quantity in the direction of the RADIUS VECTOR. In POLAR COORDINATES, the radial component of VELOCITY is dr/dt, and the radial component of ACCELERATION is

$$\frac{\partial^2 r}{\partial t^2} - r\left(\frac{\partial\theta}{\partial t}\right)^2 .$$

radian (abbrev. **rad**), *n.* a unit of measurement of angles; the angle between two radii that cut off an arc on the circumference of a circle equal in length to the radius. One radian is approximately 57.296 DEGREES, and $\pi/2$ radians equals a right angle; Fig. 306 shows an angle of one radian at the centre of a circle.

Fig. 306. The marked angle is one **radian**.

radical, *n.* **1.** (*Arithmetic*) another word for ROOT (sense 2). See also RADICAL SIGN, SOLUTION BY RADICALS.

2. Jacobson radical. (*Algebra*) **a.** (of a left module). the intersection of all maximal (proper) submodules of the TORSION MODULE.

b. (of a ring) the intersection of all maximal left IDEALS, or equivalently of all maximal right ideals; this is a two-sided ideal. In a commutative BANACH ALGEBRA, it is the intersection of all maximal ideals.

3. Noetherian radical. (*Algebra*) (of a ring) the union of all two-sided nilpotent IDEALS (for which $J^n = 0$ for some natural number n). In a left NOETHERIAN RING, this is the largest two-sided ideal all of the elements of which are nilpotent. In a left ARTINIAN RING, the Noetherian and Jacobson radicals coincide.

4. (in a commutative BANACH ALGEBRA) the set of GENERALIZED NILPOTENT elements.

5. *adj.* of, constituting, or pertaining to a RADIX.

radical axis, *n.* the locus of points of equal POWER with respect to two circles; when the circles intersect, it is the line joining the points of intersection of the circles. The radical axes are the bold lines in Fig. 307 below.

radical centre, *n.* (of three circles) the point of intersection of the respective RADICAL AXES of each pair of circles selected from the given three, as in Fig. 307. The point is finite when the three circles have non–collinear centres.

Fig. 307. The bold lines intersect at the **radical centre**.

radical fraction, *n.* another term for RADIX FRACTION.

radical sign, *n.* the symbol, '$\sqrt{\ }$', placed before a number or expression to indicate that its ROOT is to be extracted; it is often extended over a complex expression, as in

$$\sqrt{x^2 y} = x\sqrt{y} .$$

The ORDER of the root is indicated by a superscript figure (the *index*) before or above the sign; for example, $\sqrt[3]{5}$ is the third root, or *cube root*, of 5. When no such figure appears, the sign denotes the square root.

radicand, *n.* a number or quantity from which a ROOT is to be extracted, usually preceded by the RADICAL SIGN.

radices, *n. plural of* RADIX.

radius (abbrev. **rad**), *n.* **1a.** a straight line joining the centre of a circle or sphere to any point on its circumference.

b. the length of this line, usually denoted *r*.

2. either a LONG RADIUS or a SHORT RADIUS of a regular polygon.

radius of convergence, *n.* the RADIUS of the largest circle (or interval in the real case) around a given point such that a given POWER SERIES converges (absolutely) at all points strictly inside the circle. The series diverges at all points strictly outside the circle, and may do either at points on the circle. See CIRCLE OF CONVERGENCE. See also ROOT TEST.

radius of curvature, *n.* the absolute value of the reciprocal of the CURVATURE of a curve at a given point, taken as the rate of change of the tangent to the curve with respect to arc length. For $y = f(x)$, it is

$$\frac{d\left(\frac{df}{dy}\right)}{ds}$$

In particular, the radius of curvature of a plane curve at a point is the radius of a circle with curvature equal to that of the given curve at that point; Fig. 308 shows the radius and circle of curvature for the curve $y = (x-1)^2$. See CENTRE OF CURVATURE.

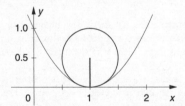

Fig. 308. **Radius of curvature.**
See main entry.

radius of gyration, *n.* (*Mechanics*) the distance from a given axis that a PARTICLE of the same MASS as a RIGID BODY must be placed in order to have the same MOMENT OF INERTIA; the square root of the quotient of the moment of inertia of the rigid body about the axis divided by the mass of the body.

radius of normal curvature, *n.* (of a surface in a given direction) the absolute value of the reciprocal of the CURVATURE of the normal SECTION of the surface at the point in the given direction.

radius of torsion, *n.* the reciprocal of the TORSION.

radius vector, *n.* a DIRECTED LINE from the origin of a coordinate system to a point in space, regarded as variable and sweeping out a curve. If it is the vector $\overrightarrow{OP}$, where the coordinates of P are $(x, f(x))$, as in Fig. 309 overleaf, it can also be represented as $\langle x, f(x) \rangle$, that is, as the *n*-tuple of its components in the directions of, or projections onto, the coordinate axes. See also POSITION VECTOR.

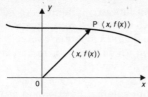

Fig. 309. $\overrightarrow{OP}$ is the **radius vector** of the curve described by P.

radix (abbrev. **rad**), *n.* another word for the BASE of a PLACE-VALUE NOTATION or system of LOGARITHMS.

radix fraction or **radical fraction,** *n.* the generalization of DECIMAL FRACTIONS to other BASES of PLACE-VALUE NOTATION.

radix point, *n.* the generalization of the DECIMAL POINT to other BASES of PLACE-VALUE NOTATION.

Radon measure, *n.* another name for RIEMANN-STIELTJES MEASURE.

Radon–Nikodym derivative, *n.* the function *f*, denoted $d\lambda/d\mu$, that possesses a LEBESGUE INTEGRAL and is unique up to measure μ, such that for each measurable set E

$$\lambda(E) = \int_E f d\mu,$$

where λ and μ are MEASURES that satisfy the conditions of the RADON-NIKODYM THEOREM.

Radon–Nikodym theorem, *n.* the theorem that, given a SIGMA-FINITE MEASURE μ and a signed measure λ that is ABSOLUTELY CONTINUOUS with respect to μ, there is a function *f* that possesses a LEBESGUE INTEGRAL, such that, for each measurable set E,

$$\lambda(E) = \int_E f d\mu,$$

that is unique (up to a set with μ-measure zero); this function is called the *Radon–Nikodym derivative* of λ with respect to μ, and is denoted $d\lambda/d\mu$. These notions are used in the general definitions of CONDITIONAL EXPECTATION and CONDITIONAL PROBABILITY.

Radon's theorem, *n.* the theorem that any $n + 2$ points in an *n*-dimensional vector space can be partitioned into two non-empty sets whose CONVEX HULLS are disjoint. This is equivalent to HELLY'S THEOREM. (Named after the Austrian algebraist, analyst, and geometer, *Johann Karl August Radon* (1887–1956).)

raise (to a power), *vb.* to multiply (a number, expression, or quantity) by itself a specified number of times; for example, 2 raised to the third power is 8.

Ramanujan, Srinivasa (1887–1920), Indian mathematician who was largely self-taught; he contributed greatly to the theory of numbers and the theory of functions using mainly intuitive methods, and is famed for his skills in manipulation of series. He lost a scholarship to Madras University by concentrating on mathematics to the exclusion of other subjects, and for some time existed on private charity before obtaining a clerical post. After corresponding with G.H. Hardy, he obtained a research scholarship at Cambridge in 1914, and in 1919 was the first Indian to be elected a Fellow of the Royal Society, before returning to India much weakened by a mysterious disease then thought to be tuberculosis. Ramanujan is often regarded as one of the romantic figures of mathematical history, and his work is only now being fully digested.

ramified theory of types, see TYPE.

Ramsey numbers, see RAMSEY THEOREM.

Ramsey theorem, n. the theorem that, for any two positive integers k and l, there is an integer $R(k, l)$ such that given any simple GRAPH with $R(k, l)$ vertices, either the graph contains a CLIQUE of k vertices (with all vertices adjacent) or an independent set with l vertices (with no vertices adjacent). These integers $R(k, l)$ are called *Ramsey numbers*. Determination of these numbers is hard, but it is known that

$$R(k, k) \geq 2^{k/2},$$

that

$$R(k, l) \leq \binom{k + l - 2}{k - 1},$$

that $R(4, 4) = 18$, and that $R(3, 3) = 6$. Hence out of six people there must be either three mutual friends or three mutual strangers.

random, *adj.* (*Statistics*) **1.** having a value that cannot be determined before the value is taken, but only described probabilistically, such as a RANDOM VARIABLE.

2. chosen without regard to any characteristics of the individual members of the population, so that each has an equal chance of being selected. See RANDOM SAMPLE.

See also RANDOM NUMBER, PSEUDO–RANDOM.

random error or **noise,** n. the difference between predicted and observed values, regarded as a RANDOM VARIABLE.

randomize, *vb.* to select or order data, individuals, etc., in a deliberately RANDOM way, usually in order to enhance the reliability of any statistical results obtained.

random number generator, n. a part of the software of most computers and many calculators, and used, for example, in statistical testing, to produce a sequence of apparently RANDOM NUMBERS. However, since computer simulation of random numbers requires that they be reproducable for purposes of comparison and checking, it does not produce truly random numbers, but only a PSEUDO–RANDOM sequence, often by taking remainders modulo a large prime (*congruential methods*), or by periodically sampling some physical quantity such as a voltage. In the *mixed congruential method*, each number n_{i+1} is generated from the preceding by the formula

$$n_{i+1} = [an_i + c] \pmod{m}$$

where a and c are non-negative integers, m is a large positive integer relative to the word size of the specific computer, and there is an initial SEED, n_0.

random numbers, n. a sequence of numbers with the property that no member can be predicted from the preceding elements; in particular, these cannot form a progression or follow any regular or repetitive pattern. Tables of random numbers are used in statistics to facilitate unbiased sampling of a population, and may be selected to approximate to a given distribution, often UNIFORM; in this case, for example, given the nth member of the sequence, all numbers in the relevant range have an equal probability of appearing as the $(n+1)$th member. See also RANDOM NUMBER GENERATOR. Compare PSEUDO–RANDOM.

random sample, *n.* a SAMPLE devised to avoid any interference from any shared property of, or relation between, the elements selected, so that its distribution is affected only by that of the whole population, and can therefore be taken to be representative of it.

random variable (abbrev. **rv**), **stochastic variable,** or **chance variable,** *n.* (*Statistics*) **1.** a quantity that may take any of a range of values (either continuous or discrete) that cannot be predicted with certainty but only described probabilistically. The RELATIVE FREQUENCY with which the random variable takes a given value or lies in a given interval is the (empirical) PROBABILITY of that value or interval.

2. formally, a measurable function defined on a PROBABILITY SPACE and having range in the interval [0, 1].

random vector, *n.* an n-TUPLE of RANDOM VARIABLES, often representing successive outcomes of a repeated experiment.

random walk, *n.* (*Statistics*) a route consisting of successive and unconnected steps in which each step is chosen by a random mechanism uninfluenced by any previous step. See GAMBLER'S RUIN.

range, *n.* **1.** the set of values that a given function takes as its argument varies through its DOMAIN; the IMAGE of the domain. This must be a subset of, but may or may not be identical with the CODOMAIN.

2. the set of values that a given variable can take in an equation, identity, etc., that is, for which the expression is defined.

3. (of points) all points on a given line.

4. the set of objects, values, etc. that are relevant to the truth conditions of a CLOSED SENTENCE. See QUANTIFIER, INTERPRETATION.

5. (*Statistics*) a measure of dispersion obtained as the absolute difference between the smallest and largest values of a variable in a sample.

range of significance or **domain of definition,** *n.* the set of subjects for which a given predicate is intelligible.

rank, *n.* **1a.** the largest number of LINEARLY INDEPENDENT rows or columns of a given matrix; the number of rows or columns in the highest order non-zero MINOR of the matrix.

b. the rank of the matrix associated with a given QUADRATIC FORM.

2. the number of elements in the basis of a FREE MODULE.

– *vb.* **3.** (*Statistics*) to array a set of objects as a sequence, especially in terms of the natural arithmetic ordering of some measure of their elements. For example, one may rank students by their test scores, by their dates of birth, or by their heights. See WILCOXON TEST, MANN–WHITNEY TEST, NONPARAMETRIC TEST.

rank correlation coefficient, see SPEARMAN'S RANK ORDER COEFFICIENT.

Rao Blackwell theorem, *n.* (*Statistics*) the theorem that if T(X) is a COMPLETE SUFFICIENT STATISTIC for a PARAMETER θ, and W(X) is an UNBIASED

ESTIMATE for $\phi(\theta)$, then $E[W \mid T]$ is the minimum variance unbiased estimate for $\phi(\theta)$.

rate of change, *n.* **1.** the ratio of the difference between values of a variable quantity at different times to the difference between the times; the change per unit time, that is,

$$\frac{f(t) - f(t_0)}{t - t_0}.$$

If the quantity is continuously variable and this ratio tends to a limit as both endpoints of the time interval tend to the same point, the *instantaneous rate of change* is the DERIVATIVE of the function with respect to time, at the point that represents that quantity.

2. more generally, any derivative: dy/dx is the rate of change of y with respect to x.

rate of convergence or **order of convergence,** *n.* any estimate of the speed with which a given sequence or iteration approaches its limit, often measured by the number of terms or evaluations involved in obtaining a given accuracy. Convergence of a sequence subject to the condition, for $p > 1$, that

$$\frac{\mid x_{n+1} - x \mid}{\mid x_n - x \mid^p} = O(1)$$

as n increases, is called p^{th}-order convergence; for example, *quadratic convergence* when $p = 2$. One similarly speaks of logarithmic convergence or exponential convergence. See also LINEAR CONVERGENCE.

ratio, *n.* a QUOTIENT or PROPORTION of two numbers, magnitudes, quantities or expressions, such as a measure of the relative size of two classes; for example, the ratio of the side of a square to the diagonal is $1:\sqrt{2}$.

ratio of similitude, see SIMILITUDE.

rational, *adj.* **1.** (*Arithmetic*) expressible as a RATIO of two integers.

2a. involving or consisting of RATIONAL NUMBERS.

b. (*as substantive*) a rational number.

3. (of an expression, equation, function, etc.) containing no variable either in an irreducible RADICAL form or raised to a fractional power; expressible as a ratio of polynomials. See also INTEGRAL RATIONAL.

rational element, *n.* a DEDEKIND CUT corresponding to a RATIONAL NUMBER in Dedekind's construction of the REAL NUMBERS.

rational form, *n.* a square matrix in which the only non-zero entries are in SUBMATRICES arranged down the MAIN DIAGONAL, that are COMPANION MATRICES of non-scalar monic polynomials such that each polynomial divides the next.

rational function, *n.* a POLYNOMIAL or ratio of polynomials.

rationalize, *vb.* to eliminate RADICALS from an expression or equation without changing its value or roots respectively. For example, the fraction

$$\frac{1}{\sqrt{a} - \sqrt{b}}$$

may have its denominator rationalized by multiplying top and bottom by $\sqrt{a} + \sqrt{b}$, producing

$$\frac{\sqrt{a} + \sqrt{b}}{a^2 + b^2}.$$

An INTEGRAL is said to be rationalized by a substitution that makes the INTE-GRAND rational.

rational number or **rational,** *n.* any number that can be expressed as a ratio, a/b, where a and b are integers of which the latter may not be zero; for example, $\frac{7}{3}$ or $-\frac{14}{35}$ (which is the same rational number as $-\frac{2}{5}$). The INTEGERS are not technically a subset of the rationals, but the rationals, generally denoted $\mathbb{Q}$, are formally equivalence classes of ordered pairs of integers, any two pairs, $\langle a, b \rangle$ and $\langle c, d \rangle$ being equivalent if $ad = bc$. Unlike the integers, the rationals are closed under division, but the integers are ISOMORPHIC to, and may be identified with, RATIONALS of the form $\langle \alpha, 1 \rangle$, that is $\alpha/1$. The rationals are DENUMERABLE, but are DENSE in the reals, and are first category with measure zero. See also REAL NUMBER.

rational polynomial, *n.* any ratio of two POLYNOMIALS. See also POLYNOMIAL RING.

rational root theorem, *n.* the number-theoretic result that if a RATIONAL number p/q, with p and q relatively prime, is the root of an integral polynomial equation, then p divides the coefficient of the constant term and q divides the coefficient of the leading term. In particular, if the polynomial is MONIC, the number is integral.

ratio scale, *n.* (*Statistics*) a scale of measurement of data that has a fixed zero value and permits the comparison of differences. For example, although time itself cannot be measured on a ratio scale, differences in time can be, since it makes sense to talk of one pair of events being twice as far apart as another. Compare ORDINAL SCALE, INTERVAL SCALE, NOMINAL SCALE.

ratio test, *n.* **1.** also called **Cauchy's ratio test.** the test for whether a complex series $\sum a_n$ is ABSOLUTELY CONVERGENT by testing whether the limit of the ratio of the absolute values of successive terms,

$$\lim_{n \to \infty} \frac{|a_{n+1}|}{|a_n|}$$

is less than or greater than unity, in which cases the series respectively converges absolutely or diverges. This test is strictly weaker than the ROOT TEST. **2. generalized ratio test** or **D'Alembert's ratio test.** the test for the convergence of an infinite series by testing whether there is a $k < 1$ such that the ratio of the absolute values of a term to its predecessor is less than k for all terms after some term; if this ratio is always greater than unity, then the series diverges. This allows one to replace the limit in the preceding entry by an appropriate limit inferior or limit superior.

ratio theorem, *n.* another term for SECTION FORMULA.

ray, *n.* **1.** (*Euclidean geometry*) a straight line extending from a point; a half-line, especially one of a PENCIL of lines radiating from a single point as in Fig. 310.

Fig. 310. A pencil of **rays**.

2. (*Algebraic geometry*) any one-dimensional subspace of a finite-dimensional vector space over the given field.

Rayleigh quotient, *n.* the ratio

$$r_A(x) = \frac{\langle x, Ax \rangle}{\langle x, x \rangle}$$

of which, for a positive definite matrix A, the supremum produces the maximum eigenvalue and the infimum produces the minimum eigenvalue. See NUMERICAL RANGE.

Rayleigh–Ritz method, *n.* a method of obtaining approximate solutions to functional equations or variational problems, by replacing the functions by appropriate finite combinations of basic elements, and finding the minimum solution within this class.

reachable set or **reachable points,** *n.* **1.** the set of state values that can be attained at a given time in a problem in CONTROL THEORY.

2. the set of vertices that can be connected to a given VERTEX in a DIGRAPH.

ready reckoner, *n.* a set of TABLES, especially one for applying rates of interest, discount, etc. to different sums.

real, *adj.* **1.** involving or consisting of REAL NUMBERS alone; having no IMAGINARY PART, or equivalently, having imaginary part equal to zero.

2. (*as substantive*) a REAL NUMBER.

real analysis, *n.* that part of modern mathematics that has its roots in the study of functions of a REAL VARIABLE, including MEASURE and INTEGRATION, some parts of TOPOLOGY (sense 1), and the elementary theory of NORMED SPACES. The term is particularly used in contrast with COMPLEX ANALYSIS.

real analytic, *adj.* (of a real function) possessing derivatives of all orders and agreeing with its TAYLOR SERIES locally. See ANALYTIC (sense 1).

real axis, *n.* the *x*-axis of an ARGAND DIAGRAM, along which the real part of the complex number to be represented is measured.

real function, *n.* a FUNCTION whose DOMAIN and CODOMAIN are sets of REAL NUMBERS. Compare COMPLEX FUNCTION.

realism, *n.* **1.** the philosophical doctrine that words refer to entities that exist in reality, rather than merely being the signs of concepts or sets of instances, whence in particular, that mathematical entities have a real existence independent of our conception of them and of physical representations or instances. See PLATONISM.

2. the philosophical doctrine that the truth or falsity of statements or propositions depends upon some fact of the matter independent of human processes of knowledge acquisition, and so that mathematical truth is not constructed by a proof, but discovered by it.

3. the philosophical theory, relevant to the interpretation of mathematical statements, that the sense of an expression is given by a specification of its truth conditions, or that there is a reality, independent of the speaker's conception of it, that determines the truth or falsehood of every statement. Compare INTUITIONISM.

See FOUNDATIONS OF MATHEMATICS.

real line, *n.* the REAL NUMBERS regarded as the points on a line; the NUMBER LINE or CONTINUUM.

real number or **real,** *n.* any RATIONAL or IRRATIONAL NUMBER. The reals, de-noted ℝ, are defined in terms of CAUCHY SEQUENCES of, or DEDEKIND CUTS on, the rationals. Technically, the rationals are not a subset of the reals but they are ISOMORPHIC to the subset of sequences that contain their SUPREMUM and are usually identified with those sequences. Whereas the rationals are DENUMERABLE, the reals are not; there are thus strictly more reals than there are integers or rationals. The reals are DENSE and form a COMPACT CONNECTED set (a continuum); they are often referred to as *the continuum.* The sum of a real and an IMAGINARY NUMBER is a COMPLEX number. See also CANTOR'S DIAGONAL THEOREM, CONTINUUM HYPOTHESIS.

real part, *n.* the term of a COMPLEX number, function, etc., that is not a multiple of i (the square root of -1). If $z = a + ib$, where a and b are REAL NUMBERS, then the real part of z, denoted $\operatorname{Re} z$ or $\operatorname{re} z$, is a.

real-symmetric, *adj.* (of a real matrix) equal to its TRANSPOSE. See ADJOINT.

real-valued, *adj.* (of a function) taking only REAL values, such as the numeri-cal TRACE of a self-adjoint matrix.

real variable, *n.* a variable that ranges over a set of REAL NUMBERS, often an OPEN INTERVAL.

rearrangement, *n.* a series of which the terms are a permutation of the terms of a given series. A result due to Riemann shows that a CONDITIONALLY CONVERGENT real series may be rearranged so that any EXTENDED REAL NUM-BER can be obtained as the limit of the rearranged series. For example,

$$1 - \tfrac{1}{2} + \tfrac{1}{3} - \tfrac{1}{4} + \tfrac{1}{5} - \tfrac{1}{6} + \ldots \;=\; \ln 2,$$

and

$$1 + \tfrac{1}{3} - \tfrac{1}{2} + \tfrac{1}{5} + \tfrac{1}{7} - \tfrac{1}{4} + \ldots \;=\; \ln\left(2\sqrt{2}\right),$$

while

$$1 - \tfrac{1}{2} + \tfrac{1}{3} + \tfrac{1}{5} - \tfrac{1}{4} + \tfrac{1}{7} + \tfrac{1}{9} + \tfrac{1}{11} - \tfrac{1}{8} + \ldots$$

is divergent. *Steinitz' theorem* shows that the set of limits of the convergent rearrangements of a series in Euclidean space always forms an AFFINE MANI-FOLD, which is a singleton if and only if the series converges absolutely.

reciprocal, *adj.* **1a.** of or relating to a multiplicative inverse; for example, x^n and x^{-n} are reciprocal functions.

b. forming a multiplicative inverse. For example, the *reciprocal function,* $y = 1/x$ takes as its value for any argument the element whose product with the given argument is unity.

2. of or relating to the result of dividing 1 by a given number or quantity: for example, the reciprocal fraction of a/b is b/a.

3. (*as substantive*) **a.** an expression of the form $1/x$.

b. any function, expression, number or quantity that is reciprocal to an-other; for example, the reciprocal of a/b is b/a.

reciprocal polar curve, *n.* a pair of curves such that the POLAR of each point of one is tangential to the other. See POLE AND POLAR.

reciprocal polar figure, *n.* the figure related to a given configuration of points and lines in the plane in such a way that each point of either figure is the POLE of a line in the other (or equivalently, each line of either figure is the POLAR of a point in the other) with respect to some given conic; for example, a pair of RECIPROCAL POLAR CURVES.

reciprocal variation, *n.* another term for INVERSE PROPORTION.

reciprocation, *n.* (*Geometry*) the transformation of a configuration of points and lines into its RECIPROCAL POLAR FIGURE.

reciprocity law, *n.* the law of QUADRATIC RECIPROCITY.

rectangle or **oblong,** *n.* a PARALLELOGRAM with four right angles. An EQUI-LATERAL rectangle is a SQUARE.

rectangular, *adj.* **1.** shaped like a RECTANGLE; having right angles.
2. mutually perpendicular.

rectangular coordinates, see CARTESIAN COORDINATES.

rectangular hyperbola, *n.* a HYPERBOLA of which the asymptotes are perpendicular; if $xy = c^2$, then the asymptotes are the coordinate axes, as in Fig. 311.

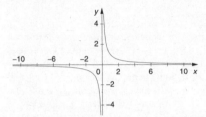

Fig. 311. **Rectangular hyperbola**.

rectangular number, *n.* any number that is not PRIME, and so is expressible as $a \times b$, for *a* and *b* greater than 1; hence, equal to the number of unit-spaced points in a RECTANGULAR array, since the total number of points in such an array equals the product of the numbers of points in the sides. If these factors are equal, the number is a *square number*. See also FIGURATE NUMBER.

rectifiable, *adj.* (of a curve) possessing a well-defined ARC LENGTH; that is, if $a_1,\ldots, a_n$ are a set of points in order along the curve, the sum of the lengths of the chords from each a_i to a_{i+1} tends to a limit as *n* tends to infinity.

rectify, *vb.* to determine the LENGTH of a curve.

rectilinear, *adj.* consisting of, bounded by, or formed of straight lines.

recur, *vb.* (of a digit or sequence of digits) to be repeated an infinite number of times at the end of the decimal expression of a fraction. For example, 3 recurs in the expansion of 241/300 as 0.83333...; this is read as 'point eight three–recurring' or 'point eight three–repeating', and is written 0.8$\dot{3}$. See RECURRING DECIMAL. Compare TERMINATE.

recurrence relation or **difference equation,** *n.* an equation of the form

$$x_{n+p} = f(n, x_n,\ldots,x_{n+p-1}),$$

that gives a RECURSIVE DEFINITION for the entire sequence, given *p* initial values, where *p* is the *order* of the recurrence relation. Often there is no explicit dependence of *f* on *n*. For example, the *n*th FIBONACCI NUMBER satisfies the recurrence relation

$$a_0 = 0, \quad a_1 = 1, \quad a_n = a_{n-1} + a_{n-2}.$$

recurring decimal, **repeating decimal** or **circulating decimal**, *n.* a RATIONAL NUMBER whose representation as a DECIMAL (or RADICAL) FRACTION contains a pattern of digits that repeats indefinitely after the decimal point. If the sequence of digits $a_1 a_2 \ldots a_n$ repeats, then the rational number represented by the recurring decimal is usually written

$$0 \cdot b_1 \, b_2 \ldots b_m \, \dot{a}_1 \, \dot{a}_2 \ldots \dot{a}_n$$

(where $b_1 \ldots b_m$ are digits that do not recur);

$$0 \cdot \dot{a}_1 \dot{a}_2 \ldots \dot{a}_n \ = \ \frac{a_1 \, a_2 \ldots a_n}{10^n - 1}$$

so that, for example,

$$0 \cdot 1\dot{4}2\dot{8}5\dot{7} \ = \ \frac{142857}{999999} \ = \ \frac{1}{7} \ ;$$

and

$$0 \cdot 1\dot{3} \ = \ 10^{-1}(1 + \tfrac{3}{9}) \ = \ \frac{4}{30}.$$

Compare TERMINATE.

recursion, *n.* the application of a function to its own values to generate an infinite sequence of values. The term may refer either to a RECURSION FORMULA, or else to RECURSIVE DEFINITION.

recursion formula, **recursion clause**, or **inductive step,** *n.* the part of a RECURSIVE DEFINITION (or ALGORITHM) that specifies the step by which each element of the sequence is generated from the preceding; for example, given the BASE CLAUSE $f(0) = 5$, the schema

$$f(n+1) \ = \ f(n) + 3$$

specifies the successive terms of the arithmetic progression 5, 8, 11, 14, ...

recursive or **iterative,** *adj.* **1.** of, involving, or expressed in terms of RECURSION. **2.** (of a function, algorithm, etc.) able to be specified in terms of a RECURSIVE DEFINITION. See CHURCH'S THESIS.

recursive definition or **inductive definition,** *n.* the definition of a sequence by specifying its first term (by the BASE CLAUSE) and an algorithm (the RECURSION CLAUSE) by which to derive any term from its predecessor; for example, a^n is defined by

$$a^0 = 1, \quad a^{k+1} = a \times a^k;$$

and $n!$ is defined by

$$0! \ = \ 1, \quad n! \ = \ n(n-1)!$$

recursive function, *n.* **1.** informally, a function defined in terms of the repeated application of a number of simpler functions to their own values, by specifying a BASE CLAUSE and a RECURSION FORMULA.
2. more formally, any of the class of functions generated from the four operations of addition, multiplication, selection of one element from an ordered *n*-tuple, and determination whether $a < b$ (which are given to be RECURSIVE) by the two rules:

if F and $G_1, \ldots, G_n$ are recursive, then so is $F(G_1, \ldots, G_n)$; and

if H is a recursive function such that for each **a** there is an x with $H(\mathbf{a}, x) = 0$, then the least such x is recursively obtainable.

This has been shown to correspond to what may be computed by a TURING MACHINE, and according to CHURCH'S THESIS, it is also equivalent to being EFFECTIVELY COMPUTABLE.

recursively enumerable, *adj.* (of a predicate, P) such that there is a RECURSIVE PREDICATE, Q, satisfying

$$(\exists x)Q(\mathbf{a}, x) \leftrightarrow P(\mathbf{a}).$$

recursively generated, *adj.* (of a SEQUENCE) able to be given a RECURSIVE DEFINITION; generated by an ALGORITHM.

recursive partial function, *n.* a function of which the GRAPH (the set of points $\langle x, f(x) \rangle$) is RECURSIVELY ENUMERABLE.

recursive predicate, *n.* a predicate that possesses a REPRESENTING FUNCTION that is RECURSIVE.

reduce, *vb.* to modify or simplify the form of an expression, especially by substituting one term for another equivalent term.

reduced complexity method, *n.* any algorithm that performs with a complexity below that of the usual or naive method. An example is afforded by multiplication using the FAST FOURIER TRANSFORM.

reduced echelon form, *n.* an ECHELON FORM satisfying the additional condition that, in a *row-reduced echelon form,* the first non-zero element in each non-zero row is 1 and is the only non-zero element in its column; in a *column–reduced echelon form* the conditions attaching to rows and columns are interchanged. (These are sometimes referred to as the *reduced row* and *reduced column echelon form* respectively.) An *m*-by-*n* matrix or array may be transformed to an echelon form by a sequence of at most *mn* ELEMENTARY ROW OPERATIONS. When the original matrix is SQUARE and NON–SINGULAR, its row-reduced and column-reduced echelon forms are identical and are an IDENTITY matrix. See also GAUSSIAN ELIMINATION.

reduced form of a cubic, see CARDANO'S FORMULA.

reduced fraction, *n.* a fraction in its LOWEST TERMS.

reduced residue system, see RESIDUE CLASS.

reduced residue class, see RESIDUE CLASS.

reduced word, *n.* a WORD on a non-empty set that contains no consecutive pair of letters of the form xx^{-1}, or $x^{-1}x$.

reducible, *adj.* **1.** (of a surface or curve) deformable within a given region to a point. See also HOMOTOPY.

2. (of a polynomial) factorizable, over a given field, into polynomials of lower degree; otherwise the polynomial is *irreducible.* The polynomial is *fully reducible* if all irreducible factors are linear.

3. (of an IDEAL) being the intersection of two ideals that are both distinct from it; otherwise the ideal is *irreducible.* Every PRIME IDEAL is irreducible.

4. (of a LINEAR TRANSFORMATION on a vector space) fixing two complementary subspaces; in the case of a HILBERT SPACE, these are orthogonal.

5. (of a set of LINEAR TRANSFORMATIONS on a vector space) fixing a non-trivial subspace.

reduct, *n.* a structure derived from another by removing all occurrences of some set of symbols; A is thus a reduct of B if B is an EXTENSION of A.

reductio ad absurdum, *n.* (*Logic*) a method of disproving a proposition, or proving its negation, by showing that it has absurd, especially self-contradictory, consequences. The Euclidean proof of the infinitude of primes may be viewed as of this form, since it argues that, were any list $p_1, p_2, ..., p_n$ to exhaust the primes, then the number $(p_1 p_2 ... p_n) + 1$ would be neither prime nor composite. See also INDIRECT PROOF.

reduction, *n.* **1.** the expansion of a FRACTION in decimal form.

2. the expression of a fraction as a ratio in which the NUMERATOR and DENOMINATOR are RELATIVELY PRIME; such a ratio is obtained by CANCELLATION.

reduction formula, *n.* a formula that expresses a desired value in terms of more easily or previously calculated values. In particular, *trigonometric reduction formulae* give the value of a trigonometric function of any angle greater than 90° ($\pi/2$) in terms of some function of an acute angle; for example,

$$\sin(90° + \theta) = \cos\theta.$$

Integral reduction formulae often occur with the use of INTEGRATION BY PARTS as when integrating $\cos^n x$.

redundant, *adj.* **1.** (of a member of a system of equations, inequalities, or axioms) implied by the remaining members of the system, and so unnecessary in the sense that nothing derivable from the redundant element would fail to be derivable without it.

2. spurious or VACUOUS; for example, when both sides of an equation containing roots are squared *extraneous roots* are introduced.

re-entering angle, *n.* another term for a RE-ENTRANT angle.

re-entrant, *adj.* **1a.** (of an angle) having a vertex that points inwards in a polygon; REFLEX; greater than 180°. Compare SALIENT.

b. (*as substantive*) a re-entrant angle, such as the angle at D in Fig. 312; a re-entrant angle is between 180° and 360°.

2. (of a polygon) having a re-entrant angle; Fig. 312 shows a re-entrant pentagon.

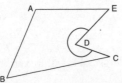

Fig. 312. D is a **re-entrant** angle in a re-entrant polygon.

reference, *n.* (*Logic*) **1.** the relation between a word, phrase, individual constant, or other expression and the entity it refers to or picks out. Compare SENSE.

2. the entity referred to by an expression; the REFERENT.

reference angle, *n.* the acute angle for which the trigonometric functions have the same absolute values as for some given angle, so that $\phi = n\pi \pm \theta$, where ϕ is the given angle, n is an integer, and θ is the reference angle, in radians. See also PRINCIPAL VALUE, REDUCTION FORMULA.

reference configuration, *n.* (*Continuum mechanics*) an arbitrary fixed CONFIGURATION of a BODY with respect to which its MOTION is described. See also MATERIAL DESCRIPTION.

referent, *adj.* (*Logic*) the entity to which a formal or linguistic expression refers. Compare REFERENCE, SENSE.

refinement, *n.* **1.** (*Topology*) (of a COVER) another cover such that each member of the second lies inside a member of the first.

2. (of a PARTITION of an interval on the line) a new partition constructed by further division of members of the original partition.

3. (of a NORMAL SERIES) a normal series containing every member of the given normal series. See also SCHRIER REFINEMENT THEOREM.

reflection, *n.* a planar TRANSFORMATION in which the direction of one axis is reversed, or that changes the polarity of one of the variables, so that it is of either of the forms

$$x' = x, \quad y' = -y,$$

or

$$x' = -x, \quad y' = y.$$

Each of the irregular figures and pair of axes in Fig. 313 is the reflection of the other about the central line.

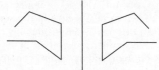

Fig. 313. The figures are **reflections** of one another in the bold line.

reflection principle of Schwarz, *n.* (*Complex functions*) the theorem concerning ANALYTIC CONTINUATION that if a function is analytic on a domain that contains one or more admissible segments of a straight line (or circle) in its boundary, while it is continuous thereon and these segments are mapped to similar segments, then the function can be continued analytically across the segments by REFLECTION (or INVERSION). The reflection is most easily viewed for the case of a function that is real on a real segment. There is a corresponding principle for HARMONIC functions.

reflex, *adj.* (of an angle) between 180° and 360°; the larger of the two angles between two line segments that meet at the vertex, as shown in Fig. 314.

Fig. 314. A **reflex** angle.

reflexive, *adj.* **1.** (of a relation) holding between any member of the domain and itself. For example, ... *is a member of the same family as* ... is reflexive since everyone is a member of his or her own family. Compare IRREFLEXIVE, NON–REFLEXIVE. See also EQUIVALENCE RELATION.

2. (of a normed space) canonically identifiable with its second normed dual space, possessing a unit ball that is weakly compact. Compare NON–REFLEXIVE. See also JAMES' THEOREM.

region, *n.* **1.** a CONNECTED subset of two-dimensional space, such as the set of points (x, y) such that $x > 0$, $y \geq 0$ and $x^2 + y^2 < 1$.

2. any set that is the union of an OPEN CONNECTED SET (DOMAIN) with some, none, or all of its BOUNDARY POINTS.

regress, *n.* a supposed definition or explanation, each stage of which requires to be similarly defined or explained; often a RECURSION FORMULA without a BASE CLAUSE. For example, if a natural number is defined simply as whatever is the successor of a natural number, then in order to show *n* to be one, we have to first show this of *n* − 1, and to show that we first need to consider *n* − 2, and so on; this never terminates since there is no base clause to specify the first element, as in PEANO'S AXIOMS.

regression, *n.* (*Statistics*) the analysis or measure of the association between a DEPENDENT VARIABLE and one or more INDEPENDENT VARIABLES, usually formulated as an equation in which the independent variables have parametric coefficients, which may enable future values of the dependent variable to be predicted.

regula falsi, *n.* another term for FALSE POSITION.

regular, *adj.* **1a.** (of a geometric figure) having all sides and all angles equal, as in a regular polygon such as the octagon of Fig. 315.

b. (of a geometric solid) having regular polygons as bases. A *regular polyhedron*, such as the regular tetrahedron shown in perspective in the second figure of Fig. 315, has identical regular polygons as bases, making equal angles with one another: a *regular prism* has regular polygons as bases; and a *regular pyramid* has a regular polygon as its base and its vertex perpendicularly above the centre of the base.

Fig. 315. **Regular** octagon and regular tetrahedron.

2. (of a complex function) another word for ANALYTIC.

3. (of a topological space) such that a point and a disjoint closed set can be SEPARATED by open sets; equivalently, for every NEIGHBOURHOOD of a point there is another neighbourhood of the point whose CLOSURE is contained in the given neighbourhood. *Complete regularity* is characterized by separation by continuous FUNCTIONALS: if $p \notin V$ with V closed in X, there is a continuous function $f: X \to [0, 1]$ such that $f(V) = 1$ and $f(p) = 0$. Compare NORMAL, T–AXIOMS, TIETZE EXTENSION THEOREM.

4. (of a curve) possessing no SINGULAR POINT; having only ORDINARY POINTS.

5. (of a summation method) giving the correct sum to a convergent series or sequence, as opposed to *conservative* methods that maintain convergence but may change the value of the limit. See TAUBERIAN CONDITION. See also ABEL SUMMATION, CESARO SUMMATION.

6. (of an OUTER MEASURE) such that every set, E, is contained in a MEASURABLE subset, A, of the same measure: $\mu(A) = \mu^*(E)$.

7. (of a BOREL MEASURE on a locally compact HAUSDORFF SPACE) assigning a finite measure to every COMPACT set and such that

(i) the measure of any Borel set is the INFIMUM of the measures of OPEN measurable sets containing the given set, and

    (ii) the measure of every open set is the SUPREMUM of the measures
       of COMPACT sets contained in the given set.

When X is compact, (i) implies (ii), and if each open subset of X is SIGMA–COMPACT (as when X is compact metrizable), it suffices for each compact set to have finite measure.

8. (of a GRAPH) having every VERTEX of the same degree.

9. (of an ACTION of a group on a set) such that the set has precisely one ORBIT under the group action, and the STABILIZER of every element of the set is trivial.

10. (of an element of a RING) such that its left or right product with any non-zero element of the ring is non-zero; that is, x is regular if and only if, for all $r \in R$, either $rx = 0$ or $xr = 0$ only if $r = 0$. For example, every non-zero element of an INTEGRAL DOMAIN is regular.

regular approximating sequence, $n.$ (*Measure theory*) a strictly increasing sequence of real-valued functions that are all BOUNDED and MEASURABLE, that converge to a given function almost everywhere.

regularity condition, $n.$ any condition imposed on a problem to guarantee that it meets the demands of a theorem or method, as with a CONSTRAINT QUALIFICATION.

regular point, $n.$ (of a HOLOMORPHIC function, f, analytic on an OPEN DISK) a point on the BOUNDARY of the open disk that can be surrounded by another disk on which there is an analytic function, g, that agrees with f at all points on the latter disk. A boundary point that is not a regular point is a SINGULAR POINT.

regular singular point, $n.$ a point a of a second order DIFFERENTIAL EQUATION

$$y'' + P(x)y' + Q(x)y = 0$$

for which a is a *singular point* (either P or Q is not REAL ANALYTIC at a), and

$$(x-a)P(x) \text{ and } (x-a)^2 Q(x)$$

are real analytic around a. See FROBENIUS METHOD.

related angle, $n.$ an angle that has the same absolute value for its TRIGONOMETRIC FUNCTIONS as a given angle, and in particular its acute REFERENCE ANGLE.

relation or **relationship,** $n.$ **1.** an association between, or a condition satisfied by, ordered pairs of objects, numbers, etc., such as $ab = 1$, ... *is greater than* ... or ... *is the father of* ... .

2. formally, any set of ordered pairs. Such a set defines a relation between the first member of each pair and its corresponding second member. If each first member is associated with only one second member, the relation is a FUNCTION. See also CORRESPONDENCE, EQUIVALENCE RELATION, SET-VALUED FUNCTION.

3. any analogous association of three or more members; a set of ordered n-tuples; an n-place PREDICATE.

4. (*Group theory*) one of a number of conditions subject to which a set generates a PRESENTATION of a given group.

relationship, $n.$ a less formal word for RELATION, especially one with a natural interpretation.

relative, *adj.* **1.** subject to some stated assumption; with respect to some chosen value.

2. another word for LOCAL; thus a relative maximum is a point at which the value of a function is greatest within a neighbourhood of that point.

3. (*Statistics*) proportional to a total; for example, RELATIVE FREQUENCY or RELATIVE ERROR.

relative acceleration, *n.* (*Mechanics*) the rate of change of RELATIVE VELOCITY.

relative angular momentum or **moment of relative momentum,** *n.* (*Mechanics*) (of a PARTICLE about a point with position vector **p**) the quantity

$$m(\mathbf{x} - \mathbf{p}) \times (\dot{\mathbf{x}} - \dot{\mathbf{p}}),$$

where *m* is the MASS and **x** is the position vector of the particle.

relative automorphism, *n.* an AUTOMORPHISM of an EXTENSION FIELD that leaves the BASE FIELD fixed. See also NORMAL EXTENSION FIELD.

relative compactness, *n.* (of a set in a TOPOLOGICAL SPACE) having a CLOSURE that is COMPACT. Every BOUNDED subset of Euclidean space is relatively compact. Compare TOTALLY BOUNDED.

relative complement, *n.* the members of a set lying outside another set. The relative complement of A in B, written B \ A and shown in Fig. 316, is the intersection of B with the COMPLEMENT of A if there is a universal set, but the notion is well-defined even without this. For example,

$$\{1, 2, 3\} \setminus \{2, 3, 4\} = \{1\}.$$

See also SYMMETRIC DIFFERENCE.

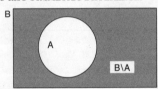

Fig. 316. The **relative complement** of A in B is shaded.

relative condition number, see CONDITION NUMBER.

relative consistency, *n.* the property of being RELATIVELY CONSISTENT.

relative displacement, *n.* (*Mechanics*) the difference between the position vector **x** of a PARTICLE and a chosen point with position vector **p**; the quantity **x** − **p**.

relative error, *n.* a measure of the difference between a number *b* and an estimate *a* given by the ratio of $|a - b|$ to $|b|$. Compare ABSOLUTE ERROR.

relative frequency, *n.* (*Statistics*) **1.** the ratio of the actual number of favourable events (successes) to the total number in a sample, often taken as an estimate of PROBABILITY.

2. the proportion of the values of a RANDOM VARIABLE taking a given value or lying in a given interval; that is, the proportion of favourable occurrences to all possible occurrences in a given SAMPLE SPACE.

relative identity, *n.* a relation of qualitative IDENTITY that defines a partition of its domain, so that the elements are equivalent for the purposes of the theory. It may then be possible to select one from each equivalence class as its CANONICAL element; alternatively, the equivalence class may itself be taken to be the element of the FACTOR SPACE. For example, the rationals may be defined either as the sets of ordered pairs $\langle kn, km \rangle$ for integral k, m, and n, or else may be taken to be the unique members of these sets for which m and n are RELATIVELY PRIME. The rationals are thus *identical relative to*, or UNIQUE UP TO, common factors.

relative interior, *n.* the INTERIOR of a CONVEX set in the INDUCED TOPOLOGY determined by the AFFINE SPAN of the set, denoted relint(A) or ri A. In infinite dimensions, one distinguishes this from the interior relative to the closed affine span. Every finite-dimensional convex set A has a non-empty relative interior.

relatively complete, *adj.* (of a PARTIALLY ORDERED SET) another term for CONDITIONALLY COMPLETE.

relatively consistent, *adj.* able to be shown to be CONSISTENT with respect to some larger theory; that is, having a model within that larger theory. This is the case, for example, for PEANO ARITHMETIC within SET THEORY, and for RIEMANNIAN and LOBACHEVSKIAN GEOMETRIES within EUCLIDEAN GEOMETRY.

relatively prime or **coprime,** *adj.* (of a pair of integers, or polynomials) not having any common divisors other than unity; for example, 8 and 9. When this relation holds, one is said to be *prime to* the other. Compare PRIME.

relative momentum, *n.* (*Mechanics*) (of a PARTICLE about a point with position vector **p**) the quantity $m(\dot{\mathbf{x}} - \dot{\mathbf{p}})$, where m is the MASS and **x** is the position vector of the particle.

relative topology, *n.* another term for the INDUCED TOPOLOGY on a subspace.

relative velocity, *n.* (*Mechanics*) the rate of change of RELATIVE DISPLACEMENT.

relativity theory, *n.* the mathematical formulation of Einstein's general theory of relativity in the setting of four-dimensional RIEMANNIAN GEOMETRY.

relatum, *n.* one of the entities related by a RELATION.

relaxation method, *n.* (*Numerical analysis, optimization*) a method in which solution of a problem is sought by initially relaxing some of the constraints that are then reimposed during the solution process. For example, when seeking to solve a problem in INTEGER PROGRAMMING by first solving the underlying LINEAR PROGRAMMING problem, one ignores the integrality restrictions.

reliability, *n.* (*Statistics*) **1.** sampling VARIANCE.

2. a measure of the precision of some quantitative method, for example, by computation of the variance of repeated measurements of the same quantity.

remainder, *n.* **1.** the amount left over when one quantity (number, polynomial, etc.) cannot be exactly divided by another; the difference between a given dividend and the largest multiple of the divisor less than (or of lower degree than) the dividend. For example, the remainder of $10 \div 3$ is 1; the remainder when $x^3 - 2x^2 + 7$ is divided by $x^2 + 2x$ is $8x + 7$. If $a = bq + r$, for non-negative r less than b, then a/b has quotient q and remainder r. See CHINESE REMAINDER THEOREM, EUCLID'S ALGORITHM, DIVISION ALGORITHM.

2. the difference between an infinite series and one of its partial sums, such as the LAGRANGE FORM OF THE REMAINDER for a Taylor series.

remainder theorem, *n.* **1.** the theorem that the remainder upon dividing a polynomial P(x) with coefficients in a FIELD by $x - a$ is equal to P(a).
2. see CHINESE REMAINDER THEOREM.

removable, *adj.* (of a DISCONTINUITY, SINGULARITY, etc.) able to be removed by redefining the function concerned; that is, by finding another function that has the same values except at a set of such points that are, depending on the setting, isolated or of MEASURE or CATEGORY zero. For example,

$$f(x) = \frac{x^2 - 1}{x - 1}$$

has a discontinuity at $x = 1$ since at this point the denominator is 0, but the discontinuity is removable either simply by setting $f(1) = 2$, or by cancelling to give $f(x) = x + 1$. See also INDETERMINATE.

renorm, *vb.* to construct an EQUIVALENT NORM (called a *renorm*) for a given norm on a NORMED SPACE. Every SEPARABLE space can be renormed so that the renormed unit ball is both (Gateaux) SMOOTH and strictly CONVEX.

repeat, *vb.* (of a sequence of digits) to RECUR infinitely often in a decimal expression of a fraction. See RECURRING DECIMAL.

repeated integral, *n.* another name for MULTIPLE INTEGRAL, particularly when evaluated as an ITERATED INTEGRAL.

repeated integration, *n.* another term for MULTIPLE INTEGRATION.

repeated root, *n.* another name for MULTIPLE ROOT.

repeated series, *n.* another term for DOUBLE SERIES.

repeating decimal, *n.* another name for RECURRING DECIMAL.

repetend, *n.* the digit or digits that repeat in a RECURRING DECIMAL.

replicable, *adj.* (*Statistics*) (of an experiment) able to be repeated under conditions that maintain some or all of the CONTROL CONDITIONS.

represent, *vb.* (of a series of functions) to converge to the value of the given function at each point of a specified set.

representation, *n.* a HOMOMORPHISM, R, from a (usually finite) GROUP, G, into GL(n; K), the GENERAL LINEAR GROUP of degree n over the FIELD, K, where the integer $n \geq 1$ is also called the *degree* of R. Two representations, R and S, are said to be *equivalent* if they have the same degree and there is a fixed c in GL(n; K) such that R(x) = c^{-1}S(x)c for all x in G. A representation over the complex numbers is an *ordinary representation*, and one over a field of prime CHARACTERISTIC is a *modular representation*. An ISOMORPHIC copy within a particular class of groups such as permutation or matrix groups is a *faithful representation*.

representing function, *n.* the function with the same domain as a given PREDICATE that, for each point in the domain, takes the value 1 if the predicate holds at that point, and 0 if it fails; the CHARACTERISTIC FUNCTION of the EXTENSION of the predicate.

repunit, *n.* a natural number all of whose digits are 1s, such as 1111. In base 10, the repunits with 2, 19, 23, 317, and 1031 digits are known to be prime.

research, see PLAGIARISM.

residual, *adj.* (*as substantive*) any difference between observed and predicted values after a MODEL has been fitted to a POPULATION on the basis of a SAMPLE.

residual set, *n.* a set that is the complement of a set of the first BAIRE CATEGORY.

residual spectrum, see SPECTRUM.

residue, *n.* **1a.** one of the integers $0, \ldots, n-1$, regarded as the remainders on division of any given integer by n (the *base*), and defining MODULAR ARITHMETIC modulo n. The set of residues mod n constitute a *residue class* RING, denoted $\mathbb{Z}_n$, under the arithmetic operations; it is a FIELD when n is prime. **b.** any member of a RESIDUE CLASS taken as the CANONICAL member of that class, and therefore sometimes identified with the residue class itself.

2. (*Number theory*) **power residue** of *m* of order *n*. a number *a* for which the congruence $x^n = a \pmod{m}$ is solvable. A number is a *non-residue* otherwise. See also PRIMITIVE ROOT, QUADRATIC RESIDUE.

3. (*Complex analysis*) the coefficient of the $(z - a)^{-1}$ term in the LAURENT EXPANSION of an analytic function at POLE, *a*.

residue class, *n.* **1.** one of the EQUIVALENCE CLASSES of numbers with identical RESIDUES modulo a given integer; each equivalence class may be represented by any of its elements, usually the least non-negative member. The residue classes mod *n* are the sets

$$\{ m : \ m = a + kn \},$$

where $0 \leq a \leq n - 1$ is a residue mod *n*, and *k* is non-negative. A *complete residue system modulo n* is a set of integers that contains one element from each class; the class of *least residues* is such a system. The *reduced residue system modulo n* contains one element from each class prime to *n*; thus $\{1, 3, 5, 7\}$ is a reduced residue system modulo 8. See also EULER'S PHI FUNCTION. **2.** one of the equivalence classes that is a COSET of an element of an IDEAL; an element of a FACTOR RING. See also TRANSVERSAL.

residue class ring, *n.* another term for FACTOR RING.

residue theorem of Cauchy, *n.* (*Complex analysis*) the theorem that if a function *f* is ANALYTIC in a SIMPLY CONNECTED domain *D* except for finitely many ISOLATED SINGULARITIES, then the CONTOUR INTEGRAL of *f* over any SIMPLE CLOSED CURVE in *D* that misses the singularities is equal to $2\pi i$ times the sum of the RESIDUES of *f* at singularities inside the contour; equivalently, the integral of a complex function around a JORDAN CONTOUR is the sum of the residues of that function inside the curve multiplied by $2\pi i w$, where *w* is the WINDING NUMBER of the curve about that point. This is of great use in evaluating CURVILINEAR INTEGRALS of MEROMORPHIC functions, and so certain classes of definite integrals of real functions.

resolve, *vb.* (of a vector) to find two or more other vectors, its COMPONENTS, usually perpendicular or in given directions, such that the RESULTANT of these other vectors is the given vector.

resolvent, *n.* the inverse of the matrix $t\mathrm{I} - \mathrm{A}$ where A is a given matrix or operator and *t* is not in the SPECTRUM of the matrix A. When operators in NORMED SPACE are considered, it is required that $t\mathrm{I} - \mathrm{A}$ have DENSE range and possess a BOUNDED inverse on that range. The set of such *t* is the *resolvent set* of A. In a BANACH SPACE setting, all numbers with $|t| > \|A\|$ lie in the resolvent set and

$$(t\mathrm{I} - \mathrm{A})^{-1} = \sum t^n \mathrm{A}^{n-1}.$$

resolvent equation, see CUBIC RESOLVENT EQUATION, CARDANO'S FORMULA.

resolvent kernel, see KERNEL.

resolvent set, see RESOLVENT.

response function, *n.* (*Continuum mechanics*) the symmetric TENSOR-valued function which describes the stress in an ELASTIC BODY.

response variable, *n.* (*Statistics*) a modern term for DEPENDENT VARIABLE.

restricted quantifier, *n.* (*Logic*) a QUANTIFIER regarded as ranging over the EXTENSION of a PREDICATE rather than over the entire domain of a logical theory. For example, PREDICATE CALCULUS standardly treats

all ravens are black

 as equivalent to

if anything is a raven it is black,

and writes $(\forall x)\, Rx \to Bx$ (with the obvious translation scheme); however, HEMPEL'S PARADOX suggests that non-ravens are irrelevant to the truth conditions of such a statement, and so that it might better be treated as quantifying only over those entities that satisfy the subject term. We might then write this as $(\forall_R x)\, Bx$. Similarly, it is clear that

most As are Bs

is not equivalent to

it is true of most things that if they are A they are B,

since the latter will be true if an absolute majority of the domain are not *A*, no matter what their relation to the *B*s; thus logics of PLURALITY require restricted quantification.

restriction, *n.* **1.** a condition that imposes a constraint on the possible values of a variable, on the domain of definition of an expression, or on the range of arguments of a function.
2. a function defined on a subset of the domain of a given function and taking the same values as the given function for those arguments. The restriction $f_E(x)$ of the function $f(x)$ to the set E is the set of pairs $\langle x, y \rangle$ such that $y = f(x)$ and x is a member of E; the restriction itself is sometimes written $f \,|\, E$. Compare EXTENSION.

result, *n.* the outcome of performing a mathematical operation or solving a mathematical problem, especially (*in plural*) a compilation of data from a statistical or other analysis.

resultant, *n.* **1.** a single vector or vectorial quantity that is the sum of two or more given vectors or quantities, especially the diagonal of a parallelogram of forces; in Fig. 317, $\overrightarrow{AC}$ is the resultant of $\overrightarrow{AB}$ and $\overrightarrow{BC}$. See PARALLELOGRAM RULE.

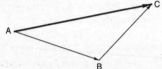

Fig 317. $\overrightarrow{AC}$ is the **resultant** of $\overrightarrow{AB}$ and $\overrightarrow{BC}$.

2. also called **eliminant.** given two polynomials, p, of degree n, and q, of degree m, with leading coefficients a_0 and b_0 respectively, the quantity

$$R(p, q) = a_0^n b_0^n \prod_{i=0}^{n} \prod_{j=0}^{m} (r_i - s_j)$$

$$= a_0^n q(r_1) q(r_2)...q(r_m) = (-1)^{mn} R(q, p),$$

where $r_1,..., r_m$ are the zeros of p, and $s_1,..., s_m$ are the zeros of q. See also DISCRIMINANT.

retract, see RETRACTION.

retraction, *n.* (of a TOPOLOGICAL SPACE onto a subspace A) a continuous EXTENSION to the entire space of the identity mapping on the subspace. The subspace is then said to be a *retract* of the space. The retract is an *absolute retract* if, whenever B is a closed subspace of a NORMAL space S and B is HOMEOMORPHIC to A, then B is actually a retract of S. The *n*-cube is an absolute retract.

reverse, *adj.* (of a MODULE, RING, or other structure) having a non-commutative multiplication operator that is the REVERSE MULTIPLICATION of the operator of a given structure. A reverse ring is also called an *opposite ring*.

reverse lexical order, see LEXICAL ORDER.

reverse multiplication, *n.* the non-commutative operator defined by reversing the order of multiplication in some given non-commutative setting; that is, defining

$$a**b = b*a.$$

reverse ordering, see ORDERING.

reverse Polish notation or **postfix notation,** *n.* (*Computing*) a notation that dispenses with the need for brackets by writing operators (such as LOGICAL CONSTANTS) after their arguments. For example, *P or Q* is written '*pq*A', and $3 + 5$ would be written '$3\ 5\ +$'; thus,

$$(3 \times (6 + 5)) \times ((2 \times 4) + 3)$$

can be written unambiguously as the string

$$3\ 6\ 5\ +\ \times\ 2\ 4\ \times\ 3\ +\ \times,$$

and

$$(P \to (Q \vee R)) \to ((P \to R) \vee S)$$

is written in reverse Polish notation

$$pqr\Lambda C pr C s A C.$$

If one uses a BINARY TREE diagram with the operators at the nodes to represent the structure of an expression, as in Fig. 318, the reverse Polish representation is obtained by starting at the bottom left-most branch, and working upwards, at each node reading the right-hand branch before the

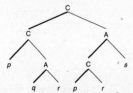

Fig. 318. **Reverse Polish notation**.
See main entry.

node itself, and iterating. Computers use this notation since it enables them to store the operator last, so that it is retrieved first and can determine the next operation of the computer. See also POLISH NOTATION. Compare INFIX NOTATION.

reversion, *n.* the formal process of computing the POWER SERIES of the inverse function to that represented by a given power series. Thus one reverts the power series for arctan to obtain a recurrence for the coefficients of the power series for tan.

revolve, *vb.* to rotate around an axis or point. See SURFACE OF REVOLUTION, SOLID OF REVOLUTION, VOLUME OF REVOLUTION.

Rhind papyrus, *n.* an early papyrus, about 6 yards long and one foot wide, containing 85 mathematical problems transcribed by Ahmes around 1650 BC. Discovered by A. Henry Rhind in 1858, it is one of the two principal sources of knowledge of ancient Egyptian mathematics, along with the smaller *Moscow papyrus* which dates from around 1850 BC.

rhomb, *n.* another name for RHOMBUS.

rhombic, *adj.* having the shape of, or pertaining to, a RHOMBUS.

rhombohedron, *n.* a six-sided PRISM of which the sides are parallelograms, as in Fig. 319.

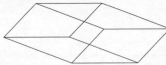

Fig. 319. **Rhombohedron**.

rhomboid, *n.* a PARALLELOGRAM of which the adjacent sides are unequal in length, as the second figure in Fig. 320b.

rhombus, rhomb, diamond, or **lozenge,** *n.* an oblique-angled PARALLELOGRAM whose four sides are equal, as in Fig. 320a. Compare SQUARE.

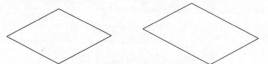

Fig. 320. (a) **Rhombus**. (b) **Rhomboid**.

Riccati equation, *n.* a non-linear DIFFERENTIAL EQUATION of the form

$$y' = f(x) + P(x)y + Q(x)y^2,$$

the solutions of which are not generally obtainable by integration. (Named after the Italian geometer and analyst, *Count Jacopo Francesco Riccati* (1676–1754).)

Richardson extrapolation, Richardson improvement, or **deferred approach to the limit,** *n.* an EXTRAPOLATION, using two computed values, h_{LARGER} and h, that is of the form

$$E(h) = \frac{F(h_{\text{LARGER}}) - f(h)\,r^n}{1 - r^n},$$

where r is the ratio of h_{LARGER} to h, and F is an $O(h^n)$ approximation to some quantity. If the TRUNCATION ERROR for F is of higher order $O(h^m)$, then the error in the extrapolation $E(h)$ will be of this order. The process can then be repeated with m replacing n.

Richard's paradox, n. the semantic paradox generated by supposing it possible to list all real numbers between 0 and 1 that can be defined by a finite condition. It is possible, by a DIAGONAL PROCESS, to define a number that differs from every number on this list (for example, by taking the nth digit of the new number to be 1 greater (mod 10) than the nth digit of the nth number on the list); however, that condition would itself clearly be a finite definition of this new number, so that it would satisfy the condition for being a member of the list, and yet differ from every member of it. Russell's solution for such paradoxes lay in his theory of TYPES, which denied sense to any expression that quantified over a domain of which it was itself a member, as occurs here. Compare BERRY'S PARADOX.

Riemann, Georg Friedrich Bernhard (1826–66), German mathematician who became Professor at Göttingen in 1859, where he had studied under, and won the support of, Gauss. His major achievements include work in function theory, the development of DIFFERENTIAL GEOMETRY from its origins in the work of GAUSS, the description of non-Euclidean RIEMANNIAN GEOMETRY, and the discovery of the RIEMANN INTEGRAL; he also postulated the RIEMANN HYPOTHESIS. He was elected a Fellow of the Royal Society shortly before his death.

Riemann condition, n. the condition that a function is INTEGRABLE on an interval if, for every $\varepsilon > 0$, there is a partition of the interval for which the UPPER SUM and LOWER SUM differ by less than ε.

Riemann hypothesis or **Riemann zeta hypothesis,** n. the conjecture that the ZETA FUNCTION has no non-trivial zeros except on the line with $\mathrm{re}(z) = \frac{1}{2}$. The trivial zeros occur at negative even integers. It is known to be true for the first 1 500 000 000 zeros, and its establishment would have many consequences for the PRIME NUMBER THEOREM and related theory, Riemann having observed that the distribution of prime numbers amongst the whole numbers, while not regular, approximates closely to the zeta function. This was one of HILBERT'S PROBLEMS and is a MILLENNIUM PRIZE PROBLEM.

Riemannian geometry, n. **1.** (*Differential geometry*) a generalization of the intrinsic geometry of a surface in which an n-dimensional MANIFOLD is endowed with a differential QUADRATIC FORM that is interpreted as its element of arc, in terms of which analogues of length, angle, curvature, etc. can be defined. Riemannian geometry in which the quadratic form need not be POSITIVE DEFINITE has applications in relativity theory, and it can be regarded as a distortion of EUCLIDEAN GEOMETRY.

2. another name for ELLIPTIC GEOMETRY.

Riemannian manifold, n. a MANIFOLD endowed with a METRIC TENSOR.

Riemann integrable, *adj.* (*Analysis*) having a RIEMANN INTEGRAL; having limits of UPPER and LOWER SUMS equal.

Riemann integral, n. the DEFINITE INTEGRAL of a real-valued bounded function, defined on a bounded interval as the value to which all RIEMANN SUMS,

$$m(i)\left[t_{i+1} - t_i\right],$$

converge as their MESH–FINENESS tends to zero, where $m(i)$ is the supremum (for the UPPER SUMS) or the infimum (for the LOWER SUMS) of the given function on the subinterval $[t_i, t_{i+1}]$; this exists if the function is continuous. Compare LEBESGUE INTEGRAL, IMPROPER INTEGRAL.

Riemann–Lebesgue lemma, n. the result that

$$\lim_{t \to \infty} \int_I f(x) \exp(itx) \, \mathrm{d}x = 0$$

for any interval I of the real line, any real variable t, and any Lebesgue integrable function f.

Riemann mapping theorem, n. (*Complex analysis*) the theorem that any SIMPLY–CONNECTED complex domain, whose boundary contains at least two points, can be mapped in a CONFORMAL fashion onto the open UNIT DISK.

Riemann sphere, n. the representation of the COMPLEX PLANE by the STEREO-GRAPHIC PROJECTION, where the pole is the POINT AT INFINITY.

Riemann–Stieltjes integration, n. an extension of the RIEMANN INTEGRAL allowing integration of a function, f, with respect to a function, g, defined as the limit of the sums

$$m_f(i)\,[g(t_{i+1}) - g(t_i)],$$

where $m_f(i)$ is the supremum (for the UPPER SUMS) or the infimum (for the LOWER SUMS) of f on the subinterval $[t_i, t_{i+1}]$; these limits are equal, and the integral exists, when f is continuous and g is of BOUNDED VARIATION.

Riemann–Stieltjes measure or **Radon measure,** n. a MEASURE on a SIGMA–ALGEBRA on a general TOPOLOGICAL SPACE such that every point has a neighbourhood of finite measure, and the measure of any set equals the supremum of the INNER MEASURE of all its COMPACT subsets. It may also be defined in terms of the infimum of the OUTER MEASURE of all open sets containing the given set.

Riemann sum, n. (for a real function f on an interval $[a, b]$) any sum of the form

$$\sum_{i=0}^{n} f(c_{i+1})\,\Delta_i,$$

where $\Delta_i = t_{i+1} - t_i$, for any PARTITION with

$$b = t_n > t_{n-1} > \ldots > t_1 > t_0 = a,$$

where $t_{i+1} \geq c_i \geq t_i$.

Riemann surface, n. a device whereby a complex SET–VALUED FUNCTION each BRANCH of which is ANALYTIC is converted into an analytic function on a more general surface by associating each branch with a separate plane or *sheet* interconnected in a consistent fashion.

Riemann zeta function, see ZETA FUNCTION.

Riesz–Fischer theorem, n. the theorem that the SQUARE–INTEGRABLE functions on a set form a COMPLETE NORMED SPACE, denoted L_2. Equivalently, every SQUARE–SUMMABLE sequence is the sequence of FOURIER COEFFICIENTS of some square-integrable function.

Riesz representation theorem or **Riesz–Kakutani theorem,** n. the theorem showing that all continuous LINEAR FUNCTIONALS on the space C(S) of real-valued continuous functions, where S is a compact HAUSDORFF SPACE, may be identified ISOMETRICALLY with differences of REGULAR BOREL MEASURES on S:

$$\psi(f) = \int_S f \, d\mu,$$

and $\| \psi \| = \| \mu \|$, the TOTAL VARIATION of μ. Moreover, the measure μ is non–negative exactly when the functional ψ is non-decreasing. In the case where S is a bounded interval $[a, b]$ on the line, the measure may be further identified with a function g of bounded variation that is continuous from the right and vanishes at a. This gives a Riemann–Stieltjes integral

$$\psi(f) = \int_a^b f \, dg,$$

and $\| \psi \|$ agrees with the TOTAL VARIATION of g on $[a, b]$.

right *adj*. **1.** also called **right-angled.** possessing a RIGHT ANGLE; for example, a right circular cone, as shown in Fig. 321, has a right angle between the AXIS and any DIAMETER of the BASE.

Fig. 321. A **right** circular cone.

2. (of an operator in a non-COMMUTATIVE theory) acting on the right: I_r is a *right identity* if $xI_r = x$ for all x; r_x is a *right inverse* of x if $xr_x = I_r$. Compare LEFT.

right angle, *n*. **1.** the angle between two PERPENDICULAR lines; an angle of $90°$ or $\pi/2$ radians; the angle between two intersecting lines when all the angles between them are equal, such as the angle ACB in the triangle of Fig. 322 below. This is a primitive concept of EUCLIDEAN GEOMETRY.
2. at right angles. perpendicular.

right-angled triangle or **right triangle,** *n*. a triangle one of whose angles is a RIGHT ANGLE, such as that in Fig. 322. See PYTHAGORAS' THEOREM.

Fig. 322. **Right-angled triangle**.

right-handed, *adj*. (of a coordinate system) given the orientation corresponding to a RIGHT–HANDED TRIHEDRAL, as shown in Fig. 323.

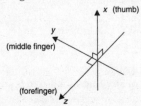

Fig. 323. A **right-handed** coordinate system.

right-handed trihedral, *n.* a configuration of three non-coplanar directed lines of which the TRIPLE PRODUCT is positive. This is so called because the thumb and first two fingers of the right hand have this orientation, as shown in Fig. 323 above; if the thumb is placed in the positive direction of the first line then the angle between the positive directions of the other two fingers is less than π. The other possibility yields a LEFT–HANDED TRIHEDRAL.

right-hand limit, *n.* the ONE–SIDED LIMIT of a function defined on an interval FROM ABOVE or *from the right*; the limit where x is restricted to values greater than a, written

$$\lim_{x \to a+} f(x) = f(a+).$$

Compare LEFT–HAND LIMIT.

right-invariant, see HAAR MEASURE.

right triangle, *n.* another term for RIGHT–ANGLED TRIANGLE.

rigid body, *n.* (*Mechanics*) a BODY in which the distance between the constitutive PARTICLES remain fixed under all possible MOTIONS.

rigid body motion, *n.* (*Continuum mechanics*) a MOTION of a BODY that consists only of ROTATIONS and TRANSLATIONS.

rigid motion, *n.* (*Euclidean geometry*) a TRANSFORMATION leaving size and shape of a configuration unchanged; the effect of a ROTATION composed with a TRANSLATION, in either order; an ISOMETRY of Euclidean space. SUPERPOSITION of plane figures is performed by a rigid motion.

rigorous, *adj.* (of a proof) making completely explicit the validity of the successive steps, usually with reference to some underpinning formal system.

ring, *n.* **1.** the area between two concentric circles, an ANNULUS.

2a. (*UK usage*) a non-empty set endowed with two binary operations, usually called addition and multiplication, such that the set is an ABELIAN GROUP under the addition and a SEMI–GROUP under the multiplication, the latter being both left- and right-DISTRIBUTIVE over addition. If, furthermore, the ring has a multiplicative IDENTITY ELEMENT, it is said to be a *ring with identity*; thus, the integers are a ring with identity, but the even integers are not. The possibility of a ZERO RING is not excluded.

b. (*North American usage*) as above, with a non-zero identity element.

See also COMMUTATIVE RING, DIVISION RING, INTEGRAL DOMAIN. Compare GROUP, FIELD.

ring homomorphism, see HOMOMORPHISM.

ring of sets, *n.* a BOOLEAN RING of sets that is closed under finite union and relative complementation. Compare SIGMA–RING.

rise, *n.* the difference between the values of the ORDINATES of a pair of points. The ratio of rise to RUN for a given pair of points gives the slope of the line segment joining the points.

rising factorial, *n.* another term for POCHHAMMER SYMBOL.

R-module, see MODULE.

rms, *abbrev. for* ROOT MEAN SQUARE.

Rodrigues' formula, see LEGENDRE POLYNOMIALS.

Rogers–Ramanujan identities, *n.* the pair of identities, subject to both terms of the equation being defined:

$$1 + \sum_{k=1}^{\infty} \frac{x^{(k^2)}}{(1-x)(1-x^2)\dots(1-x^n)} = \left[\prod_{m=1}^{\infty}(1-x^{5m-4})(1-x^{5m-1})\right]^{-1}$$

$$1 + \sum_{k=1}^{\infty} \frac{x^{k(k+1)}}{(1-x)(1-x^2)\dots(1-x^n)} = \left[\prod_{m=1}^{\infty}(1-x^{5m-3})(1-x^{5m-2})\right]^{-1}.$$

These identities were first stated by Ramanujan without proof in a letter to Hardy; however, Rogers had earlier given a proof as a consequence of some more general identities which had up till then been overlooked.

Rolle's theorem, *n.* the elementary result in mathematical analysis that if a real function is continuous at and between, and differentiable between, two points for which it has the same value, there is some intermediate point at which its derivative is zero. The MEAN VALUE THEOREM follows from this result. (Named after the French analyst, algebraist, and geometer, *Michel Rolle* (1652–1719).)

Roman numerals, *n.* the letters used by the Romans for the representation of CARDINAL NUMBERS: 1 is represented by I, 5 by V, 10 by X, 50 by L, 100 by C, 500 by D and 1000 by M. Multiples of 1000 are indicated by a superior bar; so $\overline{V} = 5000$, $\overline{X} = 10\,000$, $\overline{D} = 500\,000$, etc. Other numbers are represented by the shortest sequence of these letters with the required total value: their values are added except when a letter of lower denomination precedes one of higher in which case it is deducted from it; for example, IV = 4, IX = 9, CD = 400, but VI = 6, XI = 11, and DC = 600, etc. Note that $X\overline{D} = 499\,990$, $\overline{XD} = 490\,000$ but $\overline{X}D = 10\,500$. Compare ARABIC NUMERALS.

rook polynomial, *n.* the GENERATING FUNCTION of the number of ways of placing k mutually non-taking rooks on a chessboard C (of arbitrary shape and size). The number of ways of placing n rooks on an $n \times n$ chessboard, with none on the main diagonal, corresponds to the number of DERANGEMENTS of n objects. This suggests the utility of looking at more general rook polynomials to study permutations involving more forbidden positions. See MONTMORT MATCHING PROBLEM.

root, *n.* **1.** a value that SATISFIES or SOLVES a given equation. The roots of a polynomial or functional equation are the ZEROS of the corresponding polynomial or function, but this distinction is not always observed. See also DESCARTES' RULE OF SIGNS, NEWTON'S METHOD.

2. also called **radical.** in particular, a number or polynomial, a given integral power of which equals a given number or polynomial; if the required power is n, it is the nth root. See RADICAL SIGN.

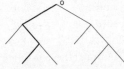

Fig. 324. O is the **root** of the tree (sense 3).

3. a vertex or node of a TREE that is uniquely distinguished as the ORIGIN; the unique point that is the initial member of the ANCESTRAL of every vertex and of every maximal chain of elements of the tree. The node O in the fragment of a tree shown in Fig. 324 is its root, since under the relation that

491

generates the tree, every node can be traced back to O as shown by the bold line.

root of unity, *n.* any n^{th} ROOT of the number 1. These may be computed in trigonometric terms by DEMOIVRE'S FORMULAE with $x = 2\pi/n$. See also PRIMITIVE ROOT OF UNITY, CYCLOTOMIC.

rooted, *adj.* (of a TREE) having a unique ROOT.

root mean square (abbrev. **rms**), *n.* the square root of the average of the squares of a set of numbers or quantities; for example, the standard deviation of a sample is the root mean square of their deviations from the mean, and is therefore sometimes known as the *root mean square deviation*.

root test or **Cauchy's root test,** *n.* the test for whether a complex series $\sum a_n$ is ABSOLUTELY CONVERGENT by considering

$$L = \limsup \left[a_n \right]^{1/n};$$

if L is less than unity, the series converges absolutely, while if it is greater than unity, the series diverges. This test is strictly stronger than the RATIO TEST. The RADIUS OF CONVERGENCE of a POWER SERIES is the reciprocal of L when the $\{a_n\}$ are the coefficients of the series.

rose, *n.* a curve shaped like a collection of petals with a common origin; its polar equation is $r = a\sin n\theta$ or $r = a\cos n\theta$; if n is odd it has n petals, while if n is even it has $2n$; for example, in Fig. 325, $n = 2$.

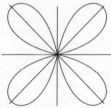

Fig. 325. A **rose** of order 2 showing axes of symmetry.

rot, *abbrev. and symbol for* ROTATION (of a function).

rotating frame of reference, *n.* (*Mechanics*) a FRAME OF REFERENCE in which the basis vectors are rotating with some ANGULAR VELOCITY ω in three-dimensional EUCLIDEAN POINT SPACE.

rotation, *n.* **1.** a circular motion of a configuration about a given point or line (the AXIS) without a change in shape; a RIGID MOTION leaving fixed the given point or line. It is a TRANSFORMATION resulting from turning the entire plane about a fixed point in the plane; this is effected by a change of variables of the form

$$x' = x\cos\theta + y\sin\theta,$$
$$y' = -x\sin\theta + y\cos\theta.$$

The positive direction is taken to be anti-clockwise.

2. (*Euclidean geometry*) a TRANSFORMATION in which the coordinate axes are turned about the origin.

3. (abbrev. **rot**) (*Vector analysis*) another name for CURL. See also IRROTATIONAL.

rotational symmetry, *n.* the property of a figure of being identical to some rotational transform of itself. The *order of symmetry* of a figure is the number of its transforms that are identical figures but are distinct in orientation, so that an equilateral triangle has rotational symmetry of order 3, since each of its three sides may be taken as the base.

Roth's theorem, *n.* another name for the THUE–SIEGEL–ROTH THEOREM.

Rouché's theorem, *n.* (*Complex analysis*) the result that when *f* and *g* are ANALYTIC in a SIMPLY–CONNECTED domain containing a SIMPLE CLOSED CURVE upon which $|g|$ strictly dominates $|f|$, then *g* and *f* + *g* have the same number of zeros inside the contour. (Named after the French analyst, algebraist, probabilist, and geometer, *Eugène Rouché* (1832–1910).)

roulette, see CYCLOID.

round, *adj.* **1.** shaped like a CIRCLE or SPHERE.

2a. forming or expressed by an integer or whole number, with no fractional part.

b. whence, expressed to one significant digit, *in round numbers*.

round angle or **perigon,** *n.* an angle of 360°; the angle swept out by a line when it returns to its initial position, such as the angle POP in Fig. 326.

Fig. 326. The angle POP is a **round angle**.

round down, *vb.* to approximate a number to a certain number of significant digits or to a whole number or number of tens, hundreds, etc. by replacing the remaining digits by zeros. For example, 432.25 can be rounded down to 432.2, 432, 430, or 400 according to circumstances. Compare ROUND UP. See ACCURACY.

rounding error, *n.* (*Numerical analysis*) the difference between the precise value of some quantity and the result of rounding up or down. See ROUND UP, ROUND DOWN.

round-off error, *n.* the error accumulated during a calculation as a combined effect of *inherent round-off* resulting from working on a fixed precision device, and the *propagated round-off* attributable to the level of precision used and cancellation or other errors. This is contrasted with the TRUNCATION ERROR intrinsic to a given APPROXIMATION, and is estimated in both relative and absolute terms.

round up, *vb.* to approximate a number to a certain number of significant digits or to a whole number or number of tens, hundreds, etc. by increasing the relevant digit by one and replacing the remainder by zeros: 486.75 can be rounded up to 486.8, 487, 490 or 500 according to requirements. Com-pare ROUND DOWN. See ACCURACY.

row, *n.* **1.** a horizontal linear array of numbers or terms, especially in a matrix; a $1 \times n$ matrix such as

$$[a \ b \ c]$$

or the row $[a \ b \ c]$ in a matrix such as

$$\begin{bmatrix} a & b & c \\ d & e & f \\ g & h & j \end{bmatrix}.$$

2. (*as modifier*) operating on, or relating to, the rows of a matrix, and contrasted with a corresponding COLUMN operation; for example, ELEMENTARY MATRIX OPERATIONS on the rows of a matrix are *elementary row operations*.

row equivalence, *n.* the relation that holds between a pair of matrices when one is obtainable from the other by a finite sequence of ELEMENTARY MATRIX OPERATIONS on its ROWS. Compare COLUMN EQUIVALENCE.

row rank, *n.* the RANK of the ROW SPACE of a matrix. This coincides with the COLUMN RANK and the RANK of the matrix.

row reduced, see ECHELON FORM.

row-reduced echelon form, see REDUCED ECHELON FORM.

row space, *n.* the VECTOR SPACE generated by the ROWS of a matrix. This space has DIMENSION equal to the ROW RANK of the matrix.

row-stochastic, see STOCHASTIC.

row vector, *n.* an *n*-TUPLE of quantities written as an $1 \times n$ matrix.

RSA code, *n.* a CODE which data are encrypted using the product of two very large prime numbers. It can be easily decrypted if one factor is known, but because it is extremely difficult to factorise such a number it is almost unbreakable otherwise.

rule, *n.* **1.** a fixed procedure for solving some problem or carrying out a procedure, such as the RULE OF THREE, or transformation rules.

2a. also called **ruler.** a straight-edge, used for drawing lines or measuring linear distance.

b. any linear scale.

ruled surface, *n.* a SURFACE that can be generated by a moving line, called its *generator* or *ruling*. A quadric surface may be generated by two distinct sets of generators, and so may be referred to as a *double ruled surface*. The *conjugate ruled surface* is constructed so that its rulings are tangent to those of the given surface.

rule of detachment, *n.* another name for MODUS PONENS.

rule of false position, see FALSE POSITION.

rule of inference, *n.* (*Logic*) a syntactic rule that is part of the definition of a FORMAL CALCULUS and by which theorems are derived from axioms and other theorems. The rules of inference are the RECURSION FORMULAE in the definition of the set of theorems of the calculus (the THEORY) in which the axioms are the BASE CLAUSES.

rule of signs, see DESCARTES' RULE OF SIGNS.

rule of three, *n.* the rule that the product of the extremes of a PROPORTION is equal to the product of its means, enabling an unknown quantity to be found: if $2 : x :: 4 : 6$, then $2 \times 6 = 4x$, so $x = 3$.

ruler-and-compass constructions, see CONSTRUCTIBLE.

ruling, see RULED SURFACE.

run, *n.* **1.** the difference between the ABSCISSAS of two points; the ratio of RISE to run gives the slope of the line segment joining the points.

2. *vb.* (of a sequence of data from a DISTRIBUTION) to be divisible into successive strings of observations with a common property; for example, the sequence 12, 16, 8, 14, 17, 11, 3, 5, 9, 18, 10, 8 runs with respect to parity.

Runge–Kutta methods, *n.* (*Numerical analysis*) a class of methods of approximately solving a DIFFERENTIAL EQUATION by approximating the TAYLOR POLYNOMIAL of a given degree. The *Runge–Kutta mid-point method* solves

$$y' = f(y, t), \quad y(a) = A$$

for $a < t < b$ by setting

$$w_0 = A, \quad h = \frac{b-a}{N}, \quad t_i = a + ih,$$

and iteratively solving

$$w_{i+1} = w_i + hf(t_i + \tfrac{h}{2}, \ w_i + \tfrac{h}{2}f(t_i, w_i)),$$

for $i < N$, to estimate the solution over the interval. Compare SIMPSON'S RULE.

Russell, Lord Bertrand Arthur William (1872 – 1970), English philosopher, logician, and mathematician, noted for his ground-breaking work in mathematical logic and the foundations of mathematics. He discovered RUSSELL'S PARADOX in the axiomatization of set theory proposed by Frege, and communicated it to him just as the second volume of his major work was about to go to press. He obtained a lectureship at Cambridge in 1910, but was dismissed from the post, and later jailed, for making pacifist statements during the First World War. Later he taught at Harvard, the National University of Peking, the University of Chicago, and UCLA, and won many prizes, including the Nobel Prize for Literature.

Russell's paradox, *n.* (*Logic*) the paradox in NAIVE SET THEORY that the class of all classes that are not members of themselves is a member of itself only if it is not, and is not only if it is; this undermines the intuitive belief in an all-inclusive universal class. The paradox was discovered by Bertrand Russell in the axiomatization of set theory proposed by Gottlöb Frege.

rv, (*Statistics*) *abbrev. for* RANDOM VARIABLE.

S

s, (*Mechanics*) *symbol for* SECOND.

saddle function, *n.* a function of two vector variables that is CONVEX in one variable and CONCAVE in the other variable; more generally, a function for which a MINIMAX THEOREM is obtainable.

saddle point, *n.* **1.** a point on a surface that is a MAXIMUM in one planar cross-section and a MINIMUM in another, such as the point X in Fig. 327. For example, $z = x^2 - 3xy - y^2 + 8xy^2$ has a saddle point at the origin.

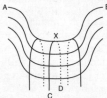

Fig. 327. **Saddle point**.

2. a point at which a function of two variables has first partial derivatives zero but which is not a local optimum; this occurs if the DETERMINANT of the HESSIAN is negative. The tangent plane at the point is horizontal but lies partly above and partly below the surface as with a saddle.
3. an entry in a matrix that is simultaneously maximal in its column and minimal in its row.
4. (*Game theory*) a point that minimizes in one variable, and maximizes in the other, the SADDLE FUNCTION associated with a MINIMAX THEOREM, and thus a point attaining the VALUE of an appropriate game.

salient, *adj.* **1.** (of an angle) less than 180°; an interior angle of a polygon is salient if its vertex points outwards, as every angle except C in the polygon of Fig. 328. See also WEDGE. Compare RE-ENTRANT.

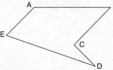

Fig. 328. Only C is not a **salient angle** (sense 1).

2. (of a point on a curve) such that two branches of a curve meet and have different tangents at the point, as happens with $2y = |x|$ at the origin, as shown in Fig. 329.

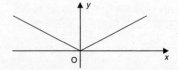

Fig. 329. The origin is a **salient** point (sense 2).

saltus, *n.* **1.** another word for a JUMP of a function.

2. another word for the OSCILLATION of a function on an interval.

sample or **sampling,** *n.* (*Statistics*) **1.** a set of individuals or events selected from a population for analysis to yield ESTIMATES of, or to test hypotheses about, PARAMETERS of the whole population. A BIASED sample, as contrasted with a RANDOM SAMPLE, is one in which the objects selected share some property that influences their distribution. See ESTIMATOR.

2. (*as modifier*) denoting a SAMPLE STATISTIC such as the sample mean, sample variance, etc.

sample point, *n.* (*Statistics*) any of the observed values of a RANDOM VARIABLE; a member of the SAMPLE SPACE of an experiment.

sample space, *n.* (*Statistics*) the set of possible outcomes of an experiment; the domain of values of a RANDOM VARIABLE. See also SAMPLE POINT.

sample statistic or **sampling statistic,** *n.* (*Statistics*) any function of observed data, especially one used to estimate the corresponding PARAMETER of the underlying distribution of the entire population, such as the sample mean, sample variance, etc. See also ESTIMATE, ESTIMATOR.

sampling, *n.* (*Statistics*) another word for SAMPLE.

sampling frame, (*Statistics*) see FRAME.

sandwich result or **squeeze rule,** *n.* one of a number of inequalities, useful in analysis, concerning the limits of sequences or functions whose terms are bounded above and below by ('sandwiched between') two others. For example, if $f(x) \leq g(x) \leq h(x)$ for all x greater than some N, and if $f(x)$ tends to A and $h(x)$ tends to A as x tends to infinity, then $g(x)$ tends to A as x tends to infinity. Another example is afforded by KATETOV'S INTERPOLATION THEOREM, and the HAHN–BANACH THEOREM can also be cast in this form.

satisfiable, *adj.* (*Logic*) (of an expression or set of expressions of a formal calculus) possessing a MODEL in which the given expressions are all true; CONSISTENT.

satisfiability problem, *n.* (*Logic*) the problem of determining whether there is an assignment of values to its variables that will SATISFY any statement of a logical calculus. Every sentence of SENTENTIAL CALCULUS is truth-functionally equivalent to a conjunction of clauses that are disjunctions of LITERALS, and even 3-satisfiability, in which exactly three literals are used in each clause, is an NP–COMPLETE PROBLEM.

satisfy, *vb.* **1.** to fulfil the conditions of a given theorem, assumption, etc. For example, $x = 3$ satisfies the equation $x^2 - 4x + 3 = 0$.

2. (*Logic*) to yield a truth by the substitution of the given value or sequence of values in a PREDICATE. For example, *x killed y* is satisfied by the ordered pair ⟨Cassius, Caesar⟩, but not by the pair ⟨Caesar, Cassius⟩; it is also defined to be satisfied by any longer sequence, including infinite sequences, of which the initial segment is identical. This enables a uniform semantic account to be given of relations and predicates, and by an extension due to Tarski, also to closed sentences regarded as zero-place predicates. Semantics for the existential and universal quantifier can be given in terms of satisfaction by sequences which agree everywhere except in the position corresponding to the bound variable.

scalar, *n.* **1.** (*Vector analysis*) **a.** a quantity with magnitude but not direction, such as speed as opposed to velocity.

b. (*as modifier*) having magnitude but not direction; for example, 2 is a scalar coefficient of the VECTOR **v** in the vectorial expression 2**v**.
Compare VECTOR, TENSOR.

2. (*Algebra*) an element of the array constituting a matrix, or of the field over which a VECTOR SPACE is defined.

3. an element of the ring over which a commutative group is a MODULE.

scalar field, *n.* a function that maps a connected domain in EUCLIDEAN SPACE into the real numbers. Compare VECTOR FIELD, TENSOR FIELD.

scalar matrix, *n.* a DIAGONAL MATRIX whose diagonal entries are all equal scalars. Multiplication by a scalar matrix is equivalent to SCALAR MULTIPLICATION by the constant scalar. A *scalar operator* is a multiple of the identity.

scalar multiplication, *n.* the multiplication of a VECTOR by a SCALAR to yield another vector. For example,

$$3\langle 1, 2, 3\rangle = \langle 3, 6, 9\rangle.$$

See VECTOR SPACE. Compare SCALAR PRODUCT.

scalar operator, *n.* a LINEAR OPERATOR that is a scalar multiple of the identity operator.

scalar product, **inner product**, or **dot product**, *n.* **1.** the defined product of an INNER PRODUCT SPACE.

2a. in particular, in (real) Euclidean space or a (complex) Hermitian vector space, where it is given by

$$\langle x, y\rangle = \sum_{i=1}^{n} x_i \, \overline{y}_i,$$

where $\mathbf{x} = \langle x_i\rangle$, and $\mathbf{y} = \langle y_i\rangle$.

b. (*Vector analysis*) a binary product of two VECTORS, written **v.w** or **vw**, that is a SCALAR equal to the arithmetic product $|\mathbf{a}|\,|\mathbf{b}|\cos\theta$, where $|\mathbf{a}|$ and $|\mathbf{b}|$ are the magnitudes of the given vectors, and θ is the angle between their directions. If the vectors are expressed in terms of coordinates, this can be calculated as above as the sum of the products of the corresponding coordinates. For example,

$$\langle 1, 2, 3\rangle . \langle 4, 5, 6\rangle = (1 \times 4) + (2 \times 5) + (3 \times 6) = 32.$$

Compare SCALAR MULTIPLICATION, VECTOR PRODUCT, TRIPLE PRODUCT.

scalar projection, *n.* (*of a vector on a vector*) a scalar derived from the given vectors equal to the length of the projection of a line segment representing the first vector onto a line segment representing the second. Thus the scalar projection of **a** onto **b** equals $|\mathbf{a}|\cos\theta$, where θ is the angle between the given vectors. It is independent of the magnitude of **b**, and is positive when the VECTOR PROJECTION is in the same direction as **b**, and negative when it is in the opposite direction.

scalar triple product, *n.* another term for TRIPLE PRODUCT.

scalar-valued, *adj.* (of a mapping) taking values in a field of scalars, as contrasted with a *vector-valued* mapping taking values in the corresponding vector space. See LINEAR FUNCTIONAL.

scale, *n*. **1a.** a sequence of *collinear* marks, usually either at regular intervals or else representing equal steps, that are used as a reference in making measurements. A *linear scale* is one in which equal distances represent equal amounts; in a *logarithmic scale*, distances are proportional to the logarithm of the amounts represented.
b. a measuring instrument having such a scale.
2a. also called **scale factor.** the ratio between the size of a representation of something and its actual size.
b. (*as modifier*) produced to some scale, such as a scale model.
3. a PLACE–VALUE NOTATION, such as the decimal scale.

scaling, *n*. (*Numerical analysis*) changes of SCALE undertaken during a computation such as GAUSSIAN ELIMINATION or a fixed point iteration in order to improve numerical performance.

scalene, *adj*. (of a triangle) having all sides unequal. Compare ISOSCELES, EQUILATERAL.

scatter diagram or **scattergram,** *n*. (*Statistics*) a graphical representation of the distribution of two RANDOM VARIABLES as a set of points whose coordinates represent their observed paired values; for example, Fig. 330 represents the observed distribution of salary and years of service in a small manufacturing company.

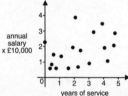

Fig. 330. **Scatter diagram**.

scattered, *adj*. (of a set in a TOPOLOGICAL SPACE) containing no non-empty PERFECT SET as a subset.

Schauder basis, *n*. another term for BASIS (sense 3).

Schauder basis problem, *n*. the problem of whether every separable BANACH SPACE has a SCHAUDER BASIS. This is now known to be generally false although all classical Banach spaces have such bases.

schema, *n*. (*Logic*) an expression using METAVARIABLES that may be replaced by expressions of the OBJECT LANGUAGE to yield a well-formed formula; thus A = A is an *axiom schema* for identity, yielding the infinite set of axioms $x = x$, $y = y$, $z = z$, etc.

schlicht, **simple**, or **univalent,** *adj*. (of a complex function on some domain, often the unit circle) ANALYTIC and taking no value more than once in the domain. A schlicht function mapping the finite complex plane onto itself is linear.

Schlömilch's form of the remainder, *n*. a form of the REMAINDER in a Taylor series that includes the Cauchy and LAGRANGE FORM OF THE REMAINDER as special cases. (Named after the German analyst, *Oskar Xaver Schlömilch* (1823–1901).)

Schnirelmann density, *n*. the INFIMUM, d(S), of the ratio $S(n)/n$ where $S(n)$, for $n \geq 1$, is the number of elements in a given sequence S of non-negative

499

integers that are no greater than n. Then $d(S)$ is 1 if and only if S equals $\mathbb{N}$. The LIMIT INFERIOR of $S(n)/n$ defines the *asymptotic density* of the set, and is called the *natural density* of the set if it exists as a limit; the square-free integers have natural density $6/\pi^2$. Schnirelmann density is sensitive to changes in the initial segment of the set and gives rise to the *alpha-beta theorem*. This result states that if $d(S) = \alpha$ and $d(T) = \beta$, for any two sets S and T, then

$$d(S + T) \geq \min\{1, \alpha + \beta\}.$$

See also UNIFORM DISTRIBUTION.

Schrier–Nielson theorem, see FREE GROUP.

Schrier refinement theorem, n. the theorem that any two NORMAL SERIES of a group have isomorphic REFINEMENTS. See also JORDAN–HOLDER THEOREM.

Schröder–Bernstein theorem or **Cantor–Bernstein theorem,** n. the theorem that establishes that two sets are EQUIPOLLENT if there is an INJECTIVE mapping from each to the other.

Schrödinger's equation, n. (*Mechanics*) the PARTIAL DIFFERENTIAL EQUATION

$$\nabla^2\psi + k\,(E - V)\psi = 0,$$

where E is the total energy, and V is the potential energy.

Schur complement, n. the quantity related to a given partitioned matrix by

$$D = B_4 - B_3 B_1^{-1} B_2,$$

when the original matrix is given as

$$\begin{bmatrix} B_1 & B_2 \\ B_3 & B_4 \end{bmatrix},$$

with B_1 invertible and B_4 square. (Named after the German algebraist and number theorist, *Issai Schur* (1875 – 1941).)

Schur's lemma or **Schur's theorem,** n. the result that any square matrix is in UNITARY EQUIVALENCE with an upper TRIANGULAR MATRIX of which the diagonal entries are the EIGENVALUES of the original matrix. Hence it follows easily that a NORMAL MATRIX is unitarily equivalent to a DIAGONAL MATRIX.

Schwartzian derivative, n. **1.** the quantity, given any thrice-differentiable function g, defined by

$$S(g) = \frac{2g'g''' - 3(g'')^2}{2(g')^2}.$$

2. see DISTRIBUTION.

(Named after the French functional analyst, mathematical physicist, topologist, and Fields medalist, *Laurent Schwartz* (1915 –).)

Schwarz inequality, n. the CAUCHY–SCHWARZ INEQUALITY, and in particular, its complex integral version. (Named after the German analyst and complex function theorist, *Hermann Amandus Schwarz* (1843 – 1921).)

Schwarz' lemma, n. the consequence of the MAXIMUM PRINCIPLE that an analytic function that maps the set of complex numbers z such that $|z| < 1$ into itself, and is zero at zero, either is a rotation or satisfies $|f(z)| < |z|$ in the punctured disk and has $|f'(0)| < 1$.

Schwarz principle, see REFLECTION PRINCIPLE OF SCHWARZ.

scientific notation, **exponential notation**, or **standard form**, *n.* (*Computing*) the expression of numbers in FLOATING–POINT notation, as multiples of the largest power of the BASE less than the given number. Thus, 123.45 is written 1.2345×10^2; many electronic calculators represent this as 1.2345E2.

scope, *n.* (*Logic*) (of an operator in an expression) that part of the expression that is governed by the given operator; that is, the operator itself together with its arguments. For example, the scope of the negation in 'P & –(Q ∨ R)' is '–(Q ∨ R)'.

sd, (*Statistics*) *abbrev. for* STANDARD DEVIATION.

se, (*Statistics*) *abbrev. for* STANDARD ERROR.

sec, *abbrev. and symbol for* the SECANT function.

sec⁻¹, *symbol for* the inverse SECANT function, ARC–SECANT.

secant, *n.* **1.** (of an angle) a trigonometric function that, in a right-angled triangle, is the ratio of the length of the hypotenuse to that of the adjacent side; the reciprocal of the COSINE function. It is usually written sec$x$, and its graph is shown in Fig. 331. The derivative of sec$x$ is sec$x$tan$x$, and an antiderivative (or indefinite integral) is

$$\ln|\sec x + \tan x|.$$

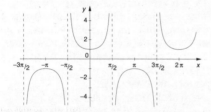

Fig. 331. Graph of the **secant** function.

2. a line that intersects a curve, usually at more than one point

secant method or **method of linear interpolation**, *n.* a variation of NEWTON'S METHOD for finding a zero of a real function that replaces the derivative by the slope of the SECANT through the two previously computed points on the graph. This demands two initial estimates, and exhibits local convergence of ORDER $(\sqrt{5}+1)/2$, but it may diverge if poor initial estimates are chosen. See also FALSE POSITION.

sech, *symbol for* the HYPERBOLIC FUNCTION hyperbolic secant, the reciprocal of the hyperbolic cosine COSH. Its derivative is – sech$x$tanh$x$, and an antiderivative (or indefinite integral) is

$$\tan^{-1}|\sinh x|;$$

its graph is shown in Fig. 332.

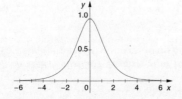

Fig. 332. **Sech**. Graph of the hyperbolic secant function.

sech⁻¹, *symbol for* the inverse HYPERBOLIC SECANT function, ARC–SECH.

second, *n.* **1. second of arc.** one 60th part of a MINUTE of arc; one 360th of a DEGREE.

2. (symbol **s**) the standard unit of TIME; one of the fundamental units of the SYSTEME INTERNATIONAL. This is defined as the time occupied by 9 192 631 770 oscillations of a molecule of cæsium.

second-category set, *n.* see BAIRE CATEGORY.

second-countable, *adj.* (of a TOPOLOGICAL SPACE) such that the topology has a countable BASE. A METRIC SPACE is second-countable if and only if it is SEPARABLE. For example, the usual topology on the reals is second-countable. Compare FIRST COUNTABLE.

second curvature, *n.* another term for TORSION.

second derivative, *n.* a DERIVATIVE derived from a given function by differentiating its first derivative. Correspondingly one speaks of *higher derivatives* such as *third derivatives*.

second derivative test, *n.* a test for the OPTIMALITY of a CRITICAL POINT of a function that uses SECOND ORDER information. Thus, for a function of one variable one checks whether the second derivative at the point is positive (a local minimum), negative (a local maximum), or zero (indeterminate). For a function of several variables one checks whether the HESSIAN at the point is positive definite (a local minimum) or negative definite (a local maximum), indefinite (a SADDLE POINT), or singular (indeterminate); if the DETERMINANT of the Hessian at the point is negative, the point is a saddle point. Compare FIRST DERIVATIVE TEST.

second diagonal, *n.* another term for SUPERDIAGONAL.

second isomorphism theorem, see ISOMORPHISM THEOREMS.

second-kind induction, *n.* another term for COMPLETE INDUCTION, by contrast with FIRST–KIND INDUCTION. See INDUCTION.

second mean value theorem, see MEAN VALUE THEOREM.

second-order, *adj.* **1.** (of a DERIVATIVE) of or involving second derivatives. This is consistent with calling a quadratic term in a polynomial second-order if one considers the polynomial as a Taylor series.

2a. (of an ORDINARY DIFFERENTIAL EQUATION) involving the first and second derivatives, but no derivative of a higher ORDER, of the dependent variable with respect to the independent variable. LINEAR, AUTONOMOUS and HOMOGENEOUS second-order equations have standard methods of solution. Equations in which the dependent variable does not explicitly occur can be regarded as FIRST–ORDER equations in the first derivative. See VAN DER POL EQUATION.

b. (of a PARTIAL DIFFERENTIAL EQUATION) such that no partial derivative of ORDER greater than 2 occurs in it. See LINEAR DIFFERENTIAL EQUATIONS, MONGE'S METHODS.

3. metatheoretical.

4. (of a LOGICAL THEORY) admitting quantification over some classes as well as individuals; for example, *second-order arithmetic* or *second-order set theory*. Compare FIRST–ORDER.

5. see TENSOR.

second species, see SPECIES.

secretary problem, *n.* another name for the INTERVIEW PROBLEM.

section, *n.* **1a.** the intersection between a plane and a surface or solid; a plane figure formed by cutting through a solid. The section is a *normal section* if the plane contains a NORMAL to the surface.

b. the shape or area of such a plane figure.

2. see GOLDEN SECTION.

3. a FACTOR GROUP of a subgroup of a given group.

sectionally, *adv.* another term for PIECEWISE.

section formula or **ratio theorem,** *n.* (*Geometry*) the theorem that if a point P divides a directed line segment, $\overrightarrow{AB}$, in the ratio $m : n$, then the POSITION VECTOR, **p**, of P, can be expressed in terms of those of A and B as

$$\mathbf{p} = \frac{m\mathbf{a} + n\mathbf{b}}{m + n}.$$

sector, *n.* the portion of a circle bounded by two radii and an arc. Any pair of radii divide a circle into two sectors; in Fig. 333, area AOBX is the MINOR sector, and AOBY is the MAJOR sector. The area of a sector is $\frac{1}{2} r^2 \theta$, where r is the length of the radius, and θ is the central angle (in radians) subtended by the arc. Compare SEGMENT.

Fig. 333. **Sector.**

see, *vb.* (of a pair of points of a set in a Euclidean vector space) to be such that the line segment between the two points lies entirely within (the given set); the convex set of points that sees the entire set is the STAR of the set.

seed, *n.* the initial state of a DYNAMICAL SYSTEM or of a RANDOM NUMBER GENERATOR.

segment, *n.* **1.** the portion of a circle bounded by an arc and a chord; any chord divides a circle into two segments; in Fig. 334, the area AXB is the MAJOR segment, and AYB is the MINOR segment. (It should be noted, however, that in ordinary usage, the cross-section of a 'segment' of a fruit such as an orange is in fact a SECTOR of the whole cross-section.)

Fig. 334. **Segment.**

2. more generally, a part of a plane or solid figure cut off by one or more lines or planes that intersect the figure.

3. a part of a line or curve lying between two of its points.

selection, *n.* (from a CORRESPONDENCE) a single-valued mapping whose value lies inside the image under a SET–VALUED FUNCTION at each argument. *Michael's continuous selection theorem* is that a lower SEMI–CONTINUOUS multifunction with closed non-empty convex images admits a continuous

503

selection through any point in its graph as soon as the domain is PARA-COMPACT and the range is a BANACH SPACE: $f(x) \in F(x)$ and $f(a) = b$ for any $b \in F(a)$. A measurable *selection theorem* asserts the existence of a measurable selection out of a suitably measurable multifunction.

self-adjoint, *adj.* **1.** (of a MATRIX or LINEAR OPERATOR on HILBERT SPACE) HERMITIAN; equal to its own HERMITIAN CONJUGATE or ADJOINT; thus

$$\langle Ax, y \rangle = \langle x, Ay \rangle$$

for all x and y in the Hilbert space. The concept makes sense for a mapping from a REFLEXIVE NORMED SPACE to its DUAL.
2. (of an algebra) having the complex conjugate of any member of the algebra in the algebra. See STONE–WEIERSTRASS THEOREM.

self-conjugate, *adj.* (of a LINE or POINT) lying on the POLAR, or, dually, running through its POLE. The only self-conjugate point on a self-conjugate line is its pole.

self-contradictory, *adj.* CONTRADICTORY, usually of a single sentence; not able to be true under any INTERPRETATION.

self-inverse, *adj.* (of an element of a GROUP, RING, etc.) being its own INVERSE, so that $xx = I$, the identity element of the structure, such as an element of a group of ORDER 2; for example, the function $f(x) = 1/x$, defined on the interval $]0, \infty[$.

self-polar, *adj.* **1.** (of a TRIANGLE) such that each vertex is the POLE of the opposite side.
2. (of a CONVEX set) equal to its POLAR set.

self-reference, *n.* (*Logic*) the property of an expression of referring to itself, which gives rise to such SEMANTIC paradoxes as that of determining the truth-value of the sentence

this sentence is false,

which is true if it is false and false if it is true. Russell proposed his VICIOUS CIRCLE principle and the THEORY OF TYPES in part to rule out the possibility of such expressions being WELL FORMED. Certainly, just as CANTOR'S PARADOX showed that there can be no all-inclusive cardinal number, and RUSSELL'S PARADOX that there can be no all-inclusive universal set, so the paradoxes of self-reference show that there can be no all-inclusive language or conception of truth; indeed the distinction between OBJECT LANGUAGE and METALANGUAGE was introduced by Tarski to resolve these paradoxes. See also LIAR PARADOX, LAWYER PARADOX, GRELLING'S PARADOX, BERRY'S PARADOX.

self-similar, *adj.* (of a set in a Euclidean geometry with positive HAUSDORFF MEASURE) INVARIANT under SIMILITUDES $\psi_1, ..., \psi_n$, such that

$$\psi_i(E) \cap \psi_j(E) = \varnothing$$

for all $i \neq j$, where E is the given set.

semantic, *adj.* (*Logic*) concerned with the meaning and truth of expressions of a FORMAL LANGUAGE, as opposed to their structure in an uninterpreted FORMAL CALCULUS. For example, truth-tables are given as an account of the meaning of the sentential connectives in terms of the contribution they make to the truth-value of compound expressions in which they occur. Compare SYNTACTIC.

semantics or **model theory,** *n.* (*Logic*) **1.** the study of INTERPRETATIONS and MODELS of formal theories.

2. the study of the relationship between the structure of a theory and its subject matter.

3. the principles that determine the truth or falsehood of sentences and the references of terms within a given formal theory.

Compare SYNTAX.

semantic tableau, *n.* (*Logic*) a TREE diagram constructed in order to demonstrate the consistency or otherwise of a set of statements by successively breaking down the given statements into simpler components; as soon as a contradiction is obtained, that branch of the tableau need be considered no further and is *closed*. If every branch is closed, the entire tableaux is closed, and the original set of formulae is thereby shown to be inconsistent. This method can be used to show the validity of an argument by testing the consistency of the set consisting of all its premises and the negation of its conclusion; if the tableau of this set is closed then the argument is valid. The tableau in Fig. 335 shows the validity of the argument form

$$P \ \& -Q \ \vdash \ -(-P \vee Q);$$

first the sentences to be tested for consistency are listed, then the simpler consequences of the conjunction and double negation are added to the list of supposed truths; the disjunction does not entail simpler consequences, but permits branching to consider the two distinct possibilities; in this example, however, each of these possibilities contradicts what is already known, and so the tableau closes. This is equivalent to a 'reverse' TRUTH–TABLE test.

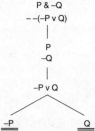

Fig. 335. A closed **semantic tableau**.

semi-axis, *n.* half the length of an AXIS of a conic.

semicircle, *n.* **1.** a plane figure bounded by the diameter of a circle and one of its arcs.

2. an arc of a circle that is half of its circumference.

semicontinuous, *adj.* **1.** (of a real-valued function, *f*) such that either *f* or –*f* satisfies

$$\liminf_{y \to x} f(y) \ \geq \ f(x).$$

If *f* itself satisfies this condition, it is *lower semicontinuous*, and this corresponds to lower LEVEL SETS of the form

$$\{ x : f(x) \ \leq \ r \}$$

being closed, or to the EPIGRAPH

$$\{(x, r) : f(x) \leq r\}$$

being closed. A function f is *upper semicontinuous* if $-f$ is lower semicontinuous. A function is continuous if and only if it is both upper and lower semicontinuous.

2. (of a SET–VALUED FUNCTION) such that the lower or upper INVERSE IMAGE of any open set is open; the multifunction is then *lower* or *upper semicontinuous* respectively. A multifunction with both properties is sometimes said to be a *continuous multifunction*. There are a multitude of competing definitions for continuity and upper semicontinuity, many involving HAUSDORFF DISTANCE. A multifunction between compact spaces that has closed images is upper semicontinuous exactly if it has a closed graph. For a single-valued correspondence, both coincide individually with continuity.

semidefinite, *adj.* see POSITIVE SEMIDEFINITE.

semi-elliptical, *adj.* shaped like one half of an ellipse, especially one divided along its major axis.

semi-group, *n.* a set endowed with an ASSOCIATIVE binary operation, usually called addition, under which it is closed. Compare MONOID, GROUPOID.

semi-interquartile range, *n.* (*Statistics*) one of several measures of the spread of a distribution, equal to half the difference between the first and third QUARTILES.

semi-inverse, see PSEUDO–INVERSE.

semi-metric, see METRIC.

semi-norm, *n.* a generalization of the concept of NORM that does not demand that only the origin has value zero. Thus a norm is a semi-norm with trivial KERNEL.

semi-prime, *adj.* (of a RING) such that if $A^n = 0$ for an IDEAL, A, and any positive integer, n, then $A = 0$; any PRIME ring is semi-prime.

semi-regular, *n.* (of a POLYHEDRON) such that all its faces are REGULAR polygons, though not all congruent, and the different kinds of face occur in the same order around each vertex. Right-regular PRISMS with square lateral faces and right-regular ANTIPRISMS whose equilateral triangular lateral faces are semi-regular.

semi-ring of sets, *n.* a FAMILY of subsets containing the EMPTY set, closed under finite INTERSECTION, and having the property that if E is a subset of F with E and F in the family, then $F \setminus E$ is a COUNTABLE union of DISJOINT members of the family. Compare RING OF SETS. See also BOOLEAN ALGEBRA.

semi-simple, *adj.* **1.** (of a MODULE) being generated by, or being the DIRECT SUM of, SIMPLE submodules.

2. (of a RING) **a.** SEMI–PRIME, often with the additional requirement that it be a right or left ARTINIAN RING. A ring is semi-simple if and only if it is the direct sum of (finitely many) MINIMAL left IDEALS.

b. such that a given RADICAL is zero.

3. (of a commutative BANACH ALGEBRA) such that the intersection of maximal two-sided ideals is zero.

semi-transcendental function, *n.* the general solution of a non-linear second

order DIFFERENTIAL EQUATION for which the general solution is not an ALGE-
BRAIC function of two constants of integration but the equation admits of a
first integral that is an algebraic function of one constant of integration.
For example, the first integral of $w'' + 2ww' = q(z)$ is

$$w' + w^2 \; = \; \int q(z) \, dz + A;$$

the general solution is therefore at worst a semi-transcendental function of
A and the second constant of integration.

sense, *n.* **1a.** one of the two opposite directions measured on a directed line,
the sign of the MEASURE as contrasted with the MAGNITUDE of a vector, so that
$\overrightarrow{AB}$ and $\overrightarrow{BA}$ have opposite sense but the same direction. See also POLARITY.
b. one of the two opposite directions of ROTATION, that is, either CLOCK-
WISE or ANTICLOCKWISE.
2. (*Logic*) **a.** the import of an expression as contrasted with its REFERENT.
Thus *the morning star* and *the evening star* have the same reference, the
planet Venus, but different senses.
b. the property of an expression by virtue of which its referent is determined.
c. that which one grasps in understanding an expression.
Compare REFERENCE.

sentential calculus or **propositional calculus,** *n.* (*Logic*) the formal theory of
which the intended interpretation concerns the logical relations between
sentences treated only as a whole and without regard to their internal
structure. Its primitive terms are the LOGICAL CONSTANTS and an unlimited
supply of sentential symbols (sometimes called propositional variables),
usually capital or lower case letters with indices if necessary. Compare
PREDICATE CALCULUS.

sentential function, *n.* (*Logic*) another term for OPEN SENTENCE.

separable, *adj.* **1.** (of a TOPOLOGICAL SPACE) containing a countable DENSE
subset. Every COMPACT METRIC SPACE or SECOND COUNTABLE SPACE is separable, as is Euclidean space since it contains the rational *n*-tuples, which are
countable and dense.
2. (of a function) able to be written so that variables are *separated*, additively
or multiplicatively, as, for example, if

$$f(x, y, z) \; = \; f_1(x) + f_2(y) + f_3(z).$$

This is very useful in computational optimization since minimization can
be performed term by term.
3. (of a polynomial) such that the irreducible factors have no repeated
roots.
4. (of an EXTENSION FIELD) such that every element in the extension has a
MINIMUM POLYNOMIAL that is separable. Every extension of a field of characteristic zero is separable.
5. (of a first order ORDINARY DIFFERENTIAL EQUATION) such that it can be
written in the form $y' = g(y) \, h(t)$ and hence can be integrated directly to
yield a solution of the form

$$\int \frac{1}{g(y)} \; = \; \int h(t) \, dt + A.$$

See SEPARATION OF VARIABLES.

separated, *adj.* **1.** (of two sets in a TOPOLOGICAL SPACE) such that neither intersects the CLOSURE of the other. A space is CONNECTED if and only if it can not be written as a union of two non-empty separated sets.

2. constituting a HAUSDORFF SPACE.

3. (of variables) see SEPARABLE (sense 2).

separate points, *vb.* (of an algebra) to satisfy the condition that given any two distinct points of the set there is a member of the algebra for which the value at the points differs. See STONE–WEIERSTRASS THEOREM.

separate variables, *vb.* to solve a DIFFERENTIAL EQUATION by SEPARATION OF VARIABLES.

separating, *adj.* (of a set of real-valued functions) such that for any x and y in the domain, there is a function, f, in the set for which $f(x) \neq f(y)$.

separation axiom, *n.* any of a number of possible additional axioms for a TOPOLOGICAL SPACE that assert at least the existence of open sets that contain one but not the other of every pair of points. See HAUSDORFF SPACE, REGULAR, NORMAL, T–AXIOMS.

separation of variables, *n.* the process of solving a DIFFERENTIAL EQUATION by rewriting it as an equation each side of which can be directly integrated with respect to one of the variables; in the simplest case, the equation is of the form

$$y' = g(x)/h(y),$$

so that one may cross-multiply. For example, cross-multiplying in

$$\frac{dy}{dx} = \frac{x}{\ln y}$$

separates the variables to give

$$\ln y \, dy = x \, dx,$$

which can be integrated directly, yielding the solution

$$y \ln y - y = \tfrac{1}{2} x^2.$$

See EXACT.

separation theorem of Mazur or **geometric form of the Hahn–Banach theorem,** *n.* the theorem that disjoint CONVEX sets lie on opposite sides of some closed HYPERPLANE. This requires one of the sets to have a non-empty topological interior. See HAHN–BANACH THEOREM. See also SUPPORT FUNCTION.

sept-, *prefix denoting* seven; for example, a *septilateral* is a plane figure with seven sides, a *septuple* is an ordered set of seven members, and a *septic* function is of seventh order.

sequence, *n.* an ordered set of objects, especially numbers, either finite or DENUMERABLE, and so capable of being indexed by the natural numbers or an initial segment of them. Compare SERIES.

sequent, *n.* a formal representation of an argument, in a logical calculus, as a set of premises and a conclusion; for example, the inference of A from A & B is written

$$A \,\&\, B \vdash A.$$

The sequent '⊢ A' represents the derivation of A from no assumptions and thus indicates that A is a theorem. The symbol '⊢' is usually called *turnstile* or *gatepost*. See also NATURAL DEDUCTION.

sequent calculus, *n.* any logical calculus presented in terms of SEQUENTS; for example, the INTRODUCTION RULE for conjunction

$$\frac{\Gamma \vdash A, \ \Delta \vdash B}{\Gamma \cup \Delta \ \vdash \ A \ \& \ B}$$

is a rule of a sequent calculus.

sequential convergence, *n.* the CONVERGENCE of a SEQUENCE, as contrasted with NET CONVERGENCE.

sequentially compact, *adj.* (of a set in a TOPOLOGICAL SPACE) such that every sequence contains a convergent subsequence with limit in the set. If the limit is not necessarily in the set, one speaks of *relative sequential compactness*. In a METRIC SPACE or in the weak topology of a BANACH SPACE, sequential compactness and compactness coincide for closed sets; so that, in a metric space such as the reals, a set is sequentially compact if and only if it is COMPACT. See NET CONVERGENCE.

serial, *adj.* (of a relation) CONNECTED, TRANSITIVE and ASYMMETRIC, thereby imposing an ORDER on all the members of the domain, such as *less than* on the natural numbers. See also ORDERING.

serial correlation, *n.* the property of a sequence of RANDOM VARIABLES, that adjacent members are correlated.

series, *n.* **1.** the sum of a finite or infinite SEQUENCE of terms; the series $a_0 + a_1 + a_2 + ... + a_n$ is often abbreviated

$$\sum a_i \ \text{or} \ \sum_{i=0}^{n} a_i$$

An infinite series has a sum if and only if the sequence of PARTIAL SUMS of its initial segments, that is, the sequence

$$\langle \ a_0, \ a_0 + a_1, \ a_0 + a_1 + a_2, \ ... \ \rangle,$$

CONVERGES.

2. see NORMAL SERIES.

serpentine, *n.* a curve that is symmetric about the origin and asymptotic to the *x*–axis, as shown in Fig. 336; it has canonical equation

$$x^2y + b^2y - a^2x \ = \ 0.$$

Fig. 336. **Serpentine**.

Serret–Frenet formulae, *n.* another name for the FRENET FORMULAE.

sesquilinear, *adj.* (of a function of two variables on a complex vector space) linear in the first variable and *conjugate linear* in the second variable; this holds for the inner product.

set, *n.* **1.** also called **class.** a collection, possibly infinite, of distinct numbers, objects, etc., that is treated as an entity in its own right, and with identity dependent only upon its members. For example,

$$\{3, \text{the moon}\}$$

is the set with two members, namely the number 3 and the moon; it is the same set as

$$\{\text{the moon}, 3\}$$

and

$$\{\text{the only known natural earth satellite, the smallest odd prime}\}.$$

2. in some formulations, a class that can also be a member of other classes.

set function, *n.* a function of which the domain is a CLASS of sets. See also MEASURE

set-theoretic paradoxes, *n.* a number of PARADOXES, such as RUSSELL'S PARADOX, CANTOR'S PARADOX, and the BURALI–FORTI PARADOX, that arise within the intuitive theory of sets, or within certain axiomatizations of SET THEORY that seek to capture that intuitive notion. Russell observed that they share a common structure: if P is a property and f a function on sets such that if $P(x)$ for all x in a set S, then $P[f(S)]$ and $f(S) \notin S$. However, if we then consider the set of all instances of P, that is $W = \{x : P(x)\}$, it follows that $P[f(W)]$ and $f(W) \notin W$, but by the definition of W, since $P[f(W)]$ it follows that $f(W) \in W$, a contradiction. The solutions to these paradoxes must deny the existence of either the function f or the set W; usually the latter device is adopted, for example, by imposing restrictions that distinguish between sets that may be members of others and PROPER CLASSES.

set theory, *n.* **1.** the elementary study of the properties of finite SETS or classes and their relations.

2. the extension of this study to include the properties of infinite sets.

3. (*Logic*) a theory constructed within first-order PREDICATE CALCULUS that yields the mathematical theory of classes, especially one that distinguishes sets from proper classes as a means of avoiding certain paradoxes. In AXIOMATIC SET THEORY the consequences of various sets of axioms are studied in the abstract, while NAIVE SET THEORY seeks to model the intuitive properties of sets as the consequences of a set of interpreted axioms. See also BOOLEAN ALGEBRA.

set-valued function, multivalued function, multifunction, carrier, or **point-to-set mapping,** *n.* a mapping that associates a number of different elements of the second set with the same element of the first; a mapping from a set into the POWER SET of another set. A ONE–MANY relation is thus regarded as a FUNCTION under which the image of a given argument is the set of its distinct images under the relation. See also CORRESPONDENCE.

sex-, *prefix denoting* six; for example, a *sextuple* is an ordered set of six members, and a *sexual* function is of the sixth degree.

sexagesimal, *adj.* related to, or based on, the number 60, or a system of MENSURATION to BASE 60, such as the division of time into hours, minutes and seconds, or the Babylonian system of counting.

sexagesimal measure, *n.* the measurement of angles in degrees, minutes, and seconds, each unit being $1/60$ of the preceding. Compare CIRCULAR MEASURE.

sextile, *n.* (*Statistics*) one of five values of a variable that divide its distribution into six equiprobable intervals; for example, the fifth sextile is the value of the variable below which lie $\frac{5}{6}$ of the population. See also PERCENTILE.

sfield, *n.* another term for SKEW FIELD.

sg or **sgn,** *abbrev. for* SIGNUM.

sh, *symbol for* the hyperbolic sine function, SINH.

sh⁻¹, *symbol for* the inverse hyperbolic sine function, ARC–SINH.

shadow prices, *n.* another name for the variables of a dual linear program in the DUALITY THEORY OF LINEAR PROGRAMMING, so called because of the economic interpretation of the dual program as determining equilibrium prices if the primal linear program models a production process.

sheaf, *n.* a family of planes that all pass through a single point. See also BUNDLE.

shear, *n.* a transformation in which all the points in one line or plane remain fixed while all other points move parallel to the fixed line or plane by a distance proportional to their distance from the fixed line or plane; for example, a shear transformation of a rectangle produces a parallelogram, as in Fig. 337.

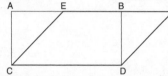

Fig. 337. ABCD and CDEF are related by a **shear** transformation.

shearing force, *n.* (*Mechanics*) the internal force perpendicular to the length of a thin beam.

shear stress, *n.* the component of the STRESS VECTOR, **t**, tangential to a given surface. Thus if **n** is the unit outward normal to the surface, then the shear stress is $\mathbf{t}^2 - (\mathbf{t} \cdot \mathbf{n})^2$. Compare HYDROSTATIC.

sheet, *n.* **1.** (*Euclidean geometry*) any of the maximal continuous parts of a surface on which it is possible to draw a curve from any point to any other point without leaving the surface. A HYPERBOLOID OF TWO SHEETS is a surface of two sheets separated by a finite distance between the vertices.

　2. (*Complex functions*) a portion of a RIEMANN SURFACE.

Sheffer's stroke, *n.* (*Logic*) a TRUTH FUNCTION of two sentences, equivalent to the negation of their conjunction, and written P | Q, where P and Q are the arguments. P | Q is false only when P and Q are both true, and it is possible to construct all truth functions out of this one alone; for example, P | P is equivalent to –P, and (P | Q) | (P | Q) is equivalent to P & Q, as shown in the truth table of Fig. 338.

P	Q	P | Q	P | P	(P | Q) | (P | Q)
T	T	F	F	T
T	F	T		F
F	T	T		F
F	F	T	T	F

Fig. 338.　Truth table for **Sheffer's stroke**.

shift, see UNILATERAL SHIFT.

short division, *n.* the method of dividing one number, usually an integer or decimal, by another, usually a small integer, by subtracting multiples of the divisor that have been calculated mentally and carrying the remainders, rather than by recording each step in the calculation as in LONG DIVISION.

short exact sequence, *n.* (*Category theory*) an EXACT SEQUENCE with five terms of which the first and last are trivial; that is,

$$0 \to a \overset{f}{\to} b \overset{g}{\to} c \to 0$$

when the image of *f* is the kernel of *g*, *f* is MONIC, and *g* is EPI.

short radius, *n.* the distance between the CENTRE of a regular polygon and any of its sides; the bold line in Fig. 339 is a short radius of a square. Compare LONG RADIUS.

Fig. 339. The **short radius** of a square.

shrinking, *n.* a HOMOTHETIC TRANSFORMATION of the form

$$x' = kx, \quad y' = ky,$$

where $0 < k < 1$.

SI, *abbrev. for* SYSTEME INTERNATIONAL.

side, *n.* **1.** one of the line segments that form a POLYGON.

2. one of the FACES of a POLYHEDRON.

side-condition, *n.* another term for CONSTRAINT.

sieve of Eratosthenes, *n.* the algorithm that obtains all the prime numbers less than any given integer *n*, by deleting from the set of all integers less than *n* the multiples of each of the primes up to $\sqrt{n}$ in turn. For example, to determine that 2003 is prime, one need only check that it is not divisible by 3, 5, 7, 11, 13, 17, 19, 23, 29, 31, 37, 41, and 43. Many more sophisticated sieves are used in prime number theory.

sigma, *n.* **1.** the symbol, Σ, denoting a sum, often written

$$\sum_{i=a}^{b} x_i = x_a + x_{a+1} + \dots + x_b$$

for the sum of the elements x_i from $i = a$ to b, or similarly over any other index set. If the sequence is infinite, it is written

$$\sum_{i=a}^{\infty} x_i.$$

See SERIES. Compare PI.

2. the symbol, σ, used to denote a countable property, such as F_σ.

3. the symbol, σ, for the SIGMA FUNCTION.

4. (*Statistics*) the symbol, σ, for STANDARD DEVIATION.

sigma-algebra or **σ-algebra,** *n.* (*Measure theory*) a collection of subsets of a set that contains the set itself, the empty set, the complements in the set of all members of the collection, and all countable unions of members.

sigma-compact or **σ-compact,** *adj.* (of a subset of a TOPOLOGICAL SPACE) expressible as the countable union of COMPACT sets.

sigma-finite or **σ-finite,** *adj.* (of a MEASURE) such that every measurable set is the countable union of sets of finite measure, as with Lebesgue measure on the line.

sigma-field (or **σ-field) of sets,** *n.* (*Probability theory*) a term for a SIGMA-ALGEBRA.

sigma function or **σ function,** *n.* (*Number theory*) **1.** the function, $\sigma(n)$, that sums the distinct divisors of n, including 1 and n. The sum of the PROPER FACTORS of n is therefore $\sigma(n) - n$. When p is prime,

$$\sigma(a) = \frac{p^{n+1} - 1}{p - 1},$$

and since σ is MULTIPLICATIVE, the value for any other argument can be easily computed from its prime factorization. In terms of this function, a PERFECT NUMBER is one with $\sigma(n) = 2n$, and a pair of AMICABLE NUMBERS have $\sigma(a) = \sigma(b) = a + b$.

2. more generally, the function $\sigma_k(n)$ that sums the kth powers of the divisors of n. In this notation, $\sigma_1(n)$ is $\sigma(n)$, and $\sigma_0(n)$ is the DIVISOR FUNCTION, $d(n)$.

sigma-ring or **σ-ring,** *n.* (*Measure theory*) a collection of subsets of a set that is closed under SYMMETRIC DIFFERENCE and countable union.

sign, *n.* **1.** any symbol indicating an operation, such as the plus sign or implication sign.

2. the positivity or negativity of a number, quantity or expression. Thus subtraction from zero changes the sign of an expression. See also SENSE, POLARITY.

signature, *n.* **1.** a number,

$$\varepsilon_{i_1, i_2, \ldots i_k},$$

denoting whether a PERMUTATION $(i_1 \, i_2 \ldots i_k)$ differs from the natural order by an odd or even number of steps: $\varepsilon = +1$ if the permutation is EVEN, and $\varepsilon = -1$ if it is ODD. See EPSILON (sense 3).

2. (of an HERMITIAN matrix or quadratic form) the surplus of positive over negative coefficients in any real DIAGONAL MATRIX or form similar to the given one; this equals the excess of positive over negative EIGENVALUES. See SYLVESTER'S LAW OF INERTIA.

signed, *adj.* able to take either SIGN; for example, *signed numbers,* or SIGNED MEASURES, as are treated by a JORDAN DECOMPOSITION.

signed measure, *n.* a countably additive set function that may take either sign. See MEASURE. See also JORDAN DECOMPOSITION.

signed minor, *n.* another term for COFACTOR.

signed number, *n.* a less common term for INTEGER.

signed-ranks test, (*Statistics*) see WILCOXON TEST.

significance, *n.* (*Statistics*) a measure of the confidence that can be placed in a result not being merely a matter of chance. The term is especially applied to the reliability of the rejection of a substantive causal hypothesis in HYPOTHESIS TESTING.

significance level, *n.* (*Statistics*) the probability in a test of wrongly rejecting the NULL HYPOTHESIS; thus a significance level of 5% or 0.05 means that there is no more than this probability of such an error (a TYPE I ERROR). Compare CONFIDENCE LEVEL, POWER. See HYPOTHESIS TESTING.

significance test, *n.* (*Statistics*) in HYPOTHESIS TESTING, a test of whether the ALTERNATIVE HYPOTHESIS achieves the predetermined SIGNIFICANCE LEVEL required for it to be accepted in preference to the NULL HYPOTHESIS.

significant, *adj.* (*Statistics*) (of a difference between an observation and a prediction) too large to be attributed to chance.

significant digits or **significant figures,** *n.* **1.** the digits of a number that express a quantity to some specified degree of accuracy, rounding the last figure up if the next would be 5 or greater. For example, 3.14159 to four significant digits is 3.142. See ACCURATE.

2. the digits of a number from the left-most non-zero digit to the rightmost non-zero digit, that is, from the largest to the smallest PLACE VALUE of which the coefficient is non-zero. In some usages, zeros with a high place value are regarded as significant; this difference in usage corresponds to the difference between FLOATING–POINT and FIXED–POINT notations.

sign test, *n.* (*Statistics*) a statistical test particularly used to analyse the direction of differences of scores between the same, or matched pairs of, subjects under two experimental conditions.

signum or **signum function,** *n.* the real function, denoted sgn(x) or sg(x), that assigns the SIGN of a non-zero number to that number; alternatively, sgn(x) can be defined as +1, 0, or –1, according as x is positive, zero, or negative. The term is also used for the corresponding function that sends x to $x/|x|$ and zero to zero, in any NORMED SPACE, and especially the complex numbers.

similar, *adj.* **1a.** (*Euclidean geometry*) (of two plane figures) having corresponding angles equal, whence having all corresponding pairs of sides in proportion; for example, in the quadrilaterals shown in Fig. 340, the ratios AB : KL and CD : MN are equal.

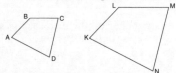

Fig. 340. **Similar** irregular quadrilaterals.

b. (of two sets of points) related by HOMOTHETY, without translation.

2. (of two classes) EQUIPOLLENT.

3. (of two matrices or operators A and B) such that there exists an invertible transformation, C, with A = C⁻¹BC; A and B then represent the same linear transformation with respect to bases related by C. Compare UNITARY EQUIVALENCE, EQUIVALENT (sense 6).

4. (of polynomial terms in several variables) containing the same powers of each indeterminate.

similarity or **similarity transformation,** *n.* (*Euclidean geometry*) a transformation preserving similarity, composed of some or all of TRANSLATION, ROTATION, and HOMOTHETY.

similitude or **transformation of similitude,** (*Geometry*) a HOMOTHETY that leaves the origin fixed; in vector terms, a transformation $\mathbf{x} \to k\mathbf{x}$ where k is a positive number (the *ratio of similitude*) and the origin is the *centre of similitude*. Two figures related by such a transformation are homothetic.

simple, *adj.* **1.** (of a root of an equation) occurring only once; not multiple.

2. (of a GROUP) containing no proper non-trivial NORMAL subgroup. One of the triumphs of 20th-century mathematics was the complete classification of finite simple groups largely achieved by Feit and Thompson. They divide into various regular classes (e.g., alternating groups, cyclic groups of prime order, Lie-type groups (five sub-varieties), and 26 *sporadic* exceptions, the largest of which is the MONSTER GROUP. The only commutative simple groups are cyclic groups of prime order; the smallest non-commutative simple group is the ALTERNATING GROUP A_5, and A_n is simple for $n > 4$.

b. (of a MODULE) having only itself and the zero module as submodules.

c. (of a non-ZERO RING) such that its only two-sided IDEALS are itself and zero; every simple ring is PRIME, every ARTINIAN simple ring is NOETHERIAN, and every commutative simple ring is a FIELD.

3. (of an equation) linear; having variables only of the first power.

4. (of a graph) having no LOOPS or multiple paths between the same pair of vertices.

5. (of an analytic function) another word for SCHLICHT.

6. (of an integration or summation) not ITERATED.

7. (of a MEASURABLE function) taking only finitely many values.

simple closed chain, *n.* a GRAPH whose initial and final NODES are identical, and in which no other node occurs more than once.

simple closed curve or **Jordan curve,** *n.* (*Complex analysis, Euclidean geometry*) a continuous curve (an ARC) in the complex plane that does not cross itself (is *simple*), but meets at the ends; if its equation is given by $z = z(t)$, then $z(t_1) = z(t_2)$ if and only if $t_1 = t_2$. See JORDAN'S CURVE THEOREM.

simple continued fraction, *n.* a CONTINUED FRACTION with unit numerators and integral denominators. See CONVERGENTS.

simple false position, see FALSE POSITION.

simple field extension, *n.* a subfield of a given EXTENSION FIELD generated from a given BASE FIELD by a single element; it is referred to as algebraic or transcendental over the base field according as the element is algebraic or transcendental.

simple fluid, *n.* (*Continuum mechanics*) a NEWTONIAN FLUID for which the variation of the DEVIATORIC part of the STRESS TENSOR, as a function of the VELOCITY GRADIENT, is independent of direction.

simple fraction, **common fraction,** or **proper fraction,** *n.* a fraction whose numerator is an integer of lower ABSOLUTE VALUE, or a polynomial of lower DEGREE, than its denominator, such as

$$\frac{(3 + x)^2}{x^4 + 2x^3 - 5x^2 + 3}.$$

simple harmonic approximation, *n.* the approximation of the motion of a particle by SIMPLE HARMONIC MOTION.

simple harmonic motion, *n.* motion governed by an equation of the form $y'' = -\omega^2 y$, which has solution $y = a\cos(\omega t + b)$, where a, b, ω are constants.

This describes motion in which acceleration is directed towards, and is proportional to distance from, the rest position; for example, it approximates to the motion of a simple pendulum provided its angle from the vertical is small.

simple interest, *n.* the interest accumulated over a given period when successive interest payments are not aggregated with the principal sum for the purpose of calculating the next interest payment. Under simple interest, capital of $\$C$ invested at $i\%$ *per annum* yields $\$i\% \times C$ every year in perpetuity, unlike COMPOUND INTEREST under which the interest is added to the principal so that the interest increases annually. After n years of simple interest, the total sum becomes $\$(1+ni/100)C$.

simple ordering, *n.* a linear ORDERING.

simple pole, see POLE.

simple root, see MULTIPLE ROOT.

simplex, *n.* (*pl.* **simplices.**) **1.** the most elementary geometric figure of a given dimension: the line in one dimension, the triangle in two dimensions, the tetrahedron in three, etc.

2. *n*-dimensional simplex. a POLYTOPE with $n+1$ affinely independent vertices; thus a triangle is a two-dimensional simplex. See also BARYCENTRE.

simplex method, *n.* the standard method of solving a LINEAR PROGRAMMING problem that proceeds by PIVOTING to produce a finite sequence of *basic feasible points* corresponding to vertices or extreme points of the feasible set and to linearly independent columns of the associated *simplex tableaux* or *simplex schemata* (these being notational schemes for keeping the updated information, especially in hand calculation). The variables corresponding to these columns are called *basic variables*. In the absence of *degeneracy* (the condition in which basic variables may be zero) the value strictly improves with each iteration and finite termination results. Otherwise *cycling* may occur (although it rarely does in practice) in that the same vertex is visited repeatedly. Variations of this method have been hugely successful in handling even enormous practical linear programming problems. See DUALITY THEORY OF LINEAR PROGRAMMING. See also HUNGARIAN METHOD, LINEAR PROGRAMMING, TRANSPORTATION PROBLEMS.

simplicial complex, *n.* a set consisting of a finite number of SIMPLICES with the property that any two intersect in a shared FACE or are disjoint. The *dimension* of the complex is the largest dimension of its component simplices. The term is also used for the underlying *skeleton* of vertices.

simplicial mapping, *n.* a mapping between two SIMPLICIAL COMPLEXES with the property that the image of any component SIMPLEX is a simplex.

simplicial triangulation, see TRIANGULATION (sense 3).

simplify, *vb.* **1.** to reduce an expression to a simpler form by cancellation of common factors, regrouping of terms in the same variable, etc.

2. to be reducible in this way; for example, $2x + 3y = y$ simplifies to $x = -y$.

simply connected, *adj.* (of a region of the complex plane) having no holes in it, so that its complement in the extended plane is also connected. For example, a circle is simply connected but an annulus is not, since its complement consists of two unconnected regions. More generally, connectivity of a surface is defined in terms of the EULER CHARACTERISTIC for the surface. A region in three-space is simply connected if every simple closed curve in the region bounds a surface of which the graph is in the region. This fails for the interior of the TORUS but holds for the sphere.

Simson line or **simson,** *n.* the line into which the PEDAL TRIANGLE of a point P on the circumcircle of a given triangle ABC degenerates.

Simpson's paradox, *n.* the statistical paradox described by the British statistician *E.H. Simpson* in 1951, in which two sets of data that separately confirm an hypothesis may support the opposite conclusion when considered together. For example, consider two tests of the comparative effectiveness of two drugs: in the first test, drug A is effective on 100 of 1000 patients (10%) while drug B is effective on 2000 of 10 000 (20%), and in the second test, A is effective on 4000 out of 10 000 (40%), and B on 600 out of 1000 (60%); it seems obvious that B is more effective in each test, yet when the results are aggregated it is apparent that A cured 4100 out of 11 000 (37%), while B cured only 2600 out of 11 000 (24%). Similarly, evidence may support two conclusions separately, but undermine their conjunction: two-thirds of detective stories may be by women, and so may two-thirds of novels about Paris, but fully two-thirds of detective stories about Paris may nonetheless be by men. See also VOTING PARADOX.

Simpson's rule, *n.* a method of approximating to an integral as a sum of quadratic terms; whence

$$\int_a^b f(x) \, dx \sim$$

$$\frac{\delta}{3} \left[f(a) + 4f(a+\delta) + 2f(a+2\delta) + 4f(a+3\delta) + 2f(a+4\delta) + ... + f(b) \right]$$

where $\delta = (b-a)/2n$. This is considerably more accurate than the TRAPEZOIDAL RULE, with an error of

$$\frac{M(b-a)^{2n}}{180 \, n^4}$$

where M is the maximum absolute value of the fourth derivative on the interval. The formula for cubics is exact and is known as the *prismoidal formula*. (Named after the English analyst, geometer, algebraist, and probabilist, *Thomas Simpson* (1710–61).)

simulation, *n.* (*Statistics, Computing*) the construction of a mathematical model for some process, situation, etc., in order to estimate its characteristics or solve problems about it probabilistically in terms of the model.

simultaneous differential equations, *n.* a set of DIFFERENTIAL EQUATIONS that are to be simultaneously satisfied. A set of simultaneous LINEAR DIFFERENTIAL EQUATIONS that occurs frequently (for example, in finding streamlines) is of the form

$$\frac{dx}{P} = \frac{dy}{Q} = \frac{dz}{R},$$

where P, Q and R are functions of the three variables x, y, z. The solution of these equations is sought by transforming them to a TOTAL DIFFERENTIAL

EQUATION; the method is to find constants a, b, c such that

$$\frac{a\,dx + b\,dy + c\,dz}{aP + bQ + cR}$$

has either zero denominator and an EXACT numerator, or a numerator that is the differential of the denominator.

simultaneous equations, n. a set of equations in several unknowns, especially when the equations are linear and algebraic, and the number of unknowns equals the number of independent equations; a unique solution can then be found by GAUSSIAN ELIMINATION. Such a solution is a set of values for the unknowns that satisfy all the equations simultaneously.

sin, *abbrev. and symbol for* the SINE function.

sin⁻¹, *symbol for* the inverse SINE function. See ARC–SINE.

sine, n. abbreviated **sin.** the TRIGONOMETRIC FUNCTION that is equal in a right-angled triangle to the ratio of the side opposite the given angle to the hypotenuse. If θ is the angle, measured in radians, swept out in an anticlockwise direction from the positive direction of the x-axis of a coordinate system by a radius of length r centred on the origin, then $\sin\theta = y/r$ where y is the ORDINATE of the end of the radius; in Fig. 341 this is the ratio XP/OP, and in general, $\sin\theta$ is the y-coordinate of the point on the unit circle round the origin whose position vector has an inclination of θ to the x-axis.

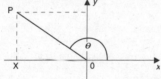

Fig. 341. **Sine**. $\sin\theta = $ XP/OP.

It is an ODD function of which the graph is the SINE CURVE, shown in Fig. 342. Its derivative is $\cos\theta$, the COSINE function and an antiderivative (or indefinite integral) is $-\cos\theta$; together they satisfy

$$\cos^2 z + \sin^2 z = 1$$

$$\sin(2z) = 2\cos z \sin z.$$

It is best defined, as a complex function, by the power series:

$$\sin z = \sum_{i=0}^{\infty}(-1)^n \frac{z^{2n+1}}{(2n+1)!}$$

See also DE MOIVRE'S FORMULAE.

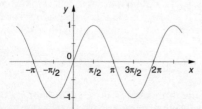

Fig. 342. Graph of the **sine** function.

sine curve, *n.* **1.** a curve with equation $y = \sin x$, the graph of the sine function. This lies between $y = -1$ and $y = 1$, is continuous, has maxima at $\pi/2 + 2n\pi$, minima at $3\pi/2 + 2n\pi$, and is zero at $n\pi$ for all integers n. (See Fig. 342).

2. another term for SINUSOID.

sine law or **sine rule,** *n.* the theorem that the sides of a triangle are proportional to the sines of the opposite angles. For a SPHERICAL TRIANGLE the sines of the lengths of the sides, measured in radians, are proportional to the sines of the opposite angles.

sine series, *n.* **1.** the SERIES expansion for the sine function:

$$\sin x = x - \frac{x^3}{3!} + \frac{x^5}{5!} - \frac{x^7}{7!} + \dots.$$

This is valid for all x.

2. any SERIES in which the terms are sine functions. See FOURIER SERIES.

single precision, see PRECISION.

singleton, *n.* any set containing precisely one member.

singular, *adj.* **1.** (of a square matrix) having DETERMINANT equal to zero; not possessing an INVERSE.

2. (of a continuous linear operator) **a.** not invertible.

b. in some usages, either not invertible or having a discontinuous inverse.

3. (of a solution to a differential equation) a solution not arising as a particularization of an attempted parametrization of a GENERAL SOLUTION. This is the case with $(y')^2 = 4y$ where the parametric family $y = (x + c)^2$ misses the singular solution zero.

4. (of a measure ν with respect to a measure μ) such that there exists a measurable set E with $\mu(E) = 0$ and $\nu(F) = \nu(F \cap E)$, for all measurable sets F; this relationship is symmetric. If two measures ν and μ are finite, then there is a *Lebesgue decomposition* $\nu = \nu_1 + \nu_2$ where ν_1 is singular with respect to $\mu (\nu_1 \perp \mu)$ and ν_2 is ABSOLUTELY CONTINUOUS with respect to μ. See also RADON–NIKODYM THEOREM.

singular point, *n.* **1.** also called **singularity**. any point at which a curve does not have a unique SMOOTH tangent, for example because it is isolated or the curve crosses itself. Compare ORDINARY POINT, DOUBLE POINT. See also CUSP, CRUNODE, SPINODE, TACNODE.

2. a point on the boundary of an open disk that is not a REGULAR POINT.

3. (of a second-order differential equation) see REGULAR SINGULAR POINT.

singularity, *n.* **1.** (*Complex analysis*) a point at which a function is not DIFFERENTIABLE although it is differentiable at points in any neighbourhood of that point; such a point may, however, be a REMOVABLE singularity. See also POLE, ISOLATED SINGULARITY, PICARD'S THEOREMS.

2. a DISCONTINUITY that is not removable.

3. another term for SINGULAR POINT (sense 1).

singulary, *adj.* (of an operator, etc.) another word for MONADIC.

singular solution, *n.* a solution of an ORDINARY DIFFERENTIAL EQUATION that cannot be obtained from the GENERAL SOLUTION by choosing suitable values of the arbitrary constants.

singular value, *n.* (of a real matrix A) any of the square roots of the eigenvalues, usually listed in decreasing order, of the product of a real matrix, A, with its transpose; for a square symmetric matrix these values can be obtained from the SPECTRAL DECOMPOSITION of A.

singular value decomposition, *n.* the representation of a real NORMAL MATRIX A as USU*, where U is UNITARY, U* is the ADJOINT of U, and S is a diagonal matrix of which the entries are the SINGULAR VALUES of A.

sinh or **sh,** *symbol for* the HYPERBOLIC FUNCTION hyperbolic sine; this is related to the SINE function, for any complex number z, by the identity

$$\sinh z = -i \sin iz$$

where $i = \sqrt{-1}$. It can be defined in terms of the EXPONENTIAL FUNCTION as

$$\sinh z = \frac{e^z - e^{-z}}{2}.$$

It is an ODD function of which both the derivative and an antiderivative (or indefinite integral) are COSH, the hyperbolic cosine function; its graph is shown in Fig. 343. Together with cosh it satisfies

$$\cosh^2 z - \sinh^2 z = 1$$
$$\sinh(2z) = 2 \sinh z \cosh z.$$

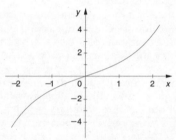

Fig. 343. **Sinh**. Graph of the hyperbolic sine function.

sinh^{-1}, *symbol for* the inverse hyperbolic sine function, ARC–SINH.

sink or **terminal,** see NETWORK.

sinusoid, *n.* any curve derivable from the SINE CURVE by multiplication by or addition of a constant; a curve of the same shape as the sine curve but with possibly different amplitude, period, or intercepts with the axis. Any such curve plots the perpendicular distance from a fixed diameter of a point moving with constant speed round a circle.

sinusoidal, *adj.* relating to, and especially shaped like, a SINE CURVE.

Sion minimax theorem, see MINIMAX THEOREM.

sistroid, *adj.* contained between the convex sides of two intersecting curves. Compare CISSOID.

skeleton, see SIMPLICIAL COMPLEX.

skew, *adj.* **1a.** also called **agonic.** neither parallel nor intersecting, such as two lines not lying in the same plane in a three-dimensional space; for example, the non-intersecting diagonals of adjacent faces of a cuboid, as shown in Fig. 344 opposite, are skew.

b. (of a curve) also called **twisted.** not lying in a plane.

2. (of a matrix) SKEW–SYMMETRIC or SKEW–HERMITIAN.

3. (*Statistics*) (of a distribution) not symmetrical.

Fig. 344. The bold lines are **skew** (sense 1).

skew field or **sfield,** *n.* a DIVISION RING, a mathematical system that satisfies all the FIELD axioms except commutativity of multiplication, such as, for example, the system of quaternions.

skew-Hermitian, *adj.* (of a matrix) equal to the negative of its ADJOINT.

skew-metric, see METRIC.

skewness, *n.* (*Statistics*) a measure of the symmetry of a distribution about its mean, especially the statistic

$$B_1 = \frac{m_3}{(m_2)^{3/2}},$$

where m_2 and m_3 are respectively the second and third MOMENTS of the distribution around the mean; in a normal distribution, $B_1 = 0$. Compare KURTOSIS.

skew-symmetric, *adj.* (of a matrix) equal to the negative of its TRANSPOSE.

Skolem form, *n.* (*Logic*) a PRENEX NORMAL FORM of a formula in which all universal quantifiers precede any existential quantifier and that contains no function symbols.

Skolem paradox, *n.* (*Logic*) the corollary of the LÖWENHEIM–SKOLEM THEO-REM that even although real arithmetic is provably uncountable, it has a countable model. The resolution is that we require to distinguish what is true in the model from what is true of the model: while the set representing the real numbers in the model is indeed countable, it is not countable in the model; the mapping between the model of the reals and the natural numbers is not itself in the model, so that within the countable model there is no contradiction to the theorem that the real numbers are un-countable in terms of the model. This dramatizes the fact that even cardinality is relative to the theory within which it is defined. More generally, the intuition that all models of a complete theory are isomorphic is false.

slack variable, *n.* a variable that is added to replace an inequality of the form $g(x) \leq 0$ by the equality $g(x) + y = 0$ and the inequality $y \geq 0$. This is often done in linear programming to enable one to place the linear program in *standard form*, which involves only equality constraints and no negativity requirements on the variables. The corresponding replacement of $g(x) \geq 0$ by the equality $g(x) - y = 0$ and the inequality $y \geq 0$, leads to a *surplus variable*. In linear programming one also encounters *artificial variables*, that is, variables that are introduced to ease a computation but are disposed of ('driven out') during the computation.

slant height, *n.* the length of a generating segment of a right circular cone

Fig. 345. **Slant height.**

or frustum; the distance from the vertex to the base (or in a frustum, between bases) measured along the surface. The slant height is shown as h in Fig. 345 above.

Slater's condition, n. the CONSTRAINT QUALIFICATION placed on a set of inequalities

$$g_1(x) \leq 0, \ldots, g_n(x) \leq 0$$

(usually with convex functions), that the inequalities should have a simultaneous strict solution: a vector z with

$$g_1(z) < 0, \ldots, g_n(z) < 0.$$

slide rule, n. a device for multiplying and dividing numbers, consisting of two strips, one usually sliding in a central groove of the other, and each marked with LOGARITHMIC SCALES of numbers, trigonometric functions, etc. Two numbers are multiplied by placing 1 on the second scale opposite one of the multiplicands on the first scale, and reading off the number on the first scale that is opposite the second multiplicand on the second scale; the effect of this is to take the antilogarithm of the sum of the logarithms of the multiplicands. The principle is illustrated by Fig. 346, using a slide rule to base 2, with the multiplicands indicated by the small arrows and the product by the large arrow.

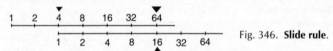

Fig. 346. **Slide rule**.

slope, n. **1a.** (of a line in a coordinate system) the tangent of the angle between the direction of the line and the x-axis; in Fig. 347 below, this is $\tan\theta$, that is, the ratio Dy/Dx. See also GRADIENT.
b. (of a straight line or segment) the ratio of the RISE to the RUN of any two distinct points on the line or segment.
2. the first DERIVATIVE of the equation of a curve at a given point. This is equal to the limit of $\delta x/\delta y$ as δx tends to zero, the slope of the tangent to the curve at that point, and Fig. 347 shows the equivalence of the three definitions.

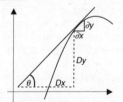

Fig. 347. **Slope**. See main entry.

slope–intercept equation, see LINE.

Slutsky's theorem, n. (*Statistics*) the result that if $X_1,\ldots, X_n$ is a sequence of RANDOM VARIABLES such that

$$\lim_{n \to \infty} P[X_n \leq x] = P[X \leq x]$$

for some random variable X for which $P[X \leq x]$ is continuous everywhere, then for any continuous function, g,

$$\lim_{n \to \infty} P[\,g(X_n) \leq y\,] \;=\; P[\,g(X) \leq y\,].$$

small or **in the small** (*im kleinen*), see LOCAL.

small circle, *n.* a circular SECTION of a sphere by a plane that does not contain the centre of the sphere. Compare GREAT CIRCLE.

smooth, *adj.* **1.** (of a function or curve) differentiable at every point; usually, continuity of the derivative is demanded.

2. (of a NORM) linearly GATEAUX DIFFERENTIABLE everywhere except at zero.

3. (of a MANIFOLD) differentiable of class greater than or equal to 1; having a $C^{(r)}$-ATLAS for $r \geq 1$.

sn, see JACOBIAN ELLIPTIC FUNCTIONS.

snowflake, see FRACTAL.

solenoidal, *adj.* (of a vector function in a region) having DIVERGENCE of zero throughout; being the curl of some potential.

solid, *n.* **1.** having, or relating to, three dimensions. For example, a solid figure is a three-dimensional diagram. See SOLID GEOMETRY.

2. (*as substantive*) a bounded volume in three-dimensional space, or the closed surface that bounds it.

solid angle, *n.* a geometrical surface consisting of lines originating from a common point (the vertex) and passing through a closed curve or polygon, as in Fig. 348. See STERADIAN.

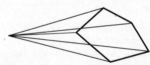

Fig. 348. **Solid angle.**

solid figure, *n.* a figure in three-dimensional Euclidean geometry.

solid geometry, *n.* the branch of geometry concerned with the properties of three-dimensional geometric figures.

solid of revolution, *n.* the solid figure generated by revolving a given curve around a line. If the axis of revolution is the x-axis, then the volume generated by the segment of $y = f(x)$ between $x = a$ and $x = b$ is

$$\pi \int_{a}^{b} f^{2}(x) \; \mathrm{d}x;$$

Fig. 349 shows this solid and an element of the integration. See also SURFACE OF REVOLUTION, PAPPUS' THEOREM.

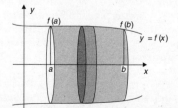

Fig. 349. **Solid of revolution.**

523

soluble or **solvable,** *n.* able to be solved. See also SOLVABLE PROBLEM.

soluble group or **solvable group,** *n.* a group with a NORMAL SERIES in which every NORMAL FACTOR is ABELIAN; equivalently, for some positive integer, the DERIVED SERIES terminates in the TRIVIAL SUBGROUP, or, for a finite group, the COMPOSITION FACTORS have prime order. For $n \geq 5$, the symmetric group S_n is insoluble, while for $1 \leq n \leq 4$, S_n is soluble. This enables the methods of Galois to be used to prove the celebrated result that the quintic cannot be solved by radicals; a polynomial has SOLUTION BY RADICALS if and only if its GALOIS GROUP is soluble.

solution, *n.* **1.** a unique set of values that yield a true statement when substituted for the unknowns in an equation or set of equations.
2. a member of a set of assignments of values to variables under which a given statement is satisfied, that is, a member of a TRUTH SET; for example, the minimum of a linear program.

solution by radicals, *n.* the possibility of obtaining an expression for the roots of a polynomial equation involving only rational operations and radicals, or, more precisely, as the end product of a *tower of radicals*: a finite sequence of numbers each of which is the radical of a polynomial in the previous members of the sequence. This is possible for all equations of degree less than 5, but not in general for QUINTIC or higher-degree polynomials, as a consequence of results of Abel and Galois. See QUADRATIC EQUATION, CARDANO'S FORMULA, GALOIS THEORY.

solution curve, *n.* the curve

$$\{(t, \mathbf{y}(t)) : t \in I\}$$

where $\mathbf{y}$ is the solution of a system of ORDINARY DIFFERENTIAL EQUATIONS, and I is the INTERVAL OF EXISTENCE of $\mathbf{y}$. Compare TRAJECTORY.

solution set, *n.* **1a.** the set of all solutions of a system of algebraic equations.
b. another term for TRUTH SET.
2. the set of optimal values in an optimization problem.

solvable, *adj.* another term for SOLUBLE. See also SOLUBLE GROUP.

solvable problem, *n.* (*Computing, logic*) a decision problem that possesses an algorithm that, given a suitably presented instance of the problem, returns the answer for that instance. If no such algorithm exists, the problem is *unsolvable*.

solve, *vb.* **1.** to find the value or set of values of the variables that SATISFY (an equation or system of equations).
2. (*Trigonometry*) to find the lengths of all the sides and sizes of all the angles of (a triangle), given information only about some of them, using trigonometric formulae such as the SINE LAW.

sound, *adj.* (*Logic*) **1.** (of a DEDUCTIVE argument) another word for VALID.
2. (of an INDUCTIVE argument) according with whatever principles ensure a high probability of the conclusion being true, given that the premises are.
3. (of a formal system) another word for CONSISTENT.

source, see NETWORK.

Souslin set or **analytic set,** *n.* the continuous image of a POLISH SPACE. (Named after the Russian analyst and topologist, *Michael Jakovlevich Souslin* (1894–1919).) See also UNIVERSALLY MEASURABLE.

space, *n.* **1.** a set of points endowed with a structure that is usually defined by specifying a set of axioms to be satisfied by the points. See BANACH SPACE, EUCLIDEAN SPACE, HILBERT SPACE, METRIC SPACE, N–SPACE, NORMED SPACE, TOPOLOGICAL SPACE, VECTOR SPACE. See also STRUCTURE (sense 2).

2. (*Mechanics*) a primitive concept that in Newtonian mechanics is assumed to be a EUCLIDEAN SPACE with a set of CARTESIAN COORDINATES with fixed axes, distance measured in METRES, and angles measured in RADIANS.

space curve, *n.* a curve in three-dimensional space; the boundary of a bounded surface.

space-filling curve, *n.* a pathological curve that passes through every point in a space of two or more dimensions; for example, a PEANO CURVE.

space–time, *n.* a four-dimensional SPACE used in relativistic physics to represent space and time and their interrelationship. See LORENZ GROUP.

span, *n.* **1.** also called **hull.** the CLOSURE of a set under some operation; the smallest set containing a given set and having a specified property. In particular, the *linear span* of a set in a vector space is the smallest linear subspace containing the set; the *closed span* and *affine span* are similarly defined.

– *vb.* **2.** to have (a given set) as its span; to contain all elements of (the given set) in the set of linear combinations of its elements. For example, the vectors $(0, 1)$ and $(1, 0)$ span the real plane.

spanning tree, *n.* (*Graph theory*) a TREE linking a given set of NODES of a graph. The tree then *spans* those nodes.

sparse, *adj.* (of a matrix or an equation system) having a high proportion of its entries zero, as frequently occurs in application. Matrices with a high proportion of non-zero entries are called *dense.* See STAIRCASE STRUCTURE.

sparse matrix technique, *n.* any technique that exploits the properties of SPARSE matrices, particularly those that are well structured, in order to reduce dramatically the work involved in solving, storing or manipulating systems of equations.

spatial equation of continuity or **continuity equation,** *n.* (*Continuum mechanics*) the consequence of the conservation of MASS that

$$\dot{\rho} + \rho \operatorname{div} \mathbf{v} = 0,$$

where **v** is the VELOCITY of the particles of a BODY with DENSITY ρ, and the DIVERGENCE is taken with respect to its current CONFIGURATION.

spatial description or **Eulerian description,** *n.* the description of physical phenomena associated with the deformation of a body in terms of FIELDS defined over the current, rather than the reference, CONFIGURATION. Compare MATERIAL DESCRIPTION.

Spearman's rank order (**correlation**) **coefficient,** *n.* (*Statistics*) a statistic that measures the extent to which two sets of discrete data place the distinct items in the same order, such as the rankings of a group of people by height and by weight, given by

$$r_s = 1 - \frac{6 \sum d^2}{n(n^2 - 1)},$$

where $\sum d^2$ is the sum, over all the items, of the squares of the differences of the ranks of each item in the two orderings, and n is the number of items.

special function, *n.* any specially identified non-ELEMENTARY TRANSCENDEN-
TAL function, among the most important of which are the BETA FUNCTION,
GAMMA FUNCTION, ZETA FUNCTION, ELLIPTIC FUNCTIONS, BESSEL FUNCTIONS,
THETA FUNCTION, and HYPERGEOMETRIC FUNCTIONS.

special induction, *n.* another term for FIRST-KIND INDUCTION, by contrast
with general induction. See INDUCTION.

special integral, *n.* a solution of a PARTIAL DIFFERENTIAL EQUATION that cannot be obtained from the GENERAL SOLUTION by substituting suitable functions for the arbitrary functions.

special linear group, *n.* the NORMAL SUBGROUP of the GENERAL LINEAR GROUP
consisting of all matrices with determinant equal to 1, denoted $SL(n, F)$. In
some cases the base field is replaced by the integers.

special orthogonal group, *n.* the NORMAL SUBGROUP of the ORTHOGONAL
GROUP consisting of all matrices whose determinant is equal to 1; the special orthogonal group is denoted $SO(n)$.

special theta function, see THETA FUNCTION.

species, *n.* a classification of a set in a TOPOLOGICAL SPACE in terms of whether
the process of computing the sequence each member of which is the DERIV-
ED SET of the preceding, starting with the given set, terminates finitely with
the empty set. If this does happen the set is of the *first species*; if it does not
happen the set is of the *second species.*

spectral decomposition, *n.* the expression of a NORMAL MATRIX A as UDU*
where U is unitary and D is diagonal; U may be taken to be real if A is real
and symmetric.

spectral form, *n.* the representation

$$\mathbf{s} = \sum_{i=1}^{n} \lambda_i \mathbf{u}_i \otimes \mathbf{u}_i$$

of a symmetric second order CARTESIAN TENSOR **s** over an n-dimensional
space, where λ_i are the EIGENVALUES and $\mathbf{u}_i$ the EIGENVECTORS of **s**.

spectral integral, see SPECTRAL THEOREM.

spectral radius, *n.* the maximum modulus of members of the SPECTRUM of a
given matrix or operator; this coincides with

$$\lim_{n \to \infty} \| T^n \|^{1/n}.$$

spectral theorem, *n.* the theorem asserting that a bounded linear operator
can be reconstructed in a canonical fashion as a *spectral integral* with respect
to a family of commuting self-adjoint projections defined on the SPECTRUM
of the given operator. For NORMAL MATRICES or for compact NORMAL OPERA-
TORS, this becomes $T = \sum \lambda_i P_i$ where P_i is the orthogonal projection on the
null space of $\lambda_i I - T$. See also SPECTRAL DECOMPOSITION.

spectrum, *n.* the set of complex numbers, denoted $\sigma(T)$, for which the RE-
SOLVENT of a matrix or (bounded) linear operator on a normed space fails
to exist as a bounded linear operator, by virtue either of lying in the *point
spectrum* where $\lambda I - T$ is not one-to-one, or of lying in the *continuous spectrum* where $\lambda I - T$ is one-to-one with proper dense range, or of lying in the
residual spectrum, that is, the remainder of the spectrum, which is the set of
numbers for which the range of $\lambda I - T$ is not dense. For a matrix, only the

point spectrum occurs and is the set of eigenvalues. For a bounded linear operator in BANACH SPACE, the spectrum is bounded and non-empty, while the residual spectrum is empty if T is NORMAL.

Sperner's lemma, n. the result that, given a triangulation of an n–SIMPLEX $\{s_0, s_1, ..., s_n\}$, and a labelling mapping $f : T \to N$ from the vertices of the triangulation T into the integers $N = \{0, 1, ..., n\}$ such that if I is any subset of N and v lies in the face generated by $\{s_i : i \in I\}$ then $f(v) \in I$, then there is a simplex in the triangulation of which the image is $\{0, 1, ..., n\}$.

sphere, n. **1a.** (*Euclidean geometry*) a three-dimensional closed surface every point of which is equidistant from a given point (its centre). Its equation is

$$(x - a)^2 + (y - b)^2 + (z - c)^2 = r^2,$$

(in Cartesian coordinates) where r is the radius, and (a, b, c) is the centre. The surface area of a sphere is $4\pi r^2$.

b. the solid figure bounded by this surface, or the space enclosed by it, with volume $\frac{4}{3}\pi r^3$.

2. (in a METRIC SPACE) **a.** the set of points equidistant in metric from a given point.

b. less commonly, another word for BALL.

3. (in a TOPOLOGICAL SPACE) another word for NEIGHBOURHOOD.

4. (in ALGEBRAIC TOPOLOGY) the bicontinuous image of the unit sphere in Euclidean n-space, written S^{n-1}.

sphere-packing problem, n. any of the class of problems involving the arrangement, in a region of n-space, of rigid, uniform, disjoint SPHERES in such a way as to optimize the volume of the spheres. The LEECH LATTICE is a 24-dimensional integer lattice that produces a particularly good packing.

spherical, *adj.* of, relating to, or shaped like a sphere. See SPHERICAL GEOMETRY, SPHERICAL COORDINATES, SPHERICAL TRIANGLE.

spherical angle, n. the angle formed at the intersection of two GREAT CIRCLES of a sphere, equal to the angle between their tangents at that point.

spherical cap, n. a portion of a sphere lying on one side of a plane that intersects the sphere. Compare ZONE.

spherical coordinates, n. the system of representing a point in three dimensional space in terms of its POSITION VECTOR: the point is identified by a triple (r, ϕ, θ), where r is the length of the position vector, $\theta \in [0, \pi]$ is the angle between it and one of the coordinate axes, and $\phi \in [0, 2\pi)$ is the angle from the plane in which both that axis and the position vector lie to either of the coordinate planes including that axis; as in Fig. 350, θ is generally taken to be the angle between OP and the z-axis, and ϕ the angle between the OPz plane and the x–z plane. Thus θ is the polar angle of the

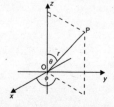

Fig. 350. **Spherical coordinates**. See main entry.

between the OPz plane and the x–z plane. Thus θ is the polar angle of the projection of OP on the x–y plane. Spherical coordinates are related to CARTESIAN COORDINATES by

$$x = r\sin\theta\cos\phi$$
$$y = r\sin\theta\sin\phi$$
$$z = r\cos\theta;$$

the JACOBIAN of the transformation to rectangular coordinates is then $r^2\sin\phi$. Compare CYLINDRICAL COORDINATES.

spherical excess, *n.* the amount by which the sum of the angles of a SPHERICAL TRIANGLE exceeds two right angles, or, more generally, the amount by which the sum of the angles of a SPHERICAL POLYGON exceeds $\pi(n-2)$ radians, where *n* is the number of sides of the polygon. This difference is related to the area of the figure by the formula $A = \pi r^2 E$, where E is the ratio of spherical excess to two right angles, and *r* is the radius of the sphere.

spherical geometry, *n.* **1.** the branch of geometry concerned with the properties of figures formed on the surface of a sphere, especially by the intersection of GREAT CIRCLES.

2. a NON–EUCLIDEAN GEOMETRY that is a RIEMANNIAN GEOMETRY, and has a model on the surface of a sphere.

spherical harmonic, *n.* a special solution of degree *n* to LAPLACE'S EQUATION in polar form, that is homogeneous of degree *n* and polynomial in three variables, denoted H_n. The spherical harmonics are built out of LEGENDRE POLYNOMIALS and associated functions, and are such that any solution analytic around the origin may be expressed as an infinite sum of such functions, one of each degree.

spherical polygon, *n.* a closed geometric figure formed on the surface of a sphere and bounded by intersecting MINOR ARCS of a number of GREAT CIRCLES; the area of the polygon is proportional to its SPHERICAL EXCESS. A SPHERICAL TRIANGLE such as that of Fig. 351 below, is a spherical polygon.

spherical surface, *n.* a surface of constant and positive TOTAL CURVATURE. Compare PSEUDOSPHERE.

spherical triangle, *n.* a closed geometric figure formed on the surface of a sphere and bounded by intersecting MINOR ARCS of three GREAT CIRCLES as in Fig. 351. The SPHERICAL ANGLES between these arcs may be all acute, all right, or all obtuse, and their sum must be strictly between π and 3π radians. The area of the triangle is proportional to its SPHERICAL EXCESS. There are eight triangles formed by three distinct great circles, but only one is formed from minor arcs and referred to as a spherical triangle; the properties of the others can be deduced from it. The study of these properties is SPHERICAL TRIGONOMETRY. Compare CIRCULAR TRIANGLE.

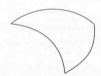

Fig. 351. **Spherical triangle**. See main entry.

spherical trigonometry, *n.* the branch of TRIGONOMETRY concerned with the measurement of the angles and sides of SPHERICAL TRIANGLES. This has importance in observational astronomy.

sphericity, *n.* the state or fact of being spherical.

spherics, *n.* the geometry and trigonometry of figures on the surface of a sphere.

spheroid, *n.* another name for an ELLIPSOID, especially an ELLIPSOID OF REVOLUTION.

spheroidal, *adj.* shaped like a SPHEROID; approximately spherical.

spheroidicity, *n.* the state or fact of being SPHEROIDAL.

spherometer, *n.* an instrument for measuring the CURVATURE of a surface.

spinode, *n.* another term for CUSP. Compare TACNODE, CRUNODE.

spin tensor, *n.* another term for BODY SPIN.

spiral, *n.* **1.** any plane curve formed by a point winding around a fixed point at an ever-increasing distance from it, such as the curves in Fig. 352. The polar equation of an *Archimedes spiral* is $r = a\theta$; of a *logarithmic spiral*, $\log r = a\theta$; and of a *hyperbolic spiral*, $r\theta = a$, where a is a constant.

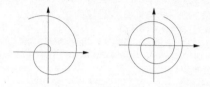

Fig. 352. Archimedean and logarithmic **spirals**.

2. a HELIX in Euclidean space.

spiral similarity, *n.* a dilatative rotation, the product of a dilatation and rotation in either order.

spline-fitting, *n.* (*Numerical analysis*) a popular type of piecewise APPROXIMATION by POLYNOMIALS of degree n (or more general functions) on an interval; these are fitted to the function at specified points on the interval (*nodes* or *knots*) where the polynomial used can change but the derivatives of the polynomials are required to match up to degree $n - 1$ at each side of the nodes, or to meet related interpolatory conditions. In addition, boundary conditions are imposed at the ends of the interval. For the *natural cubic spline*, one uses cubic polynomials and requires that the second derivative of the spline vanishes at the endpoints (a *natural boundary condition*). The *clamped boundary condition* demands that the first derivatives of the function and the spline agree at the end points.

split exact sequence, *n.* a SHORT EXACT SEQUENCE in which the second non-trivial mapping, g, has a right inverse, g', with $g \circ g' = 1$ (or equivalently, the first non-trivial mapping has a left inverse).

splitting field, *n.* the smallest EXTENSION FIELD in which a given polynomial over a given field *splits* into linear factors, this being unique up to isomorphism. Splitting fields correspond to finite NORMAL EXTENSION FIELDS.

sporadic, *adj.* (of a finite group) see SIMPLE (sense 2).

spread, *n.* a TREE whose paths are infinite, and whose nodes are sequences of natural numbers corresponding to the initial segments of the infinite sequences that are generated in accordance with some *spread law*.

spur, *n.* (*German*) another term for TRACE.

sq, *abbrev. for* SQUARE.

square (abbrev. **sq**), *n.* **1.** a plane geometric figure with four sides of equal length and four right angles; an equilateral rectangle or an equiangular rhombus.

2a. the product of two equal factors; for example, 9 is the square of 3, written 3^2. See SQUARE NUMBER.

b. the second power.

– *adj.* **3.** possessing or forming a right angle, perpendicular.

4a. (*prenominal*) denoting a measure of two-dimensional extent that is derived from a linear measure by raising it to the second power. For example, a square metre is the area enclosed by a square each side of which is one metre long.

b. (*postpositive*) denoting the extent of a square shape, each side of which is of the stated length. For example, an area three metres square is nine square metres in area.

– *vb.* **5.** to raise to the second power.

square bracket, *n.* either of the pair of BRACKETS, [and], used to indicate that the expression between them is to be evaluated before the remainder of the formula and treated as a single unit in the evaluation of the whole. In some conventions, these are only used in expressions that already contain PARENTHESES, and have lower priority than them, but higher priority than BRACES.

squared, *adj.* raised to the second power, having exponent 2.

square-free or *quadratfrei,* *adj.* (of an integer) containing no repeated prime factors.

square-integrable, *adj.* (of a measurable function on a set) such that the square of the modulus of the function has a finite integral. The set of all Lebesgue square-integrable functions defined on an interval comprises the HILBERT SPACE L_2, when functions that differ on sets of zero measure are identified and their common square-integral value is used as the norm. Compare CONVERGENT IN MEAN, RIESZ–FISCHER THEOREM, L_p SPACE.

square matrix, *n.* a matrix with the same number of rows and columns. Such a matrix has an inverse if and only if its DETERMINANT is non-zero. This should not be confused with a matrix square that is a matrix expressible as the product of another matrix with itself.

square number, *n.* an integer that is a PERFECT SQUARE of another integer, such as 1, 4, 9, 16, 25, etc. Compare FIGURATE NUMBER.

square root (abbrev. **sqrt** or **sqr**), *n.* a number or quantity that when multiplied by itself is equal to a given number or quantity, usually written $\sqrt{x}$ in arithmetic expressions, and $x^{1/2}$ in algebraic expressions. Every POSITIVE DEFINITE operator, A, has a unique positive definite square root, B, for which A = BB, and every non-singular matrix over the complex field possesses square roots.

square root theorem, *n.* the theorem that if H is a POSITIVE DEFINITE HERMITIAN matrix then there exists a unique positive definite Hermitian matrix G such that $H = G^2$.

square-summable, *adj.* (of a sequence) such that the series of the squares of its terms converges to a finite sum.

square the circle, *vb.* to construct a square of the same area as a given circle using a straight-edge and compasses only; this traditional geometrical problem has long been believed to be impossible, but this was not proved until 1882 as a consequence of the transcendence of π. See LINDEMANN THEOREM.

sqrt or **sqr,** *abbrev. for* SQUARE ROOT, common in computer programming.

squeeze rule, *n.* another term for SANDWICH RESULT.

stabilizer, *n.* the subgroup of elements of a GROUP of PERMUTATIONS of a non-empty set, under which the image of a given subset is itself.

stable, *adj.* (*Numerical analysis*) **1.** (of a problem or computational method) not overly sensitive to marginal PERTURBATIONS in the underlying data, generally meaning that the output should be continuous in some sense as a function of the perturbation. The term is used both numerically and theoretically.
2. (of a system of differential equations) such that any solution that starts close enough to a STATIONARY POINT returns over time. The solution is *totally stable* if it returns from all perturbations.
3. (of an EQUILIBRIUM POINT, $\mathbf{y}^\varepsilon$, of a system of LINEAR ORDINARY DIFFERENTIAL EQUATIONS) such that, for all positive ε, there exists a δ > 0 such that

$$\text{if } \|\mathbf{y}(0) - \mathbf{y}^\varepsilon\| < \delta, \text{ then } \|\mathbf{y}(t) - \mathbf{y}^\varepsilon\| < \varepsilon$$

for all non-negative *t*. If in addition, there is a positive R such that given positive ε, there exists a positive T, such that

$$\text{if } \|\mathbf{y}(0) - \mathbf{y}^\varepsilon\| < R, \text{ then } \|\mathbf{y}(t) - \mathbf{y}^\varepsilon\| < \varepsilon$$

for all $t \geq T$, then $\mathbf{y}^\varepsilon$ is *asymptotically stable*. If $\mathbf{y}^\varepsilon$ is not stable it is *unstable*.
4. Liapunov stable. (of an initial value of a differential equation) such that any solution that starts close enough to the initial value remains close over time. The solution is *asymptotically stable* if, in addition, as time tends to infinity the solution also converges to the initial value. Stability may be verified by construction of a LIAPUNOV FUNCTION.

stadium paradox, *n.* the classical paradox derived from the suppostion that there arc smallest indivisible units both of time and of distance. In essence, it considers two objects moving from a given point in opposite directions, for example runners in a stadium, at a constant rate for one unit of time; relative to each other they move one unit of space in half a unit of time, and so there must be a unit smaller than the originally supposed unit. See ZENO'S PARADOXES.

staircase structure, *n.* the SPARSE structure of many matrices arising in linear programming problems that model multistage production processes. The non-zero entries are clustered in blocks around the main diagonal and so resemble a staircase.

standard basis, another term for CANONICAL BASIS.

standard deviation, *n.* (*Statistics*) **1a.** a measure of the dispersion of a distribution, given by

$$\sqrt{E[(X - E[X])^2]},$$

the square root of the VARIANCE. The standard deviation is written σ as a population PARAMETER, where σ^2 is the population variance.

b. (abbrev. **sd**) hence, the unit whose multiples are used to describe the divergence of a RANDOM VARIABLE from its mean. One standard deviation from the mean, μ, of a normal distribution is either μ – σ or μ + σ; the interval within one sd of the mean contains approximately 68% of the population. See STANDARDIZE, STANDARD NORMAL DISTRIBUTION.

2. the corresponding sample STATISTIC, written *s* and used to estimate σ. This is given by

$$s^2 = \frac{\sum (x_1 - \overline{x})^2}{n-1}.$$

Compare MEAN DEVIATION.

standard equation, *n.* the canonical form of the equation of a curve derived by a suitable change of variables.

standard error (abbrev. **se**), *n.* (*Statistics*) the STANDARD DEVIATION of an ESTIMATOR of a population parameter.

standard form, *n.* another term for SCIENTIFIC NOTATION.

standard form of a linear program, see SLACK VARIABLE.

standard index form, *n.* another term for SCIENTIFIC NOTATION.

standard infinitesimal, see STANDARD PART.

standardize, *vb.* (*Statistics*) to derive a DISTRIBUTION from a given distribution, especially a NORMAL DISTRIBUTION, by a change of variables, so that the MEAN becomes zero and the VARIANCE becomes unity. Thus standardizing any normal distribution yields the STANDARD NORMAL DISTRIBUTION.

standard normal distribution, *n.* (*Statistics*) a NORMAL DISTRIBUTION with MEAN 0 and VARIANCE 1, with PROBABILITY DENSITY FUNCTION

$$\frac{\exp(-x^2/2)}{\sqrt{2\pi}},$$

that is derived from any normal distribution by an appropriate change of variables.

standard part, *n.* the function applicable only to finite NON–STANDARD REALS, returning the unique 'nearest' STANDARD REAL NUMBER. As every finite non-standard real number is a standard real number plus an INFINITESIMAL, this function is readily defined. A *standard infinitesimal* is one of the lowest possible order. See also HYPER–REAL.

standard position, *n.* the position of a plane angle with the vertex at the origin and the initial side along the positive *x*-axis.

standard real number, *n.* any NON–STANDARD REAL NUMBER that corresponds to an element of the ordinary REAL NUMBERS.

standard score, *n.* a score expressed in units of STANDARD DEVIATIONS from the mean of the distribution of such scores.

star, *n.* (*Geometrical topology*) **1.** the collection of sets containing a given member of a family of sets as a subset. The star of a SIMPLEX in a SIMPLICIAL

COMPLEX is the set of all simplices in the complex containing the given simplex as a face.

2. the set of points of a given set in a EUCLIDEAN VECTOR SPACE such that any segment from such a point to another point of the set lies in the set. The members of the star SEE the set, which is *star-shaped* if the star is non-empty. A set is CONVEX precisely if it is its own star. The *Krasnoselskii theorem* is that, in an n-dimensional Euclidean space, a compact set such that every $n+1$ points of the set can be seen from within the set is star-shaped.

star curve, see HYPOCYCLOID.

star-like region, *n.* (*Complex analysis*) a REGION, R, that contains a point z_0 such that if z_1 is any other point in R, then the segment $[z_0, z_1]$ is wholly in R; a STAR-shaped region.

state, *n.* **1a.** any of the indexed random variables of a STOCHASTIC PROCESS.
b. any of the possible outcomes of a MARKOV CHAIN.
c. (*as modifier*) denoting, or pertaining to, a STATE VARIABLE; for example, a *state equation* or *state constraint* is a restriction that involves only the state variables.
2. (*Classical thermodynamics*) a description of a system that may be specified by any three of the macroscopic variables, PRESSURE, VOLUME, ABSOLUTE TEMPERATURE, and the number of particles in the system. This is often referred to as the *macrostate*. In statistical physics the states of a system will be the solutions of the time-independent SCHRÖDINGER EQUATION for that system, these being referred to as *microstates*. The state is often represented by a point in a $2s$ dimensional space, the PHASE SPACE, where s is the number of DEGREES OF FREEDOM of the system. See also DEGENERATE STATE.

statement, *n.* **1a.** an assertion, as contrasted with a command, question, etc.
b. what is asserted, usually conceived of as being determined in part by the reference, rather than by the sense of the subject term and the sense of the predicate. Consequently, *the queen is unwell* and *the monarch is unwell* make the same statement if they are intended to refer to the same queen, say the Queen of the United Kingdom, while, if the same sentences are uttered with the intention refer to the Queen of Belgium, they make a different statement. However, to retain identity of the statement, any substituted predicate must be necessarily equivalent, and not merely coincident, so that *the Queen is not well* still makes the same statement, but *the Queen has cancelled her engagements* does not, even if the only time she cancels her engagements is when she is ill, and she always does so then. The identity of a statement is thus determined by what it is about, and (in a natural sense) what it says about it. Compare PROPOSITION.
c. an expression that makes a statement.
2. (*Computing*) one of the sequence of commands that constitute a PROGRAM.

state variable, **state**, **output variable**, or **behavioural variable,** *n.* a DEPENDENT VARIABLE as opposed to an independent or CONTROL VARIABLE of a differential equation or control system.

static friction, see FRICTION.

statics, *n.* the branch of mechanics concerned with the forces that produce a state of equilibrium in a system of bodies. Compare DYNAMICS.

stationary point, *n.* **1a.** also called (*North American usage*) **critical point.** a point on a curve at which its first DERIVATIVE is zero, so that the tangent is parallel to the axis of the independent variable (that is, in the usual two-dimensional Cartesian coordinate system, it is horizontal, as at P in Fig. 353); a MAXIMUM, MINIMUM or POINT OF INFLECTION.

b. more generally, a point at which the GRADIENT or similar variation of a function vanishes.

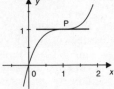

Fig. 353. P is a **stationary point**.

2. a set of values of the STATE VARIABLES of a system of AUTONOMOUS differential equations $y' = f(y)$ that make the system $f(y_0)$ zero; these are characterized by having constant solution $y(t) = y_0$. Since $dy/dt = 0$, these points are also *singular*, all other points being *regular*.

statistic, *n.* (*Statistics*) **1.** any function of a number of RANDOM VARIABLES, usually identically distributed, that may be used as an ESTIMATOR for a population parameter.

2. see SAMPLE STATISTIC. Compare PARAMETER.

statistical equilibrium, *n.* (*Statistical physics*) the STATE of a system in which the COEFFICIENT OF PROBABILITY, P, is constant with respect to time; that is, $\partial P/\partial t = 0$.

statistical inference or **inferential statistics,** *n.* (*Statistics*) the theory, methods, and practice of forming judgements about the parameters of a population, usually on the basis of random sampling. Compare HYPOTHESIS TESTING, DESCRIPTIVE STATISTICS.

statistically dependent, *adj.* (*Statistics*) (of two random variables) not INDEPENDENT. In particular, X and Y are *positively dependent* if the CONDITIONAL PROBABILITY, $P(X|Y)$, of X given Y is greater than the probability, $P(X)$, of X alone, or equivalently, if

$$P(X \& Y) > P(X) . P(Y);$$

they are *negatively dependent* if the inequalities are reversed, and these relations may depend upon the specific values of Y. In the case of equality for all values, the variables are INDEPENDENT.

statistical tables, *n.* tables showing the values of the CUMULATIVE DISTRIBUTION FUNCTIONS, PROBABILITY DENSITY FUNCTIONS or PROBABILITY FUNCTIONS of certain common distributions for different values of their parameters, used especially to determine whether or not a particular statistical result exceeds some required SIGNIFICANCE LEVEL. See also HYPOTHESIS TESTING.

statistics, *n..* **1.** quantitative data on any subject, especially data comparing the distribution of some quantity for different subclasses of the population, such as government publications of birth and death rates, often called *vital statistics.*

2. (*functioning as singular*) **a.** the classification and interpretation of these data in accordance with PROBABILITY theory and the application of methods such as HYPOTHESIS TESTING to them.

b. the mathematical study of the theory of such distributions and tests.

See also DESCRIPTIVE STATISTICS, STATISTICAL INFERENCE.

steady motion, *n.* (*Continuum mechanics*) a MOTION in which the PARTIAL DERIVATIVE with respect to time of the velocity of the BODY at fixed positions in the current CONFIGURATION is zero.

steepest descent or **gradient method,** *n.* (*Numerical analysis*) an iterative method, due to Fermat, of finding the minimum of a real valued differentiable function of n variables by moving towards the minimum of the function (an exact LINE SEARCH METHOD) along the line in the direction of the steepest negative gradient (a *steepest descent direction*) at successive points on the curve or surface. This is the prototype of non-linear OPTIMIZATION methods. The corresponding maximization technique is called the *steepest ascent method.* See also DESCENT METHODS. Compare NEWTON–RAPHSON METHODS.

Steiner point, *n.* the point of a COMPACT CONVEX set C in Euclidean n-space constructed as

$$s(C) = n \int_S x \delta_C^*(x) \, \sigma(dx)$$

where S is the $n-1$ sphere, δ_C^* is the SUPPORT FUNCTION of C, and σ is normalized LEBESGUE MEASURE. This produces an element of C, and $s(.)$ is Lipschitz in the metric defined by HAUSDORFF DISTANCE.

Steiner's problem, *n.* another term for FERMAT'S PROBLEM.

Steiner triple system, *n.* (*Combinatorics*) a BLOCK DESIGN consisting of a collection of m-element subsets of an n-element *base set* in such a way that each l-element subset of the base set lies in exactly one of the m-element subsets, denoted $S(l, m, n)$. The seven-point FINITE PROJECTIVE PLANE is an example of an $S(2,3,7)$; that is, this set of 7 points has lines of 3 points determined by any pair of points. These systems have uses in group theory and in SPHERE–PACKING PROBLEMS.

Steinitz' exchange theorem, *n.* the theorem that if, for $k < m$, $\mathbf{u}_i$ $(1 \le i \le k)$ and $\mathbf{v}_j$ $(1 \le j \le m)$ are linearly independent subsets of a VECTOR SPACE, then there exists a permutation, π, of $\{1, ..., m\}$ such that

$$\mathbf{u}_1, ..., \mathbf{u}_k, \mathbf{v}_{\pi(1)}, ..., \mathbf{v}_{\pi(m-k)}$$

are linearly independent. Consequently, any finite-dimensional vector space has a unique dimension, and any subspace of the same dimension is the space itself.

Steinitz' theorem, see REARRANGEMENT.

stem-and-leaf diagram, *n.* (*Statistics*) a HISTOGRAM in which the data points falling within each class interval are explicitly listed in order. The class intervals are visualized as the stem of a plant, and the data points as its leaves; the prime numbers in successive intervals shown in the first table of Fig. 354 overleaf would normally be written as a stem-and-leaf diagram as in the second table.

1–10	2	3	5	7		0*	2	3	5	7
11–20	11	13	17	19		1*	1	3	7	9
21–30	23	29				2*	3	9		
31–40	31	37				3*	1	7		
41–50	43	47				4*	3	7		

Fig. 354. **Stem-and-leaf diagram.** See main entry.

step function, *n.* a function that takes different constant values on each of a succession of disjoint intervals whose union is its domain, such as, for example, $[x]$, the integral part of any real x, the graph of which is shown in Fig. 355. Such functions are crucial in the definition of some forms of INTEGRATION; see LOWER SUM.

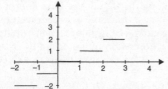

Fig. 355. **Step function**.

step-length (or **step-size) method,** *n.* an approximate DESCENT METHOD that is based on finding the appropriate length of step to take in the descent direction, as opposed to a TRUST REGION METHOD.

stepping stone method, *n.* a method of solving TRANSPORTATION PROBLEMS that adapts the SIMPLEX METHOD by exploiting the special structure of the problem.

steradian, *n.* a unit of measure of a SOLID ANGLE that is equal to the solid angle subtended at the centre of a unit sphere by a unit area on its surface.

stereographic projection, *n.* the projection from a point on a sphere (the *pole*) onto a plane that is a tangent to the sphere at the antipodal point or onto a parallel plane; this is a CONFORMAL mapping. Fig. 356 shows the stereographic projection of a circle onto a plane; P is the pole of the projection. The representation of the complex plane by this projection is known as the RIEMANN SPHERE. Compare GNOMONIC PROJECTION.

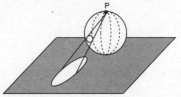

Fig. 356. **Stereographic projection**.

Stickelberger's theorem, *n.* the result that, given a monic polynomial g of degree d over the integers modulo p (p an odd prime) without multiple factors, the number of irreducible factors r satisfies $r \equiv d$ (mod 2) if and only if the DISCRIMINANT, $D(g)$, which is non–zero, is a square in $\mathbb{Z}_p$.

Stieltjes integral, see RIEMANN–STIELTJES INTEGRAL. (Named after the Dutch-born naturalized French analyst and number theorist, *Thomas Jan Stieltjes* (1856–94).)

Stieltjes moment problem, see MOMENT PROBLEM.

stiffness, see HOOKE'S LAW.

Stirling, James (1692–1770), Scottish mathematician known as S*tirling the Venetian,* who was expelled from Oxford in 1715 for his contact with the Jacobites, and pursued his studies in Venice. When he returned to Britain he was elected Fellow of the Royal Society, corresponded with Newton and Maclaurin, and published works on infinite series and gravitation. He later became manager of the Scots Mining Company.

Stirling numbers, *n.* **1. Stirling numbers of the first kind.** the integers $s(n, k)$ generated by the recursive definition:

$$s(0, 0) = 1; \quad s(n, 0) = 0 \quad (n > 0),$$

and, for $0 < k < n$,

$$s(n + 1, k) = s(n, k - 1) - ns(n, k);$$

the first terms of this sequence are shown in Fig. 357.

$k =$	0	1	2	3	4	...
$n = 0$	1					
1	0	1				
2	0	–1	1			
3	0	2	–3	1		
4	0	–6	11	–6	1	
:						

Fig. 357. **Stirling numbers of the first kind**.

2. Stirling numbers of the second kind. the natural numbers $S(n, k)$ generated by the recursive definition:

$$S(n, n) = 1 \quad (n > 0),$$
$$S(n, 0) = 0 \quad (n \geq 0),$$

and, for $0 < k < n$,

$$S(n + 1, k) = S(n, k - 1) + kS(n, k);$$

the first terms of this sequence are shown in Fig. 358. The Stirling numbers of the second kind count the number of partitions of an *n*-element set into precisely *k* parts.

$k =$	0	1	2	3	4	...
$n = 0$	0					
1	0	1				
2	0	1	1			
3	0	1	3	1		
4	0	1	7	6	1	
:						

Fig. 358. **Stirling numbers of the second kind**.

Stirling's formula, *n.* the asymptotic formula for the GAMMA FUNCTION or FACTORIAL. In its simplest form it is

$$\frac{\Gamma(s + 1)}{\left(\frac{s}{e}\right)^s \sqrt{2\pi s}} = 1 + o(1)$$

as $s \to \infty$. The relative error between $\Gamma(s + 1)$ and the denominator tends to zero like $1/(12s)$ but the absolute error is huge. There are higher-order asymptotic expansions available from Stirling's formula for $\log\Gamma(s)$. (Stirling's formula was discovered by Abraham de Moivre, though named for James Stirling.)

stochastic, *adj.* (*Statistics*) **1.** constituting a RANDOM VARIABLE; having a probability distribution, usually with finite VARIANCE. See STOCHASTIC PROCESS.

2. (of a matrix) square with non-negative elements that add to unity in all rows (when it is *row stochastic*) or all columns (*column stochastic*); it is *doubly stochastic* if it satisfies both conditions.

stochastically closed, *adj.* (*Statistics*) (of a set of STATES of a MARKOV CHAIN) a subset of states such that there is a zero probability of leaving the subset once it is entered. Compare ERGODIC.

stochastic process, *n.* a process that can be described by a RANDOM VARIABLE (*stochastic variable*) that depends on some parameter, which may be discrete or continuous, but is often taken to represent time; precisely, an indexed family of random variables, called *states*, on a probability space. A stochastic process is *finite* if the index family is countable and each state is a step function. A MARKOV CHAIN is a discrete-parameter stochastic process in which future probabilities are completely determined by the present state.

stochastic variable, *n.* another term for RANDOM VARIABLE.

Stokes' theorem, *n.* **1.** the culminating theorem of the classical theory of differential forms, stating that the integral of the derivative of a DIFFERENTIAL FORM over an appropriate volume equals the integral of the given form over the boundary of the volume.

2. the particular result that for a smooth vector field **u** defined on a domain D containing a piecewise-smooth oriented surface S with boundary contour C:

$$\int_C \mathbf{u} \cdot d\mathbf{r} = \iint_S \mathbf{n} \cdot \text{curl } \mathbf{u} \, dS,$$

where the left-hand integral is the CURVILINEAR INTEGRAL of the tangential component **u** . **T** of **u** with respect to arc length, and the right-hand integral is the SURFACE INTEGRAL of the component of **u** in the direction of the outer normal to the surface. See also GREEN'S THEOREM, DIVERGENCE THEOREM.

(Named after the British analyst and physicist, *Sir George Gabriel Stokes* (1819–1903).)

Stone–Cech compactification, *n.* the COMPACTIFICATION of a completely regular HAUSDORFF SPACE S that may be constructed as the closure in the weak-star topology of the point evaluation mappings in the dual space of the continuous bounded real-valued functions on S, written $\beta(S)$. This is a maximal compactification and is characterized (up to homeomorphism) as the unique compactification W with the property that any continuous mapping from S into a compact space T uniquely extends to a continuous mapping from W into T.

Stone–Weierstrass theorem, *n.* the extension of the WEIERSTRASS APPROXIMATION THEOREM that gives conditions for an ALGEBRA of continuous complex-valued functions on a compact set to be uniformly DENSE in the entire space. It suffices that it SEPARATES POINTS, VANISHES

NOWHERE (as happens if 1 is in the algebra), and is SELF-ADJOINT. The theorem holds for real-valued functions without the third condition.

stopping rule, see TERMINATION CRITERION.

straight, *adj.* (of a LINE) **1.** (*Euclidean geometry*) having the property that the lines passing through any pair of points on the given line are all identical. This is the primitive property of a line in EUCLIDEAN GEOMETRY, which distinguishes lines from other curves; straightness is the property of a line that yields its constant GRADIENT.

2. more generally, consisting of points that satisfy the same linear equation.

straight-edge, *n.* an instrument used for drawing straight lines, but not for measurement, referred to in the statements of the construction problems of traditional geometry, such as TRISECTING THE ANGLE. See also SQUARE THE CIRCLE.

strain, *n.* the change in relative positions of points in a medium as the result of a STRESS-produced deformation of the medium.

strange attractor, *n.* (of a DYNAMICAL SYSTEM) an ATTRACTOR set whose HAUSDORFF DIMENSION is non-integral, or else dependent on initial conditions. There is no generally accepted definition. See FRACTAL.

strategy, *n.* (*Game theory*) a particular choice of moves for a player in a game (a *pure strategy*) or a probabilistic mixture of choices to be used in repeated playing of the game (a *mixed strategy*).

stratified sample, *n.* (*Statistics*) a sample that is not drawn at random from the whole population, but is drawn separately from a number of disjoint strata of the population in order to ensure a more representative sample. See also FRAME.

stream function, *n.* (*Continuum mechanics*) a function describing the STREAMLINES of a BODY. See also COMPLEX VELOCITY POTENTIAL.

streamline or **line of flow,** *n.* (*Continuum mechanics*) a curve in the current CONFIGURATION of a BODY whose TANGENT vector is everywhere parallel to the VELOCITY.

stress, *n.* the FORCE per unit area transmitted across a surface, determined by the STRESS TENSOR; the standard unit is the PASCAL.

stress–power, *n.* the difference between the rate of change of the KINETIC ENERGY and the POWER of a SUB-BODY; that is, the integral

$$\int \mathrm{tr}(\sigma \Sigma)\, dv$$

over the volume of the current CONFIGURATION of the sub-body, where σ is the STRESS TENSOR and Σ is the EULERIAN STRAIN RATE. The quantity $\mathrm{tr}(\sigma \Sigma)$ is the *stress–power per unit volume*.

stress tensor, *n.* (*Continuum mechanics*) the second order symmetric TENSOR, σ, such that the STRESS VECTOR, **t**, at a point on a surface is given by $\sigma \mathbf{n}$, where **n** is the unit outward normal to the surface at the point. See CONSTITUTIVE EQUATION.

stress vector, *n.* the CONTACT FORCE density of a body.

stretching, *n.* a HOMOTHETIC TRANSFORMATION of the form

$$x' = kx, \quad y' = ky,$$

where $k > 1$.

strict, *adj.* (of a relation, etc.) **1.** distinguished from some other relation of the same name by the fact that it applies more restrictively, specifically by excluding the possibility of the identity of its two relata. For example, a *strict inequality* such as $x < y$ holds only between pairs of distinct numbers, while the WEAK inequality $x \leq y$ permits its arguments to be identical. Similarly, if $x < y$ implies that $f(x) > f(y)$, f is a *strictly decreasing* function; a function is *strictly convex* if the chord joining any two points on the graph lies strictly above the graph; a NORM is *strictly convex* if the corresponding unit ball contains no line segments in its boundary. See PROPER. See also ORDERING.

2. distinguished from a relation of the same name that is not the subject of formal study; for example, *strict identity* is governed by a set of axioms and is distinguished from various uses of *same* in ordinary language.

strict implication, *n.* the connective of MODAL LOGIC usually defined in terms of the impossibility of the truth of its antecedent at the same time as the falsehood of its consequent; that is,

$$P \Rightarrow Q \equiv -\lozenge(P \,\&\, -Q),$$

where $\lozenge$ is the POSSIBILITY operator. It is the relation that holds between two sentences precisely when one is validly deducible from the other, and is not TRUTH–FUNCTIONAL. Compare MATERIAL IMPLICATION.

strict inclusion, see INCLUSION.

string, *n.* a sequence of elements, often merely concatenated, as in a WORD or a RUN.

stroke, (*Logic*) see SHEFFER'S STROKE.

strong, see ORDERING.

strong completeness, *n.* the property of a logical THEORY that the addition to its AXIOMS of any WELL–FORMED formula that is not a THEOREM yields an INCONSISTENT theory.

strong convergence, see STRONG TOPOLOGY.

strong duality, *n.* the relationship between two constrained mathematical programs

$$(P) \quad p = \inf_x f(x)$$

and

$$(D) \quad d = \sup_Y g(y),$$

as, for example, a dual pair of LINEAR PROGRAMS, for which one can assert not only that $p \geq d$ (which is called *weak duality*), but actually that $p = d$ and that one or both optimal values is feasibly attained. In a case of strong duality, one says there is no *duality gap*, where the difference $p - d$ is taken as this measure. See DUALITY THEORY OF LINEAR PROGRAMMING.

stronger, *adj.* (of one of a pair of TOPOLOGIES) strictly containing the other. See COARSER.

strong ergodic theorem, *n.* another name for BIRKHOFF ERGODIC THEOREM.

strong inverse image set, *n.* see INVERSE IMAGE SET.

strong law of large numbers, *n.* (*Probability*) a precise formulation of the LAW OF LARGE NUMBERS in terms of a POINTWISE CONVERGENT sequence, as contrasted with the associated WEAK LAW OF LARGE NUMBERS that concerns

CONVERGENCE IN MEASURE. In one form it states that if a sequence of independent random variables has variances σ_n such that

$$\sum_n \sigma_n^2 / n^2$$

is finite, then the sequence of averages of the given sequence converges almost everywhere.

strong topology, *n.* the original topology on a NORMED SPACE as contrasted with the associated WEAK TOPOLOGY. Hence *strong convergence* means convergence in norm.

strophoid, *n.* the locus of points, two on each line in a pencil of lines through a fixed point, whose distance from the intercept of the line with the *y*-axis is equal to the *y*-coordinate of that intercept; in Fig. 359, P_1 and P_2 are points such that $OX = XP_1 = XP_2$. In standard form the equation is

$$y^2 = \frac{x^2(x+a)}{a-x}$$

when $(-a, 0)$ is the vertex of the pencil.

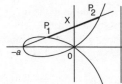

Fig. 359. **Strophoid.**

structure, *n.* **1.** (*Logic*) an assignment to a first-order language of a non-empty set (the *universe*) of which the elements are *individuals*, and of functions, predicates, and constants in that universe to the corresponding symbols of the language with the exception of identity. A structure for a theory in which its non-logical axioms are true is a MODEL for the theory.
2. a set endowed with some prescribed functions, predicates, or relations, usually of an algebraic nature. Compare SPACE.

Student's *t*, *n.* a STATISTIC often used to test the hypothesis that a random sample of NORMALLY DISTRIBUTED observations, drawn from a population with unknown parameters, has a given MEAN, μ; the statistic t is given by

$$ts = (\bar{x} - \mu)\sqrt{n},$$

where $\bar{x}$ is the mean of the sample, s is its STANDARD DEVIATION, and n is the size of the sample. (Named after *Student*, the Englishman William Sealy Gosset (1876 – 1937), who published the result under this pseudonym.)

Student's *t*-distribution, *n.* (*Statistics*) the distribution of the ratio of a STANDARDIZED NORMAL DISTRIBUTION to the square root of the quotient of a CHI–SQUARED DISTRIBUTION by the number of its DEGREES OF FREEDOM; thus one also speaks of the degrees of freedom of a *t*-distribution.

Sturm–Liouville equation, *n.* a parametric differential equation of the form

$$\frac{d}{dx}\left[p(x)\,\frac{dy}{dx}\right] + \left[\lambda\rho(x) - q(x)\right]y = 0,$$

where p is strictly positive and p, q, and ρ are continuous. Parameter values λ for which the system is solvable are *eigenvalues* and the corresponding solutions are *eigenfunctions* that form a complete orthogonal set.

Sturm sequence, *n.* the sequence, $w(x)$, defined for a given polynomial, p, as the sign variation in the sequence $p_0(x), ..., p_k(x)$, where $p_0 = p$, $p_1 = p'$, and each $p_i = -r_i$, where $r_2, r_3, ..., r_k$ are the successive remainders computed by EUCLID'S ALGORITHM for the highest common factor of p and p'. (Named after the Swiss-French analyst and physicist, *Jacques Charles François Sturm* (1803–55).)

Sturm's theorem, *n.* the theorem that if a real polynomial is non-zero at the end points of an interval, the number of roots in that interval, counting multiplicity, is the difference between the numbers of sign changes of the STURM SEQUENCE at the two endpoints. Compare DESCARTES' RULE OF SIGNS.

sub-, *prefix denoting* a SUBSTRUCTURE that is contained within a given structure and shares its structural properties.

subadditive, *adj.* **1.** (of a function with a SEMI–GROUP as its domain) such that the value at the sum of two elements is less than the sum of the values at the separate elements, that is,

$$f(x + y) \leq f(x) + f(y);$$

this makes sense whenever the range is an ordered semi-group. A function, f, is *superadditive* if its additive inverse, $-f$, is subadditive.

2. (of a SET FUNCTION on a class) such that the value for a union of two elements (that is again in the domain) is less than the sum of the values for the respective elements:

$$S(A \cup B) \leq S(A) + S(B).$$

If this holds for all finite (countable) unions that lie in the class the set function is *finitely* (or *countably*) *subadditive.* This is the case for Lebesgue OUTER MEASURE.

sub-base, *n.* (for a TOPOLOGY) a collection of open sets of which the finite intersections form a BASE for the topology.

sub-base theorem, see ALEXANDER'S SUB–BASE THEOREM.

sub-body, *n.* a subset of a BODY that is itself a body.

subclass, *n.* another word for SUBSET, especially as contrasted with a PROPER CLASS.

subcontrary, *adj.* (*Logic*) **1a.** (of a pair of statements) unable both to be false at the same time, under the same circumstances, or in the same interpretation. For example, *x is non-negative* and *x is non-positive* are subcontrary if x is restricted to the reals, since at least one must be true, but both statements are true when $x = 0$. Compare CONTRADICTORY, CONTRARY.

b. (of a single statement) not able to be false when a given statement is false.

2. (*as substantive*) a statement that is subcontrary to a given statement.

subdesign, *n.* a BLOCK DESIGN of which the sets of blocks and varieties are subsets of those of a given design.

subdiagonal, *n.* the line of entries lying immediately below the MAIN DIAGONAL of a matrix; the entries of the form $a_{i,i-1}$.

subdiagonal matrix, *n.* a matrix all the entries of which are zero except in the SUBDIAGONAL.

subfield, *n.* a SUBRING of a FIELD (or RING) that is itself a field.

subgeometry, *n.* (*Algebraic geometry*) a GEOMETRY formed by all the points that have HOMOGENEOUS COORDINATES that are LINEAR COMBINATIONS of those of a given set of points. For example, lines and planes are subgeometries of a three-dimensional Euclidean geometry.

subgradient, *n.* the set of linear functionals, denoted $\partial g(x)$, defined in terms of a given convex function, g, at a point, as

$$\partial g(x) = \{ \varphi : \varphi(y-x) \leq g(y) - g(x) \}.$$

If the function is continuous at x, each subgradient is necessarily continuous and $\partial g(x)$ is a non-empty, convex, weak-star compact set that satisfies the maximum formula

$$g'(x; h) = \max\{\varphi(h) : \varphi \in \partial g(x)\}.$$

Here $g'(x;)$ is the DIRECTIONAL DERIVATIVE of g at x. In particular, the subgradient set is singleton exactly if the function has a GATEAUX DIFFERENTIAL at x. The subgradient defines a MONOTONE MULTIFUNCTION.

subgraph, *n.* a GRAPH whose sets of vertices and edges are subsets of those of a given graph, and that contains all the vertices joined by any of its edges.

subgroup, *n.* a subset of another GROUP that is also a group under the same binary operation; for example, the integers are a subgroup of the reals under addition, but the integers modulo n are not a subgroup of these since the operations are differently defined.

subharmonic, *adj.* (of a function of two real variables on a domain) in essence, such that whenever it is dominated by an HARMONIC FUNCTION on the boundary of a subdomain, it is dominated throughout the subdomain. It follows from POISSON'S INTEGRAL formula that HARMONIC FUNCTIONS have this property. If the function has continuous second partial derivatives, it is subharmonic precisely when it has a non-negative LAPLACIAN throughout the given domain. A function, f, is *superharmonic* if its additive inverse, $-f$, is subharmonic.

subitization, *n.* the ability to recognise small numbers of objects without consciously counting them. This ability seems pervasive for one, two, or three objects, and for some individuals may go much further.

subjective probability, see PROBABILITY.

sublinear convergence, *n.* any RATE OF CONVERGENCE worse than linear.

sublinear function, *n.* **1.** a MINKOWSKI FUNCTION.

2. a function whose domain is a VECTOR SPACE that is SUBADDITIVE and positively HOMOGENEOUS, as makes sense whenever the range is an ordered vector space. A function, f, is *superlinear* if its additive inverse, $-f$, is sublinear.

submatrix, *n.* a matrix derived from a given matrix by deleting all the elements in some of its rows or some of its columns, or both, as shown in Fig. 360.

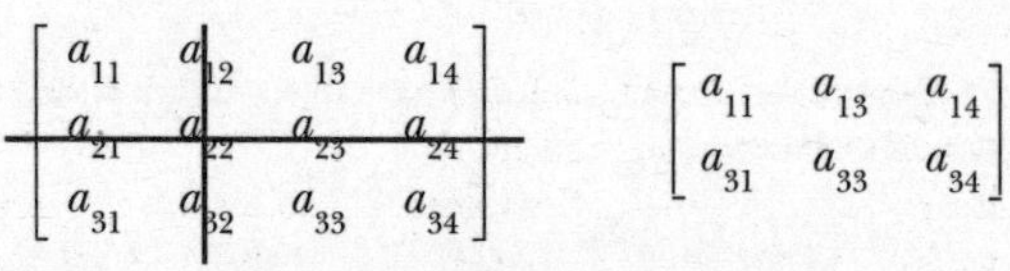

Fig. 360. **Submatrix.** See main entry.

submodule, *n.* a MODULE over a RING that is contained in another module over the same ring, and has the same addition.

submultiple, *n.* a less common term for FACTOR.

subordinate, see PARTITION OF UNITY.

subpopulation, *n.* (*Statistics*) a subset of a given POPULATION.

subring, *n.* a subset of a RING that is itself a ring under the same binary operations of addition and multiplication.

subsequence, *n.* a SEQUENCE derived from a given sequence by selecting certain of its terms and retaining their order. For example, $\langle a_2, a_3 \rangle$ is a subsequence of $\langle a_1, a_2, a_3 \rangle$, but $\langle a_3, a_2 \rangle$ is not.

subset or **subclass,** *n.* a set all the members of which are members of some given set; a *proper subset* is one strictly contained within a larger set and excluding some of its members. The relation is written

$$A \subseteq B, \ A \subset B, \ A \subsetneqq B,$$

of which the first denotes the WEAK and the last the STRICT (or strong) relation, but there are different conventions about the use of the middle notation.

subspace, *n.* a SPACE of which the elements are a subset of those of a given space and that is endowed with the same properties as the given space.

substitute, *vb.* to replace one expression by another in the context of a third; for example, substituting $x = 3y$ in $2x - 4y = k$ gives $2y = k$.

substitution, *n.* **1.** the replacement of a term of an equation by another that is known to have the same value in order to simplify the equation; for example, the simultaneous equations

$$x = 2y - 4, \quad 2x = 3y - 5$$

can be solved by substituting for x in the second the expression with which it is equated in the first, yielding $4y - 8 = 3y - 5$, so that $y = 3$.
2. (*Logic*) the uniform replacement of all occurrences of one expression by another in a given context; for example, substituting 'P & R' for 'P' in 'P $\vee$ Q' yields '(P & R) $\vee$ Q'.

substitution group, *n.* another term for PERMUTATION GROUP.

substitution instance, *n.* (*Logic*) an expression derived from another by uniform SUBSTITUTION. In a formal theory in which the given expression is a theorem, so is every substitution instance of it.

substitution rule, *n.* the rule for INTEGRATION that allows the evaluation of an integral by SUBSTITUTION. In indefinite form, if

$$\int f(x) \, dx = F(x) + C$$

then

$$\int f(g(t)) \, g'(t) \, dt = F(g(t)) + C,$$

as follows from the CHAIN RULE. The FUNDAMENTAL THEOREM OF CALCULUS allows this to be recast as

$$\int_a^b f(g(t)) g'(t) \, dt = F(g(t)) \Big|_a^b = F(g(b)) - F(g(a)).$$

For example, to evaluate

$$\int \frac{x}{1 + x^2} \, dx,$$

let $u = x^2$, so that $du/dx = 2x$; then the integral becomes

$$\int \frac{x}{1 + u} \, \frac{du}{2x} \; = \; \tfrac{1}{2} \int \frac{1}{1 + u} \, du \; = \; \tfrac{1}{2} \ln(1 + u) \; = \; \tfrac{1}{2} \ln(1 + x^2).$$

substitution theorem, *n.* (*Logic*) the theorem that a universally quantified statement implies any INSTANCE of it, and that an existentially quantified statement is implied by any instance of it. Given the DEDUCTION THEOREM, this is equivalent to the INSTANTIATION RULE for the UNIVERSAL QUANTIFIER, and the INTRODUCTION RULE for the EXISTENTIAL QUANTIFIER.

substitutivity, *n.* the principle that terms with the same reference may be substituted for one another in a sentence without changing its truth value; thus since the morning star is the evening star, if

the morning star is visible in the morning

is true, then so is

the evening star is visible in the morning.

This does not hold in OPAQUE modal contexts, or of DE DICTO modalities.

substructure, *n.* a STRUCTURE of which the elements are a subset of those of a given structure and that is closed under the appropriate operations, and so is endowed with the same properties as the given structure. In particular, SUBGROUP, subalgebra, SUBDESIGN, SUBFIELD, SUBGRAPH, sublattice, SUB-MODULE, SUBRING are all substructures of the eponymous structures.

subtangent, *n.* the projection of a tangent to a curve onto the *x*-axis in the two-dimensional Cartesian plane; a segment of the *x*-axis lying between the *x*-coordinate of the point at which a tangent is drawn to a curve and the intercept of the tangent with the axis. In Fig. 361, the tangent at P to the curve shown intersects the *x*-axis at X, and F is the foot of the perpendicular from P to the axis; FX is then the subtangent to the curve at P.

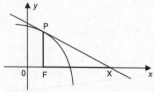

Fig. 361. FX is the **subtangent** at P.

subtend, *vb.* (of a line or curve) to define an angle at some specified point as that included between the lines drawn to the point from the endpoints of the given line or curve. For example, a chord of a circle subtends an angle at the centre that is enclosed between the radii drawn to the ends of the chord and is double the angle subtended at the circumference; in Fig. 362, overleaf, the arc AB subtends the angle 2θ at the centre C and the angle θ at any point P on the circumference.

subtract, *vb.* to calculate the difference between a pair of given quantities by SUBTRACTION.

Fig. 362.   Angles **subtended** by an arc at C and P.

subtraction, *n.*   **1.** the mathematical operation in which the difference between two numbers or quantities is calculated; the inverse operation of ADDITION, so that $a - b = c$ if and only if $a = b + c$.

2. any analogous operation, such as RELATIVE COMPLEMENT.

subtractive, *adj.*   indicating or requiring subtraction; having negative sign, such as a subtractive quantity.

subtrahend, *n.*   the number to be subtracted from another (the MINUEND).

succedent, *n.*   (*Logic*) another word for the CONSEQUENT of a sequent.

success, *n.*   (*Statistics*) an outcome of an experiment, or element of a sample space, that is in the class of which the probability is in question.

successive, *adj.*   following one after the other. This is often used to mean iterative or sequential, as with successive linear programming (SLP) or successive quadratic programming (SQP).

successive approximation, *n.*   an iterative process for finding an approximate value of a quantity such as the ROOT of a given number, by starting from a first estimate and then deriving from each approximation another that is more accurate.

successor, *n.*   (*Logic*) the element directly related to a given element of a SERIAL ORDERING, especially the natural number following a given one; the successor of n is $n + 1$, usually written Sn or n' in this context. The successors of a NODE of a TREE are those nodes related to the given node by the relation that generates the tree; graphically, these are the nodes accessible from the given node by a path away from the ROOT of the tree.   Compare PREDECESSOR.

sufficient condition, *n.*   **1.** anything that entails the truth of some statement or the obtaining of some state of affairs; having this other statement or state of affairs as a consequence without any other conditions; thus if P is a sufficient condition for Q, then P IMPLIES Q, that is *if P then Q* is true. Although a sufficient condition may also be a NECESSARY CONDITION, this is not generally the case; for example, it is a sufficient condition for x to be non-negative that it be positive, but it is not necessary. However, if P is a sufficient condition for Q, then Q is a necessary condition for P. For example, it is a sufficient condition for $x \geq 4$ to be composite that it be divisible by 3.

2. (*Optimization theory*) a condition ensuring that a previously computed solution to necessary conditions is actually optimal. Thus the FIRST DERIVATIVE TEST and SECOND DERIVATIVE TEST give sufficient conditions for a STATIONARY POINT to be optimal, and convexity of the functions in a constrained minimization problem makes KUHN–TUCKER CONDITIONS sufficient.

sufficient statistic, *n.* (for a PARAMETER) a STATISTIC T(X), such that the conditional distribution of X, given T(X), does not depend upon the given parameter. It follows that if the distribution of a sample is known, only sufficient statistics are necessary to estimate the parameters, without further reference to the data.

sum, *n.* **1a.** the result of the addition of numbers, quantities, etc.
b. formally, the number derived from a pair of given numbers such that if the given numbers are the numbers of elements in two disjoint sets, then the sum is the total number of elements in either set.
2. the limit of the SEQUENCE of partial sums of the first *n* terms of an infinite SERIES, as *n* tends to infinity. For example, the series

$$\sum_{n=0}^{\infty} \frac{1}{2^n} = 1 + \frac{1}{2} + \frac{1}{4} + \frac{1}{8} + \dots$$

has sum 2, since that is the limit of the sequence of PARTIAL SUMS

$$1, \ 1\tfrac{1}{2}, \ 1\tfrac{3}{4}, \ 1\tfrac{7}{8}, \ \dots.$$

This is in general the most convenient choice for the sum of a series, but others are possible; see CESARO SUM. See also SIGMA, REARRANGEMENT.
3. (*informal*) any arithmetical problem or calculation.
4. logical sum. another name for DISJUNCTION or UNION.
5. set-theoretic sum. another name for UNION. Compare DISJOINT UNION.

summability theory, *n.* the study of SUMMABLE quantities, and especially of methods of assigning values to DIVERGENT series and integrals. Compare ABEL SUMMATION, CESARO SUMMATION.

summable, *adj.* able to be added or integrated. See also ABSOLUTELY SUMMABLE, SQUARE SUMMABLE.

summand, *n.* a number or quantity to be added to others; one term of a SUM or SERIES.

summation convention or **dummy suffix convention,** *n.* a short-hand notation used in manipulating components of VECTORS and TENSORS, in accordance with which the summation sign Σ is omitted and the sum indicated by the repetition of an index; for example, the scalar product

$$\mathbf{a} \cdot \mathbf{b} = a_1 b_1 + a_2 b_2 + a_3 b_3$$

may be more compactly written as $a_i b_i$.

sum of squares, *n.* (*Statistics*) any quadratic sum of RANDOM VARIABLES. See SUM OF SQUARES THEOREM.

sum of squares theorem, *n.* the result that if the matrix of the QUADRATIC FORM of a SUM OF SQUARES of NORMAL random variables is IDEMPOTENT of RANK *r*, then the sum of squares is distributed proportionally to a CHI–SQUARE DISTRIBUTION with *r* degrees of freedom. Furthermore, if the mat-rix associated with a second sum of squares is orthogonal to the first, then the two sums of squares are distributed independently.

sup, *abbrev. and symbol for* SUPREMUM.

super-, *prefix denoting* a structure of which some other structure is a SUBSTRUCTURE, such as a SUPERSET.

superadditive, see SUBADDITIVE.

superdiagonal or **second diagonal**, *n.* (of a matrix) the line of entries lying immediately above the diagonal, of the form $a_{i,\,i+1}$.

superdiagonal matrix, *n.* a matrix all the entries of which are zero except in the SUPERDIAGONAL.

superharmonic, see SUBHARMONIC.

superior limit, see LIMIT SUPERIOR.

superlinear convergence, *n.* any RATE OF CONVERGENCE better than linear.

superlinear function, see SUBLINEAR FUNCTION.

superposable, *adj.* (of a pair of geometric figures) such that the image of one is CONGRUENT to the other, and so can be transposed to coincide with it.

superpose, *vb.* **1.** to transpose one geometric figure to coincide with another.
   2. (of Fourier series) to add two series to obtain a third.

superposition, *n.* **1.** the act or result of SUPERPOSING figures or Fourier series.
   2. a less common name for COMPOSITION of functions.

superposition principle, *n.* the principle that any linear combination of solutions to a homogeneous linear differential equation is also a solution. This corresponds to physically SUPERPOSING solutions.

super-reflexive, see UNIFORM CONVEXITY.

superset, *n.* a set containing the given set as a SUBSET.

superspace, *n.* a space of which some other space is a SUBSPACE, so that they have the same structure.

supertask, *n.* any hypothetical task that would require the completion of an infinite sequence of separate tasks within a finite time. See THOMSON LAMP, ZENO'S PARADOX.

sup norm, *abbrev. for* supremum norm. See CHEBYSHEV NORM.

supplement, *n.* **1.** a SUPPLEMENTARY ANGLE of a given angle.
   2. a SUPPLEMENTARY ARC of a given arc.

supplemental chords, *n.* (of a circle) the chords joining a point on the circle to two other diametral points, so that they join the endpoints of SUPPLEMENTARY ARCS; in Fig. 363, AC is a diameter, so that AB and BC are supplemental chords.

supplementary angle, *n.* an angle of which the sum with a given angle is 180°. The adjacent angles of two intersecting lines *supplement* each other.

supplementary arc, *n.* an arc of a circle that, together with a given arc, forms a semicircle, so that they sit on SUPPLEMENTAL CHORDS; in Fig. 363, AC is a diameter, so that AXB and BYC are supplemental arcs.

Fig. 363. **Supplemental chords**; **supplementary arcs**.

support, *n.* **1.** the closure of the set of arguments for which a real-valued or complex-valued function has non-zero value.
   2. also called **kernel**. (with respect to a regular BOREL MEASURE) the unique smallest closed set of which the complement has measure zero.

support function, *n.* the MINKOWSKI FUNCTION of a CONVEX set C in a real NORMED SPACE, denoted $\delta_C^*(f)$, $s_C(f)$, or $s(f, C)$, and defined by

$$\delta_C^*(f) = \sup\{ f(x) : x \in C \}$$

on the DUAL space (or, analogously, on the original space if an inner product is available). This function is everywhere finite if the set is bounded, and it will produce a NORM if the set is a symmetric convex body. The support function δ_C^* is CONJUGATE in the sense of Fenchel to the INDICATOR function δ_C.

supporting, *adj.* (of a set) constituting a SUPPORT, or consisting of SUPPORT POINTS.

support point, *n.* a point of a CONVEX SET at which there exists a non-zero linear functional (sometimes required to be continuous) that takes its supremum over the set. Geometrically, this corresponds to the existence of a (closed) *support* or *boundary hyperplane* that contains the point and has the set entirely in one SUPPORTING half-space; the *Bishop–Phelps theorem* establishes that in a BANACH SPACE the support points of such a set are dense in the boundary of the set, and are themselves norm dense in the dual space. This may fail in an incomplete normed space if the set is merely closed and bounded. See SUPPORT THEOREM.

support theorem, *n.* the consequence of the SEPARATION THEOREM OF MAZUR that a closed convex body is the intersection of its closed SUPPORTING half-spaces, and every boundary point of a convex set with non-empty interior in normed space is a support point.

supremal, *adj.* relating to, or constituting, a SUPREMUM.

supremum or **least upper bound** (abbrev. **sup**, **lub**), *n.* the unique smallest member of the set of UPPER BOUNDS for some given set, equal to its MAXIMUM if the given set has a greatest member. The supremum of τ of a set may be defined as satisfying the two conditions $\tau \geq t$ for all t in T, and for all $t < \tau$ there is a $t' > t$ in T. For example, the sequence $\frac{1}{2}, \frac{2}{3}, \frac{3}{4}, \ldots$ has as upper bound every real number greater than or equal to 1; it has no maximum, but its supremum is 1. Compare INFIMUM.

supremum norm, *n.* another term for CHEBYSHEV NORM.

surd, *n.* a numerical expression containing one or more irrational roots of numbers, such as $2\sqrt{\sqrt{3}}$, or $3\sqrt{5} + 4\sqrt{3}$. The *conjugate surd* of the latter is $3\sqrt{5} - 4\sqrt{3}$. It is an *entire surd* if it has no rational factors or terms and a *mixed surd* otherwise; it is a *pure surd* if every term is an entire or mixed surd.

surface, *n.* **1.** the complete boundary of a geometric solid.

2a. any continuous two-dimensional figure.

b. a two-dimensional graph in a three-dimensional space corresponding to a function $z = f(x, y)$, or to an implicit function $F(x, y, z) = 0$; the analogous object in higher dimensions.

surface area, *n.* the AREA of part of a SURFACE; this can be defined by the SURFACE INTEGRAL with the constant integrand 1.

surface integral, *n.* **1.** the double integral of a SCALAR FIELD, F, in three-dimensional real Euclidean space, with respect to the area of the region of the surface, denoted $\iint F \, dS$; if $\mathbf{x}(u, v)$ for real u and v is a parametric representation of S, then this is equal to

$$\iint_S F(\mathbf{x}(u, v)) \left| \frac{d\mathbf{x}}{\partial u} \times \frac{d\mathbf{x}}{\partial v} \right| du\, dv.$$

In terms of the parametrization in terms of the x- and y-coordinates, this reduces to

$$\iint_S F[x, y, f(x, y)] . \sqrt{1 + \left(\frac{\partial f}{\partial x}\right)^2 + \left(\frac{\partial f}{\partial x}\right)^2}\ dx\, dy.$$

b. the double integral of a SCALAR FIELD, F, in $\mathbb{R}^3$, with respect to the area of the region of the surface in the direction of one of the coordinate axes. It is denoted $\iint_S F\, dS_i$ and is equal to $\iint_S F n_i\, dS$, where n_i is the ith component of the unit outward normal to the surface.

c. linear combinations of such integrals; these surface integrals are analogous to the CURVILINEAR INTEGRALS of the second kind.

2. less commonly, the double integral of a function of three variables over a region S of a surface; where $F(x, y, z)$ is the function, and $z = f(x, y)$ for $\langle x, y \rangle \in$ D is the region of the surface, the surface integral

$$\iint_S F(x, y, z)\ dx\, dy$$

is evaluated as the double integral

$$\iint_D F[x, y, f(x, y)]\ dx\, dy.$$

surface of revolution, *n.* the surface of a solid of revolution; if the arc of the curve $y = f(x)$ between $x = a$ and $x = b$ is rotated around the x-axis, as in Fig. 364, the area of the resulting surface, of which an element is shown, is

$$2\pi \int_a^b y \sqrt{1 + \left(\frac{dy}{dx}\right)^2}\ dx.$$

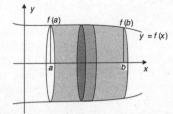

Fig. 364. The shaded surface is the **surface of revolution** of the curve.

surjection, *n.* a SURJECTIVE mapping.

surjective, *adj.* (of a function, relation, etc.) associating two sets in such a way that every member of the CODOMAIN, T, is the IMAGE of at least one member of the DOMAIN, S, although there may be members of the domain that are not mapped into any element of the codomain, as in the diagram of Fig. 365 opposite. The RANGE of a surjective mapping is thus its entire codomain; that is, $f(S) = T$. For example, the mapping from the set of all men to the set of married women that takes each man into his own wife is a *surjection.* Compare INJECTIVE, BIJECTIVE. See also EPIMORPHISM.

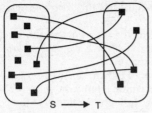

Fig. 365. A **surjective** mapping.

surplus variable, see SLACK VARIABLE.

Swinnerton-Dyer, see BIRCH AND SWINNERTON-DYER CONJECTURE.

syllogism, *n.* (*Logic*) **1.** a deductive inference consisting of two premises and a conclusion, all of which are CATEGORICAL, that is, relate two PREDICATES. The premise in which the predicate of the conclusion (the *major term*) occurs is the *major premise*, and that in which the subject of the conclusion (the *minor term*) occurs is the *minor premise*; the *middle term* occurs in both premises but not the conclusion. There are 256 syllogisms categorized in four FIGURES, but only 24 are valid; for example,

> *some men are mortal.*
> *some men are angelic.*
> so *some mortals are angelic.*

is invalid, while

> *some temples are in ruins.*
> *all ruins are fascinating.*
> so *some temples are fascinating.*

is valid. Here *fascinating, in ruins,* and *temple* are respectively the major, middle and minor terms.
2. deductive inferences of certain other forms with two premises, such as the HYPOTHETICAL SYLLOGISM.

syllogistic, *n.* the study of SYLLOGISMS.

Sylow's theorems, see SYLOW SUBGROUP

Sylow subgroup or **Sylow *p*-subgroup,** *n.* a subgroup of a given finite GROUP of ORDER a maximal prime power; that is, if p divides $|G|$ and α is the largest integer such that p^α divides $|G|$, then a Sylow *p*-subgroup is a subgroup, H, of order $|H| = p^\alpha$. By *Sylow's first theorem*, if p divides $|G|$, G has a Sylow *p*-subgroup; whence every finite group possesses subgroups of order p^s for any prime p for which p^s divides the order of the group. *Sylow's second theorem* states that in a finite group, for some fixed p, all the Sylow *p*-subgroups are conjugate and so isomorphic. *Sylow's third theorem* states that the number of Sylow *p*-subgroups for a given p is congruent to 1 mod p; consequently, any maximal *p*-subgroup of a finite group is a Sylow *p*-subgroup. (Named after the Norwegian group theorist, *Peter Ludvig Sylow* (1832–1918).)

Sylvester, James Joseph (1814–97), English analyst, number theorist, and geometer, who trained as an actuary and a barrister, and also published poetry. While he was pursuing his other careers, he gave private tuition and amongst his pupils was Florence Nightingale. He held chairs at Johns

Hopkins University and Oxford, and founded the *American Journal of Mathematics.*

Sylvester's law of inertia, *n.* (*Linear algebra*) the theorem that the RANK and SIGNATURE of a QUADRATIC FORM over the field of real numbers are independent of any non-singular change of variable.

Sylvester's theorem, *n.* the result, conjectured by Sylvester, and proved more recently by Gallai and then by Erdos, that given a finite set of non-collinear points in the plane, there is a line passing through exactly two of the points.

symbol, *n.* a letter or sign used to represent a number, quantity, function, relation, variable, etc.

symbolic logic, *n.* another term for FORMAL LOGIC.

symbolic manipulation, *n.* (*Computing*) the use of computer programs or languages (such as MACSYMA, MAPLE, REDUCE) that allow one to manipulate quantities symbolically rather than merely numerically. For example, symbolic integration of

$$\int_1^2 x^{-1} \, dx$$

would produce ln2 rather than 0.693....

symmetric or **symmetrical,** *adj.* **1.** (of a figure or configuration) identical with its own reflection in an AXIS OF SYMMETRY or CENTRE OF SYMMETRY; having pairs of points identically placed, except for sense, with respect to some line, point, plane, etc.

2. (of a relation) holding between a pair of arguments x and y when and only when it also holds between y and x. For example, ... *is a sibling of* ... is symmetrical, since one must be one's brother's or sister's brother or sister; but ... *is a brother of* ... is not symmetrical, since a female is not the brother of her brother. Compare ASYMMETRIC, ANTISYMMETRIC, NON–SYMMETRIC. See also EQUIVALENCE RELATION.

3a. (of a function, f, with respect to a point, c) such that, for all x,

$$f(c + x) = f(c - x).$$

b. in particular, if c is the origin, EVEN.

4. (of a binary operator) having the property that the order of arguments is immaterial; COMMUTATIVE.

5. (of a TENSOR) such that its components satisfy either $T^{ab} = T^{ba}$ or $T_{ab} = T_{ba}$.

symmetric design, *n.* a BLOCK DESIGN in which the number of blocks is equal to the number of varieties or points; equivalently, the number of points belonging to each block is equal to the number of blocks in which each point is contained.

symmetric difference, *n.* the set of elements that belong to one but not both of two given sets; the union of their RELATIVE COMPLEMENTS; the relative complement of their intersection in their union. The symmetric difference of A and B is written $A \ominus B$, $A + B$, or $A \nabla B$; for example,

$$\{1, 2, 3\} \ominus \{2, 3, 4\} = \{1, 4\}.$$

In Fig. 366 opposite, if the circles represent the sets S and T respectively, then the shaded region is their symmetric difference.

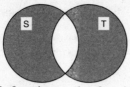

Fig. 366. **Symmetric difference**.

symmetric function, *n.* **1.** a function that is SYMMETRIC about the origin; an EVEN function.

2. a function of several variables that is unchanged by any permutation of the variables. Such functions are *absolutely symmetric* as contrasted with *cyclosymmetric* functions that are left unchanged only by cyclic permutations. The kth *elementary symmetric function in n variables* is the the sum over all k-fold products of the variables and occurs up to sign as the coefficient of x^k in the expansion of the polynomial

$$p(x) = (x + x_1)(x + x_2)\ldots(x + x_n).$$

All symmetric polynomials of degree n have unique representations as polynomials in the elementary symmetric functions. See NEWTON'S IDENTITIES.

symmetric group, *n.* the group consisting of all PERMUTATIONS of a given set; where the given set is finite of order n, the symmetric group has order $n!$. See PERMUTATION GROUP.

symmetric matrix, *n.* a square MATRIX whose entries are symmetric around its MAIN DIAGONAL, and that is therefore equal to its own TRANSPOSE; for example,

$$\begin{bmatrix} 1 & 2 & 3 & 4 \\ 2 & 5 & 6 & 7 \\ 3 & 6 & 8 & 9 \\ 4 & 7 & 9 & 10 \end{bmatrix}.$$

symmetric rotation, *n.* a ROTATION of a regular polygon or polyhedron that is congruent to the original.

symmetry, *n.* **1.** the property of a geometric configuration of being SYMMETRIC around some AXIS OF SYMMETRY, CENTRE OF SYMMETRY, or PLANE OF SYMMETRY. The *order of symmetry* of a figure is the number of distinct orientations of the figure that are indistinguishable; for example the order of symmetry of a cube is 6.

2. a transformation that reflects a figure or configuration about an axis, centre, or plane of symmetry.

symmetry group of an elastic body, *n.* (*Continuum mechanics*) (relative to a given REFERENCE CONFIGURATION of a BODY) the set of transformations to a new reference configuration such that the RESPONSE FUNCTION of the body is invariant.

synclastic, *adj.* (of a surface) having CURVATURE at a given point of the same sign in two perpendicular directions, so that it is not a SADDLE POINT. Compare ANTICLASTIC.

syntactic, *adj.* **1.** relating to or determined by SYNTAX.

2. (*Logic*) describable wholly in terms of the grammatical structure of an expression or the rules of well-formedness for a formal theory, without regard to their sense. See PROOF THEORY.

syntax, *n.* **1.** the study of the rules of well-formedness of the expressions of FORMAL CALCULI.

2. the set of rules, usually in the form of an algorithm, stated wholly in terms of structure and without regard to meaning or truth, that determines all and only the WELL-FORMED formulae of a formal calculus.

Compare SEMANTICS.

synthetic, *adj.* (*Logic*) (of a proposition) neither true nor false by virtue of meaning alone. Whether or not all such statements are A POSTERIORI (EMPIRICAL) is a matter of dispute; so it has been held that while *all effects have causes* is ANALYTIC, *all events have causes* is synthetic but not *a posteriori*.

synthetic division or **synthetic substitution,** *n.* a simplified method of recording long division of one polynomial into another.

synthetic geometry or **pure geometry,** *n.* the study of geometry (usually projective geometry) by the SYNTHETIC METHOD.

synthetic proof or **synthetic method,** *n.* the DEDUCTION of the properties of some entity from a set of AXIOMS, as contrasted with the ANALYTIC PROOF by algebraic construction.

Syracuse problem, see THREE *x* PLUS ONE PROBLEM.

system, *n.* **1.** a set of abstract entities endowed with structure by a set of axioms, and regarded as an UNINTERPRETED CALCULUS, such as groups, rings, fields, Boolean algebras, etc.

2. a set of equations or inequalities that are to be solved simultaneously or treated together.

systematic error, *n.* (*Statistics*) an error that is not random and that introduces BIAS into a statistic.

Système International (abbrev. **SI**) or *Système International d'Unités,* *n.* the METRIC SYSTEM, adopted by international agreement in 1960, and based on the METRE (m), KILOGRAM (kg), and SECOND (s) as the fundamental units of length, mass, and time respectively. Derived units such as the NEWTON (N), JOULE (J), WATT (W), and PASCAL (Pa) are defined in terms of these. Multiples and fractions of the basic units are defined in multiples of 1000, and are denoted by the following prefixes (and symbols):

10^3	kilo–	(k)	10^{-3}	milli–	(m)
10^6	mega–	(M)	10^{-6}	micro–	(μ)
10^9	giga–	(G)	10^{-9}	nano–	(n)
10^{12}	tera–	(T)	10^{-12}	pico–	(p)
10^{15}	peta–	(P)	10^{-15}	femto–	(f)
10^{18}	exa–	(E)	10^{-18}	atto–	(a).

In addition, the following 'customary' prefixes are used:

10	deka–	(da)	10^{-1}	deci–	(d)
10^2	hecto–	(h)	10^{-2}	centi–	(c).

systems analysis, *n.* the application of mathematical methods to the analysis of a task such as a production process in order to determine the most efficient method of performing it.

t

t, *symbol.* **1.** the independent real variable of a function of time.

2. an independent variable in PARAMETRIC EQUATIONS, often non-angular, as contrasted with the angular parameter θ.

3. (*Statistics*) See STUDENT'S T.

T, *symbol.* **1.** (*written superscript*) denoting the TRANSPOSE of a matrix.

2. (*with numerical subscript*) see T–AXIOMS.

3. (*Logic*) also written ' **1** ' (in contrast with **0**). the TRUTH–VALUE truth, especially in TRUTH–TABLES.

4. *abbrev. for* TERA–, used in symbols for multiples of the physical units of the SYSTEME INTERNATIONAL.

$\perp$ (*inverted*) **1.** (*Logic*) *symbol for* falsehood, especially in TRUTH–TABLES; also written **F**, or **0**.

2. *symbol for* ORTHOGONAL, PERPENDICULAR, or SINGULAR (sense 4).

tableau, 1. see SEMANTIC TABLEAU.

2. see SIMPLEX METHOD.

tables, *n.* **1.** any compilation of the values of a function for a range of arguments, such as log tables, trigonometric tables, and statistical tables.

2. a set of arrays showing the results of elementary arithmetical operations for low integral arguments, especially multiplication tables learned by rote by school-children.

tabular differences, *n.* the differences between successive function values when *tabulated* or recorded in a table.

tacnode, *n.* another name for a point of OSCULATION. Compare SPINODE, CRUNODE.

tacpoint, *n.* a point at which two curves from a family intersect with a common tangent; for example, Fig. 367 shows a tacpoint of $\sin x$ and $\frac{1}{2}\sin(2x)$ at the origin. See also TWO–POINT CONTACT.

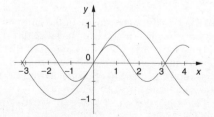

Fig. 367. The origin is a
tacpoint of the curves.

tail, *n.* the set of points in a DIRECTED SET that are greater than some given point; for example, if the natural numbers are the directed set, then the set of natural numbers greater than a GOOGOL is a (very distant) tail.

555

tail event, *n.* (*Probability*) an event that does not depend on any finite INITIAL SEGMENT of a sequence of independent RANDOM VARIABLES. See ZERO–ONE LAW.

tan, *abbrev. and symbol for* the TANGENT function.

tan⁻¹, *symbol for* the inverse TANGENT function. See ARC–TANGENT.

tangency point or **point of contact,** *n.* (of a curve or surface) the point at which a TANGENT line or plane is tangent to a given curve or surface; in Fig. 368, P is the tangency point of the line PT to the curve.

tangent, *n.* **1a.** (*Euclidean geometry*) a line that touches a curve at a point and has the same GRADIENT as that of the curve at the point; a line that has TWO–POINT CONTACT with the curve at the point; the limiting position of a chord PQ as Q approaches P, as shown in Fig. 368. See also DERIVATIVE.

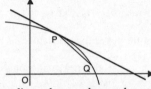

Fig. 368. **Tangent** (sense 1).

b. any line, plane or hyperplane that touches a curve or surface at a point and has the same NORMAL at that point. See TANGENT PLANE, OSCULATION.

c. (*Algebraic geometry*) a line that has an intersection with a given curve, surface, etc. at which the defining equations have at least a double ROOT.

d. (*as modifier*) constituting a tangent; tangential; especially, having a point of osculation with.

2. (*Trigonometry*) abbreviated **tan. a.** a trigonometric function that in a right-angled triangle is the ratio of the length of the side opposite the given angle to that of the adjacent side, where the lengths are taken to be positive; in Fig. 369, the tangent of angle XOP is $|y|/|x|$.

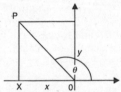

Fig. 369. **Tangent** (sense 2). $\tan\theta = y/x$.

b. more generally, the ratio of the *y*- to the *x*-coordinate of the endpoint of a line from the origin and making the given angle in a clockwise direction with the positive direction of the *x*-axis; in Fig. 369 the tangent of θ is the negative ratio y/x. More properly, the tangent function is the ratio of SINE to COSINE, of which the graph is shown in Fig. 370 opposite; its derivative is $\sec^2 x$ and an antiderivative (or indefinite integral) is $\ln|\sec x|$.

tangent bundle, *n.* (*Differential topology*) the set of TANGENT VECTORS to a MANIFOLD, M, usually denoted TM.

tangential, *adj.* of, forming, related to, or in the direction of a TANGENT.

tangent plane, *n.* a plane that is tangent to a surface at a point, in the sense that every line of the plane that passes through the point is a tangent to the surface; the DIRECTION NUMBERS of the NORMAL to this plane are the PARTIAL DERIVATIVES of the equation of the surface evaluated at the tangency point.

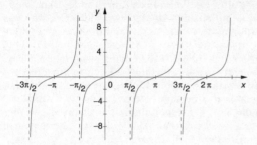

Fig. 370. Graph of the **tangent** function (sense 2b).

tangent rule, *n.* the rule

$$\tan \frac{B-C}{2} = \frac{b-c}{b+c} \cot \frac{A}{2},$$

that is used in solving plane triangles, where a, b, c are respectively the sides opposite the vertices A, B, C of the triangle.

tangent space, *n.* the set $T_x(M)$ of all vectors that are tangent to a differential MANIFOLD at a point x in M. For a surface, it is the TANGENT PLANE at a point.

tangent vector, *n.* **1.** (to a smooth space curve at a point) the rate of change of the position vector when parametrized by arc length:

$$\mathbf{T}(x) = \frac{d\,\mathbf{x}(s)}{ds}.$$

The tangent vector is, by construction, a unit vector. See FRENET FORMULAE.
2. (*Differential topology*) an equivalence class, $[x, \lambda, a]$, of triples, (x, λ, a), where a is a real n-tuple, and x is an element of a given n-dimensional MANIFOLD, M, with ATLAS

$$\{ (\phi_\lambda, U_\lambda) ; \lambda \in \Lambda \},$$

where $[x, \lambda, a] = [y, \mu, b]$ if and only if the derivative of the COORDINATE CHANGE at $\phi_\lambda(x)$ sends a to b.

tanh or th, *symbol for* the HYPERBOLIC FUNCTION hyperbolic tangent, the ratio of the hyperbolic sine function SINH to the hyperbolic cosine function COSH. Its derivative is $\operatorname{sech}^2 x$, and an antiderivative (or indefinite integral) is $\ln(\cosh x)$; it has asymptotes $y = \pm 1$, as shown in the graph in Fig. 371.

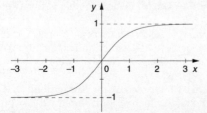

Fig. 371. **Tanh**. Graph of the hyperbolic tangent function.

tanh⁻¹, *symbol for* the inverse hyperbolic tangent function, ARC–TANH.

Tarski (or **Knaster**) **fixed-point theorem,** *n.* the theorem that an ISOTONE mapping on a set with an ORDERING that is order-COMPLETE possesses a FIXED POINT. Indeed, the infimum of the elements in $\{\, x : f(x) \leq x \,\}$ is the least fixed point. (Named after the Polish-born American logician, meta-mathematician, algebraist, and analyst, *Alfred Tarski* (1902–83), who was influential in the development of MODEL THEORY and the theory of DECIDABLE propositions.)

Tauberian, *adj.* (of a SUMMABILITY method) such that a series summable by that method actually converges to the same value. Results to this effect are called *Tauberian theorems*, the simplest being Tauber's original result that a series, $\{\, c_n \,\}$, that is summable to *s* by ABEL SUMMATION and for which nc_n tends to zero, actually sums to *s*. Correspondingly, any theorem asserting that a given method is REGULAR is called an *Abelian theorem*. (Named after the Austrian analyst, *Alfred Tauber* (1866–1933).)

tautological consequence, *n.* (*Logic*) **1.** a statement or FORMULA that is true whenever a given finite set of statements or formulae are true, especially when the relation is TRUTH–FUNCTIONAL.

2. the relationship of VALID deducability that holds in such cases.

tautology or **logical truth,** *n.* (*Logic*) a statement that is always true; especially a TRUTH–FUNCTIONAL expression that takes the value *true* for all combinations of values of its components, such as

either it is raining or it is not raining.

Compare INCONSISTENCY, CONTINGENCY.

T-axioms or **Tychonoff conditions,** *n.* (*Topology*) any of a hierarchy of increasingly restrictive SEPARATION AXIOMS that a TOPOLOGICAL SPACE may satisfy; in particular:

0. T_0-space or **Kolmogorov space.** a topological space in which one of any pair of distinct points lies in an open set excluding the other.

1. T_1-space or **Fréchet space.** a topological space in which each of any pair of distinct points lies in an open set excluding the other.

2a. T_2-space. a topological space in which distinct points lie in disjoint open sets; a HAUSDORFF SPACE.

b. $T_{5/2}$-space or **Urysohn space.** a topological space in which distinct points lie in open sets with disjoint closures.

3a. T_3-space. a topological space in which each of any pair of points lies in an open set excluding the other, and in which every neighbourhood of a point contains the closure of another neighbourhood of that point; a REGULAR T_1-space.

b. $T_{7/2}$-space or **Tychonoff space.** a COMPLETELY REGULAR T_1-space, so that the preceding separation can be achieved by a continuous function from the space into [0, 1], under which the image of the given point is 0 and of the given closed set is 1.

4. T_4-space. a topological space in which each of any pair of points lies in an open set excluding the other, and in which disjoint closed sets lie within disjoint open sets; a NORMAL T_1-space.

5. T_5-space. a topological space in which each of any pair of points lies in an open set excluding the other, and in which sets disjoint from the closure of one another lie within disjoint open sets; a COMPLETELY NORMAL T_1-space.

Taylor, Brook (1685 – 1731), British analyst, geometer, painter, and philosopher, who pioneered the INFINITESIMAL CALCULUS and wrote two works on perspective. Because he did not publish his results, some were claimed by Johann Bernoulli, and the importance of TAYLOR'S THEOREM was only recognized some 60 years later, by Lagrange. He became a Fellow of the Royal Society, and sat on the committee that adjudicated between the claims of Newton and Leibniz to have first invented infinitesimal calculus.

Taylor polynomial, *n.* a finite initial segment (that is a partial sum) of a TAYLOR SERIES that approximates to the value of a function in a small interval around a given argument.

Taylor series, *n.* a POWER SERIES for an infinitely differentiable function of the form

$$\sum_{n=0}^{\infty} \frac{1}{n!} (x - a)^n f^{(n)}(a),$$

where $f^{(n)}(a)$ is the nth derivative of f at a. See RADIUS OF CONVERGENCE, MACLAURIN'S FORMULA, TAYLOR'S THEOREM.

Taylor's theorem, *n.* the theorem of mathematical analysis that states that if a function has an $(n+1)^{\text{th}}$ derivative on an interval $[a, b]$, then its value at $(b, f(b))$, can be written as a TAYLOR POLYNOMIAL of the form

$$f(a) + (b - a)f'(a) + ... + \frac{1}{n!} (b - a)^n f^{(n)}(a),$$

plus an error term, where $f^{(n)}(a)$ is the nth derivative of $f(x)$. When the function is infinitely differentiable, the TAYLOR SERIES will REPRESENT the function at the points for which the error (see LAGRANGE FORM OF THE REMAINDER) goes to zero as n increases. Corresponding forms when f is a real-valued function of a vector variable are easily derived by considering

$$g(t) = f(a + t(b - a)),$$

provided that all partial differential coefficients of order $n + 1$ are continuous about the centre.

Tchebyshev, see CHEBYSHEV.

t-distribution, *n.* (*Statistics*) the distribution of the STUDENT'S T statistic.

technology matrix, see INPUT–OUTPUT MODEL.

telescoping series, *n.* a series of which the terms may be expressed in the form

$$a_n = b_n - b_{n+1}$$

allowing one to obtain the sum through cancellation. The series

$$\sum_{n=1}^{\infty} \frac{1}{n(n+1)}$$

is of this form since

$$\frac{1}{n(n+1)} = \frac{1}{n} - \frac{1}{n+1},$$

and the sum is found to be

$$\lim_{n \to \infty} \left[1 - \frac{1}{n+1} \right] = 1.$$

tend to, *vb.* **1.** to have as a limit, especially (of a dependent variable) as the independent variable itself tends to a limit or infinity.
2. tend to infinity. to increase without UPPER BOUND. The value of an expression tends to infinity if it is such that for any number N, however large, a value of the expression can be found that is greater still. A limit of the function $f(x)$ as x tends to infinity is a value to which the function becomes arbitrarily close as the independent variable increases without bound; this is written

$$\lim_{n \to \infty} f(x) = a.$$

See LIMIT.

tense logic, *n.* the study of the logical properties of tense operators such as *past*, *present*, and *future*, and of the logical relations between tensed sentences, by means of consideration of appropriate formal systems.

tensile normal stress, *n.* a NORMAL STRESS that is in the same direction as the outward normal at a point on a surface. Compare COMPRESSIVE NORMAL STRESS.

tension, *n.* (*Mechanics*) the internal FORCE in the longitudinal direction in a narrow body such as a rope or a thin beam.

tensor, *n.* a multilinear differential form invariant with respect to a group of permissible coordinate transformations in n-space; an element of a TENSOR PRODUCT. A tensor of type (r, s) is a member of the product

$$T_s^r = T^* \otimes \ldots \otimes T^* \otimes T \otimes \ldots \otimes T,$$

of the VECTOR SPACE T with itself r times and with its DUAL, T^*, s times, and has r superscripts and s subscripts. Different BASES of T lead to different bases of T^*, and hence to different bases for the tensors; however, a COMPONENT TRANSFORMATION LAW exists. If we choose the basis of T to be ORTHONORMAL we obtain *Cartesian tensors*. A *zero-order tensor* is a SCALAR, and has no superscript or subscript. A *first-order tensor* is a member of T or T^* (according as it is COVARIANT or CONTRAVARIANT), and corresponds to a VECTOR; it has one subscript or superscript. A *second-order tensor* can be represented by a matrix, and has a total of two subscripts and superscripts, that is, it is a member of T^2, T_2, or T_1^1.

tensor field, *n.* a TENSOR-valued function defined on a connected domain in Euclidean space. Compare VECTOR FIELD, SCALAR FIELD.

tensor product or **dyadic product,** *n.* **1.** any formal expression of the form

$$\mathbf{v} \otimes \mathbf{w} = \sum x_i y_j (\mathbf{v}_i \otimes \mathbf{w}_j),$$

where

$$\mathbf{v} = \sum x_i \mathbf{v}_i \quad \text{and} \quad \mathbf{w} = \sum y_j \mathbf{w}_j$$

are the representations of **v** and **w** with respect to the respective bases of finite dimensional VECTOR SPACES, V and W, and, for the basis vectors,

$$\mathbf{v}_i \otimes \mathbf{w}_j = t_{ij}$$

for distinguished symbols t_{ij}. The tensor product of two linear functionals is then computed by the formula

$$\langle \mathbf{v}' \otimes \mathbf{w}',\ \mathbf{v} \otimes \mathbf{w} \rangle = \langle \mathbf{v}', \mathbf{v} \rangle \langle \mathbf{w}', \mathbf{w} \rangle.$$

The tensor product is similarly defined for MODULES.
2. the VECTOR SPACE of all such expressions, denoted V ⊗ W; the vector space of all BILINEAR FUNCTIONALS on the CARTESIAN PRODUCT, V* × W*, of the DUALS of two given vector spaces, V and W. There is an isomorphism between the bilinear mappings from V × W into a third space U and the linear mappings from V ⊗ W into U.

tera- (symbol **T**), *prefix* denoting a multiple of 10^{12} of a unit of the SYSTEME INTERNATIONAL.

term, *n.* **1.** any expression that forms a separable part of some other expression; in particular, either of the expressions separated by the identity or inequality sign in an equation or inequality; either of the expressions the ratio of which is a fraction or proportion; any of the separate elements of a sequence; or any of the individual addends of a polynomial or series.
2. (*Logic*) in PREDICATE CALCULUS, a name or variable, as opposed to a predicate; that which a predicate qualifies.
3. one of the relata of a RELATION.

terminal, *n.* another word for the SINK of a NETWORK.

terminal side, *n.* the exiting side a DIRECTED angle.

terminate, *vb.* (of a decimal expansion) to have only a finite number of digits. Thus a terminating simple RADIX FRACTION in the place value system to base b can be expressed as ab^{-n}, where a is the integer whose digits are those following the RADIX POINT, and n is the number of digits following the point; to base 10, a terminating decimal is thus equal to $a \times 10^{-n}$, and a VULGAR FRACTION is expressible as a terminating decimal only if its denominator has no prime factors other than 2 and 5. Compare RECUR.

termination criterion or **stopping rule,** *n.* (*Numerical analysis*) a criterion that determines when to terminate an approximative procedure, whether because enough accuracy has been obtained (measured variously), because a solution can be ruled out, or because too much effort has been expended.

ternary, *adj.* **1.** related to, or expressed in, PLACE–VALUE NOTATION to BASE 3.
2. (of a function, relation, etc.) having three arguments.

tessellation, *n.* (*Euclidean geometry*) a covering of the plane by identical shapes; literally, a paving. A triangle or hexagon will tessellate the plane; an octagon will not. Fig. 372 shows part of a tessellation of the plane by the shaded scalene triangle.

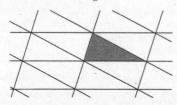

Fig. 372. **Tessellation**.

tesseract, *n.* a four-dimensional equivalent of a cube, a regular four-dimensional HYPERCUBE.

test, see HYPOTHESIS TESTING.

test statistic, *n.* a STATISTIC that has a known distribution under the NULL HYPOTHESIS of a test, and a different distribution under an ALTERNATIVE HYPOTHESIS; for example, a test statistic may have a small numerical value under the null hypothesis, but be large under the alternative.

test rule, *n.* a TEST STATISTIC, T, together with a function, δ_A, onto the set $\{0, 1\}$, such that the hypothesis is accepted if $\delta_A(T) = 0$, and rejected otherwise.

tetra-, *prefix denoting* four. For example, a *tetrahedron* is a polyhedron with four faces; a *tetravalent* predicate is one with four argument places.

tetrad, *n.* **1.** a set or sequence of four elements.

   2. a power of 10 000 (itself the fourth power of 10).

tetragon, *n.* a less common term for QUADRILATERAL.

tetrahedron, *n.* a solid figure with four plane faces; a SIMPLEX. These faces are all triangles, and if they are equilateral, it is a *regular tetrahedron*. A *tetrahedral angle* is a polyhedral angle with four faces.

t-formulae, *n.* a set of trigonometric identities that are useful as a change of variables in integration, that express functions in terms of $t = \tan(\theta/2)$; in particular,
$$\sin\theta \;=\; \frac{2t}{1 + t^2}\,, \qquad \cos\theta \;=\; \frac{1 - t^2}{1 + t^2}\,, \qquad \tan\theta \;=\; \frac{2t}{1 - t^2}\,.$$

th, *symbol for* the hyperbolic tangent function, TANH.

th^{-1}, *symbol for* the inverse hyperbolic tangent function, ARC–TANH.

the, see DEFINITE DESCRIPTION.

theorem, *n.* **1.** a statement or formula that can be deduced from the AXIOMS of a formal system by recursive application of its RULES OF INFERENCE; in a system of NATURAL DEDUCTION it is a sequent relying on no ASSUMPTIONS.

   2. less formally, a proposition derived from previously accepted results in some fragment of mathematics that may not be fully axiomatized.

theory, *n.* **1.** a systematic statement of the principles underlying some fragment of mathematics. It is often not clear whether what is described as a theory is an informal, intuitive description of a subject or its subsequent mathematization. Often, the formalization is merely partially undertaken, with no more than a tacit presumption of the possibility of its completion. Sometimes, however, the prior conception of what should constitute such a formal theory proves unrealizable, or the process of mathematical modelling itself shows the *pre-theoretical* notions to be untenable; see, for example, NAIVE SET THEORY.

   2. rigorously, a FORMAL LANGUAGE together with its AXIOMS and RULES OF INFERENCE. Such a system generates a set of truths, the THEOREMS, but it cannot itself refer to the truth of its own sentences.

theory of games, *n.* another name for GAME THEORY.

theory of types, see TYPE.

theta function, *n.* any of a class of special functions that are important in topology, number theory, and analysis. The basic such function from which all others can be derived is the ENTIRE function

$$\Theta_3(z,q) = 1 + 2 \sum_{n=1}^{\infty} q^{n^2} \cos(2nz),$$

where $q = e^{\pi i t}$ for $\mathrm{im}\, t > 0$. When analytic properties are concerned, the dependence on q is suppressed. The *special theta function*

$$\Theta_3(q) = \Theta_3(0,q) = 1 + 2 \sum_{n=1}^{\infty} q^{n^2}$$

is the GENERATING FUNCTION of the sequence of SQUARE NUMBERS. It satisfies the remarkable *theta transformation formula*:

$$\sqrt{s}\,\Theta_3[\exp(-\pi s)] = \Theta_3[\exp(-\pi/s)].$$

Compare TRIPLE PRODUCT IDENTITY.

theta notation or **Θ-notation,** see ORDER NOTATION.

theta transformation formula, see THETA FUNCTION.

thin, *vb.* (*Logic*) to weaken a form of argument, especially by adding to its premises; for example, from $A \vdash B$ we derive $A, C \vdash B$ by thinning.

third curvature, *n.* another term for TOTAL CURVATURE (sense 2).

third isomorphism theorem, see ISOMORPHISM THEOREMS.

Thomson lamp, *n.* a paradoxical machine devised as a thought-experiment by British philosopher *James Thomson* in 1970 to highlight the difficulty in ZENO'S PARADOXES of deciding whether a SUPERTASK, an infinite number of tasks in finite time, has been completed. The lamp is switched on, and after a minute switches off; after a further half minute it comes on again, goes off after another quarter minute, and so on, changing its state after each term of a convergent series of time intervals. Since this series of intervals has a sum of two minutes, the process must have terminated after that time, so the lamp must be either on or off; yet it cannot be either on or off, since each time it entered either state it was immediately switched again. This does not show, however, as Thomson argued, that Zeno's paradoxes are unresolved by taking limits, since these paradoxes concern convergent rather than oscillating series of tasks.

three-circle theorem, *n.* the theorem, due to Hadamard, that a complex function that is ANALYTIC in an annulus is such that $\log m(r)$ is a CONVEX function of $\log r$, where $m(r)$ is the maximum modulus of the function on a ring of radius r. This result is so named as it involves circles of three radii. See also ORDER (sense 9).

three n plus one $(3n + 1)$ **problem,** see THREE x PLUS ONE PROBLEM.

three-point contact, *n.* (*Algebraic geometry*) the relation between two curves, surfaces, etc., at a point at which the defining equations of the intersection have at least a triple ROOT; in Euclidean geometry this is equivalent to them touching and having the same CURVATURE at that point. Compare TWO-POINT CONTACT.

three series theorem, see KOLMOGOROV'S THREE SERIES THEOREM.

three-space property, *n.* a property that is inherited by a space, S, whenever it is possessed both by a subspace, T, and by the FACTOR SPACE, S/T; for example, being NOETHERIAN is a three-space property. Such a property of groups is called a *polyproperty*.

three-valued logic of Lukasiewicz, *n.* an early MANY-VALUED LOGIC allowing for a third and intermediate truth value.

three *x* plus one (3*x* + 1) problem or **three *n* plus one (3*n* + 1) problem,** *n.* also known as the **Syracuse problem** or **Hasses's algorithm,** the unsolved DECISION PROBLEM, attributed to Collatz, Kakutani, and Ulam, concerning the behaviour of the iterates of the function that sends odd integers *n* to 3*n*+1 and even integers *n* to *n*/2. The 3*x*+1 conjecture asserts that, for any positive starting value, repeated iteration of this function always eventually terminates with 1. The conjecture is simple to state but is apparently intractable like the existence of odd PERFECT NUMBERS, or celebrated DIOPHANTINE EQUATIONS such as FERMAT'S LAST THEOREM. Paul Erdös commented that "Mathematics is not yet ready for such problems."

Thue–Siegel–Roth theorem or **Roth's theorem,** *n.* the culminating theorem of rational APPROXIMATION that for any ALGEBRAIC irrational number α, and for any κ > 2, there are only finitely many rational numbers *p/q* solving

$$|\alpha - p/q| < Cq^{-\kappa}.$$

Compare HURWITZ' THEOREM.

T$_i$-axioms, see T-AXIOMS.

Tietze's extension theorem, *n.* (*Topology*) the theorem that a topological space is NORMAL if and only if every continuous mapping from a closed subset of the space into the unit interval has a continuous EXTENSION with range in the unit interval. (Named after the Austrian topologist and analyst, *Heinrich Franz Friedrich Tietze* (1880–1964).)

time, *n.* (*Mechanics*) a primitive term of Newtonian mechanics, represented by a single real variable and measured in SECONDS.

time algorithm, see POLYNOMIAL TIME ALGORITHM.

time series, *n.* (*Statistics*) a sequence of data indexed by time, often comprising uniformly spaced observations.

Toeplitz matrix, *n.* **1.** a MATRIX with entries of the form $a_{ij} = b_{i-j}$ and so constant on lines parallel to the MAIN DIAGONAL. Compare HANKEL MATRIX. **2.** more particularly, an infinite matrix corresponding to a REGULAR summation method.

Tonelli's theorem, *n.* (*Measure theory*) the theorem that if (X, Σ, μ) and (Y, T, ν) are SIGMA–FINITE MEASURE SPACES, and *F* is a non-negative $(\Sigma \times T)$-measurable function, then

$$\iint F(x, y)\,\mu(\mathrm{d}x)\,\nu(\mathrm{d}y) = \iint F(x, y)\,\nu(\mathrm{d}y)\,\mu(\mathrm{d}x) = \iint F\,\mathrm{d}(\mu \times \nu).$$

Compare FUBINI'S THEOREM.

tonne, *n.* (*Mechanics*) a MASS of 1000 KILOGRAMS.

topological degree, see DEGREE (sense 7).

topological dimension, *n.* an integer that measures the size of a set and is preserved by homeomorphism. The dimension may be defined for a metric space as the least integer, *n*, such that, for each ε > 0, there is an EPSILON NET of order less than or equal *n* + 1 (at most *n* + 1 of the covering sets intersect). All arcs have dimension one and Euclidean *n*-space has dimension *n*. Compare HAUSDORFF DIMENSION. See also INVARIANCE OF DOMAIN THEOREM.

topological group, *n.* a GROUP, such as the set of all real numbers, that constitutes a TOPOLOGICAL SPACE, and in which multiplication and inversion are continuous operations. For example, a Hausdorff space with a group operation such that the map

$$(x, y) \mapsto xy^{-1}$$

is continuous is a topological group.

topological invariant, *n.* another term for TOPOLOGICAL PROPERTY.

topological property or **topological invariant,** *n.* a property definable in terms of OPEN SETS and therefore invariant under HOMEOMORPHISM. For example, compactness is a topological property.

topological space, *n.* a set with an associated family of subsets, the OPEN SETS, including the whole set and the empty set, that is closed under set union and finite intersection; the family of subsets is usually denoted τ. See TOPOLOGY (sense 3).

topological vector space or **linear topological space,** *n.* a VECTOR SPACE, such as a normed space, that constitutes a TOPOLOGICAL SPACE in which addition and SCALAR MULTIPLICATION are continuous.

topology, *n.* **1. point set topology.** the branch of mathematics that is concerned with the generalization of the concepts of continuity, limits, etc. to sets other than the real and complex numbers.

2. algebraic topology or (*formerly*) **analysis situs.** a branch of geometry describing the properties of a figure that are unaffected by continuous distortion such as stretching or knotting. See also KNOT.

3. a family of subsets of a given set that constitute a TOPOLOGICAL SPACE. The *discrete topology* consists of the entire power set, while the *indiscrete topology* contains only the empty set and the entire space. The *relative* or *induced topology* on a subset is the topology constructed by taking intersections of the original topology with the subset. A topology, τ_1, is *finer* than another, τ_2, if τ_1 is a REFINEMENT of τ_2, and τ_2 is then said to be *coarser* than τ_1. Thus on any given set, the discrete topology is the finest topology and the indiscrete topology is the coarsest.

toroid, *n.* a surface or solid generated by rotating a closed plane curve around a coplanar line that does not intersect the curve. See TORUS.

torque or **moment of a force,** *n.* (*Mechanics*) the cross product, $\mathbf{x} \times \mathbf{F}$, of a force $\mathbf{F}$ on a PARTICLE with its position vector $\mathbf{x}$, and the sum of such products for a system of forces. See PRINCIPLE OF ANGULAR MOMENTUM.

Torricelli point, see FERMAT'S PROBLEM.

torsion or **second curvature,** *n.* (of a space curve at a point) the rate at which a curve leaves its OSCULATING PLANE. Explicitly, the quantity

$$\tau = -\mathbf{N}(s)\,\mathbf{B}'(s)$$

where $\mathbf{N}$ and $\mathbf{B}$ are respectively the unit principal and binormal vectors to the curve, parametrized by arc length. The quantity $1/\tau$ is the *radius of torsion*.

torsion element, *n.* **1.** an element of an R-MODULE whose product with some non-zero element of the RING is zero; this is equivalent to being NIL-POTENT for the group operation of $\mathbb{Z}$-modules. The set of such elements forms the *torsion submodule*; all other elements are *free elements*. The QUOTIENT of a module by its torsion submodule is a TORSION–FREE MODULE.

2. (*Group theory*) an uncommon term for a PERIODIC element.

torsion-free module, *n.* a MODULE with no non-zero TORSION ELEMENTS; it is isomorphic to a SUBMODULE of a FREE MODULE. Torsion-free modules over principal ideal domains are free.

torsion module, *n.* a MODULE all of whose elements are TORSION ELEMENTS.

torsion submodule, *n.* the set of all the TORSION ELEMENTS of a MODULE.

torus or **anchor ring,** *n.* a ring-shaped surface or solid obtained by rotating a circle about a coplanar line that does not intersect it, as in Fig. 373. Its volume is $2\pi^2 r^2 R$, where r is the radius of the rotating circle, and R is the distance of its centre from the axis of revolution; its surface area is $4\pi^2 rR$.

Fig. 373. **Torus**.

total boundedness, *n.* the property of being TOTALLY BOUNDED.

total curvature, *n.* **1.** the GAUSSIAN CURVATURE of a two-dimensional surface at a point. However, in higher dimensions these concepts diverge.

2. also called **third curvature.** the quantity

$$\sqrt{\tau^2 + \kappa^2}$$

where κ is the CURVATURE of a space curve at point and τ is the TORSION.

total derivative, *n.* the DERIVATIVE of a function of two or more variables with respect to a single parameter in terms of which these variables are expressed: if $z = f(x, y)$ with PARAMETRIC EQUATIONS

$$x = u(t), \quad y = v(t),$$

then, under appropriate conditions, the total derivative is

$$\frac{dz}{dt} = \frac{\partial z}{\partial x}\frac{dx}{dt} + \frac{\partial z}{\partial y}\frac{dy}{dt}.$$

Compare PARTIAL DERIVATIVE.

total differential or **exact differential,** *n.* the DIFFERENTIAL of a function of two or more variables with respect to a single PARAMETER in terms of which these variables are expressed, equal to the sum of the products of each PARTIAL DERIVATIVE of the function with the corresponding increment. If

$$z = f(x, y), \quad x = u(t), \quad y = v(t),$$

then, under appropriate conditions, the total differential is

$$dz = \frac{\partial z}{\partial x} dx + \frac{\partial z}{\partial y} dy.$$

total differential equation, *n.* a DIFFERENTIAL EQUATION of the form

$$\sum_{i=1}^{n} P_i \, dx_i = 0,$$

where each P_i is a function of the variables $x_1, x_2, \ldots, x_n$. When $n = 2$, a solution of the total equation can be found by means of a solution of the LINEAR DIFFERENTIAL EQUATION

$$P_1 + P_2 \frac{dx_2}{dx_1} = 0.$$

When $n = 3$, this equation is integrable if and only if $\nabla \times \mathbf{V} = 0$, where $\mathbf{V} = (P_1, P_2, P_3)$. See CHARPIT'S METHOD.

totally bounded, *adj.* (of a set in a METRIC SPACE) able to be enclosed in the union of a finite number of ε-BALLS round elements of the set; such that, for every $\varepsilon > 0$, there is a finite EPSILON–NET; that is, there is a finite set F in the space such that each point of the given set is within epsilon of F. For example, in the reals, the interval $(0, 1)$ is totally bounded, but the rationals are not. A set in a metric space is COMPACT if and only if it is both complete and totally bounded. Compare PRECOMPACT.

totally disconnected, *adj.* (of a TOPOLOGICAL SPACE) such that every two distinct points can be SEPARATED. The rationals are totally disconnected.

totally finite, *adj.* (of a MEASURE) FINITE, and such that the space itself is of finite measure.

totally multiplicative, see MULTIPLICATIVE.

totally ordered, *adj.* having a TOTAL ORDERING.

totally sigma-finite, *adj.* (of a MEASURE) SIGMA–FINITE, and such that the space itself is of sigma-finite measure. An example of a sigma-finite but not totally sigma-finite measure is provided by counting measure on the ring of countable sets of an uncountable set.

totally stable, see STABLE.

totally unimodular, see UNIMODULAR.

total moment, *n.* (*Mechanics*) the total TORQUE of a system of forces.

total ordering, *n.* a RELATION that orders a set in such a way that every element is related to every other either by the relation or its converse; a relation R such that, for all x and y, either xRy or yRx. Some usages demand that the order be ANTISYMMETRIC. For example, *less than* is a total ordering on the reals, as contrasted with the PARTIAL ORDERING of set inclusion.

total probability theorem, *n.* the result for a PROBABILITY SPACE (X, Σ, P), that for a PARTITION, $[E_n]$, of X, by elements of the SIGMA–ALGEBRA, Σ,

$$P(A) = \sum_{n=1}^{\infty} P(A \mid E_n)\, P(E^n).$$

total variation, *adj.* **1.** a measure of the oscillation of the function h:

$$V_h(a, b) = \sup\left\{ \sum \mid h(x_{i+1}) - h(x_i) \mid \right\}$$

over all PARTITIONS of the interval $[a, b]$, which is finite if and only if the function is of BOUNDED VARIATION on the interval. If the function is decomposed as $f - g$, with f and g monotone increasing, by setting

$$2f(x) = V_h(a, x) + h(x) - h(a)$$
$$2g(x) = V_h(a, x) - h(x) + h(a),$$

for x between a and b, then the total variation between a and b equals $f(b) + g(b)$.

2. the measure arising by adding the members of the JORDAN DECOMPOSITION of a signed measure.

totative, *n.* a number less than and relatively prime to a given *n*; the number of such totatives is the TOTIENT of *n*.

totient, *n.* the value of EULER'S PHI FUNCTION, $\phi(n)$, for an integer *n*; the number of TOTATIVES of *n*.

tour, *n.* a HAMILTONIAN CIRCUIT of a graph.

tournament, *n.* a DIGRAPH containing precisely one of the two possible edges between any pair of vertices. Intuitively, this corresponds to a graph of winners and losers in a round-robin tournament.

tower of radicals, see SOLUTION BY RADICALS.

towers of Hanoi, *n.* the ancient puzzle in which there are three poles with *n* disks of decreasing diameter all initially on one pole as in Fig. 374. The goal is to move rings one at a time until all rings sit on another pole, with the proviso that at no stage may a larger disk be placed above a smaller one. The puzzle has no known connection with Hanoi, but may be Indian.

Fig. 374. **Towers of Hanoi**.

trace, *n.* **1.** also called **spur.** the sum of the EIGENVALUES of a finite-dimensional linear transformation, or square matrix. This agrees with the sum of the diagonal entries of any representing matrix. There is a class of infinite-dimensional operators (*trace class*) to which one can meaningfully extend the concept.

2a. the projection of a curve on a given plane. The trace on the horizontal plane of a standard HELIX is a circle.

b. also called **piercing point.** a point at which a line in space cuts a given coordinate plane.

3. (of an ALGEBRAIC NUMBER) the sum of the CONJUGATES of the given number. The trace of $\sqrt{2} + 1$ is

$$(\sqrt{2} + 1) + (1 - \sqrt{2}) = 2.$$

trace class, see TRACE.

trace norm, *n.* another term for FROBENIUS NORM.

traction, *n.* the CONTACT FORCE density of a body.

tractrix, *n.* an INVOLUTE of a CATENARY of which the equation in standard form is

$$x = \operatorname{arccosh}(a/y) \pm \sqrt{(a^2 - y^2)}.$$

The solid formed by revolving this curve, shown in Fig. 375, around its asymptote is a PSEUDO–SPHERE.

Fig. 375. **Tractrix**.

trail or **Eulerian walk,** *n.* a WALK in a graph in which all EDGES are distinct. Compare PATH.

trajectory, *n.* **1.** a PATH, especially when parametrized by time.

2. a curve that cuts a family of curves at a constant angle. An ORTHOGONAL trajectory is a curve that cuts a family of curves at right angles. For example, the involutes of a curve, such as I in Fig. 376, are orthogonal trajectories to the tangents of the original curve, C.

Fig 376. **Trajectory** (sense 2).
See main entry.

3. also called **orbit.** the curve $\{\mathbf{y}(t) : t \in \mathrm{I}\}$, where $\mathbf{y}$ is the solution of a system of ORDINARY DIFFERENTIAL EQUATIONS, and I is the INTERVAL OF EXISTENCE of $\mathbf{y}$; this is a curve in PHASE SPACE. Compare SOLUTION CURVE.

transcendental, *adj.* not ALGEBRAIC; thus, a TRANSCENDENTAL NUMBER is one that is not the root of an algebraic equation with rational coefficients. A *transcendental extension field* contains a number that is not algebraic over the base field. Amongst the real numbers, there are uncountably many transcendental numbers but only countably many algebraic numbers.

transcendental function, *n.* a function that cannot be constructed in a finite number of steps from the ELEMENTARY FUNCTIONS and their inverses, such as $\sin x$. Compare ALGEBRAIC FUNCTION, ELEMENTARY FUNCTION.

transcendental number, *n.* a number that is real but not ALGEBRAIC; that is, not a root of any polynomial equation with rational coefficients. For example, e and π are transcendental numbers, while the status of γ and $\zeta(3)$ is not settled. The transcendental numbers form a set of which the complement is denumerable and so of measure zero. See APERY'S THEOREM, LINDEMANN'S THEOREM, LIOUVILLE NUMBER, THUE–SIEGEL–ROTH THEOREM.

transfinite, *adj.* having CARDINALITY that is, or indexed by a sequence whose ORDINAL NUMBER is, a TRANSFINITE NUMBER.

transfinite induction, *n.* a form of INDUCTION on ORDINAL NUMBERS that is equivalent to the AXIOM OF CHOICE, and that has the form of complete INDUCTION. Thus one proves that whenever some property $P(\alpha)$ holds for for all ordinals $\alpha < \beta$, then $P(\beta)$ holds, and one may then conclude that $P(\alpha)$ holds for all α.

transfinite number, *n.* a CARDINAL or ORDINAL NUMBER used in the comparison of infinite sets, the smallest of which are respectively the cardinal $\aleph_0$ (ALEPH–NULL) and the ordinal ω (OMEGA). The set of rationals and the set of reals have different transfinite cardinality.

transform, *vb.* **1.** to change the form of an expression by a TRANSFORMATION.
– *n.* **2.** the result of a transformation, especially a CONJUGATE element of a group or a SIMILAR matrix.
3. see INTEGRAL TRANSFORM.

transformation, *n.* **1.** a change in the position or direction of the axes of a coordinate system without changing their relative angles.

2. an equivalent change in an expression resulting from the uniform substitution of one set of variables for another.

3a. a MAPPING between two spaces, especially one in which the expression for the value is uniformly derived from that for its argument. See FUNCTION.

b. see **linear transformation.**

transformation group, *n.* a GROUP of TRANSFORMATIONS on a set with COMPOSITION as the binary operation.

transformation of similitude, *n.* another term for SIMILITUDE.

transformation of the plane, *n.* an INVERTIBLE MAPPING of the plane to itself.

transformation rules, *n.* (*Logic*) the set of rules that specify in purely syntactic terms the methods by which theorems may be derived from the axioms of a formal system; its RULES OF INFERENCE.

transition matrix, *n.* (of a MARKOV CHAIN) a STOCHASTIC matrix that gives the *transition probabilities* of moving from one state to another.

transitive, *adj.* **1.** (of a relation) having the property that if the relation holds between a first object and a second, and between the second and a third, then it holds between the first and the third; for example, ... *is greater than* ... is transitive, since for any *a*, *b*, and *c*, if *a* > *b* and *b* > *c* then *a* > *c*. Compare INTRANSITIVE, NON–TRANSITIVE. See also EQUIVALENCE RELATION.

2. (of a graph) such that the adjacency relation on the vertices is transitive. See ADJACENT.

3. (of an ACTION of a group on a non-empty set) such that the set has precisely one ORBIT.

transitive closure, *n.* the unique minimal RELATION that is TRANSITIVE on a given set and contains a given relation on the set.

transitive set, *n.* a set of which every member is also a subset. Equivalently, a set *x* is TRANSITIVE if and only if set membership is transitive, so that every element of an element of the set is an element of the set; that is, if

$$(\forall y)(\forall z)(y \in x \;\&\; z \in y \rightarrow z \in x).$$

See also ORDINAL.

translate, *vb.* **1.** (*Euclidean geometry*) to move (a figure or body) laterally, without rotation, dilation, or angular displacement, as in Fig. 377.

– *n.* **2.** the image of some figure or body under a TRANSLATION; in Fig. 377, KLM is a translate of ABC.

3. (*Topological groups*) the function derived from a given function *f* on a GROUP G when the argument is multiplied (on the left or right) by a given element *a* of G; that is, the *left translate* of *f* by *a*, denoted $L_a f$ or *af*, and the *right translate*, $R_a f$ or *fa*, are respectively defined by

$$L_a f(x) = f(ax) \text{ and } R_a f(x) = f(xa)$$

for all *x* in G.

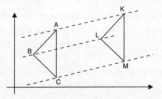

Fig. 377. **Translation.** ABC and KLM are **translates** of one another.

translation, *n.* (*Euclidean geometry*) **1. translation of axes.** a transformation in which the origin of a coordinate system is moved to another position but the new axes are parallel to the old; a change of variables of the form

$$x' = x + a, \quad y' = y + b.$$

2. equivalently, a transformation in which a figure or curve is moved so that it retains the same orientation to the axes, as shown in Fig. 377 opposite, so that lines joining corresponding points are parallel; it is a DILATATION that does not have a fixed point.

translation-invariant, see INVARIANT.

transparent, *adj.* (of a modal context) not OPAQUE; permitting quantifiers outside the scope of a modal operator to bind variables within their scope, and terms with the same reference to be substituted without change of truth value. See OPAQUE, SUBSTITUTIVITY.

transportation problems, *n.* a class of LINEAR PROGRAMMING problems about NETWORKS, modelled on the problem originally studied by Hitchcock, in which one attempts to minimize the cost of delivering integral quantities of goods produced at *n* plants to *m* outlets while balancing supply and demand. This generates a TRANS–SHIPMENT PROBLEM with no intermediate nodes and with each sink and source connected. See also MATCHING.

transpose, *vb.* **1.** to move a term from one side of an equation to the other with a corresponding change of sign; for example, transposing y in $x - y = 2$ yields $x = y + 2$.

2. to interchange the rows and columns of (a matrix); thus, transposing

$$\begin{bmatrix} a & b & c \\ d & e & f \\ g & h & i \end{bmatrix}$$

yields

$$\begin{bmatrix} a & d & g \\ b & e & h \\ c & f & i \end{bmatrix}.$$

3. *n.* the matrix derived from a given matrix by interchanging the rows and columns; the transpose of matrix M is often denoted M^T.

transpose diagonal, *n.* the OFF DIAGONAL of a matrix.

transposition, *n.* a PERMUTATION that merely interchanges two elements while leaving all others fixed; the transposition (i, j) is the permutation that only exchanges the positions of the elements i and j. For example, the transposition (b, e) takes the sequence $(a\ b\ c\ d\ e\ f)$ into the sequence $(a\ e\ c\ d\ b\ f)$.

trans-shipment problems, *n.* a class of LINEAR PROGRAMMING models for NETWORKS in which, given a network with multiple sinks and sources, and with trans-shipment costs for each arc, one wishes to minimize the cost of shipping a given quantity of material from the sources to the sinks. By adding an extra sink (a *dump*) one may presume that supply must equal demand. Compare TRANSPORTATION PROBLEMS.

transversal, *n.* **1.** also called **traverse.** a line intersecting two or more other lines; for example, in the configuration of Fig. 378, AB is a transversal.

Fig. 378. AB is a **transversal** of the other lines.

2. (*Group theory*) a set of canonical representatives of the COSETS of a subgroup of a group; a set, T, of elements such that, for a given subgroup H of the group G, for every group element, *x*, there is exactly one member, *t*, of T for which *x*H = *t*H (a transversal of the left cosets of H in G), or else H*x* = H*t* (a transversal of the right cosets of H in G).

transversality conditions, *n.* (*Calculus of variations, control theory*) orthogonality conditions that must hold at the boundary of a problem. See PONTRYAGIN'S MAXIMUM PRINCIPLE. A curve satisfying a transversality condition is called a *transversal*.

transverse axis, *n.* the axis of a hyperbola that passes through the foci. In Fig. 379 this is the *x*–axis; the *y*–axis is then the CONJUGATE AXIS.

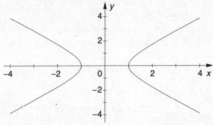

Fig. 379. **Transverse axis**.

transverse component, *n.* (*Mechanics*) the COMPONENT of a vector that is perpendicular to the RADIUS VECTOR. In POLAR COORDINATES, the transverse component of VELOCITY is $r(d\theta/dt)$, and the transverse component of ACCELERATION is

$$r \frac{d^2\theta}{dt^2} + 2 \frac{dr}{dt} \frac{d\theta}{dt}.$$

trapezium, *n.* **1.** (*mainly UK usage. North American term*: **trapezoid**.) a quadrilateral with two parallel sides of unequal length. Compare PARALLELOGRAM.
2. (*mainly North American usage. UK term*: **trapezoid**.) a quadrilateral with neither pair of sides parallel.

Fig. 380. **Trapezium** and **trapezoid** (or *vice versa*).

trapezoid, see TRAPEZIUM.

trapezoidal rule or **trapezium rule,** *n.* a method of approximating to an integral as the limit of a sum of areas of trapezia:

$$\int_{a}^{b} f(x) \ \mathrm{dx} \ \sim \ \frac{\delta}{2}[f(a) + 2f(a+\delta) + 2f(a+2\delta) + ... + f(b)],$$

where $\delta = (b-a)/n$. The rule is exact only for linear functions, the error being of the form

$$\frac{(b-a)^3 f''(c)}{12 \, n^2}$$

for some point c on the interval $[a, b]$. Compare SIMPSON'S RULE.

travelling salesman problem, *n.* the hard combinatorial problem of finding the least-distance (or least-cost) HAMILTONIAN CIRCUIT (or TOUR) of a graph. See also CHINESE POSTMAN PROBLEM.

traversable, *adj.* (of a NETWORK) forming an EULERIAN CHAIN; able to be traced without lifting one's pencil from the page, or retracing an edge.

traverse, *n.* another name for TRANSVERSAL.

tree, *n.* **1.** also called **tree diagram.** (*Graph theory*) a connected GRAPH of which the diagram is tree-shaped in that there are no loops or paths leading from any vertex back to itself. It is a ROOTED tree if one vertex is distinguished as the ROOT or origin, as in Fig. 381; otherwise it is a FREE tree; in a LABELLED TREE, a unique element is associated with each node. A finite set of node-disjoint trees comprises a *forest*.

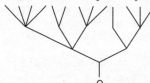

Fig. 381. A rooted **tree**.

2. a COMPACT CONNECTED set in which every pair of points is joined by a unique RECTIFIABLE path.

trefoil, see MULTIFOIL.

trend, *n.* (*Statistics*) a functional relationship between observed data and an independent variable, usually time.

trend line, *n.* (*Statistics*) a line fitted to observations made over time, often by a LEAST SQUARES approximation.

tri-, *prefix denoting* three; for example, a *trilinear* or *trilateral* configuration is one containing three lines or sides.

trial, *n.* (*Statistics*) a single experiment or observation.

triangle, *n.* (*Euclidean geometry*) a closed plane figure bounded by three straight lines meeting at three vertices. Triangles may be classified by their angles as ACUTE, OBTUSE, RIGHT–ANGLED or EQUIANGULAR, or by their sides as ISOSCELES, SCALENE or EQUILATERAL; Fig. 382 shows examples of an acute scalene, an obtuse scalene, a right isosceles, and an equilateral triangle.

Fig. 382. **Triangle**. See main entry.

The sum of angles of a plane triangle is 180°, and its area is half the product of base and height; the ratios of sides define the TRIGONOMETRIC FUNCTIONS. See also POLYGON. Compare SPHERICAL TRIANGLE.

triangle inequality, *n.* **1.** the proposition that the sum of any two sides of a triangle is greater than the third.

2. the requirement that

$$d(x, y) + d(y, z) \geq d(x, z),$$

(for a binary real-valued function *d*), or that

$$|x| + |y| \geq |x+y|.$$

See METRIC, NORM.

triangle of forces, *n.* a triangle whose sides represent the magnitudes and directions of three forces of which the resultant is zero and that are therefore in equilibrium, as in the diagram in Fig. 383. Compare PARALLELOGRAM OF FORCES.

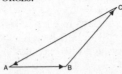

Fig. 383. **Triangle of forces**.

triangle of reference, *n.* a set of three LINEARLY INDEPENDENT points in two-dimensional ALGEBRAIC GEOMETRY that are chosen together with a UNIT POINT in order to determine a system of HOMOGENEOUS COORDINATES for the geometry.

triangulable, *adj.* (of a TOPOLOGICAL SPACE) able to be mapped by a (simplicial) TRIANGULATION, as is the sphere.

triangular, *adj.* **1.** shaped like, or relating to, a triangle.

2. (of a three-dimensional figure) having a triangular base; for example, a tetrahedron is a triangular prism.

triangular matrix, *n.* having only zero entries below (*upper triangular*) or above (*lower triangular*) the MAIN DIAGONAL. See also SCHUR'S LEMMA, CHOLESKY DECOMPOSITION. Compare HESSENBERG FORM.

triangular number, *n.* the number of dots in a triangular array of unit spaced dots, and so the sum of the first *n* integers for some *n*. See also FIGURATE NUMBERS.

triangulate, *vb.* **1.** to calculate trigonometrically, especially by TRIANGULATION.

2. to divide (an area, or region) into triangles or SIMPLICES.

triangulation, *n.* **1a.** a method of surveying in which an area is divided into triangles, and one line (the *base line*) and all angles are measured, and thence the lengths of all the other lines are calculated trigonometrically.

b. a method of fixing an unknown point (for example in navigation) by making it a vertex of a triangle whose other vertices and angles are known.

2. the network of triangles formed in triangulation.

3. also called **simplicial triangulation**. a HOMEOMORPHISM of a TOPOLOGICAL SPACE onto a polyhedron comprising the points of a SIMPLICIAL COMPLEX.

triaxial, *adj.* having three axes.

trichotomy, *n.* the property of the real line, or more generally of any strong TOTAL ORDERING, that given elements *a* and *b*, exactly one of

$$a < b, \quad b < a, \quad a = b$$

is true.

trident of Newton, *n.* a plane curve with an equation of the form

$$xy = ax^3 + bx^2 + cx + d.$$

tridiagonal matrix, *n.* a matrix with zero entries except on the MAIN DIAGONAL and the rows immediately above and below it.

trig, *abbrev. for* TRIGONOMETRY or TRIGONOMETRICAL.

trigon, *n.* an archaic word for TRIANGLE.

trigonometric function, **circular function**, or **cyclometric function,** *n.* any of a group of functions expressible naively in terms of the ratios of the sides of a right-angled triangle containing an angle equal to the argument of the function in radians. More generally for real arguments, they can be expressed in terms of the ratios of the coordinates of the points on the circumference and radius of a circle centred on the origin as the radius sweeps out an angle. Fig. 384 shows this for an angle in the second quadrant. These functions are SINE, COSINE, TANGENT, SECANT, COSECANT, and COTANGENT, and may be defined by power series or otherwise as complex functions. Compare HYPERBOLIC FUNCTION.

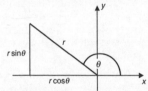

Fig. 384. **Trigonometric function**. See main entry.

trigonometric moment sequence, see MOMENT SEQUENCE.

trigonometric reduction formulae, see REDUCTION FORMULAE.

trigonometric series, *n.* any series of the form of a FOURIER SERIES whether or not the coefficients are the Fourier coefficients of any function; such a series is

$$\sum_{i=1}^{\infty} \frac{\sin(nt)}{\log(n+2)},$$

which converges almost everywhere.

trigonometric tables, *n.* tables showing the values of the TRIGONOMETRIC FUNCTIONS for values of their arguments between 0° and 90°, and hence enabling their values for any argument to be inferred by means of the trigonometric REDUCTION FORMULAE. These were much used to facilitate computation, but are rendered largely obsolete by calculators; however, it is important to note that while the trigonometric functions are defined mathematically in terms of RADIANS, most calculators default to computation in degrees, and so give incorrect results unless expressly reset to compute in radians.

trigonometry (abbrev. **trig**), *n.* the branch of mathematics concerned with the properties of the TRIGONOMETRIC FUNCTIONS and their application to the determination of the sides and angles of triangles, used in surveying, navigation, etc. See TRIANGULATION.

trihedral, *adj.* **1.** having or formed by three plane faces meeting at a point. **2.** (*as substantive*) a figure formed by the intersection of three lines in diffe-rent planes, as illustrated in Fig. 385. See RIGHT–HANDED TRIHEDRAL, FRAME OF REFERENCE.

Fig. 385. **Trihedral**.

trihedron, *n.* a figure determined by the intersection of three planes.

trilateral, *adj.* having three sides.

trilemma, *n.* a classically valid argument one of whose premises is the dis-junction of three statements from each of which the same proposition may be derived, and whose conclusion is that common consequence; such as

> *Egbert is in Paris, Bonn, or Rome.*
> *If he's in Paris, he's in Europe.*
> *If he's in Bonn, he's in Europe.*
> *If he's in Rome, he's in Europe.*
> *So he's in Europe.*

See DILEMMA.

trilinear, *adj.* consisting of, bounded by, or relating to three lines.

trillion, *n.* **1.** (*in Britain*) one million cubed, 10^{18}.

2. (*in USA*) one million million, 10^{12}.

trilogarithm, see POLYLOGARITHM.

trim, *vb.*(*Statistics*) to discard extreme observations in a sample; the *trimmed mean* is thus regarded as characterising the 'typical' values of the variable.

trinomial, *adj.* **1.** consisting of three terms.

2. (*as substantive*) a polynomial with three terms, such as the expression $a + b + c$, or the quadratic $ax^2 + bx + c$.

triple, *n.* a set or sequence with three members.

triple product or **scalar triple product,** *n.* (*Vector analysis*) the product of three vectors in three-space defined in terms of the SCALAR PRODUCT and the VECTOR PRODUCT as the scalar quantity $x \cdot (y \times z)$ of which the absolute value equals the volume of a PARALLELEPIPED with the three vectors as gen-erators. This may be computed as the determinant of the matrix of which the rows are the coefficients of the vectors in the given order. Thus

$$x \cdot y \times z = x \times y \cdot z,$$

and there is no need for parentheses. Compare TRIPLE VECTOR PRODUCT.

triple-product identity of Jacobi, *n.* the identity

$$\sum_{n=-\infty}^{\infty} x^n q^{n^2} \prod_{n=1}^{\infty} (1 + xq^{2n-1})(1 + x^{-1}q^{2n-1})(1 - q^{2n}).$$

Compare THETA FUNCTIONS, Q–BINOMIAL THEOREM.

triple vector product, see VECTOR TRIPLE PRODUCT. Compare TRIPLE PRODUCT, VECTOR PRODUCT.

trisect, *vb.* to divide into three equal parts.

trisecting the angle, *n.* the traditional problem of how to CONSTRUCT an angle equal to one-third of a given angle using only straight-edge and compasses; this was not proved generally insoluble until 1847, although it is soluble if $4t^3 - 3t - \cos\theta$ is reducible over the rationals. The trisection is possible with aid of a PROTRACTOR, or using the LIMACON OF PASCAL or a TRISECTRIX.

trisectrix of Maclaurin, *n.* the locus of the equation

$$x^3 + xy^2 + ay^2 - 3ax^2 = 0;$$

the curve, as shown in Fig. 386, is symmetric around the x-axis, contains the origin, and has a vertical asymptote at $x = -a$. If a line with angle of inclination 3α drawn through $(2a, 0)$ meets the trisectrix at P, then the line passing through the origin and P has angle of inclination α.

Fig. 386. **Trisectrix of Maclaurin.**

trisoctahedron, *n.* a solid figure with 24 identical triangular faces, each group of three being constructed on one face of an underlying octahedron.

Tristram Shandy paradox, *n.* the paradox of the infinite derivable from Lawrence Sterne's 1760 novel *Tristram Shandy,* which purports to be part of the hero's autobiography; since it has taken him two years to describe his first two days, Shandy concludes that his autobiographical aspirations are doomed to failure, but Russell pointed out that were he immortal he would be able to complete his goal even at the same rate of progress. See also HILBERT'S PARADOX.

trivial, *adj.* **1.** (of a solution of a system of equations) setting the value of all the variables to zero.
2a. obvious in a given context. Thus $x^n + y^n = z^n$ always has trivial solutions with $x = z = 1$ and $y = 0$.
b. of no interest in the given context.
c. in some way pathological, usually because the existence of the entity concerned is guaranteed at a high level of generality; for example, the trivial subsets of any set are the empty set and the given set itself.
(From Latin TRIVIUM.) Compare NON–TRIVIAL.

trivial ring, *n.* a RING in which the product of any pair of elements is defined to be zero.

trivial subgroup, *n.* the SUBGROUP of a given group whose sole member is the IDENTITY ELEMENT.

trivial vector, *n.* another term for ZERO VECTOR.

trivial ultrafilter, see ULTRAFILTER.

trivium, *n.* the lower division of the ancient and medieval liberal arts, comprising grammar, rhetoric, and logic. The higher arts – arithmetic, geometry, astronomy, and music – constituted the *quadrivium*. (from Latin for "three ways", where three roads meet, whence TRIVIAL).

trochoid, *n.* the curve described by a fixed point on the radius of a circle, or its extension, as it rolls along a straight line, as shown in Fig. 387. In some usages the term is synonymous with CYCLOID, while others use the latter term only for the special case of the locus of a point that lies on the circumference of the rolling circle. The EXTENDED CYCLOID and CONTRACTED CYCLOID (or trochoid) are sometimes also known as the *prolate trochoid* or *curtate trochoid*, but there is no consistency about which is which.

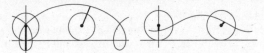

Fig. 387. Extended and contracted **trochoids**.

true, *adj.* **1a.** one of the TRUTH-VALUES assigned to a sentence in any two-valued logic.

b. the unique DESIGNATED truth-value in a MANY-VALUED LOGIC.

2. (of a proposition in a mathematical theory) validly deducible from the axioms of the theory.

truncate, *vb.* to approximate an infinite series by a finite number of terms; for example, for small x the TAYLOR SERIES for $\cos x$ can be truncated after the second term to yield

$$\cos x = 1 - \frac{x^2}{2}.$$

This technique is often useful in numerical work and evaluation of limits.

truncated, *adj.* **1.** (of a cone, pyramid, prism, etc.) having an apex removed by an intersection with a plane that is usually not parallel with the base, such as the pyramid of Fig. 388. See ARCHIMEDEAN SOLID. See also FRUSTUM.

Fig. 388. A **truncated** pyramid.

2. (of a series) constituting a finite initial segment of an infinite series to which it approximates; in particular, a truncated decimal fraction is a finite initial segment of the terms of a non-terminating decimal.

truncation error, *n.* (*Numerical analysis*) the error implicit in using a given approximation to a quantity being computed. See also ROUNDING ERROR.

trust-region method, *n.* a class of approximate DESCENT METHODS that replaces the STEP–LENGTH METHOD computation by an estimation of the positive definiteness of the approximate Hessian. If this is found 'trustworthy', a step size of 1 is used, as in NEWTON'S METHOD; otherwise a search direction is generated, based on a measure of the trustworthiness itself updated through the calculation.

truth, *n.* in general, deducibility of a proposition from the axioms of a theory in accordance with the rules of inference of that theory (*truth in a theory*). Many philosophers of mathematics do not regard any wider notion of truth as intelligible. However, a Platonist would hold that mathematical truth consists in correspondence with a realm of facts that are independent of human knowledge of them. On the other hand, an intuitionist or constructivist has a more austere view that regards truth not in terms of provability in a theory, but of the actual existence of a proof. On the former view it is still possible to hold that every proposition is either true or false, in that it is either provable or refutable, although we do not know which; for the constructivist that is unintelligible, and propositions that have neither been proved nor refuted can not even be said to satisfy the LAW OF EXCLUDED MIDDLE.

truth-function, *n.* (*Logic*) **1.** a function that determines the TRUTH–VALUE of a complex sentence solely in terms of the truth-values of the component sentences without reference to their meaning, such as NEGATION, CONJUNCTION, DISJUNCTION, and IMPLICATION.

2. a complex sentence of which the truth-value is so determined, such as a negation or a disjunction.

truth-functional, *adj.* (*Logic*) pertaining to, consisting of, or able to be represented by, TRUTH–FUNCTIONS.

truth set or **solution set,** *n.* **1.** the set of values of the variables that satisfy an OPEN SENTENCE or a set of equations or inequalities, especially one that has no unique solution.

2. (*Logic*) the set of POSSIBLE WORLDS in which a given statement is true.

truth-table, *n.* (*Logic*) a diagram that sets out all the possible combinations of TRUTH–VALUES of the atomic sentences in a set of statements, and uses TRUTH–FUNCTIONS to determine the possible combinations of the truth-values of the given set of statements themselves, and in particular whether any of them is a TAUTOLOGY or a CONTRADICTION, and whether any of them is a LOGICAL CONSEQUENCE of the remainder; Fig. 389 shows the successive steps in the calculation of the truth-table for $(P \& Q) \rightarrow (P \lor Q)$ in order to show that it takes the value *true* in every possible circumstance, that is, is a tautology.

P	Q	P & Q	P ∨ Q	(P & Q)→(P ∨ Q)
T	T	T	T	T
T	F	F	T	T
F	T	F	T	T
F	F	F	F	T

Fig. 389. A **truth-table** for a tautology.

truth-value, *n.* (*Logic*) **1.** either of the values *true* and *false* that may be assigned to a statement.

2. by analogy, any of the values that a semantic theory may accord to a statement. See THREE–VALUED LOGIC, MANY–VALUED LOGIC, VALUATION SYSTEM.

truth-value gap, *n.* (*Logic*) the possibility in certain semantic systems of a statement being neither true nor false while not being determinately of any third truth value, so that the LAW OF EXCLUDED MIDDLE fails; for example,

> *all my children are asleep,*

uttered by a childless person, would turn out to be true if the universal quantifier were interpreted standardly, as it would be equivalent to

> *if anything is my child it is asleep,*

which is true because the antecedent is never satisfied. However, this is clearly a misleading statement, but to say it is false suggests that

> *all my children are not asleep*

is true; that is, that they are awake. The solution proposed by some philosophers and logicians such as Frege is that propositions whose subject fails to refer are neither true nor false. There is a different truth-value gap in INTUITIONIST logic; for the intuitionist, only what has been proved may be asserted so that a proposition which has neither been proved nor falsified is a counterexample to the law of excluded middle.

T-space, see T-AXIOMS.

t-test, *n.* (*Statistics*) a test whose TEST STATISTIC has a STUDENT'S T DISTRIBUTION, especially a test of the hypothesis that the NORMAL DISTRIBUTION from which a given random sample is drawn has a given mean.

-tuple, *suffix* (*also used as substantive*) *denoting* a SEQUENCE or ORDERED SET; for example, a *5-tuple* or *quintuple* is an ordered set of five elements; the term *n-tuple* is often used for a finite ordered set with an unspecified number of members.

Turing machine, *n.* an ABSTRACT MACHINE that provides the generally accepted model of serial computation, which, given CHURCH'S THESIS, corresponds to what is recursively computable (see RECURSIVE). A deterministic Turing machine has a finite control, an indefinitely long input tape that is divided into units or cells of which a finite number contain a symbol drawn from a finite vocabulary, and a moving tape head. In each move the machine scans a tape cell, and, as determined by its present state and that symbol, overprints or deletes a non-blank symbol on the tape cell scanned, moves its head one cell to the left or right, and changes state. The machine can be fully described by a sequence of ordered quintuples: $(a, 0, 1, R, b)$ can be read as the instruction 'in state a, if the tape cell contains a 0, then replace it with a 1, move one cell right, and go into state b'. Some authors use ordered quadruples to describe the machine, regarding as separate the instructions to write and to move. Among the finite set of states is the initial state and a subset consisting of final states. The machine continues its operation until one of these states is encountered and the machine *halts*. The *halting problem* is that of determining whether a Turing machine will halt when presented with a given input string, and is one of many UNSOLVABLE PROBLEMS. For a DECISION PROBLEM the final states may be taken to consist of *yes* and *no* and the machine is said to accept the input string if it terminates finitely with *yes*. To correspond to a DECISION PROCEDURE the machine must halt for all possible input strings. This model can be shown to be equivalent to most other formulations proposed for serial computation, and Turing proved that such a machine with only the symbols 0 and 1 has the power of any machine devised to compute a particular algorithm.

(Named after the English mathematician and logician, *Alan Mathison Turing* (1912–54), who built some of the earliest digital computers. His death of cyanide poisoning while conducting electrolysis experiments in connection with his work on the development of cells was claimed to be accidental, but is now generally regarded as suicide.)

turning point, *n.* a STATIONARY POINT at which the FIRST DERIVATIVE of a function changes sign, so that, typically, its graph does not cross a tangent parallel to the axis of the independent variable (that is, the horizontal axis in the usual two-dimensional Cartesian system). If the second derivative is strictly negative at a stationary point, it is a local maximum of the function; if it is strictly positive, it is a local minimum. Fig. 390 shows the tangent to a curve at its turning point.

Fig. 390. **Turning point**.

turnpike theorems, *n.* a class of theorems popular in economic growth models that assert that an optimal or near optimal strategy always approaches the optimal growth rate and remains there, by analogy with highway driving.

turnstile or **gatepost,** *n.* (*Logic*) *informal term for* a symbol of the form '⊢', used to represent logical consequence when inserted between expressions to form a SEQUENT, or when prefixed to a single expression indicating that it is a THEOREM. Different variants of the symbol are often used to distinguish syntactic and semantic notions, as in

$$\Gamma \Vdash B, \models A,$$

but there is no uniform notation.

twin primes, *n.* a pair of PRIME NUMBERS that differ by 2, such as 17 and 19, or 1001 and 1003. It is an unproved conjecture that there are infinitely many such pairs.

twisted, *adj.* (of a curve) another term for SKEW.

two-dimensional, *n.* **1.** having or relating to two DIMENSIONS, usually described in terms of length and breadth, or length and height.
2. lying on a surface, especially a plane; having an area but not a volume; for example, a sphere has a two-dimensional surface in three-dimensional space.

two-person zero-sum game, see ZERO–SUM GAME.

two-point contact, *n.* (*Algebraic geometry*) the relation between two curves, surfaces, etc., at a point at which they touch and have a common TANGENT; a TACPOINT. Compare THREE–POINT CONTACT.

two-sample problem, *n.* (*Statistics*) any problem requiring a test to be applied to two independent samples. Compare PAIRED–SAMPLE PROBLEM.

two-sided, *adj.* (of a LIMIT) equal to both the ONE-SIDED limits from above and from below, as the independent variable approaches a given value.

two-tailed, *adj.* (*Statistics*) (of a SIGNIFICANCE TEST) concerned with the hypothesis that an observed value of a TEST STATISTIC differs significantly from a given value, where an error in either direction is relevant. For example, in testing the fairness of weighing scales, an inspector will seek to exclude both overweight and underweight goods. Compare ONE-TAILED.

Tychonoff condition, *n.* another term for T-AXIOM.

Tychonoff space, *n.* a completely REGULAR T_1-space. See T-AXIOMS.

Tychonoff's theorem, *n.* the theorem that the (infinite) Cartesian product of COMPACT TOPOLOGICAL SPACES is compact. (Named after the Russian topologist and physicist, *Andrei Nikolaevitch Tychonoff* (1906– 93).)

type, *n.* (*Logic*) a class of expressions, or of the entities they represent, that can all enter into the same syntactic relations. In Russell's *theory of types*, the type of a function, including a predicate, is determined recursively by the types of its arguments and values. In his *ramified theory of types*, the types of variables occurring bound in the expression are also taken into account. Both theories are advanced to avoid paradoxes such as the LIAR PARADOX and RUSSELL'S PARADOX.

type I error, *n.* (*Statistics*) the error of rejecting the NULL HYPOTHESIS when it is in fact true, the probability of which is the SIGNIFICANCE LEVEL of the test.

type II error, *n.* (*Statistics*) the error of not rejecting the NULL HYPOTHESIS when in fact it is false, the probability of avoiding which is the POWER of the test and is a function of the ALTERNATIVE HYPOTHESIS.

typical instance, see INSTANTIATION.

U

ultrafilter, *n.* a maximal proper FILTER on a set, and as such, for each subset, containing either it or its complement. The family of all sets containing a given point comprises the *trivial* or *principal ultrafilter*, all others being called *free* or *non-principal.*

ultrapower, *n.* the ULTRAPRODUCT of a set with itself.

ultraproduct, *n.* the QUOTIENT of a CARTESIAN PRODUCT of an infinite family of sets $\{A_i : i \in I\}$ with respect to the equivalence relation induced by an ULTRAFILTER U on I:

$$\{a_i\} \equiv \{b_i\} \text{ if } \{i \in I : a_i = b_i\} \text{ belongs to U.}$$

If the sets $A_i = A$ are identical this is called an *ultrapower* of the set A. In general the construction is used with a free ultrafilter.

umbilical point or **umbilic,** *n.* (*Euclidean geometry*) a point on a surface that is either *planar* (when normal curvature vanishes) or *circular* (when the principal radii of normal curvature are equal). All points where an ellipsoid of revolution cuts its axis of revolution are umbilical.

umv, *abbrev. for* UNIFORM MINIMUM VARIANCE.

unary, *adj.* another term for MONADIC.

unbiased, *adj.* (*Statistics*) **1.** (of a sample) not affected by any extraneous factors, conflated variables, or selectivity that influence its distribution; random.

2. (of an ESTIMATOR) having an EXPECTED VALUE equal to the parameter being estimated; having zero BIAS.

3. (of a SIGNIFICANCE TEST) having a POWER greater than the predetermined SIGNIFICANCE LEVEL.

unbounded, *adj.* **1.** (of a set) not having a BOUND.

2. (of a function on a set) having values that increase without bound in modulus or norm for arguments in the given set; that is, for all large N there is a value of the function of which the modulus or norm is greater than N. If MEASURE is involved, an unbounded function is one that is not ESSENTIALLY BOUNDED.

uncertainty, see INFORMATION.

unconditional, *adj.* (of an inequality) universally true; true for any values assigned to the variables. For example, $x + 1 > x$ is an unconditional inequality, while $x^2 > x$ is a conditional inequality.

unconditional convergence, *n.* the requirement that a series converge in any reordering of the terms; for a complex series this coincides with ABSOLUTE CONVERGENCE. For example,

$$1 - \frac{1}{4} + \frac{1}{9} - \frac{1}{16} + \dots$$

is unconditionally convergent, as it is absolutely convergent, while

$$1 - \frac{1}{2} + \frac{1}{3} - \frac{1}{4} + \ldots$$

is not.

uncountable or **uncountably infinite,** *adj.* not corresponding ONE-TO-ONE with the NATURAL NUMBERS; neither finite nor DENUMERABLE.

undecagon, *n.* a polygon with eleven sides.

undecidable, *adj.* (*Logic*) **1.** (of a formal system) lacking a DECISION PROCEDURE; not DECIDABLE.

2. (of a well-formed formula of a given theory) not DECIDABLE; neither provable, nor having a provable negation, within the given theory, so that neither it nor its negation is a theorem. For example, the continuum hypothesis is undecidable in Zermelo–Fraenkel set theory since both it and its negation are consistent with the axioms of the theory.

undefined element, *n.* a non-logical constant; a PRIMITIVE member of some axiomatically defined structure. For example, points and lines are undefined elements in axiomatic geometry, and '∈' is an undefined predicate in set theory.

underdetermined, *adj.* (of a system of equations, usually linear) involving fewer equations than variables. Compare OVERDETERMINED.

underlying set, *n.* the set on which a TOPOLOGY or other structure is defined.

undetermined, *adj.* (of a parameter) of an as yet unspecified form; for example, constants of integration are undetermined.

unexpected examination paradox, *n.* the paradox implicit in the statement that a group of students will be examined one day next week, but will not know in advance on which day this will occur. Obviously, the examination cannot be on Friday, since by Thursday evening they would be able to infer when it will be. But then, since Thursday is the last possible day, they would be able to draw the same inference on Wednesday; so it cannot be then either, and by iteration every day is excluded – so the students are genuinely surprised when the examination takes place on Tuesday, or, for that matter, on Friday! The paradox was described by Lennard Ekbom, a Swedish mathematician, in 1948, and is based on an actual announcement of a civil defence exercise on Swedish radio; the *hangman paradox* has the same structure and concerns an unexpected execution. It is clearly an EPISTEMIC paradox, and has been resolved by recalling that a conclusion is only justified if validly inferred from known premises rather than from true premises, and distinguishing what is known to the teacher from what is known to the students.

unicity, *n.* uniqueness, especially of BEST APPROXIMATIONS. The *unicity theorem* states that the best approximation by *generalized polynomials* (functions of the form

$$\sum_{i=1}^{n} c_i p_i,$$

for fixed continuous p_i) is unique for all real continuous functions on a compact interval if and only if the HAAR CONDITION is met.

unicursal, *adj.* (of a curve) closed but not having any part retraced; drawn in a single sweep.

uniform bound, *n.* a BOUND that holds uniformly, usually for a set of functions. See UNIFORM BOUNDEDNESS PRINCIPLE.

uniform boundedness principle or **Banach–Steinhaus theorem,** *n.* the theorem that a pointwise-bounded family of continuous linear operators between a BANACH SPACE and a NORMED SPACE is EQUICONTINUOUS (UNIFORMLY BOUNDED): if for each *x* in the unit ball,

$$\sup_I \| T_i(x) \|$$

is finite, then actually $\sup_I \| T_i \|$ is finite.

uniform continuity, *n.* the property of a function between METRIC SPACES, that CONTINUITY is uniform, so that for every $\varepsilon > 0$, there is a single $\delta > 0$ such that $d(f(x), f(y)) < \varepsilon$ whenever $d(x, y) < \delta$, whereas in general δ would depend on both ε and *x*. If the domain is compact, uniform continuity follows automatically from continuity. See also UNIFORMLY CONTINUOUS. Compare EQUICONTINUOUS.

uniform convergence, *n.* the property that all of a family of functions or series on a given set CONVERGE at the same rate throughout the set; that is, for every $\varepsilon > 0$ there is a single *N* such that for all points in the set,

$$| f_m(x) - f_n(x) | < \varepsilon$$

for all $m, n > N$, and similarly for uniform convergence as *x* tends to a value *a*. This is convergence in the *uniform* or CHEBYSHEV NORM. A power series converges uniformly within any disk of strictly smaller radius than its radius of convergence. The uniform limit of a sequence of continuous functions is continuous, as opposed to the limit in POINTWISE CONVERGENCE which may well not be. This is often the case for FOURIER SERIES. See DIRICHLET'S CONDITION, ASCOLI'S THEOREM.

uniform convexity or **uniform rotundity,** *n.* the property of a NORM or ball that for every $\varepsilon > 0$ there is a single $\delta > 0$ such that $\| x - y \| < \varepsilon$ whenever

$$\left| \frac{x + y}{2} \right| > 1 - \delta,$$

and *x* and *y* have norm one. A normed space on which such an equivalent norm may be placed is *super-reflexive*, a stronger property than reflexivity, possessed by L_p SPACES for $1 < p < \infty$. In finite dimensions such norms coincide with strictly convex ones. *Local uniform convexity* arises if *x* is fixed throughout the definition.

uniform distribution, *n.* **1.** the property of an infinite sequence of numbers $\{a_n\}$ in an interval $[a, b]$, that, for any $a < \alpha < \beta < b$, the probability that the numbers lie in $[\alpha, \beta]$ is the ratio

$$\frac{\beta - \alpha}{b - a};$$

precisely, one requires that this ratio is equal to

$$\lim_{N \to \infty} \frac{\{n < N: a_n \in [\alpha, \beta]\}}{N}.$$

2. (*Statistics*) **a.** the DISTRIBUTION of a random variable on an interval $[a, b]$ when its PROBABILITY DENSITY FUNCTION is zero outside this interval, and is $1/(b-a)$ at any point of the interval. Its mean is the midpoint of the interval, and its variance is $(b-a)^2/12$.

b. formally, the distribution associated with normalized LEBESGUE MEASURE on a set in n-space, all sets of the same size being equally likely.

uniformity, *n.* (on a set S) a FILTER of subsets F of $S \times S$, each member of which contains all points of the form (x, x), such that if V lies in F so does the inverse

$$V^{-1} = \{ (y, x) : (x, y) \in V \},$$

and such that, for each V in F, there exists a W in F with the composition

$$W \circ W = \{ (x, z) : (x, y) \in W, (y, z) \in W \}$$

lying in V. Each element of the uniformity is called a *vicinity*. Every metric space is a UNIFORM SPACE with a base for the uniformity consisting of sets of the form

$$\{ (x, y) : \rho(x, y) < \varepsilon \} \text{ for } \varepsilon > 0.$$

uniformly bounded, *adj.* (of a set of functions) having a UNIFORM BOUND.

uniformly continuous, *adj.* (of a real function on a set) such that for every $\varepsilon > 0$, there is a $\delta > 0$ such that

$$| f(x) - f(y) | < \varepsilon \text{ whenever } | x - y | < \delta,$$

for all x, y in the given set. Compare EQUICONTINUOUS.

uniform minimum variance (abbrev. **umv**), *n.* the property of a STATISTIC of having the minimal VARIANCE of all statistics of a certain class, often the class of unbiased estimators, for every value of the parameters. See GAUSS–MARKOV LEAST SQUARES THEOREM.

uniform norm, *n.* another term for CHEBYSHEV NORM.

uniform rotundity, *n.* another term for UNIFORM CONVEXITY.

uniform space, *n.* a topological space S in which the topology is induced by a UNIFORMITY (a class of subsets of $S \times S$), with U open if and only if for any $x \in U$, there exists V in the uniformity with $\{ y : (x, y) \in V \} \subseteq U$.

uniform substitution, *n.* the replacement of every occurrence of some well-formed part of a given expression by some other expression, to yield a SUBSTITUTION INSTANCE.

unilateral, *adj.* **1.** (of a limit) evaluated on one side of the limiting value of the independent variable. *Unilateral analysis* is the study of one-sided properties such as lower semi-continuity, convexity, minimality etc. See LEFT–HAND LIMIT, RIGHT–HAND LIMIT.

2. (of a surface) having only one side. See MÖBIUS STRIP, KLEIN BOTTLE.

unilateral shift, *n.* the linear operator defined on a space of (square-summable) sequences,

$$\{ x \}_{n=0}^{\infty} ,$$

by $(Sx)_n = x_{n-1}$, where $x_{-1} = 0$. The *bilateral shift* is similarly defined for doubly infinite sequences

$$\{ x \}_{n=-\infty}^{\infty}.$$

unimodal, *adj.* (of a real function f defined on an interval) possessing a single MAXIMUM or MINIMUM in the interval; that is, there is a unique point c in the given interval, $[a, b]$, such that the function is MONOTONE in opposite senses on $[a, c\,[$ and on $]c, b\,]$. Thus either

$$\text{if } x < y < c \text{ then } f(x) < f(y),$$

and

$$\text{if } x > y > c \text{ then } f(y) > f(x),$$

or vice versa. This means that LINE SEARCH METHODS based on shrinking the interval in which the minimum lies can be used since the minimum can be diagnosed to lie either in $[a, y]$ or in $[x, b]$.

unimodular matrix, *n.* a square matrix having a DETERMINANT of ±1. An integral unimodular matrix has an integral INVERSE MATRIX because of the adjoint formula. A *totally unimodular* matrix has all MINORS equal to 1, and so has an inverse of the same form.

uninterpreted, *adj.* (of a formal theory) considered solely in terms of its syntactic structure rather than in terms of any assignment of significances to its terms; lacking an INTERPRETATION.

union or **sum,** *n.* **1.** the set of elements that belong to either of a given pair of sets, written $S \cup T$, and often called CUP; if the two circles in Fig. 391 represent S and T respectively, the shaded area represents their union.

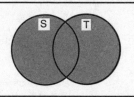

Fig. 391. **Union.** The shaded area is $S \cup T$.

2. the binary operation that forms such a set from two given sets.

3. more generally, over any COLLECTION, **C**, of subsets C_α of a given set X (indexed by $\alpha \in A$), the set each of whose elements lies in at least one member of the collection. This is denoted

$$\bigcup_{\alpha \in A} C_\alpha \quad \text{or} \quad \bigcup C;$$

in particular, $\bigcup \varnothing = \varnothing$.

See also INTERSECTION.

unique, *adj.* **1.** being the only value satisfying some condition. For example, the unique positive square root of 4 is 2, although −2 is another square root of 4. See also DEFINITE DESCRIPTION.

2. unique up to (some relation). (of elements within some structure) related by a RELATIVE IDENTITY relation; identical relative to that relation; equivalent within a given structure. For example, to say the positive integers have prime factorizations that are unique up to order, is to say that the index of each prime is unique, but that their order is irrelevant. Common phrases of this kind are *unique up to isomorphism, unique up to permutation, unique up to a constant.*

unique factorization domain, *n.* another term for a GAUSSIAN DOMAIN.

unique factorization theorem, *n.* **1.** another name for the FUNDAMENTAL THEOREM OF ARITHMETIC.

2. any theorem asserting unique factorization, as obtains in, for example, a EUCLIDEAN DOMAIN.

uniqueness theorem, *n.* **1.** any theorem proving that all solutions to a given problem, or all entities that have a given property, are identical; that is, that the solution or element with that property is unique.

2. another name for the IDENTITY THEOREM.

unique quantifier, *n.* (*Logic*) the strengthened EXISTENTIAL QUANTIFIER, used to assert that a predicate is uniquely instantiated. It is written $(\exists! x)Fx$, and defined contextually as

$$(\exists! x)Fx \equiv (\exists x)(Fx \;\&\; (\forall y)(Fy \to x = y)).$$

It is the first member of the sequence of strong NUMERICAL QUANTIFIERS. See also DEFINITE DESCRIPTION.

unit, *n.* **1.** (usually plural) the first position in a PLACE–VALUE counting system, representing a single-digit number. For example, in the decimal system, the number 27 has 7 units and 2 tens.

2a. (*as modifier*) having a value defined as 1 for the system, such as a unit vector, unit square, or unit interval; an IDENTITY ELEMENT.

b. a physical magnitude used as the basis of a measuring system, so that other magnitudes are expressed as multiples and fractions of the basic unit. See SYSTEME INTERNATIONAL.

3. (*Number theory*) an algebraic integer of which the inverse is also an integer. There are four units in the GAUSSIAN INTEGERS, namely ± 1 and $\pm i$. Since $\sqrt{2} + 1$ is an integer in $\mathbb{Q}(\sqrt{2})$ and since $(\sqrt{2} - 1)(\sqrt{2} + 1) = 1$, it is a unit.

4. (*Algebra*) a multiplicatively invertible element of a RING, an INTEGRAL DOMAIN, or other algebraic structure; an element u is a unit of R if and only if $uR = R$. The constant polynomials are units in a polynomial ring over a field. Compare UNITY.

unital or **unitary,** *n.* possessing a UNITY or IDENTITY, such as a unital semi-group or unitary ring.

unitary equivalence, *n.* the property of two operators or matrices of being SIMILAR with respect to a UNITARY MATRIX. Hence, for U unitary, A = UBU*, and the operators are also CONGRUENT. Any HERMITIAN MATRIX is unitarily equivalent to a real diagonal matrix, and any NORMAL MATRIX is unitarily equivalent to a complex diagonal matrix. If the matrix is real the unitary matrix may be supposed real and so orthogonal. See SCHUR'S LEMMA.

unitary group, *n.* the group, denoted $U(n)$, of all $n \times n$ complex UNITARY MATRICES.

unitary matrix, *n.* a matrix of which the HERMITIAN CONJUGATE is its INVERSE. For a real matrix this coincides with an ORTHOGONAL MATRIX.

unitary module, *n.* a MODULE over a RING with identity in which the product of the ring identity with each element is that element.

unitary space or **Hermitian vector space,** *n.* a complex VECTOR SPACE on which an INNER PRODUCT is defined. Compare HILBERT SPACE, INNER PRODUCT SPACE.

unitary transformation, *n.* a linear operator on Hilbert space of which the ADJOINT is its INVERSE. A finite-dimensional operator is unitary if and only if the associated matrix is a UNITARY MATRIX; an ISOMETRY of the underlying Hilbert space.

unit disk, *n.* any NEIGHBOURHOOD in a METRIC SPACE, of which the radius is unity; especially that centred on the origin in the complex plane, namely $\{x : |x| < 1\}$.

unit point, *n.* a point that, together with a given TRIANGLE OF REFERENCE, determines a system of HOMOGENEOUS COORDINATES for a two-dimensional ALGEBRAIC GEOMETRY; the unit point is chosen to be LINEARLY INDEPENDENT of any two vertices of the triangle of reference.

unit set, *n.* a set having a single member, a SINGLETON.

unit square, *n.* a square each side of which has unit length.

unit vector, *n.* a VECTOR that has unit magnitude, especially the vectors **i, j, k** in the positive directions of the coordinate axes of a CARTESIAN COORDI-NATE system.

unity, *n.* **1.** the number or numeral 1.

2. any quantity taking or assigned the value one.

3. also called **neutral element**. the element of a set of which the product with any other element under a multiplicative operation is that other element, its multiplicative IDENTITY.

4. the greatest member of a LATTICE or PARTIALLY-ORDERED set, written ∨, such as the universal set.

Compare ZERO.

univalent, *adj.* another term for SCHLICHT.

universal, *adj.* (*Logic*) **1.** (of a statement or proposition) affirming or denying something of every member of a class of objects; containing a UNIVERSAL QUANTIFIER. For example, *all men are wicked*, and *no pigs can fly* are universal statements. Compare EXISTENTIAL QUANTIFIER.

2. (*as substantive*) **a.** a universal proposition, statement, or formula.

b. a universal quantifier.

universal algebra, *n.* the study of relational structures on sets.

universal elimination, *n.* another name for UNIVERSAL INSTANTIATION. See also ELIMINATION RULE.

universal generalization, see GENERALIZATION.

universal gravitational constant, *n.* the constant, γ, determined only by the units, that occurs in Newton's law of GRAVITY; in standard units, its value is 6.673×10^{-11}. Compare LOCAL GRAVITATIONAL CONSTANT.

universal instantiation or **universal elimination,** *n.* (*Logic*) the rule of PREDI-CATE CALCULUS in accordance with which any INSTANCE may be validly in-ferred from a universally quantified statement; the ELIMINATION RULE for the UNIVERSAL QUANTIFIER.

universal introduction, see INTRODUCTION RULE.

universally measurable set, *n.* a set E that is MEASURABLE for each BOREL MEASURE on a topological space. Thus, given a measure μ, there exist $G \subset E \subset F$ with $\mu(F \setminus G) = 0$, where G and F are BOREL SETS, these being taken as the sigma algebra generated by all closed subsets. In a separable metric space, any SOUSLIN SET is universally measurable.

universal quantifier, *n.* (*Logic*) an operator containing a variable, written '(x)' or '$(\forall x)$', that indicates that the OPEN SENTENCE that follows is true of every member of the relevant domain; that is, that every replacement of that variable by a name yields a true statement. For example,

$$(x)(Fx \rightarrow Gx)$$

is read as *for all x, if x is an F then it is a G,* that is, *all Fs are Gs,* and is true if and only if $Fa \rightarrow Ga$ is true of every member of the domain.

universal set or **universe,** *n.* the domain with reference to which complementation in set theory is defined; the union of any set and its complement. RUSSELL'S PARADOX proves that this set cannot be all-inclusive, and in particular cannot include itself. Different formulations of set theory have adopted different devices to avoid this difficulty, but for practical purposes, such as VENN DIAGRAMS, it is sufficient to take the universal set to be some specific class large enough to include all the elements of any set relevant to the subject matter; for example, in a diagram of the relations between fish, mammals and aquatic animals, it may be sufficient to take animals as the universal set.

universe, *n.* **1.** another term for UNIVERSAL SET.

2. (*Statistics*) another word for POPULATION.

3. universe of discourse, universe of interpretation, or **domain of discourse.** (*Logic*) the complete set of individuals able to be referred to, or quantified over, in an interpreted theory.

unknown, *n.* a variable, or the quantity it represents, whose value is to be discovered by solving an equation; a variable in a conditional equation. For example, $3y = 4x + 5$ is an equation in two unknowns. See INDETERMINATE.

unordered arrangement, *n.* (of a set) another term for COMBINATION.

unsolvable, insolvable, or **insoluble,** *adj.* **1.** not possessing a solution.

2. provably not able to be solved. See also IMPOSSIBILITY THEOREM, TRISECTING THE ANGLE.

unsolvable problem, see SOLVABLE PROBLEM.

unstable, *adj.* (of an EQUILIBRIUM POINT of a system of LINEAR ORDINARY DIFFERENTIAL EQUATIONS) not STABLE.

update, *vb.* (*Numerical analysis*) to modify or reset parameters or variables during the course of a computation. For example, when using QUASI-NEWTON METHODS, during each step of the iteration one may perform a rank-one update of the approximate HESSIAN.

upper bound, *n.* a value greater than or equal to all of a set of given values. For example, in the LATTICE of subsets of $\{1, 2, 3, 4, 5\}$ ordered by set inclusion, the least upper bound of $\{1, 2, 3\}$ and $\{2, 3, 4\}$ is $\{1, 2, 3, 4\}$. Compare LOWER BOUND. See also SUPREMUM, MAXIMUM.

upper Darboux integral, see UPPER INTEGRAL.

upper Darboux sum, see UPPER SUM.

upper Hessenberg form, see HESSENBERG FORM.

upper integral or **upper Darboux integral,** *n.* the limit, as the MESH–FINENESS of its subintervals tends to zero, of the UPPER SUMS of a function on that interval; if this exists and is equal to the LOWER INTEGRAL, then the function is RIEMANN INTEGRABLE.

upper inverse image set, see INVERSE IMAGE SET.

upper level set, see LEVEL SET.

upper limit, *n.* **1.** (of integration) the larger of two endpoints over which a definite integral is taken. Compare LOWER LIMIT.

2. another term for LIMIT SUPERIOR.

upper semicontinuous, see SEMICONTINUOUS.

upper sum or **upper Darboux sum,** *n.* the weighted sum of the products of the supremal values of a function on a succession of subintervals of a given interval with the lengths of the subintervals; hence, the area under the step function of which the value is the supremum of the given function on each subinterval, as shown in Fig. 392. The limit of this sum of products, as the lengths of the subintervals tend to zero, is the UPPER INTEGRAL of the function. Compare LOWER SUM. See RIEMANN INTEGRAL.

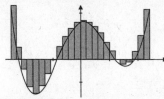

Fig. 392. **Upper sum**. See main entry.

upper triangular, *adj.* (of a square matrix) having only zero entries below the MAIN DIAGONAL. Compare UPPER HESSENBERG FORM.

up to, *prep.* differing only by the specified relation, and so permitting non-identical equivalence under that relation. Thus, to say that a function determines an antiderivative up to a constant is to say that the antiderivative is unique except for that constant, that all its antiderivatives are identical to within a constant, or that they differ by only a constant.

urelements, *n.* (*Set theory*) objects that are not sets and do not involve sets in their construction, out of which sets are then built; *pure sets* are obtained if no urelements are involved. This is the case in standard ZERMELO–FRAENKEL SET THEORY. [*from German* ur (= *primitive*) + element.]

Urysohn's lemma, *n.* (*Topology*) the result that a space S is NORMAL if and only if disjoint closed sets A and B can be functionally separated in the sense that one can find a continuous function $f: S \to [0, 1]$ with $f(A) = 0$ and $f(B) = 1$. (Named after the Russian analyst and topologist, *Paul Samuilovich Urysohn* (1898–1924).) See also METRIZABLE.

Urysohn's metrization theorem, see METRIZABLE.

Urysohn space, *n.* another name for $T_{5/2}$-SPACE. See T–AXIOMS.

utility, *n.* (*Statistics*) a measure of the total benefit or disadvantage attaching to each of a set of alternative courses of action, of which the EXPECTED VALUE is the *expected utility*. See also DECISION THEORY.

utility function, *n.* (*Decision theory*) an increasing, often continuous, real-valued function defined on a set and inducing or corresponding to a PREFERENCE ORDER, by having $u(x) \leq u(y)$ exactly when y is preferred to x. Often one requires that u be QUASI–CONCAVE as this corresponds to the 'law of diminishing returns'.

v or **V,** *symbol for* the number 5 in ROMAN NUMERALS.

vacuous, *adj.* (of an operator or expression) idle; having no import. For example, in

$$(y)(\exists x)(\text{John loves } x),$$

the universal quantifier (y) is vacuous.

valid, *adj.* (*Logic*) **1.** (of an inference or argument) **a.** also called **sound.** having premises and conclusion so related that whenever the former are true the latter must also be true.

b. often **formally valid.** so related where the inference is justified by the form of the premises and conclusion alone. Thus

> *Tom is a bachelor.*
> therefore *Tom is unmarried.*

is valid but not formally valid, while

> *today is hot and dry.*
> therefore *today is hot.*

is formally valid.

2. (*informally*) correct. Often 'valid' is used of the conclusion of an argument, but this is misleading, as the same conclusion can be both validly and invalidly inferred. Thus *today is hot* can be validly inferred as above, or invalidly inferred from *today is hot or dry*; consequently, validity cannot be said to attach to the conclusion by itself. Nor should the classification of arguments as valid or invalid be confused with the classification of component statements as true or false; all combinations of such classifications are possible with the sole exception that a valid argument cannot have both true premises and a false conclusion.

3. generally, (of a sentence in a FORMAL LANGUAGE) true in every INTERPRETATION; SATISFIED by every ASSIGNMENT of values to the variables in every interpretation, so that every interpretation is a MODEL for the statement. A sentence is valid in a theory if it is SATISFIABLE in every model for the theory.

valuation, *n.* **1.** (*Logic*) a function in an INTERPRETATION of a formal calculus that assigns individual elements of the universe of discourse to each variable of the language of the calculus; an ASSIGNMENT of denotations or *values* to the variables. It is then possible to give a recursive definition of the semantic value of other expressions of the calculus under that valuation. Thus, closed sentences will be assigned values that are the analogues of the TRUTH–VALUES *true* and *false* in classical two-valued logic; of these values some will be DESIGNATED, that is analogous to truth.

2. another word for GAUGE.

valuation system, *n.* (*Logic*) a set of values that are assigned to sentences by a VALUATION, together with a set of DESIGNATED values; thus in bivalent logic the valuation system is $\langle\{T, F\}, \{T\}\rangle$.

value, *n.* **1.** the particular quantity that is the result of applying a function or operation, for some given argument. For example, the value of the function $y = x^2$ for $x = 3$ is 9.

2a. an assignment of a significance to a variable; for example, for any value of x, $x = x$; the value of x for which $3x = 6$ is 2.

b. in particular, a value assigned to a statement by a VALUATION; a sentential variable in classical logic may take either of the values *true* and *false*.

3. (*Game theory*) see MINIMAX THEOREM.

4. the total NETWORK FLOW reaching the terminal node of a network.

Vandermonde determinant, *n.* the determinant of the square matrix of which each row consists of the zeroth to $(n-1)^{\text{th}}$ powers of one of n numbers:

$$\begin{vmatrix} 1 & x_1 & \cdots & x_1^{n-1} \\ 1 & x_2 & \cdots & x_2^{n-1} \\ \cdots & & & \\ 1 & x_n & \cdots & x_n^{n-1} \end{vmatrix} = \prod_{i<j} (x_j - x_i).$$

Van der Pol equation, *n.* the differential equation

$$u'' + \alpha(u^2 - 1)u' + \beta u = 0,$$

which has exactly one periodic solution.

van der Waerden's conjecture, *n.* the celebrated proposition, conjectured in 1926 but only recently proved, that the unique DOUBLY STOCHASTIC $n \times n$ matrix of minimum PERMANENT (of which the value is $n! \times n^{-n}$) is the CONSTANT MATRIX with all entries $1/n$.

vanish, *vb.* to become, or tend to, zero.

vanish at infinity, *vb.* (of a continuous complex-valued function on a locally COMPACT space) to be such that for every $a > 0$ there is a compact set K_a such that $|f(x)| < a$ for all x not in K_a.

vanish nowhere, *vb.* (of an algebra) to satisfy the condition that given any point of the set, there is a member of the algebra for which the value at the point is non-zero; this happens if 1 is in the algebra. See STONE–WEIERSTRASS THEOREM.

variable, *n.* **1a.** an expression that can be assigned any of a set of values.

b. (*as modifier*) able to take any of a range of values: a *variable sum*.

2. a symbol, such as x, y, or z, representing an unspecified member of a class of objects, numbers, etc. Variables may be used either existentially or universally: in elementary algebra variables occur in CONDITIONAL equations representing unknown quantities of which the values are to be found. For example, $x^2 + x = 6$ has the solutions $x = -3$ or 2. However, in an IDENTITY such as

$$(x + y)(x - y) = x^2 - y^2,$$

the stated relationship holds for all values of the variables; in the functional notation $y = f(x)$, each value of the argument, the INDEPENDENT VARIABLE, x, is associated with a unique value of the DEPENDENT VARIABLE, y. See also INDETERMINATE, UNKNOWN.

variance, *n.* (*Statistics*) **1.** a measure of the dispersion of the DISTRIBUTION of a RANDOM VARIABLE, obtained by taking the EXPECTED VALUE of the square of the difference between the random variable and its MEAN, written

$$\text{Var}(X) = E[(X - E(X))^2].$$

The variance of a random variable is usually written σ^2, and is the square of its STANDARD DEVIATION. Compare COVARIANCE.

2. sample variance. the UNBIASED estimator of the variance of a population,

$$s^2 = \sum_{i=1} \frac{(x_i - \bar{x})^2}{n-1},$$

where $\bar{x}$ is the mean of the sample x_i.

variance–covariance matrix or **covariance matrix,** *n.* (*Statistics*) (for a sequence, $\{X_i\}$, of RANDOM VARIABLES) the $n \times n$ matrix, usually denoted Σ, of which the i,j^{th} entry is $\text{Cov}(X_i, X_j)$, the COVARIANCE of X_i and X_j; the matrix is then symmetrical and non-NEGATIVE DEFINITE, and the diagonal entries are the VARIANCES, $\text{Var}(X_i)$.

variate, *n.* (*Statistics*) a RANDOM VARIABLE or a numerical value taken by it.

variation, *n.* **1.** the SUPREMUM of the OSCILLATIONS of a function over all finite partitions of a given interval. See TOTAL VARIATION.

2a. direct variation. another term for DIRECT PROPORTION.

b. inverse or **reciprocal variation.** another term for INVERSE PROPORTION.

3. admissible variation. (*Calculus of variations*) in the simplest case, a differentiable function v that vanishes at the endpoints of an interval so that for all scalars t, $x + tv$ agrees with x on the boundary.

4. variation of an integral. the DIRECTIONAL DERIVATIVE of the integral

$$I(x) = \int_a^b f(x, x', t) \, dt$$

in the direction of an admissible variation h; explicitly,

$$\delta I(x, h) = \lim_{\lambda \to 0} \frac{I(x + \lambda h) - I(x)}{\lambda}.$$

Second- and *higher-order variations* are defined by taking higher derivatives of $I(x + \lambda h)$.

variational, *adj.* pertaining to the CALCULUS OF VARIATIONS.

variational calculus, *n.* another name for CALCULUS OF VARIATIONS.

variational inequality, *n.* an inequality system in HILBERT SPACE: given a nonlinear operator, T, and a closed convex set, C, one seeks an x in C such that

$$\langle T(x), y - x \rangle \leq 0 \text{ for all } y \text{ in } C.$$

Such problems arise in the theory of partial differential equations and in optimization. If the set is the entire space one finds a solution to $T(x) = 0$. If the set is the positive ORTHANT in EUCLIDEAN SPACE, this is a COMPLEMENTARITY PROBLEM.

variation of parameters, n. **1.** the method of determining a solution to an INHOMOGENEOUS LINEAR DIFFERENTIAL EQUATION, $L(x) = f$, by first finding a FUNDAMENTAL SET OF SOLUTIONS $\{x_1, ..., x_n\}$ for the HOMOGENEOUS equation $L(x) = 0$, and then attempting to solve the equation

$$L\left(\sum_{i=1}^{n} c_i x_i\right) = f,$$

for the undetermined functions c_i. If one imposes the condition that for $0 \le k < n-1$, the derivatives

$$\sum_{i=1}^{n} c_i' x_i^{(k)}$$

vanish, the differential equation becomes

$$\sum_{i=1}^{n} c_i' x_i^{(n-1)} = f,$$

and one has obtained n equations in n unknowns. The determinant of this system is the WRONSKIAN of the solutions. Thus there is a unique solution for $dc_i(t)/dt$ and integration will provide the desired solution.
2. for INHOMOGENEOUS systems of LINEAR ORDINARY DIFFERENTIAL EQUATIONS,

$$\mathbf{y}' = A(t)\mathbf{y} + b(t),$$

the solution

$$\mathbf{y} = \Omega(t)\mathbf{c} + \int^t \Omega^{-1}(s)\, b(s)\, ds,$$

where $\Omega(t)$ is the PRINCIPAL SOLUTION MATRIX of $\mathbf{y}' = A(t)\mathbf{y}$.

variety, n. **1.** an element of the underlying set of a BLOCK DESIGN.
2. see ALGEBRAIC VARIETY.

vector, n. **1. vector quantity.** any quantity that has both a magnitude and a direction, such as velocity as opposed to speed.
2. an n-tuple of real or complex numbers viewed as a member of an n-dimensional EUCLIDEAN SPACE. These are used to represent vector quantities, where the LENGTH of the vector is the magnitude of the quantity, and they have the same DIRECTION NUMBERS; they are represented in diagrams by arrows in the same direction, of which the length is proportional to the magnitude. Vectors are usually symbolized by bold typeface letters or by arrows or bars above them:

$$\overrightarrow{AB}, \ \overrightarrow{F}, \ \overline{F}, \ \boldsymbol{F}, \ \mathbf{v}.$$

Compare SCALAR, TENSOR.
3. an element of a VECTOR SPACE.

vector analysis, n. the application and extension of the methods of DIFFERENTIAL and INTEGRAL CALCULUS to VECTOR-valued functions. See VECTOR PRODUCT, SCALAR PRODUCT, SCALAR MULTIPLICATION, GRADIENT, DIVERGENCE, CURL, STOKES' THEOREM.

vector basis or **Hamel basis,** n. BASIS for a VECTOR SPACE.

vector field, n. a mapping from a connected domain in Euclidean space into Euclidean space, especially when the values are written vectorially, as, for example,

$$\mathbf{v} = v_1(x, y, z)\mathbf{i} + v_2(x, y, z)\mathbf{j} + v_3(x, y, z)\mathbf{k}.$$

Compare SCALAR FIELD, TENSOR FIELD.

vector function, *n.* a function of which the domain is a subset of *n*-dimensional EUCLIDEAN SPACE.

vectorial angle, *n.* (*Cartesian geometry*) the angle between the POSITION VECTOR of a point and the *x*- or polar axis.

vector measure, see LIAPUNOV CONVEXITY THEOREM.

vector processing, *n.* (*Computing*) a type of parallel processing.

vector product or **cross product,** *n.* (*Vector analysis*) the product of two real VECTORS in three dimensions that is itself a vector, of which the magnitude is the product of the magnitudes of the given vectors and the sine of the angle between their directions, and of which the direction is perpendicular to the plane of the given vectors, forming a RIGHT–HANDED system. It is denoted $\mathbf{v} \times \mathbf{w}$, or $\mathbf{v} \wedge \mathbf{w}$, and is equal to the determinant

$$\begin{vmatrix} \mathbf{e}_1 & \mathbf{e}_2 & \mathbf{e}_3 \\ v_1 & v_2 & v_3 \\ w_1 & w_2 & w_3 \end{vmatrix} = \sum_{i=1}^{3} \sum_{j=1}^{3} \sum_{k=1}^{3} v_i\, w_i\, \varepsilon_{ijk}\, \mathbf{e}_i,$$

that is,

$$(v_2 w_3 - v_3 w_2)\mathbf{e}_1 + (v_3 w_1 - v_1 w_3)\mathbf{e}_2 + (v_1 w_2 - v_2 w_1)\mathbf{e}_3.$$

Thus $|\mathbf{v} \times \mathbf{w}|$ is the area of a parallelogram with sides $\mathbf{v}$ and $\mathbf{w}$. The vector product is special to three dimensions but can be identified with the EXTERIOR PRODUCT. Compare SCALAR PRODUCT, VECTOR TRIPLE PRODUCT.

vector projection, *n.* (*of a vector on a vector*) a vector derived from the given vectors which has the direction of the second vector and magnitude equal to the SCALAR PROJECTION of the first onto the second. Thus the vector projection of $\mathbf{a}$ on $\mathbf{b}$ has magnitude $|\mathbf{a}| \cos\theta$, where θ is the angle between the given vectors and the same direction as $\mathbf{b}$.

vector space, *n.* a mathematical structure consisting of two sets with their operations, such that one constitutes an ABELIAN GROUP of which the elements are called VECTORS, and the other is a FIELD, of which the elements are called SCALARS. (The term 'scalar' is still used in a MODULE, although they are elements of a ring.) A further operation, SCALAR MULTIPLICATION, is defined, yielding a vector as the product of a scalar and a vector. This operation distributes over addition of both scalars and vectors, and is associative with multiplication of scalars; that is

$$\lambda(\mathbf{v} + \mathbf{w}) = \lambda\mathbf{v} + \lambda\mathbf{w}; \quad (\lambda + \mu)\mathbf{v} = \lambda\mathbf{v} + \mu\mathbf{v}; \quad \lambda(\mu\mathbf{v}) = (\lambda\mu)\mathbf{v}.$$

In analysis, the field is typically the real or the complex numbers, and a complex vector space can be identified with a real vector space by COMPLEXIFICATION.

vector sum, *n.* **1.** the binary vector operation that yields a vector of which the length and direction are represented by the diagonal of a parallelogram of which the sides represent the given vectors; this operation is associative and commutative. In Fig. 393 opposite, if $\overrightarrow{OA}$ and $\overrightarrow{OB}$ represent the magnitude and direction of two vector quantities acting at the same point, then $\overrightarrow{OR}$ is their sum. See also PARALLELOGRAM RULE, RESULTANT.

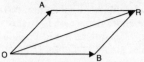

Fig. 393. **Vector sum.** See main entry.

2. the result of applying this operation to a sequence of vectors.

vector triple product or **triple vector product,** *n.* the product of three vectors in three-space, defined by

$$a \times (b \times c) = (a \cdot c)b - (a \cdot b)c;$$

this is not equal to $(a \times b) \times c$. Compare TRIPLE PRODUCT, VECTOR PRODUCT.

vector-valued, *adj.* (of a mapping) taking values in an appropriate vector space as contrasted with a *scalar-valued* mapping taking values in the corresponding scalar field.

vel, *conj.* (*Logic*) or; INCLUSIVE DISJUNCTION, written ∨. Compare AUT.

velocity, *n.* **1.** rate of change of position or DISPLACEMENT, either instantaneous or average. It is a vector quantity, unlike speed, and if unqualified is taken to be linear. The standard unit of velocity is metres per second (ms^{-1}). See also ANGULAR VELOCITY.

2. (*Continuum mechanics*) a generalization of the foregoing; the MATERIAL DERIVATIVE of a MOTION of a BODY, evaluated at a given point of the body.

velocity gradient, *n.* (*Continuum mechanics*) the GRADIENT of the VELOCITY of a BODY with respect to position in the CURRENT CONFIGURATION. See also BODY SPIN, EULERIAN STRAIN RATE.

velocity potential, *n.* **1.** a scalar function, ϕ, such that the VELOCITY is ∇ϕ in an IRROTATIONAL MOTION.

2. see COMPLEX VELOCITY POTENTIAL.

Venn diagram, *n.* a diagram in which mathematical sets or the terms of categorial statements are represented by overlapping circles within a boundary representing the universal set, so that all possible combinations of the relevant properties are represented by distinct areas of the diagram. It is possible to demonstrate the validity of an argument by showing its conclusion to be already represented in a diagram of the premises. For example, Fig. 394 shows a Venn diagram being used to test the validity of the argument

> *all logicians are mathematicians.*
> *some philosophers are logicians,*
> so *some philosophers are mathematicians.*

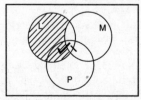

Fig. 394. **Venn diagram.** See main entry.

Here the circles L, M, P represent the three classes, the shading represents the subset given to be empty by the first premise, and the tick marks the area given to have a member by the second premise; the conclusion requires a member in either of the areas joined by the line; in fact this is satisfied, since the exclusion of the shaded part of L ensures that the member represented by the tick is in M as well as L and P. This is a more flexible method of testing validity than EULER'S CIRCLES. (Named after the Cambridge logician, probabilist, and writer, *John Venn* (1834–1923).)

vers, *abbrev. for* VERSED SINE.

versed sine (abbrev. **vers**), *n.* the trigonometric function equal to 1 minus the COSINE function.

versiera, *n.* another name for WITCH OF AGNESI.

vertex, *n.* **1a.** any point of intersection of two sides of a polygon or plane surfaces of a solid, especially that opposite the BASE in a given orientation. **b.** an extreme point of a POLYHEDRON.
2. the point of intersection of a PENCIL of lines.
3. (of an ellipse) either of the points at which the major axis intersects the curve.
4. (in GRAPH THEORY) one of the NODES that, together with associated EDGES, constitutes a graph.

vertex form, *n.* an expression for a CONIC, obtained by a suitable change of variables, in which the VERTEX is taken as the origin of the coordinate system, and the axis of the conic lies along the *x*–axis. In general,

$$y^2 = 2px - (1 - \varepsilon^2)x^2$$

where $2p$ is the PARAMETER and ε the NUMERICAL ECCENTRICITY of the conic.

vertical angles, *n.* the pair of equal angles between two intersecting straight lines; OPPOSITE ANGLES.

vibrating–string equation, *n.* the equation

$$\frac{\partial^2 f}{\partial x^2} = \frac{T}{\rho}\frac{\partial^2 y}{\partial x^2},$$

where x is the direction in which a vibrating string is stretched, y is the displacement, t is time, T is the tension in the string, and ρ is its density. Standard boundary conditions are $y = f(x)$ and $\partial y / \partial t = 0$ at time zero.

vicenary, *adj.* having or using BASE 20.

vicinity, *n.* a member of a UNIFORMITY.

vicious circle, *n.* (*Logic*) **1.** a futile form of reasoning in which a conclusion is inferred from premises of which the truth cannot be established independently of that conclusion.
2. an explanation given in terms that cannot be understood independently of what was to be explained. See INFINITE REGRESS.
3. a situation arising from SELF–REFERENCE in which some statement is shown to entail its negation and vice versa, such as *this statement is false* which is true only if false and false only if true.

Viète, François or **Franciscus Vieta** (1540–1603), French algebraist and geometer, who introduced the use of literals to algebra, but rejected the existence of negative numbers. He made original contributions to trigonometry and the theory of equations, and decoded a complex code used by Philip II of Spain in his war against the French, being accused of witchcraft for his pains.

Viète's formula or **Vieta's formula,** *n.* the formula for π, derived from the INFINITE PRODUCT for $2/\pi$, namely

$$\sqrt{\frac{1}{2}} \times \sqrt{\frac{1}{2} + \frac{1}{2}\sqrt{\frac{1}{2}}} \times \sqrt{\frac{1}{2} + \frac{1}{2}\sqrt{\frac{1}{2} + \frac{1}{2}\sqrt{\frac{1}{2}}}} \times \dots ,$$

published in 1593, and generally regarded as the first use of an infinite product.

vigesimal, *adj.* related to, based on, or proceeding by intervals of 20.

Vinogradov's theorem, *n.* (*Number theory*) the result that all odd numbers, except at most finitely many, are the sum of three odd primes. See also GOLDBACH'S CONJECTURE. (Named after *Ivan Matveevich Vinogradov* (1891–1983), Russian number theorist.)

vinculum, *n.* a horizontal line placed above a group of terms in an expression as an alternative to BRACKETS to indicate that the terms so marked are to be treated as a unit in the evaluation of the expression. For example,

$$\overline{x + y} - z = (x + y) - z$$

This notation is uncommon outside elementary contexts, as it may be confused with that for the COMPLEX CONJUGATE.

virtual work, *n.* (*Mechanics*) the total WORK done by a mechanical system in an infinitesimal displacement, subject to physical restraints. If the restraints do no work, for example, by acting at right angles to a possible motion, the system is in equilibrium if and only if the virtual work is zero.

viscous fluid, *n.* (*Continuum mechanics*) a BODY in which the STRESS TENSOR σ is given by

$$\sigma = -p\mathbf{I} + \sigma^{\mathrm{E}},$$

where p is the PRESSURE and σ^{E} is a DEVIATORIC tensor arising from the motion of the body.

viscosity, *n.* (*Continuum mechanics*) the constant, μ, in the CONSTITUTIVE EQUATION of a simple NEWTONIAN VISCOUS FLUID,

$$\sigma = -p(\mathbf{x}, t)\mathbf{I} + 2\mu[\Sigma - \tfrac{1}{3}(\mathrm{tr}\Sigma)\mathbf{I}],$$

where σ is the STRESS TENSOR, p is the PRESSURE function, and Σ is the EULERIAN STRAIN RATE.

Vitali covering, *n.* a COVERING of a set E in Euclidean *n*-space by hyper-cubes (where the ratio of largest to smallest side remains bounded) with the property that, for every element *e* of E, there is a member of the cover that contains *e* and has arbitrarily small positive measure. (Named after the Italian set theorist and analyst, *Giuseppe Vitali* (1875–1932).)

Vitali covering theorem, *n.* the result that if a class of closed hypercubes is a VITALI COVERING of a set E in *n*-space, then there is a countable sequence of pairwise-disjoint members of the covering of which the union J has Lebesgue OUTER MEASURE equal to that of E; that is, $\mu^*(E \setminus J) = 0$.

Vitali set, *n.* the (Lebesgue) NON–MEASURABLE subset of the real line constructed by taking one element from each EQUIVALENCE CLASS of real numbers where the EQUIVALENCE RELATION is differing by a rational.

vital statistics, see STATISTICS.

void intersection property, *n.* the property of a collection of sets that their intersection is empty.

Volterra, Vito (1860–1940), Italian analyst and mathematical physicist who was a pioneer of functional analysis. He became a Senator of the Italian kingdom, and joined the air force in the First World War where he proposed the military use of dirigibles and the replacement of hydrogen by helium. He was forced by the Fascists to resign his post and honours for his refusal to swear allegiance to Mussolini's regime.

Volterra's integral equation, *n.* the equation

$$f(x) = \int_a^x K(x,t)\,y(t)\,\mathrm{dt},$$

where f and the KERNEL $K(x, t)$, of the integral (defined on the triangle $a \le x \le t \le b$) are given, and y is sought. This is referred to as *Volterra's integral equation of the first kind*, by contrast with *Volterra's integral equation of the second kind*, the equation

$$y(x) = f(x) + \lambda \int_a^x K(x,t)\,y(t)\,\mathrm{dt}.$$

When both f and K are continuous the second equation has a unique continuous solution. The first equation may be reduced to the second by differentiation when $K_x(x, t)$ exists continuously.

volume, *n.* **1.** the extent of a three-dimensional space enclosed within or occupied by a solid.

2. the analogous quantity in Euclidean space, usually defined as the LEBESGUE MEASURE of a measurable set, and, by the BANACH–TARSKI THEOREM, not in fact well-defined.

volume of revolution, *n.* the volume of a SOLID OF REVOLUTION.

von Neumann, John (1903–1957), Hungarian-born American mathematical logician. He was a child prodigy who could converse in classical Greek at the age of six, and memorised sections of the telephone directory as a party trick. He studied chemistry at university, but also achieved outstanding results in mathematics despite not attending any classes, published the standard definition of ORDINAL NUMBER in 1924, and in 1926 was awarded a doctorate for a thesis on SET THEORY. Von Neumann taught in Berlin and Hamburg, before moving to Princeton, where he became, with EINSTEIN, one of the founders of the Institute of Advanced Study. He laid the mathematical foundations of quantum theory, ERGODIC theory, and his most celebrated creation, GAMES THEORY, and his recognition of the value of numerical methods led him to make significant contributions to the development of computer science. From 1940 to 1955 was a consultant to the US armed forces, and contributed to the development of the hydrogen bomb.

von Neumann architecture, *n.* a term used to describe the conceptual design of the standard serial DIGITAL COMPUTER, first described by von Neumann. All present day computers accord to this general architecture. Compare QUANTUM COMPUTER.

von Neumann minimax theorem, see MINIMAX THEOREM.

von Neumann set theory, *n.* an alternative to the orthodox axiomatization of set theory which seeks to avoid the SET-THEORETIC PARADOXES by making a radical distinction between sets and classes. Every set in the system is a class, but not all classes are sets; sets may be members of other sets, but PROPER CLASSES may not be members of other classes.

vortex line, *n.* (*Continuum mechanics*) a CURVE of which the TANGENT vector at any point is in the direction of the VORTICITY.

vorticity, *n.* (*Continuum mechanics*) the CURL of the VELOCITY of a BODY with respect to position in the current CONFIGURATION.

vorticity tensor, *n.* another term for BODY SPIN.

voting paradox, *n.* the paradox that there may be no consistent ordering of public preferences: a moderate candidate, for example, may win in two-way polls against both the conservative and the radical, yet come bottom of a three-candidate election; that is, even if every voter's PREFERENCE ORDERING is transitive, their combined ordering is not. ARROW'S IMPOSSIBILITY THEOREM proves that the existence of a consistent public preference ordering is inconsistent with certain other reasonable conditions on a democratic electoral system. See also SIMPSON'S PARADOX.

vulgar fraction, *n.* an expression of a RATIONAL NUMBER as a RATIO rather than as a DECIMAL FRACTION. Such a fraction is a PROPER FRACTION if its DENOMINATOR is greater than its NUMERATOR; that is, if its absolute value is strictly less than 1. See also MIXED FRACTION.

W, (*Mechanics*) *symbol for* WATT.

Wald's equation, *n.* (*Statistics*) the identity, for a sequence of INDEPENDENT IDENTICALLY DISTRIBUTED RANDOM VARIABLES, $X_1, ..., X_N$, and a random positive integer, N, that

$$E[X_1 + ... + X_N] = E(X_1)\ E(N).$$

Wallis formulae, *n.* the formulae

$$\int_0^{\pi/2} \cos^{2n+1}(t)\,dt = \frac{2.4.6. .... .2n}{1.3.5. .... .(2n+1)}$$

and

$$\int_0^{\pi/2} \cos^{2n}(t)\,dt = \frac{1.3.5. .... .(2n-1)}{2.4.6. .... .2n} \cdot \frac{\pi}{2},$$

from which WALLIS'S PRODUCT FOR PI is computed by repeated integration by parts. (Named after the English algebraist, logician, and theologian, *John Wallis* (1616–1703), whose work influenced Newton's development of the calculus and laws of motion; he was also party to the meetings that led to the establishment of the Royal Society in 1662.)

Wallis's product for pi, *n.* the infinite product

$$\frac{\pi}{2} = \frac{2 \times 2 \times 4 \times 4 \times 6 \times 6 \times ...}{1 \times 3 \times 3 \times 5 \times 5 \times 7 \times ...}$$

which Wallis discovered by an inspired interpolation process, but which is now established by comparing the ratio of the WALLIS FORMULAE.

walk, *n.* an alternating sequence of edges and vertices in a GRAPH. If the first and last vertices are identical it is a *closed walk*, if all edges are distinct it is a *trail*, and if all vertices (except possibly the endpoints) are distinct it is a *path*; a closed path is a *cycle* or *loop*.

Waring's problem, *n.* the problem, solved by Hilbert, of proving that every natural number can be written as the sum of a fixed smallest number, $g(n)$, of nth powers of integers. Thus LAGRANGE'S THEOREM shows that $g(2) = 4$. The exact value of $g(n)$ is known for all but finitely many n; for $n < 20\ 000$, $g(n)$ is the integer part of $(3/2)^k + 2^k - 2$. The corresponding smallest number, $G(n)$, that works with finitely many exceptions is less well understood. In particular, $g(3) = 9$, $g(4) = 19$, $g(5) = 37$, and $G(2) = 4$, $G(4) = 16$.

watt (symbol **W**), *n.* (*Mechanics*) the standard unit of POWER, being the rate of one JOULE per SECOND.

wave equation, *n.* the second order PARTIAL DIFFERENTIAL EQUATION, of importance in many areas of physics such as the study of electromagnetic, water and sound waves, describing the propagation of a classical wave. The equation is

$$\frac{\partial^2 f}{\partial t^2} = c^2 \nabla^2 f,$$

where f is the wave function, $\nabla^2 f$ is the three–dimensional LAPLACIAN, and c is the speed of propagation of the wave .

wcg, *abbrev. for* WEAKLY–COMPACTLY GENERATED.

weak, *adj.* (of an inequality, ordering etc.) permitting the possibility of identity. For example, $x \leq y$ is a *weak inequality*; if $x < y$ implies that $f(x) \geq f(y)$, f is a *weakly decreasing* function. When there is no explicit contrast with a STRICT sense, the qualifier is usually omitted.

weak duality, see STRONG DUALITY.

weak ergodic theorem, *n.* another term for the MEAN ERGODIC THEOREM.

weak inclusion, see INCLUSION.

weak inverse image set, see INVERSE IMAGE SET.

weak law of large numbers, *n.* various results concerning the CONVERGENCE IN MEASURE of the sequence of averages of a sequence of RANDOM VARIABLES to their MEAN. If $\{X_k\}$ is a sequence of random variables with the same mean, μ, and bounded VARIANCE, then for every small δ and $\varepsilon > 0$, there is an N beyond which

$$P\left(\left|\frac{\sum_{i=1}^{N} X_i}{N} - \mu\right| > \varepsilon\right) < \delta \, ;$$

equivalently,

$$\lim_{n \to \infty} P\left[\frac{\sum_{i=1}^{n} X_i}{n} = \mu\right] = 1.$$

These differ from the STRONG LAW OF LARGE NUMBERS in which the convergence is pointwise.

weakly compact, *adj.* COMPACT in the WEAK TOPOLOGY.

weakly-compactly generated (abbrev. **wcg**), *adj.* (of a BANACH SPACE) containing a weakly COMPACT set, the closed linear span of which is the entire space. Separable and reflexive spaces are weakly-compactly generated.

weakly convergent, *adj.* (of a sequence or net $\{x_n\}$) CONVERGENT in the WEAK TOPOLOGY; $\{x_n\}$ converges weakly to x if and only if, for every continuous linear functional on the space, $f(x_n)$ tends to $f(x)$.

weak-star, see WEAK TOPOLOGY.

weak topology, *n.* **1.** (of a NORMED space) the topology imposed on the underlying vector space by taking as a SUB–BASE all open half spaces containing zero. This gives the weakest topology in which all the norm continuous linear functionals are continuous. A space is REFLEXIVE if and only if the unit ball is WEAKLY COMPACT, or equivalently, by the EBERLEIN–SMULIAN THEOREM, if and only if it is *weakly* SEQUENTIALLY COMPACT. See also WEAKLY CONVERGENT.

2. weak-star topology. the corresponding topology placed on the dual normed space by requiring that a sequence or net $\{f_n\}$ converge to f (*weak-star convergence*) if and only if, for every point x of the PREDUAL SPACE, $f_n(x)$ tends to $f(x)$. In this topology the dual ball is *weak-star compact*. See BANACH–ALAOGLU THEOREM.

Wedderburn's theorem, n. the theorem that a finite DIVISION RING is a FIELD.

Wedderburn structure theorem or **Wedderburn–Artin theorem,** n. the fundamental structure theorem for SIMPLE and SEMISIMPLE rings, that states that every semisimple right ARTINIAN RING is the DIRECT SUM of a finite number of simple right Artinian rings, and a simple right Artinian ring is isomorphic to the ring of $n \times n$ matrices over some DIVISION RING, K, for some positive integer, n. (Named after the Scottish algebraist and number theorist, *Joseph Henry Maclagan Wedderburn* (1882–1948), who became Chairman of the Institute of Advanced Studies at Princeton.)

wedge, n. a convex CONE based at the origin, especially when the cone contains no full lines (as opposed to half lines). Such a convex cone is called *pointed* or *salient*.

Weierstrass, Karl Theodor Wilhelm (1815–97), German analyst, who contributed especially to the theories of complex variables, power series, elliptic functions, continuity, quadratic forms, and the calculus of variations. He was sent to Bonn University to study law, but left without a degree after four years of drinking and fencing; he then trained as a mathematics teacher and taught for 14 years. During this time, without contact with the mathematical world, he developed an entirely original rigorous approach to analysis that enabled him to describe continuous but nowhere differentiable functions and so completely undermine the intuitive approach to these concepts. After the appearance of a monograph developing the work of Abel on theory of functions, he was awarded an honorary doctorate and was found an academic post; although he did not write much more, his extremely influential lectures were published.

Weierstrass approximation theorem, n. the theorem that the polynomials are uniformly dense in the continuous functions on a closed bounded interval. This may be derived as a special case of the STONE–WEIERSTRASS THEOREM. It is shown by *Müntz' theorem* that one can achieve density precisely, using all polynomials involving the constants and any infinite sequence of powers of x of which the reciprocals diverge; thus prime powers suffice.

Weierstrass elliptic function, n. the fundamental ELLIPTIC FUNCTION

$$\mathbf{p}(z) = \frac{1}{z^2} + \sum_{m,n} \left\{ \frac{1}{\left[z - \Omega(m, n) \right]^2} - \frac{1}{\left[\Omega(m, n) \right]^2} \right\}$$

summed over non-zero integers, where

$$\Omega(m, n) = 2n\omega_1 + 2m\omega_2,$$

for independent PERIODS $2\omega_1$ and $2\omega_2$.

Weierstrass–Erdman corner conditions, see EULER–LAGRANGE EQUATIONS.

Weierstrass function, *n.* the function

$$f(x) = \sum_{i=1}^{\infty} \lambda^{(s-2)i} \sin(\lambda^i x),$$

for *s* strictly between 1 and 2, and $\lambda > 1$; it is continuous but nowhere differentiable, by reason of its recursively constructed repeated oscillation. The Hausdorff dimension of the graph has lower bound *s*; although equality seems likely, this has never been proved.

Weierstrass M–test, *n.* the test for UNIFORM CONVERGENCE of a series based on the result that, for a family $\{f_n(x)\}$ of complex functions defined on a set E,

$$\sum_{n=1}^{\infty} f_n(x)$$

converges uniformly on E as soon as one can find a summable real series,

$$\sum_{n=1}^{\infty} M_n < \infty,$$

such that

$$|f_n(x)| \le M_n$$

for all *x* in the set E. For example,

$$\sum_{k=1}^{\infty} -\frac{z^k}{k^2}$$

is convergent on the closed unit disk, since $M_k = 1/k^2$ is convergent. The analogous result holds true for integrals.

Weierstrass product expansion, see GENUS (sense 2).

weight, *n.* **1.** a value put on EDGES of a GRAPH; typically used in applications such as the TRAVELLING SALESMAN PROBLEM where the weights are the distances between adjacent vertices.

2. see ORTHONORMAL FUNCTIONS.

3. (*Mechanics*) the FORCE of the Earth's gravity on a body, equal to the product of the MASS of the body and the LOCAL GRAVITATIONAL CONSTANT.

weighted average or **weighted mean,** *n.* **1.** (*Statistics*) an average calculated by taking into account not only the frequencies of the values of a random variable, but also some other factor such as their standard deviation; the weighted average of observed data, where each of the values v_i respectively occurs n_i times, and each v_i has a weighting of w_i, is

$$\frac{\Sigma_i v_i n_i w_i}{\Sigma_i n_i w_i}.$$

2. a similar quantity computed for any integral or sum.

weighting, *n.* (*Statistics*) a factor by which some quantity is multiplied in order to make it comparable with others. See also WEIGHTED AVERAGE.

well-conditioned, *adj.* (*Numerical analysis*) **1.** (of a problem) having a small CONDITION NUMBER.

2. (of a computation) numerically stable.

Compare ILL-CONDITIONED.

well-formed, *adj.* (*Logic*) (of a formula, expression, etc.) grammatically correct; constructed in accordance with the FORMATION RULES of a particular formal system. Often abbreviated WFF (for *well-formed formula*).

well-ordered, *adj.* **1.** (of a RELATION) having the property that every non-empty subset of its field has a least member under the relation, and so is INDUCTIVELY (whence TOTALLY) ORDERED; *less than* is well-ordered on the natural numbers, but not on the reals, since an open set has no least member. See ORDERING.

2. (of a set) ordered by such a relation.

well-ordering, *n.* a WELL-ORDERED relation.

well-ordering principle, *n.* **1.** (*Logic*) also called **well-ordering theorem.** the non-CONSTRUCTIVE set theoretic result, equivalent to the AXIOM OF CHOICE or ZORN'S LEMMA, that for any set there is a binary relation under which it is WELL-ORDERED.

2. the principle that the positive integers are well-ordered.

well-posed problem, *n.* a problem that is well formulated; especially one for which, under appropriate conditions, the solution can be shown to exist, to be unique, and to vary continuously with perturbation of the data. If these three conditions do not hold, the problem is said to be *ill-posed*, although it may still be soluble. See also STABLE.

wff, *abbrev. for* WELL-FORMED formula.

whole number, *n.* a non-negative INTEGER, either a NATURAL NUMBER or zero. Usage, however, varies, and the term may be used for all INTEGERS, or only positive integers, or only for positive whole numbers.

Wiener process, *n.* (*Probability*) a STOCHASTIC PROCESS that models Brownian motion; a family of real-valued random variables X_t ($t \geq 0$) with $X_0 = 0$ almost everywhere, such that each $X_{t+s} - X_t$ (s, $t \geq 0$) is NORMALLY DISTRIBUTED with mean zero and variance s, while for $0 \leq t_0 < t_1 < \ldots < t_n$, the random variables

$$X_{t_{i+1}} - X_{t_i}$$

are independent for $0 \leq i < n$. (Named after the American analyst, applied mathematician, and pioneer of cybernetics, *Norbert Wiener* (1894–1964).)

Wilcoxon test or **Wilcoxon signed ranks test,** *n.* (*Statistics*) **1.** a one-sample test upon the RANKS of some given data, used in examining whether or not the population from which the data are drawn has a given MEDIAN; for example, a test for the relative level of the scores of the same subjects under two experimental conditions.

2. Wilcoxon Mann–Whitney test. see MANN–WHITNEY TEST.

Wilson's theorem, *n.* (*Number theory*) the result that a natural number n is prime if and only if n divides $1 + (n-1)!$. (Named after the English number theorist, *John Wilson* (1741–93).)

winding number or **index,** *n.* (*Complex analysis*) the number of times that a given closed curve, γ, passes anticlockwise around a point (that is, as the angle between the radius vector and the polar axis increases); denoted $n(\gamma, z)$. For a piecewise smooth curve this may be computed from

$$2\pi i \, n(\gamma, z) = \int_{\gamma} \frac{dw}{w - z}.$$

witch of Agnesi or **versiera,** *n.* the locus of the points of intersection of the legs of all right-angled triangles of which the hypotenuse lies on a given line through the origin, one side parallel to the *x*-axis and passing through the point of intersection of the given line with a circle of radius *a* tangential to the axis at the origin, and the third side parallel to the *y*-axis and passing through the point of intersection of the given line with the line $y = 2a$. This curve has equation

$$x^2 y = 4a^2(2a - y),$$

and is symmetrical about the *y*-axis with the *x*-axis as asymptote, as shown in Fig. 395.

Fig. 395. **Witch of Agnesi**.

within-subjects design, *n.* (*Statistics*) (of an experiment) concerned to measure the values of the dependent variable for the same subjects under the various experimental conditions. Compare BETWEEN-SUBJECTS DESIGN, MATCHED–PAIRS DESIGN.

without repetitions, see NORMAL SERIES.

word, *n.* **1.** a sequence of symbols the length of which, the *word-length*, is taken to be a unit for some purpose, such as the number of BYTES that together constitute a single MESSAGE or element of a CODE; in particular, the basic unit of storage on a digital computer.

2. (*Group theory*) an expression of the form

$$x_1^{\pm 1} x_2^{\pm 1} \dots x_n^{\pm 1}.$$

With the product of such NON–EMPTY WORDS defined as concatenation, they form a SEMI–GROUP, and when this is extended to EMPTY WORDS, 1, by $1u = u$ for all words *u*, the set of all words is a MONOID. See also FREE GROUP.

work or **work done,** *n.* (*Mechanics*) (for a force, **F**, moving along a curve, C) the negative of the integral of **F** along C; that is, the integral

$$-\int_C \mathbf{F} \cdot \mathbf{dx} \, ;$$

if **F** is CONSERVATIVE this is independent of the choice of C. The work done by all the forces in a system equals the change in KINETIC ENERGY. The standard unit of work is the JOULE.

world, (*Logic*) see POSSIBLE WORLD.

wrench, *n.* (*Mechanics*) a FORCE together with a COUPLE with axis parallel to the force. Any system of forces is equivalent to a wrench.

Wronskian, *n.* (of *n* functions on an open interval) the DETERMINANT of the matrix of which the i,j^{th} entry is the $(j-1)^{\text{th}}$ derivative of the ith function evaluated at *x*. Given that the functions have $n-1$ continuous derivatives on $]a, b[$, the functions are linearly independent if the Wronskian is not identically zero. Conversely, if the Wronskian vanishes at even one point, and the *n* functions solve an nth-order linear differential equation with continuous coefficients of which the leading coefficient is never zero, then the functions are LINEARLY DEPENDENT, and the Wronskian vanishes everywhere. (Named after the Polish-born French analyst, combinatorialist, physicist, and philosopher, *Josef Maria (Hoëné-) Wronski* (1778–1853).) See also FUNDAMENTAL SYSTEM OF SOLUTIONS.

wrt, *abbrev. for* with respect to.

X

x or **X,** *symbol for* the number 10 in ROMAN NUMERALS.

x-axis, *n.* one of the axes of a CARTESIAN COORDINATE system; conventionally, the horizontal one pointing from left to right on a graph, as shown in Fig. 396. Compare Y–AXIS, Z–AXIS.

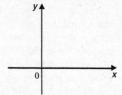

Fig. 396. The bold line is the **x–axis**.

x-intercept, *n.* in a CARTESIAN COORDINATE SYSTEM, the distance from the origin to the point at which a line, curve, or surface intersects the X–AXIS.

y-axis, *n.* one of the axes of a CARTESIAN COORDINATE system; conventionally, the vertical one in a two-dimensional graph, as in Fig. 396 above, or that pointing from back to front in a representation of three-dimensional space, as in Fig. 397 opposite. Compare X-AXIS, Z-AXIS.

y-intercept, *n.* in a CARTESIAN COORDINATE SYSTEM, the distance from the origin to the point at which a line, curve, or surface intersects the Y-AXIS.

Yang–Mills theory, *n.* (*Quantum theory*) a framework for describing elementary particles using geometrical structures, now the foundation of most of elementary particle theory. Its predictions about the strong interactions of elementary particles and the existence of a subtle quantum mechanical property called the "mass gap" – that quantum particles have positive masses, even though the classical waves travel at the speed of light – have been experimentally tested and confirmed by computer simulations, but its mathematical foundation is unclear, and providing such a foundation is one of the MILLENNIUM PRIZE PROBLEMS.

Young's inequality, see CONJUGATE (sense 9).

Young's modulus, *n.* (*Mechanics*) a constant that measures the extent to which a material is ELASTIC, and that varies with the material of a body and with the units.

Young tableaux, *n.* another formulation of FERRAR'S GRAPH.

Z

$\mathbb{Z}$ or $\mathcal{Z}$, *symbol for* the set of INTEGERS. Also called $\mathbb{I}$. Compare $\mathbb{N}$, $\mathbb{Q}$, $\mathbb{R}$.

$\mathbb{Z}^+$, *symbol for* the set of positive integers.

$\mathbb{Z}_n$, *symbol for* the ring of MODULAR ARITHMETIC MODULO n.

z–axis, *n.* one of the axes of a CARTESIAN COORDINATE system; conventionally, the vertical one in a representation of three-dimensional space, as in Fig. 397. Compare X-AXIS, Y-AXIS.

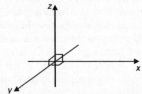

Fig. 397. The bold line is the **z–axis**.

z–intercept, *n.* in a CARTESIAN COORDINATE SYSTEM, the distance from the origin to the point at which a line, curve, or surface intersects the z-AXIS.

Zariski topology, *n.* a TOPOLOGY on an infinite set whose open sets are those with finite complements; it is not METRIZABLE as it is not Hausdorff.

Zeno's paradoxes, *n.* a collection of paradoxes, primarily regarding time, motion, and plurality, attributed to *Zeno of Elea* (*c.* 490 – 435 BC), Greek philosopher and mathematician, and known to us through Aristotle. The four most troublesome are DICHOTOMY, the ACHILLES PARADOX, the ARROW PARADOX, and the STADIUM PARADOX. The first two both argue that motion is impossible if one makes the assumption that time and space are infinitely divisible, while the last two argue that motion is impossible if one makes the contradictory assumption that time and space are only finitely divisible.

Zermelo's theorem, *n.* another name for the WELL–ORDERING PRINCIPLE. (Named after the German set theorist and analyst, *Ernst Friedrich Ferdinand Zermelo* (1871 – 1953).)

Zermelo–Fraenkel set theory (abbrev. **ZF**), *n.* the most standard axiomatization of set theory. The theory with the addition of the AXIOM OF CHOICE is denoted *ZFC*.

zero, *n.* **1a.** also called **nought.** the symbol 0 representing the absence of any magnitude.

b. the cardinality of the empty set.

c. the number whose sum with any other number is that other number.

2a. the IDENTITY ELEMENT for any additive operation, so that its sum with any other element is that other element, such as the matrix all of the elements of which are 0.

b. an element, often denoted 0, of a RING, of which the product with any other is that zero element.

3. **zero of a function**. a value of the argument of a function at which the value of the function is zero. For example $x^2 + 2x$ has a zero at $x = -2$. See also ROOT.

4. the least element of a lattice or partially ordered set, such as the empty set in the poset of subsets of a given set, written $\wedge$. Compare UNITY.

zero or **zeroize,** *vb.* to set equal to zero, especially to INITIALIZE the values of the variables in an algorithm or computation.

zero divisors, *n.* non–zero elements in a RING of which the product is zero, such as the matrices

$$\begin{bmatrix} 0 & 0 \\ 1 & 1 \end{bmatrix} \text{ and } \begin{bmatrix} 0 & 1 \\ 0 & -1 \end{bmatrix}.$$

See also DIVISION RING, INTEGRAL DOMAIN.

zero measure, *n.* **1.** another term for NULL MEASURE.

2. a MEASURE, μ, such that $\mu(E) = 0$ for every MEASURABLE set, E.

zero-one law, *n.* (*Probability*) the result, due to Kolmogorov, that the probability of a TAIL EVENT of a sequence of independent random variables is either 0 or 1. See also BOREL–CANTELLI LEMMA.

zero-order, see TENSOR.

zero ring, *n.* a RING consisting of only a single element, denoted 0, with multiplication and addition defined by

$$0 + 0 = 0 = 0 \cdot 0;$$

it is a COMMUTATIVE RING with an identity.

zero set, *n.* the set of values for which a given non-zero analytic function is zero; this set is at most countable.

zero-sum game, *n.* a GAME in which total pay-off is zero, especially a *two-person zero-sum game* in which the PAY OFF for one player is the negative of the pay off for the other. For such two person games the MINIMAX THEOREM establishes the existence of optimal mixed STRATEGIES, which simultaneously maximize the minimum expected gain of the one player and minimize the maximum expected loss of the other.

zeroth, *adj.* first in a sequence of terms indexed by the natural numbers, such as a_0 in the sequence $a_0, a_1, a_3, \ldots$

zero vector or **trivial vector,** *n.* a VECTOR with no magnitude, and hence no direction.

zeta function, *n.* a SPECIAL FUNCTION of great importance in number theory; the function

$$\zeta(s) = \sum_{n=1}^{\infty} \frac{1}{n^s} \, ;$$

in particular, $\zeta(2) = \pi^2/6$, $\zeta(4) = \pi^4/90$. While the distribution of PRIME NUMBERS is unpredictable, Reimann observed that it approximates to the zeta function. Also called *Riemann zeta function* to distinguish it from more general related functions. See also APERY'S THEOREM, RIEMANN HYPOTHESIS.

zeta hypothesis, *n.* another name for the RIEMANN HYPOTHESIS.

ZF, *abbrev. for* ZERMELO–FRAENKEL SET THEORY.

ZFC, *abbrev. for* ZERMELO–FRAENKEL SET THEORY together with the AXIOM OF CHOICE.

zigzagging, *n.* (*Numerical analysis*) poor behaviour often observed in performance of numerical methods, such as STEEPEST DESCENT, near the optimum, when small, nearly ORTHOGONAL steps are taken and a zigzag path is followed. See also JAMMING.

zone, *n.* a portion of a sphere between two parallel planes intersecting the sphere, as shown in Fig. 398.

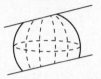

Fig. 398. A **zone** of a sphere.

Zorn's lemma, *n.* the NON-CONSTRUCTIVE set-theoretic result that, in an ordered set in which every CHAIN has an upper bound, there is a maximal element. This is important in mathematical practice and equivalent to the AXIOM OF CHOICE, the WELL-ORDERING THEOREM, and the HAUSDORFF MAXIMALITY THEOREM. (Named after the German-born American group theorist, algebraist, and analyst, *Max August Zorn* (1906–).)

Zoutendijk's method, *n.* a prototype FEASIBLE directions method of solving CONSTRAINED OPTIMIZATION problems by generating feasible descent directions. To minimize a differentiable function $f(x)$ in n-space subject to linear constraints

$$\langle a_i, x \rangle \leq b_i \text{ for } 1 \leq i \leq m,$$

one produces a new search direction by considering the LINEAR PROGRAM

$$\min \langle \nabla f(x), d \rangle$$

subject to

$$\langle a_i, d \rangle \leq 0 \quad (i \in I(x))$$
$$\sum_{i=1}^{n} | d_i | \leq 1.$$

Here $I(x)$ is the index set of binding constraints at x. One then produces a direction such that $f(x + td)$ decreases and $x + td$ remains feasible for small positive t. These methods are subject to ZIGZAGGING and other numerical problems.

APPENDICES

Symbols and Conventions

Greek Alphabet

Where letters have well-defined mathematical significance, cross-references are shown. In some cases, however, the Greek letter functions as a prefix, and the reader should check all the appropriate entries.

α	A	alpha	
β	B	BETA	
γ	Γ	GAMMA	γ : EULER'S CONSTANT, GRAVITY
δ	Δ	DELTA	δ or Δ : INCREMENT, DIFFERENCE QUOTIENT, DIFFERENCE SEQUENCE
			δ: d, DERIVATIVE, KRONECKER DELTA, DIRAC DELTA FUNCTION, FRÉCHET DIFFERENTIAL, G–DELTA, POINT EVALUATION, SUPPORT FUNCTION, TEST RULE, VARATION of an integral
			Δ, ∇, or ∂ : LAPLACIAN
			∇ : DIFFERENTIAL OPERATOR
			∂ : PARTIAL DERIVATIVE, GRADIENT, JACOBIAN
ε	E	EPSILON	ε : NEIGHBOURHOOD, ECCENTRICITY, SIGNATURE
	ς	DIGAMMA	
ζ	Z	ZETA	
η	H	eta	
θ	Θ	THETA	Θ : ORDER NOTATION
ι	I	iota	
κ	K	kappa	κ : CURVATURE
λ	Λ	lambda	
μ	M	MU	μ : MICRO-, MEAN, MÖBIUS FUNCTION
ν	N	nu	
ξ	Ξ	xi	
o	O	omicron	
π	Π	PI	π : RADIAN
			Π : PRODUCT, INFINITE PRODUCT, HOMOTOPY
ρ	P	rho	ρ : CURVATURE, CORRELATION
σ	Σ	SIGMA	σ : STANDARD DEVIATION, VARIANCE, F–SIGMA
			Σ : SUM, SERIES, INFINITE SERIES
τ	T	tau	τ : TORSION
υ	Y	upsilon	
φ	Φ	phi	φ : EULER PHI FUNCTION Φ : FRATTINI SUBGROUP
χ	X	CHI	χ : CHARACTERISTIC FUNCTION, CHROMATIC NUMBER
ψ	Ψ	PSI	
ω	Ω	OMEGA	ω : ANGULAR VELOCITY, MODULUS of continuity

Hebrew Alphabet

ℵ	ALEPH	CONTINUUM HYPOTHESIS

Conventional usages

There are a number of well-established conventions for the use of letters of the Greek and Roman alphabets as either variables or dummy terms. These are, however, no more than conventions, and context should be sufficient to determine whether or not they are relevant.

α, β, γ	DIRECTION ANGLES
α, β, γ or λ, μ, ν	SCALAR coefficients, especially in LINEAR COMBINATIONS
α, β	TRANSCENDENTAL NUMBERS
γ	a PERMUTATION or CYCLE
Γ	a general INDEX SET, and γ for a dummy member of the set
Γ	a curve or CONTOUR, especially in CURVILINEAR INTEGRALS
δ	a METRIC or DISTANCE function
θ, ϕ, ψ	angles
θ	the parameter of a system of PARAMETRIC EQUATIONS
θ	a MAPPING, especially a HOMOMORPHISM
(r, θ)	POLAR COORDINATES
λ	a RATIO
λ	a LEBESGUE MEASURE
λ	a LATENT ROOT (eigenvalue)
Λ	a general INDEX SET, and λ for a dummy member of the set
μ, ν	MEASURES
μ, ν	PARAMETERS of certain statistical distributions
ν	the NATURAL EPIMORPHISM
ξ, η, ζ	coordinate variables under a TRANSFORMATION
π, σ	PERMUTATIONS
ρ	a METRIC
ρ	the DENSITY of a BODY
σ, τ	TOPOLOGIES
ϕ, ψ	MAPPINGS, especially HOMOMORPHISMS
ϕ, ψ	FUNCTIONALS
ϕ, ψ	PREDICATES (especially variables over predicates)
$\phi(x)$	a SCALAR FIELD
(r, ϕ, θ)	SPHERICAL COORDINATES
χ	a CONFIGURATION of a BODY
(u, ψ)	a CHART
ω	an ANGULAR quantity, such an ANGULAR VELOCITY
ω	the PERIOD of a periodic function
Ω	the volume of a BODY, and $\partial\Omega$ its surface
a, b	arbitrary CONSTANTS
$\mathbf{B}$	a BODY
d	a DISTANCE function or METRIC
$f, g, h; F, G$	FUNCTIONS; the capitals are often used for the integrals of the functions represented by the corresponding lower case letters
F, G, H	PREDICATES

G, H	GROUPS
i, j, k	members of an INDEX SET of INTEGERS, for example, in a SUM or PRODUCT, or indexing the rows and columns of a MATRIX or DETERMINANT; the index set is often denoted $\mathbb{N}$ or $\mathcal{N}$ in the general case
I	an INTERVAL
I	an INTEGRAL
k	a CONSTANT
m	the GRADIENT of a LINE
m, n	INTEGERS, and N for a large integer
n	a NORMAL vector
N	a NEIGHBOURHOOD
p	a PRIME NUMBER
p	a PROBABILITY
$p, q, r,$ or P, Q, R	STATEMENTS or PROPOSITIONS
R	a RING
R	a RELATION
S, T	SETS
T^r_s	a TENSOR
u, v	functional components of a function, such as its REAL and IMAGINARY PARTS
U, V	NEIGHBOURHOODS
u, **v**, **w** or **x**, **y**, **z**	VECTORS
x, y, z	VARIABLES, especially REAL variables
z	a COMPLEX variable

Symbols   The following have entries in the body of the dictionary:

[ ]	BRACKETS
( )	PARENTHESES (commonly called BRACKETS)
{ }	BRACES
⟨ ⟩	ANGLE BRACKETS
$\bar{\ }$	BAR
′	PRIME
.	DOT
=	EQUALS SIGN
(as $\overline{x-y} \times z$)	VINCULUM

Arithmetic

+	('plus') ADDITION
−	('minus') SUBTRACTION
±	'plus or minus'; see ERROR
×	('times') MULTIPLICATION, PRODUCT
*	(*Computing*) MULTIPLICATION
÷ , /	DIVISION
≠	not EQUAL

$<, \leq, >, \geq$	INEQUALITY
$\cong$	APPROXIMATE
$\propto$	PROPORTIONAL
$\equiv$	IDENTITY
$\sqrt{}$	RADICAL SIGN, ROOT
$:, /$	PROPORTION, RATIO
$::$	MEAN
$\%$	PER CENT
$^0/_{00}$	PER MIL
x^y	EXPONENT
$^{-1}$ (*superscript*)	RECIPROCAL. This symbol is commonly also used for the INVERSE of a function, but as these very different concepts are often confused, we have followed the practice of those mathematicians who prefer to distinguish the notations.
$^{-\mathsf{I}}$ (*superscript*)	INVERSE; see also ARC–.

Geometry

$\angle$	ANGLE
$\wedge$ (as in $A\hat{B}C$)	ANGLE
Δ	TRIANGLE
$\odot$	CIRCLE
$\perp$	PERPENDICULAR
$\parallel$	PARALLEL
$(\ ,\ )$	COORDINATES
$^\circ$ (*superscript*)	DEGREE of arc
$|\ |$	DISTANCE, LENGTH
$\equiv$	CONGRUENT

Combinatorics

$!$	FACTORIAL
$^nC_k,\ C_k^n$	COMBINATION
$^nP_k,\ P_k^n$	PERMUTATION
$\binom{n}{k}$	BINOMIAL COEFFICIENT
$\binom{n}{n_1 \dots n_m}$	MULTINOMIAL COEFFICIENT
$\begin{bmatrix} n \\ m \end{bmatrix}_q$	Q–BINOMIAL

Number theory

$\equiv$	CONGRUENCE
$|$	exactly DIVISIBLE by
$(\ |\ )$	LEGENDRE SYMBOL
$d(\ )$	DIVISOR FUNCTION
$p(\ )$	PARTITION FUNCTION

Abstract Algebra	$\langle\ ,\ \rangle$	SCALAR PRODUCT
	$\rightarrow$	ARROW
	$\oplus$	DIRECT SUM
	$\otimes$	DIRECT PRODUCT, TENSOR PRODUCT
	$\wedge$	EXTERIOR PRODUCT
	$**$	REVERSE MULTIPLICATION
	$\perp$	ORTHOGONAL
	$*$	DUAL

Matrices and Operators	$[\ ]$	MATRIX; $[a_{ij}]$ is the matrix with a_{ij} at the intersection of the ith row and j^{th} column
	$*,\ ^{\text{T}}$ (as A^{T})	TRANSPOSE
	$*,\ ',\ ^{\perp}$ (as A^{*})	ADJOINT
	$\|\ \|$	DETERMINANT
	$\|\ \|$	FROBENIUS NORM
	† (as $A^{\dagger}$)	PSEUDO-INVERSE

Group Theory	$\|\ \|$	ORDER of an element
	$\|\ :\ \|$	INDEX
	$/$ (as F/G)	FACTOR SPACE
	$^{\wedge}$ (as $\hat{A}$)	COFACTOR
	$<$	NORMAL, IDEAL
	$\langle\ ,\ \rangle$	PRESENTATION
	$[\ ,\ ]$	COMMUTATOR
	A_n	ALTERNATING GROUP
	S_n	PERMUTATION GROUP
	$C_{\text{G}}(\ )$	CENTRALIZER
	$GL(\ ,\ )$	GENERAL LINEAR GROUP
	$x^{\text{G}},\ \text{orb}_{\text{G}}(\ )$	ORBIT
	$N_{\text{G}}(H)$	NORMALISER
	$O(\ )$	ORDER IDEAL
	$O(\ )$	ORTHOGONAL GROUP
	R_{G}	GROUP RING
	$R[\ ],\ R(\ )$	POLYNOMIAL RING
	$SL(\ ,\ )$	SPECIAL LINEAR GROUP
	$SO(\ )$	SPECIAL ORTHOGONAL GROUP
	$S^{\text{o}},\ S^{0},\ S^{\text{oo}},\ S^{00}$	POLAR SET
	$\text{stab}_{\text{G}}(\ )$	STABILISER
	$Z(G)$	CENTRE

Vectors	$\rightarrow$ (as $\overrightarrow{BA}$)	VECTOR, DISPLACEMENT
	$\langle\ ,\ \rangle$	VECTOR, especially a POSITION VECTOR
	$\times,\ \wedge$	VECTOR PRODUCT

621

Analysis and Topology

Real and Complex	(,) or] , [	OPEN INTERVAL
	[,]	CLOSED INTERVAL
	[, [or [,)	
	] ,] or (,]	HALF–OPEN INTERVALS
	| |	ABSOLUTE VALUE, MODULUS
	[ ]	INTEGRAL PART
	{ }	FRACTIONAL PART
	rez	REAL PART
	imz	IMAGINARY PART
	$^{-}$(as $\bar{a}$)	CONJUGATE

Functions	$\rightarrow$	FUNCTION
	$\mapsto$	FUNCTION
	$\circ$	COMPOSITION
	$f^{[n]}$	COMPOSITION of f with itself n times
	$f_E, f\|E$	RESTRICTION
	$\rightarrow$	LIMIT
	$a+$	RIGHT–HAND LIMIT
	$a-$	LEFT–HAND LIMIT
	$\uparrow$	ABOVE, INCREASING
	$\downarrow$	BELOW, DECREASING
	∞	INFINITY
	$\sim$	ASYMPTOTIC
	O, o	ORDER NOTATION
	$F(, ; ;)$	HYPERGEOMETRIC FUNCTION
	$(a)_n$	POCHHAMMER SYMBOL
	$\mathrm{Li}_n( )$	POLYLOGARITHM

Differentiation	f', D_x	DERIVATIVE
	$f^{(n)}$	n^{th} DERIVATIVE of f
	$f_x, \mathrm{D}_x f$	PARTIAL DERIVATIVE
	∂	PARTIAL DERIVATIVE, JACOBIAN
	dF	DIFFERENTIAL
	$\dot{x}$	FLUXIONS

Integration	$\int$	ANTIDERIVATIVE, INTEGRAL
	$F(x)\big|_a^b = \big[F(x)\big]_a^b = F(b) - F(a)$	
	$\iint$	MULTIPLE INTEGRATION
	$\iint_S$	SURFACE INTEGRAL
	$\int_E$	RIEMANN INTEGRATION
	$\square$	CONVOLUTION

Vector Analysis	$\int_\Gamma$	CONTOUR INTEGRAL
	∇	DIFFERENTIAL OPERATOR, DIVERGENCE, GRADIENT, CURL, FRECHET DIFFERENTIAL
Measure Theory	$\int_E d\mu$	LEBESGUE INTEGRATION
	$*$ (as $\mu*$)	OUTER MEASURE
	$<<$	ABSOLUTELY CONTINUOUS
	$\| \|_p$	L_p SPACE, l_p SPACE
	$\| \|_\infty$	CHEBYSHEV NORM
	$s_C(\ ), s(\ ,\ )$	SUPPORT FUNCTION
	$V_h(\ ,\ )$	TOTAL VARIATION
Topology	$\circ$ (as A°)	INTERIOR
	(as $\bar{A}$), $Cl(\ )$	CLOSURE
	$Fr(\ )$	FRONTIER
	ri	RELATIVE INTERIOR
	$N(\ ,\ ), N'(\ ,\ )$	NEIGHBOURHOOD
	$B_\varepsilon(\ ), B(\ ,\ )$	BALL
	$\| \|$	NORM
	$\mathfrak{F}$	FILTER
Applied mathematics	∇	DEL
	∇^2	NABLA SQUARED
	∂R	SURFACE of BODY R
	$^-$ (as $\bar{x}$)	CENTRE OF MASS
	$\frac{\partial}{\partial t} T(\mathbf{X}, t)\big|_\mathbf{X}$	MATERIAL DERIVATIVE

Set Theory and Logic

Set Theory	$\{\ \}$	SET
	$\langle\ \rangle$	SEQUENCE, ORDERED SET
	$\in$	MEMBER
	$\notin$	NON–MEMBER
	$\subset, \subseteq$	SUBSET
	$\supset, \supseteq$	INCLUSION
	$\cup$	('cup') UNION
	$\cap, \cap*$	('cap') INTERSECTION
	$\cup*$	DISJOINT UNION
	$\varnothing, \wedge$	EMPTY SET
	$'$ (as A'), $C(\ )$	COMPLEMENT
	$\setminus$	RELATIVE COMPLEMENT
	$\ominus$	SYMMETRIC DIFFERENCE
	$\|\ \|$	CARDINALITY
	$\times$	CARTESIAN PRODUCT
	$\mathcal{P}(S), 2^S$	POWER SET of S

623

symbols and conventions

Mappings	$\rightarrow$	MAPPING
	$\mapsto$	MAPPING
	$f(S)$	IMAGE
	$\mathrm{dom}(f)$	DOMAIN
	$^{-1},\ ^{-1}$ (*superscript*)	COUNTERIMAGE
	$1-1$	ONE-TO-ONE
	$\equiv$	EQUIVALENCE RELATION
	[ ]	EQUIVALENCE CLASS
	$F^{-}(B),\ F^{w}(B)$ $F^{+}(B),\ F^{s}(B)$	INVERSE IMAGE SET

Sentential Calculus	&, $\wedge$	AND, CONJUNCTION
	v	DISJUNCTION
	$\underline{v}$	EXCLUSIVE DISJUNCTION
	$\not\equiv$	non-EQUIVALENCE
	$\rightarrow,\ \supset$	IMPLICATION
	$\leftrightarrow,\ \equiv$	EQUIVALENCE
	$\sim, \neg, -,\ ^{-}$ (as $\overline{P}$)	NEGATION
	|	SHEFFER'S STROKE

Predicate Calculus	$\exists$	EXISTENTIAL QUANTIFIER
	$\forall$	UNIVERSAL QUANTIFIER
	ι	DEFINITE DESCRIPTION
	$\exists!$	UNIQUE QUANTIFIER
	$\exists n$	NUMERICAL QUANTIFIER
	$\hat{x}Fx$	ABSTRACT

Modal Logic	$\square$	NECESSITY
	$\Diamond$	POSSIBLE
	$\rightarrow$	('fish-hook') ENTAILMENT

Metalogic	$\perp$	FALSEHOOD
	$\vdash$	('gatepost', 'TURNSTILE') SEQUENT
	$^\wedge$	CONCATENATE
	$\ulcorner\ \urcorner$	QUASI-QUOTATION
	(,)	APPLICATION

Orderings and Lattices	$<, \leq, >, \geq$	ORDERING
	$\wedge$	MEET
	v	JOIN
	$\wedge$	ZERO
	V	UNITY

Statistics

$^{-}$ (as $\bar{x}$)		ARITHMETIC MEAN
B_1		SKEWNESS
B_2		KURTOSIS
Bi(,)		BINOMIAL DISTRIBUTION
E( )		EXPECTED VALUE
E(|)		CONDITIONAL EXPECTATION
Ga(,)		GAMMA DISTRIBUTION
N(,)		NORMAL DISTRIBUTION
P( )		PROBABILITY
P(|)		CONDITIONAL PROBABILITY
Po( )		POISSON DISTRIBUTION
Un(,)		UNIFORM DISTRIBUTION

Table of Derivatives and Integrals of Common Functions

$f(x)$	$f'(x)$	$\int f(x)\,\mathrm{dx}*$
x^n	nx^{n-1}	$\dfrac{x^{n+1}}{n+1}\quad(n \neq -1)$
x^{-1}	$-x^{-2}$	$\ln\lvert x\rvert$
e^x	e^x	e^x
$\ln x$	x^{-1}	$x(\ln\lvert x\rvert -1)$

Trigonometric functions

$\sin x$	$\cos x$	$-\cos x$
$\cos x$	$-\sin x$	$\sin x$
$\tan x$	$\sec^2 x$	$-\ln(\cos x)$
$\sec x$	$\sec x \tan x$	$\ln(\sec x + \tan x)$
$\operatorname{cosec} x$	$-\operatorname{cosec} x \cotan x$	$\ln(\operatorname{cosec} x - \cotan x)$
$\cotan x$	$-\operatorname{cosec}^2 x$	$\ln(\sin x)$

Hyperbolic functions

$\sinh x$	$\cosh x$	$\cosh x$
$\cosh x$	$\sinh x$	$\sinh x$
$\tanh x$	$\operatorname{sech}^2 x$	$\ln(\cosh x)$
$\operatorname{sech} x$	$-\operatorname{sech} x \tanh x$	$\sin^{-1}(\tanh x)$
$\operatorname{cosech} x$	$-\operatorname{cosech} x \coth x$	$\ln(\tanh\frac{x}{2})$
$\cotanh x$	$-\operatorname{cosech}^2 x$	$\ln(\sinh x)$

Inverse trigonometric functions

$\sin^{-1} x$	$\dfrac{1}{\sqrt{1-x^2}}\quad(\lvert x\rvert < 1)$	$x\sin^{-1}x - \sqrt{1-x^2}$
$\cos^{-1} x$	$\dfrac{-1}{\sqrt{1-x^2}}\quad(\lvert x\rvert < 1)$	$x\cos^{-1}x - \sqrt{1-x^2}$
$\tan^{-1} x$	$\dfrac{1}{1+x^2}$	$x\tan^{-1}x - \frac{1}{2}\ln(1+x^2)$
$\sec^{-1} x$	$\dfrac{1}{x\sqrt{x^2-1}}$	$x\sec^{-1}x - \ln(x+\sqrt{x^2-1})$

table of derivatives and integrals of common functions

$f(x)$	$f'(x)$	$\int f(x)\,dx*$
$\operatorname{cosec}^{-1}x$	$\dfrac{-1}{x\sqrt{x^2-1}}$	$x\operatorname{cosec}^{-1}x+\ln(x+\sqrt{x^2-1})$
$\operatorname{cotan}^{-1}x$	$\dfrac{-1}{1+x^2}$	$x\operatorname{cotan}^{-1}x+\dfrac{1}{2}\ln(1+x^2)$

Inverse hyperbolic functions

$\sinh^{-1}x$	$\dfrac{1}{\sqrt{1+x^2}}$	$x\sinh^{-1}x-\sqrt{1+x^2}$
$\cosh^{-1}x$	$\dfrac{1}{\sqrt{x^2-1}}\quad(x>1)$	$x\cosh^{-1}x-\sqrt{x^2-1}$
$\tanh^{-1}x$	$\dfrac{1}{1-x^2}\quad(\lvert x\rvert<1)$	$x\tanh^{-1}x+\dfrac{1}{2}\ln(1-x^2)$
$\operatorname{sech}^{-1}x$	$\dfrac{-1}{x\sqrt{1-x^2}}\quad(0<x<1)$	$x\operatorname{sech}^{-1}x+\sin^{-1}x$
$\operatorname{cosech}^{-1}x$	$\dfrac{-1}{\lvert x\rvert\sqrt{1+x^2}}\quad(x\neq0)$	$\operatorname{cosech}^{-1}x+\dfrac{x}{\lvert x\rvert}\ln(x+\sqrt{x^2+1})$
$\coth^{-1}x$	$\dfrac{1}{1-x^2}\quad(\lvert x\rvert<1)$	$x\operatorname{cotanh}^{-1}x+\dfrac{1}{2}\ln(1-x^2)$

Functions of $(x^2\pm a^2)$

$\sqrt{x^2\pm a^2}$	$\dfrac{x}{\sqrt{x^2\pm a^2}}$	$\dfrac{1}{2}\left(x\sqrt{x^2\pm a^2}\pm a^2\ln(x+\sqrt{x^2\pm a^2})\right)$
$\sqrt{a^2-x^2}$	$\dfrac{-x}{\sqrt{a^2-x^2}}$	$\dfrac{1}{2}\left(x\sqrt{a^2-x^2}+a^2\sin^{-1}(\dfrac{x}{a})\right)$
$\dfrac{1}{x^2+a^2}$	$\dfrac{-2x}{(x^2+a^2)^2}$	$\dfrac{1}{a}\tan^{-1}(\dfrac{x}{a})$
$\dfrac{1}{x^2-a^2}$	$\dfrac{-2x}{(x^2-a^2)^2}$	$\dfrac{1}{2a}\ln\left(\dfrac{x-a}{x+a}\right)$
		$=\dfrac{1}{a}\coth^{-1}(\dfrac{x}{a})\quad\text{if }x^2>a^2$

$f(x)$	$f'(x)$	$\int f(x)\,dx$*
$\dfrac{1}{\sqrt{x^2 \pm a^2}}$	$\dfrac{-x}{(x^2 \pm a^2)^{3/2}}$	$\ln(x + \sqrt{x^2 \pm a^2}$
$\dfrac{1}{\sqrt{a^2 - x^2}}$	$\dfrac{x}{(x^2 - a^2)^{3/2}}$	$\sin^{-1}\left(\dfrac{x}{a}\right)$
$\dfrac{1}{x\sqrt{a^2 \pm x^2}}$	$\dfrac{\pm 1}{(a^2 \pm x^2)^{3/2}} - \dfrac{1}{x^2\sqrt{a^2 \pm x^2}}$	$\dfrac{1}{a}\ln\dfrac{a + \sqrt{a^2 \pm x^2}}{x}$
$\dfrac{1}{x\sqrt{x^2 - a^2}}$	$\dfrac{-1}{(x^2 - a^2)^{3/2}} - \dfrac{1}{x^2\sqrt{x^2 - a^2}}$	$\dfrac{1}{a}\sec^{-1}\left(\dfrac{x}{a}\right)$
$\dfrac{1}{\sqrt{2ax - x^2}}$	$\dfrac{x - a}{(2ax - x^2)^{3/2}}$	$\cos^{-1}\left(1 - \dfrac{x}{a}\right)$

*Note that CONSTANTS OF INTEGRATION are omitted.

Hilbert Problems

1. The CONTINUUM HYPOTHESIS. In 1938, Gödel showed that if the Zermelo-Fraenkel axioms for set theory are consistent, then the generalized continuum hypothesis cannot be disproved from these axioms. In 1963, Cohen showed that its negation cannot be disproved.

2. Whether the axioms of arithmetic are CONSISTENT. See GÖDEL'S THEOREM.

3. Whether two tetrahedra of equal base and altitude necessarily have the same volume. Proved false by Max Dehn in 1900.

4. To construct all the METRICS in which straight lines are GEODESICS.

5. How far Lie's conception of continuous groups of transformations is approachable without assuming that the transformations are differentiable. Solved by Gleason in 1952 and Montgomery–Zippen in 1955 in the form 'Every locally Euclidean group is a LIE GROUP'.

6. To axiomatize mathematical physics. Some progress has been made.

7. Whether α^β is TRANSCENDENTAL where α is ALGEBRAIC and β is IRRATIONAL (for example, $2^{\sqrt{2}}$, e^π, etc.). Not yet solved; important work has been done by Gelfond, Schneider and Baker. See GELFOND–SCHNEIDER THEOREM.

8. The RIEMANN (zeta) HYPOTHESIS. Notoriously unsolved.

9. To find the most general law of reciprocity in an ALGEBRAIC NUMBER FIELD. (See QUADRATIC RECIPROCITY) Artin obtained it in 1927 for abelian extensions of $\mathbb{Q}$; the non-abelian case is still open.

10. To find a method to determine whether a given DIOPHANTINE EQUATION is soluble. In 1970, Matijasevich showed that no such method exists.

11. The study of QUADRATIC FORMS with algebraic coefficients. Incomplete.

12. The study of any ALGEBRAIC NUMBER FIELD extensions. Incomplete.

13. To show that the general equation of the seventh degree cannot be solved by means of functions of only two arguments. Partly solved.

14. Whether the ring $K \cap k [x_1, \ldots , x_n]$ is finitely generated over K, where K is a field, $k [x_1, \ldots , x_n]$ is a POLYNOMIAL RING, and $k \subseteq K \subseteq k (x_1, \ldots , x_n)$. Proved false by Nagata in 1959.

15. The rigorous foundation of 'Schubert's Enumerative Calculus'.

16. The investigation of the topology of algebraic surfaces.

17. The expression of a definite rational function as a quotient of sums of squares. In 1927 Artin showed that a POSITIVE DEFINITE rational function is a sum of squares.

18. Whether there exist non-regular space-filling polyhedra.

19. Whether the solutions of LAGRANGIANS are always ANALYTIC.

20. Whether every VARIATIONAL problem has a solution, provided suitable assumptions are made about boundary conditions.

22. 'To show that there always exists a LINEAR DIFFERENTIAL EQUATION of the Fuchsian class, with given singular points and monodromic group'. Solved by Deligne (1970).

23. Development of the CALCULUS OF VARIATIONS.

Millennium Prize Problems

1. **Birch and Swinnerton–Dyer Conjecture**: The conjecture that, when the solutions in whole numbers to algebraic equations are the points of an abelian variety, the size of the group of rational points is related to the behaviour of an associated ZETA FUNCTION $\zeta(\sigma)$ near the point $\sigma = 1$. In particular it asserts that if $\zeta(1)$ is equal to 0 then there are an infinite number of rational points (solutions), and conversely, if $\zeta(1)$ is not equal to 0, then there is only a finite number of such points.

2. **Yang–Mills Theory**: In quantum theory, a framework for describing elementary particles using geometrical structures, now the foundation of most of elementary particle theory. Its predictions about the strong interactions of elementary particles and the existence of a subtle quantum mechanical property called the "mass gap" – that quantum particles have positive masses, even though the classical waves travel at the speed of light – have been experimentally tested and confirmed by computer simulations, but its mathematical foundation remains unclear.

3. **P versus NP problem**: In computer science, the problem of determining whether questions exist whose answer, if known, can be quickly checked, but which require a much longer time to solve if the answer is not already known. There certainly seem to be many such questions: for example whether a given large integer is the product of two primes, but it may just be that we have not yet devised a method for solving them quickly. Stephen Cook formulated the P versus NP problem in 1971.

4. **Hodge conjecture**: The conjecture that for particular types of spaces called projective algebraic varieties, when approximating the shape of a given object by gluing together simple geometric building blocks, the pieces called Hodge cycles are actually (rational linear) combinations of geometric pieces called algebraic cycles.

5. **Riemann hypothesis**: The conjecture that the ZETA FUNCTION has no non-trivial zeros except on the line with $\mathrm{re}(z) = \frac{1}{2}$. Riemann observed that the distribution of prime numbers amongst the whole numbers, while not regular, approximates closely to the zeta function, so that and its establishment would have many consequences for the PRIME NUMBER THEOREM and related theory. This was one of HILBERT'S PROBLEMS, and is known to be true for the first 1 500 000 000 zeros.

6. **Poincaré conjecture**: The conjecture that a simply connected three-dimensional compact manifold is topologically equivalent to a three-sphere. The four-dimensional analogue of this conjecture has recently been shown false by Michael Freedman and, in April 2002, Martin Dunwoody of Southampton University announced a proof of the three-dimensional case, although this remains to be confirmed.

7. **Navier–Stokes equation**: In continuum mechanics, the identity, for a SIMPLE NEWTONIAN VISCOUS FLUID with density ρ,

$$\rho \mathbf{a} = \rho \mathbf{b} - \nabla p + \mu \nabla^2 \mathbf{v} + \tfrac{1}{3} \mu \nabla (\nabla \cdot \mathbf{v}),$$

where $\mathbf{a}$ is the acceleration, $\mathbf{b}$ the BODY FORCE DENSITY, $\mathbf{v}$ the velocity, ρ the density, p the pressure function, and μ the viscosity. Such equations are ubiquitous in modern computational science. The challenge is to develop a sophisticated mathematical theory of the solutions to these equations and their applications to complex fluid motion.

Forty-four Useful Constants

Mathematical Constants

1. $\sqrt{2}$ = 1.414 213 562 373 095 048 8
2. $\sqrt{3}$ = 1.732 050 807 568 877 293 5
3. $\sqrt{5}$ = 2.236 067 977 499 789 696 4
4. Golden mean $\frac{\sqrt{5}-1}{2}$ = 0.618 033 988 749 894 848 20
5. π = 3.141 592 653 589 793 238 5
6. $1/\pi$ = 0.318 309 886 183 790 671 53
7. $\pi/2$ = 1.570 796 326 794 896 619 2
8. e = 2.718 281 828 459 045 235 4
9. $1/e$ = 0.367 879 441 171 442 321 60
10. e^{π} = 23.140 692 632 779 269 007
11. $\log 2$ = 0.693 147 180 559 945 309 42
12. $\log 10$ = 2.302 585 092 994 045 684 0
13. $\log_2 10$ = 3.321 928 094 887 362 347 8
14. $\log_{10} 2$ = 0.301 029 995 663 981 195 22
15. $\log_2 3$ = 1.584 962 500 721 156 181 5
16. $\zeta(2)$ = 1.644 934 066 848 226 436 5
17. Apéry's constant $\zeta(3)$ = 1.202 056 903 159 594 285 4
18. $\zeta(5)$ = 1.036 927 755 143 369 926 3
19. Catalan's constant G = 0.915 965 594 177 219 015 05
20. Euler's constant γ = 0.577 215 664 901 532 860 61
21. $\Gamma(1/2) = \sqrt{\pi}$ = 1.772 453 850 905 516 027 3
22. $\Gamma(1/3)$ = 2.678 938 534 707 747 633 7
23. $\Gamma(1/4)$ = 3.625 609 908 221 908 312 1
24. Elliptic integral of the first kind $K(1/\sqrt{2})$ = 1.854 074 677 301 371 918 4
25. Elliptic integral of the second kind $E(1/\sqrt{2})$ = 1.350 643 881 047 675 502 5
26. Chaitin's universal halting constant (in binary)

    Ω = 0.000 000 100 000 010 000 100 000 100
       001 110 111 001 100 100 111 100 010 010 011 100

27. Feigenbaum's first bifurcation constant α = 4.669 201 609 102 990 ...

28. Feigenbaum's second bifurcation constant δ = 2.502 907 875 095 892 ...

29. Khintchine's continued fraction constant
    K = 2.685 452 001 065 306 445 3

30. Madelung's electrochemical constant $M_3 = \sum' (-1)^{i+j+k}/\sqrt{i^2 + j^2 + k^2}$ =
    1.747 564 594 633 182 190 3

Physical Constants

31. Avogadro's constant N_A = 6.022 141 99(47) × 1023 mol^{-1}

32. Boltzmann's constant k = 1.380 650 3(24) × 1023 $J K^{-1}$

33. Fine structure constant α = 7.297 352 533(27) × 10^{-3}

34. Newton's constant g = 6.67310 × 10^{-11} $m^3 kg^{-1} s^{-2}$

35. Planck's constant h = 6.626 068 76(52) × 10^{-34} $J\ s$

36. Speed of light in vacuum c = 299 792 458ms^{-1}

37. Mass of an electron m_e = 9.109 381 88(72) × 10^{-31} kg

38. Electron volt eV = 1.602 176 462(63) × 10^{-19} J

39. Electron radius r_e = 2.817 940 285(31) × 10^{-15} m

40. Proton/electron mass ratio m_p/m_e = 1836.152 667 5(39)

41. Diameter of earth $\approx$ 12.7 × 10^6 m

42. Astronomical unit (average distance to Sun) $AU \approx$ 150 × 10^9 m

43. Light year = 9.460 528 405 106 × 10^{15} m

44. Parsec = 3.085 677 587 679 31 × 10^{16} m

Wherever possible mathematical constants are given to 20 signficant places. Bracketed numbers for physical constants indicate uncertainty.

Definitions not in the text may be found at:

http://pauillac.inria.fr/algo/bsolve/constant/constant.html and
http://physics.nist.gov/cuu/Constants/index.html

Internet links

Finding mathematics resources on the internet

This is a selection of addresses for mathematical websites, loosely grouped by subject matter. Each site is listed with its title, address and a brief description of its content. There is a huge volume of online material available: for this reason, topic-specific sites have only been provided for mathematics likely to be encountered by secondary or early tertiary students and, in general, neither journals nor lecture notes have been explicitly included. For higher level or more specific topics, further websites may be found via the portal sites given in the first group of addresses.

General Resources

The Math Forum@Drexel
http://mathforum.org
This site is part of the The Math Forum, an online mathematics education community centre, and provides an extensive list of annotated links to other sites. Includes a well-designed search engine.

MathWorld
http://mathworld.wolfram.com
MathWorld is a comprehensive and interactive mathematics encyclopedia intended for students, educators, math enthusiasts, and researchers. MathWorld is hosted and sponsored by Wolfram Research, Inc., makers of Mathematica.

Mathematics WWW Virtual Library
www.math.fsu.edu/Virtual/
This collection of mathematics-related resources is maintained by the Florida State University Department of Mathematics as a free service to the online community.

PSU Math – Mathematics Websites
www.math.psu.edu/MathLists/
Contents.html
Searchable, browsable, list of mathematics websites maintained by Penn State University.

The Prime Pages
http://primes.utm.edu/index.html
Exhaustive index of prime number resources. Includes list of the 5000 largest prime numbers.

A Catalog of Mathematics Resources
http://mthwww.uwc.edu/wwwmahes/
files/math01.htm
List of mathematics websites organized by topic. Also includes newsgroups, publications, and software websites. Maintained by M. Maheswaran, Department of Mathematics, University of Wisconsin Marathon County.

Core Mathematics

S.O.S. Math
www.sosmath.com/index.html
Explanations and problems aimed at secondary school to first- or second-year university students; pages for algebra, calculus, trigonometry, matrix algebra, complex variables, differential equations and mathematical tables.

Mathematical Atlas
www.math-atlas.org/
This is a collection of short articles designed to provide an introduction to the areas of modern mathematics and pointers to further information, as well as answers to some common questions.

Algebra Homework Help

www.purplemath.com/
Definitions and examples for
secondary and early tertiary level
algebra. Includes basic problem
solving and algebraic manipulation,
linear algebra and introductory level
abstract algebra.

Math Nerds Free Math Help and Tutoring

www.mathnerds.com/mathnerds/
MathNerds provides free, discovery-
based, mathematical guidance via an
international, volunteer network of
mathematicians.

Calculus Solutions

www.jtaylor1142001.net/
This site consists of about 300 sample
problems and solutions on various
topics in calculus, with some review
of precalculus.

Finite Mathematics and Applied Calculus Resource Page

http://people.hofstra.edu/faculty/Stefan_
Waner/RealWorld/tutindex.html
Includes links to online text and
interactive tutorials for elementary
calculus (precalculus, differentiation,
integration and an introduction to
multivariable calculus) and finite
mathematics (linear systems,
matrices, linear programming, sets,
introductory probability and
statistics).

Statistics and Probability

Current Index to Statistics

www.statindex.org/CIS/query
A bibliographic index to publications
in statistics and related fields.
References are drawn from 162 core
journals, from which articles are
selected that have statistical content,
proceedings and edited books, and
other sources. The Current Index to
Statistics is a joint venture of the
American Statistical Association and
the Institute of Mathematical
Statistics.

Probability Abstract Service

www.economia.unimi.it/PAS//
Archive of research article abstracts
which publishes a bi-monthly
newsletter.

Statistics.com

www.statistics.com/
A site primarily offering online
courses but also containing links to
online textbooks, glossaries and other
reference material, commercial and
free statistical software and sites of
statistical data.

The Statistics Homepage

www.statsoft.com/textbook/stathome.
html
An award-winning online statistics
text containing techniques likely to
be required by science, engineering,
medical, finance and business
students.

The WWW Virtual Library: Statistics

www.stat.ufl.edu/vlib/statistics.html
Catalog of statistics resources
includes university departments,
government departments, research
groups, journals, software, and
newsgroups.

Probability Web

www.mathcs.carleton.edu/probweb/
Collection of probability resources on
the World Wide Web designed to be
especially helpful to researchers,
teachers, and people in the
probability community.

Advanced Mathematics

Abstract Algebra

www.math.uiuc.edu/~r-ash/
The home page of Professor Robert B.
Ash contains a selection of online

graduate level books on abstract algebra.

Multivariable Calculus

www.math.gatech.edu/~cain/notes/calculus.html
An online text for a course in multivariable calculus used at Georgia Institute of Technology.

Discrete Mathematics

www.maths.mq.edu.au/~wchen/lndmfolder/lndm.html
Online text covering mathematical and computational aspects of discrete mathematics including logic and sets, finite state machines, graph theory, integer programming and codes. Maintained at Macquarie University.

Number Theory Web

www.numbertheory.org/
Catalog of online information of interest to number theorists everywhere. Includes lists of number theorists, university departments, journals, theses, lecture notes, biographies, conferences and job postings.

Experimental Mathematics

www.experimentalmath.info
List of experimental mathematics resources, experimental mathematicians, and other relevant websites.

Academic Mathematics

Mathematics Web Sites

www.math.psu.edu/MathLists/Contents.html
Maintained by Pennsylvania State University's Mathematics Department; contains links to university mathematics departments worldwide, journals and preprints, societies, and so on.

History

MacTutor History of Mathematics

www-groups.dcs.st-and.ac.uk/~history
Offers mathematics history, chronologies, and more than 1600 biographies of mathematicians. The MacTutor History of Mathematics archive was created and is maintained by John O'Connor and Edmund F. Robertson at the University of St. Andrews, Scotland.

The Cornell University Library: Historical Mathematics Monographs

http://historical.library.cornell.edu/math/
A collection of selected monographs with expired copyrights chosen from the mathematics field.

British Society for the History of Mathematics

www.dcs.warwick.ac.uk/bshm/index.html
The aims of the British Society for the History of Mathematics are to promote research into the history of mathematics and its use at all levels of mathematics education.

The Mathematical Museum: The History Wing

www.math-net.org/links/show?collection=math.museum.hist
The 'History' wing of The Mathematical Museum contains pointers to selected exhibitions, hyperbooks, information systems, museums and pages related to the history of mathematics and adjacent fields.

Institute for History and Foundations of Science

www.phys.uu.nl/~wwwgrnsl/indexi.html

The Institute for History and Foundations of Science (Utrecht University) is part of the Faculty of Physics and Astronomy. The Institute consists of two distinct Sections: the History of Science Section and the Foundations of Science Section.

Slates, Sliderules, and Software: Teaching Math in America

http://americanhistory.si.edu/teachingmath/
An exhibit of objects used by teachers and parents to help students master abstract mathematical concepts. From the Smithsonian's National Museum of American History.

Careers

The Mathematical Association of America: Careers in Mathematics

www.maa.org/students/career.html
FAQs regarding mathematics careers and places to find the answers.

Mathematical Sciences Career Information

www.ams.org/careers/
The American Mathematical Society, the Mathematical Association of America, and the Society for Industrial and Applied Mathematics are dedicated to providing career information and services to the mathematics community.

Organizations

International Mathematical Union (IMU)

www.mathunion.org/index.html
IMU is an international non-governmental and non-profit scientific organization, with the purpose of promoting international cooperation in mathematics. It is a member of the International Council for Science.

American Mathematical Society

http://e-math.ams.org/
Founded in 1888 to further mathematical research and scholarship, the American Mathematical Society fulfills its mission through programs and services that promote mathematical research and its uses, strengthen mathematical education, and foster awareness and appreciation of mathematics and its connections to other disciplines and to everyday life.

Australian Mathematical Society

www.austms.org.au/
The Australian Mathematical Society (AustMS) is the national society of the mathematics profession in Australia.

Canadian Mathematical Society

http://cms.math.ca/
The Canadian Mathematical Society (CMS) was originally conceived in June 1945 as the Canadian Mathematical Congress. The founding members hoped that "this congress [would] be the beginning of important mathematical development in Canada."

European Mathematical Society

www.emis.de/
The European Mathematical Society (EMS) was founded in 1990. The purpose of the Society is to further the development of all aspects of mathematics in the countries of Europe. In particular, the Society aims to promote research in mathematics and its applications.

South African Mathematical Society

www.cam.wits.ac.za/sams/
The South African Mathematical Society is a national association of mathematicians, conscious of its African and international context as it seeks to promote the discipline of mathematics in all its facets, locally, regionally and internationally.

The Association for Women in Mathematics (AWM)

www.awm-math.org/
The Association for Women in Mathematics (AWM) is a non-profit organization founded in 1971. Its continuing goal is to encourage women in the mathematical sciences.

European Women in Mathematics

www.math.helsinki.fi/EWM/
EWM organization is an affiliation of women bound by a common interest in the position of women in mathematics.

Isaac Newton Institute for Mathematical Sciences

www.newton.cam.ac.uk/
The Isaac Newton Institute is the UK's national and international visitor research institute for mathematics.

Institute for Mathematics and its Applications (IMA)

www.ima.umn.edu/
The primary mission of the IMA is to increase the impact of mathematics by fostering research of a truly interdisciplinary nature, linking mathematics of the highest caliber and important scientific and technological problems from other disciplines and industry.

Mathematical Sciences Research Institute

www.msri.org/
The Mathematical Sciences Research Institute (MSRI) exists to further mathematical research through broadly based programs in the mathematical sciences and closely related activities.

Online Textbooks and Tutorials

Elementary Linear Algebra

www.numbertheory.org/book/
An online text for linear algebra, with lecture notes and solutions to problems.

Journals

Geometry and Topology

www.maths.warwick.ac.uk/gt/
Geometry and Topology is a fully refereed international journal dealing with all aspects of geometry and topology and their applications.

AMS Journals: Print and Electronic

www.ams.org/journals/
List of the American Mathematical Society's collection of print and electronic journals.

Data/Sources

MathDL: The MAA Mathematical Sciences Digital Library

www.mathdl.org/
MathDL, published by the Mathematical Association of America, provides online resources for both teachers and students of mathematics.

arXiv.org e-Print archive

http://arxiv.org/
arXiv is an e-print service in the fields of physics, mathematics, non-linear science, computer science, and quantitative biology. arXiv is owned, operated and funded by Cornell University.

EULER – Your Portal to Mathematics Publications

www.emis.de/projects/EULER/
European-based real virtual library for mathematics. In particular, EULER provides a world reference and delivery service, transparent to the end user and offering full coverage of the mathematics literature world-wide, including bibliographic data, peer reviews, and abstracts.

Zentralblatt MATH

www.emis.de/ZMATH/
Zentralblatt MATH is the world's
most complete and longest running
abstracting and reviewing service in
pure and applied mathematics. The
Zentralblatt MATH Database
contains more than two million
entries drawn from more than 2300
serials and journals and covers the
period from 1868 to the present by
the recent integration of the Jahrbuch
database (JFM).

AMS Books Online

www.ams.org/online_bks/online_
subject.html
Online books published by the
American Mathematical Society are
available for free download and
organized by topic; consists mainly of
higher level mathematics.

**On-Line Encyclopedia of Integer
Sequences (Look-Up)**

www.research.att.com/~njas/sequences/
index.html
The main table is a collection of
number sequences arranged in
lexicographic order.
At the present time the table contains
over 80,000 sequences.

Mathematical Software

Mathematica

www.wolfram.com/products/
mathematica/index.html
Mathematica integrates a numeric
and symbolic computational engine,
graphics system, programming
language, documentation system, and
advanced connectivity to other

applications. Student version
available.

Maple

www.maplesoft.com/
Standard mathematical packages
which provide, amongst other things,
numeric and symbolic computation,
data import and analysis,
mathematical programming,
mathematical wordprocessing, and
two- and three-dimensional graphics.
Student version available.

MATLAB

www.mathworks.com/products/
matlab/
MATLAB is a high-level technical
programming language with an
interactive environment for
algorithm development, data
visualization and analysis, and
numeric computation.

NTL

http://shoup.net/ntl/
NTL is a portable C++ library
providing data structures and
algorithms for computational
number theory.

Glossaries

Mathematical Programming Glossary

http://carbon.cudenver.edu/~hgreenbe/
glossary/index.php
This contains terms specific to
mathematical programming, and
some terms from other disciplines,
notably economics, computer science,
and mathematics, that are directly
related.